PRODUCTION SYSTEMS:
PLANNING, ANALYSIS, AND CONTROL

Wiley Series in Production / Operations Management

FOURTH EDITION

PRODUCTION SYSTEMS: PLANNING, ANALYSIS, AND CONTROL

JAMES L. RIGGS
Formerly Oregon State University

JOHN WILEY & SONS
New York   Chichester   Brisbane   Toronto   Singapore

Library of Congress Cataloging in Publication Data:
Riggs, James L.

  Production systems.

  (Wiley series in production/operations management)
  Includes bibliographies and index.

  1. Production management. I. Title. II. Series.

TS155.R45 1987   658.5   86-28998
ISBN 0-471-84793-3

Printed in the United States of America

10 9 8 7 6 5 4 3 2

ABOUT THE AUTHOR

Dr. James L. Riggs was the founder and head of the Oregon Productivity Center and was concurrently department head and professor of the department of industrial engineering at Oregon State University. He received a B.S. degree in forest engineering, an M.S. in mechanical engineering, and a Ph.D. in industrial and mechanical engineering, all from Oregon State. He spent five years as an engineering officer in the U.S. Marine Corps before entering the teaching profession. After that, he was a consultant on a wide variety of production activities for several government agencies and companies, such as AT&T, Weyerhaeuser, Tektronix, FMC, and General Foods.

In the early 1970s, Dr. Riggs conducted productivity studies in hospitals and served on the steering committee of the National Center for Productivity and Quality of Work Life. At the 1977 World Productivity Congress in Sydney, Australia, he was elected vice-president of the World Confederation for Productivity Science. He was general chairman of the 1981 World Productivity Congress and was president of the confederation from 1981 to 1984.

Articles written by Dr. Riggs have appeared in leading technical journals, and he frequently made presentations at conferences in the United States and abroad. He was a Fulbright Scholar to Yugoslavia in 1975 and was a visiting professor at universities in Mexico, Costa Rica, Greece, Yugoslavia, Hawaii, Australia, and Israel. He was also a faculty affiliate at the Japan-American Institute of Management Science.

In addition to being the author or coauthor of 19 books on engineering economics, production, operations research, and management, Dr. Riggs was the consulting editor for the McGraw-Hill series in industrial engineering and management science. He also wrote over 45 technical papers and 120 articles.

Dr. Riggs' professional activities included serving on the boards of two corporations and as a consultant or expert witness to over 40 companies and government agencies. He was a registered professional engineer and in 1978 was elected a vice-president of the Institute of Industrial Engineers. He also served on many state and national boards and committees.

Dr. Riggs is cited in *Who's Who in America, Who's Who in Engineering, Who's Who in the West, Who's Who in the World,* and *American Men and Women in Science.*

Dr. Riggs died unexpectedly in May 1986, after completing this fourth edition of *Production Systems.*

PREFACE

A decade has passed since the initial publication of *Production Systems*. The first edition heralded the 1970s; the third edition addressed the challenges and opportunities of the 1980s; and the fourth edition looks ahead to the late 1980s and beyond. Many time-tested techniques for planning, analysis, and control remain unchanged, but most have benefited from new technology and recent developments. The updated version of this book presents the newest concepts and explores the current problems facing production analysts: inflation, energy conservation, limited resources, preservation of the environment, computer-aided design and manufacturing, consumer and worker protection, productivity improvement, and many others.

PRODUCTION STUDIES

Current events confirm the vital, dynamic, and practical role of production studies. These studies embrace economics, engineering, management, and many other disciplines to confront the difficulties encountered in converting inputs to outputs. The conversion processes yield a bewildering variety of products, ranging from manufactured goods to public services, from crops to information.

The vast number of activities that comprise the collective subject "production" make its study particularly demanding. Even laypersons must recognize the complex interacting forces that surround all significant production decisions: labor, resources, laws and regulations, capital, distribution, consumers, and the like. Many techniques have been devised to assist production decision makers. Some tools have little more rigor than do rules of thumb. Others rely on sophisticated mathematics. Most are applicable to a variety of decision situations.

Although it would be impractical to hope to achieve full proficiency in the use of all the analytical methods, a working familiarity with them, including where, when, and how to apply them, is a worthwhile goal. If battles for better production management are won with the weapons of planning, analysis, and control, then this book is an arsenal.

CONTENT

The subjects and techniques included in this book provide a substantial introduction to production concepts. The diverse topics are integrated into a conceptual framework that should make the material easier to comprehend and the interrelationships more visible. A balance is sought between an exclusively descriptive coverage and a strictly analytical approach. Although such a balance may dismay purists at either extreme, it allows quantitative methods to be presented in the more qualitative language appreciated by most practitioners—and most aspiring practitioners.

Depth of presentation is essentially limited to an elementary level. Mathematical rigor is sacrificed for conceptual clarity; reasons for technical procedures are substituted for punctilious proofs. A university sophomore level of mathematical maturity is sufficient for most of the material. However, if math preparation is a bit shaky from disuse or underexposure, some sections in the text can be skipped. These optional sections are marked at the beginning and end.

FORMAT

The text can be used in class or for independent study. Alternatively, it is appropriate for a review either after a course is completed or on the job. The format is designed to provide a livelier, student-oriented text. Informal prose is intended to appeal to readers without insulting academicians. In addition, to facilitate comprehension and retention, the following features are included as learning aids:

Learning objectives and key terms: Principal concepts and important terms. They are placed at the beginning of each chapter to guide the reader in his or her study of the material and to help ensure that key points will not be overlooked. Each learning objective can be checked off once one is familiar with the concept. The key terms will enable the reader mentally to note significant terms.

Margin notes: Anecdotes, amplifications, and annotations. Short quips in the margin add explanatory extensions or commentaries without interrupting the main flow of the text.

Examples: Illustrative experiences that supplement and complement text descriptions. These minicases show the details of technique applications and cast the procedures in a practical, modern perspective.

Close-ups and updates: High-tech developments and recent changes in managerial and economic thinking. This section addresses one or two current interest topics in depth.

Summaries: Brief reviews of the highlights of each chapter. After struggling through new material, it is handy to have a digested version for a security check; esoteric vocabulary and unfamiliar concepts are warnings for a more conscientious study effort.

Self-test review: True/false questions. Each question relates to a key word in the chapter. After taking the self-test, the reader can test his or her knowledge by checking the answers in Appendix A.

Discussion questions: Queries that emphasize thoughtful rather than rote answers. They go beyond the material presented yet relate directly to the principles covered in the chapter. The questions can be used for student review, class discussion, or homework assignments.

Problems and cases: Challenging problems and scenarios that require numerical calculations, analytic thought, and educated opinions. They develop confidence in applying analytical methods, serve as homework assignments, and test the students' ability to solve typical exam problems.

ORGANIZATION

The text is organized into three major sections, as suggested by the title: planning, analysis, and control. Each one develops the atmosphere for ensuing topics.

The first seven chapters, the planning section, start with the design and development of products and services, so as to set the perspective for subsequent presentations of quantitative methods for forecasting, allocating, scheduling, and financially evaluating operations.

The next six chapters encompass the analysis section, which emphasizes qualitative aspects and the interlocking factors affecting human performance and the utilization of facilities, machines, and materials; system objectives are stressed over component optimization.

The control section is covered in the final three chapters and reinforces systems thinking by evaluating the quantity and quality of output and the productivity of processes. The last chapter combines concepts drawn from the entire book to examine pressing problems of productivity and to present productivity-improvement techniques.

NEW FEATURES

Although the style and organization of previous editions have been retained, the content has been updated. The biggest change is in the text's usefulness to students through its learning objectives, key words, self-test reviews, answers to selected problems, and comparisons of new and conventional approaches.

New topics and refinements for existing subjects suggested by past readers, as well as current issues and developments in production, have resulted in the following new features:

- New sections on material requirements planning II (MRP II) and just-in-time (JIT) have been added.

- An enlarged chapter addresses the latest advances in productivity-improvement techniques and practices.

- More attention is given to service industries so as to balance the emphasis on manufacturing.

- Learning objectives are listed in the beginning of each chapter and appear again as marginal notes in the summary, to reinforce key points.

- A list of key words now appears at the beginning of the chapter to facilitate retention by highlighting important terms.

- A Close-Ups and Updates section ends each chapter by presenting new tools and practices, that have recently attracted, are attracting, or will soon attract attention. Included are such current subjects as computer-aided engineering, trend spotting, decision-support systems, time and project management, flexible manufacturing systems, robots, automatic identification systems, optimal production technology, and interfirm productivity comparisons.

- Each chapter closes with a self-test review. There are 292 true/false questions that pertain to the key words in each chapter, with solutions in Appendix A.

- Answers to selected problems also appear in Appendix A.

PERSONALS

Every textbook is a synthesis of the author's associations and experiences. I have benefited from comments from many users of previous editions, ranging from critiques about the book's being too exhaustive to those about its being too confined, and from satisfied readers who have applauded both its breadth and depth. To all of these correspondents, students, and reviewers, I extend my appreciation for their guidance. To list everyone to whom I am indebted for suggestions would exhaust your patience and intimidate my memory. Therefore, I simply dedicate to these donors whichever improvements are appropriate, and meanwhile, I accept the customary responsibility for shortcomings.

To new readers I extend best wishes for pleasant and profitable experiences with production systems.

Corvallis, Oregon
May 1986

James L. Riggs

CONTENTS

PRODUCTION OPERATIONS AND OPPORTUNITIES

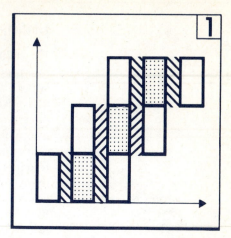

LEARNING OBJECTIVES

After studying this chapter you should

- be aware of the challenges and opportunities in the management of production systems.
- realize that service, government, and manufacturing organizations are considered to be production systems.
- be familiar with the history of production practices.
- appreciate the diversity of management theories and what constitutes the latest thinking.
- understand the organization of this text and the use of models in studying production systems.
- be conscious of the powerful impact of advancing technology on production operations, particularly the effect of computers on manufacturing.

KEY TERMS

The following words characterize subjects presented in this chapter:

scientific management

operations research

automation

management process

human relations

management science

Japanese-style management

production systems

mathematical model

satisfice

managerial excellence

computer-integrated manufacturing (CIM)

computer-aided design (CAD)

flexible manufacturing system (FMS)

electronic mail

1

1-1 IMPORTANCE OF PRODUCTION

The subject is production. It is a broad, fascinating, and timely subject. A narrow interpretation might limit it to the mass generation of commercial products in sprawling factories. Although this aspect is certainly important and dramatic, it represents only one piece of the picture. Products vary from the hardware of merchandise and machines to the nebulous properties of entertainment and information. They are produced by individuals, teams, tribes, bureaus, and corporations in offices, sheds, laboratories, and factories. Despite apparent differences in raw materials, generating processes, and ultimate output, there are many similarities. In today's society, in which so many resources are becoming scarce and environmental considerations are sensitive issues, a common concern for greater productivity is shared by industry, service organizations, and government agencies. All three production areas, though different in design and mission, use essentially the same management tools and benefit from production studies through which the resources of nature are conserved and made more useful.

1-2 SHORT HISTORY OF PRODUCTION STUDIES

No one can say when people first studied production. If we rely on written proof, the date must be set well along in recorded history, but surely some early "managers" pondered better ways to produce crude wheels, utensils, and building blocks. Maybe the Egyptians even had their own version of PERT—Pyramid ERection Technique.

Awaiting documentary evidence, we must pass by the construction marvels of the Roman Empire, the artistic masterpieces of the Dark Ages, and the craftsmanship of the medieval guilds. During the latter period, production was characterized by individual activities and muscle power instead of mechanical power.

For modern PERT, see Section 7-5.

In the 1700s, conditions changed rapidly with the introduction of steam power to replace muscle power, machine tools to reduce hand craftsmanship, and factory systems to emphasize the interchangeability of manufactured parts. These conditions heralded the Industrial Revolution and initiated many modern management headaches. Hereditary writings about how to cure the headaches also began to appear.

At the beginning of the nineteenth century, typical factory conditions were grim by today's standards. Children 5 to 12 years old put in 12- to 13-hour days six times a week. The workplace was dismal and unsafe. Management attitudes equated the sensibilities of people with that of machines and implemented cost-reduction policies by brute force. Although there were exceptions, published production guides were output oriented, concentrating on gross physical improvements, usually to the detriment of the workers' dignity. Despite the lack of social concern, production concepts inaugurated in the period included departmentalized plant layouts, division of labor for training and work study,

"[About managers] . . . By pursuing his own interests he frequently promotes that of society more effectually than when he really intends to promote it." Adam Smith, economist, 1776.

One widely admired essay was "On the Economy of Machinery and Manufactures" by Charles Babbage, mathematician and frustrated computer designer, 1832.

more orderly material flow, improved cost-recording procedures, and incentive wage plans.

Events at the start of the twentieth century shored up the foundations of production studies to make the subject more compatible with the mechanistic attitudes of the physical sciences. Significant experiments by Frederick W. Taylor characterized the new "scientific" approach. He conducted and analyzed thousands of tests to identify the relevant variables of production. From these empirical observations, he designed work methods in which person and machine were one—an operating unit inspired by an incentive wage to service a machine efficiently according to exact instructions. He segregated the planning of activities from their implementation and placed it in the province of professional management.

Taylor's work was in tune with the vaunted reputation of contemporary scientific investigations, and therefore he lodged his concepts under the title of **scientific management.** His theories received both acclaim and abuse. Critics forecast that his mechanistic views enforced by efficiency experts would completely dehumanize industry, but others saw them as logic applied to a promising new area. Whether or not people agreed with him, his beliefs and the fervor with which he expounded them stimulated industrial management.

An associate of Taylor extended his analytical methods to series of operations. Henry L. Gantt developed methods of sequencing production activities that are still in use today. His less restrictive treatment of operator–machine operations added organizational and motivational overtones to Taylor's pioneering work.

Operations-oriented thinking took new substance from the literal as well as the figurative marriage of engineering and psychology in the husband–wife team of Frank and Lillian Gilbreth: the mechanistic attitudes of engineer Frank were mitigated by the humanistic attitudes of psychologist Lillian. Together they showed that basic human motion patterns are common to many different work situations. Their analysis of micromotions to improve manual operations initiated time-and-motion studies and the use of motion pictures in work design.

The works of Taylor, Gantt, and the Gilbreths laid the foundations for the discipline of industrial engineering. Although the original developments are scarcely recognizable in the modern practices of industrial engineers, they anchored the discipline to production systems where it still serves a vital function.

In the 1920s and 1930s, things became more complicated as it was realized that people did not always behave as intuitively expected and that the complexities of emerging production processes required more controls. As demonstrated by the famous Hawthorne studies, the carrot of better wages or working conditions did not always lead to proportional increases in output; psychological factors such as morale and attention were also influential. Walter Shewhart's work provided statistical control measures to ensure the precision of interchangeable parts required for the mass production techniques initiated by Henry Ford. Perhaps even more important, when applying Shewhart's statistical controls, it became apparent that all the interacting factors of product design, plant

"'A fair day's wages for a fair day's work': it is as just a demand as governed men ever made of governing. It is the everlasting right of man." Thomas Carlyle (1795–1881) from *Past and Present*, Book 1.

The Scientific Management movement was rebuffed by a U.S. congressional investigation in 1911 which prohibited the use of stopwatches in government services.

The Gilbreths included worker fatigue, monotony, emotional reactions, and other human factors within the scope of motion studies.

The Hawthorne studies, sponsored by Harvard University, began in 1924 at the Hawthorne Works of the Western Electric Co. In one amazing case, productivity always improved regardless of changes in illumination levels. This finding led to the belief that the improvement was due entirely to a positive worker response to the attention they received in the studies.

layout, worker capacity, environmental conditions, materials, and customers' attitudes had to be considered. Such considerations naturally led to the study of entire production systems rather than isolated parts.

An interdisciplinary approach to system studies appeared in the war years of the 1940s, first in the form of British operational research teams. Members of the teams were not necessarily experts in the areas studied, because they applied accepted scientific methodologies for problems never before subjected to such analyses. That the results were favorable should not be surprising because analogies are found throughout nature and the works of human beings. Knowledge borrowed from the physical sciences and applied to management problems of similar structure offers a reservoir of decision-making techniques that is still being tapped today. From its military origin, the **operations research** (or the closely related **management science**) approach has become a fountainhead of industrial applications.

The 1940s also mark the beginning of convergent developments in automation and computerization. Although the term *automatic* has been part of the production vocabulary for many years, **automation** was coined in the forties to represent the addition of handling and control equipment to automatic machines for continuous production through a series of operations without human guidance and control. At the beginning of the automation movement, labor organizations deplored the dehumanization of the workplace and warned of future unemployment as machines replaced people. Neither of the dire predictions fully materialized, although there still are fears. Instead, automation has often made work safer and relieved workers of many tedious tasks, while having relatively little impact on employment.

"Hard" automation is typified by expensive, fixed-purpose machinery used for the long-run, high-volume production of identical items. The automotive industry is an example. "Soft" automation is associated with the computer and its peripheral apparatus. Computers are now used extensively in process industries—oil refineries and chemical plants—where they virtually run entire complexes; newer *adaptive control* computers make real-time adjustments to production machines in response to ongoing conditions, such as modifying settings to compensate for tool wear. The epitome of soft automation is the robot. Several machines with humanlike versatility are already in use. Although they bear little resemblance to humans, they do have appendages that vaguely look like arms and hands, and experimental robots demonstrate remarkable progress in recognizing their environment and reacting to changes in it. With these appendages and programmed responses, they can perform simple but arduous and/or perilous tasks.

It is almost impossible to exaggerate the effect of advances in computer technology on production systems. Across the spectrum from blue- to white-collar workers, from the bottom to the top of organizations, and in firms large or small, computers of one type or another are omnipresent. In general, the impact of computerization has been gentle, although some workers cringe at the paperwork deluge attributed to electronic data processing, and some middle

In nineteenth-century France, workers kicked newly introduced machines to pieces with their wooden shoes, or sabots, giving rise to the word sabotage.

Estimates in 1979 place the value of installed computers at $50 billion in the United States, $13 billion of which is in manufacturing industries. There are about 3000 robots in the United States, 1000 in Western Europe, and 30,000 in Japan.

managers resent a perceived loss of decision-making authority to computerized controls. Others prophesy a coming era in which computer-assisted processes will combine with servant robots to give everyone the proverbial "free lunch."

Some emotional fog about the computer issue is removed by concentrating on what has been done and what has to be done. Many mathematical techniques that we take for granted would not be feasible without the tremendous calculating speed of computers. Of course, the problems must be "programmable"—structurally adaptable to machine calculations. Therein lies the contribution of humans to the modern operator–machine partnership. Human beings must collect the necessary data (assisted by record-keeping machines), recognize the type of problem and its potential solution format, develop or select an appropriate program, and interpret or modify the machine's output. Equivalently, computers must be used if we expect to relate and evaluate the many variables in complex production systems. Both the decision makers and the helping machines must continue to develop just to keep pace with the problems and challenges they have already created.

"Man is one of the best general-purpose computers available and if one designs for man as a moron, one ends up with a system that requires a genius to maintain it. Thus we are not suggesting that we take man out of the system, but we are suggesting that he be properly employed in terms of both his abilities and limitations. Some designers have required that he be a hero as well as a genius," E. L. Thomas, *Design and Planning* (New York: Hastings House, 1967).

EXAMPLE 1.1 Development and relationship of management theories

Management theories and production practices have experienced roughly parallel development. This is not surprising, because the two are closely related. It has been suggested that production deals with *things*, whereas management is concerned with *people*, but neither can effectively yield the desired output or effects without the other. Thus it is enlightening to examine different aspects of management thinking in relation to production applications.

Long ago management positions largely resulted from an accident of birth, intrigue, or muscle. Persons occupying exalted positions were automatically accredited with the gift of astute management. It was not until near the turn of this century that management theory began to develop as a distinct discipline. There is still no agreement on a central all-purpose theorem to govern the subject, and there may never be one because the subject has so many facets.

An overview of management trends and thinking results from tracing the developments of management movements. Although the divisions are somewhat arbitrary due to infringing concepts, six schools of management thinking are delineated in the following paragraphs.

SCIENTIFIC MANAGEMENT. The theories and practices developed in this country from the 1880s to the 1920s were christened the "scientific management movement" by Louis Brandeis. The central theme was the systematic analysis and measurement of tasks to find the "one best way." The focus was on the shop level, at which the specialization of labor was coupled with economic incentives to increase productivity. With concentration devoted to the measured efficiency of individual jobs, little progress was made in broader organizational problems.

MANAGEMENT PROCESS. Early management developments in Europe centered on the problems of higher management. The writings of Henri Fayol represent the **management process** school" approach. They depicted management as a collection of processes such as planning, organizing, coordinating, directing, and motivating. With the processes thus identified, it was logically argued that professional managers could be trained to apply these processes to any organization.

When the management process school crossed the Atlantic to this country, increased attention was given to coordination. Unity of command as prototyped by the military was forwarded as a means to align efforts in one direction. Vertical authority with associated responsibilities and staff–line concepts were operational additions. The more general "processes" developed earlier were enlarged to assume the stature of management "principles." The process school is still expanding to provide guidelines for modern management functions.

BEHAVIORAL APPROACH. Attention to the effects of **human relations** on productivity was advanced by work in the behavioral sciences. The Hawthorne experiments of the 1920s stimulated questions about the authoritarian basis of the scientific management movement. The studies suggested that interpersonal relationships had as much or more effect on productivity than did work or workplace designs. Attempts to make work a more satisfying experience for employees led to several motivation theories.

Douglas McGregor depicted old-style discipline by theory X, a view that assumes that most people do not want to work, that some kind of club is needed to make them work, and that they would really rather be told what to do than think it out for themselves. The opposite view, theory Y, holds that people do not inherently dislike work, that authoritative methods are not the only way to get work done, that people do not shun responsibility, and that they will work to achieve their selected goals. A similar outlook applied to supervision divides "production-centered" and "employee-centered" management. Robert Blake suggested a "managerial grid" to rate managers on a scale of 1 to 9 for each concern: a 9, 1 supervisor (high concern for production, 9, and low concern for people, 1) follows theory X.

DECISION THEORY. A relatively recent contribution to the management movement is directed toward the power centers and communications that determine managerial action. A firm has a decision-making anatomy that may or may not conform to organizational titles and printed duty assignments. Decision theory treats both the organizational network of decison processes and the general concepts of decision-making applicable to any area of management. The intent is to improve the ability of managers to evaluate a problem thoroughly and to develop the best possible solution. Thus, effort is channeled toward the intellectual processes of individuals rather than the functional processes of systems.

QUANTITATIVE ANALYSIS. Interest in quantifying management problems has escalated rapidly in the last two decades. Leading the assault have been practitioners from engineering, mathematics, economics, and computer science. The two most familiar groups representing the mathematical approach are operations research and management science.

Quantitative analysis investigates the relationships in an organization that can be expressed symbolically and treated mathematically. The relationships can exist within one function of the organization or may connect several functions. In this respect and others, the quantitative approach is similar to the decison-making approach: both are applicable to any organizational structure. Although both have contributed significantly to problem solving, neither has as yet offered a unifying management theory.

JAPANESE-STYLE MANAGEMENT. Spectacular productivity gains recorded by Japanese industries began to attract attention in the late 1970s. By the 1980s, the techniques used by Japanese managers (so-called **Japanese-style management**) to achieve those gains became the most discussed management theory. Although many of the practices are an outgrowth of the Japanese culture and are dependent on that culture for success, several concepts are readily transferable to other settings. Among the most notable contributions to Japan's remarkable production are the following:

Lifetime employment—Employees of larger Japanese businesses are rarely laid off, even during periods of economic adversity. To maintain steady employment, there is much subcontracting during peak periods, but little when business is slow. Also, a large share of the employees' pay is in the form of bonuses, which can be deferred during downturns.

Slow promotions—Managers are rotated through an extensive series of positions so as to become familiar with many operations and employees.

Consensus decision making—Opinions are sought from everyone who will be affected by a decision. After expressing their views and hearing the views of others, everyone is expected to endorse unanimously a compatible course of action. The process can be lengthy, but it ensures complete support for a decision.

Attention to quality—Employees realize that each is responsible for the quality of his or her operations, and so together they seek ways to improve the quality of their output. Participation in quality circles (units that meet voluntarily to promote productivity) and adherence to statistical quality control principles have helped.

Long-range planning—Industrial managers concentrate on future prosperity rather than immediate profits. Government leaders cooperate by providing resources to develop new products and markets.

Supporting the visible practices of Japanese-style management is the willingness of a highly educated work force to accept and hasten technological advances. This combination has recorded such striking economic accomplishments that it has attracted a legion of believers.

It appears doubtful that a consensus or universal management theory will appear in the foreseeable future. Each movement has its champions and adherents, and each has something useful to offer in particular applications. An appreciation of all the theories with perhaps special competence in one or two areas appears to be the best preparation available for the effective management of production systems.

1-3 PRODUCTION AND PRODUCTION SYSTEMS

In our kaleidoscopic journey through history, we touched many facets of production without defining the subject. And if there is one obvious lesson to be learned from history, it is the difficulty of and the need for lucid communications. Maybe a rose, by any other name, does smell as sweet, as Shakespeare pledged, but the smell would be a lot easier to describe if everyone called the same plant a rose.

For our studies we shall say that *production is the intentional act of producing something useful.* This definition is at once liberal and restrictive. It in no way limits the method by which something is produced, but it does eliminate the accidental generation of products. The questionable property of *usefulness* is subject to individual opinions. Some might say anything saleable is useful; others would rebut that illicit drugs are certainly marketable but of uncertain worth. Even if it is agreed that "useful" implies a beneficial purpose, there is still room for debate on commodities such as armaments. Sidestepping the issue of conscience, we should recognize that a wide range of production processes have similar characteristics, regardless of the utility of the products.

Production accounts for the generation of both goods and services.

The definition of production is modified to include the system concept by stating that *a production system is the design process by which elements are transformed into useful products.* A process is an organized procedure for accomplishing the conversion of inputs into outputs, as shown in Figure 1.1.

A unit of output normally requires several types of inputs. In an industrial

The workings of a production system can be collectively called operations management, a term that accommodates production systems in which the primary function is service.

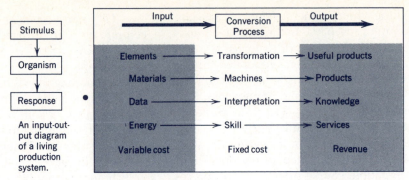

FIGURE 1.1 Block diagram and examples of a production system.

process, the inputs account for most of the variable cost of production. Conversion facilities are associated with fixed cost, and the output produces the revenue. Elementary accounting declares that profit depends on the relationship of variable and fixed costs to revenue—the interaction of input and conversion costs to output revenue.

Profit is a less apparent consideration in many service and government organizations, but these organizations still depend on balanced budgets for continued operation. As services, such as health care, have grown and government activities have proliferated in recent years, more attention is focused on the ratio of operating costs to the benefits for consumers. Efforts to reduce input and conversion costs, while maintaining or increasing output values, are using methods formerly associated with industrial production. It is now recognized that the same principles of systems analysis and work design can be successfully applied to industrial, service, and government systems.

Any system is a collection of interacting components. Each component could be a system unto itself in a descending order of simplicity. Systems are distinguished by their objectives: the objective of one system could be to produce a component that is to be assembled with other components to achieve the objective of a larger system. More sophisticated techniques are required to deal with more complex systems. It is a nip-and-tuck race between the development of ever more intricate systems and the development of capable management tools to control them. Perhaps our future will be determined by the winner.

Most organizations have an office system that supports staff systems that, in turn, support the system that produces the end product. Each is a component of the larger system, alike in mission but different in structure.

1-4 MODELS OF PRODUCTION SYSTEMS

Early efforts in production studies now appear crude, but so were the systems being studied. As the systems became more intricate, investigators naturally followed the proven path revealed by elaborate studies in the physical sciences— observe, hypothesize, experiment, and verify. This general approach has prospered from Taylor's introduction of scientific management to the currently

Even the initiated find it difficult to distinguish the disciplinary boundaries staked out by management scientists, operations researchers, industrial engineers, systems analysts, and others.

popular management science. It is best characterized by the construction, manipulation, and interpretation of *models*.

Types of models

A model is a replica or abstraction of the essential characteristics of a process. It shows the relationships between both cause and effect and objectives and constraints. Problems that defy direct solutions because of size, complexity, or structure can often be assessed through model simulations. The nature of the problem signals which of the following types of models is most appropriate.

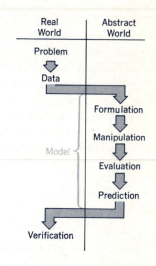

PHYSICAL MODEL

Look-alike models derive their usefulness from a change in scale. Microscopic patterns can be magnified for investigation, and huge structures can be scaled down to a manageable size. (A model of the solar system might even be confused with a model of an atom if it lacked a label.) Flow problems in a model plant are studied by easy shifts of scaled-down structures and machines, which cannot be duplicated with real items because of cost, confusion, or inconvenience. Some details are necessarily lost in models. In a physical replica, this loss may be an advantage if one factor, such as distance, is the key consideration, but it may make a study futile if the predominant influence is forfeited in the model's construction.

A three-dimensional plant layout could show the floor space assignments without recognizing possible vertical restrictions.

SCHEMATIC MODEL

Two-dimensional models are the delight of chartists. Graphics of price fluctuations, symbolic charts of activities, maps of routings, and networks of timed events all represent the real world in a digested and diagrammatic format. The pictorial aspects are useful for demonstration purposes. Some frequently encountered examples include organizational charts, flow process charts, and bar charts. Symbols on such charts can be easily rearranged to investigate the effect of reorganization. Similar experimentation with the actual work place would be crippling.

MATHEMATICAL MODEL

Quantitative expressions, the most abstract models, are generally the most useful. Formulas and equations have long been the servants of the physical sciences. In recent years they have been similarly recognized by the management sciences. When a **mathematical model** can be constructed to represent a problem situation accurately, it provides a powerful study tool; it is easy to manipulate and clearly shows the effect of interacting variables; and it is precise. Whatever faults arise from the use of mathematical models usually can be traced to the underlying assumptions and premises on which they are based. In contrast with the other types of models, what to use is harder to decide than how to use it.

Each chapter of this text presents a different schematic or mathematical model, and often both.

EXAMPLE 1.2 Construction and evaluation of a pricing model

Returns from a new product have failed to meet expectations. A study is initiated with the overall objective of increasing profit. The production system that determines whether the objective is met encompasses all aspects of the product—procurement of materials, manufacturing, marketing, and so on. A preliminary investigation narrows the area where action can be taken to the pricing policy. A model of the relationship between the number sold (N) and the price per unit (P) is expected to follow the equation

$$P = x - yN$$

where x and y are constants determined from market conditions.

If we assume that the general relationships in the equation are correct, we can gain some insights even without cost figures. First we observe that as the price decreases, the number of units sold increases ($N = (x - P)/y$). Letting $N = 0$, we have the limiting price (x) at which the units would sell if a shortage existed. Setting $P = 0$, we get the ratio x/y that identifies the total market potential or, more precisely, the amount that could be given away if the product were free.

Because our objective is to maximize profit, we must include the costs of producing the product. Under the belief that total cost (TC) is a function of fixed cost (F) and variable unit cost (v), we accept the formula

$$TC = F + vN$$

By reasoning similar to that applied to the price (and confirmed by common sense), we know that the total cost equals the fixed cost when no units (N) are produced (hence the name *fixed cost*). Now, combining all the factors with respect to the output, we have

Profit = Revenue − Total cost

$$Z = NP - TC$$

(where Z symbolizes profit)

$$Z = N(x - yN) - (F + vN)$$

(by substitution from previous equations)

$$Z = xN - yN^2 - F - vN$$

(by combining terms)

Various values of x and y can be substituted in this equation to provide a table of Z values from which the best pricing policy can be selected. A more direct approach, using differentiation, expresses the maximum value of Z. Using calculus, we get

$$\frac{dZ}{dN} = 0 = x - 2yN - v$$

(where the differential is set equal to zero to identify a maximum or minimum point)

from which

$$N = \frac{x - v}{2y}$$

Substituting the equation for the output that maximizes profit into the basic price formula, we get

$$P = x - y(N) = x - y\left(\frac{x - v}{2y}\right) = \frac{x + v}{2}$$

which is the price that provides maximum profit. Therefore, if the assumed relationships are indeed representative of the real world, the price of the product would be half the sum of the limiting amount (x) a customer would pay for the product and the variable cost (v) of producing it. The model reveals the importance of securing an accurate estimate of x and that less attention need be given to y and F. Although pricing appears to be independent of the influence of fixed cost, the actual amount of profit for a given pricing policy fluctuates with changes in F. It should also be recognized that the optimal pricing equation offers several courses of action for abiding by the preferred relationships. For instance, improved packaging would increase v and might produce a disproportionate increase in x. Thus a model is a tool to aid, not dictate, management thinking.

Use of Models

Example 1.2 illustrates that an evaluation of a problem does not depend on the availability of actual cost data. In fact, the arithmetic of handling data may obscure the significance of the underlying assumptions. Symbols are impersonal and promote the consideration of different views rather than focusing on one "answer."

The question of the amount of detail to include in a model is implied in the example. The answer lies in a balance between accuracy and simplicity. To increase accuracy, it is usually necessary to add variables and increase the complexity of relationships. Enhancing reality adds cost. An investigation can cost more than it is worth. Models that are easier to solve are also usually easier to understand and apply. However, if the model is simplified to the point that it no longer represents the real world, it will indicate erroneous or misleading outcomes.

To **satisfice** is to strive for one level of achievement while being willing to settle for another, slightly less ambitious level. Model formulation is often subject to satisficing—a sacrifice of reality for the sake of workability. The approximations may take one or more of the following forms:

1. Linear relationships are substituted in a model for actual nonlinear relationships in the system. For instance, a curve can be approximated by a number of straight-line segments that conform to the general curving pattern.

2. Variables that do not have a significant effect on the system's performance are omitted. Inventory policies are often based on the control of items that comprise the bulk of demand, under the assumption that the remaining items do not warrant the cost of attention. Such compromises between the cost of evaluation and the cost of inattention should be rigorously justified.

3. Several variables are lumped together and subsequently treated as a single variable. If the variables have essentially the same characteristics, the aggregate variable will provide a good approximation. Again using inventory policy as an illustration, items that have similar demand, storage, use, and handling properties are treated identically as one variable with little loss of reality and considerably less effort.

Many schematic and mathematical models are presented in the ensuing chapters. Most of them have "satisficing" aspects. There may be a temptation

A model offers clues to execution, but obtainment of the full effect from indicated actions depends on judgment backed by a careful appraisal of the influencing conditions in the real-world production setting.

in practice, often because of laziness or diffidence, to pick out a ready-made model and plug custom-made data into it. The outcome could be outrageous. It also could be wholly satisfactory. The difference lies in whether the inherent assumptions of the model agree with the properties of the system to which it is applied.

1-5 PLANNING, ANALYSIS, AND CONTROL OF PRODUCTION SYSTEMS

The end product of planning, analysis, and control efforts is a decision. The techniques associated with each phase of an evaluation are useful only if they contribute to that end. Mathematical tools provide a degree of confidence that is lacking with intuitive judgments, but hunch aspects will always be a part of decision making. Intuition informally embraces past experiences and current events to provide a "feeling" for a particular action. Any gambler knows that hunch players sometimes win, yet even the player cannot say why or how. Perhaps hunch explanations are just beyond the ken of our current knowledge; in the meantime it is comforting to explain the why's and how's in quantitative terms whenever possible. Therefore, though recognizing the value and need of intuitive judgment, we shall emphasize the quantitative analytical methods leading to decisions.

Planning, analysis, and control are more descriptive of the mental set of a decision maker than of a rigid problem-solving procedure. Each phase is distinguished by an objective—to anticipate, to investigate, to regulate, to design. The definition of the objective points to the most suitable quantitative technique and acts as a guide to information collection. An evaluation of an existing system might have the objective of reducing costs and would likely begin with an analysis of current operating conditions and procedures. The results of the analysis phase could lead to planned improvements for which the collected data would fuel planning and control efforts. Later the entire system could be subjected again to all three phases, starting with planning based on a new technological development. Objectives to update and to improve a system continually direct recurring studies in the pattern shown in Figure 1.2.

The organization of this book follows the planning–analysis–control sequence. The content focuses primarily on mathematical models as the source and basis of production decisions, although both time-honored and recently developed schematic models also receive attention.

A familiarity with the many types of models available to the modern decision maker necessarily includes contact with a wide range of mathematical techniques—statistics, probability, algebra, calculus, linear programming, arithmetic, and so forth. Ideally, every systems analyst would have degrees in mathematics, engineering, business, and economics; be a computer expert; have 10 years experience; and be under 30 years of age. But because very few such specimens exist, a satisfying solution is to trade depth for breadth of knowledge

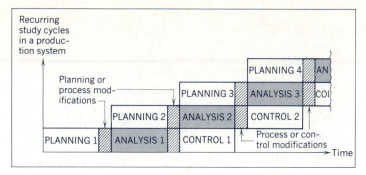

FIGURE 1.2 Cycles of production planning, analysis, and control.

in the belief that familiarity with a wide but selective range of topics will allow the investigator to know when a more penetrating study is required and where to seek the means of conducting it. This strategy is expedited by associating evaluation methods with problem areas and with the phase of study where the methods are most likely to be productive.

As intimated in Figure 1.2, the planning–analysis–control cycle can be entered at any of the three stages; the point of entry is determined by the objective of the study. Consequently, the analysis section could be encountered before the planning section. It is more important to relate the quantitative tools and the qualitative concepts *within* the sections than to be concerned about the relationships *among* the sections.

A valid reason for studying chapters in the given order is to follow the purposeful development of concepts from their introduction in the early chapters to elaboration in later pages. The validity of this approach is intuitive when you recall how difficult it was to comprehend algebra when you could barely add and subtract or how meaningless and inane an exhibition game seems unless you know the rules and have experienced the difficulties of play. In studying production, as in learning algebra and in appreciating a game, it is unrealistic to attempt to comprehend the whole before becoming familiar with the parts. Therefore, coordination and integration are stressed in the later stages of the text, *after* the integral parts have been isolated and analyzed. An equivalent approach is applicable to real-world system studies: understanding the components precedes a system study of their integrated effects.

If the planning, analysis, and control of production systems appear demanding, you have the proper impression. But they can also be exciting and rewarding. Decision making always has some traumatic aspects; they are what make the action interesting. It is the confidence created by having the right evaluation tools and knowing how to use them that transforms a potentially trying experience into one that is comfortable and satisfying. I hope that the material you are about to encounter will inspire this confidence. It is your decision. As Francis Bacon observed: "Some books are to be tasted, others to be swallowed, and some few to be chewed and digested."

The hazard of broad but shallow coverage is the creation of a self-appointed expert who knows just enough to be dangerous.

Carelessness or premeditation could cause the mirage below.

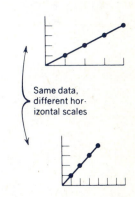

Same data, different horizontal scales

Models are only as good as the information and care put into them. Clever manipulations cannot compensate for poor data or incorrect assumptions.

1-6 CLOSE-UPS AND UPDATES

The production setting for the last half of the 1980s was created in the first half. The props are a worldwide recession and recovery, intense economic competition, and unparalleled technological advances. The international cast is composed of government planners, leery workers, uncomfortable managers, and many special-interest groups. The chorus proclaims the need for productivity. Leading roles are played by operations managers, industrial engineers, and production systems analysts. The script is still being written.

Concern for managerial excellence

As a consequence of the economic recession and miserable productivity performance by U.S. industry in the first third of the 1980s, conventional management tenets and practices were questioned. Hundreds of managers journeyed to Japan as productivity pilgrims. Production-management literature was filled with explanations of what had gone wrong and what to do about it. Suggestions for **managerial excellence** ranged from promoting more people with production experience to top leadership positions to installing more robots. Several of the most quoted recommendations follow:

Commitment to customers. Ths most successful companies are those that serve their customers courteously, efficiently, and completely. The quality of customer service should be an obsession.

Emphasis on innovation. Practical risk taking is encouraged for all employees. When they realize that they are expected to generate new and better ways of doing things and that they will not suffer if a new idea backfires, creativity flourishes. With creativity comes improvement.

Lean staffs. Most organizations are burdened with too many layers of management. Eliminating layers not only cuts overhead expense, but it also improves communications. Leaner staffs can respond faster to change. Advanced information and communication systems have reduced the need for paper-shuffling managers, thereby allowing the organization to function more efficiently with fewer moving parts.

Hands-on management. Better decisions are made by managers who are in close contact with the people they manage. Being in contact implies that the managers are listening intently to what is said and that employees feel free to say whatever they believe. Such contacts encourage a feeling of ownership in the company by letting people know what the company is doing and that they can influence the nature of their work.

Annual Rate of Change in U.S. Industrial Productivity

1979— −1.5%
1980— −0.7%
1981— +1.9%
1982— −0.1%
1983— +5.1%

Entrepreneurial spirit inside an enterprise is nicknamed intrepreneurship.

The value of direct observation is endorsed by advice to MWA, or to manage by walking around.

Sixty-second sayings. Minutes spent frugally to set goals, praise, and reprimand correctly can save management time and increase worker productivity. As recommended in *The One Minute Manager*, brief encounters of the right kind can create conditions that foster employees' full productive potential. Conditions conducive to superior performance are fashioned by communications that inform while making people feel good about themselves. Three methods of doing this are to

Blanchard, K., and S. Johnson, William Morrow, New York, 1982.

1. Guide and monitor performance by having subordinates set their performance goals and record the details of the most important ones on a single sheet of paper. Then everyone involved knows what is expected.

2. Build confidence by praising people as soon as possible after they have earned the acclaim. Catch them doing something right, and encourage them to do more of the same.

3. Deliver necessary reprimands without attacking self-esteem. When performance falls below expectation, an immediate reprimand that cites the specific difficulty should conclude with an affirmation of the chastized person's value.

An inability to do something is a training problem, whereas a refusal to do something is an attitude problem.

Most of the sermonizing about managerial competence raising in the 1980s is an elaboration of concepts that have been suggested before. But the sermon is worthwhile and noteworthy because of its intensity and ready reception. Never before have so many books aimed at productive management been best-sellers. They expose weaknesses in current management practices, providing fresh approaches to today's highly competitive business environment regarding work attitudes, physical facilities, and economic conditions. Both the advice itself and the flow of thinking that it represents deserve close attention.

Computers: Agents of change in production systems

Much of the rethinking of management practices can be attributed to the explosive computerization of both the manufacturing and the service industries. The influx of computers into even the most mundane operations has spread their effect pervasively. Neither senior managers who are reluctant to adopt new technology nor the lowest-level employees are immune to computer impact. Small and big businesses are affected. Government and not-for-profit organizations are similarly influenced. Computers must be acknowledged as the engines that have powered this decade's rapid changes in production systems.

As computers have become more accessible and powerful, more production problems have been exposed to analysis, and the results have been better.

HISTORY OF TECHNOLOGY

Chapters in the history of the human race can be titled by their associated levels of science and technology. The earliest ages are characterized by mastery of fire, language, crop cultivation, and so on. Recorded technological advances can be traced from machines designed by Archimedes through the advent of alchemy, waterwheels, firearms, compasses, clocks, and other devices that enhanced human mastery over nature.

Elementary production systems became evident in the first industrial revolution in the middle of the eighteenth century. Successive stages saw the introduction of textile production machinery, steam engines to power factories, and machine tools to broaden manufacturing. With these advances in mechanization came the social changes that are reflected in the management theories discussed in Example 1.1.

A new industrial revolution is now under way. Its technology is exemplified by microelectronics, biotechnology, and new material and energy sources. Of these, microelectronics has influenced production systems the most, being evident in the automation of factories, offices, and homes. Production managers are mainly concerned with automated machines and systems that move materials, perform operations, control processes, and handle information. These are the provinces of computers.

EVOLUTION OF COMPUTERIZED ASSISTANCE

Shaping the latest industrial revolution is the evolution of computers. The first generation of computers used vacuum tubes; the second generation, transistors; the third, integrated circuit chips; and the fourth, very large-scale integrated circuits (VLSIs). The fifth generation is not yet fully defined, but it will involve the manipulation of knowledge to make inferences in much the same way that the human brain does. This advance is expected to rely more on software than on machine refinements.

To appreciate how far computers have advanced, it is helpful to review their place in production only a few years ago. The computers of the 1950s were slow and expensive by today's standards. Data processing based on punch cards was used mostly for administrative and financial purposes. Key punches, collators, sorters, and counting machines manipulated the cards. Crude as it was, this process was a great advance over manual record keeping and found some use in manufacturing for labor reporting, job tracking, and inventory accounting.

Electronic data processing became more sophisticated when all the individual processing steps were incorporated in one machine system. Operating speeds and calculating abilities soared. By the early 1970s the minicomputer emerged, much to the benefit of engineering design, operations monitoring, and process control. As software development caught

---

The purpose of science is to explore the unknown, whereas the purpose of technology is to produce useful systems. The two fields are interdependent and synergistic.

New materials include ceramics and organic polymers, and new energy sources include atomic, solar, and geothermal power.

At the forefront of the fifth generation is artificial intelligence—the creation of computable analogues of human thought processes.

The first computer, ENNIAC, was developed in 1946. It could make calculations at a speed equivalent to 2400 desk calculators operating at the same speed. Within five years, speeds were increased 1000 times.

up with hardware capabilities, computerized automation became an integral part of the process and flow manufacturing industries: petroleum, chemicals, pharmaceuticals, and consumer goods.

In the second half of the 1970s and the first half of the 1980s, computerized production and inventory systems blossomed. When managers realized that they had to computerize in order to stay competitive, they supported the creation of hundreds of different computer applications, ranging from word processors in offices to robotics in factories. The growing availability of more powerful computers, software, and associated equipment at a modest cost brings the evolution to the threshold of **computer-integrated-manufacturing (CIM)** systems.

CIM is the amalgamation of **computer-aided design (CAD)** with **computer-aided manufacturing (CAM)** and **computer-aided testing (CAT).** This triad of computer-inspired advances was cradled in the United States and is now maturing worldwide. In the metal-working sector, a significant offshoot, **flexible machinery systems (FMS),** allows product modifications, changeovers between products, and process improvements to be made quickly and economically.

Computer-aided designing takes place on a screen. CAD substitutes computer graphics for laborious manual drafting to put pictures on a terminal that enables designers to study various views of an assemblage before it is actually built. Different designs can be tested on screen, showing how they react to changes in operating conditions such as overloads, sudden shock, and partial damage. Thus designs can be changed now in minutes instead of weeks, as before. The consequent reduction in time from product concept to hardware can go a long way toward averting time overruns that have plagued manufacturers. CAD also enables an instant exchange of design data among contractors, vendors, and subcontractors. As manufacturers more and more often ask vendors to accept parts specifications in the digital languages understood by computers, the relationship between customers and suppliers is being dramatically altered. CAD adoptions in big companies are causing a chain reaction of supporting CAD implementations in smaller companies.

CAM and CAT are less exotic than is CAD. Computer-controlled production machines are quite commonplace. In many factories the machines are served by computer-controlled material handlers, and computer-aided testing and inspections are increasing as new devices become available. Expenditures for CAM and CAT are justified by studies that indicate that a part spends 5 percent of the time being machined and 95 percent moving or waiting in a standard manufacturing process.

The linkage of CAD, CAM, and CAT into a fully integrated system is what excites production planners. Ideal computer-integrated manufacturing would allow anyone in the factory to tap the common pool of pro-

FMS replaces stand-alone tools by connecting integrated machine tool modules with a computerized delivery network; a master control computer coordinates the system production as a whole.

CAD is not inexpensive. Besides the costly equipment, programmers must write millions of lines of coded instructions to create the design images.

Group technology is a subset of CAM that groups materials and parts into families on the basis of required production operations. It enables batch production with many of the advantages of mass production, and the concept is adaptable to inventory management in service organizations.

duction data to deal with scheduling, inventory, purchasing, and other planning problems. Operational CIM reduces material waste, improves quality, shortens development time, and maximizes labor outputs and capital inputs. However, both management and labor must adjust their thinking to accommodate the benefits obtainable from CIM while minimizing the pain of its exploitation. The aches are felt by companies that lack the resources to upgrade their facilities and consequently become uncompetitive, and by people who are forced to relocate or learn new trades. Human dignity and social welfare must be considered along with operating efficiencies.

COMPUTERIZATION HERE, THERE, AND EVERYWHERE

The extravagantly advertised "office of the future" is under construction. It relies on computers as much as a high-technology factory does. The entire service industry is becoming more computer conscious, enough so to earn its own monogram: CIS—computer-integrated services. An example of one high-tech feature is **electronic mail,** a computer-based message-switching system that allows a message to be composed, edited, stored, and sent on a data network instead of through the postal service. A message can be sent to one or a number of addresses, catalogued, and stored for future reference. At the receiving end, the recipient can scan the incoming messages and then delete, print, file, forward them to others with an annotation for action, or answer them from a terminal. Huge savings are also possible from the electronic transmission of forms, which now far outnumber narrative messages in business correspondence.

The banking industry has also been the recipient of new technology. Computers have been processing funds since the mid-1950s, laying the groundwork for the introduction of word processors, electronic mail, and automatic tellers. "Big eye" machines electronically scan 70,000 pieces of paper money an hour, detecting worn and counterfeit bills while counting and bundling the good ones. These technology-driven advances have been procured mostly from ready-to-use devices developed for other applications and implanted in congenial knowledge-worker environments, making the gains affordable and convenient.

At the same time that people revel in the achievements of technology, they distrust the consequences. In general, managers look toward new technology for part of their future productivity growth, and workers have misgivings about the future effects of labor-saving machines. Neither group fully appreciates the ramifications of the rapid movement toward smart machines and computer-aided processes, because no one can accurately forecast the eventual directions the movement will take. Both groups must

Percentage of Workers by Occupational Group with Desktop Personal Computers in 1985

Technical	55.9%
Professional	39.2
Managerial	36.5
Clerical	25.0
Armed forces	20.3
Sales	16.6
Services	11.9
Farm	4.3

work together to shape events to protect employees' occupational livelihoods, preserve or enhance the quality of work life, and improve productivity.

Most of the topics introduced in this section will be explored in greater depth in later chapters of the text. The opportunities available from using advanced technologies are legion. The entire field of production and operations management, from human resources and supervisory considerations to product design and profitability, is changing rapidly to accommodate these advances: Planning methods are changing, analysis techniques are changing, and control tools are changing, all of which makes this an exciting time to become involved with production systems.

1-7 SUMMARY

A production system comprises the processes and activities required to transform elements into useful products and services. It is characterized by the input–conversion–output sequence that is applicable to a wide range of human activities. The basic principles of systems analysis and work design are applied to maximize output per unit of input within industrial, service, and government systems.

Definition of a production system

Early efforts in production were exceedingly crude. As production capabilities increased after harnessing mechanical power, the new relationships of people and machines highlighted the need for improved management techniques. Pioneering efforts to meet the need borrowed their approach from methods developed in the physical sciences. Further studies included human factors and more extensive mathematical applications based on the use of computers. As production systems became more complex, modeling techniques were developed to treat the intricate relationships.

Scope of production concerns and management practices

Models may take the form of physical images, schematic charts or templates, and mathematical representations of related variables. Mathematical models are the most abstract and generally the most useful. Model formulation is often subject to "satisficing"—a sacrifice of reality for the sake of workability. Such approximations are feasible only if the essential characteristics of the system are retained in the model.

Models for planning and analysis

Planning, analysis, and control are phases of a system study. The study may start with any phase, and over a period of time the phases tend to be repeated. The purpose of planning, analysis, and control efforts is to provide the basis for a decision, and the sections of this book follow the same sequence.

Sequence of production systems studies

Low productivity and the business turndown during the early 1980s in the United States focused attention on management and technology. Prevailing management practices are being questioned. Suggestions for improvement range from closer attention to customers' and employees' concerns to adopting practices that have been successful in Japan. Concurrently, spectacular technological advances are challenging production managers, especially in manufacturing, in which computer-aided systems known by such initials as CIM, CAD, CAM, CAT, and FMS are changing the ways in which companies operate. Computers are also having a significant impact on government and service organizations.

1-8 REFERENCES

BUFFA, E. S. *Modern Production/Operations Management*, 7th ed. New York: Wiley, 1983.

GEORGE, C. S. *The History of Economic Thought*. Englewood Cliffs, N.J.: Prentice-Hall, 1968.

MATSUMOTO, K. *Organizing for Higher Productivity: An Analysis of Japanese Systems and Practices*. Tokyo: Asian Productivity Organization, 1982.

McCLAIN, J. O., and L. J. THOMAS. *Operations Management: Production of Goods and Services*. Englewood Cliffs, N.J.: Prentice-Hall, 1980.

MEREDITH, J. R., and T. E. GIBBS. *The Management of Operations*, 2nd ed. New York: Wiley, 1984.

MILLER, D. W., and M. K. STARR. *Executive Decisions and Operations Research*, 2nd ed. Englewood Cliffs, N.J.: Prentice-Hall, 1969.

MORRIS, W. T. *Management Science in Action*. Homewood, Ill.: Irwin, 1963.

NAISBITT, J. *Megatrends: Ten New Directions Transforming Our Lives*. New York: Warner Books, 1982.

OUCHI, W. *Theory Z: How American Business Can Meet the Japanese Challenge*. Reading, Mass.: Addison-Wesley, 1981.

PASCALE, R. T., and A. G. ATHOS. *The Art of Japanese Management: Applications for American Executives*. New York: Simon & Schuster, 1982.

RIGGS, J. L., and G. H. FELIX, *Productivity by Objectives*. Englewood Cliffs, N.J.: Prentice-Hall, 1983.

SINK, D. S. *Productivity Management: Planning, Management and Evaluation, Control and Improvement*. New York: Wiley, 1985.

SMITH, A. *An Inquiry into the Nature and Causes of the Wealth of Nations*. London: A. Strathan & T. Cadell, 1793.

TAYLOR, F. W. *The Principles of Scientific Management*. New York: Harper & Row, 1911.

VAN GIGCH, J. P. *Applied General Systems Theory*, 2nd ed. New York: Harper & Row, 1978.

1-9 SELF-TEST REVIEW

Answers to following review questions are given in Appendix A.

1. T F The first *Industrial Revolution* began in the 1700s and featured the development of steam power and machine tools.

2. T F Frederick W. Taylor initiated the discipline of *operations research*.

3. T F The introduction of *automation* in the 1940s caused massive unemployment as machines replaced people in factories.

4. T F *Hard* automation is associated with computer hardware.

5. T F The *management process* approach is associated with unity-of-command and staff–line principles.

6. T F Douglas McGregor's *Theory X and Theory Y* concepts are associated with the behavioral approach to management.

7. T F The *Japanese style* of management is known for its attention to quality, lifetime employment, and reliance on quantitative decision-making techniques.

8. T F *Production* is the intentional act of producing something useful, either goods or services.

9. T F A scaled-down replica of a machine is a *schematic* model.

10. T F Mathematical models are the most abstract and generally the most useful type of models for production studies.

11. T F To *satisfice* is to strive for a goal while being willing to settle for a less ambitious one.

12. T F Studies of production systems should always follow the *planning–analysis–control* sequence.

13. T F Providing higher quality has been stressed strongly by U.S. industry in the 1980s as a means to become more competitive.

14. T F Computers are now in the fourth generation of development, which is known as the VLSI generation.

15. T F Computer-aided design (CAD) is an inexpensive substitute for manual drafting and design.

1-10 DISCUSSION QUESTIONS

1. Compare the first industrial revolution with a potential second revolution based on computer developments.

2. Look up and discuss some military applications of operations research. Do any of the military studies have counterpart industrial applications?

3. Determine several input–transformation–output production systems, and describe the process by which entering elements are made more useful. Use service as an output in at least one system.

4. A widely accepted definition of engineering is "Engineering is the art of organizing men and of directing the forces and materials of nature for the benefit of the human race." Paraphrase the definition to emphasize the affinity of the engineering function to a production system.

5. In the *Journal of Industrial Engineering* (May–June 1965), Alan J. Levy suggested that a quick view of trends in organizational theory is shown by the following evolution of management thought:

a. From: Efficiency as a mechanical process.
   To: Efficiency as a human process.
b. From: Control through command.
   To: Control through communication.
c. From: Authority from the top down.
   To: Authority from the group.
d. From: Leadership by authority.
   To: Leadership by consent.
e. From: Technological change by fiat.
   To: Technological change by consultation.
f. From: A job for subsistence.
   To: A job as a satisfying experience.
g. From: Profit from buccaneering.
   To: Profit with social responsibility.

Discuss the trends. What caused the original conditions, and why were they accepted? What present-day examples can you find for the original conditions? What has led to the current trends? Why?

6. Why should you study management when the professionals in the discipline cannot agree on the content of or even the best approach to the subject?

7. What effect on production systems can reasonably be expected from greater concern for social problems on both a local and a national level?

8. Another way of classifying types of models is to categorize them as iconic, analogue, or symbolic. Give an example of a model that would fit each classification. (*Hint:* Iconic is derived from the Greek word for image.)

9. What is meant by the statement that mathematical models may not be more accurate than are other types of models but are usually more precise?

10. What is meant by the statement that models can be used heuristically? Give an example.

11. Discuss "satisficing" in relation to studying several subjects in preparation for final examinations.

12. Discuss the following diagram of the procedures for decision making in reference to production systems. What types of problems are likely to be encountered in each step? Could the same procedures be used for decisions in physical science studies? Where does intuition fit into the steps?

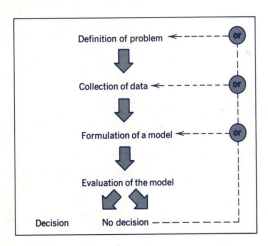

13. Consider familiar production systems such as a barbershop or gasoline station, and discuss what types of models could be used for planning, analyzing, and controlling their operations. What objectives could be appropriate to each study phase?

1-11 PROBLEMS AND CASES

1. *A Case In Which the Customer Did Not Appreciate the New Technology*

Savants of computerology stress that the setting must be ready for the insertion of new technology, or the frustration of its implementation will destroy its value. For example, many small, old-fashioned lumber mills have been modernized by computerized sawing systems, to the considerable consternation of the old-fashioned workers. Although the old hands realized the wood- and time-saving advantages of the new cutting technology and seldom forcefully opposed the change, they were not enthusiastic. After working out the bugs of installation, the company engineers expected the new machines to operate smoothly. But seldom did they do so. Breakdowns were frequent, and quality fell below expectation. Many of the failures were traced to well-intentioned mistakes. The new electronic machines were treated like the old mechanical machines they had replaced: Settings were not quite precise enough, maintenance was not strict enough, repairs were crudely made, and operating difficulties were overcome with more brute force than the sophisticated equipment could stand. Behind these difficulties was the simple fact that the newly labeled "knowledge workers" have to be given enough knowledge to squeeze productivity out of the sophisticated gadgetry.

A different kind of difficulty has frustrated the adoption of computer-assisted checkout counters in supermarkets. Electronic scanners use a tiny laser beam to read the price encoded on nearly all product labels to record and automatically total a customer's purchases. Besides speeding up the checkout lines, the scanning process eliminates the chore of marking prices on each item and keeps closer track of grocery inventories. But resistance to their wider use has come from the customers, not the employees.

Why do you think the customers might object to the electronic scanners? What effect might this displeasure have on the employees and their acceptance of scanners? How might technology come to the rescue?

2. *The Case of CIMulation*

Moves toward CIM involve more than just buying new machines. Both management and workers must be prepared to accept the new system and all the changes it entails. Organized labor has expressed its reservations about the outright acceptance of new technologies. Delegates to the 1980 UAW (United Auto Workers) convention adopted a resolution that said in part: "With their vast memories and 'real time' analysis capabilities, computers could become the 'Big Brother' watching every worker every minute; and so-called efficiency gain produced by that approach is not worth the cost in self-respect and other human value."

Suggestions made by the UAW included the following:

- Strong contract language to deal with bargaining-unit erosion, job assignments, and training.
- Prenotification to the union of plans to introduce new technologies.
- Reduction of work time without loss of pay, so that more job opportunities are created.
- Public assistance for income maintenance of victims of job displacement.
- Standards set for health and safety hazards associated with new technologies.

Discuss the suggestions from the viewpoints of the workers in a plant where CIM is being introduced, of the managers of the plant, and of the general public who will buy the products of the plant and be charged for whatever extra costs are negotiated to help the workers.

3. *The Case of Opportunities in Disguise*

The scope of responsibilities attributed to production organizations is characterized by the following charges and demands delivered by Virgil B. Day, vice-president of General Electric Company, in a 1972 speech; they are still timely:

CHARGES

1. Corporations have the ability to administer prices to achieve excessive profits.

2. The drive for greater production and lower internal costs leads to pollution and the accelerated depletion of natural resources.

3. Siting of plants detracts from the environment and negates community planning.

4. Too many products are unsafe.

5. Organizations use false and misleading advertising.

6. Corporations "regulate the regulators" and have excessive influence in communities to promote business interests.

7. International operations export U.S. jobs and exploit resources in the developing countries.

8. Production work is monotonous and dehumanizing.

9. Working conditions show minimal regard for health and safety.

10. Too many organizations are racist and sexist.

DEMANDS

1. Legislate excess profits taxes and wage/price controls.

2. Establish stringent effluents/emission controls and require that a minimum percentage of recycled materials be used.

3. Develop national, regional, and state land-use planning and environmental impact statements.

4. Institute strict product safety standards.

5. Require more technical data and have stronger FTC (Federal Trade Commission) controls on advertising.

6. Require the complete disclosure of political contributions and allow more employee leaves for community service.

7. Establish controls on overseas investments and impose restraints on imports.

8. Try participative management, job enrichment, and more flexible scheduling of work.

9. Tighten the enforcement of health and safety standards.

10. Enforce affirmative action or equal employment opportunity.

a. Some companies took a positive approach to these charges and demands. They viewed conservation measures as helping reduce costs and sought ways to better use materials to cut input costs. Similarly, equal opportunity programs were turned into new sources of employees. What additional examples can you cite that satisfied the charges and demands while improving the position of the answering organization?

b. List additional charges and demands that represent current concerns. For example, consider inflation, profiteering in service industries, and inefficiencies in government. Add an extra column to indicate what remedial actions might be suggested by a production systems specialist.

4. *The Case of the Smell of Money*

One bright Sunday afternoon Al Notz sat is his office with the shades drawn, contemplating what he should say at tomorrow's annual meeting of the company's stockholders. The facts and figures he had to deliver were discouraging. Revenue was up and operating costs had not increased proportionately, but major capital expenditures had been made to meet new clean-air standards. The end result was that the usual dividend on stock had to be halved.

Besides the angry stockholders expected to be present at the meeting, Al had heard rumors that a contingent of environmentalists from a nearby university would picket the meeting. They wanted still more controls, both to reduce further the odor from the papermaking process of the plant and to eliminate the untreated wastes being discharged into Cedar Creek. As manager of the plant, Al knew that the State Environmental Quality Department had approved a schedule for gradually controlling the remaining air and water contamination. The authorities recognized how expensive more controls would be and took that into consideration in stretching out

modifications, but the plant location alongside a busy highway made it a natural target for groups seeking to protect the environment.

The paper company had been started 22 years ago. The original owners later made it a public corporation in order to finance a major expansion that allowed the use of a new process for converting wood chips to kraft paper. The odor from the process had always been pretty strong, but the community used to be more tolerant, calling it the smell of money. Also, most of the stock had been sold locally, and the stockholders had come to expect substantial dividends from their holdings.

Al was worried that the stockholders and environmentalists would clash. In addition, he felt obliged to report that next year's dividends would also suffer from plant modifications required to meet higher safety standards in compliance with OSHA (Occupational Safety and Health Act) regulations. What depressed him most was the provable fact that the company was more productive than ever before, but the bottom line on the profit and loss statement did not reflect this achievement.

a. What approach would you recommend that Al take at tomorrow's meeting: stress productivity gains and blame the government for higher costs, or point out the benefits to the community from new pollution controls, emphasizing that these activities are the responsibilities of ownership, or what?

b. Put yourself in the role of a stockholder. Should Al be blamed for "giving in" too easily or too quickly to demands for environmental improvements and be directed to spend more effort in the future protecting dividends instead of the environment? What would be your feelings toward the picketers?

PRODUCTS AND SERVICES: DESIGN AND DEVELOPMENT

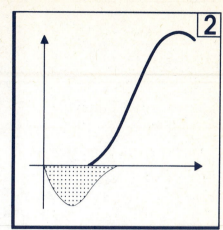

LEARNING OBJECTIVES

After studying this chapter you should

- appreciate the complexity of a production system and the relationship of functions within the system.
- understand the product/service cycle and the role of research and development in the cycle.
- be aware of ways to improve personal creativity and group approaches to ideation.
- be able to construct a cause-and-effect diagram.
- comprehend the reason for different organizational structures, their components, and how their functions interact.
- realize how new technologies and management practices are accelerating operations in both manufacturing and service organizations.

KEY TERMS

The following words characterize subjects presented in this chapter:

consumption	brainstorming
ladder of needs	synergism
product/service cycles	cause and effect (C&E)
creativity	research and development (R&D)
trigger words	project management

matrix organization

technology transfer

systems approach to management

cybernetics

feedback

product maturity

computer-aided engineering (CAE)

simultaneous engineering

age of information

expert systems

artificial intelligence

2-1 IMPORTANCE OF PRODUCT/ SERVICE DEVELOPMENT AND PRODUCTION SYSTEMS

Everyone knows that "great oaks from little acorns grow," but few people have occasion to contemplate the origin of industrial giants such as IBM, Ford, and Xerox. They, too, came from humble beginnings. Perhaps that guy tinkering with gadgets in his garage will sometime be known as the founding father of a future industrial giant, or that outgoing gal who is always organizing local activities may become the legendary founder of a powerful service agency. Such things do happen, though rarely. But people are continually starting new projects on a far more modest scale. With our rapidly changing tastes and technology, both products and services *have* to evolve just to stay current. These changes are not as dramatic as a revolutionary creation, but they are vital to continued production. You have to know *what* to start, as well as when and how.

A new product or service usually originates in response to a demand, although occasionally someone offers a new convenience so good that it creates its own demand. In either case, the originator must be astute enough to recognize the potential and competent enough to grasp the opportunity. Both attributes are enhanced by familiarity with techniques of investigation and concepts of organization.

Consumers tend to take for granted the steady parade of new products offered to them. They note the cleverness or banality of an offering without realizing the concerted efforts required to develop and produce it. Creating new designs, deciding when to drop old ones, and modifying current ones are continuing concerns of all industrial enterprises.

The success of an enterprise in developing and maintaining a product or service is influenced by the way it is organized and managed. Good ideas, skilled workers, and efficient machines are not enough. An organizational structure is needed that ensures a free flow of ideas, coordinates operations, and encourages better decision making. Different patterns are designed to fit different situations. Because production demands change with varying economic and social conditions, it is important to understand the functional relationships among the operations in a production system in order to react effectively to changes.

Innumerable folk sayings remind us of the difficulty of getting a venture off to a good start: Once begun, a task is easy; half the work is done. A good beginning makes a good ending, and so on.

2-2 PRODUCT/SERVICE CYCLES

Studies of production systems tend to dwell on ways to increase capacity, output, and internal effectiveness. In doing so, it is possible to overlook the link between production and consumption. The ability and desire to produce a product or service goes unrewarded unless there are consumers with a corresponding ability and willingness to partake.

Sources of consumption take many forms. Some products and services are considered indispensable: basic foodstuffs, health care, water supplies, protection and the like. Others are desired to maintain life-styles: air travel, beauty salons, recreational supplies, and so on. And some are perceived as needed, or at least tolerable, social benefits such as rockets for space travel, armaments, and indirect government services.

Consumption is a multistage process. Relatively few producers deal with the consumer who is satisfying his or her personal wants. Consequently, most producers produce for other producers. Recent history offers many dramatic examples of chain reaction effects than can result from a switch in the consumer's wants: downturns in housing starts that affect a multitude of suppliers, public disenchantment with the space program that rippled out to many subcontractors, and increasing demand for all types of recreational products and services that is supporting a variety of new activities.

There are time lags of varying length throughout the production–to–consumption process. Largely unavoidable delays in gearing up to produce, and sluggish changes in market demands act as buffers to provide stability for production systems. For instance, seldom do consumers all stop buying at once, except in localized situations. Sometimes a trend can be blunted by producer actions such as advertising. Unless the whole economy is suffering, a switch from one type of consumption to another usually creates a shortage in the newly favored area that causes some consumers to use the old products or services until new production capacity is ready. Consumer services in the public sector require substantial time to start up, and old projects may wither but tend to linger interminably.

All of these considerations contribute to producers' uncertainties. Very few products or services survive in the original form in which they were offered. Competition is the usual spur for change, but political and social pressures also influence the course of production. Figure 2.1 portrays the typical phases of product/service development.

The **product/service cycle** warns that production systems must be dynamic in order to survive. A static system may appear, during its zenith, to be durable. This complacent view is seldom justified. Even when activities appear healthy and vigorous, planning should be under way to identify modifications that can improve operations, and contingency plans should be constructed for major changes if internal or external conditions suddenly deteriorate. Thus, each phase in the production life cycle is subject to planning, analysis, and control.

Economist J. K. Galbraith, in *The New Industrial State,* suggested that large industrial organizations can create new demands that otherwise would never exist.

Psychologist A. H. Maslow recommended that the order of consumption key the motivation of workers. Physiological needs must be met before other wants will act as motivators. His **ladder of needs** progresses from life-sustaining needs to safety, social, ego, and self-fulfillment needs. This ladder should provide production clues to the sale of necessities and luxuries.

Products and services from pills to cars and TV repairs to education have been affected by the waves of laws designed to protect consumers. There's a message there.

Unlike nature's life cycles, in a production cycle a new, better product (progeny) replaces (ends the life of) the existing product (parent). This rapid evolution is owed largely to promises of wealth and fears of loss due to obsolescence.

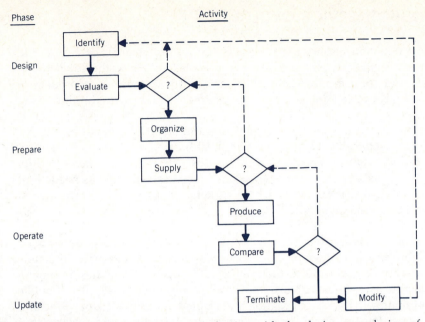

FIGURE 2.1 A product/service cycle begins with the design or redesign of the operation and eventually returns to the same state after the product/service offered is no longer sufficient. Between the end states, the operation is continually refined to accommodate changing conditions.

EXAMPLE 2-1 A business is born

After completing high school, Benny Black worked for a couple of years before enrolling as a full-time college student. His savings and odd jobs financed his first two years of study before he got the urge to be his own boss in his own business. His options were limited by meager funds and experience: he lacked the artistic talents to go into crafts, the capital to engage in manufacturing, and the education to be a licensed professional. Recognizing his limitations, he narrowed his search to finding a service he could provide. He chose lawn care and gardening. Interviews and want ads in the paper confirmed the need; the mild climate of the area allowed year-round employment; his previous work experience provided the necessary know-how; and very little investment was required to get started.

He called his business "Green Thumb Service," and it immediately was successful, but Benny was not satisfied. He saw the potential for expansion and was intrigued. Before extending his operations, he reviewed the brief history of his business to obtain an overall view of both possibilities and restraints.

Phase 1: Design. The potential for starting a lawn care and gardening service was identified. The need was verified, and the required resources were available.

Phase 2: Prepare. Necessary equipment and supplies were purchased, and Green Thumb Service was advertised. The ads were answered.

Phase 3: Operate. Initial jobs were secured, and customers appeared to be satisfied with the service. Some types of work were obviously more profitable than others. Larger jobs consumed less travel time, and those requiring some skill, such as pruning and spraying, paid better.

Phase 4: Update. Green Thumb Service could

continue as is or be modified. Alternatives include (1) limiting operations to pruning and spraying, (2) hiring help to take on more of the same types of jobs, (3) contracting to landscape new homes, or (4) combining (1) and (2) by managing a crew on routine jobs and securing mechanized equipment for specialized pruning and spraying.

Although tempted to go into landscape contracting, Benny acknowledged his limited resources by opting for the last alternative.

2-3 THE SEARCH FOR SOMETHING NEW

All new services and products originate from ideas. Some of the ideas are clever and original; others rediscover the wheel. Even good ones may be impractical when the resources required to implement them are considered. Organizations set up departments or review procedures both to foster new concepts and to protect themselves from wasting resources on impractical schemes. Individuals must rely on personal innovative talents and educated wariness in order to benefit from new pursuits. In this section, we consider the birthplace of production practices from both group and individual approaches. In doing so, it should be understood that individual and group creativity techniques complement each other and that both are enhanced by systematic search routines.

Creativity occurs when sights that everyone sees arouses thinking that nobody else has thought.

Solo search

On few occasions does an individual conceive a whole new way of doing something. When such an occasion occurs, it is most likely a reaction to a very serious problem: The present operation is in danger of collapse, or competitors have forced the issue to start anew. More frequently, ideas are sought for relatively minor modifications and improvements: a way to make an operation safer or a change in data flow that saves time.

Several books have been published suggesting ways to improve personal creativity. They are based on a core concept that creativity is an acquired, not native, talent. Different ways are offered to develop the talent. All of them rely on a positive attitude and persistence. The first few ideas are relatively easy to accumulate, but successive ones get tougher. To aid the generation of ideas, the following procedures are suggested:

1. *Trigger words.* Trigger words are a chain of related words or phrases used to stimulate imagination. The subject of the chain is established by choosing important characteristics of the problem. For example, if the problem is the *size* of an article, the chain could include bigger—smaller—lighter—heavier—rounder—flatter—thinner—fatter—taller—shorter—least—most—solid—hollow—divided—connected, and so on. Thinking about the words in the chain may lead to associations that point toward possible solutions.

2. *Checklists.* Checklists are preplanned lists of questions serving the same purpose as do trigger words. Osborn, in his *Applied Imagination*, advised key questions such as "adapt? modify? magnify? substitute? rearrange? reverse? combine?" A personally developed list of comparable questions

Characteristics of a Creative Person

Independent and strongly motivated.

Highly social.

May have college education but it is not a prerequisite.

Intelligent but not necessarily a genius; threshold IQ level is about 130.

Must be well grounded in the field where new ideas are sought.

has the advantage of being based on known similarities and contrasts that encourage recall of half-forgotten experiences.

3. *Verbalization.* Verbalization is the forced discussion of possible solutions. Everyone finds it easier to talk about doing something than to do it. If a listener is interested in the same problem, exchanged views may spark variations impossible without a dialogue. A tape recorder can be substituted for the listener, with replays acting as triggers for additional comments. Either way, the verbalization is obviously biased by the speakers' sentiments, but exposure often short-circuits personal dogmas.

Group ideation

Pioneering books on creativity include *Synectics* by W. J. J. Gordon (New York: Harper and Brothers, 1961); and *Applied Imagination* by Alex F. Osborn (New York: Scribners, 1953).

Verbalization is formalized by techniques such as **brainstorming** and **synergism:** The sum of the ideas from a group of people operating together should exceed the total number of ideas that the same people could generate operating individually.

A brainstorming group is ideally composed of 5 to 10 members, but successful sessions have been conducted with groups as large as 150. Participants are urged to think freely, mentioning any idea, no matter how wild it is, and to support one anothers' efforts. A leader tries to keep the discussion in the general problem area and guards against stopping to explore an idea deeply or criticizing the ideas of others. Ideas are stifled when a session becomes too formal, and usefulness decreases when discussions degenerate into bull sessions.

A productive brainstorming exercise should produce an idea a minute. All ideas are recorded and evaluated at a later date. The purpose of brainstorming is to promote originality through a chain reaction as one person's idea suggests a related one from a different person. It has been widely applied because it is fun and often yields interesting results, but it is not a one-shot, guaranteed survey of all alternatives.

The synectic approach to group creativity devised by Gordon resembles brainstorming, with emphasis given to *depth* rather than breadth. The leader of a synectic session plays a dominant role in keeping the discussions centered on the underlying concepts of a problem, not on the specifics of the problem itself. Because the supply of fundamentals is limited, fewer ideas are expected in such sessions than from brainstorming.

Companies try to generate a regular flow of innovations, by scheduling repeated creativity sessions, forming idea teams whose members come from different specialties, and setting up innovation centers that ensure attention is given to all new ideas.

A session, made up of 5 to 10 members chosen for their competence in the problem area, begins with a thorough briefing. Then the core problem is investigated by considering related issues. For example, the problem of how to remove garbage could lead to a discussion about the causes of waste or unwanted accumulations of anything. Promising issues act as focal points for carrying the discussion back to the specific problem. These directed discussions are particularly rewarding for technical subjects. But again, the nature of creativity precludes the possibility that any formal method will guarantee to reveal the best idea.

2-4 CAUSE-AND-EFFECT ANALYSIS

The value of visual aids in understanding a problem is tremendous. Diagrams, charts, and other graphical representations contribute to better communication by organizing data and focusing attention on specific issues. A **cause-and-effect (C&E)** diagram is a portrait of a problem. It shows the inputs that affect the problem and the results to anticipate.

Construction

The first step in constructing a C&E diagram is to develop a pithy statement of the problem or objective. This statement is entered in a six-sided box. As shown in Figure 2.2, an arrow entering from the left represents the factors that cause the problem, and an outward arrow on the right represents the composite effects a solution will provide. It is also useful to inscribe a date, usually the time at which a solution is to be implemented, that sets the point at which the causes take effect.

After the basic structure is set, the inclusive or dominant factors are identified. These are shown on the diagram by labeled ribs; the labels are entered in ovals at the *shaft* end of the cause arrows pointing to the spinal cause shaft and at the *point* of the effect arrows radiating from the spinal effect shaft. Then smaller arrows identifying subfactors run to or from the main ribs. In turn, sub-subfactor arrows can lead to or from these. Each additional class of arrows provides greater detail.

To illustrate the construction of a C&E diagram, consider a situation in which highway personnel are planning to develop a scenic viewpoint. (Although this may not seem like a production problem at first glance, it is a public-sector activity equivalent to traditional industrial activities such as plant location or facility design, and it relies on almost identical evaluation procedures.) Thus the focal point of the diagram is "develop viewpoint" with a desired completion date of, say, July 4, 1988.

With the problem stated, the key factors are identified. The *location* of the viewpoint, its *cost* and *design,* are the dominant causes of the main effects, which are *appeal* and *safety.* These five factors are the primary arrows as shown in collapsed form in Figure 2.3.

Details are usually developed by concentrating first on the cause side of

A differently structured C&E diagram that encourages the workers' involvement in its construction is described in Chapter 14.

FIGURE 2.2 First step in the construction of a C&E diagram. Spinal shafts represent composite cause-and-effect factors that pertain to the problem stated in the middle. The date establishes the time frame of action.

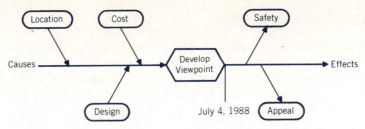

FIGURE 2.3 Basic structure of a C&E diagram, in which the main influencing factors are labeled as directed arrows joined to the spinal cause and effect shafts.

the diagram. Each of the main causes are questioned as to what components, actions, categories, or factors are involved. For example, the *design* causes include the size of the viewpoint, type of construction, materials, and the nature of the extra services to be provided. These subfactors can be further refined by identifying more detailed ingredients or alternatives. Detailing can continue to any level deemed appropriate by the analysts. A three-level detail is shown in Figure 2.4.

Use

Cause-and-effect diagrams can assist both solo and group ideation. When an individual is searching for ideas, the entries in a C&E diagram suggest additional relationships. A C&E format can also serve as a recording device for ideas generated by a group. Capturing and organizing ideas by connected arrows may inhibit a free-wheeling brainstorming session, but it can reveal relationships that might otherwise have escaped detection.

One-sided diagrams, either of cause or of effect, can be used to investigate just the origin of a problem or the expected results of a course of action. In both one- and two-sided diagrams, dependent factors or subfactors can be linked by lines to call attention to important relationships. Such modifications typify possible adaptations by users to customize C&E diagrams to their particular interests.

Another aspect of C&E diagramming is the ease with which the quality of the diagrams can be assessed. A reviewer can easily tell how thoroughly a problem was probed by the pattern of its C&E portrait. A wealth of detail, if legitimate, indicates an exhaustive effort. A bare skeleton means either the problem was negligible or the solver was negligent.

> Although C&E diagrams are usually associated with investigations, they are also valuable for presentations.

> One-sided (cause) diagrams were originated by Professor Kaoru Ishikawa in 1953, in his work on quality control. Such "cause" displays are known as Ishikawa diagrams.

EXAMPLE 2-2 Business grows

Green Thumb Service prospered. Benny Black had little trouble recruiting other students to help in his enterprise. Because almost no training was necessary for his hired help, much of his time was spent taking job orders, scheduling, checking on the quality of work done by his crews, buying supplies, billing

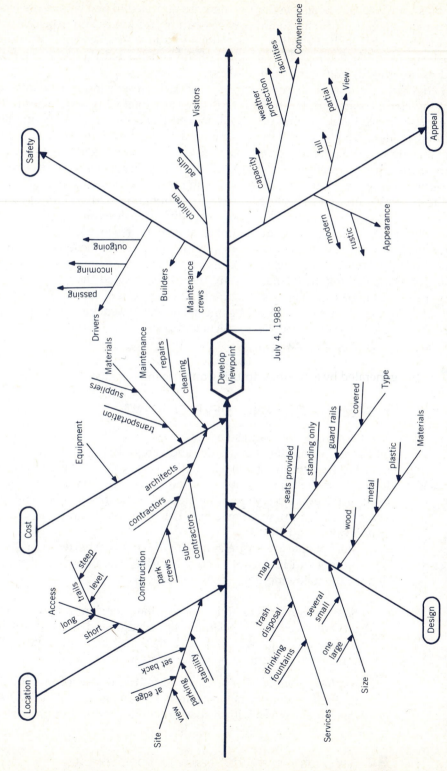

FIGURE 2.4 C&E diagram of the factors to be considered in the development of a highway viewpoint. The factors are positioned according to the areas of primary influence.

33

clients, and paying bills and wages. He also completed his graduation requirements. Now he could devote full time to practicing what he had studied.

Some of the lore he had accumulated from books and lectures worked as advertised, and some of it was disappointing. It did not take long to discover that every significant decision is snared in a web of interwoven influences. The decision-making tools he had learned were useful in identifying the primary influences and providing a feasible solution to serve as a basepoint for comparing practical alternatives. He made mistakes, but he also made enough favorable actions to build Green Thumb, Inc., into a healthy business.

In an early move he dropped routine lawn care (too much competition to allow growth) to specialize in the chemical treatment of soil and plants. After first contracting for airplane crop dusting, he learned to fly and eventually acquired his own helicopters. While flying for the U.S. Forest Service, spraying newly planted trees, he became interested in the promising future of Christmas tree farms. Such an enterprise appealed to him because he could use most of his current resources and expertise in service operations to add a product line. To evaluate the potential, he constructed the C&E diagram shown in Figure 2.5.

The C&E format helped Benny grasp the essentials of the total Christmas tree project. When it still looked promising after observing all the necessary inputs and uncertainties of future markets, he started collecting the details needed to make a firm decision. Realistic figures were developed: costs for acquiring resources, land expenses, distances, number of personnel needed, size of operations, timing of expenditures, sales volume, and so forth. Then a financial analysis (see Chapter 5) was conducted to determine the effects of the long series of preparation costs and the delayed returns from sales of the mature trees. He weighed the advice of "nothing ventured, nothing gained" against "take no risk and assume no loss." Based on his figures, he took the calculated risk.

2-5 RESEARCH AND DEVELOPMENT

The four largest industries created in the last 30 years by new technology are jet aircraft, computers, television, and xerography. These and similar advances in technology contribute heavily to the long-term increases in per-capita output in the United States. Although a few discoveries were made by individuals, most of the advances were originated by teams of scientists and engineers devoted to seeking new knowledge. This effort is broadly classified as **research and development (R&D).**

A report by the Committee for Economic Development (New York, 1962) stated that 40 percent of the total increase in income per employed person between 1929 and 1957 was attributable to "advances in knowledge."

Most people are aware generally of what is expected from R&D work, but few comprehend its breadth and intricacies. The National Science Foundation defines the following three divisions of R&D:

Basic research Research projects that represent original investigation for the advancement of scientific knowledge and that do not have specific commercial objectives, although they may be in the field of the reporting company's present or potential interest.

Applied research Research projects that represent investigation directed to the discovery of new scientific knowledge and that have specific commercial objectives with respect to either products or processes.

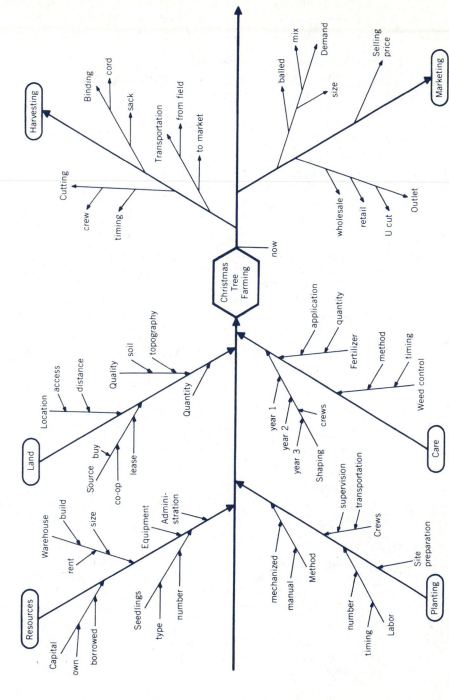

FIGURE 2.5 C&E diagram of the most critical factors affecting a decison to engage in Christmas tree farming.

Development   Technical activities concerned with nonroutine problems that are encountered in translating research findings or other general scientific knowledge into products or processes (NSF 63-7, p. 109).

Technological innovation

In the 1980s there has been a mixed attitude toward the development of new technologies. Conservationists and social critics call attention to the damage caused to our environment and way of life by some applications of technology. They stress that certain types of economic activity are more desirable than are others: Needed are basic technologies that are less wasteful of energy and materials and are more attuned to the needs of society, not "New! Improved!" consumer novelties that respond to no real demand by users.

The public also recognizes the need to combat pollution, secure more sources of energy, expand mass-transit systems, provide safer working conditions, find cures for health hazards, and attain other social goals. Some of these objectives can be met by refining discoveries made previously, but others depend on breakthroughs that are still hidden from researchers. The search for discoveries is a very uncertain process, and the lag time between discovery and application is often discouragingly long. For example, the discoveries that led to the four major industries mentioned at the beginning of this section did not result from concerted searches for commercial products and were slow in being developed:

Jet aircraft   The first jet engine came from the workshop of Frank Whittle (later knighted) who spent most of the 1930s trying to get the British aircraft industry to develop it.

Computers   The first electronic computer came from a wartime ballistics laboratory, not from R&D by the business-machine industry.

Television   TV was invented in 1919 by Vladimir Zworykin, but it was 22 years later before Westinghouse started commercial exploitation.

Xerography   In 1937 C. F. Carlson developed a xerographic process in his home workshop; in 1950 the Haloid Company made it commercially available.

Today, industrial laboratories, government task forces, and universities are engaged in multipronged investigations of nature's secrets and ways to convert them to humanity's service. One of the more difficult problems encountered is deciding which projects to fund: Would it be more valuable to discover a new source of energy or to develop a new means of mass transportation? Even after the general direction of research is known, questions still remain about which specific paths to try: Securing energy from the earth or from the sun? Providing trains without wheels that travel by magnetic suspension or on cushions of air?

Some leading candidates for technological innovations are communication satellites, waterless-web papermaking process, waste-heat utilization from nuclear power plants, coal gasification, lasers used for nuclear fusion and communications, and thermal boring devices.

The listed criteria are less applicable to basic research when what is discovered usually cannot be identified until after discovery (fundamental, physical and organic phenomena of nature).

Decisions are guided by evaluating projects according to criteria such as

1. *Promise of success* (probability that the research will yield a realizable product).

2. *Time of completion* (how long the research and/or implementation will take).

3. *Cost of project* (the direct and indirect research cost plus the cost of converting the results to production).

4. *Strategic need* (the intrinsic worth of the expected results).

5. *Market gain* (the novelty of expected results and how they will fit in with the primary purpose of the sponsoring unit).

Industrial R&D

Industrial research and development is confined mainly to large firms within a few industries and product lines in the manufacturing sector. About 85 percent of all R&D scientists and engineers are employed by firms with more than 5000 employees. The majority of all R&D funds are spent by five industries: aircraft and missiles, electrical equipment and communications, chemical and allied products, machinery, and motor vehicles. The reasons for this concentration are the great cost of R&D, the high risk that makes R&D expenditures a gamble, the opportunities to use R&D personnel for innovations in different departments within a large firm, and the ability to take advantage of process innovations through large volumes of production.

Where R&D fits into the organizational structure of a firm varies widely. It may be centralized to serve the entire corporation or be divided according to the firm's departments. A central research facility is easier to administer, but difficulties may arise in communicating problems between researchers and operating personnel. R&D is more closely linked to production in a functional organization when the work is performed under a major department such as manufacturing or when R&D is given departmental status akin to manufacturing.

After 2 years of R&D you invented THIS?!

Wider company participation in R&D can be obtained through an organizational approach known as **project management.** In this arrangement, selected people from different units in a firm are brought together to work as a team under a project manager. Those selected are chosen on the basis of the specialized knowledge they can contribute to the project. For the duration of the project, the manager has authority over the team members, budget, and resources, and he or she is accountable for their performance. Trouble may sprout from the dual nature of the undertaking. Team members must share allegiance to the project and their home units, and the project manager's needs for resources may conflict with the needs of the units that the project is supposed to serve.

The organizational structure associated with project management is called

See Section 7-9 for additional organizational considerations.

a **matrix organization.** Some say it was conceived by the Procter and Gamble Company in the 1930s, and others credit General Electric with a later introduction. Only in the last decade, however, has it received much attention. Most of that attention was created by its application to weapons production.

Developments in weaponry of almost science-fiction design are characterized by huge expenditures, accelerated R&D, highly advanced technology, complex coordination problems, and an urgency for completion. A special form of organization emerged to cope with these large-scale, diverse factors. The pattern that evolved was a two-dimensional matrix, one dimension being the different weapons under development and the other the regular line-staff components. As displayed in Figure 2.6, the rows of the matrix represent independent products or projects (weapon-systems code names), and the columns are the support units for the projects (regular operating departments).

Civilian counterparts to weapon programs are based on well-defined production or service projects such as a dam, section of a highway, educational unit, or new product development. Often the projects are of a "task force" nature to meet a pressing problem or to award special attention to a particularly promising endeavor. The following are sample applications to a government service program and a product management project:

	State Police Dept.	Motor Vehicle Dept.	Highway Division	Public Utility Commission
Traffic Safety Unit	XXXX	XX	XXX	X
Program B	XX	XXXX	XX	XXXX

X's represent participation from each department.

	Manufacturing	Facilities	Financing	Personnel	Marketing
Super Soap Manager	XXX	XXX	XXXX	XXXX	XXXX
Brand B Manager	XX	XX	X	XXX	XX

X's represent product personnel supplied from each area.

The project manager is the boss of the bosses. The power of the purse confirms and consolidates the project manager's sweeping authority.

The departure of a rectangular pattern from the traditional triangular organization design (see Figure 2.7) is more significant internally than can be shown in a chart. The most obvious difference is the premeditated denial of the one-boss decree. Under the two-dimensional arrangement each worker has a dual responsibility, first to the project or program manager for duties related to producing a special product and second to a regular department manager on matters related to functional contributions to normal departmental output. Criss-crossing lines of command invite close cooperation by opening direct communication channels between workers and staff as well as among members from

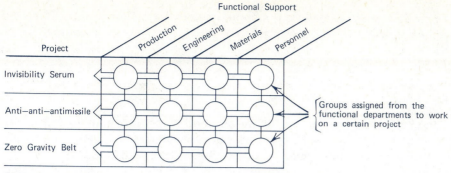

FIGURE 2.6 Matrix organization for large-scale weapon projects.

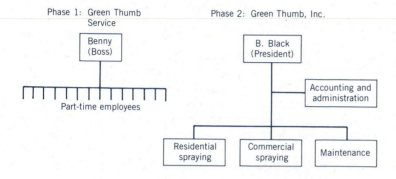

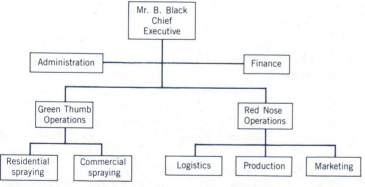

FIGURE 2.7 Organizational changes required by expansion of operations. The phase 3 chart is departmentalized according to functional areas, product lines, and customer services. Both of the line divisions are served by the staff administration and finance departments. Although it cannot be shown explicitly on an organization chart, the line departments coordinate activities by exchanging expertise, loans of equipment, and internal contracting for services, such as spraying contracted by Red Nose operations from Green Thumb operations. The reason for a formal organization structure is to give direction to departmentalized functions and at the same time to integrate all activities toward the objectives of the total system.

different departments. However, crossing patterns of authority also contribute to confusion when work orders conflict.

EXAMPLE 2-3 Business matures

Benny Black, now recognized in the business community as Mr. B. Black, was both sharp and lucky. He carefully evaluated each major decision and exercised close control over his activities, but he also enjoyed the good fortune of not being seriously hurt by factors outside his control. His venture into Christmas tree farming was successful because he managed his resources effectively and the market behaved as he hoped it would. When some of his original plans failed to materialize, he updated his strategy to move in a new direction.

"Red Nose Farms" was the name given to Benny's Christmas tree operation. Five years after the initial planting, he was ready to market his larger trees. He did so locally. The next year, as more trees were available, he sought more distant markets. Distance increased the logistic problems. After two more hectic years of individually contracting for shipping, handling, and storage, he persuaded other growers to join him in establishing their own wholesale network.

As the size and scope of operations increased, Benny Black's organizational structure went through many modifications. Three of the arrangements are shown in Figure 2.7. The first was a simple line organization, in which Benny supervised everything. In the second arrangement, he was assisted by an administrative staff and three managers, each with a defined area of responsibility. By the time Red Nose Farms were in full operation, the chain of command had lengthened until Benny was far from the people in the field. He still had to make decisions, but now the inputs to those decisions came from the line managers and the staff positions. His decisions in the mature business were of no greater personal importance than when he was the boss over his part-time crew; they just affected more people and encompassed a broader perspective. With more authority and challenges came more responsibility and opportunities.

Relationship of R&D to production

As the cost of bringing out a major new product neared $100 million in 1986, more "me-too" imitations of top sellers were introduced.

A firm engages in R&D to maintain or improve its competitive position. Discoveries may lead to new products, improved products, better production processes, and refined technical services. But competitive posture can also be improved by expenditures in marketing to improve sales and in industrial engineering to cut costs. Therefore, R&D competes with other areas of a firm for a share of the total budget.

The national patent policy that grants to an inventor exclusive control over his or her invention for 17 years influences R&D budgets. Patent incentives that promise a monopoly position make the risks in R&D more tolerable. A competitor's commanding position also leads to imitative R&D efforts so as to avoid being foreclosed from a technological field.

Hiring key people from competitors, called "industrial piracy," is of questionable ethics, and "industrial spying" is illegal.

Firms too small to support an internal R&D staff, or large firms temporarily needing extra R&D capacity, turn to external sources. Contract R&D is done by trade, industry, and private research laboratories. Patent rights can be purchased or leased. Consultants and people with the required experience can be hired to

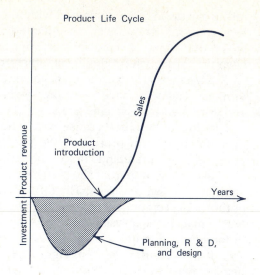

Product Life Cycle

Sometimes an old product can be repackaged to achieve the same reward of a new product without the R&D expense. Putting toothpaste in a pump and shaving cream in brushes are examples.

offer advice or to introduce new technology to a firm. Many government agencies provide free services.

An ironic aspect of R&D is the difficulty of disseminating research findings from government and nonprofit organizations to industry. With industry hungry for new technology and well aware of the cost of obtaining it, free research results should seemingly be in great demand. Actually, they are eagerly sought, but the lack of indexing, multiplicity of sources, and sheer volume of output discourages their utilization. Recently, federally funded programs have been initiated to distribute technical information developed in government-sponsored research. A few impressive spinoffs of technology from the space program offer examples of what can happen, but better means of **technology transfer** are still needed. Splendid creations that never get from prototype to production serve neither the inventors nor humanity.

Out of 17 new products that were seriously evaluated in the early 1980s, 14 were abandoned and three were marketed, two of which succeeded.

2-6 PRODUCT/SERVICE SYSTEMS

Many factors are evaluated in determining the merits of a new proposal or project. In C&E diagrams, we categorized the factors as causes and effects. Similar divisions could be made as *controllable* and *uncontrollable* factors, or *internal* and *external* considerations. Such arbitrary divisions are made to assist analyses through associations that result from classifications. The intent is to include as many influencing factors as possible in the evaluation; *to consider the whole system.*

Systems thinking

People throughout history have undoubtedly tried to anticipate the diverse effects of contemplated actions. Until the last few decades, their efforts were probably adequate in most cases because products and services tended to be rather narrow in scope compared with today's scale. They did not have so many legal concerns, pressure groups, dependent factions, and related activities with which to contend. A conscious attempt to deal with modern complexities has been the thrust behind the development of systems concepts.

A **systems approach to management** is a password to current, fashionable thinking. Today we have "weapons systems" designed by "systems engineers" and evaluated by "systems analysts." The modern emphasis is not surprising in a production context, because more complicated outputs logically require more intricate conversion processes. We also have the high-speed data-processing potential to integrate and control the refined processes.

A fascinating part of the systems approach is that a system can be almost anything you want it to be. The formal view interprets a system as a collection of functional components interacting to achieve an objective. A person working with a machine is a system. The operator–machine system is just a component in the production line system and so on, systems within systems within. . . .

If we let a triangle, as in Figure 2.8, represent the hierarchical structure of management, the system concept will be illustrated by the meandering borders of the shaded areas. Within each of these areas are the parts of the organization related by a common objective, often a particular problem to solve. The purchasing function could be a system with the objective of acquiring materials economically and on time. Another system could be a division according to shared regional interests, with the objective of serving markets within a defined geographical section. The boundaries of the systems probably would overlap. Then the ancient problem of serving two masters is created for the shared portion of the organization.

Some of the management confusion generated by dual system objectives can be avoided by the prudent allocation of divisional lines within the organizational triangle. This approach is discussed in the following sections. Two other methods to ease the problem have already been implied. One is to think big, to put the interests of the governing system above daily operational problems. This line of reasoning is presented in detail in Section 4-2. Another method is to develop information channels and decision rules to link and control subsystems.

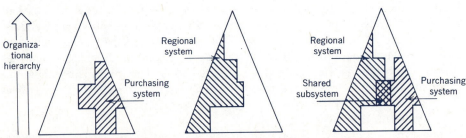

FIGURE 2.8 System boundaries within the organizational structure.

Systems theory

The study of system controls is a rapidly growing field of management research. Various labels have been pinned to the effort, with the most intriguing one called **cybernetics.** A thermostat used to control the heating system of a building is the classic example. A furnace produces the heat. The temperature of the building it heats is measured by a thermometer. A thermostat compares the actual temperature with the desired temperature to regulate the heat. The key to operation is **feedback;** information about deviations from the system's objectives feeds back to regulate inputs and thereby control the process.

Some physical mechanisms for self-regulation are easily visible, such as the old Watt governor for steam engines. Although the basic principle is the same as that shown in Figure 2.9, highly complex mechanical–electrical systems and management systems use more sophisticated feedback loops and regulators. Managers rely on the flow of information from above, below, and across for the feedback that connects input to output. When executives monitor the flow to regulate input, they do so by applying their decision rules to output feedback. A kind of self-regulating system emerges when these decision rules are passed to subordinates in the form of policies and rules that allow them to control the process input without the executive's attention.

The input–output portrayal of a system is nicely symbolic, but it does not divulge the fiercely interacting parts within the system. As we observed in Figure 2.8, divergent views within an organization can cause conflict; including outside competitive forces and regulating agencies multiplies the turmoil. A planner can be tempted to quit in dismay if he or she takes the extreme stance that everything depends on everything else. Even with the most sophisticated mathematics and unlimited computer capacity, it would be impossible to include all the factors affecting a major decision, because no data bank is all-inclusive. Fortunately, not all decisions are major, and those that are do not need *all* the related data for an adequate decision. Figure 2.10 shows how an activity combines internal and external influences but is not affected by all possible influences.

Systems theory encourages the identification of *significant* interactions and the consideration of their *combined* influence on a decision. Although it may seem like a step backward, the first step in applying systems theory is to understand the functions of the system's individual parts and thereby be able to recognize what factors could affect them. In subsequent chapters, we shall investigate the evaluation tools suggested by systems theory, but before then, it

Norbert Wiener (1947) developed the term *cybernetics* from the Greek root meaning "steersman." "Plato has used the word cybernetics in his time and Amperé has borrowed the term also as a name for the science of government; but Wiener must take the final responsibility for the currency of this ugly word, and also the credit for its aptness." S. Beer, *Cybernetics and Management.*

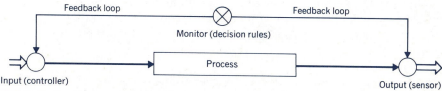

FIGURE 2.9 Input, output, and process control.

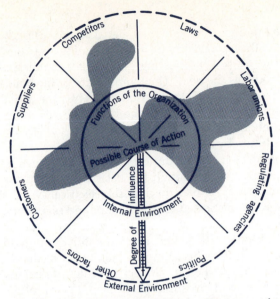

Labels in figure (clockwise from top): Competitors, Laws, Labor unions, Regulating agencies, Politics, Other factors, Customers, Suppliers

Inner labels: Functions of the Organization, Possible Course of Action, Influence, Internal Environment, Degree of, External Environment

FIGURE 2.10 Environmental factors affecting or affected by a contemplated action. Not all the factors have identical influence and consequently do not deserve the same consideration in decision making.

is necessary to become familiar with the production functions within the production system.

2-7 PRODUCTION FUNCTIONS

Production organizations are designed to generate an output. Several sequential or concurrent operations are usually involved in converting inputs to outputs. If the output is service, resources must be available to combine with professional skills to yield the desired service. A product is produced by refining resources to increase their value. In both situations the output relies on the coordinated activities of many people. Coordination is easier when the organization is divided into manageable chunks. These chunks are typically composed of the people and resources associated with a shared aspect or phase of product/service development. Depending on the size, composition, and purpose of the production system, more emphasis is given to some functions than to others.

Most functional divisions of an organization exhibit the following characteristics:

1. Regardless of the division of functional areas, they overlap. Overlapping areas often require special attention and usually provide high returns for control effort.

Slicing management by work phases leads to the planning, analysis, and control divisions described in Chapter 1. Another slicing pattern leads to work-type headings similar to the text's chapter titles.

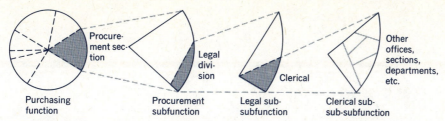

FIGURE 2.11 Operational subsections within basic functions. Multiple divisions help define responsibilities, but they increase the danger of conflicts from overlapping areas of influence.

2. Links between functional areas form a communication network by which activities of an organization are coordinated.

3. Management actions within all functional areas are basically the same. The administrative duties generally common to all subsystems include planning, organizing, staffing, directing, and controlling. Such recurring patterns give rise to the "universal truths" or "rules" of good management. They also add substance to the "generalist" role of the manager: good executives operate effectively regardless of the production system's output or organization.

The concept of "generalist" management conforms to the inclusive description of production systems in Section 1-3.

4. Each functional area can be divided, subdivided, and redivided again and again to reveal ever-smaller operating components. As shown in Figure 2.11, one section of the purchasing function is procurement. Part of procurement is the legal division. One source of legal advice is the clerical subsection. And on and on.

Functions of an industrial enterprise

The major functions of a relatively large industrial firm are depicted in Figure 2.12. Similar diagrams could be developed for smaller firms, government activities, social programs, or any organization characterized by our broad definition of a production system.

The core area of the diagram represents the organization's policymaking group. In a hierarchic triangle, this group would occupy the apex. From this lofty but core position, authority radiates outward to embrace responsibilities ranging from policy decisions for internal administrative functions to dealings with external contacts. The peripheral dealings include relations with stockholders, government agencies, competitors, and the public exclusive of customer and vendor relations handled by other sections of the firm. The commitment of an organization to social programs other than established union relations is becoming a more important policy consideration. With allegiance owed to employees, the community, and owners, policymakers have to be diplomats as well as company managers.

The ring of functions depicted as interlocking circles in Figure 2.12 represents an organization's administrative services. Broad instructions set by the

Henry Ford II once said: "To subordinate profit to broad social goals would be totally irresponsible. On the other hand, socially responsible behavior is essential to long term growth and profitability of the corporation." *Business Week*, November 2, 1968.

FIGURE 2.12 Policy and administrative functions of a large organization.

policy group are converted to overt actions by the administrative group. The overlapping portions of the circles denote the cooperation needed from the two groups in order to establish overall policy and suggest when policy interpretations are necessary to achieve desired objectives. The scope of each function and its relationship to the production process are briefly discussed next.

MANUFACTURING

A fundamental function of many production systems is to produce a physical output. Manufacturing includes the operations and direct support services for making a product. *Industrial engineering* is concerned with production scheduling, performance standards, method improvements, quality control, plant layout, and material handling. A *plant services* section typically handles shipping, receiving, storing, and transporting raw material, parts, and tools. The *plant engineering* group is usually responsible for in-plant construction and maintenance, design of tools and equipment, and other problems of a mechanical, hydraulic, or electrical nature.

PERSONNEL

The recruitment and training of the personnel needed to operate the production system are the traditional responsibilities of the personnel function. The problem of keeping people in the organization includes health, safety, and wage administration. Labor relations and employee services and benefits are increasingly important.

PRODUCT DEVELOPMENT

As described in Section 2-5, some companies include research and the development of new products as a major emphasis. Nearly all companies have at

"There is nothing more difficult to carry out nor more doubtful of success, nor more dangerous to handle, than to initiate a new order of things," said a realist named Machiavelli.

least a concern for product improvement. Design efforts vary from a search for new, basic products to the development of by-products and the economic utilization of waste products.

MARKETING

Many ideas for product development come through the marketing function. Sales forecasts and estimates of the nature of future demands are developed to aid other management functions. Selling is the prime interest of marketing. Promotional work is a highly specialized activity involving advertising and customer relations. Contact with customers provides feedback about quality expected from the firm and opinions on how well the products meet quality standards.

FINANCE AND ACCOUNTING

Internal financing includes reviewing the budgets for operating sections, evaluating of proposed investments for production facilities, and preparing financial statements such as balance sheets. The underlying activity is that of a scorekeeper, to see how well the firm and its component departments are scoring in the business competition game.

In this business-game analogy, the accounting function could be likened to the game referee. Cost data are collected for materials, direct labor, and overhead. Special reports are prepared regarding scrap, parts, and finished-goods inventories; patterns of labor hours; and similiar data applicable to production activities. In some firms the accounting function provides data-processing services for other divisions. In other cases, particularly in which computers are used to solve problems instead of to keep records, data processing is a separate function.

PURCHASING

In a narrow sense, purchasing is limited to acquiring materials from outside sources. But carrying out this basic function requires investigating the reliability of vendors, determining what materials are needed, coordinating deliveries with production schedules, and discovering new materials and processes. Because the purchasing function obviously serves the other functional areas, overlap sometimes stretches into activities such as inventory control, material inspections, shipping and receiving, subcontracting, and internal transportation.

Functions of a production process

Another way to group functions is according to their relative positions in a production process. The sequential arrangement in Figure 2.13 is admittedly oversimplified, even for a small firm. However, it indicates the material and information flows that relate management functions to the production of a product.

The functions displayed in the production cycle are fairly self-descriptive.

"As sort of a capsule observation, it could be said that the computer is the LSD of the business world, transforming its outlook and objectives. None of the existing goals of the twentieth-century business enterprise can survive the impact of the computer for even ten years." Marshall McLuhan and Quentin Fiore, *War and Peace in the Global Village* (New York: Bantam Books, 1968).

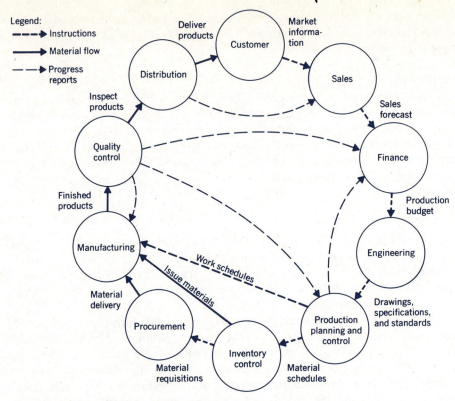

Legend:
- - - - -▶ Instructions
────▶ Material flow
- - - -▶ Progress reports

FIGURE 2.13 Cycle of production functions.

In the following chapters each will be treated in detail. Less obvious are the support functions and implied coordination of operations:

A popular acronym pertinent to information flow is KISS: Keep It Simple, Stupid!

1. The indicated information and progress report paths illustrate only the formal, usually preprinted, forms of paper flow communications. If all the verbal and informal memorandum information channels were shown, the diagram would resemble a windblown spider web.

2. The clerical work and personnel requirements are scarcely suggested by the activity descriptions for each function. But they are very real concerns.

3. One-half the cycle is complete before material begins to flow physically in the production process. The period that must be allowed (lead time) before the production actually begins is a prime consideration in planning. Each function is owed its share of lead time to prepare for a new product, order, or style. The anticipation and control of lead time are among management's major responsibilities.

TABLE 2.1
Overview of Technologies That Are Changing Manufacturing Functions.

Manufacturing Function	Selected Technological Developments
Product Design	Computer-aided Design and Designing for Manufacturability.
Manufacturing/ Engineering	Automatic Generation of Processing and Tool Requirements from Design Database, and Computer Programs to Control Machine Tools.
Industrial Engineering/ Material Requirements Planning	Applications of Computers to Capacity Planning, Labor Planning, Inventory Control, Scheduling, and Shop-floor Control.
Fabrication	Computer Control of Machine Tools and Multifunction Machine Centers.
Materials Handling	Computer-controlled Storage and Retrieval Systems.
Assembly	Robots.
Inspection/Testing	Automated Visual Inspection and Automated Testing.
Sales/Distribution	Application of Computers to Logistics Planning, Order Entry, and Delivery Date Forecasts.

SOURCE: Adapted from *The New Manufacturing: America's Race to Automate,* a Report by the Business–Higher Education Forum (Washington, D.C.: Business–Higher Education Forum, 1984).

EXAMPLE 2-4 Technological impacts on manufacturing functions

Virtually every aspect of modern manufacturing—from the design of products, planning, control of material flows, to the sales and distribution of goods—is in the process of being integrated and computerized. Selected technological developments associated with manufacturing functions are shown in Table 2.1.

Reasons for adopting new technologies vary. Companies making small batches of discrete parts automate so as to increase their efficiency without losing their ability to manufacture a diverse mix of products. Producers of custom-designed, high-performance goods, such as aerospace and electronics manufacturers, push technology to meet ever-increasing quality and reliability standards. Producers of high-volume goods, such as appliances and automobiles, expect technological advances to hold down costs and facilitate retooling for new production models. Production systems are thus being automated to secure higher quality, lower costs, and greater flexibility.

2-8 CLOSE-UPS AND UPDATES

If the managers who operated businesses a century ago could return to visit today's version of their operations, they would be overwhelmed by the modern machines and facilities, and they would be astonished by the pace of operations. Many operations that took hours only a decade ago now take minutes. Communications race at incredible speeds to satisfy information demands. R&D is prodded to conceive new products faster; computerized engineering does the designing faster; manufacturers produce faster; and sellers rely on new technology to keep up.

The shrinking product cycle

Instant product obsolescence for computers has become a familiar story. No sooner does one company get its new computer in the stores than another company announces a newer model that does the same things faster and costs less. Sometimes a company is forced to supersede its own product earlier than it had planned, in order to keep ahead of the competition. Similar experiences with truncated product-life patterns are occurring in businesses besides microelectronics, although the passage from infancy to old age is normally not as quick.

The message to production managers is that the traditional notion of conservative product development and marketing may itself be obsolete. The whole production cycle displayed in Figure 2.1 is being squeezed.

Keeping up begins with companywide attention to innovation and continues with a hefty commitment to R&D. Not only must the R&D unit have technical competence, but it also must be in close contact with customers and the competition. Both provide directions for new developments that might escape a cloistered staff.

Once a new design has been tested and found commercially attractive, it must be propelled rapidly through the manufacturing and marketing phases so as to take advantage of its uniqueness. Because competition is so intense in some markets, new products may have to sacrifice a degree of refinement for development speed. A shorter period of **product maturity**—the time when profits are expected after development costs have been recovered—means that a delayed introduction can make an otherwise worthy product unprofitable. It may even be necessary to cut prices during a product's growth phase to ward off competition, a practice deplored just a few years ago.

The race to accelerate the development of a new product cannot be a tightly regimented exercise. Innovation entails disorder. Rapid direction changes, as usually required, are disruptive. Return on the investment is a gamble. An orderly, conservative corporate culture may not be able to accommodate the high-speed product-development process without a wrenching change in managerial philosophy. Such is the challenge of keeping up.

The first microcomputers in 1977 could process only 8 bits of information at a time (a bit is one or zero in binary code). By 1981 there was a 16-bit micro, followed by a 32-bit micro in 1983.

Between 1981 and 1986, companies almost doubled their new product introductions from the previous five-year period.

The market for refrigerators took over 30 years to mature. It took about 10 years for microwave ovens. The video game market came and went in about five years.

Cooperative and computer-aided engineering

The exacting function of transforming a design concept into a producible product and then adapting it to manufacturing realities is traditionally lengthy. But as we just explained, competition is forcibly shortening the process. One response has been to integrate the design phases. Another is **computer-aided engineering (CAE).** The former is associated with old-line industries, and the latter is being developed for and by the electronic industry.

Design engineers frequently drew blueprints for a new product without consulting the manufacturing engineers charged with building the product or the marketers responsible for selling it. As a result the product often had to be altered to fit the manufacturing process because of practical considerations the designers did not foresee, and this sometimes led to a product that was not what customers had specified. The resolution of the conflict is called **simultaneous engineering,** a procedure in which design and manufacturing engineers work together while maintaining close contact with the marketing staff and customers.

CAE is a cousin to CAD but differs in two ways: (1) CAE is used only to design electronic components, and (2) it is more of an engineering than a drafting tool. CAE systems are used from the conception phase to the detailed diagrams needed for chips or printed circuits.

A CAE work station is at once an example of a highly successful new product introduction and a solution to the search for faster ways to develop new electronics, an obviously winning combination. While still in its adolescence, during which improvements continually appeared, CAE sold well because it shortened the development phase by freeing electrical engineers from having to build prototypes to check out their creation. With a CAE work station, engineers simulate prototypes to find errors and then interact with the computer to make corrections. Depending on the job, CAE systems have cut design time by up to 90 percent. When time is money, CAE is bullion.

Service providers stampede into the information age

The glue that holds together a production system is information. As displayed in Figure 2.13, all the production functions rely on the flow of instructions and reports. Although external communication linkages are not as conspicuous as are internal channels, they are equally important and may offer greater gains from improvement, particularly in service industries.

The proliferation of computers in the industrialized world affords a

The advocated switch from serialized to parallel designing results in a much more complicated process, but it can be justified by better quality as well as improved producibility.

CAE grew from nothing in 1980 to a $350 million business in four years.

CAE also contributes to better designing by allowing "what-if" explorations and saves money on testing equipment because less debugging is needed.

For the 10 years before 1986, corporate spending for information processing had an average annual increase of 16 percent.

broad spectrum of information-oriented opportunities. It is changing the way companies relate to one another, customers, and suppliers. Beyond the expected result of keeping employees better informed and supporting decision making with better data, the expanding **age of information,** as enthusiasts label it, has created a demand for new products, new services, and new technologies. Businesses are being built on the more intelligent use of information. Entirely new enterprises have grown out of systems put in place originally just to provide information. A few examples follow.

Retail stores. Until computerized cash registers replaced old-fashioned cash registers, it could take weeks before department store chains knew what was selling and what to reorder. In addition to the customary use of computers to track cash and credit purchases, clever new uses emerged. Some stores provided point-of-sale computer terminals at which shoppers could place orders that would be ready for them at the checkout counter. One progressive retailer programmed its computers to place replenishment orders directly with suppliers when stock levels dropped below specified numbers; then it sold its programs to other retailers.

Railroads. In few industries has the evolution of technology been so prolonged and fought so bitterly as in the railroads. The grudge goes back to the legendary John Henry who died winning a race against an automatic spike driver. Since then, hundreds of technological advances have been made, and most of them reduced the size of the labor force, but the remaining jobs are safer. Several railroads have leveraged their expertise in automation so as to enter related fields.

Health care. A hospital may not at first seem to be a production system. But it is. A health-care system fits the input–transformation–output model shown in Figure 1.1. It increasingly relies on technology that has close counterparts in the factory. Similarities are apparent in the following scenario:

> *A patient enters a hospital complaining of dizziness and double vision. A nurse feeds his symptoms into a computer. Alerted by the computer printout, a physician diagnoses a brain tumor and, to confirm the diagnosis, recommends tests that are conducted by highly sophisticated machines. After confirmation, microscopic brain surgery is performed; a robot assists in positioning the drill. Later a computer suggests when the patient is well enough to leave intensive care.*

Customers are often the biggest beneficiaries of information-improvement ventures. Services become more complete, faster, and cheaper.

Why inventory control is being computerized is obvious when the interest owed on the amount of money tied up in merchandise is considered.

The question of responsibility if a computer misdiagnoses an illness is still a debated issue.

Such computer programs may be **expert systems** that use **artificial intelligence** techniques to duplicate and codify the medical experts' knowledge and decision-making process. Several organizations are developing the smart computers and software that are destined to transform health care.

2-9 SUMMARY

New products and services are continuously being created to take the place of old ones that are no longer appreciated or needed. The search for improvements is directed toward new concepts for products/services and ways to better production methods and processes. Idea generation is intensified through individual and group exercises. *Cause-and-effect diagrams* are used to give order to investigations and to encourage comprehensive examinations. They are also useful for presentation purposes.

    Research and development activity is usually limited to government programs and larger industrial organizations. It encompasses efforts to advance scientific knowledge (*basic research*), to discover scientific information of commercial interest (*applied research*), and to translate research findings into products or processes (*development*). The effectiveness of R&D may be increased by organizational methods, management of projects, and improved means of technology transfer.

    A product or service typically passes through the phases of *design* (identification and evaluation of a new approach), *preparation* (setting up an organization and securing supplies), *operation* (production and control of output), and *updating* (continuation, termination, or modification of the output or process). These phases are akin to the concepts of systems theory that are based on the analysis of a system's inputs, conversion processes, outputs, and feedback links. Analyses focus on the significant factor interactions, both internal and external, and the combined influences.

    Numerous and diverse functions are required to produce a product or service. These functions, such as purchasing, manufacturing, product development, finance, and accounting, must be coordinated to achieve the overall objectives.

    The pace of production operations is accelerating as a result of advancing technology. Products are rushed to markets faster, and their periods of maturity are shorter. CAE systems hasten the design of electronic products. Development is expedited by having *design* and *manufacturing engineers* work together. Service providers also benefit from computerization through improved information flow and the introduction of new technologies such as *expert systems*.

Creativity

C&E diagrams

Research and development

Product/service cycle

System complexity

Production functions

Advanced technology

2-10 REFERENCES

ADAM, E. E., and R. J. EBERT. *Production and Operations Management: Concepts, Models, and Behavior*, 2nd ed. Englewood Cliffs, N.J.: Prentice-Hall, 1982.

ARIETI, S. *Creativity: The Magic Synthesis.* New York: Basic Books, 1976.

BLAKE, S. P. *Managing for Responsive Research and Development.* San Francisco: Freeman, 1978.

BLANCHARD, B. S. *Engineering Organization and Management.* Homewood, Ill.: Irwin, 1976.

CHASE, R. B., and N. J. AQUILIANO. *Production and Operations Management*, 3rd ed. Homewood, Ill.: Irwin, 1981.

GERSTENFELD, A. *Effective Management of Research and Development.* Reading, Mass.: Addison-Wesley, 1970.

KNELLER, G. F. *The Art and Science of Creativity.* New York: Holt, Rinehart and Winston, 1965.

ORLANS, H. *The Nonprofit Research Institute.* New York: McGraw-Hill, 1972.

RIGGS, J. L. *Productive Supervision.* Englewood Cliffs, N.J.: Prentice-Hall, 1985.

WEINER, N. *Cybernetics.* New York: Wiley, 1948.

2-11 SELF-TEST REVIEW

Answers to the following review questions are given in Appendix A.

1. T F Patterns of *consumption* indicate which products and services hold the most promise for development.

2. T F Phases of the *product/service cycle* are design, prepare, operate, and terminate.

3. T F A group of people should generate more *ideas* than should the same individuals acting independently.

4. T F *Brainstorming* can be done individually or in a group.

5. T F All labeled ribs in a *C&E diagram* point toward the boxed project statement in the middle of the diagram.

6. T F A C&E diagram is usually drawn to a *time scale*.

7. T F A *basic research* project has no specific commercial objective, but it may have commercial value.

8. T F In a *project management* approach to R&D, a team of specialists is formed for the duration of the project.

9. T F A *matrix organization* is associated with R&D project management.

10. T F An advantage of a matrix-organization over a *triangular-organization design* is that each worker has only one boss in the matrix pattern.

11. T F U.S. *patent policy* grants perpetual rights to inventors for control over their inventions.

12. T F *Technology transfer* programs attempt to disseminate technical information to the private sector about government-sponsored research.

13. T F Although a *system* can have almost any boundaries, confusion is avoided by defining production systems along organizational lines.

14. T F *Feedback* controls an operation through decision rules.

15. T F   *Systems theory* emphasizes the combined effect of the factors in production systems.

16. T F   All *policy and administrative functions* of an industrial enterprise should receive equal attention, because all of them affect performance.

17. T F   When *production functions* are arranged according to their position in the production process, they all are connected by material flows.

18. T F   A product is considered to be *mature* when most of the development costs have been recovered and few refinements are likely.

19. T F   *Computer-aided engineering* is used almost exclusively for electronics design.

20. T F   In *simultaneous engineering,* the work of industrial and mechanical engineers is integrated to improve its producibility.

21. T F   *Expert systems* use artificial intelligence techniques to assist the decision-making process.

2-12 DISCUSSION QUESTIONS

1. Assume that you have decided to set up a tutoring service for first-year mathematics students. Your decision is based on a modest survey of potential customers and agreements secured from three graduate students to cooperate with your project. Thus you are the manager of your own business.

Based on the product/service cycle in Figure 2.1, describe the activities you believe will be involved in implementing the four phases of the cycle for the tutoring service.

2. What purpose does feedback serve in a production system? What are the physical means of feedback in an industrial organization?

3. Give an example of a mechanical self-regulating system. How are the decision rules implemented? What are the physical forms taken by the sensors and controllers?

4. Where will overlap in management activities likely occur among the functions displayed in Figure 2.12? How can a definition of boundaries ease the deleterious effects of overlap? Relate the definition of boundaries to internal lines of communication.

5. What should the managers of the functional areas of a production organization know about the organization's research and development program?

2-13 PROBLEMS AND CASES

1. Systems thinking can be used in conjunction with other creativity techniques to enlist new ideas. For example, consider the question of improving the product of a company that makes egg cartons. The first inclination is to stereotype thinking by limiting ideas to the modification of traditional egg-carton designs. One way to get away from trite thinking is to consider the function involved in the input–process–output system. First, look at the manufacturing system: Why use certain materials? Why make egg cartons? and so on. Then consider the larger system that includes the manufacturing operation as only one component: Why package eggs? Why do they need protection? Who wants eggs and in what form? and so on.

Identify at least two system levels and apply solo creativity methods to solve the following problems. Do not limit yourself to practical answers; those that initially seem wild may eventually offer clues to workable solutions.

a. Design a better way to package eggs.

b. Eliminate overcrowding at an X-ray facility at a local hospital.

c. Develop a better absenteeism-reporting method.

2. The relative merits of replacing the traditional 40-hour, five day work week by one of four 10-hour days has created a great deal of debate. Develop a C&E diagram for the "adoption of a 40-hour, four-day week." Base it on the needs of a small plant that fabricates electrical subassemblies and includes an office staff. Use the following five dominant factors:

Three-day off-period (value of having three consecutive days free).

Worker 10-hour day (comparison with traditional 8-hour day).

Process 10-hour day (factors influencing physical conditions).

Work output (effects on the production from the plant).

Worker productivity (effects on the personnel in the plant).

3. Develop a one-sided cause diagram for the factors that influence employee morale. Include at least the four main factors of *financial, social, security,* and *expectations.*

4. You are given the problem of suggesting labor-intensive products that are suitable for production in cottage industries but that require no outstanding talent to produce. The products will be manufactured in homes and therefore must rely on hand tools, and these hand tools cannot be expensive. Materials must also be inexpensive and readily available. Because the finished products will be sold at street fairs, booths in supermarkets, and craft shows, the units must be easy to transport and not fragile. The units should be designed to sell for less than $35 apiece, and materials should not cost more than 25 percent of the selling price.

a. Use a solo search or group ideation method to develop creative solutions to the problem. Comment about the method used and your success in using it.

b. Identify reasonable input restrictions and desirable output features for the cottage-industry prod-

ucts, and show them on a C&E diagram. List the products that meet the conditions shown on the diagram.

5. *The Case of a Not-so-Hot Solution*
After an audit revealed that fuel costs had increased 40 percent in recent months, the facilities manager took steps to correct the situation. She figured that there would be significant savings by lowering the temperature settings while the buildings were unoccupied. Each building had a master steam control that the night janitors were told to set at the new lower reading. However, the savings failed to reach expectations. An investigation revealed that the janitors were using the lower settings *except* while they were cleaning the building; then they put the setting for a whole building at the regular working-day level, although only one or two persons occupied the building.

The facilities manager changed the temperature controls. The new controls could be modified only by using a special tool. She assigned one person to reset the heating level each morning and evening. Before long, the union filed a grievance against the new practice, claiming the janitors could not be expected to work in cold buildings.

a. Draw a cause-and-effect diagram to identify the factors affecting the fuel conservation plan.

b. What should be done now?

6. *The Case of the Squashed Triangle*
J. P. Smith is an inventive, enterprising individual.

> Mr. Smith
> designs,
> produces,
> finances,
> sells.

In his one-man organization he performs all the business functions and is successful.

As his business prospers, Mr. Smith hires employees to do the production work, freeing him to concentrate his time on designing, financing, and selling. He establishes his first supervisor–subordinate relationships.

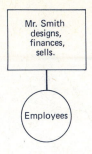

Mr. Smith designs, finances, sells.

Employees

Mr. Smith's business continues to prosper. He hires additional staff to help him design, sell, and handle finances. As he adds more production employees, he sees the need for a production manager.

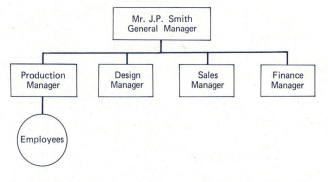

Mr. J.P. Smith
General Manager

Production Manager | Design Manager | Sales Manager | Finance Manager

Employees

J. P. S. enterprises continues to grow with the addition of new product lines. Three lines are established, each under the direction of a supervisor. A supply manager is added at the staff level to handle the greater volume of material. Smith's increasing administrative duties cause him to create a new position for himself and to hire an assistant to fill his old position as general manager.

With business still on the rise, Smith considers a reorganization. He wants to regain the close-knit, cooperative relationships he believes were responsible for his previous successes. He also seeks a more flexible arrangement for adding or deleting product lines, but he does not want to add more layers to his organization.

a. Construct a matrix organization chart as a possible new design for J. P. S. enterprises (include all the positions in the most recent organization chart for the company).

b. What difficulties could be expected in converting the existing organizational structure to a matrix pattern? (*Hint:* Consider bookkeeping, reporting, performance evaluation, and the like.)

7. *The Case of the Defective Berry Boxes*
(Adapted from "Describe Your System with Cause and Effect Diagrams," by M. S. Inoue and J. L. Riggs, *Industrial Engineering*, April, 1971).
A rehabilitation center employs handicapped workers to produce various wood products, such as pallets, glass-packing cases, novelties, and berry crates. The work is purposely designed to emphasize hand labor for training the handicapped.

A quality problem occurred when the center

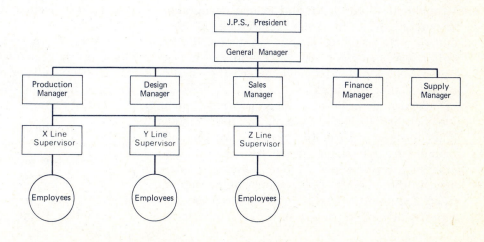

J.P.S., President

General Manager

Production Manager | Design Manager | Sales Manager | Finance Manager | Supply Manager

X Line Supervisor | Y Line Supervisor | Z Line Supervisor

Employees | Employees | Employees

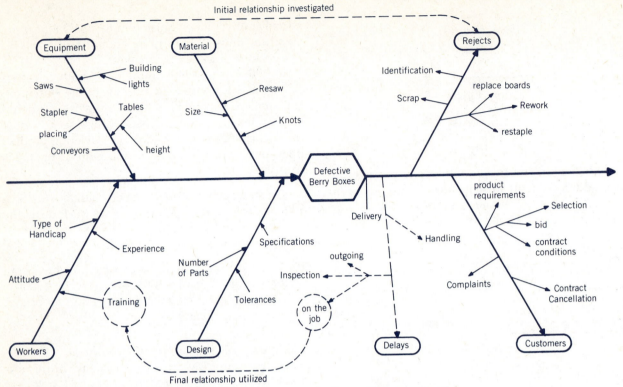

FIGURE 2.14 Initial and revised diagram of the causes and effects of poor production on a rehabilitation center project.

began work on a contract to produce berry boxes. The customer theatened to cancel if sturdier boxes were not produced.

The first pass at the problem identified the obvious causes and effects, as shown in Figure 2.14. An apparent soluton was to add an inspection station for outgoing boxes to ensure the use of good wood and adequate fastening. Then it was noticed that attention was being focused strictly on effects, not causes. Maybe it would be better to have the workers inspect as they produced.

Following the internal inspection idea, attention switched to better physical facilities: brighter lights, lower worktables for wheelchair workers, jigs for more accurate staple placing, and so on. These and other considerations led to the addition of a new effect in the diagram, *Delay* (shown by dashed arrows). To minimize delay, a suggestion was made to improve the motion patterns of workers. Finally, the

subject of training was broached. It was recognized that training was directed at general skills, but little effort was made to show each worker what was especially important to each particular contract. The problem was solved by simply informing the workers of what was needed and how to do it with existing facilities.

Perhaps the answer may appear patently obvious, but how many times have sophisticated solutions been initiated when a simple change could have produced the same results at lower cost and less loss of dignity?

a. How were the concepts of system theory used to analyze the problem? Identify the components of the berry-box production system that are comparable to the system in Figure 2.9.

b. What are the functions of the rehabilitation center as implied by the C&E diagram?

c. After the current berry-box operation is corrected, as depicted in Figure 2.14, by better training, a customer might want to change "product requirements." What factors in the *cause* side of the diagram would most likely be affected by revised customer requirements? How can this relationship of causes and effects be used in production management?

8. *The Case of the Faltering Factory*

An engineer with a flair for production management was the principal owner and founder of a company in Los Angeles. He had a special talent for working with airplane manufacturers, producing on a contract basis their designs of various small devices and subcontracting component parts. He operated a small factory where he inspected and assembled parts purchased outside and tested and shipped the completed assemblies.

Subsequently, to meet competition that sprang up, he decided to manufacture his own parts, which represented the greater part of the total value of his products. To have a favorable labor market and a climate that was good for his type of manufacturing, he located his new plant in Arizona, several hundred miles from Los Angeles. Not only did he set up the parts manufacturing there, but he moved the testing and assembling operations there as well. The owner, however, remained in Los Angeles, in his executive headquarters, with a small group of sales engineers. This permitted him to maintain the same close contacts with his customers that had made him successful.

A qualified factory manager had been put in charge of the new plant. During the starting-up period, the owner made frequent trips to the plant, keeping in touch with what was going on and providing leadership and motivation to the local management. But as time went on, these trips became less and less frequent as the complexity of his personal activities made his continuing presence in Los Angeles more and more compelling. Accordingly, the time came when full responsibility for the factory operation had been shifted to the manager, with the owner depending completely on the manager's activities and results.

Soon the owner began to hear from several of his customers that some of his prices were not competitive, some deliveries were seriously late, and quality was not up to the expected high standard. The owner began to develop a sense of disquietude. His uneasiness reached a point where he procured the services of a consulting industrial engineer.

On his introductory trip through the plant, the consultant got the impression that a rather slow working tempo prevailed. This led her to review the payroll records. Here she learned that for several months the workers' productivity had been slipping. An investigation showed that it was due to delays and slowing down of workers caused by an uneven flow of work. When this was explored more deeply, the cause of the uneven flow proved to be a combination of substandard materials and inadequate machine maintenance. Working backward, the consultant unearthed the root of this difficulty—poor control of materials. Purchased materials were not up to original standards. At first this caused low productivity, which increased labor costs. Then, to compensate for this, a drive toward overhead cost reduction ensued, which included a cutback in maintenance personnel. The result only lowered productivity still more. All this resulted in higher overall cost, lower capacity, and poorer quality, which showed up later in customer complaints.

A meeting of the owner, factory manager, and the consultant revealed that deterioration in purchased materials and parts was because of a misguided program on the part of the factory manager to reduce production costs by saving on purchases. The use of substitute materials created other cost increases that overbalanced several times the slight saving on purchases.

Discuss the case in terms of production functions and decision making in a production system.

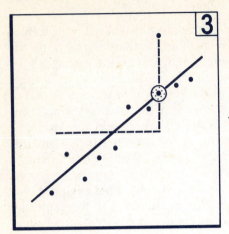

FORECASTING

LEARNING OBJECTIVES

After studying this chapter you should

- be aware of different sources of qualitative forecasts and the mathematical methods used in forecasting.
- be able to fit a straight line to time-series data by the least squares method and to construct a control chart.
- know how to apply the moving average and exponential smoothing methods to obtain single-period forecasts.
- understand the concepts of correlation analysis and how it can be used to develop predictions.
- grasp the factors that determine the cost-versus-accuracy relationships for various forecasting methods.
- appreciate the present value of anticipating events that might occur in the distant future.

KEY TERMS

The following words characterize subjects presented in this chapter:

consumer opinions   trend lines

distributor surveys   cyclic variations

executive opinions   least squares method

market trials   seasonal index

market research   moving average

time-series analysis   exponential smoothing

smoothing constant, α	Delphi method
standard error of the estimate	technological forecasting
correlation analysis	trend spotters
leading indicator	factory of the future
coefficient of correlation, r	robotization

3-1 IMPORTANCE

The impetus for forecasting comes from the necessity to plan, and equivalently, the need for planning comes from the necessity to work today on activities that are intended to meet future demands. Nothing is completely static. Changes in nature may be as gradual as wind erosion in the desert or as sudden as a tornado. Similar conditions occur in production. Demand for a stable item may be essentially constant for years, whereas demand for a different item may disappear before it leaves the design phase. Services in high demand one year may be unwanted the next, replaced by a sought-after service no one dreamed of the year before.

Any production planner would be ecstatic to have perfect information about the future, but no one can get it. Many methods have been tried, and new approaches are being developed. The search continues because the efficient use of resources depends so much on allocation and scheduling to meet coming conditions. Most forecasts are directed at estimating future demand or sales. As shown in Figure 3.1, decisions affecting a variety of activities depend on sales estimates.

"If a man takes no thought about what is distant, he will find sorrow near at hand."
Confucius

3-2 PROBLEMS

Accuracy is the measure of merit in forecasting. When a direct cause-and-effect relationship exits, an expectation of high accuracy is legitimate. The behavior of a chemical compound or maximum stress in a new design can be accurately anticipated, but when we leave the physical sciences and enter the realm of economics, relationships become obscure.

FIGURE 3.1 The sales forecast and production decisions.

A complex set of highly variable factors surrounds any significant economic question. To expect reliable answers is unrealistic. Still, some business executives expect such quality answers, and some forecasters are brave (or foolish) enough to attempt them. Instead of duping each other, the executive should realize that precise business forecasts are usually impossible with currently available techniques, and the forecaster should recognize that his or her efforts provide an indication rather than the final word on future conditions. Such attitudes would at least reduce friction and ulcers.

3-3 HISTORY

Throughout history executives have felt the need for clairvoyant counselors. Kings and merchants had their crystal gazers, palm readers, and astrologers. By mixing psychological black magic with a little court intrigue, some soothsayers did very well. Others received the witch treatment. A few still have followers.

The rhymed prophecies of Nostradamus have been popular for years. Much of the clairvoyance attributed to Nostradamus and similar prognosticators is due to the open-ended, obscure wording of their forecasts. Such wording allows an interpretation of current events to conform to the predictions. This tempting format for long-range forecasts has a special appeal even today. When few specific dates or quantities are specified, ample room is left for maneuvering. Adroit interpretations might even make the general forecast appear prescient. Though such forecasts tease the imagination, they are misfits in an industrial setting.

Early in this century, few formal measures were made to predict future business conditions. Production levels were set by managers to correspond to their estimates of demand. Workers were hired and supplies accumulated at a rate geared to the optimism of department supervisors. But few companies formulated a coordinated product–demand policy.

Before World War II, industries started to recognize that an integrated production system was necessary. Forecasts are a coordinating link in production planning. Distinct groups were established to prepare required predictions. The age of market surveys and product questionnaires began and frequently with it came a credibility gap that fostered subsequent disillusionment. A naive belief that sufficient expenditures could buy a reliable window into the future caused some impressive fiascos. Formal forecasting programs often became suspect.

Today the pendulum of forecasting acceptance appears to be rising again, but not at the fervent pitch previously experienced. The complicated interrelationships of today's economy make forecasts a vital step in operational planning. It is further recognized that the best forecasts will be made by people with special training but that even qualified specialists will make errors. This rational approach permits central forecasts to guide departmental planning with enough flexibility to compensate for prediction deviations.

3-4 SOURCES

Many companies cannot afford a staff of forecasting specialists. Although production planning may be under the auspices of a single individual, he or she is still expected to provide appropriate forecasts. To fulfill these responsibilities, he or she should be familiar with available statistics, be able to analyze and interpret statistical measures, and have a thorough knowledge of the internal and external aspects of all operations.

A surprising amount of free information is available to aid forecasters. General data about the national economic health, pricing indexes, consumer-spending trends, and the like, are offered in different magazines, newspapers, and publications from trade associations and government agencies. Even specific information about particular industries is available from various publishers.

More elaborate forecasting techniques can be employed when their expense is warranted. These methods may be used singularly, but a much better estimate results when they are used in combination with one another.

For example, the Wall Street Journal, Business Week, *and the U.S. Department of Commerce studies.*

For example, farm prices from the U.S. Department of Agriculture or construction trends from the Constructor.

Consumer opinions

The consumer can be questioned. The **consumer's opinions** are objective compared with a producer's opinions, but they may change from day to day. What a consumer intends or hopes to do and actually does may be entirely different. The relatively expensive process of determining consumers' opinions can be reduced by designing the survey to glean other useful information such as the effect of promotional campaigns.

Customer opinions

Individuals who have purchased a product can be asked why they made the purchase. Answers can be solicited when the sale is completed, and a questionnaire can be attached to the guarantee or to sales follow-up literature.

"The government is very keen on amassing statistics. They collect them, add them, raise them to the nth power, take the cube root and prepare wonderful diagrams. But you must never forget that everyone of these figures comes in the first instance from the village watchman, who just puts down what he damn well pleases." Sir Josiah Stamp, a British economist (1880–1911)

Distributor surveys

Estimates of expected sales (**distributor surveys**) can be requested from retail outlets and the company's sales force. Retailers may be more objective than salespeople, but they are less likely to devote the time necessary for conscientious estimates. Both sources are most suitable for short-term forecasts of a year or less.

Many companies rely heavily on judgments made by their sales personnel. The sales front is where the action is. Salespeople and sales managers can spot buying trends and competitor activity. But does a good observer and a hustling agent make a good forecaster? If he or she is aware of influencing economic conditions and the significance of those efforts, that person may be very good. By averaging many opinions, individual optimism and pessimism is somewhat canceled. There is little doubt that the sales force is an acceptable source of estimates, and there is great doubt that it should be the sole source.

Executive opinions

The executive–manager–supervisor level is closer to corporate policy than is the sales force and, correspondingly, is further from the consumer outlook. In common with the sales force, **executive opinions** may be colored by personal prejudices. Opinions are secured individually or from committees, and, one hopes, bias and errors are canceled by compensating views. This divergence in viewpoints, backgrounds, and interests usually provides a good cross section of estimates, but potentially conflicting forecasts may be difficult to reconcile to a consensus opinion.

Consensus executive forecasting is widely used. It can be obtained quickly and with little extra expense. What is usually lacking is the recognition of underlying economic factors. Opinions are extremely valuable for interpreting market data, but they are no substitute for a quantitative analysis of that data.

Marketing trials

The development and introduction of a new product presents special problems. If the new product is a replacement or modification of an existing line, data and opinions applicable to the old product will likely be useful in anticipating the reception of the new version. But when the product is radically different, new data must be generated.

In addition to the forecasting methods already discussed, it may be helpful to expose the new product to a very limited **market trial.** Such a trial is like a controlled experiment in which the market area and method of presentation are carefully selected and controlled. In any experiment, there is a danger of selecting the wrong variables, improper handling, and lack of control. The cost of overcoming these conditions is often enormous and consequently limits the extent and scope of marketing trials.

Market research

Another forecasting approach is an internal or contracted **market research** program. It can be used for new products or existing products for which more extensive marketing data are needed. The purpose of the research is to identify the nature of consumer consumption. After determining how general sales vary with differences in location, buyer occupation, prices, quantity, quality, consumer income, and other factors, this information is related to a specific product, and a forecast is developed. The economic forces identified in a market survey are particularly valuable when combined with other forecasting methods. The means of correlating product sales with market conditions are detailed later in this chapter.

Historical data

Basing estimates of future activity on the past performance of a product is one of the most commonly used and most reliable forecasting methods available. It has the advantage of being quantifiable and objective. Still, it is not perfect.

Inaccuracies result when the economic conditions that prevailed in the past no longer operate. Judgment remains a necessary ingredient of any forecasting method.

In the following sections we consider different methods of analyzing collected historical data. The best method to use for a given set of data is the one that most closely approximates its pattern. In essence, we are using estimates of past performances to estimate future performance; we are putting our faith in two old adages: history is a good teacher and history repeats itself. Then we must check to see whether history is living up to its reputation.

3-5 TIME-SERIES ANALYSIS

A **time-series analysis** defines how a certain production indicator varies with time. Total yearly sales over the last several years would be a production indicator. The manner in which the volume of sales changes on a year-to-year basis is formulated; the resulting expression establishes a sales–time relationship that is used to predict future sales levels.

A good analysis, like a good recipe, takes several steps to complete and includes several components. Paying attention to each step and each ingredient results in a better end product, a more reliable forecast. The most commonly used expression for a time-series forecast is

$$Y = TCSR$$

where

Y = forecasted value
T = underlying trend
C = cyclic variations about the trend
S = seasonal variations within the trend
R = residual or remaining unexplained variations

Each factor in the formula helps account for the many interacting elements that influence the variable being forecast. The relationship of the factors is amplified in Figure 3.2.

It takes a good cookbook plus expertise to be a master.

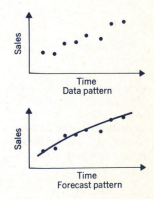

Data pattern

Forecast pattern

Trend

Returning now to our basic equation $Y = TCSR$, we can see that the **trend line** (the dashed line in Figure 3.2) is a line fitted to the long-term historical pattern. The closeness of fit is measured by the aggregate distance by which the smooth trend line misses the data points (circles). Only in the rarest cases does the trend coincide with a line connecting all the points.

Cycles

Cyclic variations account for some of the variance between the trend line and the data points. The wavelike cycles do not necessarily form a repeating pattern. They are caused by the reaction of the plotted variables to changes in general

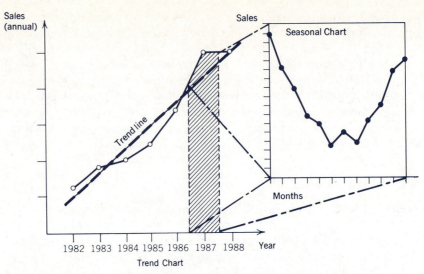

FIGURE 3.2 Cyclic, seasonal, and residual variations.
Axes: The vertical axes in both charts represent sales volume. In the trend chart, the sales per year are shown for a range of years; in the seasonal chart the amount of sales by month for a given year, 1987, is detailed.
Data: Records of past sales are indicated by circles.
Analysis: The lines connecting the circles depict the historical pattern of sales. By analyzing this known pattern, we try to anticipate where the next unknown point will occur. When this forecast value becomes known, it in turn joins the previously known data as a foundation for the next forecast.

A depressed economy would affect luxury sales differently than it would the sales of necessity items, and some luxury items would be affected more than would others.

business conditions. Each industry, and probably every company within the industry, is affected uniquely by the causes of economic fluctuations. The situation is further complicated by the inconsistent internal responses of individual firms. For instance, a signal that an areawide business cycle has bottomed out and sales are expected to rise could trigger expansion plans on one occasion, but the same indicators could lead to a "wait and see" attitude on another occasion. If expansion is planned, the same business indicators would be viewed differently the next time because of the greater capacity available.

The magnitude, timing, and pattern of cyclic fluctuations vary so widely and are due to so many causes that it is generally impractical to forecast them.

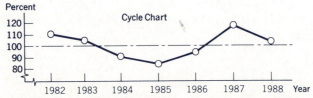

FIGURE 3.3 Comparison of actual values with calculated values (actual sales/forecast sales) $\times 100\%$.

Nevertheless, at least an awareness of the effect of cyclic variations should influence a forecaster. The percentage ratio between a calculated point and the actual point from a trend chart, as plotted in Figure 3.3, indicates any cyclic irregularities. A cycle chart is based on the implied assumption that the trend line is the "normal" condition about which the cyclic variations fluctuate. This reasoning leads to curiosity about the fit of the trend line to the data and about influences accounted for in the "normal" trend.

EXAMPLE 3.1 Unsophisticated sales forecast

An estimate is needed for the expected sales of a product during the second quarter of next year. Sales records since the product's introduction five years ago are available. An approximate forecast can be quickly made by plotting the sales data and visually fitting a straight line to the data. The "eyeballed" trend line in the sales chart of Figure 3.4a starts at $60,000 on the left and increases by annual increments of $12,500. The trend estimate for next year, 6, is then

$$T = \$60,000 + (\$12,500)(5) = \$122,500$$

From the cyclic variations of the sales chart, a forecaster might estimate that actual sales will be slightly greater than the trend indicates. A judgmental value of +8 percent, or 1.08 times the trend forecast, is selected.

A breakdown of annual sales by the percentages sold each quarter is shown in the quarterly sales chart of Figure 3.4b. A simple forecast by quarters could result from an arithmetic average or by a visual inspection of the quarterly data. By using the latter

method, which allows more weight to be given to recent sales figures, second quarter sales are estimated to be 41 percent of the annual sales estimate for next year. Based on the individually developed forecast factors and the recognition that residual errors cannot be forecast, the desired estimate is calculated as

$$Y_{(2)} = TCS$$

$$Sales_{(2nd\ quarter)} = \$122,500 \times 1.08 \times 0.41$$

$$= \$54,243$$

This example is titled "unsophisticated" because the analysis relies on inspection rather than calculation. Unsophisticated is not a synonym for worthless. Indeed, the simpler "eyeball" approach may be more appropriate than are refined methods. Depending on the data accessible, the time available, and the purpose intended, refinements could add cost without a corresponding increase in the value of the forecast.

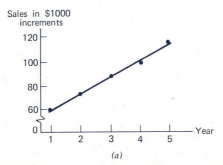

(a)

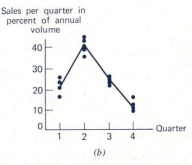

(b)

FIGURE 3.4 Sales data and forecasts by inspection: (a) annual sales and (b) quarterly sales.

Seasons

Seasonal variations are fluctuations that take place in *one* year and are annually *repetitive*. Innumerable products are in demand only a portion of each year. Most of them are associated with the weather. Because weather follows a broadly predictable pattern and because the production of seasonal products must be geared to demand, monthly or even weekly forecasts are common.

Errors

The last term in the basic forecasting equation, R, represents the random or chance variations that are not explained by trend, cyclic, or seasonal movements. By definition, residual variations *cannot* be forecast. They may arise from unforeseeable events such as "acts of God" or a sudden shift in politics. Such events are a bane to forecasters, but philosophically we have to admit that the variety they introduce saves us from a programmed existence, which would deprive life of much of its spice, interest, and alibis.

3-6 TIME-SERIES CALCULATIONS

The mathematics available for analyzing data range from quick and clean arithmetic to tedious but powerful statistical techniques. Selecting the most appropriate method is no simple task. One has to seek the age-old balance between the cost of application and the value of results. A preliminary study may not warrant a detailed analysis, whereas a major new venture would require the most exacting methods possible. Less time and money are usually devoted to forecasts for low dollar-volume products than for high ones. The most refined methods are typically employed when a company has the staff and computing capabilities to handle them, but even pros are occasionally boxed in by a stagnant policy of "this is the way we have always done it."

Data

For an unbiased start, the first step is to decide what type of data will be used in the forecast. Will planned production be a function of past production, or will it be related to some other economic index? In this section we assume that time-series data are appropriate.

The next hurdle is to collect the data. This may amount to an in-house record search or a generation of new information. After the information is collected, it should be questioned. Did certain events that will not recur affect the data? Will future events such as the emergence of new competitors or the development of improved products make previous records inapplicable? Have population shifts or fickle consumer tastes altered the applicability of earlier data? A "yes" answer to these or similar questions indicates a need for caution.

Sometimes historical data can be modified to present a truer picture of past events. For instance, a sharp decline in sales might have been created by temporarily bad publicity. After interest faded and corrective measures were taken,

Watch out for trends. Statistics show that if the current population trends for Houston, Texas, and the United States as a whole continue at their same rate, early in the next century the population of Houston will surpass that of the United States.

sales returned to their previous pattern. This period of reduced sales could be eliminated from the data used in forecasting, or records could be adjusted to compensate for the atypical period. The danger in such juggling is that subjective overtones can be inserted into otherwise objective calculations.

Calculation methods

After making the major assumption that we have representative data, we can choose a method for translating the data into a forecast. We begin by plotting the data to a convenient scale. The plot does not have to be exact because all we wish to decipher is the general pattern. This pattern gives us a good clue to the most appropriate calculation method. Further clues result from familiarity with the methods.

Some forecasting methods are very elaborate and require considerable mathematical competence. Others are of the rule-of-thumb variety and develop a prediction with simple arithmetic. Because there is no unanimous opinion as to a single "best" method, we investigate several of the ones with wide acceptance. They will be illustrated by application to the following basic data:

Generox, Inc., has been producing hand-operated nail-driving machines for five years. The plant has operated at near capacity for the last two years. Forecasts are needed to schedule production for the coming year and to provide estimates for planning future expansion of production facilities. Sales records for the five years have been tabulated by quarters, as shown in Table 3.1. Figure 3.5 shows a plot of total annual sales.

Least squares

Whenever the plotted data points appear to follow a straight line, we can use the **least squares method** to determine the line of best fit. This line is the one that comes the closest to touching all the data points. Another way of saying the same thing is that the desired line minimizes the differences between the line and each data point. This latter explanation gives rise to the origin of the name for the least squares method; it gives the equation of the line for which the sum of the squares of the vertical distances between the actual values and the line values is at a minimum. A further property of the line is that the sum of the same vertical distances equals zero.

$\Sigma(Y_a - Y_F)^2 = $ minimum
where

$Y_a = $ actual value

$Y_F = $ least squares value

$\Sigma(Y_a - Y_F) = 0$

The two equations can also be obtained by differentiating $\Sigma(Y - Y_F)^2$ with respect to a and then with respect to b.

TABLE 3.1
Quarterly Sales in Thousand of Dollars

Year	1984	1985	1986	1987	1988
Quarter 1	190	280	270	300	320
Quarter 2	370	420	360	430	440
Quarter 3	300	310	280	290	320
Quarter 4	220	180	190	200	220
Totals	1080	1190	1100	1220	1300

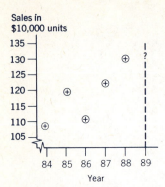

Sales in
$10,000 units

FIGURE 3.5 Pattern of annual sales.

Year	X
1984	−2
1985	−1
1986	0*
1987	+1
1988	+2
	$\Sigma X = 0$

*Base point

A straight line is defined by the equation $Y = a + bX$. For a time-series analysis, Y is a forecast value at a point in time, X, measured in increments such as years from a base point. Our objective is to determine a, the value of Y at the base point, and b, the slope of the line.

Two equations are used to determine a and b. The first is obtained by multiplying the straight-line equation by the coefficient of a and then summing the terms. With the coefficient of a equal to 1 and N as the number of data points, the equation becomes

$$\Sigma Y = Na + b \Sigma X$$

The second equation is developed in a similar manner. The coefficient of b is X. After multiplying each term by X and summing all the terms, we have

$$\Sigma XY = a \Sigma X + b \Sigma X^2$$

The two equations thus obtained are called *normal equations*.

The four sums required to solve the equations, ΣY, ΣX, ΣXY, and ΣX^2, are obtained from a tabular approach. We can simplify the calculations by carefully selecting the base point. Because X equals the number of periods from the base point, selecting a midpoint in the time series as the base makes the ΣX equal to zero. The smaller numbers resulting from a centered base point also make other required products and sums easier to handle. After the four sums are obtained, they are substituted in the normal equations, and the values of a and b are calculated. Then these values are substituted into the straight-line equation to complete the forecasting formula:

$$Y_F = a + bX$$

Exponential

Sometimes a smooth curve provides a better fit for data points than does a straight line. A smooth curve implies a uniform percentage growth or decay instead of the constant increment or decrement exemplified by a straight line. The equation for a curve may take the exponential form, $Y = ab^x$, which indicates that Y changes in each period at the constant rate b.

EXAMPLE 3.2 A least squares line fitted to the trend of the basic problem data

A straight, sloping line appears to be a reasonable fit for the data in Figure 3.5. To illustrate different versions of the least squares method, we first use 1984 as the base point, and then 1986. Using a tabular format with Y as sales in $10,000 increments to determine ΣY, ΣX, ΣX^2, and ΣXY, we have

Year	Y	X	X²	XY	
1984	108	0	0	0	base point
1985	119	1	1	119	
1986	110	2	4	220	
1987	122	3	9	366	
1988	130	4	16	530	
Sums	589	10	30	1225	

which yield the normal equations

$$589 = 5a + 10b$$
$$1225 = 10a + 30b$$

These equations are solved simultaneously to give

$$a = 108.4 \quad \text{or} \quad \$1,084,000$$
$$b = 4.7 \quad \text{or} \quad \$47,000$$

Using the same data and format but changing the base point from 1984 to 1986, we get

Year	Y	X	X²	XY	
1984	108	−2	4	−216	
1985	119	−1	1	−119	
1986	110	0	0	0	base point
1987	122	1	1	122	
1988	130	2	4	260	
Sums	589	0	10	47	

which allows a and b to be calculated as

$$a = \frac{\Sigma Y}{N} = \frac{589}{5} = 117.8 \quad \text{or} \quad \$1,178,000$$

$$b = \frac{\Sigma XY}{\Sigma X^2} = \frac{47}{10} = 4.7 \quad \text{or} \quad \$47,000$$

A forecasting equation is developed by substituting the a and b values into the straight-line equation. The forecast for 1989 is five years away from the 1984 base point of the first version which provides the equation

$$Y_F = \$1,084,000 + \$47,000X$$

and the forecast

$$F_{1989} = \$1,084,000 + \$47,000(5) = \$1,319,000$$

Similarly, using the formula derived from a base point at 1986,

$$Y_F = \$1,178,000 + \$47,000X$$

the forecast for 1989 is three years from the base point and is calculated as

$$F_{1989} = \$1,178,000 + \$47,000(3) = \$1,319,000$$

A comparison of the two forecasts and the computations in developing the equations confirms that centering the base period shortens the arithmetic without altering the forecast values.

We can determine the values for a and b by the least squares method if we convert the exponential equation to its logarithmic form:

$$\text{Log } Y = \log a + x \log b$$

The logarithmic version plots as a straight line on semilogarithmic paper: the Y scale is logarithmic and the X scale arithmetic. This property allows us to set up normal equations in the manner described in Example 3.2. Thus the normal equations

$$\Sigma (\log Y) = N (\log a) + \Sigma X (\log b)$$

$$\Sigma (X \log Y) = \Sigma X (\log a) + \Sigma X^2 (\log b)$$

can be solved by setting up a table to obtain $\Sigma (\log Y)$, ΣX, $\Sigma (X \log Y)$, and ΣX^2. When the base point is selected to make $\Sigma X = 0$, the solution reduces to calculating

$$\log a = \frac{\Sigma (\log Y)}{N}$$

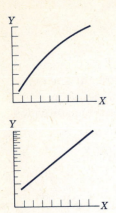

The same $Y = ab_x$ equation plotted on arithmetic scales (top) and semilogarithmic scales (bottom).

and

$$\log b = \frac{\Sigma (X \log Y)}{\Sigma X^2}$$

After solving these equations or simultaneously solving the normal equations when ΣX does not equal zero, the exponential equation is established by taking the antilogarithms of a and b.

If the curved line from the exponential equation does not represent the data adequately, forecasting equations can be based on algebraic series such as

$$Y = a + b_1 X + b_2 X^2 + \cdots + b_n X^n$$

or trigonometric functions such as

$$Y = a + b_1 \sin \frac{2\pi X}{b_2} + b_3 \cos \frac{2\pi X}{b_4}$$

Probability and composite models are also available. An understanding of these methods demands mathematics beyond the level and intent of this text. When there is a need for such sophistication, consult the books listed in the references at the end of this chapter.

Simple average

When b in the straight-line equation $Y = a + bX$ is equal to zero, the line is level. The forecast for the next period then becomes the simple average of all the Y values to date:

$$Y_F = \frac{\Sigma Y}{N}$$

The calculation of a simple average for a trend forecast is then a special case of the least squares method.

EXAMPLE 3.3 An exponential line fitted to the trend of the basic problem data

Because it is not obvious whether a straight or a curved line will fit the given data better, a smooth curve is a legitimate contender. The fitting procedure is analogous to that employed for straight lines; the variation is the use of logarithms for Y values:

Year	Y	X	X²	log Y	X log Y
1984	109	−2	4	2.0334	−4.0668
1985	119	−1	1	2.0755	−2.0755
1986	110	0	0	2.0414	0
1987	122	1	1	2.0864	2.0864
1988	130	2	4	2.1139	4.2278
Sums		0	10	10.3506	0.1719

Because $\Sigma X = 0$, we can solve for a and b as

$$\log a = \frac{\Sigma(\log Y)}{N} = \frac{10.3506}{5} = 2.0701$$

and therefore, $a = 117.5$ or \$1,175,000, and

$$\log b = \frac{\Sigma(X \log Y)}{\Sigma X^2} = \frac{0.1719}{10} = 0.0172$$

making $b = 1.0405$ or an increase of 4.05 percent each period. Thus the forecasting equation is

$$\log Y = 2.0701 + 0.0172X$$

or

$$Y_F = \$1,175,000(1.0405)^x$$

The values resulting from a straight line and a curve should be compared with the observed data. From Table 3.2, it appears that either equation fits the data fairly well. Differences between the fitted lines become more pronounced as the forecasts are pushed further into the future. The exponential formula continuously shows higher values beyond 1988.

Five observations are too few data points to recommend definitely one fitted line over the other. At the present stage, forecasters would have to rely on

TABLE 3.2

Year	X	Actual Y	Straight-Line Fit $Y_F = 117.8 + 4.7x$	Curved-Line Fit $Y_F = 117.5(1.0405)^x$
1984	−2	108	108.4	108.5
1985	−1	119	113.1	112.9
1986	0	110	117.8	117.5
1987	1	122	122.5	122.3
1988	2	130	127.2	127.2
1989	3		131.9	132.3
1990	4		136.6	137.7

judgment. If they are optimistic, they will likely choose the exponential equation that predicts a brighter future. As the historical data accumulate, the trend should become better defined, and the validity of either equation will be evident by the deviations of its forecast from actual values.

Average-value calculations are more often associated with seasonal variations, which take place *within* an overall trend. By definition, seasonal variations are limited to fluctuations during a single year. Therefore, we must collect data for several time periods within the year in order to determine seasonal patterns. Monthly or quarterly records are the most common. When data covering a substantial number of years are available, the arithmetic mean of each period *within* a year tends to damp or average out cyclic effects *among* the years.

After arranging the data according to quarters, months, or whatever time increments are appropriate, the Y values representative of each period are totaled, and the sum is divided by the number of years, N. The resulting mean for each period will be the estimate for the next period if incremental trend efforts are not pronounced. When the trend is significant, the simple average can be corrected for overall growth or decay.

A pattern of seasonal fluctuations often remains relatively constant, even though the trend may rise or fall. The pattern can be easily defined by dividing the simple average of each period by the sum of the averages. The resulting figure is a percentage estimate of the amount of activity expected during each period. This **seasonal index** is converted to sales or other demand units by multiplying the period percentage by the annual trend forecast.

EXAMPLE 3.4 Simple average applied to
the seasonal index of the
basic problem data

A perusal of the quarterly data from Generox, Inc., leads us to question the first year's sales. The introductory phase of a new product often exhibits a pattern all its own. After the initial advertising effort and consumer response have run their course, the seasonal variations tend to be more stable. Because the 1984 sales are not markedly different from those of subsequent quarterly patterns, we will include them as "clean data."

A further check on the suitability of simple averages results from plotting the periodic sales and visually fitting a trend line to each period. If the fitted lines are roughly parallel, no radical change is developing in the demand experienced in each period. When the fitted lines diverge significantly, the averages for the affected periods should be corrected. In some cases a sum amounting to the approximate

shift can be added and subtracted from the affected averages. In more extreme cases, it may be necessary to calculate a formula for the demand per period.

Lines sketched for the quarterly sales of Generox appear approximately parallel in Figure 3.6. Therefore, we can calculate without corrections the average sales for each quarter and the average for all quarters combined:

Year	Q1	Q2	Q3	Q4	Annual
1984	190	370	300	220	1080
1985	280	420	310	180	1190
1986	270	360	280	190	1100
1987	300	430	290	200	1220
1988	320	440	320	220	1300
Sums	1360	2020	1500	1010	5890
Averages	272	404	300	202	1178

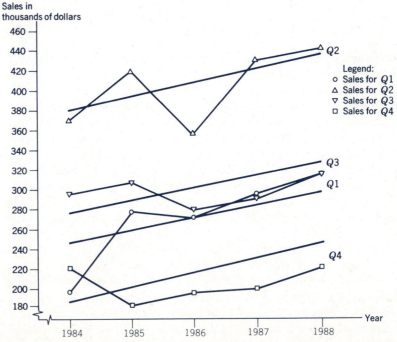

FIGURE 3.6 Trend of quarterly sales established by visual inspection.

The accepted quarterly averages are next converted to a seasonal index. A quarterly index is the quotient obtained from dividing the simple average of each quarter by the average of all quarters:

$$I_{Q1} = \frac{272}{294.5} = 0.92 \qquad I_{Q2} = \frac{404}{294.5} = 1.37$$

$$I_{Q3} = \frac{300}{294.5} = 1.02 \qquad I_{Q4} = \frac{202}{294.5} = 0.69$$

The index thus developed can be used to estimate quarterly sales for the coming year. As applied to the straight-line forecast for 1989, we have

$$F_{Q1} = \frac{\$1,319,000}{4} \times 0.92 = \$303,000$$

$$F_{Q2} = \frac{\$1,319,000}{4} \times 1.37 = \$452,000$$

$$F_{Q3} = \frac{\$1,319,000}{4} \times 1.02 = \$336,000$$

$$F_{Q4} = \frac{\$1,319,000}{4} \times 0.69 = \$228,000$$

Total $1,319,000 = trend forecast for 1989

Moving average

A **moving-average** forecast is obtained by averaging the data points over a desired number of past periods. This number usually encompasses one year in order to smooth out seasonal variations. The smoothing results because high and low values during a year tend to be canceled out. Extending the moving average to include more periods increases the smoothing effect but decreases the sensitivity of forecasts to more recent data.

A moving average is distinguished from a simple average by the requirement for consecutive calculations: each average moves forward in time to include a more recent observation while dropping the oldest point. If a given 12-month moving average is the average demand for January 1988 through December 1988, the next moving average will include the demands for February 1988 through January 1989. In the former case, the moving average represents the demand at midyear; in the latter case, it represents the demand for July 30 or August 1. An average of the two values would center the demand to July.

A moving average calculated for a number of the most recent data observations is seldom a good forecast for the next period unless the data pattern is relatively constant. A seasonal index referenced to the moving average improves the forecast. An index value is calculated by dividing the actual demand by the centered moving average for that period. A more reliable index is obtained by averaging several index values for common time periods. The forecast is thereby the product of the most recent centered moving average for a period and the index value for that period.

EXAMPLE 3.5 Moving average applied to the seasonal index of the basic problem data

Again we refer to the quarterly sales figures of Generox to develop a quarterly forecast for 1989. A four-period moving average will be used. The first moving average is one-fourth of the total sales for 1984 and represents a point in time between the end of the second quarter and the start of the third quarter. The second moving average is the sum of the last three quarters' sales in 1984 and the first quarter of 1985 divided by 4. This value is associated with the end of Q3, 1987, and the start of Q4, 1984. An average of the two numbers obtained gives a moving average centered at Q3, 1984. This procedure is repeated for all the quarters to obtain the moving averages of Table 3.3.

The last column in Table 3.3 is the seasonal index for each quarter. It is obtained by dividing the actual sales for a quarter by the centered moving average for that quarter. We can get a better estimate of a quarter's index by averaging all the values available (Table 3.4).

Before applying the average seasonal index, two checks should be made:

1. The average of the periodic indexes should total to 1.0. In the example the average is

$$\frac{0.9750 + 1.3750 + 1.0125 + 0.6675}{4}$$

$$= \frac{4.03}{4} = 1.0075$$

Therefore, the indexes must be adjusted, or else the sum of the quarterly forecasts will exceed the implied annual forecast by 0.75 percent.

2. Attention should be given to any obvious trends in a quarterly index. In Table 3.4, Q1 appears to be increasing, and Q3 has a distinct downward trend.

TABLE 3.3

Computation of Moving Averages and Quarterly Seasonal Index Values for the Sales of Generox

Year	Quarter	Sales in Units of $10,000	Four-Period Moving Average	Centered Moving Average	Seasonal Index
1984	Q1	190			
	Q2	370			
	Q3	300	270	281	1.07
	Q4	220	292	298	0.74
1985	Q1	280	305	306	0.91
	Q2	420	307	302	1.39
	Q3	310	297	296	1.04
	Q4	180	295	287	0.63
1986	Q1	270	280	276	0.98
	Q2	360	273	274	1.32
	Q3	280	275	279	1.00
	Q4	190	283	286	0.66
1987	Q1	300	300	301	1.00
	Q2	430	303	304	1.42
	Q3	290	305	307	0.94
	Q4	200	310	311	0.64
1988	Q1	320	312	316	1.01
	Q2	440	320	322	1.37 = 4.41
	Q3	320	325		
	Q4	220			

TABLE 3.4
Calculation of an Adjusted Seasonal Index

	Q1	Q2	Q3	Q4
			1.07	0.74
	0.91	1.39	1.04	0.63
	0.98	1.32	1.00	0.66
	1.00	1.42	0.94	0.64
	1.01	1.37		
Totals	3.90	5.50	4.05	2.67
Average seasonal index	0.975	1.375	1.0125	0.6675
Adjusted seasonal index	0.97	1.37	1.00	0.66

Our adjusted seasonal index reflects these considerations. A fraction (4.00/4.03) of each average index was taken, and the results were rounded off with respect to the trend in Q3. Thus calculations provide the forecasting framework, but the finishing touches are supplied by judgment.

The final step is to make a forecast. It is done by taking the product of the most recent centered moving average and its respective seasonal index. Forecasts for the first two quarters of 1989 are

$$Q1_{1989} = 316 \times 0.97 = 307 \text{ or } \$307,000$$

$$Q2_{1989} = 322 \times 1.37 = 411 \text{ or } \$441,000$$

Exponential smoothing

Any quantitative forecasting method serves to smooth out fluctuations in a demand pattern. In **exponential smoothing,** we control the smoothing characteristic by adding the **smoothing constant** called alpha (α), which directs more emphasis to recent demands. Although the exponential smoothing can be applied to any time-series forecasting technique, we shall examine it in connection with averages.

A forecast using exponential smoothing results from the equation

$$F_n = \alpha Y_{n-1} + (1 - \alpha) F_{n-1}$$

which can be rearranged as

$$F_n = F_{n-1} + \alpha(Y_{n-1} - F_{n-1})$$

where

$$F_n = \text{forecast for next period}$$

$$F_{n-1} = \text{forecast for previous period}$$

$$\alpha = \text{smoothing constant } (0 \leq \alpha \leq 1)$$

$$Y_{n-1} = \text{actual value for previous period}$$

Thus, a smoothed forecast is equal to the previous smoothed forecast plus some fraction α of the difference between the forecast and actual values during the previous period. From this description it is apparent that we must determine a previous forecast and the value of α before a new forecast can be made.

When past data are available, the initial value, F_{n-1}, can be a simple average of the most recent N observations. Determining the number of data points to employ shares the same implication as considered for moving averages: when N is large, the estimate is very stable, but it fails to reflect more recent pattern

Short-term forecasting methods such as exponential smoothing find wide application in inventory and production control. Forecasts are exploded into raw material, labor and machine time, and other requirements for production planning.

changes. If no data are available, F_{n-1} is developed from opinions as to what the process is supposed to do.

The rationale for selecting an α value is also similar to that used for selecting the number of periods in a moving average, N. In general, α is some value between 0.1 and 0.3. The response to a changing pattern improves with higher smoothing constants, just as it does with smaller values of N. This rapid response is acquired at the expense of the ability to smooth out random fluctuations. The relationship between α and N that uses the same average age for data is $\alpha = 2/(N + 1)$. Therefore, if we have a reason to be satisfied with a certain N value, we can easily calculate an α value that gives equivalent results.

That exponential smoothing is just a weighted average can be observed by following a series of forecasts. If we let the first forecast be F_0 and subsequent forecasts be indicated by F_1, F_2, and so on, a series of predictions would appear as

$$\text{period 1: } F_1 = \alpha Y_0 + (1 - \alpha) F_0$$

$$\text{period 2: } F_2 = \alpha Y_1 + (1 - \alpha) F_1$$

$$= \alpha Y_1 + (1 - \alpha)[\alpha Y_0 + (1 - \alpha) F_0]$$

$$\text{period 3: } F_3 = \alpha Y_2 + (1 - \alpha) F_2$$

$$= \alpha Y_2 + \alpha(1 - \alpha) Y_1$$
$$+ (1 - \alpha)^2 [\alpha Y_0 + (1 - \alpha) F_0]$$

or, in general,

$$F_n = \alpha Y_{n-1} + (1 - \alpha) F_{n-1}$$

$$= \alpha Y_{n-1} + \alpha(1 - \alpha) Y_{n-2} + \alpha(1 - \alpha)^2 Y_{n-3} + \cdots$$

$$\alpha(1 - \alpha)^{n-1} Y_{n-n} + (1 - \alpha)^n F_0$$

where

$$F_n = \text{forecast for period } n$$

$$Y = \text{historical data}$$

$$Y_{n-n} = \text{starting forecast}$$

$$= F_0, \text{ if an initial forecast cannot be made}$$

This function is a linear combination of all past data weighted according to the smoothing constant α. If $\alpha = 0$, no data since the original forecast should be included. When $\alpha = 1$, the next forecast is the same as the most recent actual value.

The basic exponential smoothing formula is most appropriate for a relatively constant demand pattern. When the demand follows a linear trend, the basic exponential smoothing operator can be applied again to the output of the original forecasting equation to provide a more responsive prediction. This procedure is naturally called *double exponential smoothing.* A further modification to facilitate a quadratic demand function results from essentially applying the smoothing operator a third time to the data, and it is known as *triple exponential smoothing.* Such extensions of basic forecasting models to more sophisticated applications are the result of continuing mathematical efforts generated to yield ever more accurate forecasts, a search doomed never to be completely successful but one that will always be enticing.

EXAMPLE 3.6 Exponential smoothing applied to forecasts for the basic problem data

The exponential smoothing formula limits a forecast to a single period in the future. We can apply this method to the trend forecast (Table 3.1) as a weighted average of the collected data to estimate total sales for 1989. First we apply a smoothing constant of 0.2 to get

$$F_{1989} = (0.2)130 + (0.2)(0.8)122 + (0.2)(0.8)^2 110$$
$$+ (0.2)(0.8)^3 119$$
$$+ (0.2)(0.8)^4 108 + (0.8)^5 \ 118*$$
$$= (0.2)130 + (0.16)122 + (0.13)110$$
$$+ (0.10)119 + (0.08)108 + (0.33)118$$
$$= 119 \text{ or } \$1,190,000$$

Next we can let $\alpha = 0.8$, which makes the formula more responsive to recent demands, as shown by

$$F_{1989} = (0.8)130 + (0.8)(0.2)122 + (0.8)(0.2)^2 110$$
$$+ (0.8)(0.2)^3 119$$
$$+ (0.8)(0.2)^4 108 + (0.2)^5 118*$$

*The initial forecast is taken as the average of all five available annual demands so far experienced.

$$= 0.8)130 + (0.16)122 + (0.03)110$$
$$+ (0.01)119 + (0.0)108 + (0.0)118$$
$$= 128 \text{ or } \$1,280,000$$

From these significantly different forecasts we can easily see the importance of a careful selection for α.

The initial forecast, F_n, for exponential smoothing can also be obtained from other averaging methods. We can forecast the sales for the first quarter of 1989 by starting from either a simple average or a moving average for $Q1$, 1988. We assume that an α value of 0.15 strikes a balance between stability and sensitivity.

The predictions in Table 3.5 differ greatly. This marked difference illustrates how the choice of an initial forecast can affect the subsequent forecast. Regardless of the initial prediction method, if exponential smoothing is consistently applied for several periods, the forecasts will eventually approach the same value. It is those intermediate periods that delight the second-guessers.

TABLE 3.5
Exponential Smoothing Applied to Quarterly Forecasts Based on Simple and Moving Averages

Forecasting equation: $F_n = F_{n-1} + 0.15(Y_{n-1} - F_{n-1})$

	Simple Average	Moving Average
$F_{n-1} = F_{Q1,\ 1988}$	$\dfrac{190 + 280 + 270 + 300}{4} = 260$	316 (unadjusted center moving average from Table 3.3)
$Y_{n-1} = Y_{Q1,\ 1988}$	320	320
$F_n = F_{Q1,\ 1989}$	$260 + 0.15(320 - 260) = 269$ or \$269,000	$316 + 0.15(320 - 316) = 316.6$ or \$316,600

Comparison and control of time-series forecasts

Two basic types of forecasting methods have been considered. Line-fitting equations provide a means to extrapolate forecasts several periods in the future. Iterative procedures rely on timely data to make the next forecast.

A simple average can fit a constant trend and is considered a line-fitting method.

LINE-FITTING METHODS

The extent to which an equation fits a trend is evaluated by the **standard error of the estimate,** S_y. This measure of dispersion of actual data points about the line of forecast points is calculated as

S_y is used appropriately only where $\Sigma(Y - Y_F) = 0$.

$$S_y = \sqrt{\frac{\Sigma\,(Y - Y_F)^2}{\nu}}$$

where

$$Y = \text{historical data points}$$

$$Y_F = \text{calculated fit or forecasted points}$$

$$\nu = \text{number of degrees of freedom}$$

In our examples only one Y has been given for each X, but this is not a required condition.

Before we attempt to use S_y, the assumptions included in the calculations should be explored. First, we are assuming that the forecast value Y_F is the mean of all the Y observations associated with a certain point in time, X. Then we assume that the Y values associated with each X have the same distribution and equal variances as depicted in Figure 3.7. The significance of these conditions will become more apparent when we consider **correlation analysis** in this chapter.

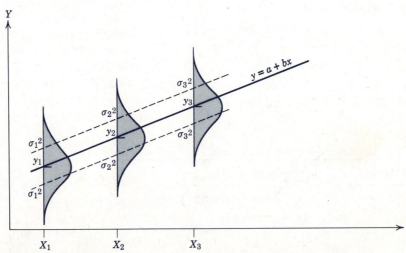

FIGURE 3.7 Line-fitting assumptions: Mean values fall on the predicted line; therefore, $Y_n = Y_{Fn}$. Variances are equal; therefore, $\sigma_1^2 = \sigma_2^2 = \sigma_3^2$.

Our estimate of the true variance (σ^2) is S_y^2. The more observations we have, the better will be our estimate. When the number is about 30, we assume that Y values are normally distributed. Under this assumption, we expect 95 percent of the observations to fall between plus or minus two standard errors of the mean, $Y_F \pm 2S_y$. With fewer than 30 data points, we usually assume that Student's t distribution is more representative of our data. This assumption broadens the band containing 95 percent of the observations. For instance, if a sloping-line prediction is based on 13 observations, 11 degrees of freedom will be used in the calculation of S_y: From the Student's t distribution table, we find that the 95-percent band is $Y_F \pm 2.2S_y$ units wide.

The loss of 13 − 11 = 2 degrees of freedom is due to the two constants, *a* and *b*, in the prediction equation.

The standard error of the estimate gives a reasonable measure of the relationship of the fitted line to past data. By itself, it does not reliably measure how well the line will fit the future. Judgment must be added. To make the application of judgment easier, we can plot existing and new data on charts. The resulting pattern, interpreted with respect to standard error control limits, offers a clue to the sustained value of the predicting equation.

EXAMPLE 3.7 Calculation of standard errors from the basic problem data

The preparation and use of charts are demonstrated by application to the simple average and least squares trend forecasting equations. The standard error of the estimate for each method is calculated as follows:

$$\text{Level line: } Y_F = \frac{\Sigma Y}{N} = 117.8$$

Year	Y	Y_F	$Y - Y_F$	$(Y - Y_F)^2$
1984	108	117.8	−9.8	96.04
1985	119	117.8	1.2	1.44
1986	110	117.8	−7.8	60.84
1987	122	117.8	4.2	17.64
1988	130	117.8	12.2	148.64
		Sums	0.0	324.80

$$S_y = \sqrt{\frac{\Sigma(Y - Y_F)^2}{N - 1}} = \sqrt{\frac{324.8}{4}} = 9$$

Note that the degrees of freedom for the level-line forecast number $5 - 1 = 4$, and that the sloping-line forecast has $5 - 2 = 3$ degrees of freedom. The difference is accounted for by the number of constants in each equation; the slope constant, b, does not occur in the level-line equation. The lower standard for error for the sloping-line prediction tends to confirm the visually obvious: a sloping line fits the data better than a level one does.

Two charts are illustrated in Figures 3.8 and 3.9. Both are based on limits expected to include 95 percent of the data. The range that accounts for a given percentage of the data is a function of the type of distribution and its standard deviation. We have cal-

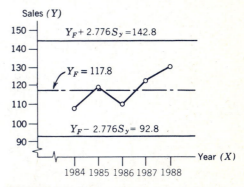

FIGURE 3.8 Y plot chart: control chart for the simple average forecast.

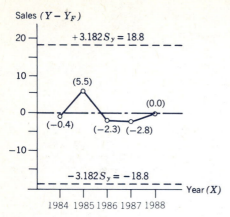

Sales $(Y - Y_F)$

$+3.182 S_y = 18.8$

20

10

(5.5)

(0.0)

0

(−0.4)

(−2.3) (−2.8)

−10

$-3.182 S_y = -18.8$

Year (X)

1984 1985 1986 1987 1988

FIGURE 3.9 Cumulative $(Y - Y_F)$ chart: control chart for the least squares forecast.

culated our estimate of the standard deviation, S_y, and have assumed that Student's t is the appropriate distribution because of the limited number of data points. Though recognizing that the prediction equations have different degrees of freedom, the respective ranges from the table of Student's t values are

$\pm 2.776 S_y = \pm 25$ for the simple average forecast

$\pm 3.182 S_y = \pm 18.8$ for the least squares forecast

The statistical control limits based on the standard error computations can be used in different control chart formats. In Figure 3.8 the actual observations are plotted directly as they occur. If the fitted line is not level (sloping or curved), the control limits are set parallel to the slope or curve. The second version, shown in Figure 3.9, tracks the *cumulative*

deviations $(Y - Y_F)$ of the actual values from predicted ones. In the example, the last plotted point falls on the 0 line because the forecasting equation was developed by using all the data shown in the chart to make $\Sigma(Y - Y_F) = 0$. Unless the equation is absurdly revised each period, future total deviations will likely stray to either side of the 0 line.

In either chart the plotted points should randomly fall above or below the center line but within the established limits as long as the forecasting equation characterizes the data. When points accumulate on either side of the center line or if one falls beyond the control limits, the forecaster should be on the alert. Such patterns may be just chance happenings, or they may signal a change in the trend. In the shadowland of prediction, a forecaster cannot ignore any signals or clues.

Sloping line: $Y_F = 117.8 + 4.7X$

Year	Y	Y_F	$Y - Y_F$	$(Y - Y_F)^2$
1984	108	108.4	−0.4	0.16
1985	119	113.1	5.9	34.81
1986	110	117.8	−7.8	60.84
1987	122	122.5	−0.5	0.25
1988	130	127.2	2.8	7.84
		Sums	0.0	103.90

$$S_y = \sqrt{\frac{\Sigma(Y - Y_F)^2}{N - 2}} = \sqrt{\frac{103.9}{3}} = 5.9$$

Sometimes a forecast of accumulated demand for several periods is required. An approximate measure of the standard error for the cumulative demand is calculated as $S_y \sqrt{n}$, where n is the number of periods over which the forecast demand is accumulated. As applied to the least squares forecast already developed, the standard error for a two-year cumulative demand is calculated as follows:

Year	Period, n	Forecast Demand, F_y	Cumulative Forecast Demand, ΣF_y	Standard Error, $5.9 \sqrt{n}$
1988	now	127.2	—	
1989	+1	131.9	131.9	5.9
1990	+2	136.6	268.5	8.3

Thus, about five times in a hundred, the cumulative actual demand through 1990 would fall by chance outside the range of

$$268.5 \pm 3.182(8.2) = 268.5 \pm 26.1$$

$$= 294.6 \text{ to } 242.4$$

A chart of control limits for accumulated demand would obviously show increasing divergence for extended forecasts.

In general, control charts are simple and useful tools. They serve primarily as easily maintained warning devices. Plotting historical data gives an indication of their statistical stability. Adding new data points as they become available shows whether current activities are following the historical pattern. To some extent an emerging new pattern suggests how forecasting equations can be suitably altered. A subtle but important aspect of charting is the forced continual comparison of actual and forecast values during updating. The time to recognize change is while it is occurring.

When there are enough observations, say 30 or more, a normal distribution is a legitimate assumption; the confidence level of 95 percent would then be established as $1.96 S_y$.

OTHER METHODS

All forecasting systems should follow the idealized system sequence shown in Figure 2.9: input–conversion–output–feedback. Data are cranked into a forecasting procedure and translated into a prediction that *must* be checked against actual outcomes. Different methods perform better under certain circumstances. Unless estimates are continually monitored, a forecaster cannot properly evaluate performance.

Statistics-based control charts can be used for iterative methods as well as for line-fitting methods. For example, a moving-average forecast is charted by plotting the difference between actual data and forecast values. Control limits for these plotted points are set by multiplying the average of the differences (called the *moving range*) by ± 2.66. These limits are then interpreted in just the same way as are those established for line-fitted forecasts.

The analysis of qualitative forecasts is less formal. Comparison of estimated values with actual amounts reveals which opinions are most reliable and the direction if there is bias. A survey may yield a forecast with a large error, but it may still be a fine predictor if the error is consistent. Simply plotting the forecast against the corresponding actuality provides a visual check and a reminder of the success of each forecasting method.

3-7 CORRELATION

A correlation analysis examines the degree of relationship between variables. Our work with the standard error of the estimate in Section 3-6 was a step in this direction. We calculated S_y to determine how well our prediction equation fitted the dependent variable, Y sales, to the independent variable, X years. From this beginning we can delve deeper into the correlation to obtain other useful insights.

Simple correlation expresses the relationship between two variables and is

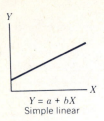

$Y = a + bX$
Simple linear

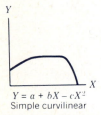

$Y = a + bX - cX^2$
Simple curvilinear

FIGURE 3.10 Regression lines for one independent variable.

Leading indicators, regression, and econometric models are collectively classified as "causal models" to indicate that the causes of activity are analyzed rather than just the past patterns.

associated with regression lines typified in Figure 3.10. Multiple correlation measures relationships among more than two variables. The surfaces shown in Figure 3.11 represent the relationship between two independent variables and one dependent variable. As in previous sections, we shall focus our attention on the most direct application, simple linear correlation, to understand the principles without being hobbled by excessive calculations. Many computer programs are available to relieve computational effort when it appears advisable to include a number of variables in regression equations.

Correlation is not limited to time-series analysis. It can be applied to the investigation of any "regression line" relating variables. Our predicting equations were regression lines relating sales to time. Sales could also be related to the gross national product, housing starts, population growth, or other economic indicators. Relationships of this kind are known as **leading indicators.**

A search for a leading indicator starts with assumptions as to the causes of demand. A house builder could hypothesize that housing demand depends on population growth, prices for houses, personal incomes, demolition of old homes, housing starts by competitors, and availability of financing. After data regarding the possible indicators have been collected, correlation studies are conducted to see whether they are indeed related to the housing data. Ideally, an indicator will be identified that has a pattern similar to the historical demand but leads it by several periods. For example, it might be discovered that the housing demand is correlated with the interest rates: as the rates climb, the demand declines, but the change in demand follows a change in interest rates by six months. In this case, a builder could observe today what is happening to interest rates and use the observation as a leading indicator to predict with some confidence what the demand for houses will be in six months.

Coefficient of determination

This dispersion of data points about regression lines is characterized by three sums-of-squares:

$$\Sigma(Y - \bar{Y})^2 = \Sigma(Y_F - \bar{Y})^2 + \Sigma(Y - Y_F)^2$$

total variation — explained variation — unexplained variation

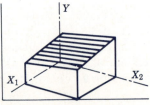
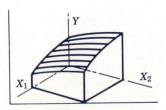

FIGURE 3.11 Regression lines for two independent variables.

TOTAL VARIATION

The deviation $(Y - \overline{Y})$, as shown in Figure 3.12, is the vertical distance between a data point and the mean of all observations, $\overline{Y} = \Sigma Y/N$. This term is a measure of the total variation of the dependent variable and is divided into two parts: the explained and the unexplained variations.

EXPLAINED VARIATION

The variation explained by the regression line is represented by $(Y_F - \overline{Y})$. In Figure 3.12, this deviation appears as a vertical distance between the line of regression and a horizontal line at $\overline{Y}$. Thus, a b value of zero in a regression line for two variables makes $Y_F = \overline{Y}$, and the sum-of-squares of the explained variation equals zero. This makes the total variation equal the unexplained variation, as was the case for the level-line forecast, $Y_F = 117.8$, in Example 3.7. Including a slope factor b in the forecasting equation, $Y_F = 117.8 + 4.7X$, improves the prediction because the explained part of the total variation is increased.

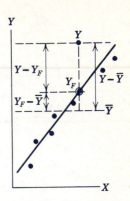

FIGURE 3.12 Deviation of a data point.

UNEXPLAINED VARIATION

The sum-of-squares of the unexplained variation is the familiar term used in calculating the standard error of the estimate. Because S_y measures the fit of the regression line to the data, it is logically based on the deviations not explained by the fitted line. This term will equal zero only if all the data points fall directly on the regression line. Then, of course, the regression line will explicitly define the relationship of the variables. In the more common case in which the data do not coincide with the prediction equation, each deviation appears as the vertical distance from a point (Y) to the regression line.

The ratio of the sum-of-squares of the unexplained variation to the sum-of-squares of the total variation, $\Sigma(Y - Y_F)^2/\Sigma(Y - \overline{Y})^2$, measures the proportion of the total variation that is not explained by the regression line. Therefore,

$$1 = \frac{\Sigma(Y - Y_F)^2}{\Sigma(Y - \overline{Y})^2}$$

measures the proportion of the total variation explained by the regression line. This expression is called the *coefficient of determination*.

Coefficient of correlation

The square root of the coefficient of determination,

$$r = \sqrt{1 - \frac{\Sigma(Y - Y_F)^2}{\Sigma(Y - \overline{Y})^2}}$$

is the more commonly recognized **coefficient of correlation, r.** The value under the radical can never be greater than 1 nor less than 0. However, because the radical has both positive and negative roots, the value of r is between $+1$ and -1. The plus or minus is indicative only of the slope of the regression line as depicted in the side charts. When $r = +1$, all the data points fall on an upward

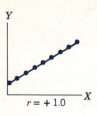

$r = +1.0$

$r = +0.8$

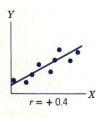

$r = +0.4$

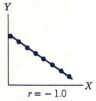

$r = -1.0$

sloping regression line. When r is between $+1$ and 0, the regression line still slopes upward, but the data points fall on either side of the line. The closer they cluster around the line, the closer r approaches 1.

COMPUTATION

When sufficient data are available, we can calculate r using the sums-of-squares as indicated or, more directly, from the formula

$$r = \frac{N \, \Sigma XY - (\Sigma X)(\Sigma Y)}{\sqrt{N \, \Sigma X^2 - (\Sigma X)^2} \, \sqrt{N \, \Sigma Y^2 - (\Sigma Y)^2}}$$

This equation is appropriate when the degrees of freedom do not have to be considered because there are enough data.

When the data are limited, the sums-of-squares each are divided by their associated degrees of freedom. In calculating r for the sales figures in Example 3.7, we have

$$r = \sqrt{1 - \frac{\Sigma(Y - Y_F)^2/(N - 2)}{\Sigma(Y - \overline{Y})^2/(N - 1)}}$$

$$= \sqrt{1 - \frac{103.9/3}{324.8/4}} = 0.757$$

The variation explained by the regression line, $Y_F = 117.8 + 4.7X$, is about $(0.757)^2 = 57$ percent of the total variance between actual sales and average sales.

CONFIDENCE

Now we have a coefficient of correlation equal to 0.757, but what can we say about our faith in the relationship it implies? We know that a greater number of observations increases our confidence in a statistic, but how many does it take to be, say, 95 percent confident? There are tests that allow us to make probability statements about r values obtained from a certain number of observations, N. Some critical values of r and N for 95-percent and 99-percent confidence levels are provided in Table 3.6. This table shows the value of r that must be exceeded for a sample size N to ensure that the correlation coefficient is not actually zero.

The values of r in Table 3.6 were developed by a statistical test of the hypothesis that the actual value of r is zero. The hypothesis is rejected for each confidence level when r is above the values shown. Before we can be reasonably sure of some relationship, $r \neq 0$, the calculated value of r must exceed a critical value based on the number of observations from which it is developed. Thus, for a simple size of 10, $N = 10$, r must exceed ± 0.632 if we are to be 95 percent confident that the correlation is not just a chance occurrence. For the sample size, r must be larger (± 0.765) so as to increase our confidence from 95 to 99

TABLE 3.6

Critical Values of r for 95-percent and 99-percent confidence levels

N	95%	99%	N	95%	99%	N	95%	99%
10	0.632	0.765	30	0.361	0.463	50	0.279	0.361
12	0.576	0.708	32	0.349	0.449	60	0.254	0.330
14	0.532	0.661	34	0.339	0.436	70	0.235	0.306
16	0.497	0.623	36	0.329	0.424	80	0.220	0.287
18	0.468	0.590	38	0.320	0.413	100	0.197	0.256
20	0.444	0.561	40	0.312	0.403	150	0.161	0.210
22	0.423	0.537	42	0.304	0.393	200	0.139	0.182
24	0.404	0.515	44	0.297	0.384	400	0.098	0.128
26	0.388	0.496	46	0.291	0.376	1000	0.062	0.081
28	0.374	0.479	48	0.284	0.368			

SOURCE: By permission from W. J. Dixon and F. J. Massey, *Introduction to Statistical Analysis,* 2nd ed. (New York: McGraw-Hill, 1957).

percent. As we would expect, critical *r* values decrease for the same confidence level as the sample size increases.

Interpretation of correlation analysis

The interpretation of a regression and correlation analysis deserves respect—even suspicion. It is said that figures can lie as well as liars can figure. Recent history provides numerous examples of forecasting fiascos created by acts bordering on statistical larceny. History also records spectacular accolades earned by accurate predictions. The difference between end results is often the judgment that accompanies an analysis.

The *purpose* of forecasting is to predict the future. In doing so, we may or may not reveal a fundamental economic truth. Although it helps to know the basic foundations from which demand originates, even emanations from the source can lead to acceptable predictions. Because universal truths are seldom known, we have to inspect our estimates judiciously for shaky relationships and spurious intimations.

A value of *r* approaching 1 is an encouraging sign. It indicates that the relationship being investigated is potentially a useful predictor. But it is not a binding guarantee of an accurate prediction. It could be misleading because

1. The data used to develop the regression equations are unreliable or unrepresentative.

2. An apparent correlation is a result of random chance factors.

3. A high correlation exists between two variables that both are functions of a third variable, which has not been identified or included in the analysis.

Sales of English and math textbooks could show a high correlation, but they both are a function of an independent variable: number of students.

EXAMPLE 3.8 Regression and correlation analysis

The management of Generox, Inc., think that sales of nail drivers should logically be related to the amount spent on construction. If a relationship does exist, published government and construction industry forecasts of anticipated building levels can be used as an additional sales predictor. First a check is made to confirm that the building-level forecasts are relatively accurate. Next the national figures are broken down to conform to the Generox marketing areas. Then the records of monthly building volume and nail-driver sales for corresponding months are collected. The resulting data are tabulated as shown in Table 3.7, in which the construction volume is in $100 million units and product sales are in $10,000 units. The column headings correspond to the values needed for the calculation of r and a linear regression equation.

From this reasonably large sample, we can calculate the coefficient of correlation as

$$r = \frac{N \Sigma XY - (\Sigma X)(\Sigma Y)}{\sqrt{N \Sigma X^2 - (\Sigma X)^2} \sqrt{N \Sigma Y^2 - (\Sigma Y)^2}}$$

$$= \frac{20(519.5) - (50.4)(202.4)}{\sqrt{20(130.5) - (50.4)^2} \sqrt{20(2076.62) - (202.4)^2}}$$

$$r = \frac{10,390 - 10,201}{\sqrt{2610 - 2540.16} \sqrt{41,532.4 - 40,965.7}}$$

$$= \frac{189}{(8.36)(23.81)}$$

$$= 0.95$$

TABLE 3.7
Computations Required to Calculate r and Y_F

Nail-driver Sales × 10^{-4} Y	Construction Volume × 10^{-8} X	Y^2	X^2	XY
7.1	1.8	50.51	3.24	12.78
9.9	2.3	98.01	5.29	22.78
9.0	1.9	81.00	3.61	17.10
10.4	2.6	108.16	6.76	27.04
11.1	3.1	123.21	9.61	34.41
10.9	2.8	118.81	7.84	30.52
10.5	2.9	110.25	8.41	30.45
9.8	2.4	96.04	5.76	23.52
11.1	2.8	123.21	7.84	31.08
10.2	2.5	104.04	6.25	25.50
9.7	2.3	94.09	5.29	22.31
10.9	2.8	118.81	7.84	30.52
8.8	2.1	77.44	4.41	18.48
8.6	1.9	73.96	3.61	16.34
12.3	3.2	151.29	10.24	39.36
11.4	3.0	129.96	9.00	34.20
11.2	2.8	125.44	7.84	31.36
10.2	2.6	104.04	6.76	26.52
10.7	2.7	114.49	7.29	28.89
8.6	1.9	73.96	3.61	16.34
202.4	50.4	2076.62	130.50	519.50

Comparing the calculated $r = 0.95$ with the critical value of $r = 0.561$ in Table 3.6, we can be at least 99 percent confident that some relationship exists between the sales of nail drivers and construction volume. The coefficient of determination

$$r^2 = (0.95)^2 = 0.9$$

specifies that about 90 percent of the variation in sales is explained.

Although the correlation is not exact, it is definitely worthy of consideration for prediction purposes. From the sums developed in Table 3.7, we can apply the least squares method to obtain a forecasting equation. The normal equations are

$$\Sigma Y = Na + b\Sigma X \qquad 202.4 = 20a + b(50.4)$$

$$\Sigma XY = a\Sigma X + b\Sigma X^2 \qquad 519.5 = a(50.4) + b(130.5)$$

which are solved for a and b to obtain

$$a = 3.37 \text{ or } \$33,700 \qquad b = 2.68 \text{ or } \$26,800$$

The forecasting equation is then

$$Y_F = 3.37 + 2.68X$$

Thus a projected construction volume of $275,000,000

for the next month would suggest a sales volume of

$$Y_F = 3.37 + 2.68(2.75) = 10.74$$

Therefore, the sales of nail drivers during the next month should be in the vicinity of $107,400.

The need for caution does not mean that regression and correlation analysis should be avoided. Every method of forecasting requires caution. It will be a long, long time before we can turn the crank to a collection of equations and expect foolproof answers about the future. In the meantime we must link our equations to a thorough knowledge of influencing economic elements and a careful diagnostic evaluation of each step in the forecast development.

END OPTIONAL MATERIAL _____

3-8 SELECTION OF A FORECASTING METHOD

A single organization may use several different forecasting methods to anticipate the future of its various activities. It also will likely use different methods during the life cycle of a single product. The selection may depend on any or all of the following factors:

1. Availability and accuracy of historical data.
2. Degree of accuracy expected from the prediction.
3. Cost of developing the forecast.
4. Length of the prediction period.
5. Time available to make the analysis.
6. Complexity of factors affecting future operations.

Cost versus benefits is always a critical issue to management, and forecasting is no exception, although the dollar values are less certain than in most other management decisions. First, a manager has to estimate the losses that might accrue from inaccurate forecasts; this is essentially a forecast of the forecasting results—uncertainty compounded. Then the manager must evaluate forecasting methods in terms of practicality and cost. A balance is sought between making the best use of data to meet real needs and applying costly techniques that promise potentially greater accuracy but may require more information and competence than are available. These relationships are implied in Figure 3.13. (adapted from "How to Choose the Right Forecasting Technique," by J. C. Chambers, S. K. Mullick, and D. D. Smith, *Harvard Business Review*, July–August, 1971).

The least sophisticated forecasting model is a "persistence prediction": the next period will be the same as this one. It has the intuitive appeal of sticking with a winner and works well for stable conditions, but it is worthless for predicting when a change will occur.

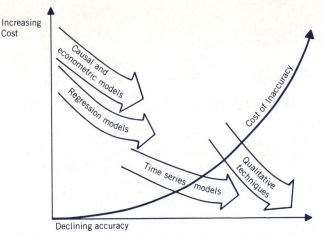

FIGURE 3.13 Comparison of forecasting costs for different forecasting methods and the cost of inaccuracy. Relative values are based on a medium-range (three months to two years) forecast. Differences in cost and accuracy for each type of forecast reflect expected results from the application of various methods. For example, qualitative techniques range from the expensive and accurate market research method to the less costly but also usually less reliable method of relying on one individual's prophecy.

Because forecasts attempt to predict future levels of activity, different forecasting methods are associated with product/service life cycles based on demand levels. Three stages are shown in Figure 3.14: note the similarity to the life cycle divisions portrayed in Figure 2.1 to represent general production interests.

In addition to the forecasting techniques already described, many other methods have been developed to treat the special requirements of product/service life-cycle states. Some of the more prominent methods are discussed in the following paragraphs.

BOX–JENKINS TECHNIQUE

The **Box–Jenkins technique** is an especially good method for short-term forecasting that is a major refinement of the exponential smoothing method. The special property of Box–Jenkins models is that successive observations in a time series are assumed to be *dependent* on the behavior of prior data in the series. Because 50 to 100 observations are desirable for the development of a Box–Jenkins model, this method is most frequently applied to time series in which the interval between observations is small, so that a relatively long history can be readily obtained.

X-11

The **X-11** is a method developed by Julius Shishkin of the Census Bureau, which decomposes a time series into trend, cycles, seasons, and irregular elements, to

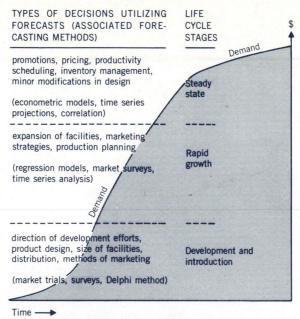

TYPES OF DECISIONS UTILIZING FORECASTS (ASSOCIATED FORECASTING METHODS)

LIFE CYCLE STAGES

promotions, pricing, productivity scheduling, inventory management, minor modifications in design

(econometric models, time series projections, correlation)

Steady state

expansion of facilities, marketing strategies, production planning

(regression models, market surveys, time series analysis)

Rapid growth

direction of development efforts, product design, size of facilities, distribution, methods of marketing

(market trials, surveys, Delphi method)

Development and introduction

Demand

$

Time ⟶

FIGURE 3.14 Types of decisions that rely on forecasts during life cycle stages of a product or service.

predict turning points and to time special events. It works well for forecasts in the medium range, three months to a year.

DELPHI METHOD

The **Delphi method** is a qualitative forecasting method developed at the Rand Corporation, which solicits and collates opinions from experts to arrive at a reliable consensus. A series of questionnaires is sent to a panel composed of experts from selected technological specialties. Each questionnaire solicits written opinions about specific subjects and requests supporting reasons for the opinions. These reasons are summarized in each iteration and returned for inspection by the whole panel. Through this series of exchanged arguments and transfer of knowledge, a consensus prediction is gradually forged. Advocates of the Delphi method claim that the anonymity of written responses from the expert panel preserves the desirable features of a committee of specialists while reducing the bandwagon and dominant-personality effects that unduly sway group opinions. Delphi forecasts may stretch several decades into the future.

TECHNOLOGICAL FORECASTING

Technological forecasting is the concept of attempting to identify, far in the future, technological developments as guides for current activities. Several forecasting techniques are used to anticipate coming innovations. The Delphi method is frequently employed to develop time-scaled "maps" of future inventions,

"State of the art" forecasts are made for subjects such as computer developments, aircraft designs, health-care needs, sources of food, use of lasers, labor requirements, and so on.

supporting technology, and social or economic developments. Very sophisticated models of interrelated factors have been programmed for computer simulation of possible future conditions for companies, cities, and even the world. These predictions may stretch decades into the future and are consequently suspect, but they provide some evidence and thought-provoking scenarios for long-range planning.

ECONOMETRIC MODELS
Econometric models are a system of interdependent equations that describe the factors influencing economic activity. Detailed models are relatively expensive to develop, but they yield commendable long-range forecasts and are very useful for anticipating turning points in activity levels.

3-9 CLOSE-UPS AND UPDATES

One thing that the tidal wave of technology has not delivered is an infallible sales forecaster. More information about the buying public is available than ever before, and computers can digest the data faster, but predictions have not improved commensurately. Even technological forecasting, which is less subject to purchasers' whims, is not significantly more reliable today. The faster pace of change and the rapid expansion of interrelated factors that sway consumer behavior are given as reasons. The same reasons are behind the unending drive to improve forecasting reliability: producers must keep pace with the mushrooming market forces and be able to anticipate their future impact.

The trend toward trend watching
Although everyone is fascinated with the future, planners are obsessed with it. They pursue all promising clues. Newspapers are dissected to count how many times certain key phrases appear or how many inches of ''help wanted'' ads are printed. People are polled to determine their optimism regarding future economic conditions. Census data are checked to detect demographic shifts. All such information is collected and analyzed to discern patterns that might reveal incipient trends. The early detection of even a minor shift in demand can provide a bountiful harvest of sales.

The economic turbulence in recent years has caused some companies to turn to **trend spotters**—organizations that eschew conventional data and analysis techniques for forecasting, relying instead on secondary sources from which intuitive connections are gleaned. For instance, the Naisbitt group studies local newspapers in search of emerging patterns. The Business Intelligence Program of the Stanford Research Institute International

Trends with marketing implications in the early 1980s included

high concern for health and fitness.

greater buying power of the elderly.

slower growth in the Southwest and higher growths in the Northeast.

shift from brown to white alcoholic beverages.

Naisbitt, J., and P. Aburdene, *Reinventing the Corporation,* Warner Books, New York, 1985.

gathers about 150 experts from diverse fields in monthly meetings to interact in a quest for major and minor trends. Other companies draw on repeated surveys of selected people and perceptions obtained from reading many magazines, journals, and books.

Trend spotters claim to be more adept at detecting changes than are number crunchers, who have to extrapolate from old data to envision the future. By examining events as they occur and creatively analyzing them for business implications, trends supposedly can be detected early enough to take advantage of them. Skeptics claim that any group of informed people could make comparable predictions by simply discussing current events in terms of their future impact. However, the services provided by trend-spotting organizations will continue to be attractive to the insecure, who fear that viewing the future as a projection of the past may miss discontinuities that foretell changes.

Trend watching supposedly involves *right-brain thinking* to supplement *left-brain linear reasoning* with intuition and imagination.

Near-now and far-out futures for production

Heady visions of a few years ago are often commonplace events of today: nuclear power, satellite communications, lasers, gene splicing, better mousetraps, and the like. Futurologists are peering ahead to identify prospective changes that should be considered by today's strategic planners. They foresee such things as cars that spot their own parking places, household appliances that talk back, computers powered by protein cells, wristbands that continually monitor health, and factories in orbit, to name only a few.

Sensational predictions are made about scientific discoveries and future life-styles. Gloomy forebodings of overcrowded, dismal cities and scarcities of food and energy are countered by visions of sea farms, new energy sources, and more socially responsive governments. Behind all such scenarios are production systems that are expected to produce goods and services with ever-increasing efficiency.

The much talked about **factory of the future,** totally unmanned and automatic, is surprisingly near at hand. Necessary technologies have begun to gel—drives and manipulators to shape products, controllers to turn on machines and robots to tell them what to do, sensors to monitor the work, and interactive communication networks to tie everything together. Small-scale unmanned factories have been operated on a trial basis. They are proving to be feasible. The not-so-distant prospect of the complete automatization of continuous-process industries and high-volume assembly-line operations is focusing attention on what will happen to displaced workers.

The future effects of **robotization,** for example, have alarming social implications. Robots will certainly replace many workers, causing a still-undetermined amount of unemployment. An indication of the magnitude

Products from off-earth, gravity-free factories would include ball bearings, pharmaceutical drugs, and computer components.

Overpopulation and nuclear war are generally conceded to be the most frightening societal issues.

At the Fujitsu Funac factory, robots make parts for robots. Only two humans, confined to a control room, are in the factory. The work continues 24 hours a day.

Past waves of labor dislocation, such as factory mechanization in the 1950s, were solved by shorter workweeks, earlier retirement, and new jobs in the expanding service sector.

of job losses can be seen from a comparison with the mechanization of agriculture. In a little over 75 years the number of U.S. farmers was reduced from 30 percent of the labor force to less than 4 percent, whereas agricultural output increased more than three times. Some pessimists predict that robot-displaced workers will create a permanently higher level of joblessness. Others disagree.

Although predictors of the far-out future often miss their mark, they attract early attention to factors that might otherwise be ignored. Perhaps some of their dire predictions miss because of preventive actions taken as a result of their warnings. Returning to the robot issue again, research is under way to find out the effects of social isolation in a robotic factory with very few workers, of retraining older workers whose jobs no longer exist, and of realigning the work-place hierarchy in which a new "elite" may be evolving to serve robots. Such preparations, based on advanced warnings from forecasters, will facilitate the transition from today's to tomorrow's production systems.

On the positive side, robots increase safety and contribute to lower product prices.

A "rainbow-collared" worker is neither a manager nor an operator but a combination of both, managing and operating smart machines.

3-10 SUMMARY

Modern production is heavily dependent on forecasts of demand or activity. Forecasts can be developed from *opinions* of consumers, distributors, experts, and executives; by marketing trials and *market research;* or through analysis of *historical data and causes.*

Qualitative forecasts

Quantitative forecasts

A *time-series analysis* defines how production indicators vary with time. The factors affecting a time-series forecast, $Y = TCSR$, are

T = underlying long-term trend of the indicator

C = cyclic variations about the trend

S = seasonal variations within the trend

R = residual or remaining unexplained variations

Predictions extrapolated from a line fitted to time-series data

Best-fit trends lines, either straight or curved, can be developed by the *least squares method.* The resulting line equation is extrapolated to estimate future demand. A simple average or a *moving-average* forecast corrected by a seasonal index damps out seasonal fluctuations to indicate future activity. *Exponential smoothing* emphasizes more recent data according to the selection of α, the smoothing constant, in the forecasting equation $F_n = \alpha Y_{n-1} + (1 - \alpha)F_{n-1}$.

Single-period forecasts

All forecasts should be checked for reliabiliity by comparing predicted with actual values. Statistical control charts are used to monitor time-series forecasts. Control limits are calculated from the standard error of the estimate, $S_y = \sqrt{\Sigma(Y - Y_f)^2/\nu}$, which measures the fit of the forecasting equation to the historical data.

A *correlation analysis* examines the degree of relationship between variables. It starts with the calculation of a regression equation which mathematically relates two or more variables. The coefficient of determination indicates the proportion of the data dispersion explained by the regression equation. The coefficient of correlation is the square root of r^2 and measures the amount of association between variables on a scale from $+1$ to -1.

The most appropriate forecasting method depends on the type and quantity of data available, the allowable analysis cost and time, and the desired predicting range and accuracy. Many methods are available, ranging from elaborate, mathematical (econometric) models to the sophisticated, qualitative *Delphi method*. The effective use of any forecasting technique requires an intimate knowledge of the economic forces affecting the prediction and a searching evaluation of the data sources, mechanics of the model, and interpretation of results.

Trend spotting to detect patterns by studying printed media is a recent addition to forecasting methods. Predictions about events far in the future allow planners to conduct current activities in a manner likely to avoid future difficulties.

Correlation concepts

Accuracy–cost relationship

Anticipation of future events

3-11 REFERENCES

ARMSTRONG, S. J. *Long-Range Forecasting*. New York: Wiley, 1985.

BOWERMAN, B. L., and R. T. O'CONNEL. *Time Series and Forecasting*. Belmont, Calif.: Duxbury Press, 1979.

BOX, G. E. P., and G. M. JENKINS. *Time-Series Analysis, Forecasting, and Control*. San Francisco, Holden-Day, 1970.

BROWN, R. G. *Smoothing, Forecasting and Prediction of Discrete Time Series*. Englewood Cliffs, N.J.: Prentice-Hall, 1962.

GRANGER, C. W. J. *Forecasting in Business and Economics*. New York: Academic Press, 1980.

MONTGOMERY, D. C., and L. A. JOHNSON. *Forecasting and Time Series Analysis*. New York: McGraw-Hill, 1976.

MORRISON, N. *Introduction to Sequential Smoothing and Prediction*. New York: McGraw-Hill, 1969.

THOMOPOULOS, N. T. *Applied Forecasting Methods*. Englewood Cliffs, N.J.: Prentice-Hall, 1980.

WHEELWRIGHT, S. C., and S. MAKRIDAKIS. *Forecasting Methods for Management*, 4th ed. New York: Wiley, 1985.

3-12 SELF-TEST REVIEW

Answers to the following review questions are given in Appendix A.

1. T F Many free data, such as *customer opinions,* are available to aid forecasters.

2. T F *Forecasts by salespeople and executives* are subject to personal prejudices and must therefore be judged accordingly.

3. T F A *market trial* is conducted to evaluate customer acceptance of a new product.

4. T F In the formula for a time-series forecast, $Y = TCSR$, the last factor (R) represents variations not explained by the other factors.

5. T F   Sales data typically have repeated patterns of identical *cyclic variations* that are impractical to forecast.

6. T F   Two *normal equations* are used in the least squares method to determine Y_a and Y_f.

7. T F   In a straight-line equation, *a* is the value of the line at its base point, and *b* is its slope.

8. T F   A trigonometric function converted to its logarithmic form plots as a straight line, the formula for which can be established by the *least squares method*.

9. T F   A *moving average* provides a forecast for only one time period beyond the most recent data point.

10. T F   The *smoothing constant, alpha,* can have a value from -1 to $+1$.

11. T F   Forecast values plotted in a *control chart* should fall inside statistical control limits, which are based on standard error computations.

12. T F   *Correlation analysis* examines the relationship among variables.

13. T F   The coefficient of determination is the square root of the *coefficient of correlation, r.*

14. T F   There is no correlation between two variables when $r = -1$.

15. T F   A *leading indicator* is a known variable that indicates the future level of another variable.

16. T F   Experts are polled in the *Delphi method* to forecast technological developments.

17. T F   *Trend spotting* relies on the statistical analysis of data from the printed media to determine a trend.

18. T F   Potential unemployment from future *robot* installations is a current concern of production systems planners.

3-13 DISCUSSION QUESTIONS

1. Assume that you are manager of a small sawmill and logging operation. You have no formal forecasting personnel or procedures. It is now time to prepare your broad production plans for next year. An estimate of the total board feet required from the logging operation and a rough breakdown of the sizes of finished lumber are required. From what sources could estimates be obtained? Evaluate the sources.

2. Why are consumers' opinions likely to be more objective than distributors' opinions?

3. Two firms are considering market trials for new products they have developed. One firm has perfected a portable, heavy-duty, laser metal cutter; the other firm has a new educational game for preschoolers. Both plan to use the trial to determine the acceptance and potential demand for their new products. Is this plan logical for both firms? What considerations should be included in designing the market trial?

4. Why should the historical data used in forecasting be questioned if you know the information is correct? Give some examples.

5. Does the sum-of-squares of $(Y - Y_F)$ for the exponential line of Example 3.3 equal zero? Why?

6. Why bother to center a moving average?

7. What limits the use of the most recently calculated moving average as a forecast for the next period? (For instance, in Example 3.5 the most recent centered moving average is 322 and might be used as the forecast for the next period, Q_{1989}.)

8. Explain the following statement: The response to a changing pattern improves with higher smoothing constants (α) in exponential smoothing, just as it does with smaller values of N in moving-average forecasts.

9. Could a standard error of the estimates, S_y, be calculated for exponential smoothing predictions? Why?

10. How is the standard error of the estimate related to the coefficient of determination?

11. Persistence predictions may seem too naive to be classified as a forecasting method, but they can be surprisingly successful under some conditions. Comment on the following observations:
a. In sporting events the usual forecast is for the current champion to win again.
b. A decision not to make a decision is a persistence prediction.

12. Suggest possible leading indicators of readily available data for the following products and services:
a. Industrial production in a developing country.
b. Enrollment in a private business school specializing in data-processing training.
c. The number of convicts for whom space must be provided in penitentiaries in less populous states.
d. Demand for prepared baby foods.

13. What are some of the major difficulties encountered in developing technological forecasts?

3-14 PROBLEMS AND CASES

1. The sales figures for two products first marketed six months ago are shown in the following table:

Month	Product 1	Product 2
January	$110,000	$ 54,000
February	102,000	63,000
March	95,000	80,000
April	85,000	98,000
May	78,000	112,000
June	70,000	133,000

a. What July forecasts for each product are obtained by using a six-month moving average?

b. How do you explain these results from part a with respect to the dissimilar patterns?

2. Monthly sales in thousands of dollars for the past two years are as follows:

Month	2 Years ago	1 Year ago
Jan.	253	250
Feb,	236	252
Mar.	245	248
Apr.	246	241
May	260	247
June	251	244
July	249	249
Aug.	242	251
Sept.	234	238
Oct.	244	249
Nov.	246	252
Dec.	257	

a. Fit a line to the data, and determine a forecast for the next month (January).

b. Calculate the coefficient of correlation, and interpret your answer.

c. Establish the cumulative demand through June of

next year. What is the range within which the cumulative demand would be expected to fall 95 percent of the time?

d. Establish a forecast for January of next year by using a six-month moving average.

e. Select an initial forecast from part a, and use $\alpha = 0.2$ to determine the forecast for next January by exponential smoothing.

f. Compare the forecasts from a, d, and e. Which one would you select? Why?

3. The current sales for the product depicted in Problem 2 are as follows:

Month	Sales	Month	Sales
Jan.	251	July	257
Feb.	261	Aug.	266
Mar.	258	Sept.	280
Apr.	256	Oct.	271
May	262	Nov.	293
June	258	Dec.	278

a. Plot current sales on a control chart (95-percent limits) developed from the equation used in Problem 2. What conclusions can you draw from the chart?

b. What forecast would you make for January of the following year? Explain your reasons.

4. Given:

Year:	1	2	3	4	5	6	7	8
Demand:	90	100	107	113	123	136	144	155

a. Plot the data and establish a forecast for year 9.

b. Establish the 95-percent confidence limits, and plot them on the graph developed in part a.

c. Compare forecasts using exponential smoothing with $\alpha = 0.15$ and $F_1 = 85$, with the Y_F values

obtained by the regression equation established in part a.

5. Annual sales for a company in the steel industry closely followed the regression line $Y = 422(1.025)^X$ for a period of 12 years. During the following 5 years, the sales followed $Y = 567 - 0.116X$. In the last 2 years, sales increased by 3 percent each year. What forecast would you make for the next year?

6. Quarterly unit demands for a product are as follows:

Year	Winter	Spring	Summer	Fall
1	81	64	73	83
2	80	70	84	74
3	86	59	71	73
4	98	72	74	64
5	106	68	75	60

a. Using a four-period moving average, determine a seasonal adjusted index, and establish a forecast for each quarter of next year.

b. Use line-fitting methods to determine a forecast for each period of next year.

7. Private aircraft sales in a three-state marketing area and the number of students enrolling each year to take flying lessons are shown in the following table:

Year	Aircraft sales	Student starts
1	900	8,000
2	800	9,000
3	900	11,000
4	1000	10,000
5	1100	9,000
6	1200	12,000
7	1000	13,000
8	900	11,000
9	1100	12,000
10	1200	13,000

a. Using just the past history of aircraft sales, estimate the number of sales to expect in years 11 and 12. How comfortable are you with the forecast?

b. Using the number of students starting flying lessons as a leading indicator for aircraft sales, develop the equation

$$\text{Sales}_{(\text{year } t)} = a + b(\text{Starts}_{(\text{year } t\text{-}2)})$$

to indicate the expected sales in years 11 and 12 ($t = 11$, $t = 12$) from the student starts in years 9 and 10 ($t - 2$).

8. *The Case of the Market Trial*

The average large supermarket stocks about 7000 items, counting different sizes of the same product as separate items. About 30 percent of these items were stocked a year earlier; thus 20 percent are "new products." Yet only one in seven of the new products will still be stocked after a year has passed.

Most of the new products either are undergoing a market trial or have already passed preliminary trials and are receiving full promotion. Obviously very few survive. But the stakes are high enough in the marketing game to keep companies playing, even with the ante fee of an expensive market trial.

In addition to the risks inherent in all forecasting, market trials are subject to special vagaries. One might be called luck: unusual weather that affects the outcome, or localized economic strife in the area that inserts bias. Another is competition. During the typical three-month period, a competitor might initiate price cutting or, just the opposite, withdraw a competing product from stores. Both tactics make the trial results difficult to interpret: "Do they want us to believe our new product is great because they know they can beat it, or are they afraid that we'll find out we have a real winner?"

Such considerations are perhaps more pronounced in market trials, but similar vagaries can influence other forecasting methods. Discuss special conditions that could affect and thereby bias historical data and qualitative input to forecasting techniques. How can these be compensated?

9. *The Case of a Factory Refitted to Face the Future.*

Adapted from "Building on the Past," by Jeffrey Zaslow, *Wall Street Journal*, Sept. 16, 1985). Fierce competition is a powerful stimulant for change. When General Motors (GM) saw its market being eroded by imports, the company realized that its outdated factories were undermining its competitive position in terms of both cost and quality of output. In 1982 GM decided to modernize a plant that was built in 1935. The challenge was to rebuild, retool, rethink practices, and to do it all without shutting down the plant.

Less than two years later the plant had been converted into a prototype factory of the future, and the assembly line had run for all but four months of the transition. This conversion ran counter to conventional thinking that high-tech factories had to be built from scratch. Few thought that a plant could stay in operation during such a retrofit.

A four-month shutdown was necessary for two purposes. It allowed the 2600-person retrofit team to work day and night to finish the project while GM's auto workers attended retraining classes to learn how to work in a computerized, roboticized environment. The classes were designed to win over workers to automation and prepare them for the new responsibilities that come with machine-paced processes.

The start-up was awkward. Old work teams were broken up and their members scattered throughout the plant. Robots banged into one another. Workers scrambled to keep up with the swift line speed. But gradually they adapted, and the equipment downtime receded. Eventually productivity rose.

Robots replaced about 1000 hourly positions. For example, automated equipment and 48 robots handled 95 percent of the welding, formerly a manual operation. Seven robots did all the painting. Although the loss of jobs hurt, the labor unions gave their approval for machines that performed undesirable and unsafe jobs. It also became apparent to employees that the machines supported huge inventory and material savings plus consistently higher quality.

The resulting competitive edge for the plant promised job security.

a. What are the advantages and disadvantages of retrofitting an old plant to accommodate new technology rather than building a new state-of-the-art plant?

b. List the types of forecasts that might have been sought, using forecasting methods from Sections 3-3 and 3-8, to help decide whether to close or refit the plant. For instance, a time-series forecast of car sales by type and volume would have had to support the plant's planned output.

SYSTEM ECONOMICS

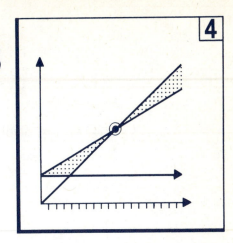

LEARNING OBJECTIVES

After studying this chapter you should

- appreciate the scope of economic factors that affect production systems.
- understand the concepts of suboptimization and sensitivity of decisions.
- be able to construct and interpret break-even charts for both single and multiple products.
- comprehend linear and nonlinear economic relationships with respect to marginal revenue and cost, average fixed and variable cost, and profit.
- realize what factors are involved in determining a product's life cycle and how life cycle costing encourages trade-off analyses for investment decisions.
- know how fixed and variable task times can be used in capacity planning for labor requirements.
- be aware of the support provided by decision support systems and the information available from the Bureau of Economic Analysis for system decisions.

KEY TERMS

The following words characterize subjects presented in this chapter:

tactics	unit contribution	multiproduct planning
strategy	dumping	make-or-buy decision
suboptimization	nonlinear relationships	life cycle analysis
decision sensitivity	marginal revenue or marginal cost	capacity planning
break-even point		life cycle costing
break-even chart	average unit cost	design-to-cost

fixed and variable tasks

decision support system (DSS)

management information system

Bureau of Economic Analysis

underground economy

gross national product (GNP)

4-1 IMPORTANCE

Money is the philosophers' stone of production systems. It is the substance that converts one type of resource into another. It is the subject of economics, a "science concerned chiefly with description of the production, distribution, and consumption of goods and services."

Economic considerations impinge on all aspects of production. They appear in every chapter of this book and are always present in the minds of production managers and analysts. We know that "time is money," and this influences scheduling. Materials, machines, products, and personnel are money, too, in the sense of the philosophers' stone. Therefore, economic effects underlie most of the decisions made in a production system.

In this chapter we examine the relationships among revenue, costs, capacity, product life, and intangible factors that influence production planning. These relationships will be viewed from a system perspective to observe potential strategies. In Chapter 5 the scope is narrowed to specific operations that support production.

4-2 TACTICS AND STRATEGIES

"Quantities derive from measurements, numbers from quantities, comparisons from numbers, and victory from comparisons." Sun Tsu, *Art of War.*

A popular retort by harassed managers is "You don't understand the big picture!" Not only is the retort popular; it also is often accurate. Too much attention to the elements of production can obscure the purpose of production. At the other extreme, a preoccupation with production is wasted if day-to-day operating problems are neglected and they subsequently damage output. Profitable production depends on both efficient operations and an effective policy, or equivalently, efficient tactics and effective strategy.

There is no sharp demarcation between tactics and strategy. The military version of the terms distinguishes **tactics** as the art of handling troops, from **strategy** as the employment of means on a broad scale. But when does the handling of troops cease to be the employment of means? Consider a military objective to be the capture of a hill by several combat platoons. The strategic plan is for one platoon to make a diversionary frontal assault while the remaining platoons perform an enveloping movement. To the leader of the assault platoon, a direct attack might seem an absurd plan. He could lead his men on a gradual advance that would certainly reduce the platoon's casualties. This tactic is more efficient from his viewpoint, but it would ruin the effectiveness of the strategic plan and would likely increase total casualties. This is an example of **suboptimization**—an attempt to optimize a tactical segment of a problem with little or no regard for the solution's strategic effectiveness.

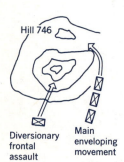

Hill 746

Diversionary frontal assault

Main enveloping movement

System suboptimization

Less brutal but equivalently damaging effects of suboptimization continually plague managers. In a production setting, tactics could be the daily handling of people and machines. Strategy is the coordination of the entire production function with other functions of the firm. Suboptimization might develop from the commendable effort of a supervisor to increase the length of production runs. This tactic would decrease the setup costs and probably increase the efficiency of the involved persons and machines. It would also increase the storage costs by enlarging the inventory on hand. Perhaps the capital tied up in inventory is urgently needed elsewhere in the organization. Thus an action designed to cut costs in one area could actually decrease total profit—suboptimization.

Organizational suboptimization is logically attacked by improving communications. The approach is far easier to recognize than it is to implement. Traditionally, tables of organization are established to outline the areas of responsibility and authority of personnel. Information is ideally channeled vertically or horizontally on a "need to know" basis. Decisions are made on one level subject to review by an upper level and carried out by a lower level. Years of operation have proved that this orthogonal arrangement works quite well when messages are transmitted freely and accurately. Modifications such as internal information centers are designed to take advantage of new high-speed data-processing machines.

Problem area crossing organizational boundaries.

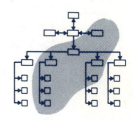

Orthogonal organization and information flow.

The advent of computers and automated data processing has certainly increased the timeliness, quantity, and quality of information, but it has not necessarily solved the problem of suboptimization. Figure 4.1 symbolically shows an organization's flow of information and decisions concerning a tactical portion of the production system. In theory, such a flow should make each production decision compatible with the total organizational policy. The weak link in the system is people. Until all messages are originated expeditiously, interpreted as intended, and acted upon when expected, suboptimization will remain a danger.

Sensitivity

Another aspect of tactics and strategies deals with the **sensitivity of a decision,** the vulnerability of an optimal solution to changes in controlling conditions. As tentative strategies are formulated, they are rated as to their effectiveness. The effectiveness scale may be in dollars, tons of production, or some nondimensional rating such as percentage points that reflect a combination of tangible and intangible factors. There are often several ways to pursue each strategy. These alternative means of executing a certain policy are then rated as to their tactical efficiency. When the tactical alternatives of a less attractive strategy rate with or above the tactics associated with the preferred strategy, we have a sensitive decision situation, as shown in Figure 4.2*b*.

The degree of sensitivity in a decision situation indicates the need for additional planning. In Figure 4.2*a*, the tactics required to implement strategy *A* in both plants show a higher efficiency than does any tactic available for strategy *B*. This condition confirms the superiority of strategy *A* over *B*. In the

An economic sensitivity study is conducted by varying the value of estimates used in making a decision. A sensitive decision is one in which a small change in an estimate switches the choice among alternatives.

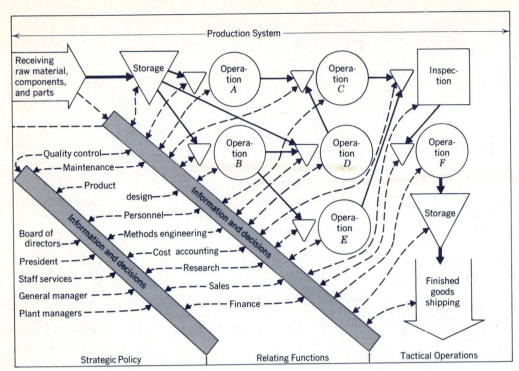

FIGURE 4.1 Information, decision, and product flow in a production system.

sensitive decision situation depicted in Figure 4.2*b*, the best tactic for strategy *B* rates with or above the highest tactic for strategy *A* in each plant. A slight miscalculation of efficiencies could reverse either of the top ratings: The need for more study to confirm the ratings should be apparent to the decision maker. The amount of study warranted is a function of the analysis cost and the significance of the decision. In the following sections, we consider several types of planning studies.

4-3 BREAK-EVEN ANALYSIS

Theoretical notions of strategies and tactics are translated into realities when an organization commits its resources to specific production objectives. The relationship of revenue to production cost and capacity is a critical factor in production planning. The level of output at which a course of action begins to yield a profit is its **break-even point**. A break-even analysis considers the economic implications of production plans to determine preferred pricing, servicing, manufacturing, and scheduling policies.

A break-even model relates fixed cost, variable cost, and revenue to the quantity of units produced. Relationships are conveniently displayed on graphs to assist communication among decision makers. Mathematical expressions of

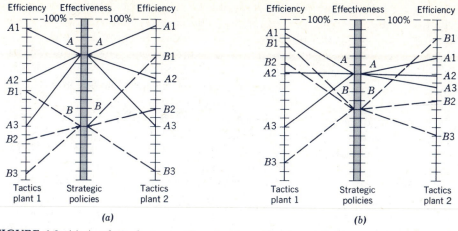

FIGURE 4.2 (*a*) A relatively insensitive decision situation and (*b*) a sensitive decision situation.

the relationships support the graphs and are used to test the sensitivity of break-even decisions. Any variable in the model can be altered to observe its effect on the other variables. In this way, alternative courses of action may be simulated and evaluated in the planning stage.

Revenue, costs, and capacity relationships

A traditional **break-even chart** shows revenue and costs as a linear function of output. As is illustrated in Figure 4.3, total cost is the sum of fixed cost and total variable cost for each level of production noted on the horizontal axis. Revenue is the product of the selling price and the number of units made and sold. To appreciate the graph's diagnostic potential, we have to consider the significance of implied assumptions and the mathematical relationships of the charted variables.

Fixed costs (*FC*) are represented by a horizontal line. The assumption that

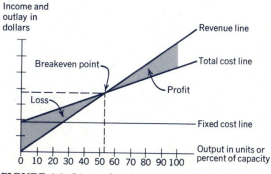

FIGURE 4.3 Linear break-even chart.

these costs are constant over a given capacity range is feasible for two reasons. The first is that by definition, fixed costs are expenses that continue regardless of production levels. Typical costs included in this category are rent, interest, property taxes, insurance, research, and indirect labor. The second reason is that the chart represents conditions expected over a relatively short time span. Therefore, no special capacity-induced procedures, such as layoffs or new equipment investments, are included in the limited planning range.

Unit variable costs (V) are also constant over the chart's period and capacity range. These costs account for the direct expenses of producing a product, such as spoilage, packaging, raw materials, and direct labor. Total variable cost (TVC) is the unit cost times the number of units produced (N), or $TVC = NV$.

A linear revenue line (R) results from the assumption that each product sells for the same price (P). When this assumption is valid, $R = NP$.

The break-even point (B) indicates the number of units that must be made and sold before costs equal revenue, $R = TVC + FC$. To solve for B mathematically, we simply find the value of N that equates costs and revenue. $B = N$ when $NP = NV + FC$, or

$$\text{Break-even volume} = B = \frac{\text{fixed cost}}{\text{contribution}} = \frac{FC}{P - V}$$

For production less than B, the **unit contribution** ($P - V$) serves to pay off the fixed cost. When N exceeds B, ($P - V$) is the *incremental profit* expected from each additional unit made and sold.

SUMMARY OF SYMBOLS

N = number of units
FC = fixed cost
V = unit variable cost
TVC = total variable cost
    = NV
TC = total cost
    = $TVC + FC$
P = unit selling price
R = revenue
    = NP
Z = profit
    = $R - TC$
B = break-even volume
    = $FC/(P - V)$
$P - V$ = unit contribution

EXAMPLE 4.1 Break-even alternatives

One type of appliance manufactured by a large electronics firm has consistently experienced a loss. The product must be continued in order to offer distributors a complete line. The most recent revenue and cost figures show an annual fixed cost of $90,000 and a total variable cost of $192,000 on sales of 12,000 units, which account for a revenue of $240,000. Costs and revenue are directly proportional to the production rate which is 25,000 units per year at 100-percent capacity. What alternatives should be evaluated in an attempt to improve the product's competitive position?

At least four alternatives are immediately apparent to relieve the current loss (negative profit):

$$Z = R - (TVC + FC)$$

$$= \$240,000 - (\$192,000 + \$90,000)$$

$$= -\$42,000$$

1. Fixed costs could be reduced. For Z to equal zero in order to break even with the other variables remaining unchanged, the new fixed cost (FC') must not exceed

$$FC' = R - TVC$$

$$= \$240,000 - \$192,000 = \$48,000$$

The required reduction, $90,000 - $48,000 = $42,000, equals the total current loss and would certainly be difficult to obtain by cutting only fixed costs. A check of the accounting procedures should at least be made to see whether the product is bearing a disproportionate share of the total factory fixed costs, such as supervisor/overhead and equipment depreciation.

2. Total variable cost is a more likely candidate for a cost-reduction program. Improved methods, materials, processes, or work procedures offer po-

tential savings. The current unit variable cost,

$$V = \frac{\$192{,}000}{12{,}000 \text{ units}} = \$16 \text{ per unit}$$

must be reduced to

$$V' = \frac{R - FC}{N} = \frac{\$240{,}000 - \$90{,}000}{12{,}000} = \$12.50$$

to break even when R, FC, and N are fixed.

3. Revenue can be increased by selling more units, raising the price per unit, or a combination of both actions. As opposed to internal cost control, pricing policies and sales volumes are largely restricted by factors outside the firm. Higher price tags are easily attached to a product, but they usually make it harder to sell. Competition tends to set a limit on price boosting. (Under rather rare circumstances, a product may be immune to competitive pressure. New inventions, geographical isolation, limited supplies, or first issues may provide a temporary monopoly.) Marketing is further restricted by including the finite consumer demand along with the pressures of competition. Therefore, it appears unreasonable to rely on a new price,

$$P' = V + \frac{FC}{N} = \$16 + \frac{\$90{,}000}{12{,}000} = \$23.50$$

or an increase of $\$23.50 - (\$240{,}000/12{,}000) = \$3.50$ to avoid the present deficit with other conditions fixed.

4. The potential of increasing output can be further explored by observing the profit picture that results from using excess capacity. Current conditions are shown by the bold lines in Figure 4.4. With a contribution per unit of $P - V = \$20 - \$16 = \$4$, the break-even point is $\$90{,}000/\$4 = 22{,}500$ units. At this level of production, the firm would no longer be losing money if they could sell all they produced, but how can sales be increased with no boost in marketing expenditures?

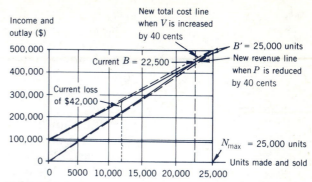

FIGURE 4.4 Current conditions of a $42,000 loss and alternative strategies to eliminate the loss by using the maximum manufacturing capacity. Sales could be increased by spending an extra 40 cents per unit on marketing or by reducing the unit selling price by 40 cents.

One possibility is to strive for a break-even condition at maximum capacity, N_{max}. This strategy is shown by the dotted line in Figure 4.4. With B' at 25,000 and P and FC unchanged, the new unit variable cost could be

$$V' = P - \frac{FC}{B'} = \$20 - \frac{\$90{,}000}{25{,}000} = \$16.40$$

or, correspondingly, a new selling price could be

$$P' = V + \frac{FC}{B'} = \$16 + \frac{\$90{,}000}{25{,}000} = \$19.60$$

The 40-cent difference from the original conditions could be applied as a price reduction or allotted to a higher advertising budget to try to increase sales to $N_{max} = 25{,}000$ units.

It is doubtful whether any of the individual measures we have considered would be applied singularly. Some combination would appear more realistic. One danger in production planning is mental tunnel vision that blocks a complete survey of cost and revenue relationships.

Step and slope changes in cost lines

Special pricing and cost arrangements can be incorporated in a linear break-even analysis by allowing discontinuous lines and slopes. Occasionally fixed or vari-

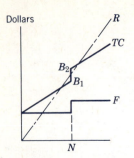

Step increase in fixed cost with two break-even points.

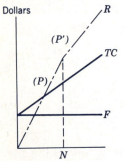

Dumping price for sales beyond N.

able costs rise or fall abruptly beyond a certain production level. A sharp rise in fixed costs could occur when a new machine or production line must be added to increase capacity beyond a certain point. Direct labor costs climb abruptly when overtime is necessary to boost output above a given level. Such variable cost changes are shown on a chart as steeper TC lines associated with affected segments of output.

Dumping

The practice of reducing the selling price on a distinct portion of output in order to use otherwise excess capacity is called **dumping.** It is graphically represented by a flatter slope of the revenue line over the extra output permitted through dumping. The basis for this strategy is that the main market for a product has only a limited maximum demand. In the belief that no additional units can be sold to this market, a secondary demand is developed by offering the same quality of units at a lower price. This lower price is necessary to make the product attractive in a market area otherwise unavailable. For instance, foreign sales are a secondary market, or the same product under a different name sold through different retail outlets can be a subordinate domestic market. The obvious limitation is that the consumers must not be aware that they can get the same product at two different prices, or the lower price will become the base price. When dumping works as intended, both total profit and factory utilization are improved.

EXAMPLE 4.2 Utilization of plant capacity

The manufacturers of "Wonderwashers" are presently operating at 75-percent capacity. A large retail chain store will sell Wonderwashers under the chain's own trade name if extra accessories are added that will increase variable production cost by 4 percent. The tentative agreement will increase production to the plant's full capacity, but the selling price for the extra 25-percent output will be only 89 percent of the usual price for Wonderwashers. To exceed the 80-percent capacity, the manufacturer will have to re-open an outdated assembly line in which fixed costs are 25 percent higher and variable costs are 6 percent greater. The present plant operating conditions are

1. Output: 91,500 units per year (75-percent plant capacity).
2. Price: $110 per unit (wholesale price to regular distributors).

3. Variable cost: $50 per unit.
4. Fixed cost: $4,000,000.

The immediate consequences of accepting the agreement allowing 100-percent utilization of facilities are apparent in a break-even chart, as shown in Figure 4.5. At 75-percent capacity, incremental revenue decreases to $P' = \$110(0.89) = \$97.90$ per unit, and the unit variable cost increases to $V' = \$50(1.04) = \$52$. A step increase of $\$4,000,000(0.25) = \$1,000,000$ in fixed cost occurs at 80-percent capacity with another increase in variable cost to $55. The total profit under present conditions is

$$Z = 91,500(\$110) - [91,500(\$50) + \$4,000,000]$$

$$= \$1,490,000$$

The expected profit obtained by dumping through

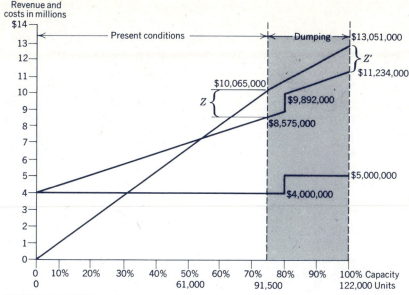

Revenue and costs in millions

FIGURE 4.5 Effect of a proposed dumping agreement.

the chain is

$$Z' = 91,500(\$110) + 30,500(\$97.90)$$
$$- [91,500(\$50) + 6100(\$52) + 24,400(\$55)$$
$$+ \$4,000,000(1.25)]$$

$$= \$1,817,000$$

Although the potential profit increase is inviting, the makers of Wonderwashers should look beyond the period depicted in the break-even chart to question possible future revenue and cost changes.

1. Will the fixed cost of the old assembly line used for the 20-percent extra production remain as estimated? If the fixed cost increases by 35 percent instead of 25 percent, the company will lose money on the agreement.

2. Will selling the same product under a different brand name affect sales of Wonderwashers? A 7-percent drop in basic sales would make total profit less than is currently earned.

3. Can the dumping agreement be extended for several years? If so, unit variable costs on the old assembly line probably can be reduced to the level of the rest of the plant.

Nonlinear relationships

It is occasionally worthwhile to fit a curve to revenue and cost functions. Straight lines are normally close enough approximations for break-even planning, even though real-world data seldom plot with such convenient directness. When realism or accuracy suggests **nonlinear relationships,** the methods discussed in Chapter 3 can be employed to develop appropriate line formulas.

Nonlinear economic models often appear quite frightening. A liberal sprinkling of differential and integral signs creates a discouraging facade. But in most instances an elementary knowledge of calculus is sufficient preparation. The graphical interpretation of curved and straight lines is similar, but calculus rather than algebra is the foundation for the quantitative evaluation of nonlinear expressions.

After line-fitting procedures are applied to develop descriptive equations, the expressions can be differentiated to determine slopes for revenue, cost, or profit lines. Being curved lines, the slope changes in infinitesimal steps for different output values. The rate of change of slope is called the **marginal revenue** or **marginal cost**—difference in income or outlay caused by the next unit of output at a specified level of production. At a point where the slope is level and equal to zero, a maximum or minimum of the function is indicated. Such points

Discussed in more detail in Section 4-4.

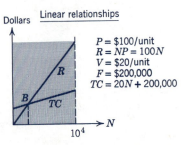

Linear relationships

Dollars

$P = \$100/\text{unit}$
$R = NP = 100N$
$V = \$20/\text{unit}$
$F = \$200,000$
$TC = 20N + 200,000$

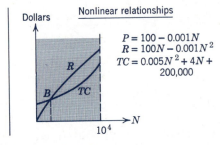

Nonlinear relationships

Dollars

$P = 100 - 0.001N$
$R = 100N - 0.001N^2$
$TC = 0.005N^2 + 4N + 200,000$

To determine break-even output, B:

$$\begin{aligned} Z = 0 &= R - TC \\ &= 100N - 20N - 200,000 \\ 0 &= 0\,80N - 200,000 \\ N = B &= 2500 \text{ units} \end{aligned}$$

$$\begin{aligned} Z = 0 &= R - TC \\ &= 100N - 0.001N^2 - 0.005N^2 \\ & \quad -4N - 200,000 \\ 0 &= -0.006N^2 + 96N - 200,000 \\ N = B &= 2462 \text{ units} \end{aligned}$$

To determine output for maximum profit:

Marginal profit = contribution
Contribution $= P - V = \$80/\text{unit}$

Total contribution increases as N increases

Maximum profit $= Z_{max}$ occurs at $N_{max} =$ 10,000 for given data

Marginal profit $= dZ/dN$
$$\frac{dZ}{dN} = \frac{d(-0.006N^2 + 96N - 200,000)}{dN}$$
$$= -0.012N + 96$$

Z_{max} occurs where $dZ/dN = 0$:
  then $0.012N = 96$
Z_{max} occurs at $N = 96/0.012 = 8000$

To determine output for minimum average unit cost:

Average cost $= AC = \dfrac{TC}{N}$

$AC = 20 + \dfrac{200,000}{N}$

AC_{min} occurs at N_{max}

$N_{max} = 10,000$ for given data

$AC = \dfrac{TC}{N} = 0.005N + 4 + \dfrac{200,000}{N}$

AC_{min} occurs where $\dfrac{d(AC)}{dN} = 0$

$0 = 0.005 - \dfrac{200,000}{N^2}$

AC_{min} occurs at $N = \sqrt{\dfrac{200,000}{0.005}} = 6325$

FIGURE 4.6 Comparison of calculations for linear and nonlinear relationships.

are useful in deciding what level of production produces the maximum profit or minimum **average unit cost.** Calculations for some critical production points based on linear and nonlinear relationships are compared in Figure 4.6.

Multiple products

Some companies grow by adding new but related products to their basic line. Many factors force such diversification. As the technology increases, a company has to run to just maintain its present position. Each new advance means at least a model change to remain competitive. Each model change means a wider range of products because some users stay loyal to older, familiar models. Product and production planning must organize or even govern the explosion of catalog offerings.

Decisions to add and drop products are subject to suboptimization. One problem, which we shall term *insular suboptimization*, deals with a tendency to focus on the solution to a particular issue at the expense of wider welfare. For instance, one product in a family of related products could be in trouble. To extract this product from its unfavorable position, it is necessary to redeploy company resources such as advertising budgets, engineering time, and capital investments. The actions could heal the ailing product but poison the other family members. A planning evaluation to avoid such problems is described in Example 4.3.

EXAMPLE 4.3 Multiproduct planning

Ador, Inc., manufactures exterior and interior house doors, wooden screen doors, and overhead garage doors. House doors were the original product, followed by garage and screen doors. Increasing popularity for aluminum screen doors with consequent sagging sales for wooden screen doors inspired a study to see whether the company should enter the metal door market. The study concluded that preformed metal door parts could be subcontracted and then fabricated in the plant using most of the personnel now engaged in making wooden screen doors. Very little new equipment would have to be purchased. Most of the equipment for the wooden doors could be retained because it is also used for the other products. The following cost and marketing figures presented a glowing endorsement for the proposed product:

	Wooden Screen Doors	Metal Screen Doors
Revenue	$325,000	$400,000
Variable cost	$105,000	$280,000
Fixed cost	$185,000	$ 60,000
Profit	$ 35,000	$ 60,000

A more comprehensive picture of product relationships is obtained by including all the economic facts in one graph. The multiproduct break-even chart of Figure 4.7 has the total fixed cost of all Ador's products entered on the vertical scale below the zero

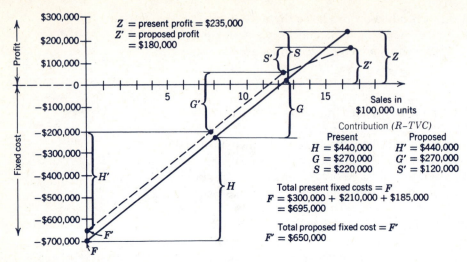

Z = present profit = $235,000
Z' = proposed profit
    = $180,000

Contribution (R−TVC)

	Present	Proposed
	$H = \$440,000$	$H' = \$440,000$
	$G = \$270,000$	$G' = \$270,000$
	$S = \$220,000$	$S' = \$120,000$

Total present fixed costs $= F$
$F = \$300,000 + \$210,000 + \$185,000$
$= \$695,000$

Total proposed fixed cost $= F'$
$F' = \$650,000$

FIGURE 4.7 Multiproduct break-even chart.

profit point. The horizontal scale is capacity in units of production or sales. Each product is represented by a line segment sloping upward to the right. The slope of each segment is set by the product's contribution; the length is set by the number sold. The break-even point for the chain of products occurs when the chain crosses the zero line.

Before and after information is needed for all related products to make a multiproduct chart comparison. Present and proposed data for the house and garage doors are as follows and are included in Figure 4.7:

	Present		Proposed	
	House Doors (H)	Garage Doors (G)	House Doors (H')	Garage Doors (G')
Revenue (R)	$800,000	$450,000	$800,000	$450,000
Total variable cost (TVC)	$360,000	$180,000	$360,000	$180,000
Fixed cost (F)	$300,000	$210,000	$350,000	$240,000

It is apparent from the cost estimates that the introduction of metal screen doors is not expected to boost sales of the other doors, but it will increase their burden of fixed costs. This logical but easily overlooked development results from a switch in materials. Previously, all three doors shared the fixed cost of a basically wood-oriented plant. If metal replaces one wood product without a corresponding deletion of fixed overhead, the remaining wood products must carry the load of existing plant facilities and operations. Thus we see that an "insular" study could show savings of $60,000 − $35,000 = $25,000 in the trouble area that actually creates an overall loss of $235,000 − $180,000 = $55,000.

Multiple comparisons

The revenue or costs expected from several alternatives also can be displayed in a break-even chart. Other than the visual presentation possibilities, there is little reason to commit the comparisons to a graph. A chart simply shows the result of algebraically equating each pair of alternatives. The break-even point, B, is the output that makes two courses of action equivalent. The conditions

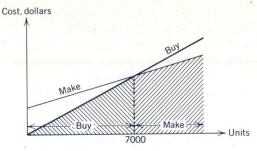

FIGURE 4.8 Make-or-buy comparison.

that determine the feasibility of a production level above or below *B* control the decision.

Many **make-or-buy decisions** fit the criteria of primary comparisons which can be displayed on breakeven charts. Figure 4.8 depicts a tactical decision between building or buying a part used in a fabricated product. When a part is made by a company, some share of fixed cost must be charged to its production. When the part is purchased, there is only the variable cost of its price per unit. If we assume that the variable cost of a produced part is less than the price of a purchased part, the decision to make or buy will rest on an estimate of the number of parts needed.

Make-or-buy decisions also illustrate another type of suboptimization. *Temporal suboptimization* is the curse of shortsighted planners, as it leads to a loss of tactical efficiency through not looking far enough ahead. Assume that the demand for the part described in Figure 4.8 is 3000 units per year. A planning horizon of 2 years yields a preference to buy the part. This choice will obviously be suboptimal if the demand for the part continues for 2.5 years or longer. With *de facto* knowledge, such errors seem foolish, but they can be avoided only by exposing the decision to the demons of uncertainty.

4-4 LIFE CYCLE ANALYSIS AND CAPACITY PLANNING

In the previous section we manipulated costs and revenue to observe what profit resulted from different output quantities. Further enlightenment about production planning is provided by a different perspective that relates cost to the life of a segment of the production system. **Life cycle analysis** forces a decision maker to look ahead and attempt to comprehend all of the factors, not just the immediate consequences, of an anticipated course of action.

A production system can be described by many characteristics. Each characteristic varies as a function of time. A successful product typically has an initial surge of sales that pushes it into a strong competitive position in the marketplace. As time passes, product sales usually level off and eventually decline as new competitors erode its market. Associated with the product life cycle are the costs

Life cycle analysis is widely used in forecasting the workforce requirements for product development. An S-shaped growth curve is applicable to most situations.

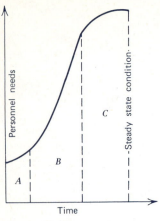

A, establishment. B, expansion. C, consolidation.

of production. These vary over time as a result of product development, start-up, operating, and termination decisions. In addition, costs attributable to marketing and overhead must be considered. And underlying all of these factors are the economic considerations for the financing of the production facilities, the control of operating expenses, and the justification for capital expenditures to continue or improve on operations.

Although no two products exhibit identical cost and revenue patterns during their lives, most pass through the same stages of maturation and decline. A typical pattern is shown in Figure 4.9. It displays both the rise and fall of cost and revenue from the original product and the potential for renewal from redesigning a promising product to launch it on a second life cycle.

The planning phase for a product starts with research to determine roughly its acceptability. If the concept appears promising, R&D efforts, as described in Chapter 2, are accelerated, and market research efforts are launched to verify the potential demand. As the basic design becomes firm and production facilities are developed, the first promotion activities commence. Successful marketing ensures that units will be sold at a minimum lag time from their production schedule. Production costs precede revenue from sales, of course, so that fi-

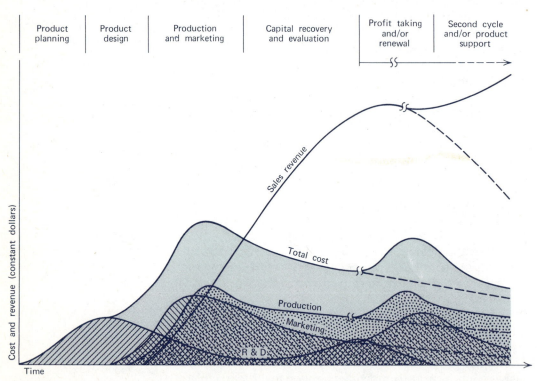

FIGURE 4.9 Phases and economic factors in a typical product life cycle. Dashed lines indicate the decline stage of the product, if it is not redesigned and promoted again.

nancial considerations pressure managers to have sales dollars flowing in as soon as possible to recover the invested capital.

Production costs level off after an initial peak caused by start-up expenses. Meanwhile, marketing costs usually decline after initial advertising and promotion expenses, and the R&D costs almost cease as attention is diverted from the original product to other products. During the period when sales growth continues, initial expenses are recovered, and profit is generated. Finally, a saturation point is reached at which sales decline because of a diminishing number of potential customers who remain unaware of the product, the replacement demand subsides, and better products or substitutes enter the market.

A mature product that has paid for its developmental costs and maintained a respectable level of sales is a candidate for a "new and improved" version. A refined model may be a major revision or merely a face-lift on the original. In either situation some R&D costs are incurred, and an upsurge in marketing cost is required to promote the new model. Production cost increases again to accommodate the changes needed to produce the modified output. If the renewal is successful, the product builds on its past record to achieve new sales records.

A decision to abandon the product is affected by and, in turn, affects many parts of a production system. The actual behavior of sales is only the surface influence for a decision to terminate production. Underlying factors include both pressure from within the firm to seek new directions and external pressure from local, national, and international sources that dampen future expectations for the product. When these factors in concert indicate termination, the closure may be abrupt or gradual, depending on the effect it will have on the community and other company operations. An abrupt cancellation often occurs when the company is ceasing operations completely or when the product is being replaced by a new line. A gradual phaseout, which meets residual product demand at minimum cost, causes less hardship for the community and is often economically preferable, even when the company is going out of business.

Internal pressures: new management, new owners, current financial considerations, and other product lines.

External pressures: domestic and foreign competition; new technology; tax, market, and supply changes; and revised government regulations.

Capacity planning—Output

Few products ever become best sellers. Production planners know this and build their projections on reasonable expectations, but there is still the threat of a major flop, a product that just does not sell regardless of the promotion effort. Cautious production scheduling can provide some protection from a disaster by limiting initial investments in production machinery, holding material purchases to absolute minimum levels, and employing temporary labor. These measures reduce the monetary commitment to a new product at the expense of larger-than-necessary unit cost.

Capacity planning attempts to integrate the factors of production so as to minimize facility costs over the life of a product or a project. Because there are a bewildering number of individual facility designs involved in introducing a major new product, an early planning step is to determine a realistic sales goal. The total number sold should be large enough to recover investment yet be well within the market potential. The goal balances selling price and production costs.

See Figure 4.6, in which a
reformulation of nonlinear
relationships reveals that

$$\frac{d(R)}{dn} = \frac{d(TC)}{dn}$$

$$\text{at } Z = \text{maximum}$$

Because profit is maximized when marginal cost equals marginal revenue, the number of units made and sold should be planned with reference to this economic relationship.

Leaving the question of selling price to market considerations, we can get a fix on preferred production quantity by analyzing production costs. As is indicated in Figure 4.10, the output quantity for minimum average total cost is spotted at the point where the marginal cost curve crosses the average cost curve. We shall next explain each curve.

AVERAGE FIXED COST

Business people say they "spread the overhead by selling more."

Fixed costs are assumed to be constant over the life of a product, although *all* costs tend to be variable over the long run. The share of fixed cost attributable to each unit produced obviously decreases as more units are produced.

AVERAGE VARIABLE COST

Unit variable costs are relatively high for the first units of output. They usually come down as the output increases because of quantity discounts on materials, more efficient use of personnel, and the like. Later in the life cycle, unless close control is maintained, unit variable costs may increase as a result of complacency, overstaffing, deteriorating facilities, and working conditions.

AVERAGE TOTAL COST

The economic *law of diminishing returns:* As amounts of a variable resource are added to fixed resources, beyond some point the marginal product will diminish.

Because average total cost is simply the sum of average fixed cost and average variable cost, it shows the combined effects of spreading the fixed charges and diminishing returns from overused variable resources.

MARGINAL COST

The cost of producing one more unit is the marginal cost of that last unit. If it costs $1000 to produce 100 units and $1012 to produce 101 units, the marginal cost will be $12.

The message of Figure 4.10 to production planners is to watch marginal cost. When it climbs to the point that it equals average total cost, continued production will lead to ever-higher unit total cost, and this condition will decrease per-unit profit if the selling price is constant. When the selling price declines with larger output, the quantity made and sold at the level at which marginal revenue equals marginal cost is the capacity that maximizes profit. Therefore, marginal cost is a barometric reading that production analysts can monitor to control a production system better.

EXAMPLE 4.4 Analysis of production costs

A manufacturing company produces several different products that are manufactured with the same equipment but that are marketed individually. An audit of one product revealed the data shown in Table 4.1; all figures are corrected to a base year to avoid distortion by inflation of the general economy.

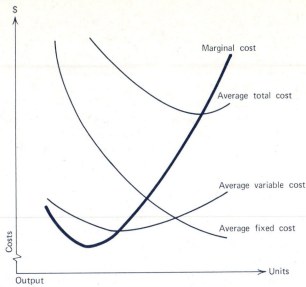

FIGURE 4.10 The relationship of average and marginal costs of production. The average total cost per unit generally follows a U-shaped curve. It is at a minimum when it equals the marginal cost of the last unit produced.

Fixed cost chargeable to the product totals $400,000 and covers the product's prorated share of plant facilities and management. Variable costs were high during the start-up period, dropped as production became more efficient, and climbed during the third and fourth years as a result of "relearning"

when production runs were not continuous, because of declining sales. Although average fixed cost drops consistently with additional output, average variable and total cost figures show the effect of increasing production costs.

Marginal costs, collected for increments of 1000

TABLE 4.1
Product Cost and Revenue Patterns over a Four-Year Product Life

Product Life (year)	Output (N units)	Fixed Cost (FC)	Total Variable Cost (TVC)	Total Cost, TC (TC = TVC + FC)	Incremental Marginal Cost (1000 units)	Average Fixed Cost (FC + N)	Average Variable Cost (TVC + N)	Average Total Cost (TC + N)	Total Revenue (R)	Incremental Marginal Revenue (1000 units)	Total Production Profit (R – TC)
	0	$400,000	0	$400,000			0		0		
					$80,000					$200,000	
Yr. 1	1000	400,000	$80,000	480,000		$400	$80	$480	$200,000		–$280,000
					32,000					200,000	
	2000	400,000	112,000	512,000		200	56	256	400,000		–112,000
					22,000					200,000	
	3000	400,000	134,000	534,000		133	45	178	600,000		34,000
					26,000					180,000	
Yr. 2	4000	400,000	160,000	560,000		100	40	140	780,000		220,000
					30,000					150,000	
	5000	400,000	190,000	590,000		80	38	118	930,000		340,000
					62,000					120,000	
Yr. 3	6000	400,000	252,000	652,000		67	42	109	1,050,000		398,000
					100,000					110,000	
	7000	400,000	352,000	752,000		37	50	107	1,160,000		408,000
					128,000					110,000	
Yr. 4	8000	400,000	480,000	880,000		50	60	110	1,270,000		390,000

units of output, focus attention on key output levels. The lowest marginal cost occurred in the first year when the annual production rate was highest. At about 5000 units, marginal cost equaled the average variable cost at its lowest point. Average total cost bottomed out near 7000 units, at which it equaled the marginal cost. At about this same level, rising marginal costs met the dropping marginal revenues to signal the point of maximum profit. The decline in marginal revenue, stated in 1000-unit increments, is attributed to pressure from competitors.

Unless variable costs can be slashed or prices increased, the product is a candidate for termination or renewal. Production factors that will influence the decision include the expected utilization of manufacturing facilities for other products, labor and material availability, and resource requirements (both money and time) for restoring productivity or producing a revamped product.

Life cycle costing

Life cycle costing is identical in concept to life cycle product analysis. Both encourage decisions based on totality: all inputs and outputs are considered for the entire period of use or activity. In life cycle costing, attention is directed to all funds expected to be expended during the useful life of a production asset, including associated activities directly linked to the employment of the asset. The asset may be a single item, such as a truck, or a system composed of several integrated items that are useful only if applied together. Stated another way, life cycle costing evaluates the total cost between the time an asset becomes recognizable, either physically or in concept, and the time it is phased out of operation.

The purpose of life cycle costing is to get all relevant costs into the decision process. These costs include not only acquisition charges but also expenses required to maintain performance at a given standard. This approach forces a trade-off analysis between one-time costs and recurring costs. An elementary version of life cycle decision making is shown in the make-or-buy chart in Figure 4.8. A more extensive analysis encompasses all the cost categories shown in Figure 4.11.

A life cycle study can be applied to service projects as well as to asset acquisition. For example, government contracts for training might be evaluated with respect to initial education provided, continued auditing of learning experiences, and individual upgrading of skills. To compare these costs, accurately, which occur at various times during the life cycle, the time value of money must be considered. Discounting procedures that do this are presented in the next chapter.

Comparisons often involve an alternative with a high purchase price and low operating cost versus one with a low first cost but high operating expenses. The length of life is critical in this analysis.

Design-to-cost *is an estimating method that reverses "buildup" estimating in life cycle costing; it starts with a known market price for an item and works back to the cost of each element.*

Capacity planning—Labor

Planning for the size and composition of the work force parallels planning based on anticipated output requirements. This type of labor-requirement planning is clearly a strategic issue, whether it is for an entirely new venture or for an expansion or contraction of existing production. In some organizations the overall life cycle pattern shown in Figure 4.11 is complicated by seasonal variations of the type described in Examples 3.4 and 3.5 for sales forecasting. Seasonal fluctuations are often more severe in service industries because firms cannot

The role of learning curves in determining labor requirements is discussed in Section 14-4.

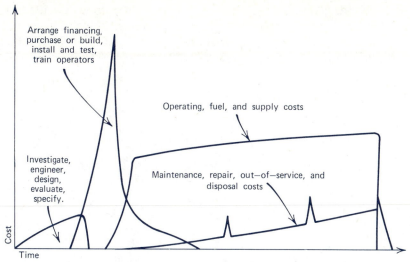

Arrange financing,
purchase or build,
install and test,
train operators

Operating, fuel, and supply costs

Investigate,
engineer,
design,
evaluate,
specify.

Maintenance, repair, out—of—service, and
disposal costs

Cost

Time

FIGURE 4.11 Typical cost profiles that collectively characterize life cycle costing. All of the cost categories must be included to provide a comprehensive evaluation of the total cost for services provided by an asset.

store server time, as they can store products, to have a reserve supply on hand when needed. Nor are service organizations as free to vary rates of output, as do many manufacturers, to maintain a relatively constant level of employment.

Capacity planning in short-run operations is based on how much labor is required to get each job done. The sum of workers' times to complete all of the necessary jobs during a specified period is the employment level for that period. Two questions are asked: How often is each task done, and how long does each task take? The answers to these questions tell supervisors the number of employees needed to produce a specified volume of output or, inversely, how much output to expect from a specified number of workers.

From the standpoint of time, all tasks fall into one of two categories—either fixed or variable. **Fixed tasks** have a constant duration and frequency of occurrence. They can be large, as for a crew working on a lengthy contract, or as short as a daily phone call. All other staffing requirements are based on the number of **variable tasks** needed and the time assigned to each one. Occurrences of variable-time tasks can be difficult to anticipate, especially for new operations. Reasonably accurate forecasts of both frequency and the pattern of time durations are obtained by referring to records of similar work and by the time-study methods described in Chapter 10. The end result of collecting these tasks is a staffing table that converts times to labor needs listed by number and type.

Organizations that know all the tasks in their tactical operations are better equipped to respond to work-load fluctuations and to predict staffing needs to accomplish their strategic mission. Even if the task times are not absolutely accurate and their correspondence to future tasks is not known with complete

Mathematical models for schedules that level employment are introduced in Chapter 6.

EXAMPLES OF FIXED
SERVICE TASKS
Write a weekly report.
Attend a monthly meeting.
Conduct an annual inspection.
Teach a daily class.

EXAMPLES OF VARIABLE
SERVICE TASKS
Prepare a lecture.
Type a report.
Investigate an accident.
Interview an applicant.

certainty, the quantified basis that they provide for staffing decisions is preferable to a "gut-feel" for future needs. Capacity planning is always beset with imponderables because future demands are undependable. Such uncertainties are characteristic of system economics.

4-5 CLOSE-UPS AND UPDATES

Good planning relies on organized information, and the information must be timely. The information system that handles a production system's economic data should have current and historical data conveniently on call, in the proper form, and indexed for easy reference. Naturally, computers do the information processing, but people still have to input the data, program the process, and analyze the output in order to make decisions. A major goal of management information specialists has been to develop computer systems that not only access the data but also interact in the decision process.

Decision support systems

Computers have now gone far beyond their initial applications, which increased business efficiency by relieving people of such tedious tasks as manually generated payrolls and inventory records. The step above a *database system* is a **decision support system (DSS).** This advance beyond routine matters is aimed at helping those who are responsible for business-unit performance to make more effective decisions.

Just as the decision-making process varies from person to person and from firm to firm, so must decision support systems vary in composition and function. They are designed to manipulate data rationally in order to develop a comprehensive set of feasible alternatives from which a manager can select the best solution. Alternatives may be for long-range planning or options for a quick reaction to an unexpected difficulty.

Common DSS applications include financial modeling, facilities planning, sensitivity analyses, and "what-if" problem solving. Most of the mathematical- and information-modeling techniques presented in this text can be used in a decision support system. Microcomputers accommodate most DSS applications, although mainframes are needed when a very large volume of data is involved.

DSS gives middle managers the ability to perform sophisticated decision analyses based on their specific needs, using data that incorporate their personal experiences, and customized to current operating conditions. Much of the information recall and deductive reasoning formerly done by decision makers is now executed by DSS with greater accuracy and ex-

Generations of computers can be associated with system developments as well as the hardware developments described in Chapter 1. The system generations are

1. Database systems.
2. Decision support systems.
3. Expert systems.
4. Knowledge based English systems.
5. ??

A **management information system (MIS)** that performs only arithmetic functions, with little ability to investigate alternative assumptions, is not a DSS.

A major advance of DSS over MIS is convenience of use by managers who are not computer experts.

pediency. Managers thus avoid tedious data engagements and are better able to observe the organization's present state, view alternative scenarios, and inspect those isolated courses of action that appear to be most promising.

A short glossary of terms associated with DSS is given in Table 4.2. The terms are listed in a functional order rather than alphabetically.

BEA: The national well for economic data

The corporation president looks at the numbers and says, "Great! Just what I need for this afternoon's conference."

An importer looks at the same numbers and sighs, "This is going to cost us a bundle unless we act quickly."

A banker frowns, "These figures mean that we're going to have trouble with some of our loans."

In a state office building, an economist thinks, "If we're lucky, state tax revenues should increase, but the data might mean a decline in investments which would. . . ."

TABLE 4.2
Terms Commonly Encountered in DSS Discussions

Information Center: A facility established within an organization, staffed by experienced data-processing personnel, for users who want to learn, develop, or customize software programs.

Database: A collection of related data organized in some predetermined format and stored in a computer file.

Database management system (DBMS): A software program to coordinate and control the accessing and modifying of a computer system's various databases.

Data Dictionary: A cross-reference file used by a DBMS to associate the data names entered by users with the actual databases in which the named data are stored.

Context Editor: A program that allows the user to view or change data stored in a file.

Full-screen Editor: A program that allows the user to modify data on the video display terminal.

Screen Format: The way in which information is presented on the terminal screen.

High-level Programming Language: A computer language that allows the programmer to use English verbs, symbols, and commands rather than machine code. Examples are FORTRAN, COBOL, PASCAL, and BASIC.

Nonprocedural language: A programming language that allows users to enter variables and data in any order they see fit, rather than requiring that variables be defined in a strict, determined order. Most traditional languages (such as FORTRAN, COBOL, PASCAL, BASIC) are *procedural languages* that require a specific order for actions.

Menu-Driven: A menu-driven program provides an option list to guide the user step by step through the program. With a *command-driven* program, the user must enter specific commands.

Spreadsheet: A program that automates financial tasks traditionally done manually. Numerical data entered into arrays or matrices are manipulated using user-defined mathematical formulas.

Array: A rectangular arrangement of mathematical data in rows and columns. Synonymous with a matrix.

Goal seeking: A common objective of DSSs, goal seeking means calculating the change in one variable to produce specified results in another variable.

What-if analysis: Analyses performed through modeling, which allow the user to see the results of changed conditions or variables. Answers are given to questions such as "What if production increases by 10 percent?"

Data security: The safeguarding of computer file data to limit access to authorized users only, and the protection of data from theft, loss, and environmental hazards. *Data integrity* is the validity of the secured data.

The president, importer, banker, and economist each could be looking at the same data just released by the **Bureau of Economic Analysis (BEA),** a branch of the U.S. Department of Commerce charged with measuring and analyzing economic activity. Data flow to BEA from many sources, including other government agencies such as the Bureau of Census and Bureau of Labor Statistics. Over 400 BEA economists, statisticians, and computer personnel work with the data to produce meaningful statistics.

Economic analysts concerned with either government or private business systems rely on BEA releases for data on the production of goods and services, income, profits, capital expenditures, consumer spending, foreign trade, and a wide variety of other measures. Forecasters search for trends in BEA statistics. Managers use regional- and county-level information for plant location- and capacity-planning decisions. Government policymakers rely on national and international measures in assessing their strategic options.

Being able to provide acceptably accurate estimates of the **gross national product (GNP),** which quarterly adds up to trillions of dollars, is by itself an immense undertaking. All of the contributing data must be checked for reliability and internal consistency. Then they must be properly combined. Both the amount of data handled by BEA and the demands for more and better measures of economic activity are increasing as the economy grows and system analysts include more factors in their studies.

4-6 SUMMARY

Scope of economic factors

Production decisions associated with operations involve tactical alternatives measured by their efficiency. Strategic alternatives pertain to broader or longer-range decisions and are rated as to effectiveness. Suboptimization occurs when a tactical segment of a problem is optimized without regard to its strategic effectiveness. Types of suboptimization include organizational (communication problems), insular (multiple-product issues), and temporal (planning-horizon questions) suboptimization.

Suboptimization

A decision situation is highly sensitive when a small change in the efficiency of a tactic can alter the preference of strategies. The degree of sensitivity is a clue to the amount of effort that should be devoted to evaluating alternative plans.

Sensitivity

Break-even analysis

A break-even analysis investigates the relationship of fixed and variable costs to revenue. The break-even point identifies the production quantity where revenue exactly meets total cost. The amount of profit obtained from each product made and sold above the break-even output is its contribution—the difference between the unit selling price and the variable cost. The practice of reducing the selling price on a distinct portion of output to use otherwise excess capacity is called *dumping*.

For nonlinear price and cost functions, the difference in income or outlay caused by the next unit of output at a specified production level is, respectively, marginal revenue or marginal cost. The point at which marginal revenue equals marginal cost is the production level for maximum profit.

Nonlinear economic relationships

Life cycle analysis considers all the factors that affect a product or project over its life. A new product typically goes through several phases, starting with a planning stage and ending with a decision for its termination or renewal. A conservative production plan limits initial expenditures and focuses on marginal costs. Life cycle costing assists the selection of the most economical choice by examining *all* outlays involved in ownership or use: it forces a trade-off analysis between one-time costs and recurring costs.

Product life cycle

Capacity planning, based on predicted output levels, determines strategic facility and labor requirements. The number of employees needed for tactical operations can be estimated from the time required for necessary fixed and variable tasks.

Capacity planning

A decision support system assists managerial decisions by means of computer programs that access data, perform calculations that define alternatives, and allow sensitivity analyses of various options. Economic measures are also available from the Bureau of Economic Analysis.

Decision support system

BEA

4-7 REFERENCES

BARISH, N. N., and S. KAPLAN. *Economic Analysis for Engineering and Managerial Decision Making,* 2nd ed. New York: McGraw-Hill, 1978.

DEGARMO, E. P., J. R. CANADA, and W. G. SULLIVAN. *Engineering Economy,* 7th ed. New York: Macmillan, 1982.

MAO, J. C. T. *Quantitative Analysis of Financial Decisions.* New York: Macmillan, 1969.

OSTWALD, P. F. *Cost Estimating,* 2nd ed. Englewood Cliffs, N.J.: Prentice-Hall, 1984.

SCHALL, L. D., and C. W. HALEY. *Financial Management.* New York: McGraw-Hill, 1978.

STUART, R. D. *Cost Estimating.* New York: Wiley, 1984.

WEBB, S. C. *Managerial Economics.* Boston: Houghton Mifflin, 1976.

4-8 SELF-TEST REVIEW

1. T F *Strategic* considerations may involve several alternative tactics to achieve an objective.

2. T F *Tactics* are more concerned with the effectiveness of policies than with the efficiency of operations.

3. T F *Organizational suboptimization* occurs when a subordinate disobeys a directive from a supervisor.

4. T F The *sensitivity* of a decision depends on the vulnerability of an optimal solution to changes in controlling factors.

5. T F A *break-even point* indicates the ideal level of output.

6. T   F   When N exceeds B, $(P - V)$ is the incremental profit.

7. T   F   With R and FC fixed, a decrease in V increases B.

8. T   F   A company is engaged in *dumping* whenever it sells the same product in distinctly differentiated markets.

9. T   F   For linear relationships, maximum *profit* and minimum *average cost* occur at maximum N.

10. T   F   For nonlinear relationships, the break-even point occurs when *marginal revenue* equals *marginal cost*.

11. T   F   Marginal cost equals *average total cost* when marginal cost is at a minimum in a nonlinear situation.

12. T   F   A make-or-buy decision illustrates *temporal suboptimization*.

13. T   F   In a chart in which the X and Y axes are, respectively, time and personnel needs, *work-force requirements* typically follow an S-shaped curve.

14. T   F   Short-run labor requirements can be calculated by summing the times for necessary *fixed* and *variable tasks.*

15. T   F   Trade-offs between one-time costs and recurring costs are featured in *life cycle cost* analysis.

16. T   F   The primary functions of a *decision support system* are to store information and facilitate arithmetic operations with the data.

17. T   F   A *database management system* is a collection of related data organized in some predetermined format.

18. T   F   A *what-if* analysis explores the sensitivity of a decision to changes in influencing factors.

19. T   F   The Bureau of Economic Analysis measures economic activity, analyzes the data, and provides policy advice on its analyses.

4-9 DISCUSSION QUESTIONS

1. Assume you have invented the proverbial "better mousetrap" and plan to set up a production facility to manufacture your invention. List some of the strategic problems you face. What tactics are associated with these strategic considerations?

2. Give an example of how a strategic decision changes to a tactical decision as a function of an organization's growth.

3. A widely used term in government–industry contractual relationships is *cost effectiveness.* Explain your interpretation of this term with respect to tactics and strategies.

4. Could an operation with a near-perfect rating in efficiency have a very low rating in effectiveness? Give an example.

5. A student's problems in getting a college education are similar to those in many production systems. Identify some courses of action that could result in

a. Organizational suboptimization.
b. Insular suboptimization.
c. Temporal suboptimization.

6. What value is there in knowing whether a decision situation is highly sensitive?

7. List several types of cost (insurance, research, depreciation, spoilage, interest, sales commissions, worker wages, executive salaries, advertising, and so on), and decide whether they are fixed or variable costs.

8. Where would a line representing income taxes be shown on a break-even chart?

9. If the goal is to increase profit, discuss the relationship of total cost and volume to revenue under the options of increasing, maintaining, or decreasing the selling price.

10. What is the purpose of life cycle analysis? What factors in a production system could be subjected to life cycle analyses?

11. What types of pressure might induce a company to terminate production of a product that has consistently yielded a modest profit?

12. Give two examples in which an abrupt termination of product production is normally the logical maneuver and two examples in which a gradual phaseout is economically preferable.

13. On three charts drawn to the same scale, graph the product life cycle that characterizes output from each of the following production organizations:
a. One film from a motion picture company.
b. A gasoline service station.
c. A cherry orchard.
Label key stages for each product. Discuss the different patterns.

14. Applications of life cycle costing are described in W. J. Kolarik, "Life Cycle Costing and Associated Models," in the *Proceedings* of the 1980 AIIE Conference:

The Columbian Rope Company of Auburn, New York, claims that by using an LCC design approach on their $7 million plant, they will save $5.5 million in terms of a 20 year life cycle for the plant. Estimated savings in buildings usually result from increasing the initial investment to curb energy consumption in the structure. However, many designers, architects, and engineers still find it difficult to convince developers and owners to consider LCC, rather than initial cost alone. Sometimes, as in the case of the 38 story Federal Reserve Bank of Boston, an LCC approach may actually reduce initial cost as well as operating and maintenance costs by forcing designers to become extremely cost conscious in all aspects of their design.

Discuss how a closer focus on all the costs associated with a product or service could lead to a more valuable economic analysis. Consider LCC with respect to resource shortages and environmental concerns that might affect "twilight" costs late in a life cycle.

4-10 PROBLEMS AND CASES

1. Variable cost to produce one unit of a product is composed of $1.22 for labor and 64 cents for material. Forty percent of the revenue from the product is allocated to fixed cost and profit. What is the selling price?

2. Costs for producing a new product are estimated at $FC = \$280,000$ and $VC = \$3000$ per unit. The demand for the product is responsive to the selling price. It is expected to show the following pattern:

Unit price	$8000	$10,000	$12,000	$15,000
Market demand	75	65	55	35 (units)

Construct a break-even chart with four revenue lines radiating from the origin. Then mark on each ray the total revenue expected at that unit price. Connect the marked revenue expectations with a "demand" curve. Analyze the resulting graph, and indicate what price should be charged for the product.

3. Prices are going up. Rather than just increase the pricetag on the present product, "Dooper Cleaner," the plan is to put the same product in a new box and call it "Super Dooper Cleaner." The new container will be introduced by an extravagant publicity program. Current sales are 1,000,000 boxes a year at $1 per box. Fixed costs are $300,000, and variable costs are $0.60 per unit. It is anticipated that neither sales volume nor variable costs will increase but that the publicity campaign will double fixed costs. If the new price is $1.29 per box, what will be the new break-even point? How many boxes above present sales would have to be sold under the "old product plus persuasion" plan to double the present profit?

4. Sales forecasts indicate that a minimum of 5000 units will be sold each year for the next three years. Two design modifications are being considered for the product. One modification will increase fixed costs by $30,000 per year, but it will reduce variable costs by $8 per unit. The other modification will increase fixed costs by $7000 and reduce variable costs by $6 per unit. Variable cost is currently $30 per unit.

a. Which design modification should be adopted?

b. At what point would you be indifferent to the two alternatives?

c. What comments could be made about the planning horizon?

5. A company now has a total sales volume of $2 million on four products produced in the same mill. The sales and production cost figures for the products (A–D) are as follows:

	A	B	C	D
Percentage of total sales	10	20	30	40
Contribution (percent of P)	45	40	45	35
Fixed cost charged	$70,000	$180,000	$210,000	$220,000
Profit	$20,000	-$ 20,000	$ 60,000	$ 60,000

Recognizing the loss incurred with product B, the company is considering dropping the product. If it is dropped, the sales volume will decrease to $1.8 million, and the sales and cost pattern will change to the following figures:

	A	C	D
Percent of total sales	15	35	50
Contribution (percent of P)	45	45	35
Fixed cost charged	$100,000	$250,000	$290,000

Should product B be dropped? How do you account for the decrease of only $40,000 instead of $180,000 in fixed costs for the revised product mix?

6. Another alternative for the product mix described in Problem 5 is to substitute a different product for B, called BB. If BB replaces B, total sales volume and the sales–cost figures for products A and C will remain unchanged. The new values for D are percent of sales = 45, contribution (percent of P) = 35, fixed cost charged = $275,000, and profit = $40,000. For an increase in total profit of $10,000 and an increase in total fixed cost of $15,000, what will the contribution rate be for product BB?

7. A small company manufactures rubber matting for the interiors of custom cars. During the past year a revenue of $202,000 from sales was earned with the given current costs:

Current Operating Costs	
Direct material	$51,000
Direct labor	42,000
Maintenance	11,000
Property taxes and depreciation	17,000
Managerial and sales expense	35,000

The forecast for the next year is a drastic drop in custom car sales, which is expected to limit mat sales to $90,000. There is insufficient time to develop new markets before next year. With a skeleton force for the reduced production, the anticipated operating costs are as follows:

Expected Operating Costs	
Direct material	$28,000
Direct labor	23,000
Maintenance	7,000
Property taxes and depreciation	17,000
Managerial and sales expense	35,000

The company can operate at an apparent loss, or mats can be purchased from a large supplier and resold to the custom car builders at a price that will just meet purchase and handling costs. Either alternative will retain the market for the company until the following year when sales are expected to be at least equal to last year.

Which of the two alternatives is the better course of action?

8. Given a nonlinear price function of

$$P = 21,000n^{-1/2} \text{ dollars per unit}$$

where V = $1000 per unit and FC = $100,000 per period, determine

a. The break-even point.

b. The production level for maximum profit.

9. Operating expenses and revenue for a manufacturing plant are closely approximated by the following relationships:

$$R = 100n - 0.001n^2$$

$$TC = 0.005n^2 + 4n + 200,000$$

(both in dollars)

a. What is the output for maximum profit?

b. What is the output at the break-even point?

c. What is the output for minimum average cost?

10. A woman with a glittering sales record was asked by a publishing house to write a book on "how to sell." Being a complete saleswoman, she decided to write, produce, and sell the book herself. Fixed costs of production would be $6000 for artwork, typesetting, plates, and so forth. Variable costs for paper, printing, binding, and similar expenses would come to $2000 for every 1000 books printed. The author believes that the number of books she will sell depends on the price she asks for the book.

a. Using regression techniques, find a relationship between price and demand using the model $P = a + (b/D)$ and the following data:

Number of books sold, D (in thousands)	Price per book, P
1	$8.00
4	3.00
9	1.33
16	0.50

b. What price will allow her just to break even?

c. What price will give her the greatest profit?

11. Life cycle is intuitive for most buyers when they are considering a major purchase. Very few customers look only at the sticker price of an automobile or the asking price for a house. They also examine the upkeep costs and resale value.

Construct a series of cost–time distributions similar to those in Figure 4.11 that represent the life cycle costs of automobile ownership. Base it on a medium-priced car ($15,000) that is driven 15,000 miles annually for six years. Assume that the car is purchased on a four-year contract with a 20-percent down payment. Discuss the value of recognizing the full range of ownership costs.

12. *The Case of Calling It Quits*
A small lumber mill (5 managers/supervisors, 40 production workers, and 8 office staff) is forced to terminate operations because it can no longer buy logs within an economical hauling distance. A decision has been made to reduce operations in steps, rather than close abruptly, because the mill can operate at a reduced capacity and its output can be sold easily. The logging plan can be coordinated to log off the remaining stand of trees to meet the scheduled reduction in mill output.

N	TVC	N	TVC	N	TVC
10,000	$320,000	50,000	$1,050,000	90,000	$2,160,000
20,000	520,000	60,000	1,260,000	100,000	2,500,000
30,000	700,000	70,000	1,540,000	110,000	2,860,000
40,000	880,000	80,000	1,840,000		

Many of the assets of the mill can be sold piecemeal. For instance, logging trucks can be sold as the need for logs decreases, and much of the equipment in the mill can be sold when it is no longer required for operations at full capacity. Similarly, personnel can be laid off as their job functions cease because of reduced operations. These two conditions will effectively reduce the fixed cost and the variable cost of production as the output declines. The revenue will naturally fall as the quantity of produced lumber drops.

a. Construct a break-even chart to analyze the termination process. Graph an abrupt closure and the planned gradual phaseout. What does the chart indicate about the optimal point to cease operation?

b. What advantages and disadvantages are associated with an abrupt termination versus a gradual phaseout?

13. *The Case of the Fish Machine*

A company is being organized to produce a new type of one-piece fishing rod and reel to be called the "Fish Machine." One million dollars has been budgeted for fixed costs over a three-year period, including the advertising budget. Sales in the first year are ex-pected to be 20,000 units and will be priced to recover one-half the fixed cost. Second-year sales are expected to double those of the first year. After all fixed costs have been recovered, the selling price of the Fish Machine may have to be reduced to encourage more sales. In the third year, production will be upped to 50,000 units, and variable costs are expected to rise sharply because production facilities will be operating over the design capacity. The expected total variable cost pattern is shown above.

a. Prepare a graph of marginal cost, average variable cost, and average total cost.

b. How much total profit would result if the price could be held at the original level for all three years?

c. Assume that the selling price holds at the original level for two years but then it must come down to meet competition from larger manufacturers that have entered the market. During the third year, prices must be dropped by 10 percent to sell each additional increment of 10,000 Fish Machines. What is the total profit for three years? Show that the maximum profit would be obtained if production were terminated at the output at which marginal revenue equals marginal cost.

OPERATIONS ECONOMY

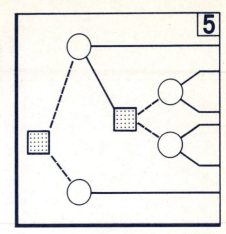

LEARNING OBJECTIVES

After studying this chapter you should

- appreciate the influence of interest charges, taxes, risk, and intangibles on decision making.
- be able to perform annual cost, present-worth, and rate-of-return comparisons for replacement studies.
- understand the function of depreciation accounting and the procedures for after-tax evaluations.
- be able to do sensitivity studies and to construct a sensitivity graph.
- be able to calculate and interpret expected values.
- be familiar with concepts and the use of discounted decision trees.
- appreciate the intent of capital rationing and the use of a payback period and a benefit–cost ratio.
- understand how intangibles can be quantified and combined with ratio-scale numbers in a priority decision table.
- realize the effect of technological advances on replacement studies, office operations, and management positions.

KEY TERMS

The following words characterize subjects presented in this chapter:

discounted cash flow comparison   equivalent present worth

time value of money   equivalent annual cost

nominal interest rate   coterminated projects

rate of return

incremental investment

depreciation

economic life

defender or challenger

sunk cost

after-tax cash flow (ATCF)

pessimistic and optimistic
estimates

sensitivity graph

expectation or expected value

payoff table

indifference probability level

value of perfect information

discounted decision tree

payback period

benefit–cost ratio

capital inventory

ordinal, interval, and ratio scales

priority decision table

office productivity

downsizing

5-1 IMPORTANCE

The name of the game is money, and the directions are to spend it wisely, at the opportune time and with an understanding of the risk involved. Economizing is a challenge shared by both private and public organizations. It is important because there are more ways to spend money for the good of operations than there is capital available. The capital must be rationed to be used for the most worthwhile projects. The challenge then becomes one of making economic evaluations that select the most advantageous expenditures. Among the questions that must be considered are

What is the pattern of future cash flows?

What is the effect of variations from expected flows?

What effect will depreciation and taxes have on proposed expenditures?

What risks are involved, and how should they be evaluated?

How should funds be rationed among several worthy projects to improve operations?

How can financial data be combined with intangible considerations in decision making?

Several ways to conduct economic comparisons are discussed in the sections that follow. Which method to apply depends on the nature of the proposals to improve operations and the type of analysis that is most appropriate.

5-2 DISCOUNTED CASH FLOW COMPARISONS

Everyone concerned with the economy of operations knows the importance of *time*. Operations managers realize that jobs have to be completed on time in order to stay within the budget. That is obvious. Less apparent is the effect that time has on investment capital.

A dollar received a year from now will be valued less than is a dollar received today. This relationship is known as the **time value of money.** The reason that today's dollar is worth more than a future dollar is that today's dollar can be invested and will have a value greater than will the dollar received in the future.

Present and future dollars are equated by interest factors. At an annual interest rate of 10 percent, $1.00 would grow in value to $1.10 in one year. From an investor's viewpoint, when risk is ignored, there is no difference between $1.00 promised today and $1.10 promised one year from now. The two sums are equivalent when we assume that future outcomes are certain to occur and inflation is disregarded.

The concept of equivalence allows us to compare alternatives realistically with different time patterns of receipts and disbursements. Knowing the amount and time of each cash transaction, we can calculate one number that is equivalent to all the transactions. Then the cash flow equivalent value for each alternative is directly contrasted for a final selection. Several equivalent time–money patterns are depicted in Figure 5.1.

Interest formulas

The task of calculating the equivalent cash flows depicted in Figure 5.1 would be very tedious if each payment were individually compounded through each time period. Fortunately, tables are available to ease the reckoning. Interest tables for six discrete interest-compounding formulas at various interest rates are provided in Appendix E. The six formulas comprise three future-value re-

Interest is the cost of using capital. Its history extends as long as humans have recorded transactions. Partly subsidized interest rates in the Greek and Roman empires were about 10 percent for the most reliable borrowers.

"The gain which he who lends his money upon interest acquires, without doing injury to anyone, is not to be included under the head of unlawful usury." John Calvin (1509–1564).

Tables for continuous compounding can also be used. We limit our attention to discrete compounding because it is more widely understood, accepted, and used by industry.

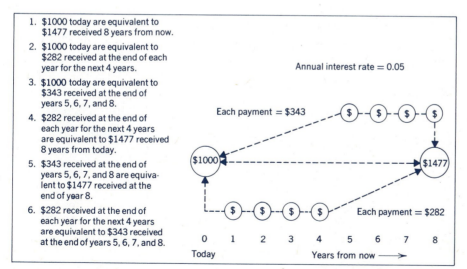

1. $1000 today are equivalent to $1477 received 8 years from now.

2. $1000 today are equivalent to $282 received at the end of each year for the next 4 years.

3. $1000 today are equivalent to $343 received at the end of years 5, 6, 7, and 8.

4. $282 received at the end of each year for the next 4 years are equivalent to $1477 received 8 years from today.

5. $343 received at the end of years 5, 6, 7, and 8 are equivalent to $1477 received at the end of year 8.

6. $282 received at the end of each year for the next 4 years are equivalent to $343 received at the end of years 5, 6, 7, and 8.

Annual interest rate = 0.05

Each payment = $343

$1000

$1477

Each payment = $282

0   1   2   3   4   5   6   7   8
Today   Years from now ⟶

FIGURE 5.1 Equivalent cash flows at a 5-percent annual interest rate.

Formula	Symbol	Function	
$(1 + i)^n$	$(f/p, i, n)$	To find the future value (F) given the present worth of a single amount (P)	
$1/(1 + i)^n$	$(p/f, i, n)$	To find the present value (P) given the future worth of a single amount (F)	
$\dfrac{i}{(1 + i)^n - 1}$	$(a/f, i, n)$	To find the value of annuity payments (A) given the future worth of the annuity (F)	
$\dfrac{i(1 + i)^n}{(1 + i)^n - 1}$	$(a/p, i, n)$	To find the value of annuity payments (A) given the present worth of the annuity (P)	
$\dfrac{(1 + i)^n - 1}{i}$	$(f/a, i, n)$	To find the future value (F) given the amount of annuity payments (A)	
$\dfrac{(1 + i)^n - 1}{i(1 + i)^n}$	$(p/a, i, n)$	To find the present value (P) given the amount of annuity payments (A)	

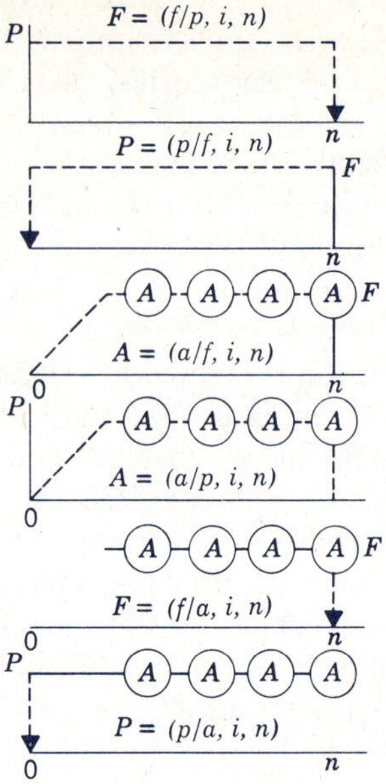

FIGURE 5.2 Interest formulas, symbols, and descriptions. Numerical values of the interest factors at rates from 0.5 to 20 percent are provided in Appendix E.

lationships and their reciprocal present-value counterparts. Mnemonic symbols for the formulas are used to make their application easier. The formulas, symbols, functional descriptions, and graphical cash flow representations are collected in Figure 5.2.

To use the interest formulas effectively, we must understand the mechanics of their application. The following five components are directly employed to produce equivalent quantities through interest–time conversions:

P = present worth, a lump-sum value at time "now."

F = future worth, a lump-sum value at a certain future point in time.

n = number of interest periods. There is often more than one interest period per year.

i = interest rate. In discrete compounding, i is the interest rate per period. Sometimes a **nominal interest rate** is quoted, such as "12 percent compounded quarterly." This statement means that interest is compounded four times a year at a rate per period (quarter) of 3 percent. Therefore, to find the value of $1 invested for 1 year at 12 percent compounded quarterly, we have

$$F = P(1 + 0.03)^4 = \$1(1.126) = \$1.13$$

or

$$F = P(f/p,3,4) = \$1(1.03)^4 = \$1(1.126) = \$1.13$$

A = one annuity payment. Our interest tables are based on an annuity composed of *equal payments* occurring at *equal time intervals* with the *first payment at the end of the first period.* Annuity factors are used to convert a series of payments to an equivalent single future or present sum and to translate single sums into a series of payments occurring in the past or future.

Many different symbols represent interest factors. The ones used in this text are mnemonic—they assist memory by associating letters with the functions to be performed.

A statement of a problem usually includes three of the P, F, i, n, and A factors and requires a solution for a fourth factor. Note that the symbol for each formula contains four of the factors. By identifying the known and unknown factors, we can identify the appropriate formula.

EXAMPLE 5.1 Use of interest formulas and symbols

The inventor of a patented production process is willing to consider a lump-sum payment in place of annual royalties for the use of his process. The firm using the process believes it will be needed for eight more years. Royalty payments will be $15,000 per year for the next five years and $10,000 a year for the last three years. Payments will be made at the beginning of each year.

If both parties agree to an annual interest of 7 percent, the present worth of the payments can be calculated in two steps corresponding to the two annuity patterns. The first five payments comprise an immediate outlay of $15,000 followed by an annuity of four $15,000 year-end payments. The present worth of these five payments, designated P_0 is

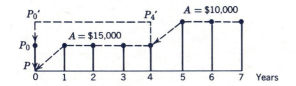

$$P_0 = \$15,000 + \$15,000(p/a,7,4)$$
$$= \$15,000 + \$15,000(3.387)$$
$$= \$65,805$$

The worth of the second annuity at the end of year 4, designated P_4' is

$$P_4' = \$10,000(p/a,7,3)$$
$$= \$10,000(2.624) = \$26,240$$

It must be discounted four years to "now" or year 0 as

$$P_0' = P_4'(p/f,7,4) = \$26,240(0.7629) = \$20,018$$

The present worth of all the royalty payments is then

$$P = P_0 + P_0' = \$65,805 + \$20,018 = \$85,823$$

which represents the equivalent amount that could be paid now to avoid future expenses. The single sum and the series of payments will be equivalent if taxes and any changes in buying power of money are ignored.

Equivalent comparison methods

Three methods are commonly applied to **discounted cash-flow comparisons:** equivalent present-worth, equivalent annual cost, and rate-of-return calcula-

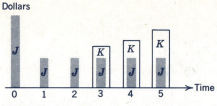

FIGURE 5.3 Coterminated projects J and K.

tions. All three methods indicate the same preferred alternative from similar data when applied correctly. Why then have three methods? The answer lies in personal preferences for the way comparisons are stated (total dollars, annual dollars, or percentages) and because some methods are better suited to certain types of problems than others are.

EQUIVALENT PRESENT-WORTH COMPARISONS

The **equivalent present worth** of an alternative is determined by converting all future receipts and disbursements to a present amount and adding the immediate cash income or outlay. The resulting single value can then be compared with similarly obtained values for other alternatives. In this way, alternatives with entirely different patterns of cash flow are reduced to a standardized present-worth index, allowing direct evaluation of expected returns.

Present-worth calculations are best suited to **coterminated projects:** comparisons in which the study periods or lives of all the alternatives end at one distinct point in time. A pattern typical of coterminated projects is shown in Figure 5.3, in which the solid and hollow bars represent the amount and time of costs expected from different alternatives, J and K. Costs associated with both projects terminate at the end of the five-year study period. A large initial investment and relatively low annual costs in J are contrasted with larger but later expenses in K. The alternative with the lower present cost is preferable in terms of total discounted outlays.

The aggregate cost figure that results from a present-worth calculation is often very large. It may appear ominous and distort thinking unless understood, but this single figure is advantageous in deciding how much can be spent in today's dollars for deferred benefits.

EXAMPLE 5.2 Present-worth comparison of coterminated projects

Special equipment to produce parts required to fulfill a contract can be developed, or the work can be subcontracted. Both alternatives provide equal quality. The contract will expire in three years, and it is doubtful whether it will be renewed or a similar one obtained. A firm bid to provide the necessary 30,000 parts each year at $2 per part has been submitted by a subcontractor. If the parts are produced by the prime contractor, annual direct costs are expected to be $22,000. Indirect costs are 50 percent of direct costs. What will be the maximum allowable developmental cost if the special equipment has only a 10-percent scrap value at the termination of the contract?

Assuming an interest rate of 10 percent and that all transactions take place at the end of a year, we can equate the first cost of the equipment (P)

less the present worth of the salvage value ($0.1P$) to the present worth of the difference between bid ($30{,}000 \times \$2 = \$60{,}000$) and production ($\$22{,}000 \times 1.5 = \$33{,}000$) costs as

$$P - 0.1P(p/f,10,3) = (\$60{,}000 - \$33{,}000)(p/a,10,3)$$

$$P[1 - 0.1(0.7513)] = \$27{,}000(2.487)$$

$$P = \frac{\$67{,}149}{0.925} = \$72{,}594$$

Therefore, if the special equipment can be developed for less than $72,594, its use will repay the amount invested plus 10-percent interest.

EQUIVALENT ANNUAL COST COMPARISONS

In an **equivalent annual cost comparison,** all receipts and disbursements occurring over a study period are converted to an equivalent yearly uniform income or outlay. Nonuniform transactions are converted to equivalent annuities through the use of appropriate interest formulas and are then added to other constant yearly transactions to provide one comparison figure. Annual costs are usually easy to collect and understand because most accounting and financial reporting periods are one year.

Annual cost comparisons are particularly suitable for *repeated projects:* studies involving physical assets that are renewed as necessary for continuing service over an indefinite time. Such cyclic patterns are based on the assumptions that the projects are initiated at one time, that the operations do not change, and that the cost figures remain constant during the foreseeable future. A common comparison is between higher-priced, longer-life items and lower-priced, shorter-life items.

EXAMPLE 5.3 Annual cost comparison of repeated projects

A material-handling problem can be solved by installing a conveyor system or by purchasing three lift trucks. The required return on invested capital is 10 percent. The cost pattern of each alternative is as follows:

	Conveyor	Lift Trucks
First Cost	$175,000	$16,000
Economic Life	18 years	6 years
Salvage Value	$ 20,000	$ 2400
Annual Operating Costs	$ 3000	$22,000
Annual Taxes and Insurance	$ 1000	$ 700

To make an annual cost comparison, we need only obtain the annuity equivalent to the capital recovery cost of the initial investment minus the discounted salvage value and add to it the other annual costs:

$$
\begin{aligned}
AC_{conveyor} &= \$175{,}000(a/p,10,18) - \$20{,}000(a/f,10,18) \\
&\quad + \$3000 + \$1000 \\
&= \$175{,}000(0.12193) - \$20{,}000(0.02193) \\
&\quad + \$4000 \\
&= \$21{,}338 - \$438 + \$4000 = \$24{,}900
\end{aligned}
$$

$$
\begin{aligned}
AC_{trucks} &= \$16{,}000(a/p,10,6) - \$2400(a/f,10,6) \\
&\quad + \$22{,}000 + \$700 \\
&= \$16{,}000(0.22961) - \$2400(0.12961) \\
&\quad + \$22{,}700 \\
&= \$3674 - \$311 + \$22{,}700 = \$26{,}063
\end{aligned}
$$

Thus the annual savings expected from installing the conveyor system rather than buying lift trucks is $26,063 − $24,900 = $1163.

The same alternative is selected by a present-worth calculation for the 18-year study period, the period that coterminates both projects:

which differs from the first present-worth calculations only by "rounding off" errors. In many problems it is necessary to combine PW and AC calculations. A frequent example is in which one repeated project starts at a date different from the one with which it is compared.

$$PW_{conveyor} = \$175,000 - \$20,000(p/f,10,18) + (\$3000 + \$1000)(p/a,10,18)$$

$$= \$175,000 - \$20,000(0.1799) + \$4000(8.201)$$

$$= \$175,000 - \$3598 + \$32,804 = \$204,206$$

$$PW_{trucks} = \$16,000 + \$13,600(p/f,10,6) + \$13,600(p/f,10,12) - \$2400(p/f,10,18)$$

$$+ (\$22,000 + \$700)(p/a,10,18)$$

$$= \$16,000 + \$13,600(0.5645) + \$13,600(0.3186) + \$2400(0.1799) + \$22,700(8.201)$$

$$= \$16,000 + \$7677 + \$4333 - \$432 + \$186,163 = \$213,741$$

The equivalence of the two methods is demonstrated by calculating the present worth of the annual cost as

$$PW \text{ of } AC_{conveyor} = \$24,900(p/a,10,18) = \$24,900(8.201) = \$204,205$$

$$PW \text{ of } AC_{trucks} = \$26,063(p/a,10,18) = \$26,063(8.201) = \$213,742$$

RATE-OF-RETURN COMPARISONS

A **rate of return** is the percentage of an investment returned each year as a gain or profit. When an asset loses value as it provides service, such as a machine producing salable products, the difference between the revenue it produces and its operation cost should meet the loss in value of the asset as well as provide a surplus or profit. This requirement is incorporated in the interest formula, $(a/p,i,n)$, which is commonly called the *capital recovery factor*. In multiplying the present worth of an investment by the capital recovery factor, we determine the minimum return that will cover depreciation expense plus a certain percentage gain. As applied to a rate-of-return calculation for a depreciable asset, we have

Depending on how the rate of return is calculated, it may indicate ROI—return on investment, ROE—return on equity, or IRR—internal rate of return. IRR is used here.

$$(a/p,i,n) = \frac{\text{annual revenue} - \text{operating cost}}{\text{amount invested in asset with a service life } n}$$

where the rate of return is the value of i in the capital recovery factor.

The rate of return is easily calculated for a single action, because the only unknown in the equation is i which is revealed by interpolation between interest rate tables for the indicated capital recovery factor. A pairwise comparison complicates the calculations because all the transactions for both alternatives must be equated with a common point in time. Several interest factors usually are included. Only one value of i will make the two alternatives equal. Therefore, the value of i that establishes the rate of return by which one alternative exceeds the other is disclosed by trial and error.

EXAMPLE 5.4 Rate of return for a single alternative

Replacing manual labor in a loading operation by automatic equipment will reduce net operating expenses by $27,200 a year. The equipment costs $100,000, and the scrap value will just cover removal cost at the end of its five-year life. Should the equipment be installed when the minimum acceptable rate of return on new investments is 10 percent?

Because the equipment is subject to physical depreciation, we calculate the rate of return as

$$\text{savings} = A \qquad \text{investment} = P$$

$$\$27,200 = \$100,000(a/p,i,5)$$

$$(a/p,i,5) = \frac{\$27,200}{\$100,000} = 0.27200$$

where the value of i is obtained by interpolating between the factors $(a/p,10,5)$ and $(a/p,12,5)$ as

$$i = 0.10 + 0.02 \frac{0.27200 - 0.26380}{0.27741 - 0.26380} = 0.112$$

The 11.2 percent rate of return means that the savings obtained by installing the automatic equipment will allow the initial investment of $100,000 to be repaid in five years plus provide annual earnings of 11.2 percent on the unrepaid balance.

Rate-of-return comparisons are useful in evaluating the effect of additional increments of investment in a single project. For instance, we know that pumping costs go down as the diameter of a pipeline increases. Pipe comes in certain increments of size. The problem of determining the most economical diameter is solved by ascertaining the rate of return from each extra increment of investment used to lower pumping cost by increasing the diameter. Stated as a gross equation, we have

$$\text{Rate of return} = \frac{\text{savings resulting from increase in pipe size}}{\text{extra investment required to increase pipe diameter}}$$

Although a rate of return is a universally understood measure of economic success, there is still some danger of misinterpretation in incremental analyses. The two most common abuses are tendencies to select automatically the alternative with the greatest rate of return on *total* investment or to select the alternative offering the largest *total* investment that meets the minimum acceptable rate of return. The correct interpretation is to select the largest **incremental investment** that exceeds minimum acceptable rate of return.

EXAMPLE 5.5 Rate of return for a pair of alternatives

Another alternative for the decision situation of Example 5.4 is to replace the manual operation with semiautomatic equipment which yields a net savings of $11,000 per year on an equipment investment of $60,000. The semiautomatic equipment will last 10 years and retain a salvage value of $20,000. Which alternative offers the higher rate of return?

A pairwise comparison can be made between the semiautomatic equipment and doing nothing or between the semiautomatic and automatic equipment. For the former comparison, we have

$$60,000(a/p,i,10) - \$20,000(a/f,i,10) = \$11,000$$

from which i is determined by trial and error to be approximately 15 percent, as

$$\$60,000(a/p,15,10) - \$20,000(a/f,15,10)$$
$$= \$60,000(0.19925) - \$20,000(0.04925)$$
$$= \$10,970$$

The second comparison, equating the annual cost of the two types of equipment under the repeated project assumption, appears as

$$\$60,000(a/p,i,10) - \$20,000(a/f,i,10) - \$11,000$$
$$= \$100,000(a/p,i,5) - \$27,200$$

or

$$\$16,200 = \$100,000(a/p,i,5) - \$60,000(a/p,i,10)$$
$$+ \$20,000(a/f,i,10)$$

and

$$\$16,190 = \$100,000(a/p,\tfrac{1}{2},5) - \$60,000(a/p,\tfrac{1}{2},10)$$
$$+ \$20,000(a/f,\tfrac{1}{2},10)$$

Therefore the rate of return on the extra increment of investment to go from semiautomatic to automatic equipment is only 0.5 percent. Because this rate is well under the required return, a preference for the semiautomatic equipment is indicated.

EXAMPLE 5.6 Rate of return for incremental investments

Three alternative methods of accomplishing a project are shown in the following table. Each investment has the same economic life, eight years, and no salvage value. The minimum acceptable rate of return is 10 percent.

The rate of return for alternative A is calculated as

$$(a/p,i,8) = \frac{\$16,000}{\$70,000} = 0.22857$$

making $i = 16$ percent. Since A is an acceptable

	Method A		Method B		Method C
Investment	$70,000		$90,000		$100,000
Incremental Investment		$20,000		$30,000*	
Net Profit	$16,000		$19,100		$ 22,300
Incremental Profit		$ 3100		$ 6300*	
Rate of Return on the Total Investment	16%		13.5%		15%
Rate of Return on Incremental Investment		5%		13.2%*	

*Based on the last acceptable increment of investment.

alternative because its rate of return is greater than 10 percent, we next evaluate the increment of investment added to A to afford the greater profit from B:

$$(a/p,i,8) = \frac{\$3100}{\$20,000} = 0.15500$$

making i = 5 percent. This rate of return is less than 10 percent for the $20,000 extra investment and makes B an *unacceptable* alternative. A comparison is always made with the *next lowest acceptable level of investment.* Therefore, alternative C is compared with A to determine the rate of return for the next increment of investment:

$$(a/p,i,8) = \frac{\$22,300 - \$16,000}{\$100,000 - \$70,000}$$

$$= \frac{\$6300}{\$30,000} = 0.21000$$

for which i = 0.132. Because this $30,000 increment shows a rate of return greater than the required minimum, C is the most attractive alternative. Note that A has a greater rate of return on the *total* investment than does C, but the selection of A instead of C would eliminate the opportunity to invest $30,000 at the attractive 13.2 percent rate.

5-3 REPLACEMENT PLANNING

It is a generally valid observation that machines lose value with age. Occasionally a machine of outstanding craftsmanship or one with antique value becomes more valuable as it gets older, but most of the tools of production depreciate in value over time. The decrease in worth, or **depreciation,** is due to a combination of the following four causes:

PHYSICAL DEPRECIATION
Normal wear and tear of operations gradually lessen the machine's ability to perform its intended function. A good maintenance program can slow decay, but only an economically questionable "rebuilding" can rejuvenate a machine to approximately its original condition. Continuously more expensive repairs are typically required to even keep an older machine operating.

"Accidental" damage is considered normal wear, but the actual amount is still a function of the care shown.

FUNCTIONAL DEPRECIATION
A change in demand or service expected from a machine makes it less valuable to the owner, even though it is still capable of performing its original purpose. If an original 0.01-inch tolerance requirement for parts is permanently changed to 0.0001-inch and the machine producing the parts is designed to meet only the first specification, the machine has functionally depreciated. It can still perform its intended function, but that function is no longer required or adequate.

Changes in capacity requirements, either up or down, are the main causes of functional depreciation.

TECHNOLOGICAL DEPRECIATION
The development of new and better methods of performing a function suddenly makes previous machine designs uneconomical. A "breakthrough" in technology is now such a routine happening that obsolescence is a major concern with any machine purchase. A decision to switch to new materials or improved product designs also can instantly make existing custom equipment obsolete.

Machine designs based on new concepts are frequently in the drawing board stage before previous designs are in production.

MONETARY DEPRECIATION

A change in the purchasing power of money causes subtle but definite depreciation. Accounting practices typically relate depreciation to the first cost of a machine, not the replacement cost. Over recent years, prices have continually increased. Therefore, the capital recovered from a machine's service is often insufficient to buy a replacement when the old one can no longer produce competitively. Thus, depreciation actually applies to the invested capital representing the machine rather than to the asset itself.

Under the influence of these largely unpredictable causes of depreciation, it is indeed difficult to estimate a machine's **economic life**—the number of years that minimizes the equivalent annual cost of holding an asset. Life estimates are needed to evaluate the relative attractiveness of new machine alternatives and to plan replacement schedules and taxes. The Bureau of Internal Revenue suggests life limits for different classes of machines. Past experiences aligned to reflect current practices and recent equipment designs also provide a basis for life comparisons.

In addition to physical and legal aspects, management philosophy plays a major part in depreciation programs. On one hand is a desire to retain machines in service as long as possible to lower the per-year capital recovery charge. Conversely, planning a shorter life reduces the chance of inferior quality and unexpected technological disruptions. Of course, managers' personal opinions color both views, such as an attraction to a particular brand of equipment or a passion to own only the most modern machines. All views are relevant to varying extents and deserve consideration, but the evaluation should be oriented toward system merits (long-range plans of corporate objectives, utilization of personnel, product development, material control, and the like) rather than concentrated on individual machines.

Depreciation accounting

Depreciation accounting serves two major purposes. It sets a pattern for recovering the capital invested in an asset, and it relates the cost of owning a machine to its output. In effect, a portion of the earnings derived from the operation of a machine is set aside in bookkeeping accounts toward the cost of replacement.

An ideal depreciation method is easy to use, provides a realistic pattern of cost, recovers all the capital invested, recognizes any tax advantages, and is acceptable to the Internal Revenue Service. Several methods satisfy these attributes to various degrees. The simplest to apply is *straight-line depreciation*, by which a constant amount is charged in each year of the asset's life to recover the invested capital. Thus the annual depreciation charge (*ADC*) is calculated as

$$\text{Annual Depreciation Charge} = ADC = \frac{P - S}{n}$$

where

P = purchase price of the asset

S = salvage value at the end of the asset's economic life

n = economic life in years

Depreciation charges are not actual cash flows, and the true decrease in the market value of an asset may not correspond to the allowable deductions. Tables and formulas approved by the Internal Revenue Service establish the pattern of depreciation. Major changes in depreciation accounting occurred with the passage of the Economic Recovery Act of 1981, which introduced the *Accelerated Cost Recovery System* (ACRS). This system specifies percentages for the recovery of investments in various property classes that determine the amount that can be deducted each year for tax purposes. Because many situation-specific conditions are regularly involved in investment recovery and the rules governing depreciation are voluminous, *appropriate IRS publications should be consulted for all depreciation decisions*.

Provisions for recovering capital invested in income-producing assets are made by charging depreciation against current income. These depreciation charges and the regular operating expenses are deductible from gross income in determining taxable income. In the section on after-tax evaluations, straight-line depreciation is applied to illustrate the effect of taxes on cash flow. This depreciation method has been the most widely used historically and is still allowed under current laws, but it does not provide the accelerated write-offs available in ACRS tables.

Economic evaluation of defenders versus challengers

Depreciation-accounting methods relate the cost of ownership to production costs, but they contribute little to replacement studies. In fact, their contribution may be misleading. A replacement study compares the operating expenses and capital costs of a machine presently owned, called a **defender,** with those of a replacement machine, called a **challenger.** A study must be made on current information. The best available estimate of what a defender is worth today is the amount it can be sold for now, not what the accounting records say it is worth.

Sunk cost is the difference between the present market value and the worth shown by depreciation accounts. As depicted in Figure 5.4, straight-line depreciation yields a book value at the end of year 2 of $3000. The actual worth of the asset is set by the amount that can actually be recovered today, $2000. The sunk cost of $3000 − $2000 = $1000 appears as a loss of capital arising from life and salvage estimates made when the present defender was a challenger.

Most replacement patterns are best compared by the discounted cash flow comparison models already introduced. The annual cost model is generally

One rationale for allowing the rapid write-offs of depreciable property is that inflation erodes the purchasing power of deducted depreciation charges. Therefore, a replacement is likely to cost more than the amount accumulated in the depreciation reserve for its purchase.

ACRS recovery percentages for three-year property in 1986 were

Year (n)	Depreciation Charge (%) in Year n
1	33
2	45
3	22

The outcomes of a decision cannot start before the moment a decision is made. Therefore, past events should not be considered unless they influence future events.

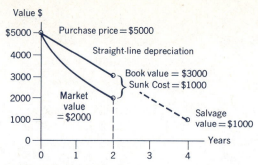

Value $

$5000 — Purchase price = $5000

4000 —

Straight-line depreciation

3000 — Book value = $3000
Sunk Cost = $1000

2000 —

Market
value
1000 — = $2000

Salvage
value = $1000

0 —

Years

0   1   2   3   4

FIGURE 5.4 Sunk cost of an asset.

Relevant replacement factors include (1) labor costs, (2) material costs, (3) taxes, (4) insurance, (5) capital costs, (6) maintenance costs, (7) cost of waste, and (8) indirect costs.

applied because cost data are normally tabulated in yearly amounts and replacements follow the "repeated projects" pattern. Part of the information needed for a comparison is the same used in depreciation accounting: purchase price, economic life, and salvage value. Operating expense and other costs associated with ownership are also needed. These factors are then discounted to equivalent bases by applying interest formulas.

The capital recovery costs for both the defender and challenger are calculated by the formula

An equivalent expression is
$AC = P(a/p,i,n) - S(a/f,i,n)$.
Both yield the same results.

$$\text{Annual cost}(AC) = (P - S)(a/p,i,n) + Si$$

where P is the present cost (purchase price for the challenger and market value for the defender) and S is the salvage value. An application of this formula is shown in Example 5.7.

EXAMPLE 5.7 Replacement study based on an annual cost comparison

An overhead crane conveyor system was installed in a warehouse assembly area 15 years ago at a cost of $40,000. The economic life was estimated at 20 years with an $8000 salvage value. Depreciation accounting is by the straight-line method. Operating costs for the conveyor system, including labor and maintenance, are $18,500 per year. Because maintenance and repairs have not been excessive, the crane is now expected to last another 10 years and then have a zero salvage value.

When the crane was installed, the products being produced were heavier and bulkier than current production. Today the same service could be provided by two lift-trucks at one-half the present operating

costs. Lift trucks can be purchased for $6000 apiece with an expected life of eight years and a salvage value equal to one-tenth their first cost. However, 200 square feet of storage space valued at $3 per square foot per year would be lost in providing space for the trucks to maneuver. If the trucks are purchased, the crane will be dismantled at a cost of $4000 and sold at a firm bid of $9000.

Based on the company's desired rate of return of 10 percent, the replacement study is conducted as follows:

CAPITAL COSTS. The defender, the crane, has a present value of $P = \$9000 - \$4000 = \$5000$. Latest estimates place its life at $n = 10$ years, with the salvage value just equal to the cost of removal, $S = 0$. It is noted that the book value, $40,000 - \frac{15}{20}($40,000 - $8000) = $16,000$ would show a sunk cost of $16,000 - $5000 = $11,000$. That this cost is irrelevant is apparent from the newer life and salvage figures; the sunk cost is derived from estimates now 15 years old, and it plays no part in the present study.

The pair of lift trucks, the challenger, has an equivalent capital recovery plus return annual cost of

$$AC_{2\ trucks} = 2[(P - 0.1P)(a/p,i,n) + 0.1Pi]$$

$$= 2[($6000 - $600)(a/p,10,8)$$

$$+ $600(0.10)]$$

OTHER COSTS:

Annual operating cost of
crane conveyor system = $18,500

Annual operating cost of
lift trucks = $18,500 × ½ = $9250

Annual cost of lost storage
if trucks are used = 2000 sq ft × $3/sq ft
= $6000

Comparison Table	Defender	Challenger
Capital cost: $5000(a/p,10,10) + 0 = $5000(0.16275)	$814	
2[$5400(a/p,10,8) + $60] = $10,800(0.18744) + $120		$2144
Other costs: operating	18,500	9250
loss of storage		6000
Total cost	$19,314	$17,394

The advantage of the challenger over the defender is $19,314 - $17,394 = 1920 per year. Although the advantage is significant, it is not dramatic. The edge given to the challenger probably stems from functional causes: the products are lighter and smaller now than in the past. Is there any chance the function will change again? Should a belt conveyor also be considered? Would the operation of lift trucks create a safety hazard? Will storage space become more valuable as a result of company expansion? Such questions surround every replacement study and deserve careful attention.

5-4 AFTER-TAX EVALUATIONS

Successful production systems are subject to income taxes, because the revenue they generate exceeds allowable tax deductions. Tax laws applicable to production vary from year to year as to rates and specific types of allowable deductions, but in general,

Taxable income = gross revenue − expenses − interest and taxes − depreciation

and

Taxes = taxable income × tax rate

As applied to cash flow analysis,

After-tax cash flow (ATCF) = before-tax cash flow (BTCF) − taxes

The reason for considering taxes in evaluating alternative proposals for capital investment is that **after-tax cash flow (ATCF)** more accurately represents operating conditions. Occasionally an after-tax analysis contradicts the order of

Investment tax credits were first allowed in 1962 to encourage corporate investments. Since then they have been canceled and reinstated twice, and credit percentages have been changed many times.

preference that resulted from a before-tax comparison, mainly when the choice is among equity funding, borrowing, and leasing.

Benjamin Franklin observed that "in this world nothing is certain but death and taxes." He might have added that attention to both increases chances of survival and prosperity.

After-tax cash flow

The basic procedure for converting before-tax to after-tax cash flow is implied by the previous relationship. Those equations assume the tabular form shown in the following for computations:

End of Year	Before-Tax Cash Flow	Cash Flow for Loan	Depreciation Charges	Taxable Income	Taxes [tax rate × (4)]	After-Tax Cash Flow
N	(1)	(2)	(3)	(4) = (1) + (2) + (3)	(5)	(6) = (1) + (2) + (5)

A tabular format represented by the table headings relies on the use of correct signs to indicate the direction of cash flow: income is positive and outflow is negative. As applied to the purchase of an asset, the purchase price is negative, as are loan repayments, interest, depreciation, and taxes. Revenue, salvage value, and loan receipts are positive.

EXAMPLE 5.8 Comparison of before- and after-tax cash flows

A productive asset can be purchased for $120,000 . It will have no salvage value at the end of its six-year life. Operating cost will be $12,000 per year, and it will produce a revenue of $40,000 annually. Straight-line depreciation is used, and the firm's tax rate is 40 percent. Compare the after-tax cash flows for a cash purchase with owned capital and acquisition when two-thirds of the purchase price is met with capital borrowed at 10 percent interest.

The cash flow pattern for equity funding is shown

TABLE 5.1
Cash Flow Pattern When the Purchase is Funded with Equity Capital

Year (N)	BTCF (1)	Loan (2)	Depreciation (3)	Taxable Income (4)	Taxes at 40% (5)	ATCF (6)
0	− $120,000					− $120,000
1	28,000		− $20,000	$8,000	− $3,200	24,800
2	28,000		− 20,000	8,000	− 3,200	24,800
3	28,000		− 20,000	8,000	− 3,200	24,800
4	28,000		− 20,000	8,000	− 3,200	24,800
5	28,000		− 20,000	8,000	− 3,200	24,800
6	28,000		− 20,000	8,000	− 3,200	24,800

TABLE 5.2
Cash Flow Pattern When the Purchase is Funded by Two-Thirds Borrowed Capital

Year (N)	BTCF (1)	Loan and Interest (2)	Depreciation (3)	Taxable Income (4)	Taxes at 40% (5)	ATCF (6)
0	− $120,000	$80,000				− $40,000
1	28,000	8,000	− $20,000	0	0	20,000
2	28,000	8,000	− 20,000	0	0	20,000
3	28,000	8,000	− 20,000	0	0	20,000
4	28,000	8,000	− 20,000	0	0	20,000
5	28,000	8,000	− 20,000	0	0	20,000
6	28,000	8,000	− 20,000	0	0	20,000
6		− 80,000				− 80,000

in Table 5.1. The first column shows the initial outlay of − $120,000 at time zero, followed by annual net revenue of $40,000 − $12,000 = $28,000. Because there is no loan when the purchase is with owned capital, column 2 is blank. Column 3 indicates straight-line depreciation at $120,000 ÷ 6 = $20,000 per year, shown as a negative value. The taxable income in column 4 is simply the algebraic sum of columns 1 and 3; these values are then multiplied by the 40-percent tax rate to obtain annual taxes in column 5. Finally, the after-tax cash flow is tabulated in column 6 as the first cost in the top line followed by taxable income minus taxes in subsequent lines.

The alternative purchasing plan, one-third equity funds and two-thirds borrowed capital, is shown in Table 5.2. Columns 1 and 3 are the same as Table 5.1 because the before-tax cash flow and the depreciation charges do not change with the type of financing. The difference is in column 2 where it is assumed that the loan of $120,000 × ⅔ = $80,000 will be repaid at the end of year 6; receipt of the loan

is a positive inflow at year 0, and its repayment is a negative cash outflow in year 6. Interest charges create a negative cash flow of $80,000 × 0.10 = $8000 per year, based on the outstanding balance of the loan. Because interest is a deductible expense, it and depreciation are subtracted from the BTCF to obtain taxable income in column 4. ATCF in column 6 lists in descending order the first cost minus the loan, annual revenue minus interest charges and taxes, and repayment of the loan.

A comparison of the two cash flow patterns indicates that no taxes are owed when borrowed capital is used, but the year-by-year cash flow is still greater when equity funds are used because there are no interest charges. It is also apparent that initial and final outflows are different; $80,000 of the purchase price is delayed by six years until the loan becomes due. The relative attractiveness of the two patterns depends mostly on the applicable interest rates for borrowed and equity capital.

Comparison methods

Tabulated cash flows can be evaluated by any of the three comparison methods discussed in Section 5-2. Using the figures from Example 5.8 and a minimum acceptable rate of return of 15 percent, the following two cash flow patterns are evaluated by the present-worth and equivalent annual worth methods to obtain the same indication of preference.

Present Worth

With no borrowed funds:	P	$= -\$120,000 + \$24,800(p/a,15,6)$
		$= -\$120,000 + \$24,800(3.784) = \underline{-\$26,159}$
With 67% borrowed funds:	P	$= -\$\ 40,000 + \$20,000(p/a,15,6)$
		$\qquad\qquad\qquad - \$80,000(p/f,15,6)$
		$= -\$\ 40,000 + \$20,000(3.784) - \$80,000(0.4323)$
		$= \underline{\$1096}$

Equivalent Annual Worth

With no borrowed funds:	A	$= -\$120,000(a/p,15,6) + \$24,800 = \underline{-\$7,509}$
With 67% borrowed funds:	A	$= -\$\ 40,000(a/p,15,6) + \$20,000$
		$= -\$\ 80,000(a/f,15,6) = \underline{\$291}$

Present worths of BTCF and ATCF comparisons are the same because taxes were 0 for 67 percent loan financing.

The advantage gained by employing money borrowed at a lower interest rate than an organization's required minimum rate of return is apparent.

It is enlightening to compare the present worths of before- and after-tax cash flows:

$$\text{No borrowed funds: } P_{(\text{BTCF})} = -\$120,000$$
$$+ \$28,000(p/a,15,6) = \underline{-\$14,048}$$

$$67\% \text{ borrowed funds: } P_{(\text{BTCF})} = \underline{\$1096}$$

The preference for using borrowed funds is again apparent, although the difference between the two financing plans is not as great as that in the ATCF analysis.

5-5 SENSITIVITY ANALYSIS

Difficulties encountered in cost estimating and the need to test the sensitivity of those estimates were introduced in Chapter 4. Regardless of the care devoted to estimating cash flows, there is always lingering doubt about their accuracy. Forecasts may be inaccurate. Data might be accurate but not applicable to a given proposal. When the data are in doubt, as is the usual case, experienced analysts consider the consequences of using faulty "facts." Considerations range from just checking the effect of a single change in one estimate to investigating the effects on a proposal of a range of changes in several factors. A sensitivity analysis measures the magnitude of change in one or more elements of an economic comparison to detect levels that will reverse a decision among alternatives.

In a sensitivity analysis of a single proposal, elements are varied to find the level at which acceptance switches to rejection.

Optimistic-pessimistic brackets

A popular form of informal sensitivity analysis is to speculate about outcomes that might evolve from slightly altered situations, called *scenarios*. Each scenario starts from the same basic structure, and changes are conceived to expose different conclusions. As applied to economic evaluations, a proposal for capital investment is the basic structure. Changes occur when original cash flows are replaced with new patterns. The conclusion of the scenario based on the introduced variations is a different present worth or rate of return than initially determined.

An elementary but revealing sensitivity study results from best- and worst-case scenarios. By bracketing the most likely condition, normally the original statement of a proposal, with data derived from **optimistic and pessimistic estimates,** analysts can observe damages that might occur when things go wrong and the greatest good that could result from everything going well. These boom-and-bust scenarios suggest what happens during extreme conditions, but no probability estimates are made for the likelihood of these conditions occurring. Therefore, they are most useful for protecting an investor from a ruinous commitment and to identify possible bonuses from an accepted proposal.

Economic risk probabilities are discussed in Section 5-7.

Optimistic-pessimistic evaluations are typically based on before-tax cash flows. The original proposal is considered to be an objective estimate of the *most likely* cash flow. An *optimistic* scenario is based on an advantageous interpretation of future events. A *pessimistic* scenario is an assessment of the future that adversely affects cash flow. Viewed from these perspectives, appraisals are made for critical decision inputs such as asset life, revenue, salvage value, and expenditures.

Neither optimistic nor pessimistic scenarios should portray ultimate extremes—once-in-a-billion happenings. A scenario should have a reasonable chance of occurrence to deserve attention.

Sensitivity graph

A **sensitivity graph** records the effects of changes in several variables on the present worth of a proposal. By displaying the effects visually, a proposal is more open to review by decision makers who did not take part in the proposal's preparation.

EXAMPLE 5.9 Optimistic—most likely—pessimistic analysis for the installation of a prototype operation

Current mail-sorting operations on one line at a mail distribution center cost about $1 million a year. A newly developed and essentially unproved but promising system for sorting mail by automatic address readers is being considered. Although the devices have been tested and approved in laboratories, there is still doubt about how they will perform in regular operations. Questionnaires describing less-than-favorable and more-than-favorable operating conditions were sent to people familiar with the devices and the mail distribution process. Consensus data from the two scenarios and the most likely estimates are as shown.

Factors Estimated	Pessimistic Estimate	Most Likely Estimate	Optimistic Estimate
First cost, including installation	$2,112,000	$985,000	$915,000
Life, years of full utilization	2 years	4 years	6 years
Annual maintenance, minor repair	$221,000	$81,000	$75,000
Annual operating cost, standbys	$929,000	$714,000	$588,000

A before-tax evaluation is conducted by calculating the annual cost for each set of estimates, using a discount rate of 10 percent.

$$AC_{(pessimistic)} = \$2,112,000(a/p,10,2) + \$212,000$$
$$+ \$929,000$$
$$= \$2,112,000(0.47619) + \$1,150,000$$
$$= \underline{\$2,155,713}$$

$$AC_{(most\ likely)} = \$985,000(a/p,10,4) + \$81,000$$
$$+ \$714,000$$
$$= \$985,000(0.21547) + \$795,000$$
$$= \underline{\$1,007,238}$$

$$AC_{(optimistic)} = \$915,000(a/p,10,6) + \$75,000$$
$$+ \$588,000$$
$$= \$915,000(0.12961) + \$663,000$$
$$= \underline{\$781,593}$$

Annual costs indicate that the potential losses are larger than the potential gains for the projected scenarios. Although these figures neither reject nor accept the proposal, they do suggest apprehension about its implementation, especially when it is compared with the $1-million annual cost for the existing operation it would replace.

Assume that a decision is to be made based on the following estimates and tentative before-tax analysis:

Economic Factors	PW at $i = 10\%$
Minimum acceptable rate of return ($i = 10\%$)	
Economic life ($N = 6$ years)	
Annual receipts ($A = \$40,000$)	$174,200
Annual disbursements ($A = -\$15,000$)	$-\$65,325
First cost ($P = -\$120,000$)	$-\$120,000
Salvage value ($S = \$20,000$)	$\underline{\$11,290}$
Before-tax present worth	$165

The present worth so calculated is akin to the most-likely estimate used in the optimistic-pessimistic bracketing and is the focal point for a sensitivity graph. In the graph, each of the economic factors is subjected to a range of changes while all other factors are concurrently held constant. The resulting pattern for each factor indicates its degree of sensitivity compared with the other factors.

Figure 5.5 shows curves for the six economic factors in the sample data as each factor deviates from its original estimate within a -30 to $+30$-percent range. The curves are generated from the following formula in which one variable at a time is altered:

$$PW = -(\text{first cost}) + \text{salvage value } (p/f,i,n)$$
$$+ (\text{receipts} - \text{disbursements})(p/a,i,n)$$

The intersection of all the curves conforms to the PW of the most likely estimate.

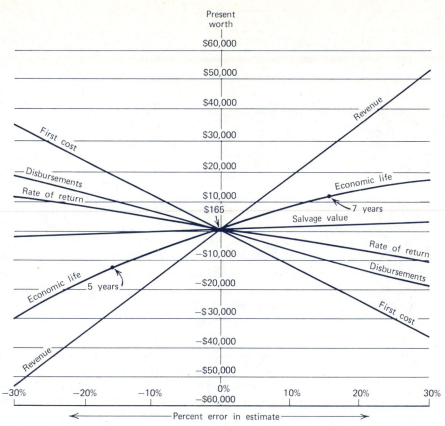

FIGURE 5.5 Sensitivity graph of the effect on a proposal's present worth when economic factors deviate from their original estimates.

Steeper curves indicate a higher degree of sensitivity to deviations from original estimates. For the given conditions, variations in first cost and revenue estimates have the greatest effect on the present worth.

5-6 CONSIDERING RISK

The next refinement beyond a sensitivity analysis is considering the probability of occurrence of each scenario for a proposal. Once the cash flows for the scenarios and their relative likelihood of happening are determined, the laws of probability can be applied to indicate a preferred alternative. Probabilities measure the "risk" that the outcome of a proposal will differ from the original expectations.

Every planner knows that cash flow estimates are indeed estimates. Variability is a recognized factor in every phase of production systems: the properties of materials vary over time and with sources; the quality and quantity of worker

Convenient categories for decision models are certainty, risk, and uncertainty. Uncertainty is distinguished from risk by its inability to assign probabilities to future states.

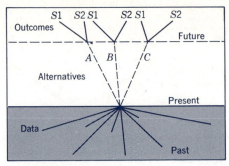

FIGURE 5.6 Decision tree.

output fluctuate widely; and even machines occasionally perform in unanticipated ways. Planners would drive themselves to distraction trying to account for the infinite number of deviations that could upset their cash flow proposals. To avoid an obsession with tactical estimating details, we can adopt a strategic plan of applying a broad safety factor or other means of overdesign. In economic studies, an obvious safety factor is to require an overly high rate of return or an unrealistically short payback period from an alternative before it is deemed acceptable. The surplus return engendered by measuring alternatives against these protective standards provides security from inaccurate estimates.

In this section our main concern is with ways to reduce the second source of estimating errors: how to include more than one future state in an evaluation. The correspondence between the past, present, and future and the data, alternatives, and outcomes is depicted in Figure 5.6. We draw on the past data to recognize alternatives and then look to the future to evaluate their worth.

Studies recognizing risk focus attention on the top section of the decision tree. They demand inspiration, perspiration, and judgment. Inspiration helps identify possible future states. Perspiration characterizes the effort required to estimate the cash flow associated with each state for each alternative. Judgment is used in grouping the states into workable units. For instance, the best method of carrying out a construction project could depend on the weather expected. If the work were sensitive to the amount of rainfall, the future states would be different possible precipitation levels. Assuming that the range were between 0 and 20 inches during the duration of the project, we could divide this range into any number of states corresponding to some fraction of the total. However, if we know that the same outcomes will be exhibited over ranges of 0 to 9, 10 to 14, and 15 to 20 inches, the effects of only three states will receive attention. Then the cash outcomes will be estimated for all three states for each feasible method of conducting the construction.

In addition to discerning future states and estimating their associated outcomes, the relative likelihood of each future must be assessed. For the construction project example, the probability of each level of rainfall could be estimated from past weather data. The sources of probability estimates are similar to those discussed for forecasting in Section 3-4. After numbers have been attached to all the branches (S1 and S2) of the decision tree of Figure 5.6, the final step is to select a comparison method for the risk evaluation.

Expected value comparison

The concept of **expectation** or **expected value** proffers a most useful and satisfying aid to risk evaluations. It is useful because it is versatile yet compels a precise statement of the problem. It is satisfying because it produces the sort of jury verdict based on an objective weighing of evidence that generates intuitive confidence.

		State	
		S1 P(1)	S2 P(2)
Alternative	1	O_{11}	O_{12}
	2	O_{21}	O_{22}

FIGURE 5.7 Payoff table.

PAYOFF TABLE

All conditions for the statement of a problem recognizing risk are included in a sample format called a **payoff table** (Figure 5.7). All alternative courses of action available in the problem situation are listed and thereby set the number of rows in the table. Each applicable future state establishes a column. Associated with each state is the probability of its occurrence. The sum of the probabilities for all the future states should equal 1.0. The cells in the table show what outcome (O) is anticipated from each alternative for each future that can occur. The outcomes can be expressed in pounds, inches, dollars, or any other units as long as the dimensions are consistent.

EVALUATION

The "expectation" from each alternative is the weighted average of all its outcomes. They are weighted according to their probability of occurrence. For the alternatives in Figure 5.7, the expected values are

$$E(A1) = [O_{11} \times P(1)] + [O_{12} \times P(2)]$$

$$E(A2) = [O_{21} \times P(1)] + [O_{22} \times P(2)]$$

If the outcomes show profit, the alternative with the largest weighted average or greater profit expectation will be preferred.

EXAMPLE 5.10 Expected value for a study recognizing risk

The image-improvement campaign initiated by the "Ipso Facto" data processors was so successful that they are now considering a major expansion. The best available probability estimates of the future demand for automated data-processing services are 0.1 to decline slightly, 0.3 to remain constant, and 0.6 to increase rapidly. If Ipso Facto launches an expansion now, it feels it can capture most of the new local demand. It would also suffer a considerable loss from unused capacity if the demand declined. The conservative alternative is to try to increase facility utilization from its present 85-percent level to 100 percent. The net average annual gains or losses over four years for each future from the alternatives are shown in the accompanying payoff table.

		Decline $P(D) = 0.1$	Constant $P(C) = 0.3$	Increase $P(I) = 0.6$
Major expansion	(E)	-$180,000	-$5000	$90,000
Increase utilization	(U)	-$10,000	$10,000	$40,000

Using the expectation criterion, we have

$$E(E) = -\$180,000(0.1) - \$5000(0.3)$$
$$+ \$90,000(0.6)$$
$$= -\$18,000 - \$1500 + \$54,000 = \$34,500$$

$$E(U) = -\$10,000(0.1) + \$10,000(0.3)$$
$$+ \$40,000(0.6)$$
$$= -\$1000 + \$3000 + \$24,000 = \$26,000$$

which indicate a preference for major expansion.

	S1 0.1	S2 0.9	E
A1	0	$200	$180
A2	$12,000	-$1000	$300

FIGURE 5.8 Expected value.

INTERPRETATION

The expected value of an alternative, such as $300 for $A2$ in Figure 5.8, is the average gain expected from repeatedly choosing $A2$ in the same decision situation. Actually, each time an outcome occurs, it will have a value of +$12,000 or -$1000. Because precisely identical decision situations can hardly be expected too often in an industrial setting, the expectation outcomes cannot be followed blindly. If a $1000 loss would be acutely painful, it might be wise to eliminate that alternative from the table, regardless of the attractiveness of other outcomes. When a conservative approach is dictated by intangible factors, a safe alternative is a prudent choice, even when the expectation of a riskier one is greater. Such decisions are subjective options accrued from factors not included in the payoff table. Expectation is a guide, not a dictate; but when followed over the long run, even when individual problems are not identical, it offers a sound, objective principle of choice.

Sometimes there is considerable disagreement or uncertainty regarding the probability of each future. One way to avoid the dilemma of pinpointing an exact probability estimate is to calculate an **indifference probability level.** Letting $P(1)$ and $P(2) = 1 - P(1)$ to be the probabilities for the two future states in Figure 5.9, the indifference probability is calculated as

$$E(A1) = E(A2)$$

$$\$0(P1) + \$200[1 - P(1)] = \$12,000(P1) - \$1000[1 - P(1)]$$

$$P(1) = \frac{\$1000 + \$200}{\$12,000 + \$1000 + \$200} = 0.09$$

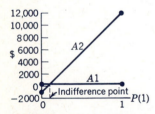

FIGURE 5.9 Indifference points can also be determined graphically as shown for the data from Figure 5.8.

indicating that when the probability of $S1 = 0.09$, the expected value of both alternatives is the same, and a decision maker would be indifferent to either course of action. Consequently, any estimate for $P(1)$ in the range of 0.091 to 1.0 would swing the preference to $A2$. It is usually easier to extract a consensus from a policymaking group when they are offered a range rather than a point to agree upon.

Spending extra effort and money can usually improve the reliability of probability estimates. When it appears that research can increase estimating confidence, the question becomes how much it is feasible to spend. An indication is obtained by calculating the **value of perfect information.** By using the data from Figure 5.9, we would always choose A2 for a gain of $12,000 if we knew S1 were *sure* to occur. Similarly, if we *knew* S2 would transpire, we would always follow A1 for a gain of $200 instead of a $100 loss from using A2. Because A1 and A2 are expected to occur, respectively, one time and nine times in 10, the expected return from prescient knowledge is

$$\$12,000(0.1) \ + \ \$200(0.9) \ = \ \$1380$$

The difference between the expectation with perfect information and the preferable alternative A2 based on blindness is the value of perfect information, $1380 − $300 = $1080. Thus we presumably would be willing to pay $1080 for perfect information each time a decision is required in the given situation. The only hitch is to find the marketplace where perfect information is sold.

OPTIONAL MATERIAL _____

Discounted decision tree comparisons

The hybrid comparison technique resulting from a combination of the expectation principle with the time value of money is a **discounted decision tree.** Its illustrative branching format is particularly useful for planning and presentation. The evaluation based on the expectation of discounted cash flow is realistically direct; inclusion of extended time-oriented decisions adds versatility.

For additional details, see J. F. Magee, "Decision Trees for Decision Making," *Harvard Bus. Rev.*, July–Aug. 1964.

FORMAT

The initial identification of future states and associated outcomes for each alternative is integral to any risk evaluation. The distinctive feature of a discounted decision-tree evaluation is the portrayal of successive decisions. Such decisions arise when the primary or immediate solution depends on which courses of action will be followed at decision points in the future.

The relationship of successive decisions is conveniently displayed by the "tree" format shown in Figure 5.10. The first decision to select a course of action at time zero is denoted by the square at the left, D1. Time flows from left to right. The dotted line connecting D1 to the circled A1 represents a possible course of action, alternative 1. The two solid lines from A1 indicate possible outcomes from the alternative; the top line O1 shows the expected returns over both time periods, and the lower line O2 represents returns for only the first period. The second square D2 indicates that a second decision will be made *if* the outcome O2 occurs. That is, the choice between A11 and A12 will be made only if the future state represented by O2 transpires. The conditional outcomes for the two alternatives from the second decision point are denoted as O2a–d.

The conditions creating the need for a second decision point originate from a desire to modify a course of action if a certain future state occurs. Let us assume that the outcomes in Figure 5.10 pertain to the following three estimates of

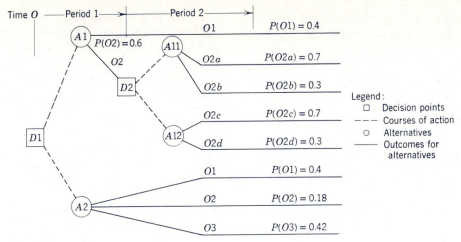

FIGURE 5.10 Decision-tree format.

future facility utilization:

State 1. Constant throughout the two periods:
$$P(C_1C_2) = 0.4$$

State 2. Increase by 20 percent during the first period, then constant for the second: $P(I_1C_2) = 0.18$

State 3. Continue to increase for both periods:
$$P(I_1I_2) = 0.42$$

From these estimates we can say that the probability of an increase during the first period is

$$P(I_1) = P(I_1C_2) + P(I_1I_2) = 0.18 + 0.42 = 0.60$$

Further, the probability of an increase during the second period, *given that* utilization increased in the first period, is

$$P(I_2|I_1) = \frac{P(I_2I_2)}{P(I_1)} = \frac{0.42}{0.60} = 0.7$$

Then the probability of constant utilization, given a first-period increase, is $1 - P(I_2|I_1) = 1 - 0.7 = 0.3$, or

$$P(C_2|I_1) = \frac{P(I_1C_2)}{P(I_1)} = \frac{0.18}{0.60} = 0.3$$

Thus, $D2$ represents a decision to modify $A1$ if future I_1 occurs.

EVALUATION
The outcomes for each alternative are first discounted to a common point in time. For $D2$, this point is the beginning of the second time period; for $D1$ it is

the beginning of the first time period or "now." After the present worth of all
the receipts and disbursements is added to any initial costs, the expectation for
each alternative is determined from the weighted outcomes. For the alternatives
at D2, we have

$$E(A11) = [PW(O2a) \times P(O2a)] + [PW(O2b) \times P(O2b)]$$
$$E(A12) = [PW(O2c) \times P(O2c)] + [PW(O2d) \times P(O2d)]$$

where PW is the present worth of the outcomes at the beginning of period 2.
From the two expectations, the most profitable is selected to designate the
preferred alternative at D2.

   After completing the evaluation at D2, we proceed to A1 because *decisions
are always made in reverse chronological order*. The preferred dependent alternative
from each decision point becomes an outcome for its related alternative from
an earlier decision point. If we assume that A11 is the preferred alternative from
D2, the expected value of A11 will be included in O2 to account for states 2 and
3 of alternative A1. Therefore the expectation of A1 is

$$E(A1) = [PW(O1) \times P(O1)] + \{PW[O2 + E(A11)] \times P(O2)\}$$

which is compared with

$$E(A2) = [PW(O1) \times P(O1)] + [PW(O2) \times P(O2)] + [PW(O3) \times P(O3)]$$

where PW is the present worth of the outcomes at the beginning of the first
period. The more valuable expectation determines which primary alternative is
preferred at D1.

EXAMPLE 5.11 Start-up plans evaluated with the discounted decision-tree procedure

All of the major airlines serving a northern coastal
region fly from a single international airport located
near a large city. Most people living within 100 miles
drive to the airport and leave their cars in a huge
parking facility during a trip by air. Having a waiting
car is convenient for their return, but it is also expensive. Usually a car has a single passenger during
the connecting rides to and from the airport, and it
is idle for the length of the trip. A calculating entre-
preneur figured that a limousine service from outlying population centers to the airport could be profitable. A survey was made in an industrial area 40
miles from the airport that confirmed the demand;
enough air travelers to support the service said they
would prefer to pay for a ride to the airport if the

service were reliable, comfortable and relatively inexpensive.

   Limoserv, the name selected for the transit company, could be started as a local or regional service.
The local route would serve only the industrial area
where the survey was taken. A regional route would
include stops at a string of towns beyond the industrial area. A larger initial investment would be
required to start the longer route and the risk of
loss would be greater, but it also would offer much
higher potential profits. A compromise between the
two start-up scales is to begin small and then to expand if the first phase is successful. The compromise
avoids part of the risk linked with the high initial
capital costs for a regional network but will require

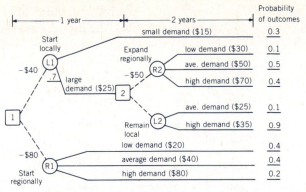

(First costs and annual returns shown in thousands of dollars)

FIGURE 5.11 Two investment alternatives for starting a limousine service with associated outcomes resulting from different demand levels over a three-year period.

a higher total investment if the larger route is followed later.

Start-up patterns for a three-year study period are shown by the decision tree in Figure 5.11. Required initial investments are indicated along the dashed lines representing the alternative courses of action. Estimated probabilities and net operating profits for all outcomes are labeled on solid lines. The compromise plan, to start locally and then expand if feasible, is displayed as the conditional decision at $\boxed{2}$ between alternatives. $\widehat{R2}$ and $\widehat{L2}$; thus Limoserv could invest $40,000 in facilities and equipment to start a local service $\widehat{L1}$ and then expand to regional routes, after one successful year, by investing another $50,000, or it could remain local for the rest of the three-year period. The other alternative is to invest $80,000 immediately to establish the full regional net work, $\widehat{R1}$.

Calculations start at decision point 2. The discounted expected values of the outcomes for alternatives R2 and L2 are computed, using an interest rate of 10 percent and a time span of two years:

$E(R2)$ = ($30)($p/a$,10,2)(0.1) + ($50)(p/a,10,2)(0.5)
   + ($70)($p/a$,10,2)(0.4) − $50

   = ($52.1)(0.1) + ($86.8)(0.5)
   + ($121.5)(0.4) − $50

   = $5.2 + $43.4 + $48.6 − $50

   = $47.2, or $47,200

$E(L2)$ = ($25)($p/a$,10,2)(0.1) + ($35)(p/a,10,2)(0.9)

   = ($43.4)(0.1) + ($60.7)(0.9)

   = $59, or $59,000

The computations indicate that Limoserv should not expand to regional service even if the demand during the first year of local operations is high. The expected value of the discounted profits from the local operation during the last two years of the study period can be entered on a revised decision tree, as shown in Figure 5.12.

With the conditional decision made to continue local operations for all three years if Limoserv starts as a local service, the primary evaluation of alternatives L1 and R1 are conducted:

$E(L1)$ = ($15)($p/a$,10,3)(0.3) + ($25 + $59)
   × (p/a,10,1)(0.7) − $40

   = ($37,3)(0.3) + (76.4)(0.7) − $40

   = $24.7, or $24,700

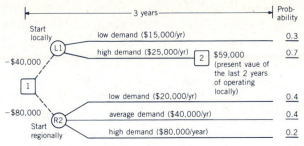

FIGURE 5.12 Revised decision tree for Limoserv operations, showing the effect of using the more profitable alternative at decision 2. The expected value of the conditional decision ($59,000) must be discounted one more year to decision point 1.

$$E(R1) = (20)(p/a,10,3)(0.4) + (\$40)(p/a,10,3)(0.4)$$
$$+ (\$80)(p/a,10,3)(0.2) - \$80$$

$$= (\$49.7)(0.4) + (\$99.5)(0.4) + (\$199)(0.2)$$
$$- \$80$$

$$= \$19.5, \text{ or } \$19,500$$

The last rollback calculation suggests that Limoserv should limit its operations to local routes. But before a final decision is made, additional questions such as the following should be considered:

1. Is the three-year study period long enough?
2. Should other alternatives be evaluated?
3. How much faith can be put in the probability estimates?
4. Are the investment and profit estimates accurate?
5. What intangible considerations should be included?
6. Should additional data such as the salvage value of purchased equipment be included in a more detailed analysis?

If answers to these queries support the discounted decision-tree evaluation, Limoserv can reasonably expect to maximize its profit by serving just the local demand.

END OPTIONAL MATERIAL

5-7 CAPITAL RATIONING

A new investment must pass three tests before any overt action is undertaken:

1. An economic evaluation compares the cost of initiating and operating a project with the benefits expected, in order to determine whether it is worth doing.

2. An intangible evaluation investigates the worthiness of a project in terms of human values, which are difficult to express quantitatively.

3. A financial evaluation compares the attractiveness of investments in different projects in relation to the quantity and quality of available investment funds.

To some extent each test is a gate to acceptance, because a project can be shut out for a decidedly subminimal showing in any one category. For the projects that survive the cutoff levels, a low rating in one test may be compensated by a strong showing in another test.

In this chapter we have considered several comparison methods applied to the economic and intangible tests. These same methods are appropriate for financial evaluations: instead of comparing alternatives to one another, the comparisons match projects to one another and to the source or amount of financing funds. This capital-rationing procedure can be roughly classified according to the scarcity of funds.

When unlimited funds are assumed to be available from borrowing, the returns from a project are measured against the cost of capital used to implement the project.

When a top limit is assumed for the amount of capital available in a given period, the projects must compete against one another in order to secure a portion of the limited funding.

Payback method

When the source of capital is limited to internally generated sums or a maximum set by management policy, the problem is to determine which of several attractive projects should be funded. This problem is prevalent in tactical situations, for example, departmental budgets. The "payback" method is used worldwide to give a rough measure of the desirability of relatively small investments; surveys consistently reveal that a majority of industrial organizations use the method in this country, and reports indicate that it is also extensively employed in Asia, Europe, and the Soviet Union. The **payback period** is simply the length of time required for returns from an investment to equal the amount invested.

$$\text{Payback period} = \frac{\text{required investment}}{\text{annual receipts} - \text{annual disbursements}}$$

$$= \frac{\text{first cost}}{\text{annual savings}}$$

Embellishments, such as a factor to account for the asset's expected life, can be added to this ratio, but the intent is still to provide a quick check that rations capital at the operating level.

Some organizations require new tools and equipment to have a payback period of one-half a year or less. As the size of the investment increases, the time allowed for the payback typically increases. Shorter payback periods decrease the investment risk and increase capital flexibility by making recovered funds available sooner for other allocations. However, the simplicity that makes payback comparisons convenient also allows deceptive interpretations, because the time value of money is excluded from the calculations.

Benefit–cost ratio

The national government is the body that comes closest to having unlimited capital, but there are so many proposals competing for a share that even the huge federal budget cannot accommodate them all. The **benefit–cost ratio** has been frequently used to rate proposed public projects. It ranks projects by a ratio of the present worth of benefits to costs:

$$\text{Benefit--cost ratio} = \frac{\text{present value of all benefits}}{\text{present value of all costs}}$$

Projects with a ratio of less than 1.0 are eliminated. Those with ratios greater than 1.0 are evaluated by the relative size of their B–C ratios and related intangible effects.

The Flood Control Act of 1936 first required federal projects to have at least a 1.0 B–C ratio to justify their funding.

Two key reservations plague the benefit–cost criterion for public works. One is the question of what interest rate to use for the present-worth calculations. Critics claim that the rate has traditionally been too low when compared with industrial rates of return, but proponents argue that public projects have different objectives and risks than do private investments. The other question concerns the equity of the comparisons. For instance, should taxes from the inner-city poor be used to fund suburban expressways, or should taxes from farmers be allocated for urban renewal projects? Such multiobjective questions cloud all public project evaluations, and no one has yet found a way to satisfy all of them in a single comparison model.

Capital inventory

A rate-of-return comparison is suitable for both economic and financial evaluations. The profit expected from each project compared with the investment sets a project rate of return. This rate is then compared with the rate that must be paid for borrowed capital. In theory, all the projects with return percentages greater than the interest charges for loans are acceptable. In practice, a "capital inventory" approach is useful. A **capital inventory** is a schedule of available capital sources listed in order of increasing interest rates. The sources may be internal funds, such as retained earnings, or external funds, such as bond issues. The least expensive funds from the capital inventory are allocated to those projects with the largest rates of return. The allocation is continued until the project rate of return is equal to or less than the cost of money to finance it.

Project Rate of Return	Capital Inventory
25%	6%
21%	7%
15%	10%
Cutoff	
12%	12%
9%	13%
8%	14%

After a comparison method is selected, there may still be difficulties. Some projects are not independent. For instance, project A could have a very high rate of return, but it is not feasible unless project B, with a lower rate of return, is also instigated. Similarly, full benefits from a project may not be realized without subsequent investments that depend on future budgets. The actual cost of capital can also be questionable. How much should retained earnings earn? These questions and a multitude of funding policies guarantee that capital budgeting will remain an occasionally bewildering and always challenging subject.

5-8 DIMENSIONS FOR DECISIONS

The end result of an economic analysis is a decision: Lease or buy? Increase capacity or not? Select machine X or machine Y? Put off the decision until more information is obtained? All of these are also subject to influences that are not readily quantifiable but still affect the decision: Is the pride of ownership worth the extra cost of buying over leasing? Would new facilities that increase capacity also create a more favorable impression for potential customers? Machine X looks more modern than machine Y, but Y could be purchased from a loyal friend of the company. Even a decision to put off a decision could result from feelings, rather than data, which suggest that the timing is wrong or a more diplomatic approach is advisable.

Difficult to measure characteristics are often called *intangibles*. The only easy evaluation involving intangibles occurs when just two alternatives are compared for a single criterion. Then it is merely a matter of judging which alternative is more satisfying. For instance, in choosing between two prints of a photograph, one might be preferred because it creates a better impression. There are no established scales by which to rate impressions, as they are intangible.

Rating scales

Most production decisions are made on the basis of **ratio-scale** measurements: weight, distance, monetary units, and so on. These are tangible values that define characteristics of importance. Ratio-scale numbers can be subjected to all the familiar mathematical operations: addition, multiplication, and so forth. Also, a ratio scale has a natural zero.

At the other extreme is an **ordinal scale.** The choice between the two photos mentioned previously was a ranking exercise by simple ordering. Values of 1 and 0 could have been assigned to the photoprints with no lessening or improvement of the comparison. However, if another print were included, a 1, 2, 3 ranking would give more information than a single choice would. Assuming that three prints are labeled X, Y, and Z with respective ranks of 1, 2, and 3, the ranker expresses preference of X over Y and Y over Z. But there is still no clue as to how much X is preferred to Y and Z or why it is preferred.

Nonetheless, ordinal-scale rankings do have uses. Personnel rating forms usually ask a rater to assign a number or to pick a proficiency level. Descriptions of performance levels are often provided to ease the chore. The satirical appraisal form in Table 5.3 is an innocent ranking example. More serious versions are widely used in industry and government. When you pick a caption, you get a number. The ranked, ranker, and especially the reviewer should be aware of what has taken place and what can be done legitimately in analyzing the numbers thus generated.

About the best measurement of intangibles we can strive for is a rating on an **interval scale.** This type of scale provides a relative measure of preference in the same way a thermometer measures relative warmth. An interval scale is a big improvement over an ordinal scale, but it still cannot be used in the same way as a ratio scale is. This limitation stems from the lack of a natural zero. A zero in a ratio scale has a universal meaning: a zero distance or a zero weight means the same thing to everyone. A zero temperature can convey different meanings according to the type of interval scaling employed, Fahrenheit or

When *X* is preferred to *Y* and *Y* is preferred to *Z* but *Z* is ranked over *X*, there is an obvious problem, called *intransivity*. An exit from this circular reasoning is found in more precise definitions. Each option must be ranked on the basis of a *single* criterion during each comparison.

TABLE 5.3
A Spoof of an Employee-ranking System That Makes Just Enough Sense to Be Frightening

	Employee Appraisal Guide				
Ranking	Phenomenal 4	Marvelous 3	Good 2	Not So Good 1	Pathetic 0
Competitiveness	Slays giants	Holds his own against giants	Holds his own against equals	Runs from midgets	Gets caught by midgets
Personal appearance	Could be a professional model	Could model but wouldn't be paid much	Could model as the "before"	Goes unnoticed in a crowd	Panics a crowd if noticed
Leadership	Walks on water consistently	Walks on water in emergencies	Wades through water	Gets caught in hot water	Passes water in emergencies
Intelligence	Knows everything	Knows a lot	Knows enough	Knows nothing	Forgets what he never knew
Communication	Talks to big shots	Talks to little shots	Talks to himself	Argues with himself	Loses those arguments

centrigrade. But once this zero value is understood, both temperature scales use standardized units of measurements that allow certain arithmetic operations, such as averaging, to be performed with the scaled values.

The first step in establishing an interval scale for intangibles is to define the criterion. This is done by identifying the ideal state or perfect solution that best satisfies the criterion. For instance, if the criterion is "reputation," the ideal state is defined by naming the most reputable object that fits the situation. Next, the least desirable state is identified. These two states, best and worst, are the end points of the interval scale, 0 and 1.0 or 10. A zero rating is the absolutely unacceptable condition, and the 1.0 or 10 rating is associated with the ideal state. In most situations it is possible to recognize an existing object or situation that epitomizes each state. For instance, when reputation is a criterion for selecting a supplier, actual suppliers with the very best and very worst histories could represent the top and bottom of the interval scale.

The rating for a specific decision alternative is obtained by comparing the alternative's feature with the best and worst situation, the top and bottom on the scale for that criterion. In the reputation example, each supplier is compared with the ideal and least desirable known supplier. If one of the suppliers being considered is neither particularly good or bad compared with the extremes, the rating will be 0.5 or 5. If a supplier is very close to the perfect supplier, the rating may be 0.9 or 9. A 10 rating indicates indifference between being served by the supplier named to represent the top of the scale and the one being evaluated.

The importance of criteria is measured on an interval scale. A 1.0 or 10 rating goes to a vital criterion, a paramount consideration. A zero rating is reserved for an irrelevant criterion, one that does not deserve consideration.

Intangible comparisons

The quantification of recreational virtues is highly important to the evaluation of public projects such as dams or parks.

Intangibles affect a decision whether or not they are formally evaluated. Sometimes they take the form of a vague feeling; at other times they create an ill-defined but urgent impression to act in a certain fashion. A step toward valid comparisons is taken by just recognizing that intangibles exist. Further progress is made by attempting to reduce intangibles to some rating system. And a major advance is achieved by including the ratings in a formal evaluation.

In comparing the intangible qualifications of alternatives, we usually need two sets of ratings: one is the alternative's rank within each quality designated, and the other measures the relative importance of the qualities. Both ratings can be completed by an individual or by the consensus of a group. Either way, the ratings will be more reliable if care is taken to establish as clearly as possible the attributes of the best possible score and what constitutes an unsatisfactory level, the bottom rating.

Priority decision table

Another model that incorporates intangibles along with objective data is described in Example 9.1.

Just as there are several ways to make quality ratings, there are different opinions concerning the best method for including the ratings in a comparison model. The model must be able to accommodate all dimensions for the various qualities considered; each alternative could be evaluated with respect to dollars of cost, weight in pounds, length in inches, efficiency in percentages, and attractiveness estimated on a subjective scale. In this section, we shall consider the popular additive model. Multiplicative and exponential models are also available, but the additive model with minor variations is apparently the most widely used. It is called a **priority decision table.**

The following five steps are a procedure for combining different outcome dimensions into a single number that represents each alternative course of action:

1. Select independent criteria with which to compare all alternatives.

2. Rate the relative importance of the criteria.

3. Determine whether there is a cutoff score for any criterion that makes an alternative unacceptable, regardless of the scores on the rest of the criteria. If so, state the cutoff ratings.

4. Assign values to the extent that each alternative satisfies each criterion. An alternative will be eliminated if any value falls below a criterion cutoff score.

5. For each alternative, multiply its criteria ratings by the respective importance factors, and add all the resulting products. The total scores thus obtained can be compared so as to determine the most attractive alternative.

EXAMPLE 5.12 Development of ratings for decision criteria

One way to develop ratings can be observed from a study conducted for a hospital to evaluate which type of thermometer should be used. Three types were considered:

1. *Standard.* A mercury thermometer that is reprocessed after each use.
2. *Wet kit.* A mercury thermometer with a disinfectant solution stand that holds the unit between readings.
3. *Electronic.* A battery-charged electronic thermometer.

Each type has qualities that recommend its use, and inherent operating disadvantages, such as

1. *Standard.* Patients expect a fresh thermometer for each reading and enjoy the personal attention during the three-minute wait for a reading, but the reprocessing cost is high.
2. *Wet kit.* Operates like the standard thermometer, but the patient is charged for the kit and keeps it after being released from the hospital.
3. *Electronic.* A temperature reading is obtained in about 20 seconds by placing a covered probe under the patient's tongue.

Criteria for evaluating the three types of thermometers were developed by hospital administrators, nurses, and doctors. Then these individuals were asked to rate the relative importance of the criteria on a 0-to-10 scale, with a 0 rating showing absolutely no importance and the 10 level indicating vital importance. Individual ratings were averaged to obtain the preceding scale:

Three of the criteria can be compared on measurable qualities, but two are based on subjective evaluations by patients and nurses. To obtain ratings for these two criteria, a large number of patients and nurses were exposed to trials of all three temperature-taking devices. Then the users were asked to make comparisons on a 0-to-10 scale. The averaged results are shown on the following scales:

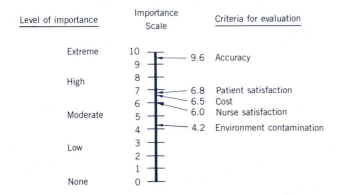

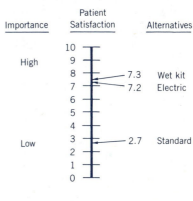

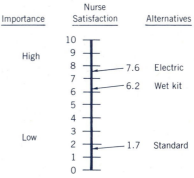

The method used to obtain the ratings by structured interview with patients and nurses is just one of many ways to quantify intangibles. For instance, difficult-to-quantify recreational benefits can be compared by measuring the travel costs that users incur to use a recreational facility, and the value an individual places on money can be rated as a function of risk, using methods suggested by *utility theory*. However, no one has yet devised a consistently reliable way to quantify intangibles. Until someone does, if ever, the imperfect methods still deserve consideration because intangibles will always influence decisions and their effect begs evaluation.

The comparison procedure is illustrated by returning to the hospital's thermometer evaluation. We have already observed the selection of the criteria, the development of importance factors, and the thermometer ratings for the two subjective criteria. Each alternative must still be rated for the other three criteria:

1. *Accuracy.* A cutoff level for accuracy was set at $\pm 0.5°F$. All three types of thermometers were tested in a water bath and found to be accurate within $\pm 0.2°F$. Therefore, all were awarded a rating of 10 on a 10-point scale.

2. *Environment contamination.* The weight of the waste products resulting from the use of each type of thermometer is a measure of environment contamination. Waste includes plastic, paper, glass, and mercury. The standard thermometer has the least waste, 0.2 grams per reading. The electric and wet kit have respective wastes per reading of 2.35 and 1.71 grams. Ratios are used to convert all values to a 10-point scale rating.

By this method the characteristic with the most desirable value (lowest cost, highest profit, and so on) is automatically ranked at the top of the scale, with all other characteristics rated proportionately lower.

Standard	*Electric*	*Wet kit*
$\dfrac{0.2}{0.2} \times 10 = 10$	$\dfrac{0.2}{2.35} \times 10 = 0.9$	$\dfrac{0.2}{1.71} \times 10 = 1.2$

3. *Cost.* Total cost includes wages (nurses and nonprofessional), breakage, first cost (0 for wet kit because the patient pays for it), and supplies (probe covers, alcohol, tissue, and the like). Again using ratios, with the top rating given to the wet kit, which had the lowest total cost during the test period ($17,255), we have

Standard	*Electric*	*Wet kit*
$\dfrac{17,255}{18,843} \times 10 = 9.2$	$\dfrac{17,255}{25,623} \times 10 = 6.7$	$\dfrac{17,255}{17,255} \times 10 = 10$

Table 5.4 summarizes the mechanics of the rating process and provides a format for calculating each alternative's overall rating. By this procedure the preferred alternative is the wet kit thermometer.

TABLE 5.4

Format for Tabulating Criterion Ratings and Comparing Qualities of Decision Alternatives. Because Higher Numbers Show a Preference in All Cases, the Wet Kit Thermometer is Preferred for the Given Criteria in the Priority Decision Table

Alternatives		Standard			Electric			Wet Kit		
Criteria	*Imp.*	*Observations*	*Rate*	*R × I*	*Observations*	*Rate*	*R × I*	*Observations*	*Rate*	*R × I*
Accuracy (min. ± 0.5°F.)	9.6	within ± 0.2°F.	10	96	within ± 0.2°F.	10	96	within ± 0.2°F.	10	96
Cost	6.5	$18,843	9.2	60	$25,623	6.7	44	$17,255	10	65
Environment contamination	4.2	02. gm.	10	42	2.35 gm.	0.9	4	1.17 gm.	1.2	5
Nurse satisfaction	6.0		1.7	10		7.6	46		6.2	37
Patient satisfaction	6.8		2.7	18		7.2	49		7.3	50
Totals				226			239			253
Decision:		Select the wet kit thermometer								

EXAMPLE 5.13 A planning decision with intangible considerations

"Ipso Facto," a thriving, independent, data-processing company, is planning an image-improvement and business-expansion campaign. Three courses of action have been proposed:

1. Tell-sell. Develop a staff to increase personal contacts with old and proposed clients; offer short courses and educational programs on the benefits of modern data-processing methods.

2. Soft-sell. Hire a staff to put out a professional newsletter about data-processing activities; volunteer data-processing services for community and charity projects.

3. Jell-sell. Hire personnel to develop new service areas and offer customized service to potential customers; donate consulting time to charitable organizations.

Only one of the three alternatives can currently be implemented, owing to budgetary limitations.

The effects or returns for each alternative are rated according to desired qualities, and the importance of each quality is ranked as follows, where effectiveness and importance are rated on a 0-to-1.0 scale.

	Annual Cost	*Immediate Effectiveness**	*Long-Range Effectiveness**
Tell-sell	$250,000	0.9	0.7
Soft-sell	150,000	0.8	0.6
Jell-sell	180,000	0.6	0.9
Importance*	0.3	1.0	0.6

*1.0 is the top rating.

Using an additive model to compare the alternatives, the overall ratings are developed as

$$\text{Tell-sell: } \left(\frac{150,000}{250,000} \times 1.0 \times 0.3 \right)$$
$$+ (0.9 \times 1.0) + (0.7 \times 0.6) = 1.50$$

Soft-sell: $\left(\dfrac{150,000}{150,000} \times 1.0 \times 0.3\right)$

$\qquad + (0.8 \times 1.0) + (0.6 \times 0.6) = 1.46$

Jell-sell: $\left(\dfrac{150,000}{180,000} \times 1.0 \times 0.3\right)$

$\qquad + (0.6 \times 1.0) + (0.9 \times 0.6) = 1.39$

The narrow edge given to Tell-sell over Soft-sell

suggests that the subjective values used in the comparison model be appraised carefully. For instance, increasing the importance rating for cost from 0.3 to 0.4, with all other figures unchanged, creates the same overall rating for the top two alternatives, 1.56. Such tests of sensitivity reveal how large a shift in a factor is required to alter the preference from one option to another.

5-9 CLOSE-UPS AND UPDATES

There are reasons to suspect that it is easier to make long-range decisions of huge proportions than it is to make short-run decisions for small investments. Strategic decisions often have vague objectives that tend to change over time, which reduces the accountability of the decision makers, assuming that they are still around for a future audit of their decisions. The reverse is true of tactical decisions. Objectives are specific—a 20-percent rate of return is required this year, or a payback must be in six months. Successes and failures are remembered for annual performance reviews. Such realities cause investment proposals to be scrutinized, especially for high-technology replacements that are too novel to warrant complete confidence.

Adventures in advanced-technology replacements

By definition, an adventure is "an enterprise involving financial risk." Investments in high-technology equipment fit the definition. Because uncertainties are associated with all new developments, estimates of receipts and disbursements are always suspect. Performance of a conceptually advanced machine does not always live up to its billing. There may even be doubt about whether to make a replacement now or wait until a still-more-advanced challenger becomes available.

Production planners must evaluate the merit of replacing individual machines with updated versions or with models that perform the same function in an entirely different way, perhaps requiring whole new support systems. Sometimes a replacement has the capacity to perform several tasks done by other machines, instantly making them redundant. An added challenge is to coordinate replacements into a schedule that maintains production capabilities during the installation, debugging, and learning phases. Dismayed planners have discovered that the whole exercise can be unexpectedly expensive.

Robots are the most controversial replacements. A robot replacement may result from a *problem-pull* condition or a *technology-push* situation. The former attempts to correct a known difficulty, and the latter is an experiment with promising technology. Solving production problems with piecemeal robotics is unlikely to yield maximum savings, because each introduction is an independent occasion. Comparably, installing a robot only because it is fashionable is hardly justifiable. An analysis should incorporate conventional cash flow categories plus reasonable estimates of difficult-to-quantify factors such as work-force morale and losses owing to inflexibility. Problem solutions can then pull in appropriate technology to push financially responsible modernization.

The paradox of office automation and productivity

Thirty years ago the service sector of the U.S. economy had half the employees of the manufacturing sector and almost equal productivity. Now service employment is double that of manufacturing, and its productivity has dropped to 61 percent of the national average. During the economic expansion of 1982–1983, the service sector gained 560,000 full-time equivalent jobs, but the rest of private industry lost 144,000 jobs. These figures underlie the determined drive to improve service operations.

Although obvious advances are being made in information processing and office automation, they have done little to boost **office productivity.** This paradox is painfully evident. Between 1977 and 1982, U.S. output rose 8 percent, yet at the same time clerical employment rose 15 percent and total white-collar employment 18 percent. American banks reinforce the paradox. Despite a heavy investment in computerized accounting systems and automated tellers, in 1982 it took more hours of work to produce a unit of output with the new equipment than it took in 1977 without it. Apparently the hardware of the new technology must be strengthened by the software of a new *office sociology.*

The estimated cost in 1986 for all U.S. office-based white-collar workers, ranging from chief executives to file clerks, is $1 trillion. Managers and professionals account for about 73 percent of the workers, clericals the other 27 percent. Conversely, about 70 percent of the expenditures to update offices is directed at clerical work, with 13 percent for professionals and 17 percent for managers. The obvious interference is that automation has missed the high-cost bull's-eye of the office target.

Smart computers, omnipresent copiers, and cleverly communicating computers have assuredly improved information processing. But the gains have not been as great as predicted. Part of the blame is laid on managers who are reluctant to run a microcomputer. Another part is attributed to the misuse of computerized accounting, which accelerates paper flow with-

A drive to improve quality may warrant a robot even if it cannot be justified on the basis of earned savings, because the experience gained might be an adequate payback.

In 1983 there were 30 robots for every employee in Sweden, 13 in Japan, 5 in West Germany, and 4 in the United States.

Annual U.S. Rate of Productivity Growth in 1948–1965 and 1977–1983

Services	1.2%	0.3%
Total private business	3.3%	0.8%

White-collar workers comprise about 50 percent of all manufacturing employees and about 70 percent of the payroll.

Office designs are examined in Chapter 9.

Old symbols of status, such as personal secretaries for executives, are being discarded in the emerging office sociology.

out increasing information flow. Bureaucrats also need to realize that they do not have to have hard copies of everything for their files. The departure from old habits for communicating and the entrance of more efficient utilization practices for office machines are milestones of office productivity.

Deciding who gets to decide by downsizing

During the first half of the 1980s, mostly a period of increasing employment, U.S. companies eliminated the jobs of half a million managers. The **downsizing** of corporate bureaucracies had been advocated for several years. It finally came about when foreign competition and domestic deregulation combined to cause an almost-universal preoccupation with cost control. Managerial ranks offered a plump target. The result is depicted in Figure 5-13.

The squeezed organizational triangle appeared in the first edition of this book in 1970 and was accompanied by the following prophetic text:

> The permanent effects of electronic assistance are more difficult to anticipate than their interim developments. The triangular organizational hierarchy may become pinched in the middle. The bottom-heavy hourglass shape could result from a garrote tightened around the traditional structure by bigger, faster, data-processing facilities and the managerial mastery required to harness the quickened data flow. The squeeze in the middle management section could take the form of programmed decisions—directives formulated automatically from structured-operating rules applied to incoming data. The bulge below the waistline accommodates a downward classification of some managers to the function of directing operations instead of planning them. At the top of the organizational structure, the upper-management group proportionately assumes enlarged functions for planning, coordinating, and innovating. The consequent challenges to maintaining morale, evaluating personnel, developing "creators," coordinating actions, and controlling the "programmed structure" would be enormous.

As anticipated in the quotation, advancing technology has made the jobs of many managers irrelevant or redundant. In some organizations, one and even two entire levels of management have been sliced out, making

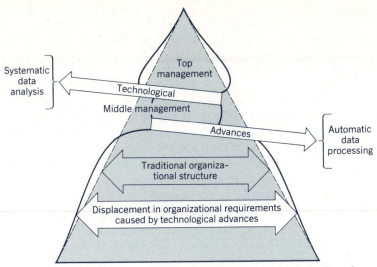

FIGURE 5.13 Distortion of traditional managerial hierarchy by advances in technology.

the triangle flatter. More often, the number of positions has been cut back, making the triangle leaner. Many of the cuts have occurred in the office area.

Three consequences of modern office technology have made some managers expendable:

1. Machines have reduced the amount of time required to conduct certain managerial duties.

2. Machines provide faster access to a vastly greater amount of information which allows managers to make better decisions over a wider range of responsibilities.

3. Machines have accelerated the flow of information both vertically and laterally, which allows a broader span of control with minimal loss of direction and feedback.

The machines are fostering what the managers themselves know should be done but too often lack the time to do. Managers' time is interrupted and confiscated by unanticipated events, which sidetrack an orderly approach to performance improvement. This accumulated neglect bloated the managerial ranks to the extent that downsizing was inevitable.

Corporate America generates enough paper each day to circle the world 40 times, and half of that comes off computer printers.

5-10 SUMMARY

Different patterns of receipts and disbursements are compared on an equivalent basis by calculating the time value of cash flow. Six interest formulas are used to convert a single amount or a series of payments to an equivalent future or present sum or annuity. Three methods of discounted cash flow comparisons are (1) equivalent present worth—for coterminated projects, (2) equivalent annual cost—for repeated projects, and (3) rate of return—for incremental investments.

After-tax evaluations are based on funds available to an organization after taxes have been deducted. Taxes are calculated by applying the organization's tax rate to the before-tax cash flow from which allowable deductions have been subtracted.

Sensitivity analyses are conducted to observe the effect of deviations from the estimated cash flow. A pessimistic–most likely–optimistic study concentrates on best-worst scenarios. Sensitivity graphs are constructed by varying one cash flow factor while all others are held constant. Resulting curves show the effect of variations on outcome measurements.

Comparisons recognizing risk require that the applicable future states be identified and their probabilities estimated. Outcomes for each state from each alternative can be displayed in a payoff table. The expected value of an alternative is the weighted average of all its outcomes when the "weights" are obtained from the probability of each outcome's occurrence.

A discounted decision tree combines the expectation principle with the time value of money. Conditional alternatives are evaluated by determining the expectation of the discounted cash flow at successive decision points. Decisions are made in a reverse chronological order, making the outcome from a preferred dependent decision an input to an earlier decision point. The primary alternative with the highest expectation at time zero is the indicated course of action.

Capital is rationed whenever there are more acceptable ways to invest funds than there is funding available. Investments should be subjected to economic, intangible, and financial evaluations. The payback method is a widely used criterion for industrial investments; only projects that return the amount invested within a certain period are acceptable. Proposals for funding public projects are frequently rated by benefit–cost ratios; any proposal with a B–C ratio less than 1.0 is rejected. A capital inventory compares the rates of return expected from investments with the cost of acquiring capital; only projects with a rate of return greater than the percentage cost of capital are funded.

Difficult-to-quantify human values that often influence decisions are termed intangible considerations. One method of including them in an economic evaluation is to identify the key characteristics of the problem, rate the relative importance of the characteristics, convert quantifiable values to a common scale (often 0 to 10), subjectively rate the remaining characteristics on the same scale, multiply each characteristic rating by its importance factor, and sum the products. The alternative with the best overall rating is the preferred solution.

Discounted cash flow comparisons

After-tax evaluation

Sensitivity studies and graphs

Expected value

Discounted decision trees

Capital rationing

Payback method

Benefit–cost ratio

Quantification of intangibles

Priority decision table

5-11 REFERENCES

CANADA, J. R., and J. A. WHITE, JR. *Capital Investment Decision Analysis for Management and Engineering.* Englewood Cliffs, N.J.: Prentice-Hall, 1980.

GRANT, E. L., W. G. IRESON, and R. S. LEAVENWORTH. *Principles of Engineering Economy,* 7th ed. New York: Wiley, 1982.

HERTZ, D. B., and H. THOMAS. *Risk Analysis and Its Applications.* New York: Wiley, 1983.

PALM, T., and A. QAYUM. *Private and Public Investment Analysis.* Cincinnati: South-Western, 1985.

OLSON, V. *White Collar Waste.* Englewood Cliffs, N.J.: Prentice-Hall, 1983.

RIGGS, J. L., and T. M. WEST. *Engineering Economics,* 3rd ed. New York: McGraw-Hill, 1986.

SHIZUO, S. T., T. FUSHIMI, and S. FUJITA. *Profitability Analysis for Managerial and Engineering Decisions.* Tokyo: Asian Productivity Organization, 1980.

WILKES, F. M. *Capital Budgeting Techniques.* New York: Wiley, 1977.

5-12 SELF-TEST REVIEW

Answers to the following review questions are given in Appendix A.

1. T F A dollar received a year from now is worth more than a dollar received today because of the *time value of money*.

2. T F A *nominal interest rate* is the average value of interest earned in a year.

3. T F The given mnemonic symbols for interest formulas are based on five letters: *P, F, a, n,* and *i*.

4. T F The first payment in the type of *annuity* used in this text occurs at the end of the first period.

5. T F When applied correctly to identical data, the *present-worth, annual cost,* and *rate-of-return* methods always indicate the same order of preference among alternatives.

6. T F In calculating the rate of return of an *increment of investment*, it is compared with the next lowest level.

7. T F Normal wear and tear causes the *functional depreciation* of a machine.

8. T F The *economic life* of a machine is the number of years that minimizes its equivalent annual cost.

9. T F Cash payments for *depreciation charges* are collected and saved to fund the purchase of a replacement.

10. T F *Sunk cost* is the difference between book value and market value.

11. T F AFTC = BFTC − (taxable income × tax rate).

12. T F A *sensitivity graph* records pessimistic, optimistic, and most likely estimates of key economic factors.

13. T F The *expected value* of an alternative is the sum of the outcomes weighted by their probability of occurrence.

14. T F The difference between correct and incorrect outcome estimates in a payoff table determines the *value of perfect information.*

15. T F Calculations in a *discounted decision tree* are made in reverse chronological order of the time flow.

16. T F A *payback period* is calculated by dividing the first cost of an investment by the discounted value of the savings it produces.

17. T F *Benefit–cost ratios* do not rely on discounted cash flow calculations.

18. T F A *ratio scale* does not have a natural zero.

19. T F *Intangibles* are usually rated on an interval scale.

20. T F The selection criterion in a *priority decision table* is the preference for the alternative that has the highest sum of $R \times I$ scores.

21. T F *Technologically advanced replacements* deserve scrutiny because their cost and performance are less certain than are those of conventional machines.

22. T F *Office productivity* has risen rapidly in recent years as a result of middle-management *downsizing.*

5-13 DISCUSSION QUESTIONS

1. Why is the present-worth comparison method associated with coterminated projects and the annual cost method with repeated projects?

2. A central air-conditioning unit has been installed in an office. Describe how the unit could lose value owing to each of the four causes of depreciation. Which cause would likely account for most of the decrease in value during the next:
a. 3 years?
b. 10 years?

3. Describe the more important causes of the retirement of
a. Steam locomotives.
b. Passenger cars.
c. Electronic computers.
d. Dump trucks.
e. Automatic screw machines.
f. Wheelbarrows.

4. Should the economic life of a machine equal the service life? Why?

5. Discuss the two major purposes of depreciation accounting with reference to assets such as a furnace,

an automobile, and land. Would the purpose be better served by using replacement cost instead of purchase price?

6. Could the salvage value of an asset be negative? How?

7. A salesman agreed to purchase a building lot by making seven annual payments of $900 each. Right after the first payment was made, the salesman was transferred to a different town. Two years and two payments later the salesman returns to find he can buy an equivalent lot for $3000 because land values decreased while he was gone. A friend advises him he has a sunk cost of $600, so he should drop the contract to buy an equivalent lot. The salesman feels he will lose $2700 if he drops the contract. Assuming the salesman will suffer no penalty for reneging on his original agreement, what would you advise? How would you explain it to the salesman?

8. Comment on the following rationalization that often accompanies a decision to keep a defender with a large sunk cost when the challenger provides lower future costs: "I've had some back luck on repairs, but it is running now, and I've got so much in it that I can't afford to sell it for what it would bring."

9. Sandy worked for several years as a general handyman in construction. He inherited $50,000 and decided to start a landscaping service. Combining his inheritance with a $40,000 loan, he bought a used tractor, truck, and hauling rig. Business was good. A gross income of $75,000 per year covered annual operating and loan-repayment expenses of $31,000. The rest he spent personally. After five years of good living Sandy found that the loan was repaid but that the equipment was worn out. Because he had made no provisions to compensate for the loss of value, he was left with almost worthless assets and no money to renew them. What did he do wrong?

10. An indifference point is often associated with an "aspiration level." How can such a point or level be used in bargaining?

11. Relate the black box input–transformation–output model to a decision tree.

12. How can you reconcile using the expected-value criterion when you believe a particular decision situation will occur only once in your lifetime?

13. What use can be made of knowing the value of perfect information for a problem?

14. Assume you are ranking your three favorite sports and, for the sake of argument, say the ranking came out skiing, golf, and jogging, in that order. Then you think again. Skiing is more thrilling than golf, and golf is more fun than jogging, but jogging is healthier, handier, and cheaper than skiing. Now what ranks where? Discuss a way out of this circular reasoning and how to arrive at a decision.

15. What is the logic behind the mathematical operation that assigns a 10 rating to the choice that has the least cost or highest return among the choices being evaluated by a mixed dimension comparison model?

5-14 PROBLEMS AND CASES

1. A proposed improvement in an assembly line will have an initial purchase and installation cost of $67,000.

Annual maintenance cost will be $3500; periodic overhauls once every three years, including the last year of use, will cost $6000 each. The improvement will have a useful life of 12 years, at which time it will have no salvage value. What is the equivalent annual expense of the lifetime costs of the improvement when interest is 8 percent?

2. A company borrowed $100,000 to finance a new product. The loan was for 20 years at a nominal interest rate of 6 percent compounded semiannually. It was to be repaid in 40 equal payments. After one-half the payments were made, the company decided to pay the remaining balance in one final payment at the end of the tenth year. How much was owed?

3. Five equal annual payments starting today will be invested to allow two payments of $3000 each to be drawn out 12 and 15 years from now. The two payments will close the account. Interest is 5 percent compounded annually. What is the amount of payment to be made today?

4. Additional parking space for a factory can either be rented for $6000 a year on a 10-year lease or purchased for $90,000. The rental fees are payable in advance at the beginning of each year. Taxes and maintenance fees will be paid by the lessee. The land should be worth at least $60,000 after 10 years. What rate of return will be earned from the purchase of the lot?

5. Machine A has a first cost of $9000, no salvage value at the end of its 6-year useful life, and annual operating costs of $5000. Machine B costs $16,000 new and will have a resale value of $4000 at the end of its nine-year economic life. Operating costs for B are $4000 per year. Compare the two alternatives on the basis of their annual costs. Which one would you select when interest is at 10 percent?

6. The annual cost of a piece of equipment is to be compared with a rental cost for comparable equipment. When the interest rate is 8 percent compounded annually, what rental charge would make the two alternatives equally attractive?

Purchase price	$12,000
Salvage value	$3500
Economic life	7 years
Annual operating costs	$2200
Taxes	2% per year of first cost
Overhauls	$1200 each at the end of years 3 and 5
Insurance	$460 for a seven-year policy payable in advance

7. "Ipso Facto" offers to handle the data-processing requirements of a company for three years at a lump-sum price of $19,000 payable in advance. An alternative offer is to charge $7000 per year payable in advance each year. What rate of return is earned by the subscribing company if it accepts the three-year contract rather than successive one-year arrangements?

8. Compare the rate of return for the following plans, and select the preferable alternative. The minimum acceptable rate of return is 6 percent.

	Plan 1	Plan 2	Plan 3
First cost	$40,000	$32,000	$70,000
Salvage value	$10,000	$6000	$20,000
Economic life	7 years	7 years	7 years
Annual receipts	$18,000	$18,000	$24,500
Annual disbursements	$11,000	$14,500	$14,500

9. A machine was purchased on January 1, 1986 for $6000. Installation costs were $300. It is expected to have an economic life of seven years and a resale value of $900, but it will cost $200 to dismantle the machine and prepare it for resale. Compute the depreciation charge during the third year by means of the straight-line depreciation method.

10. If $P = $10,000$, $S = 1000, and $n = 6$, what percentage of the anticipated replacement cost will have been accounted for by accumulated depreciation after half the economic life using the straight-line depreciation method?

11. A new contract requires a special kind of machine that the contractor believes will have no use after the project is completed in five years. Two types of machines are suitable. The service life of machine X is five years and of machine Y is three years. From the anticipated capital cost and maintenance patterns following, calculate the annual cost of each machine. The minimum acceptable rate of return is 10 percent.

Year	Machine X	Machine Y
0	First cost = $35,000	First cost = $15,000
1	——	——
2	——	Repairs = $500
3	Overhaul = $700	Cost of new machine less trade-in = $12,000
4	——	——
5	Salvage = $1000	Salvage = $3000

12. Two years ago an office duplicating machine was purchased for $1000. Its economic life was estimated at five years, at which time the salvage value would be $100. The current book value by straight-line depreciation is $640. Total material, labor, and maintenance expense is $4100 a year. A copying machine will be needed throughout the foreseeable future.

A new model produced by the manufacturers of the presently used machine sells for $1500, but its faster speed will cut annual operating costs to $3800. The manufacturer has offered $800 for the old machine as a trade-in on the new model. The expected salvage value for the new model is $200 after 10 years.

Another company leases duplicating machines at $800 a year. The lease agreement covers all maintenance and repairs, which cuts the anticipated operating cost to $3500 per year. The leasing firm will buy the present machine for $500.

Using an interest rate of 8 percent, compare the defender to each challenger.

13. A company in the 40-percent income-tax bracket is considering the purchase of equipment that will eliminate a rental expense of $9000 per year. The equipment will cost $30,000 and have a zero salvage value at the end of its eight-year life. Maintenance and operating costs are estimated at $18,000 per year. Insurance is 1 percent of the first cost. Straight-line depreciation is used for tax purposes. The company requires a 7 percent after-tax rate of return. Make a recommendation regarding the purchase of the equipment.

14. Machine A was installed 6 years ago at a total cost of $8400. At that time it was estimated to have a life of 12 years and a salvage value of $1200. Annual operating costs, excluding depreciation and interest charges, have held relatively constant at $2100. The successful marketing of a new product has doubled the demand for parts made by machine A. The new demand can be met by purchasing an identical machine that now costs $9600 installed. The economic life and operating costs for the two machines will be the same. The salvage value for the second A-type machine will be $1600.

     Machine B, a different type, costs $17,000 installed but has twice the capacity of machine A. Its annual operating costs will be about $3100, which should be relatively constant throughout its 10-year economic life. Salvage is expected to be $4000. The present machine can be used as a trade-in on the new machine B. It is worth $3000.

a. Make a before-tax comparison of the two alternatives using equivalent annual cost when the interest rate required is 10 percent.

b. Determine the after-tax annual costs for machines A and B when the required after-tax rate of return is 8 percent. Straight-line depreciation is to be applied to the challenger, and straight-line depreciation will continue to be used for the defender. The corporate effective income tax rate is 50 percent.

15. There is considerable doubt about the need for and performance to be obtained from a new process developed by the R&D department. A decision about launching a small-scale pilot project is being evaluated. Three estimates of possible outcomes of the pilot are as follows:

	Most Likely Estimate	Pessi- mistic Estimate	Opti- mistic Estimate
Income per year	$200,000	$150,000	$250,000
Expenses per year	$80,000	$90,000	$70,000
Start-up cost	$300,000	$350,000	$300,000
Life of project	3 years	1 year	4 years
Salvage value	$100,000	$50,000	$100,000

a. What is the present worth of each possible future when the minimum attractive rate of return is 15 percent?

b. What other considerations could affect the decision to launch the pilot project? Should the money already invested in research and development be a consideration? Why?

c. Compare the range of estimates method with a sensitivity graph.

16. A proposal is described by the following estimates: $P = \$20,000$, $S = 0$, $N = 5$, and net annual receipts $= \$7000$. A rate of return of 10 percent is desired on these proposals. Construct a sensitivity graph of the life, annual receipts, and rate of return for deviations over a range of ± 20 percent. To which element is the decision most sensitive?

17. Based on the given payoff table:

a. Determine the indifference point of the alternatives, and state how the values obtained can be used to aid decision making.

	S1	S2
A1	− $10,000	$60,000
A2	$35,000	$35,000
A3	$120,000	− $20,000

b. If the probabilities of S1 and S2 are, respectively, 0.3 and 0.7, what will be the value of perfect information?

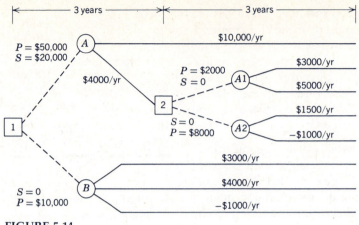

FIGURE 5.14

18. Which of the two alternatives displayed in the decision tree of Figure 5.14 is more attractive? There are two 3-year study periods and the annual interest rate is 7 percent. Assuming that positive values represent costs, seek the minimum-cost alternative.

19. It has already been decided that accounting and billing procedures will be computerized. The question now is how much equipment should be leased. Two different capacities are being considered. The larger system (B) leases for $240,000 per year on a four-year lease or $300,000 per year on a two-year lease. The smaller system (S) has an annual lease cost of $110,000 and $150,000 on leases of four and two years, respectively.

Hint: There are four primary alternatives.

A four-year study period will be used. The forecast is a 0.7 probability of high activity (*H*) for the next two years and 0.3 for low activity (*L*). If the activity is high for the first two years, it is equally likely to be high (*HH*) or low (*HL*) for the successive two years. If the first two years produce low activity, then there is a probability of 0.8 for two more years of low activity (*LL*).

Expected savings per year from the installation of either system for the different activity patterns are as follows. These savings do not include any leasing costs but do reflect the increased accuracy, quantity, and speed expected.

Which computer system should be leased now, and should it be leased for two or four years?

	First Two Years	Second Two Years Given First Two
Large system (B):	H = $300,000	HH = $350,000; HL = $200,000
Small system (S):	L = $200,000	LH = $300,000; LL = $100,000
	H = $200,000	HH = $250,000; HL = $100,000
	L = $100,000	LH = $200,000; LL = $50,000

20. Sites for a new research lab have been narrowed to three localities. The construction cost of the plant will be approximately the same, regardless of the location chosen. However, the cost of land and intangible factors largely applicable to personnel recruitment vary considerably from one location to another. Based on the following figures, in which higher ratings show a preference, which site should be selected?

Characteristics	Site 1	Site 2	Site 3	Importance
Availability of technicians	7	10	2	5
Adequacy of subcontractors	5	9	5	4
Proximity to a university	10 miles	40 miles	30 miles	4
Cost of land	$300,000	$400,000	$50,000	3
Recreational potential	7	2	10	2
Climate	6	1	9	1
Transportation	9	10	5	1

21. A life cycle cost analysis is used to select the supplier of switching units. Life cycle costs include the purchase price, shipping cost, and maintenance expense. A total of 2 million switching operations are required during the life of the project. Bids were received from three suppliers, and samples from each were subjected to laboratory tests. Shipping costs were estimated from the distribution center of each supplier. Results of the bids and tests are as shown.

a. Evaluate the switches, and select the preferred supplier on the basis of cost.

b. Suppose that cost were assigned an importance rating of 10 and "reliability" of the supplier were assigned an importance rating of 8. How much could the ratings for reliability vary before a change in preference from part a occurs when the mixed dimension comparison model is applied?

Supplier	Operations per Switch	Maintenance Cost per Switch	Shipping Cost per Switch	Bid Price per Switch
I	2850	$16.82	$84.12	$412.75
II	2190	19.46	57.70	330.50
III	2010	17.14	62.86	290.00

22. The Case of Uncertainty

Decisions based on forecasts of future events always create at least a slightly uncomfortable feeling. Sometimes they are even frightening. There is no way to eliminate all the queasiness caused by risky decisions, but some of it can be allayed by analyzing data from different perspectives. Of course, there is also the chance that the extra analysis will further confuse and complicate the question.

Consider the following table of probabilities of various rates of return that could be earned from three equal-size investments:

Alternative	−2%	0%	2%	4%	6%	8%
A	0.27	0.21	0	0.11	0	0.41
B	0	0.12	0.53	0.35	0	0
C	0.14	0.04	0.15	0.22	0.39	0.06

The expected-value principle suggests that alternative C is preferred according to the following calculations:

$$E(A) = -2(0.27) + 0(0.21) + 4(0.11) + 8(0.41)$$
$$= 3.18\%$$

$$E(B) = 0(0.12) + 2(0.53) + 4(0.35) = 2.46\%$$

$$E(C) = -2(0.14) + 0(0.04) + 2(0.15) + 4(0.22)$$
$$+ 6(0.39) + 8(0.06)$$
$$= 3.72\% \text{ (preferred)}$$

However, the varied pattern of possible returns implies that other criteria might better represent the degree of optimism of the decision maker. Any or all of the following principles could be applied to the data to conduct a more thorough analysis:

1. *Minimax principle:* Select the alternative that minimizes your maximum loss. Because alternative B is the only one for which no possible loss is foreseen, it receives the nod from this conservative principle.

2. *Most-probable-future principle:* Identify the rate of return for each alternative with the greatest likelihood of occurrence, and choose the alternative with the highest indicated rate. If two alternatives have the same highest rate, select the one with the largest probability for that rate. The most probable future for each alternative and its likelihood are as follows, in which this optimistic principle indicates a preference for alternative A:

Alternative	Most Probable Future	Probability	
A	8%	0.41	(preferred)
B	2%	0.53	
C	6%	0.39	

3. *Aspiration-level principle:* Choose the alternative that maximizes the probability of obtaining a certain level of return. For instance, if the aspiration level of the decision maker appraising the given data were 4 percent, the probabilities for all returns greater than 4 percent would be summed for each alternative to get

Alternative	Probability of Greater Than 4% Return	
A	0.41	
B	0	
C	0.39 + 0.06 = 0.45	(preferred at 4% aspiration level)

Similarly, if the aspiration level were stated as "I don't care which alternative we use as long as it makes money," alternative B would be selected because

Alternative	Probability of Greater Than 0% Return	
A	0.11 + 0.41 = 0.52	
B	0.53 + 0.35 = 0.88	(preferred)
C	0.15 + 0.22 + 0.39 + 0.06 = 0.82	

The obvious problem now is to decide which way to decide. Which principle do you prefer? Why?

23. *The Case of Cost versus Prestige versus Utility* An engineer and an accountant had two things in common. Both saw the potential for computers in production enterprises, and their favorite meal was

a prime rib dinner. The outgrowth of these shared interests was a computer service bureau named Prime Rib, Inc. (PRI) that provided customized softwear for production costing and scheduling. They started the company three years ago and now employ 35 people. Business is good and promises to get even better.

At a dinner to commemorate their first three years together, serving prime rib, naturally, Ed, the engineer, and Al, the accountant became engaged in a heated discussion about purchasing a company car. They seldom disagreed on the directions PRI should take to serve customers, but delicate questions about perquisites caused them trouble. In this instance, the question was what type of vehicle the company should buy for their use. They both agreed a vehicle was needed. They disagreed on the type it should be.

ED: You're thinking too big, Al. We need a limo like a beggar needs a tux. All we require is a nice little economical pickup truck to run errands.

AL: We gotta' think big. We're on our way and should let other people know it. A flashy set of wheels will do it. Do you want to meet a customer at the airport in a pickup?

ED: Maybe that's all we can afford. I figure the price of a fancy sedan is three times that of a pickup. Upkeep and gas will at least be double. That's expensive prestige you want to buy.

AL: It's all a matter of priorities, old buddy. We may have to stretch a bit to meet the sticker price, but we can surely afford the upkeep as we get more accounts. And that "fancy sedan," as you call it, will help us get those accounts. I say prestige should be our top priority in the decision, and my type of car would have at least five times the prestige of yours. You gotta' look successful to be successful.

ED: OK. I can go along with your rating for prestige, reluctantly. On the same scale, I'd rate cost 8 and upkeep 5. I'd also rate utility right next to prestige. We need something versatile, not just a gas hog that impresses people.

AL: I'll accept your numbers, and your argument for utility too, but I also think my big, impressive, luxurious chariot would be great to run errands in. It'll have a trunk big enough to carry our stuff in. I'd rate its utility umpteen times a pickup's.

ED: No way! We're going in circles. Let's pretend its someone else's problem and solve it the Prime Rib way. I'll set up the decision matrix.

He pulled out a pencil, grabbed a napkin, and started scribbling.

Assuming Prime Rib, Inc., uses a mixed-dimension comparison model for decision analysis, complete the matrix Ed started. Using the figures given in the dialogue, analyze the situation giving particular attention to the ratings for utility. Suggest how an interval scale could be set up (describe the 10 and 0 extremes) to determine a rating for utility. Discuss the sensitivity of the decision.

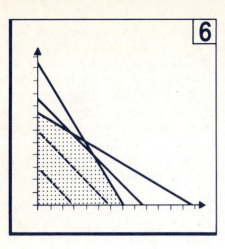

6

ALLOCATION OF RESOURCES

LEARNING OBJECTIVES

After studying this chapter you should

- understand the concepts of aggregate planning and the use of linear programming models in determining optimal resource allocations.
- be able to use the assignment method for matching supply with demand and the graphical LP method for optimizing a two-dimensional product mix.
- be aware of the simplex method and its basic operations.
- be able to apply the distribution method by using VAM to obtain an initial feasible solution and to check its optimality by the stepping-stone method.
- appreciate the strategies available for aggregate planning and the different models that can be used for decision making.
- realize the importance of time management and know how to avoid time wasters.
- know when and how to employ the nominal group technique.

KEY TERMS

The following words characterize subjects presented in this chapter:

linear programming (LP)	isoprofit lines
simplex method	feasible solution space
assignment method	distribution method
square matrix	rim conditions
row-and-column subtraction	Vogel approximation method (VAM)
opportunity costs	
objective function	stepping-stone method

dependency

degeneracy

epsilon quantity (ϵ)

aggregate planning

linear decision rule

management coefficient model

knowledge worker

blue-collar and white-collar workers

nominal group technique (NGT)

task statement

6-1 IMPORTANCE

In our planning steps thus far, we have attempted to peek into the future to decide which courses of action would be feasible. Then we have applied the dollar yardstick to see which one would be most worthwhile. Now we consider the assignment of resources to the selected plan.

After forecasts have been accepted, resources are usually allocated under the assumption of certainty. That is, for a given set of conditions deemed most realistic, the requirements of a project are analyzed to apportion available resources optimally to achieve desired objectives. Occasionally there is only one way to do things. Then allocations are anticlimactic. More commonly, there is considerable resource convertibility, and astute planning can use this flexibility to advantage.

In this chapter we consider formal procedures to assist resource planning. Linear programming is one of the most powerful tools available for optimizing resource utilization. Special attention is devoted to the distribution model, which has graphic properties that hasten the understanding of allocation procedures and is suitable for manual calculations. It is also used to demonstrate the value of aggregate planning to collectively schedule resources for production.

The mathematical techniques discussed for allocating resources are programmed for computers. Manual calculations are practical for problems of limited size, but computer competence can be a great time saver for larger applications. Yet technique familiarity and calculating expertise cannot substitute for creativity and diligence in planning. Techniques are ingenious succors in the evaluation of massed data, but they do not validate the data. Reliable information and well-advised objectives are prerequisites to an assurance that the right amounts of resources are assigned to the right places.

6-2 LINEAR PROGRAMMING

Problems of allocating scarce resources to competing activities are candidates for **linear programming (LP)** procedures. These procedures are appropriate whenever the variables in the problem are linearly related to one another. Applications have been extensive in industry and sometimes have produced dramatic savings. Although LP is by no means the only way to optimize returns for a given system, many fundamental relationships of quantities invite its utilization.

Linear means directly proportional relationships among variables. *Programming* applies to calculating procedures for solving a set of linear equations or inequations.

See, for example, the marginal analyses for nonlinear break-even data in Section 4-3.

The development of LP is closely associated with economic theory and research. Early economic studies assumed a continuous relationship of variables that could be evaluated with calculus. Although these assumptions were applicable to some systems, such as agriculture, they were not satisfactory for many industrial systems. The input–output method of analysis by economist W. W. Leontiff started the trend toward LP methods. Current developments grew from the work of G. B. Dantzig, who originated the simplex method of LP in 1947.

Studies of production systems present many opportunities for LP applications. LP can be used profitably in all three evaluation stages: planning, analysis, and control. Applicable problems include planning the location of supply facilities to minimize transportation costs, analyzing operations and methods to improve profits, and controlling machine loading to achieve maximum utilization. In the following sections, the easy-to-apply assignment method is introduced first. It is followed by a graphical LP approach for determining the most profitable product mix. This leads to the more generally applicable distribution method. Then, after a discussion of aggregate planning, the distribution model is applied to production scheduling. While observing these LP models, remember that all the problems could have been solved by the encompassing **simplex method,** which is presented in Appendix F.

EXAMPLE 6.1 Application of operations research techniques in production

Several quantitative approaches to management have become closely associated with the discipline of operations research (OR), although they are also considered to be basic tools for other disciplines such as industrial engineering, production management, decision science, and management science. A survey (by W. N. Ledbetter and J. F. Cox, "Are OR Techniques Being Used?" *Industrial Engineering,* February 1977) of the 500 largest industrial firms in the United States was undertaken to observe the actual utilization of OR techniques. The relative use of the top five techniques are shown in Table 6.1. Frequency of

TABLE 6.1
Percentage Utilization by Large Industrial Firms of the Five Most-Applied OR Techniques

| OR Technique | Degree of Utilization (% of Respondents) | | | | | Mean Score | Chapter in Which Presented |
	Never (1)	(2)	(3)	(4)	Very Frequently (5)		
Regression analysis	9.5%	2.7%	17.6%	21.6%	48.6%	3.97	3
Linear programming	15.4	14.1	21.8	16.7	32.0	3.36	6
Simulation (production)	11.4	15.7	25.7	24.3	22.9	3.31	11
Network models	39.1	29.0	15.9	10.1	5.8	2.41	7 and 14
Queuing theory	36.6	39.4	16.9	5.6	1.4	1.96	11

TABLE 6.2
Application of OR Techniques in 11 Areas of Production
Percentages indicate the proportion of responding firms that use each technique in each production area

Application Areas	Percent Utilization of OR Techniques				
	Linear Programming	Simulation	Network Models	Queuing Theory	Regression Analysis
Production scheduling	41.1%	35.6%	8.2%	12.3%	6.8%
Production planning/control	26.0	24.7	9.6	5.5	4.1
Project planning/control	13.7	12.3	38.4	2.7	0
Inventory analysis/control	15.1	37.0	4.1	5.5	16.4
Quality control	2.7	2.7	4.1	0	20.5
Maintenance and repair	0	11.0	4.1	4.1	5.5
Plant layout	17.8	26.0	6.8	6.8	2.7
Equipment acquisition	5.5	15.1	1.4	1.4	0
Blending	43.8	6.2	1.4	1.4	4.1
Logistics	37.0	32.9	11.0	4.1	8.2
Plant location	43.8	31.5	11.0	1.4	6.8

use was rated on a five-point scale from *never* (1) to *very frequently* (5); techniques are listed in order of their mean scores. The stub column on the right gives the chapters in which each of the techniques are discussed in this book.

The survey also attempted to relate OR techniques to their areas of application in production. Partial results are shown in Table 6.2. The popularity of linear programming is shown by its relatively intense utilization in most of the production functions. Simulation has the greatest breadth of utilization. Regression analysis, which received the top rating in Table 6.1 for overall use within a firm, is apparently not as extensively used as are linear programming and simulation for production activities. Lesser-used techniques mentioned in the survey include dynamic programming, game theory, heuristic programming, and material requirements planning.

6-3 ASSIGNMENT METHOD

The **assignment method** uses a clever routine, called the Hungarian algorithm, to solve a special type of LP problem in an easy, visual manner. This algorithm is applicable to any matching situation in which some type of rating can be given to the performance of each pairing and there are the same number of applicants as there are positions open. For instance, operators are matched to machines according to pieces produced per hour by each individual on each machine; teams are matched to projects by the expected cost of each team to accomplish each project. In this section we shall apply the assignment method to the allocation of employees to jobs.

Let us assume that there are four candidates to fill four different jobs. In

Named in honor of Hungarian mathematician, D. Konig, who proved the theorem for its development.

Job	1	2	3	4
Employee 1	2	6	3	5
2	1	2	5	3
3	4	3	1	5
4	2	4	1	5

FIGURE 6.1 Matrix format.

Job	1	2	3	4
Employee 1	0	4	1	3
2	0	1	4	2
3	3	2	0	4
4	1	3	0	4

FIGURE 6.2 Row subtraction.

Job	1	2	3	4
Employee 1	0	3	1	1
2	0	0	4	0
3	3	1	0	2
4	1	2	0	2

FIGURE 6.3 Column subtraction.

Job	1	2	3	4
Employee 1	0	3	1	1
2	0	0	4	0
3	3	1	0	2
4	1	2	0	2

FIGURE 6.4 Zero-covering lines.

The placement of the zero-covering lines does not affect the final solution. The only restriction is to use as few as possible.

the assignment method the number of operators must equal the number of operations, or expressed in another way, the format must be a **square matrix.** The relative ratings for each candidate for each position could be determined from test scores, trials, or subjective opinions. These ratings are then arranged in a matrix, as shown in Figure 6.1. Assuming low numbers indicate a better rating for a job, we see that there is no immediately obvious solution. It is interesting to note that there are 4! = 24 possible different assignments. In a matrix with 15 candidates and 15 jobs, there would be over a trillion unique ways to match employees and jobs. For such a situation, a time-saving algorithm to discover the best pattern is not a luxury—it is indispensable.

The first step in the solution is to obtain the opportunity costs for each row and column, through **row-and-column subtraction.** This step is accomplished by subtracting the smallest number in each row or column from the remaining values in the respective line. It makes no difference in the final solution whether the row or the column is subtracted first. Figure 6.2 shows the row subtraction. Note that one zero occurs in each row. The other values in the row are the **opportunity costs** that would result from not assigning the candidate with the best score to the most suitable job. After each operation on the matrix, a check should be made to see whether an optimal assignment has been attained. When there is a unique zero for each row and column, the best possible match is identified. In Figure 6.2, there are no zeros in the columns for jobs 2 and 4; therefore, the assignment method must be carried through at least one more step.

Column subtraction is conducted similarly to row subtraction. The lowest value in each column of the matrix resulting from the row differences is subtracted from all other values in the column. The outcome of the operation is shown in Figure 6.3. Note that columns 1 and 3 are unchanged from Figure 6.2; they already contained zeros. The zeros now reveal the opportunity costs for the employee–job interactions. Again a check for an optimal solution is due. At first glance it appears that there might be a zero for each employee–job match, but closer inspection shows that employee 2 has three of the zero opportunity costs available. Therefore, another matrix operation is needed.

The next step is two phased and serves a dual purpose. The initial phase is to cover all the zeros in the matrix from the previous step with as few straight horizontal or vertical lines as possible. If the number of lines is equal to the number of rows (or columns), a solution has already been obtained in the previous step. As shown in Figure 6.4, the sample problem requires only three lines to cover all the zeros. Because there are four rows, it means that a solution has not been obtained and thereby confirms our previous conclusions reached by inspecting the zeros independently. This optimality check is the first purpose of the lines.

The second purpose and second phase is to modify the matrix in a continuing effort toward optimization. The procedure is to choose the smallest number not covered by the recently drawn lines. This number is added to all values at line intersections and subtracted from all uncovered numbers. In Figure 6.4, the smallest uncovered number is 1 in the cells for employee 1, job 4 and employee 3, job 2. It is added to the value of each cell at a line intersection; $0 + 1 = 1$ at employee 2, job 1 and $4 + 1 = 5$ at employee 2, job 3. Then 1 is subtracted from the uncovered cells to produce the matrix shown in Figure 6.5.

Job	1	2	3	4
Employee 1	0	2	1	0
2	1	0	5	0
3	3	0	0	1
4	1	1	0	1

FIGURE 6.5 One added to line intersections and subtracted from uncovered cells.

For a quick check, the matrix is again subjected to zero-covering lines. In the sample problem there is no way to cover all the zeros with less than four orthogonal lines. Eureka, a solution! Specific assignments are identified by locating any zeros that occur uniquely in a row *or* column. The only zero in column 1 is at row 1. Therefore, employee 1 is assigned to job 1. Now we have a 3×3 matrix left: row 1 and column 1 are already taken by the first assignment. In row 4 the only assignment possible is employee 4 to job 3. The remaining two assignments are employee 2 to job 4 and employee 3 to job 2. The matchings and associated ratings are as follows. No other combination can provide better ratings per job. Expressed in another way, this is the *minimum opportunity cost* schedule.

Alternate assignments with the same minimum cost are often identified in the solution. This condition is signaled when there is no unique zero in a row or column.

Employee–job match:	E1 to J1	E2 to J4	E3 to J2	E4 to J3
Employee–job rating:	2	3	3	1

The assignment algorithm is more flexible than it might appear from the foregoing sample. It can be used when ratings are given in profits (maximization problem) by first subtracting all the values that appear in the profit matrix from the largest number in the matrix. In effect, this process converts all values to "relative costs" and allows the application of the previously described minimization algorithm. If a given employee–job match is impossible or unwieldy, such an assignment can be blocked by entering a very large cost in the appropriate cell. If there is an extra operator or operation and the intent is to identify which one should be eliminated to achieve minimum cost, a dummy with zero costs in each cell is added to make the number of rows equal the number of columns. This concept is elaborated on in Section 6-8 to account for unequal supply and demand in the transportation method. Several of these special conditions are illustrated in Example 6.2.

EXAMPLE 6.2 Employee–job matching by the assignment method

A distributor plans to penetrate a new market area. The region has been divided into four districts. Five candidates have been recommended as representatives for the new area. Because each district offers unique opportunities and each is encumbered by special difficulties, different managers are expected to perform with varying degrees of success. The anticipated annual sales by each representative in each district were estimated and are shown next (in $10,000 units).

District	I	II	III	IV
Sales managers: Browne	28	19	23	12
Greene	16	17	18	18
Redd	24	16	22	19
White	23	11	16	18
Jones	21	16	21	16

The sales figures represent profit and must be converted to relative costs in order to use the minimization assignment algorithm. In the following matrix, the values in each cell have been subtracted from the maximum value in the matrix, 28, and a dummy sales district, V_d has been added with zero costs for all assignments in its column.

	I	II	III	IV	V_d
B	0	9	5	16	0
G	12	11	10	10	0
R	4	12	6	9	0
W	5	17	12	10	0
J	7	12	7	12	0

The dummy column provides a zero for each row, making the row subtraction step redundant. Successive solution steps of column subtraction, line covering, number modification, second line-covering check, another round of number modifications, and a final line-covering check reveal the optimal assignments.

	I	II	III	IV	V_d
B	0	0	0	7	0
G	12	2	5	1	0
R	4	3	1	0	0
W	5	8	7	1	0
J	7	3	2	3	0

Column subtraction and first zero-covering lines.

	I	II	III	IV	V_d
B	0	0	0	7	1
G	11	1	4	0	0
R	4	3	1	0	1
W	4	7	6	0	0
J	6	2	1	2	0

Values modified by subtracting 1 from uncovered values and adding 1 to line intersections.

	I	II	III	IV	V_d
B	0	0	0	8	2
G	10	0	3	0	0
R	3	2	0	0	1
W	3	6	5	0	0
J	5	1	0	2	0

Values again modified according to zero-covering lines of the previous step; new zero-covering lines drawn.

The first assignment, the only unique zero in a row or column, is at $B,1$. By eliminating the B row and column I in the initial allocation, the zero at G, II is unique in column II. Now the reduced matrix appears as follows. Because there are no unique zeros in the rows or columns of the 3 × 3 reduced matrix, equally attractive alternative assignments are indicated.

The expectation of equal sales from either assignment schedule is shown next. Factors other than those included in the matrix should be used to make the management choice between the costwise equal assignments.

Assignment 1:	B to I	G to II	R to III	W to IV	Total:
Expected sales:	28	17	22	18	$850,000
Assignment 2:	B to I	G to II	J to III	R to IV	Total:
Expected sales:	28	17	21	19	$850,000

	I	II	III	IV	V_d
B	[0]	0	0	8	2	→ First assignment −, B to I
G	10	[0]	3	0	0	→ Second assignment −, G to II
R	3	2	0	0	1
W	3	6	5	0	0	Matrix reduced to 3 × 3
J	5	1	0	2	0

Alternate Schedules Indicated by Reduced Matrix

	III	IV	V_d			III	IV	V_d
R	[0]	0	1		R	0	[0]	1
W	5	[0]	0		W	5	0	[0]
J	0	2	[0]		J	[0]	2	0

6-4 GRAPHICAL METHOD

The selection of an optimal mix is an ideal introduction to LP methods. Determining the proportion of each product to produce when production resources are limited is typical of LP applications that specify how to use scarce resources to maximize profit. A two-product mix has the further advantage of simplicity, which allows the key features to be accented.

Requirements for an LP solution

One difficult task in evaluating production systems is to recognize the best method of attacking a problem. Often several computational techniques can be applied. Usually one of them will give better results or produce equivalent results with less effort. To identify this better method, you have to match the characteristics of the problem to the requirements of different problem-solving methods. An awareness of the following LP requirements should reveal whether a problem is amenable to this form of solution.

THE OBJECTIVE MUST BE STATED EXPLICITLY

The objective to get the "best equipment for our purposes" is too broad to be useful. Similarly, the goal of "obtaining the best equipment for the least money" is bewilderingly indefinite. Such contradictory or nebulous aims must be reconciled before the objective function can be expressed in the necessary equation.

ALTERNATIVE COURSES OF ACTION MUST BE AVAILABLE

If there is only one way to accomplish a necessary task, the sole recourse is to do it the given way or to do nothing. However, when faced with a Hobson's choice, at least a little doubt should be cast on the problem formulation. Perhaps a more creative effort or additional information would expose more alternatives.

A choice without an alternative is sometimes called "Hobson's choice," an allusion to the practice of Thomas Hobson (d. 1631) who let horses and required every customer to take the horse standing nearest the door.

RESOURCE LIMITATIONS MUST BE KNOWN

Rarely is there an unlimited supply of anything worth having, but commonly it is difficult to know how scarce a resource really is. Again it may take some imagination and considerable searching to set realistic limits on variables.

RELATIONSHIP OF VARIABLES MUST BE KNOWN

A corollary is that the variables must be quantifiable. Relationships among variables are expressed by inequations to show that "the allocation of resource 1 plus resource 2 cannot exceed a certain quantity" or by equations to set "the number sold of product A plus product B yields a certain profit."

Product mix—Two-dimensional case

Product mix problems occur when several products are produced in the same work centers. When only two products are involved, regardless of the number of work stations through which they pass, it is a two-dimensional case. That is, when the problem is displayed on a graph, the two products are shown on the X–Y axes. Data for linearly related product–mix restrictions and profits are shown in Table 6.3.

Note that contribution per unit, not profit per unit, is linearly related to the number of units made and sold. See Section 4-3.

The two products, A and B, must pass through all three machine centers, J, K, and L, but their passage can be in any order. The contribution per unit made and sold of each product is shown at the bottom of each product's column. These figures establish a profit equation:

$$Z = \$12A + \$8B$$

which states that the total profit is equal to the contribution of $A(\$12)$ times the number produced of A plus the contribution of $B(\$8)$ times the production of B. This profit expression is the **objective function** that we seek to maximize. It can be depicted graphically by a series of **isoprofit lines** of constant slope. A dashed line representing all the product mixes that will yield a total profit of $120 is shown in the margin sketch.

The far-right column of Table 6.3 shows the time relationships of the limiting resources. At most, only 48 hours are available in machine center L. These hours could be used to produce $^{48}\!/_2 = 24$ units of A, $^{48}\!/_4 = 12$ units of B, or

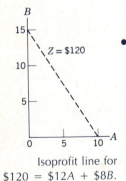

Isoprofit line for
$120 = $12A + $8B.

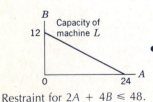

Restraint for $2A + 4B \leq 48$.

TABLE 6.3
Production Times and Profit Data for Two Products

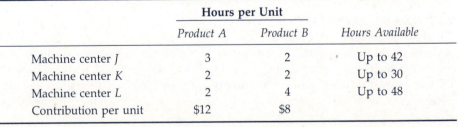

	Hours per Unit		
	Product A	*Product B*	*Hours Available*
Machine center J	3	2	Up to 42
Machine center K	2	2	Up to 30
Machine center L	2	4	Up to 48
Contribution per unit	$12	$8	

some combination of both A and B, such as 8 units of each. This condition can be expressed by the *inequation*

$$2A + 4B \leqslant 48$$

which states that the sum of the times for both products in the machine center must be less than or equal to ($\leqslant$) 48 hours. A line representing this restraint is shown in the margin sketch. Any mix on or to the left of the line meets the restriction. A mix represented by any point to the right of the line requires more time than is available in machine center L.

Combining the three machine center restrictions:

$$3A + 2B \leqslant 42 \qquad \text{for machine center } J$$

$$2A + 2B \leqslant 30 \qquad \text{for machine center } K$$

$$2A + 4B \leqslant 48 \qquad \text{for machine center } L$$

with the obvious restriction that we cannot produce negative products ($A \geqslant 0$, $B \geqslant 0$) to save time, we have the graph shown in Figure 6.6. The shaded area is the **feasible solution space.** This area represents all possible mixes that meet the restraints imposed by the machine-center time limitations. For instance, mix 6 (6 units of both A and B) is well within the time limits for all restraints, but mix 8 (8 units of A and B) exceeds the time available in machine center $K((2 \times 8) + (2 \times 8) = 32 \nleqslant 30)$ by two hours, making it a restraint violation.

With all feasible mix alternatives delineated, the remaining task is to select the one that meets the objective of maximizing profit. It is natural to expect that producing more products will increase profit. This is confirmed by the increasing magnitude of isoprofit lines as they depart farther from the origin, O. The logical

The restriction $A \geqslant 0 \leqslant B$ establishes that the graph is in the positive X–Y quadrant.

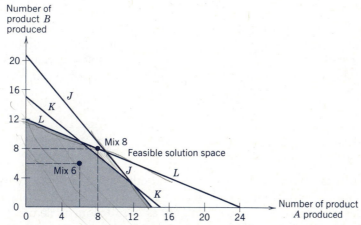

FIGURE 6.6 Feasible solution space set by machine center restraints.

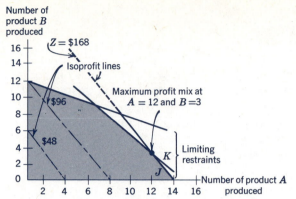

FIGURE 6.7 Most profitable mix identified by isoprofit lines.

conclusion is then that the isoprofit line the farthest from O but still within the feasible solution space identifies the mix that maximizes profit.

As shown in Figure 6.7, the last isoprofit line ($168) that touches the perimeter of the feasible solution polygon indicates the proportion of products for maximum profit. The mix can be read directly from the graph if the scale is large enough, or it can be determined algebraically from the restraint lines (J and K) that intersect at the indicated vertex. Because all the allowable time in each machine center is used at any intersection point, the coordinates can be determined from a simultaneous solution of the equations of the intersecting lines:

$$3A + 2B = 42 \qquad \text{from machine center } J$$
$$\underline{2A + 2B = 30} \qquad \text{from machine center } K$$
$$A = 12 \text{ units and } B = 3 \text{ units}$$

Thus, all the available time from machine centers J and K is used to produce 12 units of A and 3 units of B for a total profit of $168. Although the utilization of machine center L is $48 - [(2 \times 12) + (4 \times 3)] = 12$ hours short of capacity, no other combination of products will yield a greater profit.

The problem described in Table 6.3 is also solved by the simplex method in Appendix F. The presence of equally profitable solutions ($168) is shown to occur along constraint line J from $A = 12$ and $B = 3$ to $A = 14$ and $B = 0$.

EXAMPLE 6.3 Graphical solution of a mix to minimize cost

A concrete products company has two sources of sand and gravel aggregate. The material from one source is much coarser than the other. The cost from either source depends on pit cost, hauling, and refining. The amount of each grade of rock and sand required for the production of concrete pipe, ready-mix, and other products is fairly well known for one month in advance, but it changes from month to month as a function of demand. The objective of determining the proportion of material to take from each source is to *minimize* monthly procurement, handling, and storage costs. Necessary cost and

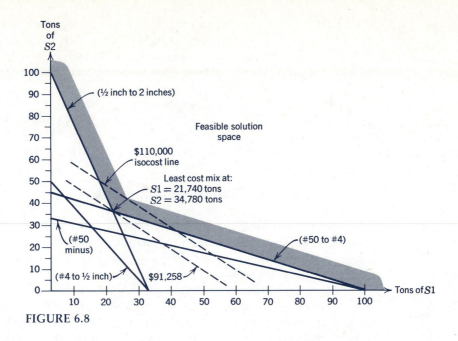

FIGURE 6.8

quantity relationships are tabulated as follows:

| Aggregate Size | Pounds per Ton | | Tons Required |
	Source 1	Source 2	
½ inch to 2 inches	600	200	10,000
#4 to ½ inch	900	600	15,000
#50 to #4	400	900	20,000
Less than #50	100	300	5,000
Cost per ton	$1.51	$1.68	

The first step in solving the problem is to convert all measures to the same dimensions. Because most units are in tons, the pounds of each aggregate size per ton can be converted to percentages of tons to establish the following restraint inequations:

$$0.30\ S1 + 0.10\ S2 \geqslant 10,000 \quad \text{for ½ inch to 2 inches}$$

$$0.45\ S1 + 0.30\ S2 \geqslant 15,000 \quad \text{for #4 to ½ inch}$$

$$0.20\ S1 + 0.45\ S2 \geqslant 20,000 \quad \text{for #50 to #4}$$

$$0.05\ S1 + 0.15\ S2 \geqslant 5,000 \quad \text{for less than #50}$$

TABLE 6.4
Quantity and Cost Summary of the Optimal Mix

| Aggregate Size | Tons per Monthly Estimate Period | | | | |
	Source 1	Source 2	Total	Required	Excess
½ inch to 2 inches	6,522	3,478	10,000	10,000	0
#4 to ½ inch	9,783	10,433	20,216	15,000	5,216
#50 to #4	4,348	15,652	20,000	20,000	0
Less than #50	1,087	5,217	6,304	5,000	1,304
Totals	21,740	34,780	56,520	50,000	6,520

Cost: $1.51 (21,740) + $1.68 (34,780) = $91,258

The function to minimize is

$$\text{Cost} = \$1.51\ S1 + \$1.68\ S2$$

These conditions can be graphically represented, as in Figure 6.8.

Because the quantity of rock and sand must be greater than or equal to given monthly demands, the solution space falls to the right of the restraint lines. Two isocost lines are shown in the figure. It is ap-

parent that cost decreases as the lines move to the left and that the lowest cost mix is defined by the intersection of the (½ inch to 2 inches) and (#50 to #4) aggregate size restraints. From the equations for the indicated restraints, the mix is determined to be 21,740 tons from S1 and 34,780 tons from S2. This mix exceeds the monthly 50,000-ton demand, as shown in Table 6.4. But no other combination will meet the requirements at a cost less than $91,258.

Product mix—Multidimensional case

A graphical solution to a product mix works well for two products, becomes difficult for three products, and is impossible for more than three products. Other solution methods are therefore required. For relatively small problems, algebraic or systematic trial-and-error solution procedures are possible, but the simplex method is usually more manageable.

A graphical solution to a three-product mix depends on the utilization of planes in *X-Y-Z* space.

Most significant applications of LP contain a formidable array of restraints and variables. In these instances, any manual approach becomes exceedingly tedious. Individual calculations are not difficult, but there are just an awful lot of them. Fortunately we can call on electronic assistance. Computer programs are widely available and extensively employed for high-capacity applications.

6-5 DISTRIBUTION METHOD OF LP

The **distribution method** is more versatile than the graphical routine and retains the succinctness that makes it a worthy tool for manual calculations. It is used to determine preferred routes for the transportation of supplies from a number of origins to different destinations. Although the name *distribution* tends to conjure up images of warehouses supplying retail outlets with produce, the method also can be used to identify the least-cost or most profitable distribution pattern for any resource.

The distribution model is also widely known as the *transportation method.*

The solution format is a matrix that defines (1) the amount and location of both supply and demand and (2) the cost or profit created by supplying one unit from every origin to every destination. There is no limit to the number of origins or destinations that can be included in the matrix. An optimal distribution is obtained by first developing an initial solution and then sequentially testing and revising improved solutions until no further improvements are available. There may be several equal cost–distribution patterns. These alternative routes are identified by the solution procedures.

Details of the solution procedures will be illustrated with a sample application. Assume a chain of bakeries, producing a complete line of packaged cakes, cookies, pies, and other baked desserts, plans to construct a new baking plant–distribution center. The company has two bakeries in metropolitan centers which are providing products to adjacent smaller communities. These plants operate

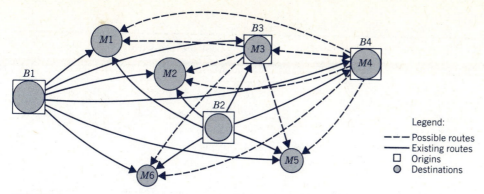

FIGURE 6.9 Existing and possible supply and demand pattern.

at full capacity but cannot meet current demand. Unless expected competition appears, the market potential should continue to grow. To meet the anticipated demand, a new bakery will be built in one of two locations. The additional capacity will satisfy the local sales of the city in which the bakery is located and the demand of nearby cities. The problem is to decide which location will minimize the cost of distributing the products.

An idea of the geographical layout and distribution pattern is available from Figure 6.9. Rectangles represent baking plant–distribution centers, and circles represent the outlying cities where the products are sold. The size of the symbols provides a rough indication of capacity. The arcs show possible distribution patterns. Solid lines in both the arcs and rectangles show existing routes and patterns. Dashed lines represent the alternative new plant locations and associated distribution patterns.

Only one alternative is to be selected. Thus, if $B4$ is the chosen bakery site, the local market $M4$ logically will be supplied by $B4$. If $B3$ is the selected site instead of $B4$, $M3$ will be the home market.

Packaged bakery goods are delivered by company-owned trucks to nearby cities. The product mix of the goods is customized to the market demand. The total contribution is a function of the mix, but the transportation cost is a function of the volume moved. Therefore, supply, demand, and transportation costs are rated in units of truckloads. The cost per truck to supply each marketing area and the capacity relationships are given in Table 6.5. The supply at each bakery indicates the balance available for outlying markets after the local demand has been met. Thus $B4$ has a total capacity of 44 truckloads per day but has only 24 loads available for daily shipments to other cities after supplying $M4$.

All the data from Table 6.5 appropriate to the bakery site $B3$ are compactly included in a transportation matrix. As shown in Figure 6.10, the available and required quantities associated with each origin and destination are entered in the bottom row and outside right column. These entries, called **rim conditions,** correspond to the restraint lines of the graphical method. Each cell in the matrix represents a possible routing of supplies from a source to a destination. Each

TABLE 6.5
Data Required for a Transportation Problem

Capacity in Truckloads per Day				Distribution Cost per Truckload						
Bakery	Supply	Market	Demand		M1	M2	M3	M4	M5	M6
B1	22	M1	16	B1	$14	$24	$30	$50	$44	$16
B2	18	M2	18	B2	$20	$14	$16	$32	$16	$14
B3	29	M3	15	B3	$16	$11	—	$18	$26	$45
B4	24	M4	20	B4	$35	$26	$18	—	$20	$50
		M5	10							
		M6	5							

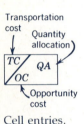

Transportation cost — Quantity allocation

TC / QA / OC

Opportunity cost

Cell entries.

route has a distinct cost of profit, which is shown in the left slice of the cell. The portion of the cell to the right of the slash is reserved for quantity or opportunity cost entries.

Reviewing the given problem conditions with respect to the requirements for an LP solution, we have

1. An explicit objective to minimize the total cost of distributing bakery products to the market areas.

2. Several alternative combinations of truck routes.

3. Limited supplies available from the bakery-distribution centers to satisfy the demand that is also limited.

4. Variables linearly related according to supply, demand, and distribution costs.

With the problem data and the solution method identified, we are ready to plunge into the solution procedures.

Origin	M 1	M 2	M 4	M 5	M 6	Supply
			Destination			
B 1	14	24	50	44	16	22
B 2	20	14	32	16	14	18
B 3	16	11	18	26	45	29
Demand	16	18	20	10	5	69 / 69

Rim conditions

FIGURE 6.10 Transportation matrix for B3 alternative.

6-6 INITIAL FEASIBLE SOLUTION—VOGEL'S APPROXIMATION METHOD

An *initial* solution is a first attempt to match supply and demand along advantageous distribution routes. For it to be a *feasible* solution, it must meet the following conditions:

1. The rim conditions of the transportation matrix must be satisfied—the sum of the quantities in occupied cells for rows and columns must equal the corresponding rim quantities.

2. The number of occupied cells must equal one less than the sum of the number of origins (O) and destinations (D): $O + D - 1$.

3. The occupied cells must be independent positions. Dependent positions allow a round trip from an occupied cell back to itself using only horizontal and vertical movements with right-angle turns at occupied cells.

The significance of the required number of occupied cells and the need for independent positions will become apparent when the initial feasible solution is checked for optimality by means of the stepping-stone technique (see Section 6-7).

A desired but not required quality of an initial feasible solution is that it be as close as possible to the optimal solution. Iterative tests for optimality are time-consuming. Spending a little extra time on the intial step often pays big dividends in time saved from subsequent steps. For this reason we determine the initial feasible solution through the use of **Vogel's approximation method (VAM).**

VAM is based on the concept of penalties charged for not making an allocation to the lowest cost route. A penalty for each row and column is calculated by subtracting the cost for transporting by the least expensive route from the next most expensive route. The cell causing the highest penalty receives the first allocation. After each cell entry, the penalties are recalculated, and successive entries are made to avoid the highest penalties. When two penalties are equal, a judgment based on a mental preview of forthcoming penalties decides the issue. The process is continued until the conditions for a feasible solution are met.

A detailed application of VAM to the matrix of Figure 6.10 is presented in Figure 6.11. For clarity, the example uses separate allocation and penalty matrices. But in actual practice, the shorthand form used in Example 6.4 is more convenient.

A quick check of the final allocations of the VAM solution reveals that the rim conditions are satisfied; there are $3 + 5 - 1 = 7$ allocations, and those allocations are in independent positions. Therefore, the solution is ready for the optimality check.

A cell is occupied if it contains an entry that represents a quantity transfer from an origin to a destination.

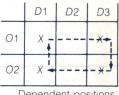

Dependent positions:
X = occupied cell.

	D1	D2	D3
O1	X	X	
O2		X	X

Independent positions:
X = occupied cell.

When two cells in the same row or column have the same minimum cost for transportation, the penalty for that row or column is zero.

Other methods of obtaining an initial solution include assignments by inspection or by the "Northwest Corner Rule," on which assignments starting at the northwest corner of the matrix progressively meet rim requirements.

FIGURE 6.11 Detailed application of Vogel's approximation method.

Operations | Penalty Matrices | Allocation Matrices

Operations: The highest penalty is circled in column M4. The lowest cost for M4 is from B3 where the required 20 units are allocated. B3 now has 9 units left.

Penalty Matrix:

	M1	M2	M4	M5	M6	Row Penalty
B1	14	24	50	44	16	2
B2	20	14	32	16	14	0
B3	16	11	18	26	45	5
Column Penalty	2	3	(14)	10	2	

Allocation Matrix:

	M1	M2	M4	M5	M6	Available
B1						22
B2						18
B3			(20)			29
Required	16	18	20	10	5	

Operations: M4 is eliminated from the penalty matrix because the demand is satisfied. Now M5 has the highest penalty, and 10 of the 18 units available in B2 are allocated to it.

Penalty Matrix:

	M1	M2	M5	M6	Row Penalty
B1	14	24	44	16	2
B2	20	14	16	14	0
B3	16	11	26	45	5
Column Penalty	2	3	(10)	2	

Allocation Matrix:

	M1	M2	M4	M5	M6	Available
B1						22
B2				(10)		18
B3			20			9
Required	16	18	0	10	5	

Operations: After eliminating M5, the penalty in row B3 is the highest. The lowest cost in the Row is M2 which requires 18 units, but only 9 are still available from B3. Therefore, these 9 are allocated to M2.

Penalty Matrix:

	M1	M2	M6	Row Penalty
B1	14	24	16	2
B2	20	14	14	0
B3	16	11	45	(5)
Column Penalty	2	3	2	

Allocation Matrix:

	M1	M2	M4	M5	M6	Available
B1						22
B2			10			8
B3		(9)	20			9
Required	16	18	0	0	5	

Operations: With row B3 eliminated and the penalties recalculated, M2 shows the highest penalty. 9 units are required but only 8 are available in the low cost row B2.

Penalty Matrix:

	M1	M2	M6	Row Penalty
B1	14	24	16	2
B2	20	14	14	0
Column Penalty	6	(10)	2	

Allocation Matrix:

	M1	M2	M4	M5	M6	Available
B1						22
B2		(8)		10		8
B3		9	20			0
Required	16	9	0	0	5	

Operations: After the allocation of 8 units to B2,M2, another penalty matrix reduction serves no purpose because there is no choice for the remaining allocations. To meet the rim conditions, 16 of the 22 units available from B1 are rationed to M1, 1 to M2, and 5 to M6.

Allocation Matrix:

	M1	M2	M4	M5	M6	Available
B1	(16)	(1)			(5)	22
B2		8	10			0
B3		9	20			0
Required	16	1	0	0	5	

6-7 OPTIMAL SOLUTION—
STEPPING-STONE METHOD

An optimal solution is the "best possible" course of action under the restrictions imposed by the problem. When the objective is to minimize cost, the optimal solution provides the lowest cost. Optimality in a transportation problem is tested by trying to locate less expensive distribution routes. Each cell left unoccupied in the initial feasible solution is a potential new route. By calculating the cost of using each new route and comparing it with corresponding allocations from the initial solution, we can determine whether any changes will provide a less expensive distribution pattern. This approach, called the **stepping-stone method,** determines whether a change is advisable, where it should be made, and how much is saved by making it.

The "stepping-stone" name follows naturally from the way the method checks for optimality. A matrix can be thought of as a number of interconnected horizontal and vertical channels. Quantities can flow either way in the channels but cannot cut across them. By further assuming that right-angle turns to change channel passages can be made only at occupied cells, we figuratively have the stepping-stone method: turning points (occupied cells) are stepping-stones, and we check the cost of using each stone as alternative distribution routes are traced.

The logic behind the stepping-stone method is apparent in its application. Starting with the initial feasible solution developed in Figure 6.11, we can select any unoccupied cell for the first check. The low transportation cost in $B2$, $M6$, ($14) appears to be a good candidate for a lower-cost route. The check is made by calculating the cost of changing the distribution pattern to accommodate the shipment of *one* unit (truckload) from $B2$ to $M6$. Adding one unit to $B2$, $M6$ violates the rim conditions. To put the supply-and-demand relationships back in balance, we have to subtract one unit from the previous allocations to $B1$, $M6$ and $B2$, $M2$. Now the $B2$ row is in balance, but column $M2$ is one unit short. By adding one unit to $B1$, $M2$, all the rim conditions are again satisfied. We have thus used the occupied cells as stepping-stones to make a complete circuit from the cell being tested. The addition–subtraction–addition–subtraction routine of closing the quantity-transfer loop is shown in Figure 6.12.

After the transfer circuit is identified, it is a simple matter to calculate the net cost of the one-unit shift. Wherever a unit is added to a cell, that cell is charged with the transportation cost for the extra unit. Conversely, a deletion of one unit from a cell creates a savings in the amount of one unit's transportation cost. The positive and negative charges can be calculated directly from the matrix or tabulated in T-accounts, as shown in Figure 6.12. By either method, the end result is one figure that tells whether it costs more or less to include the tested cell in the distribution pattern. If the extra cost exceeds the savings in the complete circuit, no changes should be made. When the negative charges (savings) are larger than the positive costs, the tested cell should be included in a revised solution.

A solution is improved by transferring as many units as possible to the

The highest cost cells in a matrix seldom provide less expensive routes and thereby deserve only a cursory check.

A negative transfer cost (saving) is a positive opportunity cost because it represents a cost incurred by not selecting the best possible alternative. Similarly, an extra cost is a negative opportunity cost.

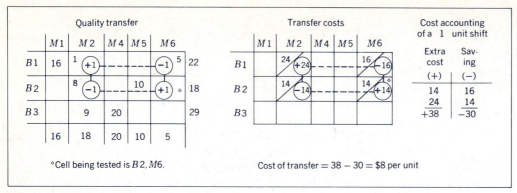

FIGURE 6.12 Stepping-stone transfer circuit and associated cost.

vacant cell that allows the greatest savings. The maximum number of units that can be transferred is limited by the smallest number in a negative cell in the transfer circuit. Using this number reduces to zero the quantity in the limiting negative cell and adds the same amount to the previously unoccupied cell. All cells in the circuit receive similar treatment, either increasing or decreasing by the limiting quantity. The total gain achieved by the shift is equal to the number of units transferred multiplied by the savings per unit.

The results from testing all unoccupied cells in the sample are given in Figure 6.13. The arbitrary order of testing unoccupied cells is shown by the circled numbers in the matrix; the corresponding cells in the test circuit are given with the tested cell first in the sequence. In this example no refinements are necessary for the solution obtained by VAM. The minimum daily transportation cost expected from building the new bakery at site $B3$ is $1059.

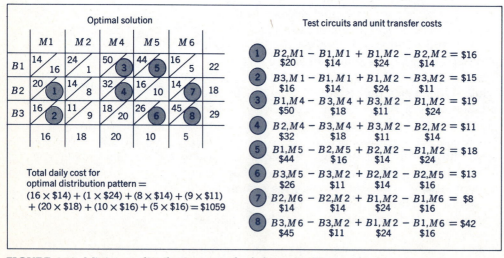

FIGURE 6.13 Minimum distribution costs for bakery site $B3$.

6-8 IRREGULARITIES IN DISTRIBUTION PROBLEMS

Very few distribution problems allow such a neat, direct solution as encountered in the previous section. Sometimes the problem cannot be formulated so compactly, or difficulties enter into the solution procedures. The more common irregularities and the means to overcome them are presented here.

Dependency

After studying the stepping-stone method and the employment of closed transfer circuits, the necessity of independent locations for initial allocations should be apparent. If a **dependent** allocation pattern slips through a check of the initial solution, it will assuredly be discovered during revisions: a transfer circuit that includes an unoccupied cell plus an *even* number of occupied cells is telltale evidence of initial dependent locations. The cure is to rearrange the allocations by inspection and then to proceed with the routine checks for unoccupied cells.

If the necessity is not clear, try to test an unoccupied cell in the margin matrix with the dependent positions in Section 6-6.

Degeneracy

When a solution has fewer than $O + D - 1$ allocations, it is **degenerate.** Both the symptoms and the cure of degeneracy are similar to those for dependency. An unsuccessful struggle to close a transfer circuit is a symptom of degeneracy; a count of occupied cells can confirm the condition. The cure is to assign an infinitesimal **quantity,** called **epsilon** (ϵ), to a promising unoccupied cell. Epsilon has the unusual but convenient properties of (1) being large enough to treat the cell in which it is placed as occupied and (2) being small enough to assume that its placement does not change rim conditions.

$O + D - 1$ allocations mean that the number of occupied cells is one less than the number of rows and columns.

$$1 + \epsilon = 1$$
$$1 - \epsilon = 1$$
$$\epsilon - \epsilon = 0$$

Degeneracy can develop in either an initial or a revised solution. In both cases it is treated in the same way. One or more cells receive epsilon allocations to make the total number of occupied cells equal to $O + D - 1$. The cells to receive epsilon allocations should be carefully chosen because a transfer circuit cannot have a negative epsilon stepping-stone. The reason, of course, is that you cannot subtract a unit from a cell that contains only an infinitesimal part of a unit. Therefore, epsilon quantities should be introduced in cells that accommodate the solution procedures.

A final optimal solution can be degenerate.

The second alternative available to the bakery chain described in the previous section illustrates degeneracy during the initial feasible solution. VAM is applied to a matrix including data for the alternative bakery site, $B4$. Instead of using reduced matrices for calculating the penalties as illustrated in Figure 6.11, the successive penalty calculations are shown outside the rim requirements in Figure 6.14. The circled numbers in the matrix indicate the order in which the allocations were made. For instance, after ③ allocations, column $M1$ has the highest penalty (21) in round ④ of the penalty calculations. Therefore, 16 of the remaining 17 units (5 units from $B1$ were allocated to $M6$ in allocation ②) are assigned to $M1$ in allocation ④ . After this allocation no further penalty calculations are needed because the remaining units must be allocated according to unfilled rim requirements.

The data for Figure 6.14 are from Table 6.5.

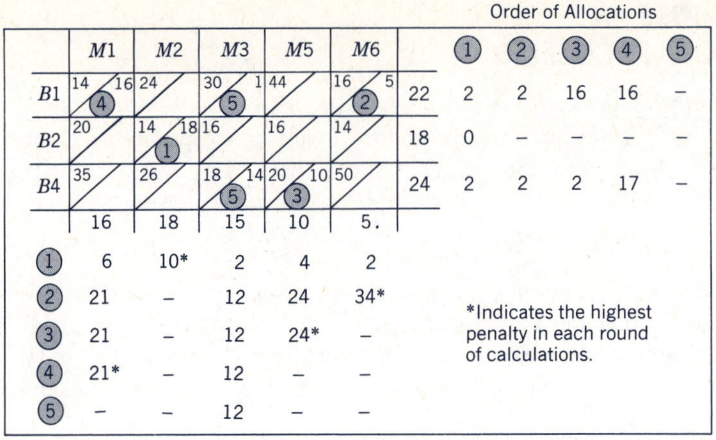

	M1	M2	M3	M5	M6		Order of Allocations ①	②	③	④	⑤
B1	14 \16 24		30 \1 44		16 \5	22	2	2	16	16	–
	④		⑤		②						
B2	20	14 \18 16	16	14		18	0	–	–	–	–
		①									
B4	35	26	18 \14 20 \10 50			24	2	2	2	17	–
			⑤ ③								
	16	18	15	10	5.						
①	6	10*	2	4	2						
②	21	–	12	24	34*						
③	21	–	12	24*	–						
④	21*	–	12	–	–						
⑤	–	–	12	–	–						

*Indicates the highest penalty in each round of calculations.

FIGURE 6.14 Degenerate initial solution by VAM.

The initial solution is degenerate owing to 6 instead of $3 + 5 - 1 = 7$ occupied cells. A logical improvement is to eliminate the high-cost $B1$, $M3$ route. The unoccupied cell $B2$, $M3$ makes a good starting point. In the existing degenerate condition, it is impossible to make a stepping-stone circuit around $B2$, $M3$. An epsilon quantity must be introduced. Epsilon could be added to $B2$, $M1$ to complete a transfer circuit, as shown in Figure 6.15a, but the epsilon stepping-stone is negative and consequently allows only an inconsequential transfer. By placing epsilon in $B1$, $M2$, it becomes a positive transfer cell in the circuit $+ B2$, $M3 - B1$, $M3 + B1$, $M2 - B2$, $M2$. A $(+ 16 - 30 + 24 - 14 =) -\$4$ cost or $\$4$ saving results from the transfer of one unit (see Figure 6.15b). This transfer further increases the number of occupied cells by one and thereby eliminates the degeneracy.

The maximum number that can be transferred in a circuit is the smallest allocation in a negative cell.

After each revision it is necessary to check for possible additional improvements. With the distribution pattern altered to include the revision from Figure

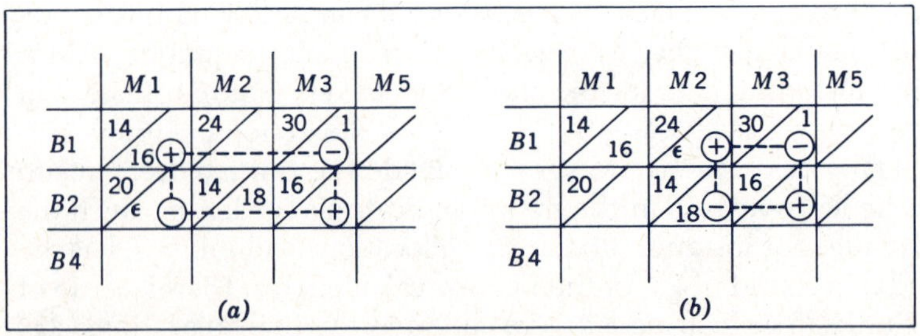

FIGURE 6.15 Introduction of ε to relieve degeneracy. (a) Incorrect ε placement. (b) Correct ε placement.

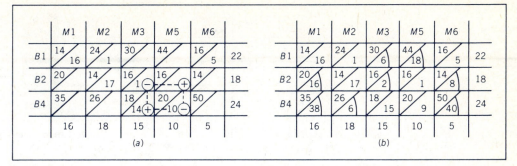

FIGURE 6.16 Revised and optimal solution to the bakery location problem. (*a*) Distribution pattern after first revision and transfer circuit for next revision. (*b*) Allocations after second revision and final, optimal distribution pattern.

6.15, one route, $B2$, $M5$, still provides further savings:

$$+ B2, M5 - B4, M5 + B4, M3 - B2, M3$$
$$+ 16 \qquad - 20 \qquad + 18 \qquad - 16 \qquad = -\$2$$

This circuit and the maximum possible one-unit shift it provides are shown in Figure 6.16.

The second and final revision reveals the optimal solution; all remaining routes exhibit greater transfer costs, as shown in the lower left corner of each unoccupied cell of the final matrix. The total transportation cost expected from locating the new bakery at site $B4$ is $\$1032 = (14 \times \$16) + (1 \times \$24) + (17 \times \$14) + (1 \times \$16) + (15 \times \$18) + (9 \times \$20) + (5 \times \$16)$. Selection of the $B4$ location over the $B3$ alternative provides daily distribution savings of $\$1059 - \$1032 = \$27$.

	M1	M2	M3	M5
B1		−1	+1	
B2		+1		−1
B4			−1	+1

Note the lopsided figure 8-shaped circuit necessary to check $B1$, $M3$ in Figure 6.16*b*.

Alternative optimal solutions

Instead of a positive transfer cost in all the unoccupied cells, as was the case in Figure 6.16, one or more cells may show zero cost. A cell with a zero transfer cost denotes an alternative, equal-cost routing. The inclusion of such a cell in the distribution pattern is decided by intangible factors not shown in the transportation matrix. A preference for a certain route could depend on loading convenience at the origin, reliability of the carrier, or service expected at the destination. Equivalent alternatives are valuable for the flexibility they allow.

Maximization problems

When origins are related to destinations by a profit function instead of cost, the objective is to maximize instead of minimize. The most convenient way to deal with the maximization case is to transform the profits to *relative costs*. The trans-

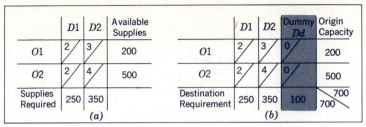

	D1	D2	Available Supplies
O1	2 /	3 /	200
O2	2 /	4 /	500
Supplies Required	250	350	

(a)

	D1	D2	Dummy Dd	Origin Capacity
O1	2 /	3 /	0 /	200
O2	2 /	4 /	0 /	500
Destination Requirement	250	350	100	700 / 700

(b)

FIGURE 6.17 Supply and demand balanced by a dummy destination. (*a*) Unbalanced rim conditions. (*b*) Balanced rim conditions.

formation is conducted by subtracting the profit associated with each cell from the largest profit in the matrix. In this way the largest profit shows a zero relative cost, and the objective changes to minimizing these relative costs. The solution procedures for relative costs are identical to those already described. After an optimal solution to the transformed minimization problem is determined, relative costs in occupied cells are restated in their original profit functions, and the total profit is calculated according to quantities in the optimal distribution pattern.

Unequal supply and demand

Actual capacity available at the origins and demand requirements of the destinations are seldom equal. There may be practical reasons other than distribution expense that certain demands should be met before others or that some supply capacity should be conserved. The reasons could evolve from strategic or tactical policy considerations and be used to balance available units politically to required units. When policy reasons are lacking, the mathematics of the solution procedures can point out where shortages or overages should exist so as to minimize the total distribution cost.

Reasoning similar to charging zero transportation cost to a dummy is behind the practice of charging an artificially huge transportation cost to a cell in a route that is not open or undesirable.

Excess supply capacity is accounted for in a transportation matrix by a dummy destination, as shown in Figure 6.17. The demand of the dummy destination is equal to the difference between the total available and total required supplies. The cost of shipping from any origin to the dummy destination is assumed to be zero. The zero cost is rational because it represents an expense-free, fictional transfer and forces an allocation to the dummy. Excess demand is treated equivalently; a dummy origin is introduced to provide fictitious supplies, and zero transportation costs are charged to supplies from the dummy. After the origin capacities are balanced with the destination requirements, the matrix is amenable to the usual solution procedures. The solution to the problem in Figure 6.17 reveals that the dummy destination is allotted 100 units from O2. The physical interpretation of the solution is that O2 ships 250 units to D1, ships 150 units to D2, and retains the last 100 units at the origin.

EXAMPLE 6.4 Distribution problem in which demand exceeds supply and the objective is to maximize profit

An area manager for three supermarkets (M1, M2, and M3) can obtain fresh apples from cold-storage lockers at four fruit cooperatives (C1, C2, C3, and C4). He needs 200 boxes for M1, 600 boxes for M2, and 400 boxes for M3. It is late in the apple season, and so supplies are limited. C1, C3, and C4 each have 400 boxes available, and C2 has 300. The quality of apples, wholesale costs, and transportation costs are reflected in the following profit he expects to make per box from each source:

	C1	C2	C3	C4
M1	$8	$ 3	$2	$7
M2	$9	$11	$5	$2
M3	$6	$10	$6	$4

From the statement of the problem and a little preliminary arithmetic, it is apparent that the profits must be transformed to relative costs and that a dummy market should be introduced to absorb the excess supply of apples. The corrective measures for the irregularities are shown on page 204, where an initial solution is developed by applying VAM.

A check of the unoccupied cells in the initial feasible solution reveals two transfer circuits which allow a $3-per-box saving in relative cost. Either circuit (as shown) could be used for the first revision, because both lead to the same final solution. A choice of M2, C3 leads to the transfer circuit and allocations shown for the second revision in the sequential solution matrices. The indicated transfer circuit for M3, C3 is the second revision which in turn leads to the distribution pattern for the third revision. After 200 boxes for M3 are switched from C4 to C3 in the third revision, the final solution is obtained; optimality is

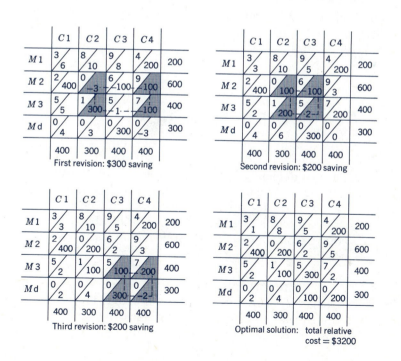

First revision: $300 saving

Second revision: $200 saving

Third revision: $200 saving

Optimal solution: total relative cost = $3200

	C1	C2	C3	C4		Penalty			
M1	3	8	9	4 / 200	200	1	1	1	5*
M2	2 / 400	0	6 / 100	9 / 100	600	2	2	4*	3
M3	5	1 / 300	5	7 / 100	400	4	4*	0	2
Md	0	0	0 / 300	0	300	0	—	—	—
	400	300	400	400	1500				

Penalty	2	0	5*	4
	1	1	1	3
	1	—	1	3
	—	—	1	3

*Indicates the highest penalty in each round of calculations.

confirmed by the extra transfer costs associated with all unoccupied cells. Finally, relative costs are restated in profit per box to disclose the maximum expected gain. Total profit (from the optimal distribution pattern and the profit function) is

200 boxes from C4 sold at M1 for $ 7 per box	$1400
400 boxes from C1 sold at M2 for $ 9 per box	$3600
200 boxes from C2 sold at M2 for $11 per box	$2200
100 boxes from C2 sold at M3 for $10 per box	$1000
300 boxes from C3 sold at M3 for $ 6 per box	$1800
	$10,000

6-9 AGGREGATE PLANNING

In a healthy organization, planners concurrently lay out detailed schedules for the next several days' activities, plan how best to balance total available resources against demands expected for weeks and months into the future, and develop strategies to follow several years hence. In this section we consider intermediate-range plans for operations a month to a year ahead. Within this range, the physical facilities are assumed to be fixed for the planning period. Therefore, fluctuations in demand must be met by varying labor and inventory schedules. **Aggregate planning** seeks the best combination to minimize costs.

Planning inputs

Aggregate planning for service organizations primarily concerns the labor force, whereas in manufacturing both the work force and the product output are involved.

Demands for products or services are seldom constant over several months. A prerequisite to aggregate planning is the development of a forecast. The predictions typically concern the overall level of demand, not broken down to a specific mix within a product line. For example, the forecast could estimate the number of gallons of a drink to produce without specifying the size of bottles to put it in, or the number of hospital beds that will be occupied without detailing whether they are in single rooms, double rooms, or wards. In most planning models, once the forecast is accepted as reasonably accurate, predicted demands are treated as certainties.

Planners need to know what options are available to meet demand variations and the costs of the options. Options vary among organizations because of different management policies. For instance, a policy might be to never run out of stock. A more lenient policy could be to allow a stockout occasionally while making sure that customers' orders are filled within two weeks. The costs of the options are not always recorded in accounting records. In the case of later deliveries to a customer, what will the customer's displeasure eventually cost the supplier in terms of extra expense to handle complaints and possible loss of future orders? Such real but difficult-to-measure costs must be collected, along with readily available data on inventory expense, overtime charges, and similar costs, in order to develop a reliable production plan.

The purpose of the production plan is to smooth out or eliminate sporadic disruptions in operations caused by fluctuations in demands. This is accomplished by scheduling operations to meet demand patterns over several periods in the future. Suppose that the demand for a product declined for six months and then rose for six months. If the production manager had no way to anticipate the demand pattern, he or she would probably decrease output each month by laying off workers or going on a reduced workweek. Then as the demand rose, the manager would likely hire, institute overtime, or fall behind in meeting orders. These short-term reactions can hurt the morale of the work force, reduce productivity, and increase personnel costs. An extended planning horizon allows the manager to weigh the effects of different ways to match production rates to sales demands, and to select the plan that minimizes disruptions.

Other options include building up inventory levels during a decline and not accepting new orders when operating at capacity.

Sometimes less-than-desirable tactics are inevitable when forecasts miss badly, but the long-term effects must still be considered.

Planning strategies

Combinations of four controllable inputs to the production process are commonly used in aggregate planning. Each strategy has pros and cons.

1. *Vary the size of the work force.* Output is controlled by hiring or laying off workers in proportion to changes in demand. Costs of hiring cover recruiting, interviewing, examinations, new records, training, and a period of low productivity while workers are becoming familiar with job conditions. Layoff costs include expenses for unemployment compensation and severance pay, plus the subtle effects of lower morale in the remaining work force and weakened public relations.

Placement problems are discussed in Section 8-3.

2. *Vary the hours worked.* Overtime and shortened workweeks raise or lower production rates. Direct costs of overtime are well known, but the effects of lower efficiency from working longer days are not so apparent. Wages saved from using less than a 40-hour week are also obvious, but the del-

Wage considerations are presented in Section 10-6.

eterious effects, such as attrition of the work force, are less visible. Undertime periods, when employees are paid but remain idle owing to lack of work, can be another major cost.

3. *Vary inventory levels.* Demand fluctuations can be met by stockpiling items during periods of lower demand and then drawing from the stockpile for higher demands. Costs of holding stock include insurance, interest charges, damage, and taxes. If the accumulated stock plus current production cannot meet demand, costs will be incurred for backordering and customer dissatisfaction.

4. *Subcontract.* Upward shifts in demand from low-level, constant-production rates can be met by using subcontractors to provide extra capacity. The apparent cost of subcontracting is the difference between the subcontracting company's price per unit and the marginal cost of manufacturing above the given production rate. Quality and delivery time problems may contribute to higher subcontracting costs.

EXAMPLE 6.5 Comparison of pure production strategies for meeting fluctuations in product demands

A "pure" strategy uses only a single alternative rather than a combination. Rarely, if ever, would pure strategies be used in practice, but insights can be gained by observing their hypothetical application.

Suppose that a company is starting to produce a new product. The company policy is to maintain a level work force. Consequently, the alternative of hiring and laying off workers during demand fluctuations is eliminated. For brevity, the analysis is limited to a six-month planning period, although the company would probably use a year because an annual plan covers the full range of sales variations. Relevant data are listed in the adjacent column.

Sales Period, Month	Demand, Units	Cost Data and Conditions
1	1300	No initial inventory is on hand
2	850	Regular manufacturing cost = $100 per unit
3	1200	Overtime manufacturing cost = $150 per unit
4	1450	Subcontracting cost = $130 per unit
5	900	Holding cost for inventory = $2 per unit per month
6	800	Backorder cost = $10 per unit per month

PLAN 1 *Use a minimum-size work force, allow no backorder or holding costs, and subcontract for additional capacity as needed.*

A policy not to use inventory accumulation could result from a lack of storage facilities. The policy of limiting the initial size of the labor force is a conservative tactic for producing a new product. Therefore, a minimum capacity of 800 units (the lowest monthly demand) is used, and the production of additional units to exactly meet the demand each month is subcontracted. It is assumed that subcontractors have the necessary capacity and can produce at less than the company's overtime cost per unit.

The total cost to produce 6500 units under Plan 1 is $530,000 + $156,000 = $686,000.

Month	Sales Demand	Regular Units*	Production Cost	Production Units	Subcontract Cost
1	1300	900	$ 90,000	400	$ 52,000
2	850	850	85,000	0	
3	1200	900	90,000	300	39,000
4	1450	950	95,000	500	65,000
5	900	900	90,000	0	
6	800	800	80,000	0	
	6500	5300	$530,000	1200	$156,000

*The number of units produced per month with a constant work force varies with the number of working days in a month.

Month	Sales Demand	Regular Units	Production Cost	Backordered Units	Backorder Cost
1	1300	1100	$110,000	200	$2000
2	850	1050	105,000	0	
3	1200	1100	110,000	100	1000
4	1450	1150	115,000	400	4000
5	900	1150	115,000	150	1500
6	800	950	95,000	0	
	6500	6500	$650,000	850	$8500

PLAN 2 *Use a work force large enough to meet the annual sales demand without overtime or subcontracting, and allow inventory-holding and backorder costs.*

The size of the work force is based on an estimated average monthly demand of 1100 units (6500/6 ≐ 1100) with actual production varying from 950 to 1150 units, depending on the number of working days available each month. It is assumed that orders not filled during a month can be carried over to succeeding months without danger of cancellation.

The cost of Plan 2 is $650,000 + $8500 = $658,500.

The apparent $686,000 − $685,500 = $27,500 advantage of Plan 2 over Plan 1 shows the potential for using stock carry-overs for the relatively stable demand pattern in the example. However, if a buffer stock of 400 units could have been produced before taking orders in Plan 2, no stockouts would have occurred, and the total inventory-holding cost would have been (400 + 200 + 400 + 300 + 250 units)($2/unit) = $3100. Unfortunately, many additional factors restrict actual production capabilities and thwart the use of pure strategies in aggregate planning.

6-10 DISTRIBUTION MODEL FOR PRODUCTION PLANNING

Trial-and-error methods relying on tables and charts were used exclusively for aggregate planning before the advent of mathematical techniques in the 1950s. Production planning to meet sales demand was structured as a distribution

"Production Planning by the Transportation Method of Linear Programming," by Bowman, appeared in the *Journal of the Operations Research Society* in February 1956.

model by E. F. Bowman. Destination columns of the familiar distribution matrix become demand periods, and the supply rows take the form of various ways to produce the demanded units. The costs of production and inventory take the place of transportation costs, and resource limits establish the rim conditions. As shown in Figure 6.18, demand fluctuations, initial and ending inventory levels, backorder and holding costs, and production capacities from the regular work force, overtime, and subcontracting can be represented in the matrix.

The symmetry of the distribution matrix quickly exposes its construction. Consider the top row, beginning inventory. If the initial supply of units (I_0) is not used to meet demand in the first period, a holding cost of h dollars will be charged. If I_0 is held for two periods, the total holding charge will be $2h$ and so on, to the planning horizon, nh. All rows are treated similarly. If the units

	Period 1	Period 2	Period 3	$\cdots$	Inventory at the end of period n	Unused capacity	capacity
Beginning Inventory	0	h	$2h$	$\cdots$	nh	0	I_0
Regular 1	r	$r+h$	$r+2h$	$\cdots$	$r+nh$	0	R_1
Subcontract 1	s	$s+h$	$s+2h$	$\cdots$	$s+nh$	0	S_1
Overtime 1	t	$t+h$	$t+2h$	$\cdots$	$t+nh$	0	T_1
Regular 2	$r+b$	r	$r+h$	$\cdots$	$r+(n-1)h$	0	R_2
Subcontract 2	$s+b$	s	$s+h$	$\cdots$	$s+(n-1)h$	0	S_2
Overtime 2	$t+b$	t	$t+h$	$\cdots$	$t+(n-1)h$	0	T_2
Regular 3	$r+2b$	$r+b$	r	$\cdots$	$r+(n-2)h$	0	R_3
Subcontract 3	$s+2b$	$s+b$	s	$\cdots$	$s(n-2)h$	0	S_3
Overtime 3	$t+2b$	$t+b$	t	$\cdots$	$t+(n-2)h$	0	T_3
$\vdots$	$\vdots$	$\vdots$	$\vdots$	$\cdots$	$\vdots$	$\vdots$	$\vdots$
	Demand 1	Demand 2	Demand 3	$\cdots$	Final inventory, I_n	Dummy demand	Grand total

FIGURE 6.18 Distribution matrix for aggregate planning, in which r = production cost per unit for regular size work force, s = subcontracting cost per unit, t = overtime cost per unit, h = holding cost per unit of inventory per pay period, b = backorder cost per unit per period of delay, and n = number of periods in the planning horizon.

produced by the regular work force (R_1) in period 1 are not assigned, they will be charged holding costs until they are allocated to a monthly demand.

When insufficient items are produced to meet demand, a backorder charge is made, b. The charge accumulates every month. For example, if the demand for period 1 is not satisfied until period 3, a backorder charge for two periods ($2b$) will be added to the production cost (r_3, s_3, or t_3) in column 1. A policy of no backorders blocks out all assignments to each column below the current production period.

The total production capacity given in the matrix is the sum of the beginning inventory (I_0), regular time $\left(\sum\limits_{i=1}^{n} R_i \right)$, subcontracting $\left(\sum\limits_{i=1}^{n} S_i \right)$, and overtime $\left(\sum\limits_{i=1}^{n} T_i \right)$. This total capacity customarily exceeds the total sales demand. Therefore, to balance the matrix, a column is added that accounts for the unused capacity. The cost for an assignment to this dummy demand is, of course, zero.

An application of the distribution method to aggregate planning is demonstrated in Example 6.6. Note that restrictions are placed on the amount to be produced in each period from each source and that additional sources (such as second and third shifts) could be incorporated in the model.

EXAMPLE 6.6 Application of the distribution method to production scheduling

The company described in Example 6.5 has recognized additional restrictions on its future production. The demand and cost data are unchanged, but the regular work force is limited to 60 percent of that anticipated in Plan 2; overtime cannot exceed 25 percent of regular time; and the capacity of the subcontractor is confined to 300 units per month for the first three months and 450 units for the last three. An initial inventory of 300 units was stocked from the subcontractor before period 1, and a stockpile of 400 units is desired at the end of period 6. These restrictions are shown as rim conditions in the distribution matrix of Figure 6.19. For instance, the top entry in the right-hand column shows the beginning inventory level of $I_0 = 300$ units and is followed by the number of units that can be produced in period 1 by the regular work force ($R_1 = 660$), subcontractor ($S_1 = 300$), and overtime ($T_1 = 165$). Capacities for successive periods are listed vertically. Sales demands per month and the final inventory, $I_n = 400$ units, are listed horizontally at the bottom.

Assignments in the matrix are calculated and checked according to the usual procedures of the distribution method. Unused capacity is allocated to the dummy demand. Matrix entries indicate the following production schedule and a total cost of $752,230.

Type of Production	Period (Month)					
	1	2	3	4	5	6
Regular	660	630	660	690	690	570
Subcontract	300	300	300	450	450	450
Overtime	40	75	165	170	0	0

As highlighted in Figure 6.20, the schedule causes a delivery delay of 60 units at the fourth month and so uses inventory from four of the sales periods to meet the 1450 demand.

	Month 1	Month 2	Month 3	Month 4	Month 5	Month 6	I_n	Unused capacity	
I_0	0 / **300**	2	4	6	8	10	12	0	300
Regular 1	100 / **660**	102	104	106	108	110	112	0	660
Subcontract 1	130 / **300**	132	134	136	138	140	142	0	300
Overtime 1	150 / **40**	152	154	156	158	160	162	0	/ 125	165
Regular 2	110	100 / **630**	102	104	106	108	110	0	630
Subcontract 2	140	130 / **220**	132 / **80**	134	136	138	140	0	300
Overtime 2	160	150	152	154 / **75**	156	158	160	0	/ 85	160
Regular 3	120	110	100 / **660**	102	104	106	108	0	660
Subcontract 3	150	140	130 / **300**	132	134	136	138	0	300
Overtime 3	170	160	150 / **160**	152 / **5**	154	156	158	0	165
Regular 4	130	120	110	100 / **690**	102	104	106	0	690
Subcontract 4	160	150	140	130 / **450**	132	134	136	0	450
Overtime 4	180	170	160	150 / **170**	152	154	156	0	170
Regular 5	140	130	120	110	100 / **690**	102	104	0	690
Subcontract 5	170	160	150	140 / **60**	130 / **210**	132	134 / **180**	0	450
Overtime 5	190	180	170	160	150	152	154	0	/ 170	170
Regular 6	150	140	130	120	110	100 / **570**	102	0	570
Subcontract 6	180	170	160	150	140	130 / **230**	132 / **220**	0	450
Overtime 6	200	190	180	170	160	150	152	0	/ 140	140
	1300	850	1200	1450	900	800	400	520	7420

FIGURE 6.19 Distribution matrix for a six-month planning period with capacity limitations on regular, subcontracted, and overtime production. Backorders are not restricted.

Minor modifications of the solution shown in Figure 6.20 can be made by switching allocations within closed loops of the distribution matrix that do not alter the minimum total cost. Alternative schedules can be analyzed by modifying the estimated costs or capacities and observing any allocation changes that result from reworking the matrix. For instance, backorders can be eliminated by assigning a prohibitive cost for b, and s can be raised to see the effect on the production pattern if subcontracting costs increase.

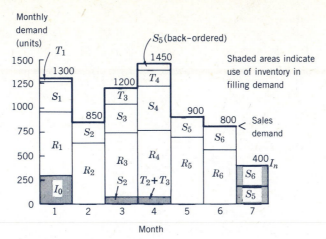

FIGURE 6.20 Production sources utilized to satisfy month-by-month sales demand.

6-11 MORE MATHEMATICAL MODELS FOR AGGREGATE PLANNING

LINEAR PROGRAMMING MODEL

The transportation model is a sample of the linear programming approach to production planning. The general LP models relax some of the restrictions imposed by the transportation format. A significant improvement is the capability of including the cost of changing the production rate, as in going from one work-force size to another. A possible extension is the inclusion of several products at several locations in a single analysis. Numerous specialized versions have been developed to better represent physical situations, but the increased realism comes at the expense of more computational effort.

Once the problem has been formulated, the simplex method described in Appendix F can be used to obtain a solution.

LINEAR DECISION RULE

The use of linear equations as **linear decision rules** was forwarded in the mid-1950s as a method of specifying optimal production levels and work-force size.[2] Separate quadratic cost curves were developed for employee hiring and layoff, overtime, and inventory holding and shortages. By differentiating these cost functions and combining them with regular payroll costs, two basic expressions result: One is a rule that defines the size of the work force for the period, and the other rule sets the production rate for the same period. The form of the expression is shown in the following equation:

The developers were C. C. Holt, F. Modigliani, J. F. Muth, and H. A. Simon. Results were first reported in "A Linear Decision Rule for Production and Employment Scheduling," *Management Science*, Vol. 2, October 1955.

$$P_t = a_0 S_t + a_1 S_{t+1} + a_2 S_{t+2} + \cdots + bW_t + c - dI_{t-1}$$

$$W_t = e_0 S_t + e_1 S_{t+1} + e_2 S_{t+2} + \cdots + fW_{t-1} + g - hI_{t-1}$$

[2]The developers were C. C. Holt, F. Modigliani, J. F. Muth, and H. A. Simon. The results were first reported on "A Linear Decision Rule for Production and Employment Scheduling," *Management Science* 2 (October 1955).

where

$$P_t = \text{production rate for period } t$$

$$W_t = \text{size of the workforce for period } t$$

$$S_j = \text{forecast of the demand in period } j$$

a's, b, c, d, and e's, f, g, and h = constants set according to statistical estimates from accounting data and subjective estimates from management

The two decision rules are tedious to develop initially, but repetitive applications are simple. In tests, the rules lowered costs significantly from those encountered under the existing management procedures. However, in situations in which quadratic cost functions were not representative of the actual cost structure, the rules did not perform so well.

MANAGEMENT COEFFICIENTS APPROACH

"Consistency and Optimality in Management Decision-Making," by E. F. Bowman, published in *Management Science,* Vol. 4, January 1963.

A procedure for modeling management decision making by applying regression analysis to past decisions was advocated by Bowman.[3] He suggested that it might be a better policy to improve on past managerial decisions rather than to introduce new, explicit models for "optimal solutions." His **management coefficient model** is supported by the logical argument that managers must be aware of what factors are important to their work and that they must have a "feel" for proper solutions, or they would not be holding their current positions. Therefore, good guidelines for the future are available from records of past performance, from which hasty or emergency actions of lower quality have been eliminated.

The feedback principle on which this approach is built makes it suitable for many types of management activities in addition to aggregate planning.

For a given repetitive situation, the first step is to express the decision factors in terms of observable or predictable values. A formula for the production rate might take the following form:

$$P_t = aW_{t-1} - bI_{t-1} + cF_{t+1} + K$$

which implies that the rate in period t depends on the size of the work force and the inventory on hand during the previous period ($t - 1$), and the demand forecast for the next period ($t + 1$) plus some constant K. The critical values for the coefficients, a, b, and c are derived from actual historical values of P, W, I, and F. In practice, the management coefficient approach has been effective and has an intuitive appeal to managers that makes its implementation easier than that of more theoretical models. But difficulties are encountered in the subjective selection of the basic formulation of the decision rule, and the tendency for the method to retain undesirable bias that weakened previous decisions.

[3] E. F. Bowman, "Consistency and Optimality in Management Decision-Making," Management Science 4 (January 1963).

OTHER TECHNIQUES

Several investigators suggest that the actual cost function of a particular production situation should be formulated as accurately as possible and that then the variables should be systematically varied in the formulation until a reasonable solution is reached. This is a *simulation* approach. Computer-assisted simulation is a respected method for obtaining a *satisficing* solution to problems for which fully acceptable optimizing methods are unavailable. Although there is no guarantee that simulation will reveal the best solution, it does allow inclusion of more complex functions than are allowed by linear programming and linear decision rule procedures.

Parametric production planning is another search routine.[4] This model assumes two linear feedback rules, one for the work force and the second for the production rate. The rules are applied to likely sequences of sales demands to generate a series of labor-force levels and production rates. Then the relevant costs are developed from the actual costs of the organization being studied. A preferred solution is identified by applying a grid search technique. Both simulation and computer search models are especially promising for tailor-made applications to unusual or unwieldy aggregate planning situations.

A few aggregate planning models include learning effects. It is not uncommon for productivity rates to increase as additional production experience is gained. Increases are quantitatively described by learning curves or manufacturing progress functions. A combination of productivity growth with aggregate planning recognizes situations in which workers become more proficient with practical experience and thereby change work-force and material requirements, as compared with situations in which worker output is unchanging over time. Two of the models that incorporate a learning factor are described next.

1. Additional constraints are added to the conventional linear programming model for aggregate planning to assign increasing productivity rates.[5] Worker productivity is assumed to increase in proportion to time spent in production activities. Thus output and material requirements per person rise as long as the same people are doing the job. Keeping track of the progress of productivity gains causes the computational load for the refined LP model to enlarge: For a 12-month plan with six classes of work productivity, 288 variables and 168 constraints are required.

2. A learning function is directly used in a model proposed by Ebert.[6] It includes mathematical expressions to represent direct labor, variable labor overhead, hiring and firing, and inventory-carrying and shortage costs. The direct labor term reflects productivity improvement as a function of the cumulative production output. A search routine is used to solve the model.

Computer simulation is examined in Section 11-7.

"Parametric Production Planning" by G. H. Jones appeared in Management Science, *Vol. 13, July 1967.*

See Section 14-4 for a discussion of learning curves.

P. M. Wolf, et al., "Aggregate Planning Models Incorporating productivity—An Overview," Proceedings, *AIIE Conference, Spring 1979.*

R. J. Ebert, "Aggregate Planning with Learning Curve Productivity," Management Science, *Vol. 23, 1976.*

[4]C. H. Jones, "Parametric Production Planning," *Management Science* 13 (July 1967).
[5]P. M. Wolferal, "Aggregate Planning Models Incorporating Productivity—An Overview," *Proceedings, AIIE Conference,* Spring 1979.
[6]R. J. Ebert, "Aggregate Planning with learning Curve Productivity," *Management Science* 23 (1976).

The time–cost trade-offs and resource-leveling techniques discussed in the next chapter for project scheduling are additional ways to meet demand fluctuations. Some network-based approaches are fundamentally similar to the popular charting and tabular methods of aggregate planning. Traditional graphic methods still are probably the most-used approach, but quantitative methods are gaining rapidly.

6-12 CLOSE-UPS AND UPDATES

The efficient allocation of physical resources has a clear effect on bottom-line profitability, but the allocation and management of human resources are more critical to continued prosperity. The group to target in the total work force is the **knowledge-worker** segment. As noted in Section 5-9, this group of managers and professionals accounts for over two-thirds of the cost of white-collar employment which, in turn, accounts for about two-thirds of the total payroll in the United States.

The rewards for the astute utilization of knowledge workers are obviously huge. Because these workers have less-structured jobs than do other **white-collar employees** and **blue-collar workers,** their effectiveness is raised by building a work environment in which they thrive intellectually and socially and by encouraging them to use their time and talents more efficiently.

Knowing the knowledge workers

The knowledge-workers' work consists largely of gathering, analyzing, and transferring knowledge. They have many titles, among which are executives, production managers, engineers, accountants, and supervisors. Their job sites are offices. The tools of their trade are computers, telephones, and meeting places. Information is their energizing force. Having the "right" information marks their power. How well this knowledge is used is reflected by the success of their organization.

Production managers and supervisors are knowledge workers themselves and normally supervise knowledge workers. Yet they may have difficulty in relating to the occupational needs of their peers and subordinates. Characteristics of knowledge workers and contemporary changes in traditional work expectations are displayed in Table 6.6.

Time—Use it or lose it

Table 6.7 displays the results of two of many studies of how managers spend their time. The constant in all of them is that an eight-hour day is always the same length. Variables include lengthening the standard workday by stealing minutes or hours from other pursuits, adjusting to uncon-

Blue- and white-collar workers are distinguished by the nature and location of their jobs—blue-collars are associated with manual work in extraction, manufacturing, and construction, and white-collars appear in clerical, managerial, and professional jobs.

Automated equipment for knowledge workers was described in Sections 2-8 and 4-5.

Adapted from "Improving Use of Discretionary Time Raises Productivity of Knowledge Workers," by Harry P. Conn, *Industrial Engineering*, July, 1984.

TABLE 6.6
Profile of Knowledge Workers and Their Changing Expectations

Characteristics of Knowledge Workers	*Expectations of Contemporary versus Traditional Knowledge Workers*
1. *Type of work.* Generally engaged in nonrepetitive jobs. Requires higher level of education to accomplish assigned tasks. Relies heavily on problem-solving and communications skills.	Traditional employees are likely to have more loyalty to and to expect less from employers. Contemporary employees stress the importance of more interesting and creative jobs.
2. *Scope of work.* Can determine own work methods within broad limits. Actions frequently affect a large number of people both inside and outside the organization. Efficiency often depends on the adoption of new technologies and a willingness to accept innovative practices.	Today's greater mobility allows contemporary knowledge workers to place less emphasis on job security and more on fulfilling work. They are not as willing to accept deferred recognition or to delay satisfaction of their material wants.
3. *Work environment.* The production rate of knowledge workers is influenced by the support and challenge of associates. Being more self-assured and self-motivated, work changes tend to be accepted more readily. Work is seldom done in isolation.	Contemporary employees show greater interest in participative management techniques, and they want more information about their work with respect to its value to the overall operation of the organization.
4. *Output.* Performance is difficult to measure because output cannot be readily quantified. Feedback on major decisions may take months or even years.	Compensation is more rigidly linked to performance in the minds of contemporary employees. They are less likely to place work before family and leisure.

SOURCE: Adapted from Harry P. Conn, "Improving Use of Discretionary Time Raises Productivity of Knowledge Workers," *Industrial Engineering*, July 1984.

trollable conditions during the workday, and better managing controllable activities. Comparisons show, with remarkable consistency and irrespective of the nationality or type of industry, that the biggest wasters of knowledge workers' time are

- Telephone and drop-in-visitor interruptions.
- Meetings and impromptu socializing.
- Crises and responses to shifting priorities.
- The necessity of doing things that should be done by others or not done at all.
- Inability to say no to special interests.

Planning, analysis, and control all have an implied dimension of time: searching for it, slicing it, saving it, and even buying it, as described in the next chapter.

TABLE 6.7
Time Consumption Patterns of Japanese and American Managers

| Management Activities* | Percentage of Time Spent on Each Activity by | |
	Japanese Managers	American Managers
Attending scheduled meetings	24.4	13.1
Conferring with others	21.6	25.2
Reading	15.7	16.6
Writing	16.3	20.0
Traveling outside headquarters	8.5	13.1
Planning or scheduling	5.7	4.7
Other	7.8	7.2

SOURCE: Adapted from 1980 surveys conducted by the Japanese Management Association and by IBM in independent studies to study time management by middle- to upper-level managers.

*Each activity category includes related operations. For example, the 20-percent allocation for "writing" by American managers includes actual writing (9.8 percent), dictating (5.9 percent), proofreading (1.8 percent), calculating (2.3 percent), and copying (0.2 percent).

- Indecision and procrastination in routine affairs.
- Receipt and composition of unnecessary or unorganized reports.
- Confused objectives, priorities, or responsibilities.

There is no order of importance for this list because different professions and positions are exposed to different conditions, but the first three are nearly universal.

One of Murphy's laws, acclaimed for its profundity and painfully obvious, is that *everything takes longer than you think*. It suggests that things will require more time than you estimate for them, and correspondingly, when something is done, you find it took more time than you thought it would. The former implication suggests more disciplined planning, and the latter suggests more disciplined performance. Both are addressed in the following techniques for personal time saving.

Set priorities. Not all tasks are equally important. Distinguish among what *absolutely* has to be done, *should* be done, and *could* be done. Observe how time is parceled out all day and ration it to the best uses. This means not only prioritizing one's own activities but also not wasting other people's time.

A famous rule for reducing the time spent on correspondence is *touch it only once*.

Set deadlines. Start each day by listing the important activities for that day and the time by which they will be completed. The trick is to divide the important things into small goals that can be accomplished one after the other: divide and conquer.

Avoid procrastination. Everyone likes to do things that are fun, easy, and interesting. But many important activities fall in the less pleasant and more complex category. Reward yourself when a difficult obstacle has been overcome by taking a break or doing one of the fun jobs.

Be flexible but disciplined. Recognize that some time obligations are beyond personal control but must still be accommodated. Leave at least 20 percent of the day unplanned.

Know how to conduct a meeting. Posting the agenda in advance, starting promptly, maintaining the pace, and summarizing at the end are respected practices for conducting group meetings. What may be forgotten is that essentially the same preparations improve one-on-one meetings.

Surveys indicate that interruptions and meetings are the most frustrating time-consumers. One way to flag meeting cost is to add up the salaries of participants and be stunned by the cost per minute.

Hundreds of books and articles have been written about *time management.* The foregoing suggestions are merely a sample of the plethora of how-to's for the allocation of time. A sustained interest in time conservation is ensured by the growing availability of ever-faster machines to assist knowledge workers. Nearly instantaneous communications and faster production processes have placed a premium on managerial response time, which has the end effect of requiring managers either to work faster or to use their time better. Because haste traditionally makes waste, time is a prime resource for utilization optimization.

Making the most out of knowledge and time by using the nominal group technique

Recent generations of managers have been labeled by their appearance or the machines with which they worked. There have been crew-cut or gray-flannel managers as part of the automation era or computer generation. Now we have knowledge workers in an information age. Moreover, different managerial techniques are associated with each generation, as described in Example 1.1. Today the **nominal group technique (NGT)** is gaining popularity.

The nominal group technique capitalizes on the strengths of knowledge workers. It involves a participative process that relies on creativity and mental discipline. It is used for group efforts to solve problems, set program priorities, develop long-range plans, and determine resource allocations. The participants' direct involvement in the process leads to a

NGT dates back to 1968 when it evolved from studies by behavioral scientists and industrial psychologists.

NGT uses the insights derived from education and motivation of knowledge workers to spot opportunities.

sense of ownership—they form a commitment to the actions forwarded by NGT.

Procedures for NGT are highly structured so as to increase the productivity of group action. They are designed to stimulate ideation, facilitate aggregation of individual judgments, and reach conclusions that leave participants feeling satisfied. There are typically six phases in the process.

1. *Group preparation.* A group of 6 to 12 members is led by an experienced facilitator. After explaining the mechanics of the technique, a **task statement** is read. This carefully conceived statement expresses simply and without bias the issue to be confronted. The statement is briefly discussed to clarify the group's thinking.

2. *Silent generation of ideas.* Each participant is asked to list ideas or possible solutions for the issue. No interaction is allowed. Approximately 20 minutes is provided for this phase.

3. *Round-robin reporting.* The facilitator calls on participants one by one to state one of the responses he or she has written. A participant may pass on one round and come in again later. The facilitator or a scribe records on a flipchart each item as it is given and numbers them sequentially. No side comments or discussion are permitted.

4. *Clarification discussion.* Each item on the flipchart is discussed in turn. The purpose of the discussion is to clarify and eliminate duplications and broad generalities. Differing points of view can be expressed, but no arguments are allowed on the merits of any item.

5. *Ranking and voting.* Each person selects a predetermined number of items from the flipcharts (five, for example) that are believed to be the best. The selected items are recorded on 3 × 5 cards by their number from the flipchart and a few key words. Then each person is asked to lay down all five cards face up. The most important item is chosen, and a score of five is given to it. After writing in the five, the card is turned over. The least important item is determined next, given a score of one, and the card is turned over. The remaining three items are scored in a similar fashion, choosing which is next most important to earn a score of four and thus deciding which of the last two is less important for its score of two. The completed cards are collected, and the scores are tallied for each item.

6. *Discussion and decision.* The total score for each item is displayed on the flipchart. The leader may initiate further discussion to examine inconsistencies in the ranking patterns and why some items may have received too few or too many points. A final vote is then made.

Traditional unstructured meetings commonly mix solutions with problems, meander excessively, suppress conflicting ideas, and lack a fitting closure.

Some NGT leaders distribute the task statement before the meeting and ask participants not to discuss it.

Participants are encouraged to add items to their list during the round-robin.

Research has shown that individuals are better at generating ideas than groups are. But a group of individuals can benefit from seeing others in the group work hard at the task.

A constructive discussion resolves conflict, promotes consensus, and provides a sense of accomplishment.

> Participants may be given just one vote or, in an alternative approach, may be given three votes which can be distributed among the items or all given to one item. The item with the largest number of votes establishes a decision or a list of ranked alternatives as the outcome of the meeting.

6-13 SUMMARY

Linear programming is an optimizing technique that is appropriate to problems whose objective is stated in linear terms, alternative courses of action are available to achieve the objective, resources are known, and variables are quantifiable. The assignment method, an LP variant and a special case of the transportation method, is a procedure used to match single candidates to single positions. A graphical solution is feasible for a limited class of LP-type problems, such as the best mix for two or three products. The transportation model is more versatile than is the graphical method and succinct enough for hand calculations.

Assignment method

Graphical LP method

A transportation matrix shows the amount of supply and demand (rim conditions) and the cost to supply one unit from each origin (O) to each destination (D). A dummy origin or destination is introduced to absorb any imbalance in supply and demand. An initial feasible solution results when allocations meet the rim conditions, are in independent positions, and number $O + D - 1$. The Vogel approximation method (VAM) for determining an initial feasible solution uses row-and-column penalties to make initial allocations as nearly optimal as possible.

Vogel approximation method

An optimal solution to a transportation problem is developed by a stepping-stone method, in which each unoccupied cell is checked to see whether its transfer cost is less than a route presently in the solution. The check is made by summing the costs of transferring one unit around a closed orthogonal circuit of occupied cells to the unoccupied cell being checked. When a new route offers a saving, as many units as possible are allocated to it. A degenerate allocation pattern, less than $O + D - 1$ occupied cells, is remedied by assigning an infinitesimal quantity (ϵ) to an unoccupied cell in a propitious position. This cell is considered occupied in a transfer circuit.

Stepping-stone method

A maximization problem in which supply and demand is related by a profit function is changed into a minimization problem by transforming profits to relative costs. The transformation is made by subtracting the profit associated with each cell from the largest profit in the matrix. After the transformation, the problem is treated by the usual minimization procedures.

The intention of aggregate planning is to find the most economical way to deploy production resources to meet fluctuating demands for output. The planning horizon is generally a year or so. Planning typically considers combinations of alternatives for changing the size of the labor force, working time, subcon-

Aggregate planning

tracting, and letting inventory levels increase or decrease. Many of the costs associated with the alternatives are not recorded in accounting data and must be estimated. Mathematical approaches include linear programming, linear decision rule, management coefficients, and simulation. (The simplex method for solving linear programming is described in Appendix F.)

Time management

Knowledge workers are managers and professionals, who account for about two-thirds of white-collar employment costs. Better time management improves their productivity. Setting priorities and deadlines, avoiding procrastination, and being flexible helps conserve managerial time.

Nominal group technique

The nominal group technique is a highly structured procedure by which a small group of selected participants identifies solutions to a given issue or problem and then prioritizes courses of action. NGT encourages creativity and consensus.

6-14 REFERENCES

BUFFA, E. S., and J. S. DYER. *Essentials of Management Science/Operations Research.* New York: Wiley, 1978.

DANTZIG, G. G. *Linear Programming Extensions.* Princeton, N.J.: Princeton University Press, 1963.

HADLEY, G. *Linear Programming.* Reading, Mass.: Addison-Wesley, 1962.

HILLIER, F. S. and G. J. LIEBERMAN. *Introduction to Operations Research,* 3rd ed. Oakland, Calif.: Holden-Day, 1980.

JOSHI, M. V. *Management Science.* North Scituate, Mass.: Duxbury Press, 1980.

LOOMBA, N. P. *Linear Programming: A Managerial Perspective.* New York: Macmillan, 1976.

PHILLIPS, D. T., A. RAVINDRAN, and J. J. SOLBERG. *Operations Research.* New York: Wiley, 1976.

RIGGS, J. L., and M. S. INOUE. *Introduction to Operations Research and Management Science.* New York: McGraw-Hill, 1975.

SILVER, E. A., and R. PETERSON. *Decision Systems for Inventory Management and Production Planning,* 2nd ed. New York: Wiley, 1985.

WAGNER, H. M. *Principles of Operations Research,* 2nd ed. Englewood Cliffs, N.J.: Prentice-Hall, 1975.

6-15 SELF-TEST REVIEW

Answers to the following review questions are given in Appendix A.

1. T   F   *Linear programming* is an allocation method developed in the last decade for use on minicomputers.

2. T   F   The *simplex method* for solving LP problems was created by G. B. Dantzig.

3. T   F   The first steps in the *assignment method* obtain opportunity costs for each row and column.

4. T   F   *Zero-covering lines* indicate either an optimal solution or the matrix modifications needed to approach a solution more closely.

5. T   F   The *objective function* for an LP problem is stated in the form of an inequation.

6. T   F   The *graphical LP method* is limited to problems with no more than two resource restrictions.

7. T   F   All possible mixes that meet restraints are shown in the *feasible solution space* of an LP graph.

8. T   F   An optimal product mix is identified by the *isoprofit line* that touches the point of the feasible solution polygon nearest the origin of the graph.

9. T   F   The *distribution method* can be used only when an LP problem can be formatted in a square matrix.

10. T   F   *Rim conditions* in a transportation matrix show the supply-and-demand data.

11. T   F   *VAM* is applied to check the optimality of a solution.

12. T   F   A "stone" in the *stepping-stone method* is always an occupied cell.

13. T   F   When a distribution solution has fewer than $O + D - 1$ allocations, it is *degenerate*.

14. T   F   A *dummy* row or column in both the assignment and distribution methods has a zero cost for any allocation made to it.

15. T   F   *Aggregate planning* includes the determination of long-range investments needed to meet future demand.

16. T   F   Four *controllable inputs* to the production process used in aggregate planning are size of the work force, inventory levels, R&D activities, and subcontracting.

17. T   F   Several conceptually different *mathematical models* have been developed to replace traditional graphical methods for aggregate planning.

18. T   F   A specific level of education qualifies a person to be classified as a *knowledge worker*.

19. T   F   The principles of *time management* are generally applicable to knowledge workers in every industry in every country.

20. T   F   To *conserve time*, realize that all tasks are equally important.

21. T   F   *NGT* encourages participants to argue openly about the merits of each idea or suggestions presented.

22. T   F   A *facilitator* guides a meeting using the nominal group technique.

23. T   F   An NGT session usually produces a prioritized list of items pertinent to the issue posed in the *task statement*.

6-16 DISCUSSION QUESTIONS

1. Show how the product-mix problem of Example 6.3 fits the requirements for an LP solution as outlined in Section 6-4.

2. With reference to a graphical solution, describe a condition in which all limiting resources would be used to their capacity.

3. Explain the visual checks that determine the acceptability of an initial feasible solution. Why are they necessary?

4. Discuss the way that you would break a tie between two equal and highest penalties in a VAM penalty matrix.

5. Why is a positive net transfer cost stated as a negative opportunity cost?

6. Describe a transfer circuit that would create a degeneracy in a revised solution.

7. Why must a cell containing ϵ in a previously degenerate solution be a positive cell in a transfer circuit?

8. Why is the smallest allocation in a negative cell of a transfer circuit the limiting quantity that can be transferred to a lower-cost route (the cell being checked)?

9. Why must a complete transfer circuit contain an equal number of positive and negative cell allocations?

10. Explain the significance of a zero transfer cost for an occupied cell. How can it be useful?

11. Assume that a transportation matrix with a supply greater than demand shows a *profit* relationship between origins and destinations. After converting the profits to relative costs and adding a dummy destination to absorb excess supply, a zero cost is assigned to the dummy transportation routes. Is it reasonable to consider the dummy zero cost equivalent to the relative zero cost obtained from the route with the greatest profit in the original matrix?

12. What is implied by the word *aggregate* in aggregate planning?

13. Which of the required inputs for aggregate planning are the most difficult to obtain and verify?

14. What business conditions encourage and discourage the use of backorders?

6-17 PROBLEMS

1. A production problem is described by the following algebraic statements and is to be solved by LP methods: Maximize $3A + 3B$ subject to $4A + 7B \leqslant 1800$ and $3A + 11B \leqslant 1300$, where $A \geqslant 0$ and $B \geqslant 0$.

2. Large orders for four parts are to be assigned to four person–machine centers. Some machines are better suited to produce certain parts, and their operators are more proficient at producing some parts than others. The costs to produce each part at each center are as follows:

Part	MC1	MC2	MC3	MC4
P1	12	9	11	13
P2	8	8	9	6
P3	14	16	21	13
P4	14	15	17	12

a. Which part (P) should be assigned to each machine center (MC)?

b. Assume that a new machine has been added to the facilities. One old machine is to be phased out. The operator on the old machine will operate the new one if it can produce one of the parts less expensively than the assignment made in part a. The estimated cost using the new machine ($MC5$) is

Part:	P1	P2	P3	P4
Production cost at $MC5$:	11	7	15	10

Should the new machine be used? If so, which part should it produce?

Hint: Add the new machine costs and a dummy part ($P5$) to the matrix used to solve part a. Use zero costs to produce $P5$ in each machine.

3. A new office building has been constructed to allow the centralization of administrative functions. Six regional managers will be moved to the new central location. All office managers are on the same floor. Each office has the same room area and furnishings, but the exposures and views differ. To please as many managers as possible, the managers were asked to rank their preferences for offices, with six being the most desirable and one the least desirable. The following rankings were submitted:

Office	201	205	209	212	216	220
Manager A	4	2	5	1	3	6
B	1	3	5	2	4	6
C	3	5	6	2	1	4
D	2	4	6	1	3	5
E	5	2	6	4	1	3
F	1	6	3	5	2	4

Determine the assignment that will provide the most overall satisfaction.

4. A small ceramic shop produces two products for which the market is essentially unlimited: small (S) indoor figurines and large (L) outdoor ornaments. The bottlenecks in production are the required oven time and the artistic hand glazing. The owner has no intention of increasing the capacity of the shop by additional investments in machines or by training more glaziers, but she does want to use her facilities as advantageously as possible. The relevant annual data on the average availability of oven time (O) and artist time (A) and the contribution per unit produced are as follows:

	Processing Time in Hours/Unit		Hours Available per Year
	Large (L)	Small (S)	
Oven time (O)	8	5	100,000
Artist time (A)	1	2	10,000
Contribution/unit	$1.50	$2.50	

a. What is the optimal product mix for these data?

b. Assume that a very foolish contract was signed disregarding the effect of inflation. A total of at least 8000 units of L and/or S is to be supplied at prices that are expected to result in a loss to the shop; the contribution is estimated at $-\$0.05$ per unit for product L and also for product S. If the owner decided to fulfill the contract and then produce no more figurines and outdoor ornaments, how should the product mix be juggled?

5. A box factory currently has contracts for berry crates and pallets. Because of the closure of nearby mills, the availability of lumber is limited. The grade and size of lumber are similar for three lots on hand. The requirements to produce one crate or pallet from the material on hand are as follows:

Lot Number	Requirement in Board Feet		Thousands of Board Feet Available
	Berry Crate	Pallet	
1	1.2	0	21
2	0.9	4.1	24
3	0.6	6.4	60
Contribution/unit	$0.17	$0.26	

a. How many of each product should be produced to maximize profit from the material on hand?

b. The contribution is based on current lumber prices. How much extra could the box company pay (and still break even) for enough additional lumber with the mix of Lot 2 to keep production going until all current lumber lots are exhausted? Why might this action be considered?

6. Transportation costs in dollars per unit and the supply–demand relationships are provided for two problems:

Problem A						Problem B					
4	8	6	2	16		94	63	71	84	56	70
7	5	4	3	24		59	84	65	77	68	47
5	6	4	4	21		85	75	60	65	70	66
12	18	15	16	61		70	68	86	72	62	50
						33	47	49	72	32	233

a. Determine an initial feasible solution by VAM.

b. Obtain an optimal solution by the stepping-stone method.

7. Assume that the supermarket manager described in Example 6.4 discovers that $M3$ needs only 300 instead of 400 crates of apples. All other conditions remain the same.

a. Show results from applying VAM to obtain an initial feasible solution. (Show the placement of ϵ for the first revision.)

b. What is the maximum possible profit? State where and what quantities are used.

8. A company has 740 out-of-style units to dispose of. The units are distributed through three company-owned warehouses that supply customers in five major geographical areas. The capacity of the warehouses and market areas are as follows:

	North	South	East	West	Central	Total
Market requirements	110	90	200	140	100	640
Warehouse capacity (units)	W1 – 240		W2 – 350		W3 – 150	

Based on the following unit transportation costs, what stock should be assigned to the warehouses, and which markets should each warehouse supply?

	North	South	East	West	Central
W1	7	40	10	35	14
W2	38	18	8	30	6
W3	35	16	30	9	6

Hint: Two dummies may be introduced to balance origin and destination capacities with the supply available.

9. A cost matrix in which the amount available exceeds the amount required follows. Identify the best allocation pattern.

				Available
8	9	3	7	130
3	6	8	4	180
7	5	4	5	240
8	7	2	6	160
5	4	2	9	90
8	4	4	3	120
Required 300	100	150	200	

10. A producer of exclusive jewelry products has a maximum capacity of 600 units per year. The company can sell as many units as it can produce. An extra touch of luxury is added by packaging each jewelry set in a handmade case. Cases are hand-crafted by artists in three locations. The sets are shipped to the artists' establishments where they are individually packaged. The maximum units each artist can handle and the cost of shipping each unit to each artist are as follows:

Artist	Capacity	Shipping Cost
A1	150	$2.00
A2	300	$3.00
A3	250	$1.50

The products are sent to five market areas after packaging. The current demand from each market area and the transportation cost per unit from the artist to the market area are as follows:

Market Capacity	M1 100	M2 200	M3 150	M4 100	M5 200
A1	$1.80	$1.50	$8.00	$6.00	$5.00
A2	$2.00	$3.00	$6.00	$2.00	$3.50
A3	$2.30	$6.00	$4.00	$2.50	$2.00

a. How many jewelry sets should be allocated to each artist, and to which markets should the packaged units be delivered?

b. How much extra compensation per unit should the jewelry company receive if it agrees to supply the entire demand of M3 while maintaining the same total transportation cost?

Observe the same hint given in Problem 8.

11. Develop a production plan for the data given in Example 6.6 that does not have any backorders.

a. Compare the cost of the no-backorder plan with the one given in Figure 6.19, which has a total cost of $752,230 when no charge is made for the initial inventory I_0.

b. What are some reasons that the more expensive no-backorder plan might be used rather than the one shown in Figure 6.19?

12. Pickup truck campers are manufactured by Cavemen Canopies, Inc. The campers are distributed regionally with fluctuations in seasonal sales. The following data have been collected for production planning:

Month:	April	May	June	July	August	September
Demand	300	600	1200	800	400	200
Working days	21	22	20	20	22	21

Labor hours required per camper = 30
Standard pay = $5 per hour for an 8-hour day
Storage costs = $15 per month per camper
Marginal cost of a stockout = $30 per month per unit
Marginal cost of subcontracting = $120 more than a

camper produced at the standard pay rate. (Materials furnished by Cavemen.)
Cost of hiring and training = $300 per worker
Layoff costs = $400 per worker

a. Compare the three production plans in (1), (2), and (3) that have no starting inventory and no limit on the number of campers subcontracted.
   (1) Exact production with a varying work force: Number of workers at beginning of April = 54.
   (2) Constant work force with varying inventory and stockouts, and no subcontracting: Constant work force = 120.
   (3) Constant low work force with subcontracting and varying inventory: Constant work force = 60.

b. Use the distribution method to develop a production plan subject to the following additional conditions:

Overtime pay rate = $7.50, limited to 25 percent of the hours of the regular work force in any month

Constant work force = 75

Subcontracting is limited to a maximum of 100 campers per month

Subcontractors absorb any holding costs for the units that they provide

Starting inventory = 10 campers

Ending inventory = 50 campers

Backorders up to one month are allowed

What is the production cost of the plan?

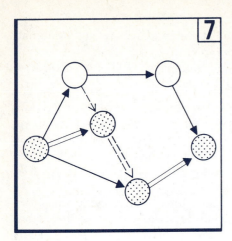

7

RESOURCE SCHEDULING

LEARNING OBJECTIVES

After studying this chapter you should

- understand what is involved in project scheduling and management.
- be able to construct an arrow diagram, calculate the critical path, and determine boundary times.
- know when it is appropriate to use CPM or PERT and be aware of the statistical properties of PERT applications.
- appreciate the concept of time–cost trade-offs and how float can be adjusted to smooth resource usage.
- realize that projects are becoming larger and more complex, which increases the difficulties and opportunities in project management.

KEY TERMS

The following words characterize subjects presented in this chapter:

Critical Path Method (CPM)	expected time (t_e)
arrow diagram	milestone
network dummy	time–cost trade-offs
activity start and finish times (ES, LS, EF, LF)	job shop
total float (TF)	project management
boundary timetable	jumbo projects
Program Evaluation and Review Technique (PERT)	S-curve
	management by walking around (MBWA)

7-1 IMPORTANCE

Even if we assume that the procedures in Chapter 6 gave us ways to allocate the right amounts of resources to the right places, we still face the problem of having them ready at the right time. After setting on a basic course of action, we must decide which acts to perform first. For a relatively small task, a mental list of materials coupled with another mental list of the order of accomplishment will normally suffice. For slightly larger tasks, it usually helps to commit plans to an informal "shopping list" schedule. When projects get too complex to analyze mentally or too large to "carry in your head," more formal and systematic procedures are required. In this chapter we investigate a network approach for integrating actions and scheduling resources.

A seldom mentioned feature of network scheduling is its basis for a common language. Once learned, the standard terminology and symbols simplify planning communications between staff and line and among planners from different disciplines.

The complexity of many of today's projects, even those of moderate size, demands not only consistent, disciplined thinking but also a method of systematically summarizing and presenting the results of this thinking. The graphical networks and associated calculation techniques assist effective thinking by sponsoring a step-by-step routine for coordinating work assignments and resource utilization with project objectives. Control criteria for the evaluation of work progress are established, and the most economical means for correcting delays are diagnosed.

Thus, network scheduling contributes to the two critical features of project planning: formulating the initial plan and monitoring progress. It operates in the delicate interface in which ideas are translated into acts.

7-2 NETWORK SCHEDULING

Coordination of operations has been a concern of production planners for hundreds of years. Consider the following excerpt from the diary of a visitor to a seventeenth-century Venetian arsenal. He described the outfitting of large fighting galleys which carried crews of 171 plus 45 swordsmen. The operation was obviously a cooperative effort requiring significant resource scheduling.

> And as one enters the gate there is a great street on either hand with the sea in the middle, and on one side are windows opening out of the houses of the arsenal, and the same on the other side, and out came a galley towed by a boat, and from the windows they handed out to them, from one the cordage, from another the bread, from another the arms, and from another the balistas and mortars, and so from all sides everything which was required, and when the galley had reached the end of the street all the men required were on board, together with the complement of oars, and she was equipped from end to end. In this manner there came out ten galleys, fully armed, between the hours of three and nine.

Venetian Ships and Shipbuilders of the Renaissance, F. C. Lane, Johns Hopkins Press, Baltimore, Md., 1934.

Because even greater coordination is required in modern production, network scheduling has evolved.

A network depicts the sequence of activities necessary to complete a project. Segments of a project are represented by lines connected together to show the

Closely related graphical techniques for operational control are presented in Section 14-4.

interrelationship of operations and resources. When a duration is associated with each segment, the model shows the time orientation of the total project and its internal operations. This information is used to coordinate the application of resources. The most celebrated versions of network scheduling are the **Critical Path Method (CPM)** and the **Program Evaluation and Review Technique (PERT).**

Few if any new management tools have received such wide acclaim so rapidly as CPM and PERT have. The U.S. Navy developed PERT in 1958 to coordinate research and development work. It was first applied to the Polaris ballistic missile program with significant success. In the same year a similar network scheduling tool was developed by the DuPont company for industrial projects. It was called the Critical Path Method in deference to the path of critical activities that control the project's duration.

Other closely related techniques included in the critical path scheduling family are PEP, IMPACT, CPA, PERTCO, SCANS, GERT, NMT, and MCX.

A network is a picture of a project, a map of requirements tracing the work from a departure point to the final completion objective. It can be a collection of all the minute details involved or only a gross outline of general functions. It can be oriented to the use of a particular physical resource, or it can define broad management responsibilities. Networks have been constructed for building dams, introducing new products, organizing political campaigns, meshing new production lines with existing facilities, initiating research, coordinating charity drives, developing complex defense systems, and a multitude of other purposes. But regardless of its use, the fundamentals of network construction are the same. You have to determine a list of necessary activities, establish a restriction list that sets the order of activity accomplishment, and then combine the two lists with a set of drawing conventions to construct a graphical network.

7-3 CONSTRUCTION OF A CPM NETWORK

An **arrow diagram** similar to CPM networks is shown in Figure 7.1. In gross terms it represents the operations required to produce a custom-made machine. The project starts with the design work to satisfy the customer's specifications and culminates in the delivery of the finished product. Each arrow symbolizes a distinct activity. The chronology of the activity arrows shows the way in which work is expected to progress. The purpose of committing a project to a network is to allow an evaluation of different modes of operation before the work is under way. The evaluation should foretell how resources can be applied most advantageously to fulfill desired objectives such as time restrictions or budgetary limitations.

It is doubtful that such a small project as shown in Figure 7.1 would be subjected to CPS, but the logic is the same for more extensive applications.

Activity list

The initial step in a CPM application is to break the project down into its component operations to form a complete list of essential activities. This task may appear easy but is usually difficult. The burden is the importance of a representative listing: without a valid list, all subsequent steps are meaningless.

GIGO—if you put Garbage In, you get Garbage Out.

An activity is a time-consuming task with a distinct beginning and end

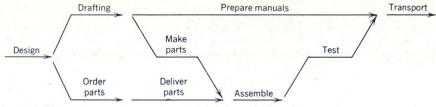

FIGURE 7.1 Arrow diagram of production activities.

point. Some easily identified characteristic should be associated with each start and finish point. For instance, in Figure 7.1 the start of the first activity could be signaled by the receipt of a specification listing from the customer and could end when detailed sketches are delivered to the drafting and order departments.

Better activity lists are ensured by using all available information. A network is a composite picture of an entire undertaking. The efforts of many people are included, and they should be consulted with regard to their expected accomplishments. Suppliers, cooperating departments, subcontractors, and other representatives should review the activity list to see whether their expectations pertaining to their respective interests are accurately and realistically detailed.

Restriction list

A restriction list establishes the precedence of activities. A list of operations commonly evolves from the general order of completion because it is natural to think of activities following a familiar sequence. This rough sequence forms a starting point for the restriction list. Each activity is bracketed by the answers to two questions: What must precede? What must follow? The answer to the first question identifies the *prerequisite*—the activity that immediately precedes another activity. The answer to the second question gives the *postrequisite*—the activity that immediately follows.

Accuracy and completeness are marks of a good restriction list. Accuracy stems from a careful appraisal of true priorities, not priorities set by habit. The inclination in dealing with familiar operations is to accept them as they have always been done. For instance, in sequencing the old caution "stop-look-listen," the temptation is to say you first stop, then look, and then listen. But a better restriction list would put "stop" as the prerequisite to both "look" and "listen" because the two senses can operate concurrently. Completeness is a function of the persistence with which the priority of each activity is questioned. It is easy to overlook a relationship between an activity occurring early in a project and one occurring near the end.

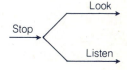

In drawing an arrow network, we are interested in the arrow that immediately follows the one just drawn—the prerequisite–postrequisite relationship. Therefore, a restriction list preferably shows only this relationship. A more extensive listing, such as all the jobs that follow each activity, is unnecessarily tedious and even contributes to errors. The activity list, prerequisite–postreq-

TABLE 7.1
Activity and Restriction Lists for the Project in Figure 7.1

Activity List		Activity Relationships		Restriction List*
Description	*Symbol*	*Prerequisite*	*Postrequisite*	
Design	A		Drafting, Order Parts	A < B,C
Order Parts	B	Design	Deliver Parts	B < D
Drafting	C	Design	Prepare Manuals, Make Parts	C < E,F
Deliver Parts	D	Order Parts	Assemble	D < G
Prepare Manuals	E	Drafting	Transport	E < I
Make Parts	F	Drafting	Assemble	F < G
Assemble	G	Deliver Parts, Make Parts	Test	G < H
Test	H	Assemble	Transport	H < I
Transport	I	Test, Prepare Manuals	—	—

*This sign < is read precedes as A precedes B and C.

uisite relationships, and a shorthand notation for the restriction list for the project depicted in Figure 7.1 are shown in Table 7.1.

Network conventions

Each management tool seems to require special definitions and conventions, and CPM is no exception. Because much of the CPM language has been introduced already, we can now associate the language with the graphical representations.

An activity is represented by a line or arrow. This line or arrow connects two events. Each event is a specific point in time, marking the beginning and/or end of an activity. A dual-purpose symbol for an event is a circle topped by a cross: the circle facilitates computer computations, and the cross aids manual arithmetic. As shown in the activity chain of Figure 7.2, the first activity, "design," can be identified by the abbreviation A or by the numbered events as activity 1,2. The latter numbering system is used for computer applications. It also clarifies the definition of an event: event 2 in the chain marks the end of activity A and the beginning of activity B.

Nodal numbers of activities are a combination of the event numbers at the beginning (i node) and end (j node) of any activity: For example, activity i,j.

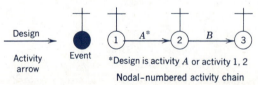

FIGURE 7.2 Activity and event conventions.

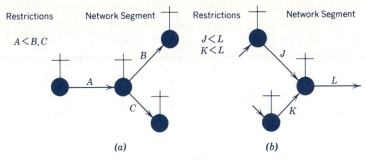

FIGURE 7.3 Activity (a) burst and (b) merge patterns.

When two or more activities end with the same event, that event is referred to as a *merge*. Similarly, when two or more activities can begin at the same time, the event denoting that time is called a *burst*. A merge or a burst can be recognized in a restriction list by the occurrence of the same prerequisites or postrequisites for two or more activities. The recognition of typical recurring patterns, such as those in Figure 7.3, speeds network drawings.

A dashed-line arrow is used in a network to show the dependency of one activity on another. It is called a **network dummy activity** and has all the restrictive properties of regular activities except that it requires zero time. The employment of dummies can be distinguished according to the purpose they serve; an *artificial* dummy is inserted to facilitate node numbering for computer applications, and a *logic* dummy is necessary to portray graphically certain restraint relationships between nodes.

The special arrangement for graphing two activities that begin and end with the same event is shown in Figure 7.4. Both versions in the figure depict the same restriction: activities A and B can start at the same time, and both must be completed before activity C can begin. This relationship is visually obvious for manual manipulations, but a computer relies only on event numbers for identifying and sequencing activities. Without a dummy, both activities A and B have the same identifying event numbers 1,2 and are thereby indistinguishable in a nodal-numbering system. To avoid this duplication, an artificial dummy is introduced after either activity. The dummy forces a unique set of numbers for each activity ($A = 1,3$; $B = 1,2$; dummy $= 2,3$) without changing the logic of the network. For consistency, we shall use artificial dummies even when the calculations are performed manually.

Some newer computer programs do not require unique node numbers.

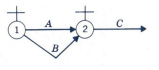

Adequate for manual
calculations

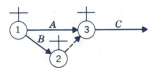

Necessary for computer
applications

FIGURE 7.4 Artificial dummy.

The dummy could also be placed in front of B without changing the logic:

$$A = 1,3$$
$$\text{dummy} = 1,2$$
$$B = 2,3$$

Case	Spillover	Cascade	Unnecessary	Redundant
Restriction	$A < C$ $B < E$ $C < D, E$	$A < D, E$ $B < E, F$ $C < F$	$A < C$ $B < D$ $C < D$	$A < C$ $B < C, D, E$ $C < E$
Incorrect network segment	*B* is not a prerequisite of *D*	*A* is not a prerequisite of *F*	Artificial dummy 3, 4 is unnecessary	*B* is a redundant restriction on *E*
Correct network segment				

FIGURE 7.5 Correct and incorrect use of network dummies.

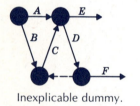

Inexplicable dummy.

The careful use of logic dummies will eliminate many beginner's errors that creep into network drawings. Four of the most common mistakes are displayed in Figure 7.5. The *spillover* error illustrates how a dummy aims a restriction in one direction only: *C* restricts *E*, but *B* does not restrict *D*. The *cascade* effect emphasizes that the dummy shows a zero time or a simultaneous relationship between events. The incorrect version neglects the fact that all activities (and a dummy is a zero-time activity) in a merge are prerequisite to all activities in a burst from the same event. *Unnecessary or redundant* dummies do not spoil the logic of a network; they just create extra calculations and network complications. Every unnecessary dummy increases the computer time required to develop a schedule and introduces another chance for misinterpretation.

Network drawing

A CPM network is merely a graphical version of the activity and restriction lists based on adopted conventions. The visual representation is a noteworthy communication device. The relationship of each part to the whole project is easier to comprehend when it is in network form, and mistakes of commission and omission are more apparent.

It is convenient to use a blackboard, which allows easy changes for a preliminary network.

Drawing techniques vary. The initial attempt to graph a project is usually an approximate sketch. Sometimes several activities are lumped into a single arrow to accentuate the general flow. After the overall sequence is checked, the compound activities can be subdivided into detailed operations. Very large projects are handled more conveniently by treating subdivisions independently in preliminary networks and then integrating them with dummy arrows to relate the segments. A similar approach is suitable for portions of a project that are repeated several times. For example, in the construction of a multistory building, several floors often have essentially the same characteristics and, consequently, identical activity lists. It is thus necessary to make only one network for the common list; dummies are used to establish the order of cyclic repetitions of this basic network.

Networks are easier to check if all arrows flow from left to right and as many arrow crossovers as possible are eliminated. Labeling each activity with a concise description rather than a letter symbol avoids continued reference to the activity list. Nodal numbers should be assigned after the entire network is drawn and checked, because some computational methods and computer programs require a numbering system in which the node at the beginning of an activity (i node) is always smaller than the ending node (j node).

EXAMPLE 7.1 Development of an arrow network for a new product

The research department reports promising developments with a refined control unit. A laboratory model functioned reasonably well and, with minor redesign, should merit field tests. Construction of a dozen prototypes for the field tests is to be put on a critical path network. Construction can commence when the final design from the research department is submitted. As indicated in the following activity list, most operations involve securing components by inventory requisition of stock on hand, from orders to outside subcontractors, or from internal fabrication work:

Activity	Symbol	Restriction
Final design	A	A < B,C
Engineering analysis	B	B < D,F,H
Prepare layouts	C	C < E
Requisition material	D	D < E
Fabricate parts	E	E < J
Requisition parts	F	F < G
Receive parts	G	G < J
Place subcontracts	H	H < I
Receive subcontracted parts	I	I < J
Assemble	J	J < K
Inspect and test	K	

A network based on the activities and restrictions is constructed and questioned. An engineer notices that the amount of material needed in "fabrication" will not be known until certain layouts are completed. This condition can be shown in the network by subdividing the activity "layout" into "layout 1" and "layout 2," in which the end event for "layout 1" is the completion of the layouts required for material takeoffs. Then the fabrication supervisor notices that no time has been allowed for delivering the requisitioned material. Material lead time is included by adding another activity, "receive requisitioned material." Alterations in the activity list to incorporate these revisions are as follows:

Activity	Symbol	Restriction
Final design	A	A < B,C1
Engineering analysis	B	B < D,F,H
Prepare layout 1	C1	C1 < C2,D1
Prepare layout 2	C2	C2 < E
Prepare material requisition	D1	D1 < D2
Receive requisitioned material	D2	D2 < E

The original network, with changes required by the revised activity list shown in the inset, is given on page 234. (Note that block numbering of the activities allows additions that do not alter the technological order of the nodal numbers.)

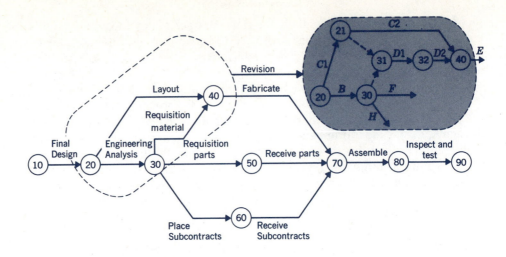

7-4 THE CRITICAL PATH

See Figure 7.6.

A network composed of activities required to develop a critical path schedule would probably label the first activity "investigate project" followed by a burst of three activities: "identify activities," "list restrictions," and "estimate durations." This arrangement shows that the three tasks could be performed almost concurrently, if it is more advantageous from a resource-utilization standpoint, they could be completed in a sequence. Thus the network pictures how a project *can be done*, whereas a schedule establishes how it is *planned to be done*. The key to network scheduling is the *critical path*, the chain of activities that determines the total project duration.

The time to conduct each activity is not required for network logic, but times are necessary to determine critical activities. Two approaches are available for estimating activity durations. A *deterministic* approach, appropriate for most industrial projects, relies on a single, most likely time estimate; a *statistical* approach uses a range of possible activity times to determine a single duration.

Most likely time estimates

A single time estimate rather than a range of estimates for an activity duration is feasible for most industrial projects because the operations required are usually similar to previous experiences.

An estimate of the elapsed time required to accomplish the objectives stated in an activity's description is called the *activity duration*. All activity durations should be expressed in the same time units, such as days, weeks, or months. Estimates do not reflect uncontrollable contingencies, such as flash floods or unexpected legal delays, but they should account for weather conditions and other factors that at least can be anticipated.

A project manager's final duration estimates should be studied and accepted by subcontractors so as to increase confidence in the resulting schedule.

The estimation of activity durations closely parallels the approaches discussed for forecasting. Objective sources of data for estimating include records of past performances on similar projects, historical data such as weather records or suppliers' literature, and applicable laws or regulations. Subjective estimates are obtained from persons responsible for doing the work, such as supervisors

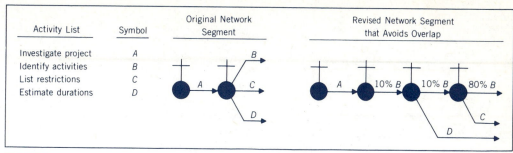

Activity List	Symbol	Original Network Segment	Revised Network Segment that Avoids Overlap
Investigate project	A		
Identify activities	B		
List restrictions	C		
Estimate durations	D		

FIGURE 7.6 Network representation of overlapping activities.

and subcontractors, or from professional assistance such as consulting firms and other project managers. The reliability and applicability of the sources are bound to vary; hence, judgment is still the vital ingredient for converting guesstimates to estimates.

As activity durations are developed, it is often necessary to revise activity descriptions. Troublesome contradictions surface when part of the work traditionally associated with one activity is included in a different one. This condition is annoying when the sources of time estimates pertain to traditional descriptions, but it can be alleviated by either changing activity breakdowns or by breaking down traditional operations to conform to a new grouping if this arrangement better fits the management objectives. A similar problem occurs when a restriction list shows that two or more activities can theoretically be done concurrently, but in practice, some portion of one must be completed before the other can start. A timely example is a network for starting a CPM analysis. As described previously, the first burst shows that activity and restriction listing and duration estimating can occur at the same time. This is almost true for large lists, but at least a few activities must be identified before restricting postrequisites can be set. To portray such overlapping conditions realistically, activities are subdivided, as shown in Figure 7.6.

Boundary-time calculations

After estimated activity performance times are secured, each duration is entered beside the appropriate arrow in the network. The network is then ready for boundary-time calculations. These calculations can be performed either manually or with computerized assistance to provide time relationships suitable for activity scheduling. The most useful boundary times are described as follows:

Earliest start (ES)—the earliest time an activity can begin when all preceding activities are completed as rapidly as possible.

Latest start (LS)—the latest time an activity can be initiated without delaying the minimum project completion time.

Earliest finish (EF)—the sum of ES and the activity's duration (D).

Free float, or activity float, is sometimes calculated. It is the difference between an activity's LF and the smallest ES of any postrequisite activity.

Latest finish (LF)—the LS added to the duration (D).

Total float (TF)—the amount of surplus time or leeway allowed in scheduling activities to avoid interference with any activity on the network critical path; the slack between the earliest and latest start times ($LS - ES = TF$).

EXAMPLE 7.2 Adjusting activity time estimates according to probable outcome

Activity times are based on the *successful* completion of the work. When an activity is known for its difficulty and there is sufficient experience to estimate the degree and likelihood of that difficulty, probability theory can be used to improve the estimates. Consider the case in which an activity is either successfully completed or has to be repeated until it is done adequately. This condition is represented below; there is an 80-percent chance that activity A will be done correctly the first time in four days and a 20-percent chance that it will have to be repeated, taking another four days, or repeated again with the same odds of success. The actual time that activity B

must wait to start is 4, 8, 12, or – days, depending on how many times A must be repeated. From probability relationships, the expected time for A to be completed is

$$EX(A) = \frac{\text{normal duration}}{\text{probability of success}} = \frac{4}{0.8} = 5 \text{ days}$$

which is the average time it would take to complete the activity if it were conducted a large number of times under the same conditions.

An activity may fall into the category of being partially successful; that is, if it is not finished successfully, it will not have to be completely repeated. The following network segment represents the application of paint to a production unit:

Nine times out of 10 a satisfactory paint job is finished in 40 minutes. The 10 percent of the units that fail inspection require 10 additional minutes for retouching. If it is assumed that retouching always produces a passable paint job, the expected duration for the PAINT activity is

$$EX(PAINT) = 40 + 10(0.1) = 41 \text{ minutes}$$

When it is assumed that the retouching has the same probability of failure as does the original painting, and the retouching time remains constant, the expected duration for PAINT becomes

$$EX(PAINT) = 40 + \frac{10(0.1)}{0.9} = 41.1 \text{ minutes}$$

Actually, the duration could be 40, 50, 60, or – minutes, but "on-the-average" it will take just over 41 minutes to complete each paint job.

Identification of the critical path is a by-product of boundary-time calculations. A critical activity has no leeway in scheduling and, consequently, zero total float.

The ES times are calculated manually by making a forward pass through the network, adding each activity duration in turn to the ES of the prerequisite

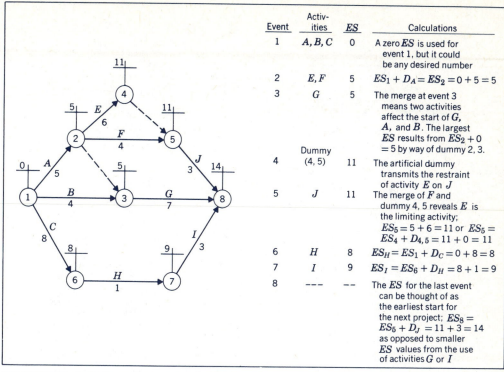

Event	Activities	ES	Calculations
1	A, B, C	0	A zero ES is used for event 1, but it could be any desired number
2	E, F	5	$ES_1 + D_A = ES_2 = 0 + 5 = 5$
3	G	5	The merge at event 3 means two activities affect the start of G, A, and B. The largest ES results from $ES_2 + 0 = 5$ by way of dummy 2, 3.
4	Dummy (4, 5)	11	The artificial dummy transmits the restraint of activity E on J
5	J	11	The merge of F and dummy 4, 5 reveals E is the limiting activity; $ES_5 = 5 + 6 = 11$ or $ES_5 = ES_4 + D_{4,5} = 11 + 0 = 11$
6	H	8	$ES_H = ES_1 + D_C = 0 + 8 = 8$
7	I	9	$ES_I = ES_6 + D_H = 8 + 1 = 9$
8	---	--	The ES for the last event can be thought of as the earliest start for the next project; $ES_8 = ES_5 + D_J = 11 + 3 = 14$ as opposed to smaller ES values from the use of activities G or I

FIGURE 7.7 Activity earliest start calculations.

activity. When a merge is encountered, the largest $ES + D$ of the merging activities is the limiting ES for all activities bursting from the event. Dummies are treated exactly the same as are other activities. Each limiting ES is recorded on the left bar of the event markers so as to keep track of the cumulative entries. The cross at the i node of an activity indicates the ES for that activity. Details of ES calculations for a small network are given in Figure 7.7.

The limiting ES is the latest ES for an event.

LS times are calculated in almost a reversed procedure from that for ES times. A backward pass is made through the network, subtracting activity durations from the limiting LS at an event. The limiting LS, the smallest one at a burst event, is entered on the right bar of the cross. Subsequent LS's are calculated by subtracting activity durations from the LS on the j-node cross.

The main difference between LS and ES calculations is that each activity from a common event can have a different LS, whereas all activities starting from the same event have the same ES. To accommodate this situation, a "shelf" is connected to each activity in a burst that has a larger LF value than the limiting one. Thus, an individual LS is recorded for each activity, as detailed in Figure 7.8.

The limiting LS is the earliest LS (smallest number) in a burst.

The initial step in LS calculations is to make the right bar of the last cross in the network agree with the left bar. This entry confirms the critical nature of

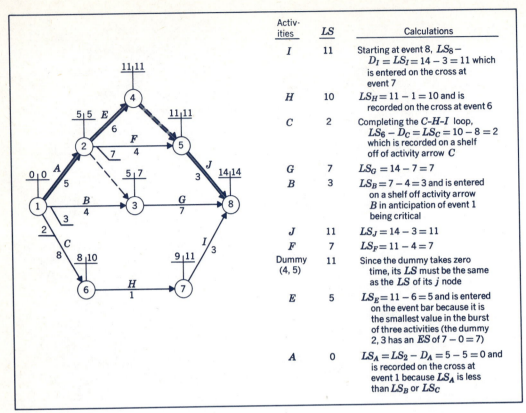

Activities	LS	Calculations
I	11	Starting at event 8, $LS_8 - D_I = LS_I = 14 - 3 = 11$ which is entered on the cross at event 7
H	10	$LS_H = 11 - 1 = 10$ and is recorded on the cross at event 6
C	2	Completing the C–H–I loop, $LS_6 - D_C = LS_C = 10 - 8 = 2$ which is recorded on a shelf off of activity arrow C
G	7	$LS_G = 14 - 7 = 7$
B	3	$LS_B = 7 - 4 = 3$ and is entered on a shelf off activity arrow B in anticipation of event 1 being critical
J	11	$LS_J = 14 - 3 = 11$
F	7	$LS_F = 11 - 4 = 7$
Dummy (4, 5)	11	Since the dummy takes zero time, its LS must be the same as the LS of its j node
E	5	$LS_E = 11 - 6 = 5$ and is entered on the event bar because it is the smallest value in the burst of three activities (the dummy 2, 3 has an ES of $7 - 0 = 7$)
A	0	$LS_A = LS_2 - D_A = 5 - 5 = 0$ and is recorded on the cross at event 1 because LS_A is less than LS_B or LS_C

FIGURE 7.8 Activity latest start calculations.

Nothing is gained by calculating an LS for logic dummies, and so no shelves should be placed on dashed-line arrows.

the last event, a distinct point in time at which a subsequent project relying on the resources employed in the current project can begin. Successive subtractions of activity durations from each limiting event LS eventually lead to a zero LS for the first node in the network. A nonzero LS for event 1 means your arithmetic needs polishing.

After the ES and LS calculations are completed, the remaining boundary times are established by routine manipulations. The most direct approach is to set up a **boundary timetable,** as shown in Table 7.2. Under the activity heading is a column of activity descriptions. (Table 7.2 lists only the activity symbols in conformance to the network of Figure 7.8.) The values for activity durations, ES's and LS's, are transferred from the network to the table. The EF column is completed by adding the activity duration to each ES. Similarly, LF times are equal to the duration plus LS. TF can be determined by the difference between the two start times or the two finish times: these differences must be the same for each activity.

The critical path is easily identified from either the network or boundary

TABLE 7.2
Boundary Timetable

Activity	Duration (D)	Earliest Start (ES)	Latest Start (LS)	Earliest Finish (EF)	Latest Finish (LF)	Total Float (TF)
A	5	0	0	5	5	0
B	4	0	3	4	7	3
C	8	0	2	8	10	2
E	6	5	5	11	11	0
F	4	5	7	9	11	2
G	7	5	7	12	14	2
H	1	8	10	9	11	2
I	3	9	11	12	14	2
J	3	11	11	14	14	0

timetable. In the network, each event with the same entries on both bars is critical. A line connecting these events along the activities responsible for the event times traces the critical path, as shown by the double-lined arrows in Figure 7.8. Critical activities in the boundary timetable are indicated by zero float.

EXAMPLE 7.3 Boundary times for the network from Example 7.1

After agreeing on the accuracy of the revised network for field testing the new product, activity durations are secured from the different departments involved: research and development, engineering, fabrication, supply, subcontractors, and the like. One person coordinates the estimating because information of an interdepartmental and intradepartmental nature is often required before realistic and practical durations can be determined. An exchange of information among the managers responsible for the activities is facilitated by routing a time-labeled network for final verification. Upon acceptance of the duration estimates, network calculations and boundary times can be developed as shown.

Activity	D*	ES	LS	EF	LF	TF
A	3	0	0	3	3	0
B	1	3	4	4	5	1
C1	2	3	3	5	5	0
C2	2	5	5	7	7	0
D1	1	5	5	6	6	0
D2	1	6	6	7	7	0
E	4	7	7	11	11	0
F	1	4	8	5	9	4
G	2	5	9	7	11	4
H	1	4	5	5	6	1
I	5	5	6	10	11	1
J	2	11	11	13	13	0
K	1	13	13	14	14	0

*Activity durations are in weeks.

THE CRITICAL PATH **239**

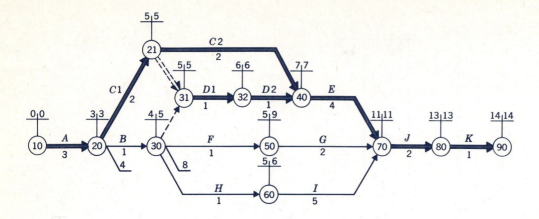

7-5 PERT CALCULATIONS

The statistical approach to activity and project durations is a characteristic of the Program Evaluation and Review Technique. This approach is appropriate for unfamiliar or experimental projects. In such projects it is difficult to agree on one most likely time estimate for each activity because there is little or no precedent for the operations involved. In research and development work, only very naive or adventurous managers will commit themselves to a definite completion date when they are unsure of what they have to do or how to do it.

    Much of the hesitancy concerning time estimations is relieved by allowing a range of estimates. Instead of a commitment to a single duration estimate, a most likely time (m) is bracketed between an optimistic (a) and a pessimistic (b) estimate of performance time. The most likely time is the duration that would occur most often if the activity were repeated many times under the same conditions. The optimistic and pessimistic estimates are the outside limits of completion time when everything goes either all right or all wrong.

    It is assumed that there is a very small probability that an actual completion time will fall outside the range a to b and that the proportion of durations within the range will follow a beta distribution. If an activity were performed many times and the relative frequency of completion times were plotted, a distribution curve as shown in Figure 7.9 would result. The mean point dividing the area under the curve into two equal parts is called the **expected time** (t_e), and it is calculated as a weighted average of the three time estimates according to the formula

$$t_e = \frac{a + 4m + b}{6}$$

PERT is recommended for planning and scheduling R&D projects as described in Section 2-5.

Extreme conditions such as "acts of God" are not accounted for in the range of estimates.

The density function of the beta distribution is $K(t - a)^\alpha (b - t)^\beta$ where t = actual completion time.

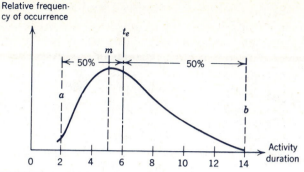

FIGURE 7.9 Relationship of a, b, and m to t_e.

where $a \leq m \leq b$. After the expected times are computed for each activity, they are treated the same as are single time estimates in determining the critical path and boundary times, as illustrated in Figure 7.10.

Each activity in a PERT network also has a variance associated with its completion time. This variance measures the dispersion of possible durations and is calculated from the formula

$$\sigma^2_{i,j} = \left(\frac{b - a}{6}\right)^2$$

where $\sigma^2_{i,j}$ = variance of an individual activity with pessimistic and optimistic performance time estimates of b and a, respectively.

When $b = a$, the duration of an activity is well known, and consequently, the variance is zero. A wide variation in the outside limits of estimated times produces a large variance and indicates less confidence in estimating how long it will actually take to complete the activity. Sample calculations of activity variances are provided in Figure 7.10.

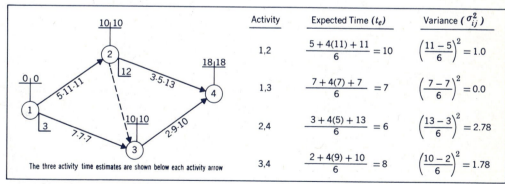

Activity	Expected Time (t_e)	Variance (σ^2_{ij})
1,2	$\dfrac{5 + 4(11) + 11}{6} = 10$	$\left(\dfrac{11 - 5}{6}\right)^2 = 1.0$
1,3	$\dfrac{7 + 4(7) + 7}{6} = 7$	$\left(\dfrac{7 - 7}{6}\right)^2 = 0.0$
2,4	$\dfrac{3 + 4(5) + 13}{6} = 6$	$\left(\dfrac{13 - 3}{6}\right)^2 = 2.78$
3,4	$\dfrac{2 + 4(9) + 10}{6} = 8$	$\left(\dfrac{10 - 2}{6}\right)^2 = 1.78$

The three activity time estimates are shown below each activity arrow

FIGURE 7.10 Expected times, variances, and critical path calculations.

Having a variance for each activity offers the interesting possibility of calculating the probability that a certain time scheduled for an event will be met.

The scheduled start (SS) is called a **milestone,** which represents the time of the planned accomplishment of an event of particular managerial importance. This significant date may originate from the need to complete a vital activity by a given date so that other portions of the project can commence, or it may result from contractual obligations outside the project.

A milestone date is compared with the ES of an event. The ES evolves from a cumulative total of activity expected durations to the milestone event. Similarly, the variance associated with that event is the sum of the variances of the activities along the path determining the milestone ES. The difference between the SS and the ES is the numerator in the equation

$$Z = \frac{SS - ES}{\sqrt{\Sigma\sigma_{ES}^2}}$$

where Z is the probability factor from which the relative likelihood is derived, and $\Sigma\sigma_{ES}^2$ is the sum of the individual variances of the activities used to calculate ES.

The actual completion times for the milestone event are assumed to be normally distributed about ES. Therefore, the probability factor is a function of the normal curve areas given in Appendix B. A Z value of zero means that the probability of completing the event on the scheduled date is 0.50. With reference to Figure 7.10, the Z factor for an SS of 17 for event 4 is

$$Z_4 = \frac{SS - ES}{\sqrt{\sigma_{1,2}^2 + \sigma_{3,4}^2}} = \frac{17 - 18}{\sqrt{1 + 1.78}} = \frac{-1}{1.67} = -0.6$$

As a rule of thumb, probabilities between 0.4 and 0.6 are considered to be a reasonable balance of risk. Above 0.6, there may be excessive resource expenditures; below 0.4 there is danger of insufficient resources causing a delay.

and indicates that the SS (17) is 0.6 standard deviations less than the ES (18). This Z value corresponds to a probability of 0.2743. Therefore, if time "now" were zero and the time units were days, we could expect to complete the project depicted in Figure 7.10 on or before 17 days with a likelihood of 0.27. This probability is applicable without any expediting measures applied to the project. When an extra day beyond the expected 18-day project duration is allowed, the probability of meeting the milestone (19) is 0.7257:

$$Z_4 = \frac{19 - 18}{\sqrt{1 + 1.78}} = \frac{1}{1.67} = 0.6$$

A backdoor approach to milestones would set the SS at a date with a desired probability of completion. Thus a manager might promise a finished product on a date that is two standard deviations beyond the ES of the last event, to give an 0.977 probability of meeting the date. Any probability could be incorporated in the formula $SS = ES + Z(\sqrt{\sigma_{ES}^2})$ to provide the desired safety margin.

END OF OPTIONAL MATERIAL _____

7-6 NETWORK APPLICATIONS IN PROJECT PLANNING

A network is an abstraction of a real project. The cost in time, money, and inconvenience is far less when manipulating a model project than a real project. Several network revisions may be required before the desired realism is developed or the preferred method of accomplishment is obtained. Regardless of the number of changes, paper alterations are still less expensive than a single major change in the physical project. Although most of our attention has been devoted to the coordination of activities and the perfected utilization of time, expenditures of other resources also can be incorporated into the network model.

Time–cost trade-offs

An initial network often reveals that a project will take longer than anticipated. The critical path exposes the group of activities from which cuts should be made to shorten the project, but it does not indicate which cuts will be least expensive. To obtain a cost priority for reducing the project duration, more information than we used in the boundary-time calculations is needed. For each activity a range of reduced completion times and associated costs (a **time–cost trade-off**) is estimated, as shown in Figure 7.11. Sometimes a cost is developed for each time unit cut from an activity's normal duration. More commonly, only a few (usually two) time–cost points are set, and the costs for completion points in between are calculated by linear interpolation.

"Crash schedules" always introduce overtime, waste, confusion, and other factors that increase an activity's cost as its duration is shortened from its normal time.

Manual methods are available for determining not only which activities should be cut to meet a deadline but also the least expensive project duration when both direct and indirect project costs are considered. The procedure is to identify an ascending order of activity-cutting costs and to make successive time reductions by using the lowest cost available. A cut always crosses the critical path; continued cutting makes more activities critical as float is reduced. The sums of direct activity cost and indirect project cost are compared for each new duration. An optimal costwise schedule is apparent when the lowest point in the total cost curve is discovered. A typical cost pattern is shown in Figure 7.12.

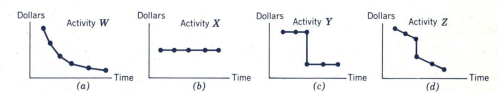

FIGURE 7.11 Common activity time–direct cost patterns. (*a*) Increasing marginal cost function. (*b*) Constant cost function. (*c*) Step increase-constant cost function. (*d*) Step increase-increasing cost function.

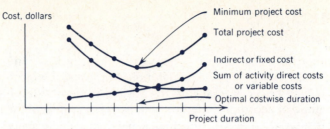

FIGURE 7.12 Direct, indirect, and total project cost patterns.

Both time and cost calculations can be conducted using LP formulations.

When a small adjustment in a project duration is required, manual methods are feasible for identifying the least expensive means. However, a manual cost analysis of a total project, even a very small one, is a long, tedious, and error-prone process. Acceptable computer programs are widely available to relieve hand computations. These cost-evaluation programs have not been used extensively because of the difficulty in developing the necessary cost data, but they will undoubtedly receive greater attention in the future.

Resource assignments

Activity costs and durations are functions of resource expenditures. These resources—skilled and unskilled labor, supervision, equipment, materials, and the like—may be limited in quantity or quality. Activity estimates should compensate for a limited supply of a resource, and adroit scheduling should ensure that the supply is applied advantageously.

In many cases the deployment of a limited resource is aided by the efficient use of float time available to noncritical activities. If only a single resource is limited, such as a one-of-a-kind machine, the activities requiring that machine can be labeled and visually checked to see whether the machine is scheduled to be in two places at the same time.

Another type of time chart for critical path networks is presented in Section 14-4.

Visual checks are easier if the network is converted to a time scale. One popular format is a bar chart, because it can be produced as a direct computer printout. The format shown in Figure 7.13 has each activity represented by a

I	J	D	Description	Day 1 2 3 4 5 10` 15
1	2	5	A	X X X X X I I
1	3	4	B	X X X X 0 0 0 I
1	6	8	C	X X X X X X X X 0 0 I
2	4	6	E	I I X X X X X X I
2	5	4	F	I I X. X X X 0 0 I
3	8	7	G	I I X X X X X X X 0 0 I
6	7	1	H	I I X 0 0 I
7	8	3	I	I I X X X 0 0 I
5	8	3	J	I I I X X X I

FIGURE 7.13 Format of a computer generated bar chart for the network of Figure 7.8 and the boundary times of Table 7.2.

bar of X's corresponding to its duration and beginning at the ES. The zeros following the X's represent the TF available to the activity. When activities are spaced to avoid overlapping assignments of a limited resource, the float in a chain of activities must be rationed judiciously because all the activities in the chain share that float. When the float is taken by an adjustment for one activity, it is no longer available for the others. For instance, Figure 7.13 is a bar graph for the network from Figure 7.8, in which the activity chain C-H-I has 2 days of TF. If activity C is assigned to start on day 3 (the end of day 2), then the start times for activities H and I will also be set at 11 and 12 days, respectively, because all the float in the chain is taken by the delayed start of C.

There are many versions of Gantt or bar charts, and other formats will be used in later chapters.

When no usable float remains for activities with excessive resource demands, a choice has to be made between acquiring more of the limiting resource and altering the scheduled times. Methods of modifying a schedule include the following choices:

1. Working overtime to shorten the duration of one or more activities requiring the limited resource.

2. Shortening the duration of prerequisite or postrequisite activities to allow float for rescheduling key activities.

3. Changing resource requirements. A substitute resource may change the duration of the activity involved.

4. Extending the total project duration.

The criterion for selecting from these choices will likely be cost, but less tangible factors such as goodwill and convenience may also influence the decision.

EXAMPLE 7.4 Expediting the project schedule for Example 7.3

A rumor that a competitor is perfecting a similar control unit has spurred the project manager to expedite the prototype field tests. A decision to reduce the project duration to 10 weeks has been made with the full realization that some confusion and considerable waste are sure to result. Through attentive network rescheduling, the disorder can be minimized.

The revised plan has "layout" preparations ($C1$ and $C2$) being done concurrently with "final design" (subdivided to $A1$ and $A2$). "Engineering analysis" (B) will start when the "final design" is partially completed ($A1$). "Assembly" is divided into two parts to allow the "requisition parts" (G) to be assembled ($J1$) before the "subcontract" (I) and "fabricate" (E) parts are received. The network resulting from these modifications is shown in Figure 7.14.

The changes reduce the project duration to the desired 10-week deadline, but the increased number of critical activities requires strenuous control measures. No activity times have been shortened, but the revised concurrent scheduling of some activities will certainly create challenging coordination problems. In general, the expedited project will probably cost more, be more frustrating, and come closer to meeting the 10-week milestone than the original schedule would have.

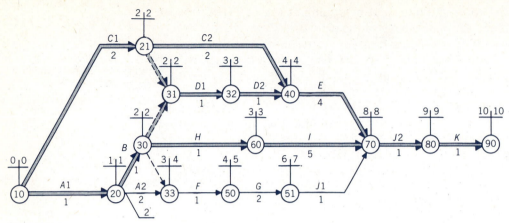

FIGURE 7.14

7-7 NETWORK APPLICATIONS IN PRODUCTION PLANNING

An example of product/project manufacturing is the production of heat exchangers. They differ in size and design, typically requiring 10,000 to 200,000 shop labor hours, and fabrication may take up to four years for completion.

The difference between project planning and production planning is fuzzy. Projects usually infer self-contained, one-of-a-kind products. Construction projects are an ideal example—seldom are two big buildings, dams, bridges, or housing projects identical. However, projects rely on the production system for accomplishment because they are a form of production. The boundary gets even fuzzier when certain types of job-shop production are considered. Because a **job shop** by definition is a place where individually different but similar jobs are worked on in the same facility, each job can be interpreted as a project—or as a product. The contrast then narrows to the degree of difference between products, both in design and resource requirements. This gray area is a fertile field for network applications.

Where CPM fits into production

Production environments that are most conducive to network planning have the following characteristics:

CPM is one of the rare quantitative management techniques that has equal or greater applicability to service operations than to manufacturing.

1. Products are complex and essentially one of a kind, yet several products are manufactured at the same time in one facility.

2. Start (go-ahead) dates and delivery schedules are difficult to predict (customers negotiate dates as their needs become known, and they tend to want delivery yesterday).

3. Design and manufacturing groups work as teams on each product/project.

4. Materials for each product are ordered individually because of the unpredictability of customers' needs for specific items.

5. Manufacturing methods and tooling vary from one product to the next, sometimes requiring significant modifications or new development.

6. Costs of work-force fluctuations, training/learning, and rework tend to be high, owing to lack of product repeatability.

How CPM helps job-shop production

Given the foregoing production characteristics, the value of network representations for job-shop production is evident for both planning and control. The inputs to the network are the usual ones spiced with a manufacturing flavor: Activities stretch from design and engineering, to shop fabrication, to test and delivery; activity times are often quoted in direct labor hours; and activity restrictions arise from facility limitations, material availability, and customer delivery requirements for components, in addition to normal sequencing constraints.

Networks are developed for individual components and tied together to portray the complete product/project. They encompass predesign consultation, design, product drawings, bills of material, tool requirements, quality assurance measures, and manufacturing procedures. Preliminary networks are reviewed for compliance with facility loadings from other products, opportunities to improve the design and manufacturing procedures, and error correction in material and fabrication requirements.

Once the bugs are out of the schedule and it has been approved for production by the different departments involved, the control phase begins. One way to track progress is to issue "start" and "completion" cards for key activities. These cards contain activity descriptions from the network *without* boundary time dates. They are dated when dispatched to the shop where actual start and completion times are recorded. Equivalent control over material is exercised by "issued" and "received" cards. Network updating is based on the returned cards; float times are adjusted and a new critical path is designated if necessary. Expediting is selectively authorized for activities to which the application of extra resources will have the strongest effect on the total project.

Continued refinement of CPM for production planning has fascinating possibilities. A library could be established for standard activity descriptions, complete with labor and material usage rate data, and standard subnetworks for segments of production that are frequently undertaken in essentially the same way. These segments would serve as building blocks to construct alternative production plans and to simulate future operations under different conditions. Networks of simulated projects built on projections furnished by sales and engineering departments could be integrated with ongoing projects in order to develop long-range plans. All of the information could be stored in computers and recalled on graphic display terminals to make product/project planning easier, faster, and more complete.

Adapted from CPM applications at Air Products and Chemicals, Inc., as reported by John E. Hughes, Jr., "Job-Shop Productivity Improvement Via Application of the Critical Path Method," *AIIE Proceedings*, Spring 1979.

Eventually managers will be able to plan individual projects on microcomputers and send them electronically to a corporate mainframe where they will be integrated into the master plan.

7-8 PRACTICE VERSUS THEORY IN PROJECT SCHEDULING

Both beginners and experienced practitioners have difficulty defining the end points of activities. In theory, all activities are independent and discrete, but in operation, they tend to overlap. For instance, as an activity nears completion, part of the crew assigned to it may commence the next activity while the rest of the crew finishes the current one; or the physical work on an activity may be completed much sooner than the associated administrative work, which is conducted by a different part of the organization, one that handles several activities at once and may even be following a different timetable. Overlapping could be nearly eliminated by making activities smaller and smaller until each potential overlap is itself an activity. However, such extreme detailing would not significantly improve the managerial contribution of networks. Both workers and managers accept the inevitability of some overlap and understand that it exists even if the network indicates otherwise.

Estimates vary on the cost of applying CPM (including updating) from 0.5 to 1 percent of the total project cost.

The predictability of activity functions, results, and completion times is a fundamental assumption of network theory. Difficulties in predictions arise from the following causes:

1. New technologies used in R&D work defy definition. It is unrealistic to expect accurate, detailed breakdowns of the steps required to conduct a project that has never been attempted before.

2. Activities are not always completed successfully the first time, but the network implies first time right, every time. Failure of an activity can create a whole string of new activities, as when the test of a prototype R&D model fails and the whole project goes in a different direction, or when the wall in activity "Build wall" collapses to cause a series of new activities: Clean-up debris, Move-in materials, Set forms, Rebuild wall (hope it stands).

3. Time estimates are largely subjective evaluations that are influenced by the nature of the estimator and the operating environment. Some people are naturally optimists, whereas others are more cautious, and their estimates reflect their nature. Degrees of optimism or pessimism soar or plunge with the unique conditions for each project. Highly uncertain R&D projects tempt conservative estimates, and cost-plus contracts may lead to shorter-than-usual estimates to increase the chances of winning the contract award. To the extent that personal moods and the estimating environment override normal judgments, the purposes of the network exercise are thwarted.

A counterargument is that network scheduling allows simulations of new conditions by simply varying times or sequences to observe the effects on other activities.

A criticism of network scheduling shared with other scheduling techniques is the way it locks management into an inflexible routine. Especially in R&D, a skillful manager needs flexibility to shift resources among activities in response to new information. In any project, unanticipated emergencies can delay activities on one path, causing the project manager to shift resources to another path while waiting out the delay. Minor modifications can be made by updating the original network, but major adjustments require a new project plan. However, the initial schedule guides adjustments and provides a basis for comparing the results of changes with the original estimate.

A multitude of computer programs have been written to handle special project conditions, such as minimizing both idle crew time on repeated projects and resource allocations when capital is limited. A continued parade of improved programs for cost and resource analyses has reduced application costs and the amount of computer memory required for the analyses of large projects. Schedulers now face the traditional information explosion problem: how to keep up with the developments that would make their jobs easier and their output more effective.

7-9 CLOSE-UPS AND UPDATES

Proliferation of project management practices

The complement of project scheduling is project management. Critical path methods are closely linked to the planning and control of projects. **Project management** is the inclusive name given to the techniques, philosophies, and practices for efficiently using resources to complete a limited but complex task. It is being applied more frequently and to larger undertakings.

A project can be of almost any size and character, but for productive purposes it is considered to be a discrete undertaking with defined objectives, limited resources, and a finite time duration. Constructing a skyscraper is a project, as are making a motion picture and conducting a political campaign. These fit the project management mold because they have designated budgets, major resource commitments, and specified timetables.

The modern era of project management is usually said to have started with the Manhattan Project which led to the atomic bomb. The Polaris project ushered in CPM. The space program to reach the moon raised technical input to a new zenith. Away from public programs, projects to develop new products proliferated as product life cycles shortened. Government agencies increasingly rely on special projects to rally support and concentrate resources on specific, limited missions. The project approach is popular.

The matrix management structure described in Section 2-5 is also associated with project management.

The penchant for creating task forces to confront problems in both the public and private sectors is evidence of the popularity of the "project" approach.

JUMBO PROJECTS AND MEGA PITFALLS

As technology advances, designs become more complex; projects increase in size; and demands on managers become correspondingly greater. For example, the production intricacies and reliability requirements for a project to produce a space vehicle are enormously more demanding than are those for most earth-bound vehicles. Recent construction projects, such as the Alaska pipeline and offshore platforms in the North Sea, dwarf previous undertakings. And many are hampered by harsh environmental conditions or unpredictable political entanglements. The nature of **jumbo projects** thus underlines the necessity for competent scheduling and management.

The concepts of network scheduling have not changed much in recent years, but the applications have. Instead of walls papered with PERT charts or cabinets full of CPM network segments, as they were a few years ago, relationships of time to resources to activities are now stored in computer memory where they can be called up as needed. PERT/CPM programs have evolved into interactive, on-line integrated information systems for planning, costing, and controlling the entire project life cycle.

Although computer integration has made massive projects manageable, many practices have had to be changed. In network scheduling, the level of detail has been sacrificed to reduce the total number of activities. Alternative sequences have had to be identified in recognition of technological and political uncertainties. More regulated feedback concerning schedule deviations has been required of subcontractors. And the weekly rebalancing of resource allocations has been instituted in high-priority projects.

One of the few unchanged aspects of project management is the shape of a project's utilization profile. The cost and level of effort still gradually climb as the project gets under way, plateau at a maximum level, and rapidly drop at the end. The familiar **S-curve** shown in the margin diagram represents the cumulative progress and expenses over the life of a typical project. On jumbo projects, represented by the dotted curve, the buildup portion tends to last longer, and the rundown may have a slight upward tilt at the end.

For such major construction projects as tunneling under the English channel, the following considerations that were not applicable a few years ago are now commonplace:

1. Decentralizing so as to move decision making closer to the work and reducing management layers to improve communications.

2. Integrating all participating parties into a governing group that plans and controls all phases of a project, regardless of their physical location.

3. Realizing the impact that governments have on start-up requirements and progress pitfalls, and the staffing needed to adjust to these influences.

Some experts say that 5000 activities is a practical limit for any project, but 10,000 to 20,000 activity projects are not uncommon.

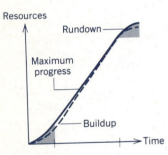

FIGURE 7.14

S-curve of project completion percentages and resource expenditures.

4. Understanding that the traditional practices for scaling up capacity and cost ratios from conventional projects to jumbo projects are seldom accurate.

TWO-FACED PROJECT MANAGERS

Having two faces is normally associated with double dealing and dishonesty. Not so for project management. Because project managers must simultaneously deal with functional and line managers, who may have opposing objectives, or with owners and contractors, whose interests often clash, they must dexterously expose the proper face to each. This dual treatment is most important when the project is besieged by time deadlines and resource shortages.

In the project buildup phase, personnel must be acquired along with the equipment and supplies needed to get started. An internal project to launch a new product is usually well supplied, whereas the start of a construction project in a remote area may be slowed by insufficiencies. Personnel are usually "borrowed" for internal projects, which may require lengthy negotiations to obtain desired skills; line managers are often reluctant to lend skilled workers. It may even be necessary to convince workers that they should accept a temporary project assignment.

No amount of planning can eliminate all the emergencies that plague projects. Project managers typically endure a series of crises from a project's inception to its conclusion. Multiple facings may be needed concurrently to reassure inspectors, motivate laggards, soothe upper management (or owners), and bully stubborn suppliers.

Project managers are advised to **manage by walking around (MBWA)** in order to discover irregularities before they become disasters. Supervisors who MBWA can detect upstream conditions that may adversely affect their operations and can pass along information that supplements regular reports. By facing both ways, observing what will soon affect their own operations and warning downstream operations of impending difficulties, the project management team can avoid many budget-busting, time-eating crises.

A new breed of manager appears to be emerging to handle projects. Besides being willing to travel and endure discomforts, contemporary project managers must relish challenges and be strongly people oriented. They must first convince everyone in the project that success relies on uninhibited cooperation and then use their technical expertise to coordinate activities so that success is possible.

Dr. H. Kerzner in "Project Management in the Year 2000," *Journal of Systems Management,* October 1981, suggests that the terms product and project management will become synonymous in the future.

The challenge and variety provided by project work attracts managers, but burnout is a career hazard.

7-10 SUMMARY

Critical Path Method

Network Construction

Program Evaluation and Review Technique

Time–cost trade-offs

Boundary times

And most important . . . . .
It self-destructs if we
fall behind schedule!

Project management

Network scheduling is a graphical approach to the sequencing and coordination of activities necessary to complete a project economically and on time. The first step in a CPM application is to break down the project into its component operations to form a complete list of essential activities. An activity is a time-consuming task with distinct beginning and end points called events. As the activity list develops, an order of completion is established by a restriction list, a statement of prerequisite–postrequisite relationships for each activity. From the two lists evolves a network drawn according to conventions, in which arrows representing activities connect nodes showing the sequence of events. Dummy arrows are included to allow distinctive nodal numbering for computer applications and to show certain event restrictions.

A single activity duration is estimated in the deterministic approach: a range of time estimates is used in the statistical PERT approach. With PERT, expected times (t_e) result from the formula $t_e = (a + 4m + b)/6$ where a and b are, respectively, optimistic and pessimistic estimates and m is the most likely duration. With activity durations estimated by either method, boundary times are calculated for all network activities to determine the float available for non-critical activities and the chain of activities that sets the total project duration—the critical path.

In PERT networks the variance of activities, $\sigma^2 = [(b - a)/6]^2$ can be used to determine the probability of meeting a certain scheduled completion time called a milestone. With deterministic durations, time–cost trade-offs can be used to identify the least-cost measures to reduce the project duration.

Computer programs for CPM applications are widely available. Manual calculations are feasible for smaller problems, but electronic assistance is almost a necessity for larger ones. In network analyses, a computer can perform the boundary-time calculations and generate bar charts for easier checks of resource assignments.

Network-theory assumptions that activities are independent, discrete, and predictable are not always met in actual applications, but this departure from reality does not seriously affect the planning and coordinating values of CPM.

Project management encompasses the skills and practices employed to conduct a discrete undertaking with limited resources and a finite time duration. When projects become larger and more complex, conventional techniques are less appropriate. Computer integration of far-flung project components has helped, but skillful management is the key to a successful project.

7-11 REFERENCES

ARCHIBALD, R. D., and R. L. VILLORIA. *Network-based Management Systems.* New York: Wiley, 1967.

BURMAN, P. J. *Precedence Networks for Project Planning and Control.* New York: McGraw-Hill, 1972.

CLELAND, D. I., and W. R. KING. *System Analysis and Project Management*, 3rd ed. New York: McGraw-Hill, 1983.

FORD, L. R., and D. F. FULKERSON. *Flows in Networks.* Princeton, N.J.: Princeton University Press, 1962.

HOARE, H. R. *Project Management Using Network Analysis.* New York: McGraw-Hill, 1973.

MEREDITH, J. R., and S. J. MANTEL, Jr. *Project Management: A Managerial Approach.* New York: Wiley, 1985.

MODER, J. J., C. R. PHILLIPS, and E. W. DAVIS. *Project Management with CPM, PERT and Precedence Diagramming,* 3rd ed. New York: Van Nostrand–Reinhold, 1983.

RIGGS, J. L., and C. O. HEATH. *Guide to Cost Reduction through Critical-Path Scheduling.* Englewood Cliffs, N.J.: Prentice-Hall, 1966.

SHAFFER, L. R., J. B. RITTER, and W. L. MEYER. *The Critical Path Method.* New York: McGraw-Hill, 1965.

WHITEHOUSE, G. E. *Systems Analysis and Design Using Network Techniques.* Englewood Cliffs, N.J.: Prentice-Hall, 1973.

WIEST, J. D., and F. K. LEVY. *A Management Guide to PERT/CPM,* 2nd ed. Englewood Cliffs, N.J.: Prentice-Hall, 1977.

7-12 SELF-TEST REVIEW

Answers to the following review questions are given in Appendix A.

1. T   F   An *arrow network* is an ordered collection of arrows that shows the sequence and times for the activities in a project.

2. T   F   The *critical path method* was created in 1956 to improve the planning and control of research and development projects.

3. T   F   No activity on the critical path ever has *float.*

4. T   F $TF = LF - LS$.

5. T   F $LS = ES$ plus the activity's duration.

6. T   F   The *critical path* indicates the longest possible duration of a project.

7. T   F   An activity with LS larger than ES always has float.

8. T   F   Arrow networks for large projects are normally drawn by *computers.*

9. T   F   When a project is falling behind schedule, consideration should be given to *reallocating resources* from noncritical activities to critical activities.

10. T   F   A one-day delay in finishing a *critical activity* will cause the project to be completed one day late unless a subsequent activity's duration is cut.

11. T   F   A *job shop* is the modern version of traditional hiring halls where workers gather to find what jobs are available.

12. T   F   A major limitation of CPM is the assumption that all *activity durations* can be accurately predicted.

13. T   F   A *project* is an understanding that has defined objectives and a distinct end point.

14. T   F   Triangular organizational structures with a strong leader at the top are associated with *project management.*

15. T   F   *Functional managers* are usually anxious to release their personnel to participate in internal projects.

7-13 DISCUSSION QUESTIONS

1. Graphical methods have been described as applicable in the areas of survey, analysis, and presentation. Explain and give an example of how CPM techniques can be used in each of the three areas.

2. Define each of the following terms:

a. Network   b. Merge   c. Earliest start
d. Activity   e. Logic dummy   f. Critical path
g. Event   h. Expected time   i. Float
j. Postrequisite   k. Milestone   l. Time–cost
                                            trade-off

3. What is the reason for putting the *earliest* of the LS times on the cross at a burst?

4. Discuss the three time estimates used in PERT networks with respect to deterministic estimates of activity durations. Which type do you think would be easier to obtain?

5. Under what conditions are m and t_e the same for an activity?

6. What is the probability of completing a project on the date set by the LF of the last activity in a PERT network? Why?

7. Will the probability of meeting a milestone SS less than the event's ES be greater when the variances of the activities leading to the event are smaller? Why?

8. What network convention would replace a situation in which the correct logic of a restriction was shown by a two-headed dummy?

7-14 PROBLEMS AND CASES

1. Complete the network for the project "Conduct a CPM Evaluation" begun in Figure 7.6. Use only gross activities such as "calculate boundary times" instead of detailed computation procedures for ES, LS, EF, LF, TF, and so on.

2. Some computer programs print out another type of float called *free float*. It is calculated by subtracting an activity's EF from the earliest start time of all postrequisite activities. Calculate the free float for the activities in the network of Figure 7.8. What value is there in knowing an activity's free float?

3. Describe the resources involved for an activity that would likely cause a time–direct cost curve similar to each of the curves shown in Figure 7.11.

4. Consider another pattern for the PAINT activity in Example 7.2. It is now assumed that when the paint job is unsuccessful, the paint must be removed before the new paint is applied. After the 10-minute paint removal, the product will undergo another 10-minute paint job. The sequence is repeated each time the paint job fails to pass inspection. The probability of a successful original painting and each successful retouching is 0.9.

a. Portray the successful-damage pattern in a network segment, and state the actual durations the PAINT activity could take.

b. Use probability theory to determine the expected duration of the successful-damage sequence.

5. Construct an arrow network segment to portray each of the following activity prerequisite–postrequisite relationships:

Set 1	Set 2	Set 3	Set 4	Set 5
$A < E$	$A < D,E$	$A < F$	$A < D$	$A < C,D,E$
$B < E$	$B < D,E$	$B < D,E$	$B < D,E$	$B < C,D,E$
$C < D,E$	$C < E$	$C < E$	$C < D$	$C < F$
$D < F$	$D < F$	$D < F$	$D < F$	$D < F$
$E < F$	$E < F$	$E < F$	$E < F$	$E < F$

6. Construct a CPM network to conform to the following relationships. Note that the activities are not in alphabetical order and that there are redundant restrictions:

Activity	Must Precede
A	C
B	A,C,F
D	C,F
E	A,C

7. Construct a network based on the following restriction list:

Activity	Duration	Restriction
A	3	A < E,F
B	5	B < C,D,E,F
C	3	C < H
D	4	D < G,H
E	6	E < G,H
F	7	
G	4	
H	5	

From the given durations, calculate the boundary times. If the *ES* for the first event is 23, what will be the *LF* of the last event?

8. The following list gives nodal-numbered activities and their durations. Each activity is described by an *i* and a *j* node with its duration in parentheses. What is the minimum time by which the project can be completed?

Activity:	1,2	1,5	2,3	2,4	2,5	3,4	4,6	5,6
Duration:	(5)	(4)	(3)	(4)	Dummy	Dummy	(2)	(7)

9. Activity descriptions, durations, and node numbers (in place of a restriction list) for a construction project follow. Complete a boundary timetable for the project.

i node	j node	Activity Description	Duration
1	2	Move in	2
2	3	Foundation	12
3	4	Lead time for parts	4
3	6	Concrete slabs and columns	25
3	14	Filling and grading	133
4	5	Dummy	
4	14	Air conditioning and plumbing	130

i node	j node	Activity Description	Duration
5	14	Electrical	130
6	7	Lead time for framing	3
6	8	Lead time for steel	20
7	9	Framing	20
8	10	Erect structural steel	15
8	14	Other concrete work	15
9	10	Walls and insulation	25
10	11	Roof deck	14
10	14	Lath and stucco	60
11	12	Carpentry	14
11	14	Roof and sheetmetal	15
12	13	Painting	14
12	14	Hardware	45
13	14	Finishing	25
14	15	Clean up	4

10. A firm seeks the opinions of the local populace about adding a new process to its manufacturing fa-

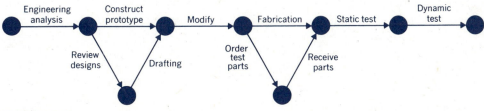

FIGURE 7.15

cilities that will create disagreeable odors. The smell should be noticeable about 20 percent of the time because of prevailing winds, but the process will provide many new jobs. A survey will be made by hiring specially trained personnel to call on 5000 selected households. The interviewers will follow a questionnaire designed for this particular situation. Set up a network to show the necessary activities and estimated times for the activities, assuming that you are the manager of the project and have one assistant to handle all details (questionnaire design, training, review, report, and so on). What estimate would you give for the completion time of the project for a 95-percent probability of meeting your milestone?

11. In what ways might the project shown in Figure 7.15 be expedited?

12. Given the following data for the activities required to launch the promotion of a new product:

Activity	Time Estimates			Postrequisites
	a	*m*	*b*	
Develop training plans	2	6	10	Conduct training course
Select trainees	3	4	5	Conduct training course
Draft brochure	1	3	4	Conduct training course, Print brochure, Prepare advertising
Conduct training course	1	1	1	
Deliver sample products	3	4	4	
Print brochure	4	5	6	Distribute brochure
Prepare advertising	2	5	7	Release advertising
Release advertising	1	1	1	
Distribute brochure	2	2	3	

a. Calculate the expected time for each activity.

b. Complete a boundary timetable.

c. Calculate the activity variances and determine the probability of completing the project in 12 weeks.

d. Calculate the probability that the advertising will be "released" by the time of its *LF*.

13. For the network shown in Figure 7.16*a*, answer these questions:

a. What is the most likely duration for the project? What is its probability of occurrence?

b. Using basic PERT assumptions, calculate the duration at which there is a 92-percent likelihood that the project will be done by that time or sooner.

c. What is the probability that activity *I* can begin after 16 days?

14. Computer assistance is a necessity for a time–cost analysis of a large project. However, a small project or a distinct portion of a larger network can be analyzed manually by using a bar chart or similar graphic aid to keep track of the trades between cost and time. Construct a network based on the following data and determine:

a. The normal duration of the project.

b. The minimum crashed duration.

c. The minimum cost to cut the project by one day,

Activity	Postrequisites	Activity Duration		Cost to Cut an Activity by One Day
		Crash	Normal	
V	—	1	1	No cut possible
F	V,M,O	6	6	No cut possible
B	F	6	7	$30
D	S	3	5	$50 per day
M	—	2	3	$40
S	V,M,O	5	6	$70
A	F	2	4	$25 per day
O	—	3	4	$85
C	D	2	2	No cut possible
K	A,D	3	4	$80

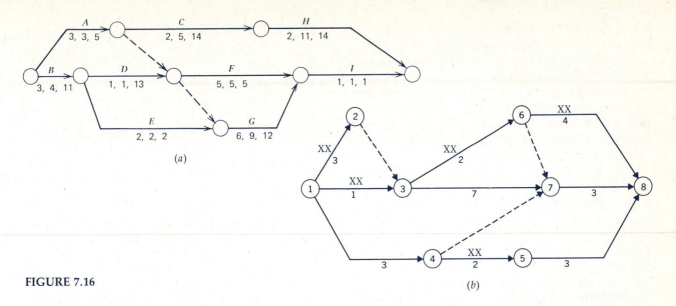

FIGURE 7.16

(a)

(b)

two days, and three days (indicate which activities are cut and their cost).

15. The activities in the network of Figure 7.16b that require the use of a special, one-of-a-kind, machine are labeled XX. Determine the minimum project duration, and show the schedule in a bar chart format.

16. *A Case of Tangled Restrictions*
You have been assigned to interview craftsmen who build specialized assemblies. You are to make a network diagram of the sequence of operations to record what must be done if the experienced craftsmen are not available when a special assembly is needed in the future. There are eight steps in the assembly process, identified as A to H. To make the exercise more realistic, have someone read to you the following directions that are attributed to the lead man responsible for the assemblies:

"OK, here's how we make it," said the lead man.
"First I start with A. I have to finish A before Ed and Bill can work on F and H. You can't do G or B before H is done. F also comes before G and C. And B has to come after D."
"C is the last thing we do, but we can't start it until G and B are done. Oh yeah, we have to get E from the shipping department in order to start C. There's another

thing too; Ed usually does G and B, but he can't do them at the same time."

As the directions are read to you, jot down your notes, and construct a rough diagram. What difficulties did you have extracting information from the monologue? Suggest improvements for recording project information obtained from an interview.

17. *A PERT Puzzle Case*
In a PERT network, all activities on the critical path have a probability of 0.5 of being completed by their earliest finish time. Occasionally a noncritical path terminating at a critical event will cause the merge node to have a lower completion probability than that derived from the critical activity ending at the same

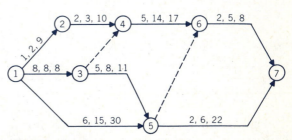

FIGURE 7.17 Three time estimates shown above each activity arrow.

event. This happens when activities on the noncritical path have very high variances. Therefore, whenever a scheduled start (SS) calculation is being conducted, it is necessary to check suspicious noncritical paths as well as the critical path, to determine which chain has the limiting (lowest) probability. For the network in Figure 7.17, calculate the limiting probability, and indicate the path from which it was taken for each of the following conditions:

a. To complete event 7 by day 25.

b. To complete event 7 by day 28.

c. To complete event 5 by day 16.

18. The Cost-cutting Case

Key production machines in a factory are overhauled twice a year. During the downtime period of the overhaul, the outage cost for an idle machine is $100 per hour. Activities involved in the overhaul were collected in a network, as shown in Figure 7.18, to see whether the repair and maintenance process could be improved. The following table indicates both crash and normal times and costs for the activities:

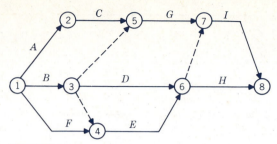

FIGURE 7.18

	Crash Schedule		Normal Schedule	
Activity	Duration (H)	Direct Cost ($)	(H)	($)
A	5	910	7	840
B	4	400	6	360
C	4	250	4	250
D	5	1100	9	1000
E	6	300	8	220
F	2	590	5	410
G	3	835	5	705
H	4	200	4	200
I	5	800	6	720

Because the project is relatively small, it is feasible to conduct a cost analysis without the assistance of a computer. One way to do so is to convert the network to a bar-chart format with the bar for each activity starting at its ES and followed by its float based on network calculations using *normal* times. The bar chart should show a total normal time duration of 22 hours. Then the cost per hour to cut each activity is computed by assuming that the normal and crash schedules are linearly related. For instance, the cost per hour to cut up to two hours off activity A is

$$\frac{\$910 - \$840}{7 - 5 \text{ (hours)}} = \frac{70}{2} = \$35 \text{ per hour}$$

To cut the total project duration, it is necessary to reduce one of the critical activities. When there is no parallel noncritical path, or if there is float available on a path parallel to the critical path, the only cost for a time reduction is the cost of cutting the critical activity. Otherwise, the cost per cut is the sum of the per-unit cutting costs for all involved activities. For the given data, the lowest-cost initial cut is through the critical activity A. The paths parallel to activity A involve activities B and G, both of which have float available. This cut reduces the project duration from 22 to 21 days and reduces the total cost of the project (indirect cost at $100 per day plus the sum of all activity costs) from

($100 per day × 22 days) + $4705 = $6905

to

($100 per day × 21 days) + ($4705
$$+ \$35) = \$6840.$$

Each additional activity cut should cost the same or more than the last time reduction but will lower the total project cost until a minimum point is identified (see Figure 7.12).

We can use the bar chart to keep track of successive time reductions by inserting a vertical line through the activity or float bar to indicate the position of each cut. No activity can be reduced in duration below the crash schedule. The procedure of selecting the next-lowest-cost cut and recording it on the bar chart is continued until the minimum total-project-cost duration is determined.

Calculate the most economical project duration and its cost. Show the revised schedule on a bar chart. Comment on the difficulty of obtaining time–cost estimates and of conducting time-cost analyses manually.

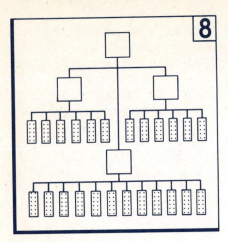

HUMAN FACTORS

LEARNING OBJECTIVES

After studying this chapter you should

- appreciate the wealth of human abilities and the factors that affect the use of those abilities in production systems.
- know how applicants are attracted to, screened for, and placed in job positions.
- be aware of the importance of proper orientation for new employees, characteristics of a successful training program, and the impact of high-technology on training.
- realize the complexity of motivation and be aware of the methods used to motivate work performance, including behavior modification and performance reviews.
- comprehend the causes of accidents and how a safety program works.
- understand how the structure of an organization affects management and why good human relations are vital to effective supervision.

KEY WORDS

The following words characterize subjects presented in this chapter:

human-stimulus interface

data identification and interpretation

job success criteria

permanent traits

employee–job matching

orientation programs

effective motivation

incentive pay

job enrichment

Occupational Safety and Health Act (OSHA)

unsafe acts

unsafe conditions

accident pyramid

organizational structure

management by exception

span of control

human relations

performance expectation

behavior modification

performance reviews

interactive training

mechatronics

8-1 IMPORTANCE

People initiate and operate production systems to serve other people. The inspiration for production is the satisfaction of human wants. The acceptance accorded a product is a measure of how well it was designed to fit the want and how well the production system functioned; not only must the design fulfill the intended purpose, the production process also must deliver a product of adequate quality at an acceptable price to the market where it is wanted. Figure 8.1 characterizes how people are benefactors, components, and beneficiaries of the production systems they design, construct, and use.

It is easy to recognize that humans are an integral part of complex systems, and it is easy to overlook that humans themselves are complex systems. There are many things they can do, and there are also many things they cannot do. An efficiently operating human–machine combination implies an efficient human operating an efficient machine. The muscle and blood component of the human–machine partnership deserves more attention than does the metal component because human beings' abilities and limitations are far more difficult to comprehend than are those of their metal partners.

In the following sections we consider the human factor. There are few exact rules to follow, but an awareness of human abilities can sharpen our daily observations of people in action and help us develop guidelines for analyzing their critical role in production. This role includes getting the right people for the right jobs, training them, maintaining a safe place to work, and providing supervision that produces effective motivation.

"Call a man a machine if you want to, but don't underestimate him when you come to do experiments on him. He's a non-linear machine; a machine that's programmed with a tape you can't find, a machine that continually changes its programming without telling you; . . . a machine that may try to out-guess you in your attempts to find out what makes him tick, . . ." A. Chapanis, *Research Techniques in Human Engineering* (Baltimore, Johns Hopkins University Press, 1957).

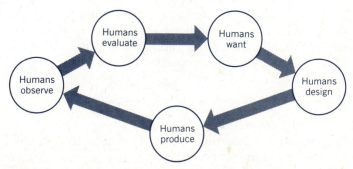

FIGURE 8.1 Human-oriented production cycle.

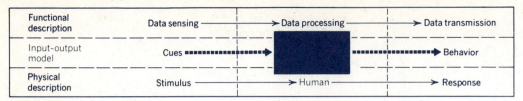

FIGURE 8.2 Black-box input–output model of a human.

8-2 HUMAN ABILITIES

Human factors should be the most fascinating and simple subject associated with production. After all, we are human and have daily contact with the factors that affect us. This exposure can lead to a false sense of security. Conversely, our familiarity can breed contempt for the subject; continued observations of unexpected behavior might even weaken faith in human disciplines. One way to avoid the skeptical and mystical aspects that clutter the interpretation of human behavior is to treat a person as a black box.

The "black box" treatment allows us to review the inputs and outputs of the "human system" unencumbered by the intricacies of the conversion process; the conversion of inputs to outputs is blanked out by the black box (Figure 8.2). By this procedure we can concentrate on observable reactions and leave the explanation of internal linkages to medical and psychological investigators.

Data sensing

It is estimated that people rely on their visual sense for 80 percent of their information about their environment.

The sources of data in any environment are tremendous in number and variety. People can detect many forms of energy such as light waves, temperature, or internal chemical reactions. They are also blind to the extreme ranges of energy such as infrared light waves or ultrasonics. Even when the energy is of sufficient intensity, it may still be unrecognized if they are not posturally set to receive it.

One advantage of an audio over a visual warning device is that it requires no postural set to be received.

The **human–stimulus interface** has been intensively studied. Although sensing ability varies between individuals and with time, it is relatively easy to anticipate which cues from an environment a person will be physically able to detect. What are more difficult and more important to anticipate are the data that will be specifically detected and thereafter will influence behavior. Several design considerations appropriate to a production environment are listed in Table 8.1.

Data processing

Humans have a miraculous ability to process the data picked up by their sense organs. They can identify the source of energies and distinguish differences in type and intensity. From these identifications and the overriding patterns of

TABLE 8.1
Design Considerations for Data Sensing

Data-sensing Factors	Production Examples
Only energies of certain types can be picked up by a human.	Dangerous radiation cannot be detected by a human's unaided senses.
Sensory organs are more sensitive to some kinds of energies than to others.	High-frequency noise is more annoying than is low-frequency noise. A bitter taste is easier to detect than is a salty taste.
The capacity of sensory organs is limited; only energy of sufficient intensity will be picked up.	Verbal communication at ordinary voice level is difficult in a noisy factory because of masking background noise.
Certain energies will not register on a human unless he or she is posturally set to receive them.	A gauge displaying vital information fails its purpose unless the operator looks at it.
The same type of energy can be received differently by each individual.	A color-blind person obtains little information from color-coded controls.
Sensory organs can be damaged by overexposure to energy.	Hearing loss may result from unusually noisy plant conditions.

attitudes and emotions, they derive meaning that leads to judgments and decisions.

DISCRIMINATION AND IDENTIFICATION OF DATA

The first data-processing step in the human–black box is the differentiation of stimuli contacting the body. Although the greatest discriminatory skills are visual, humans can make relatively fine discriminations among sounds, temperatures, pressures, odors, tastes, and the like. They can also discriminate sensations arising from within their bodies. Chemical, electrical, and mechanical activities inside the body add to and color the discriminations concerning outside stimulating energies. The aggregate discriminations are a human's basic contact with the world. They are the means by which behavior is determined and controlled.

Such **data identification** is the process of grouping differentiated sensations. In most instances, groupings rather than isolated sensations are acted upon. The ability to identify rests on the ability to discriminate and is the direct result of learning shapes, scales, and models. This knowledge is stored and used when needed. Beyond learning, the ability to identify also depends on skills of recalling, organizing, manipulating, and comparing. How far humans could go in storing and using experiences is difficult to determine. We do know that training can reinforce this ability and make them more adept in a particular function.

INTERPRETATION OF DATA

The workings of the human–black box assumes even greater complexity when we consider human social and personal abilities. Just as a person's behavior is limited by sensory and mental abilities, it is also limited by attitudes, interests, and adjustmental abilities.

An attitude is an opinion, belief, feeling, conviction, or emotional tone that colors each event a human encounters. The attitude is the background against which sensations and identifications are judged. Whether or not on a conscious level, such **data interpretation** means that people find themselves liking or disliking, approving or disapproving, wanting or not wanting, or somewhere along a scale of positive or negative feeling with regard to all they encounter. And they bring these attitudes to work with them. They may seem unreasonable or irrational, but they still influence people's behavior.

What is it that interests humans? They have been known to become interested in almost everything at one time or another. We hear statements such as "Jane could do a good job if she wanted to." What is really meant is that "Jane could do a good job if Jane *could* want to." The usual interpretation is that humans have limitations on sensory, mental, and motor abilities but not on their interests. Interests are as necessary to success on a job as is any other ability.

Many additional adjustmental abilities vary among individuals. Some people are ambitious, striving always to get ahead. Others are willing to accept things as they come. Still others do not want to get "ahead" and would refuse a promotion. There are conformists and rebels, leaders and followers, responsible persons and shirkers of responsibilities. Many times these abilities or limitations are temporary; often they are basic to a person. Obviously these traits fit a person better to one job than another.

DECISION MAKING

The ability to identify and interpret data has little value alone; it is a human's ability to use the data to make decisions for effective action that counts. This ability to reason, to manipulate ideas, and to judge is a human's most highly developed and valuable talent. In this capacity, a human excels over all machines. A human-system program for given input and output could follow the sequence shown in Table 8.2.

Individuals differ considerably in their abilities to manipulate verbal and numerical symbols and figures in space and to perform problem-solving activities. Quite often the word *intelligence* is used to refer to the level of ability in these activities. Although this term is highly ambiguous in some ways, it does provide a summary term for evaluating the general level of decision-making ability. But too often we regard this term with undeserved esteem. A high test score on a certain intelligence test does not guarantee that a worker will do well on all jobs. An individual who makes sound judgments on a written test may make very poor judgments on social problems or may panic under pressure. Some repetitive jobs may be conducted more effectively by a person with a lower

"Any attempt to build a machine that could operate at even a moron level of intelligence would call for an expenditure of billions of dollars and parts and equipment covering hundreds of square miles." A. W. Jacobsen.

TABLE 8.2
Data Input–Output Sequence of the Human System

Functional Description	Physical Description	Environmental Examples
Environmental data	Stimulus	Stop sign, warning siren
	Energy	Light waves, sound waves
Data sensing	Receptor organs	Eyes, ears
	Sensations	Light, sound
Data processing	Discriminations	Color, shape, pitch, loudness
	Identifications	Stop sign, warning siren
	Interpretations	Worthy of attention and action
	Decision	Must stop, take emergency action
Data transmission	Performance	Application of force to brake, report conditions

intelligence score. As a tongue-twisting, summarizing conclusion, we might observe that the intelligent use of intelligence tests tests the intelligence of the intelligence tester.

Data transmission

In a production setting, the overt behavior that follows data sensing and processing often takes the form of data transmission to other humans and to machines. Such behavior is observable, whereas the data processing is implied. Being easily discernible, communications among sentient beings or inanimate objects have been widely studied. In fact, we can speculate about the internal linkages and activities within the human–black box through observation of overt behavior. As is depicted in Figure 8.3, to design facilities that will improve *motor performance*, we study motor performance.

Overt behavior includes the exertion of force and other body movements.

COMMUNICATION WITH HUMANS

Data may be transmitted to another person by talking, writing, or signaling. Each device relies on symbols. This use of symbols reduces the basic sensations

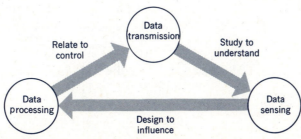

FIGURE 8.3 The motor performance cycle.

to classifications. Communication of these classifications attempts to reconstruct the original sensations of the sender in the mind of the receiver. Because the classifications are not identical to the sensations, structural distortions usually appear. Messages are further distorted by misinterpretation of symbols and by alterations of meaning attributable to the setting or the manner in which the communications take place.

COMMUNICATION WITH MACHINES

Humans have muscles, tendons, ligaments, bones, and joints, and they have neural connections between their sensory apparatus, their central nervous system, and their muscles. They operate and control machines and equipment by using all or part of this makeup. When machines or tools are not available or appropriate, they use the same makeup to act directly on their surroundings.

Humans are versatile but not infinitely so. They become fatigued; they can exert only so much force; and they can react only so quickly. In a production setting, the environmental conditions, machines, tools, and layout can be designed to enable them to function more effectively within the constraints of their abilities and the limitation imposed by their tasks. In the final evaluation, it is people's ability to communicate and their other motor performances that pay. The payoff is maximized by attention to their individual and collective abilities.

EXAMPLE 8.1 Human abilities versus machine abilities

One characteristic that typifies the present trend of production systems is the replacement of humans by machines. And machines are becoming more complex. The more ambitious automated systems tend to replace the muscle power of humans and, to some extent, usurp their decision-making responsibilities. One way to allocate and justify task assignments between humans and machines is to recognize where each excels. The following list offers a basis for the distribution of tasks.

A HUMAN IS MORE CAPABLE OF . . .

Handling unexpected events.

Profiting from experience.

Being sensitive to a wide variety of stimuli.

Using incidental intelligence with originality.

Improvising and adapting flexible procedures.

Selecting own input.

Reasoning inductively.

A MACHINE IS MORE CAPABLE OF . . .

Monitoring humans and other machines.

Exerting large amounts of force smoothly and precisely.

Consistently performing routine, repetitive tasks.

Rapid computing and handling of large amounts of stored information.

Responding quickly to signals.

Reasoning deductively.

8-3 PLACEMENT

Different workers produce different amounts, regardless of wages paid, working conditions, training, motivation, or basic abilities. The poor producer is an expensive worker. Overhead and capital investment are the same for him or her as for the good producer. The cost of hiring a poor worker is the same as for a good worker. Tenure or seniority systems may not allow the dismissal of poor workers, and often the poor worker cannot be identified until after his or her position has been established.

Special care must be exercised to avoid discriminatory selection procedures. Codes protect against discrimination because of sex, age, race, religion, and cultural beliefs.

It is imperative that a procedure for selecting employees results in some degree of accuracy in employee–job matching. Managers should never feel that they can secure perfect matchings by picking employees. It may be necessary to alter certain working conditions to fit the abilities of the people available or to attract applicants from other locations. More specifically, a job can be tailored to the employee in a reversal of the typical "employee wanted" appeal.

Establishing predictors

The objective of any employment program should be to detect good workers *before* they are hired. This requires spelling out what is expected of a good worker and how a good worker differs from a poor one. These goals are easier to state than to attain.

Before answering what makes a good worker, it must be decided what success on a job means. One way to define success is to select the most important traits or characteristics associated with specific jobs. These **job success criteria** are largely subjective opinions of management, and they vary from plant to plant for the same job as well as from job to job. Although the judgments are subjective, they should be made on the basis of objective observations of specific situations. Another approach is to study trouble spots in production. A definition of failure is perhaps a backdoor entry to success predictives, but each trouble spot should reveal at least a strong hint of common criteria associated with deficiencies. The more serious the trouble is, the more important are the criteria. Some common criteria connected with success and failure are shown in the margin notes.

Success Criteria

Quantity or quality of production
Equipment damage
Training time required
Longevity
Versatility
Promotability
Absenteeism
Accident rate
Material spoilage
Horseplay
Grievances
Company loyalty
Compatibility

Still further judgments are necessary to make the criteria meaningful. How much spoilage of material is too much spoilage? How much absenteeism is more than can be tolerated? High production is high only in that it is higher than less production. There is a point along the scale of each criterion at which performance is not good enough to justify presence on the job. This is the management-determined cutoff point. Workers with performance levels above the cutoff point are acceptable and successful employees. It is simple logic that an employer should attempt to hire only those people whose performance will be above the cutoff point on the criterion scales after they go to work.

The only sure way to find the performance level of applicants is to hire them and observe their performances. When no criteria are available, as in a new type of job, this random-selection method might be feasible; otherwise the

Permanent Traits

Physical dimensions

Age

Sex

Appearance

Race or nationality

Acuity of senses

Attitudes, sociability, and mental measures

hit-and-miss procedure could lead to an expensive game of musical chairs. A better procedure is to find the **permanent traits** of those individuals who are above the cutoff points that are different from the permanent traits of those individuals below the limits. For instance, it may be found that for a certain job, "tall" individuals (defined as over six feet) are "safer" workers (defined as having less than one lost-time accident per 5000 working hours) than are "short" workers who consistently fall in the group of "unsafe" workers. Thus height would be considered a permanent trait that could be used to predict success on that particular job. The designer of a placement program has the duty of matching such traits from a worker's historical and testing program records with his or her on-job performance records. The identified traits should, at a minimum, provide a predictive, **employee–job–matching** yardstick by which to measure candidates in the labor market.

Attracting applicants

There is little merit in developing performance predictives if the labor supply is severely restricted. Selectivity is possible only if there is a qualified pool of applicants from which to choose. It may be economically advantageous, or sometimes a necessity, to *entice* workers to apply for work. These campaigns operate according to the age-old selling principle: if it is worth buying and buyers hear of it, it will sell. They are also subject to merchandising principles regarding false advertising and tarnished reputations.

Worker Attractions

Starting salary

Potential growth

Company prestige

Geographic location

Regular salary increases

Permanent position

Paid vacations and holidays

Quick advancement

Insurance and health plans

Company size

Educational facilities

Good climate

Retirement or pension plans

Housing

Challenging work

The rule is to offer what people want. In the job market the wants may be money, prestige, location, working conditions, or many others. Again, it is management's task to determine which wants are in demand and then to weigh the cost of satisfying them against the benefits of selectivity. The task is greatly simplified by considering hiring while planning the plant. Locating a plant near a known labor supply cancels the need to entice qualified workers from other areas. Similarly, inviting surroundings and clean, safe, working areas attract available workers to specific plants.

When employers cannot offer exactly what workers want, employers can teach workers to like what they have to offer. In ancient times humans were induced to work by use of a whip. The want of freedom from the whip was taught by lashings; the means, of course, to achieve the want was work. The modern version is a Madison Avenue selling campaign that creates a want or cultivates a dissatisfaction. "Come to California where you can work and play in the sun all year." Or conversely: "Tired of the same monotonous climate? Enjoy the change of seasons and your work in the Midwest!"

EXAMPLE 8.2 A program to improve job attractiveness

Some time ago it became necessary to establish new research and testing facilities for nuclear investiga-

tions. The project was sponsored by the government but administered by private companies.

The first problem encountered was where to locate the establishment. The nature of the work decreed a large, nonpopulated area with abundant water resources. These conditions, though not conducive to recruiting employees, were unalterable. The best compromise location was a large government-owned, arid tract near a big river and a medium-sized town.

When the physical facilities were completed, the personnel problem arose. Highly trained technical workers were absolutely necessary. But high wages alone were not enough to attract and, particularly, to keep the caliber of employees required. The factors needed for efficient operations were opposed to contented workers.

Effort was directed toward the "intangibles." Instead of trying to improve the small government housing area near the reactors, modern buses were purchased to run between the test area and the nearest established town. These towns were encouraged to and aided in developing recreational outlets. Some rental homes were built in the towns, and choice building sites were made available for individual construction. With the promise of expanded business, more retail stores moved into the area. Cosponsorship of company and civic groups provided new educational and cultural activities. The local community was well informed of the nature of the work and the type of employee at the nuclear base. Whenever possible, local people were employed for routine work. Before long the atmosphere changed from a company town to a town's company.

Some industries are very adept at attracting applicants. Some also suffer from unkept promises. Unless the hiring enticement is accompanied by job satisfaction, the selling effort is a waste.

Screening applicants

Once there are enough applicants for a job to allow a selection, the screening phase begins. It is a two-sided process. The intentions are to pick the applicant whose prediction of success for a particular job is greatest and to create a favorable impression while doing so.

An applicant's physical traits can be determined from observations or questions. Some potential sources are listed in the margin notes. In reviewing the sources, a skeptical attitude is not unhealthy: recommendations reveal no more than what a selected recommender says; work histories tell only what someone did somewhere else; and even a carefully designed testing program reveals only what certain tests show.

Screening information should be obtained as quickly and painlessly as possible for the mutual benefit of the applicant and the hiring agency. A well-designed employment flow system gathers only the needed information through a series of progressively restrictive trials. For instance, an application received by mail could reveal the lack of desired physical traits and thereby save the applicant a wasted visit. An interview could eliminate a contender before he or she is subjected to an expensive battery of written tests. The number of such hurdles is largely a function of the level of employment sought, but regardless of the level the applicant deserves respect. A comfortable screening experience should leave rejected applicants feeling they have been treated fairly, and successful applicants should believe they have found a pleasant place to work.

Sources of Predictive Information

Test scores
Physical exams
Interviews
Job experiences
Recommendations
Mock-up trials
School records

Employee–job matching

When a key position is to be filled, such as that of a research scientist or plant manager, the predictives are definitive and the candidate selected by the screening process is tailored to the position. Screening approval for people hired for general production does not necessarily carry with it an assignment to a particular job. Different jobs usually require special skills, attitudes, or adjustmental abilities. Some individuals possess several skills and the versatility to suit several openings.

The allocation of individuals to particular jobs is sometimes a matter of whim or convenience: department *A* gets the first 10 people hired today, or a clerk arbitrarily sends prospects to departments. Objectivity is increased by referring only to the physical traits and predictives related to a definite job. In the case of application blanks, this objectivity is accomplished by making an overlay based on the cutoff points determined for each job. The overlay is called a *job template* and takes the physical form of a stiff, blank sheet of paper with holes cut in it. The template is placed on top of a completed application blank. Only the pertinent information for the given job shows through. The levels of traits revealed by the template windows are compared with the cutoff levels developed for the job. Because each job, and possibly the same job in different departments, has distinctive windows and varying cutoff scores, the template differentiates the applicants to provide at least preliminary assignment directions.

8-4 TRAINING

From the minute an employee sets foot on plant grounds, the training program is under way. Learning occurs whether or not production is benefited by it. By refusing to direct the learning experiences of workers, an employer does not stop the learning process—he or she merely forfeits control to chance or other factors. When workers want something, they will set about finding a means of achieving it. If what they find does work, they will seek more effective methods. The broad objective of a production training program is to help people discover what behavior is rewarding and how to develop skills that lead to goal satisfaction.

Orientation

The time to start training is the day employment begins. A worker who goes to work with little or no knowledge of company policies, regulations, and purpose starts with a severe handicap. He or she may discover these things as days pass; he or she may also learn what is not so and piece together a distorted picture.

A knowledge of what a plant has to offer a worker and what the worker is counted on to give can be achieved by a simple orientation program. Here is where morale has its start. The layout of the plant, what goes on here and there,

The assignment method discussed in Section 6-3 provides a systematic way to match candidates with position openings.

Gossip is the wild card of orientation. It can work for or against and is nearly impossible to control. As E. W. Howe observed in his *Country Town Sayings:* "What people say behind your back is your standing in the community."

office rules, and plant regulations must be known to some extent before the worker can fit comfortably into the scheme of things. Many organizations give the worker a period of indoctrination at company expense. Often this program includes a conducted tour, introduction to key personnel, a series of talks, a lecture, a movie, a question-and-answer session, a packet of literature, or perhaps a day to wander around and investigate the plant.

The extent and nature of **orientation programs** vary widely, but all should be aimed at specific objectives. A review of trouble areas is fertile ground for harvesting objectives. For instance, if accidents have been a problem, safety practices should be stressed. The message may be delivered subtly by inferences and suggestions or blatantly by shocking examples and rigid directions. The best routine depends on the message, the audience, and the skills of the orientators. Worker-oriented guidelines that accent the human factor are

1. Make the new employees feel relaxed and part of the group.

2. Explain what is expected of them.

3. Be sure they know their immediate boss.

4. Keep them occupied with meaningful activities.

5. Instruct them properly and show confidence in their abilities.

6. See that the employees have all necessary tools and materials.

7. Give credit where it is due or criticize constructively.

Training programs

Humans cannot always do today what they might be able to do tomorrow. Through training, they learn the skills that make their intrinsic abilities valuable to a production system. All humans can learn, but some are more proficient at it than others are. A training program has to be planned for someone. A general indoctrination can be aimed at everyone. Special programs are directed to individuals or segments of a working group to develop certain abilities. Individuals may be selected because of promotions or recognition of special attributes. Groups, such as new employees, supervisory personnel, individuals on express assignments, or different levels of management, receive training that should overcome present or anticipated troubles. The extent of the training program largely depends on whether the troubles are localized or widespread.

There is no one answer to the question of what training methods to use. Many types have been developed, and all have their champions. Probably the most reliable rule is to select the method that is most likely to alleviate the problem for which the training program is initiated. For instance, a program to develop specific job skills should be based on the study of errors—missing skills, steps of an operation being left out, or inefficient application of existing skills. If all workers share the same shortcomings, an inclusive demonstration and practice program is appropriate. When just a few workers make errors, a su-

Training Methods

On-the-job: task training at the work place.

Vestibule school: separate classroom area specially equipped with training aids.

Apprenticeship: working with an experienced person for craft or executive training.

Conference: preplanned discussion or problem solving, often with outside experts.

Role playing: preplanned exchange of roles.

Affiliated programs: liaison with schools to teach technical skills.

pervisor can work with them individually while they stay on the job. Several common training methods are listed in the margin notes.

Although the definition of the training objective is a prominent step in the training program, it is only the first one. Unless the sessions are well organized, they will likely fail to satisfy the objectives. Selected students must understand the importance of the program. The facilities and the teachers should be fully prepared. Nearly everyone has suffered through boring, incompetent presentations held in inferior classrooms with inadequate teaching aids. The irony is that many of the former students become teachers and perpetuate the same obstacles to learning. Good teaching requires careful attention to details and plenty of preparation time. Great teaching takes even more time plus a flair that appeals to and inspires students. Because the cost of industrial training programs is due mainly to the students' lost production and wages, every effort should be made to ensure that the presentations make the most profitable use of available time. Some characteristics of successful training programs are

Too often training concentrates on "know-how" at the expense of "know-why."

1. A well-qualified instructor with an ability to teach well.

2. Carefully planned subject matter that stresses methods designed to counter specific troubles.

3. A detailed timetable of instruction that is followed.

4. Adequate physical facilities (classrooms, mock-ups, materials, and the like) that allow instruction without disturbing interruptions.

5. Motivated students that know the purpose of the training and are eager to learn.

6. A schedule long enough to include time for practice and repetition.

7. An atmosphere that allows students to make mistakes and to learn from corrections.

8. A summary that relates all parts of the training to the overall intent.

9. A follow-up routine to reinforce learning and confidence.

10. Program evaluation to detect faults that will pave the way for future improvements.

8-5 MOTIVATION

Humans are employed in a production system for the work they do. This work is physical, mental, or both. It takes effort. There must be reasons for workers to expend the effort. Their behavior in any situation, work or play, is the result of a complex pattern of cause-and-effect relations. It is a temptation to grab the first apparent cause, usually wages, and proclaim that its effect is work. The extension of this cause–effect formula is that the more you pay, the more work you will get. Though there is some merit in the "dollars equal doing" equation, it is incomplete. The problem is to find what other elements should be included in the motivation formula.

During the World War I era, the prevailing philosophy was that people worked solely to feed and shelter themselves. The way to get more out of them was to threaten in a tough, dictatorial voice backed by the power to fire them. From the 1930s through most of the 1950s, the feeling switched to opinions that people worked out of "loyalty" to an organization; therefore the way to increase productivity was to organize company ball teams, publish chatty house organs, pump in soft music, and pile on the fringe benefits. Today we are still searching. Money retains its influence as a motivating factor up to a certain level. Beyond that point, money and fringe benefits do not automatically lead to more productivity. Even the power to fire is a diminishing threat and motivating force, owing to increased worker mobility and skill shortages. Perhaps the motivators that in the long run make better workers are psychological—feelings of responsibility and accomplishment.

Statesmanship
⇧
Participation
⇧
Paternalism
⇧
Servitude
⇧
Slavery
⇧

The management metamorphosis from manipulation to motivation.

Want strength

"Man does not live by bread alone" is a threadbare statement because it states so clearly a human's motivational needs. Employees who are paid a reasonable wage, a want no one will deny they have, usually show up for work. How much they do on the job depends a lot on how well the job provides for the whole pattern of their wants. Their productivity is influenced by their personal wants, such as the approval of those they respect or their vision of success. It takes more than money to motivate them.

A want pattern starts with the basic physiological need for food, water, shelter, exercise, stimulation, and other factors of well-being. Less tangible wants, often referred to as psychological needs, are superimposed on the basic pattern. People may want prestige and not actually need it in a physiological sense; but if they want it, then they need it psychologically to the exact extent that they want it. Humans do not always want what will benefit them. Their wants may seem completely irrational to an impartial observer, but what they want is what they believe they need. And for these wants they will work.

Personal wants are specific. It is not just recognition that is wanted; it is a certain kind of recognition from certain people. Promotions, attention, status, freedom, and security all are broad categories within which lodge the special wants of each individual.

Resistance to effort

In nature, a state of rest is the equilibrium position. To change the state takes effort. We can view the human–black box similarly. In a state of rest, humans have few of their wants satisfied. To satisfy a want takes effort. The want and the resistance to effort are opposing forces. For work to result, humans' wants must be stronger than their resistance.

Resistance to effort varies with individuals and situations. If people are sick, weak, or fatigued, their resistance will be greater even when their wants remain strong. Noisy or unsanitary conditions, personality clashes, and autocratic supervision contribute to the resistance. Some people are more mindful of particular discouragements than others are. One person can ignore distrac-

Contributors to Resistance
Fatigue
Discouragement
Lack of interest
Illness
Weakness
Lack of success
Laziness
Futility
Tension
Emotions
Drugs
Inferiority

tions, whereas another will surrender to them. Nevertheless, attention to well-established design considerations for human comfort will obviously reduce resistance for the majority. Further attention to human relations by supervisory personnel can cater to special problems of individuals and minority groups.

Effective motivation

The difference between want strength and the resistance to effort, as depicted in Figure 8.4, is **effective motivation,** the driving force to the exertion of effort on a particular task. If the task is too difficult, individuals will not attempt to satisfy the want. Thus, more difficult tasks require special incentives, or the tasks must be redesigned to allow their accomplishment with less effective motivation. Bonus compensations are an example of the former, and machine-assisted operations are an example of the latter.

Most work-for-pay situations are at a disadvantage from the start. In industrial plants the workers exert effort to earn rewards that must be translated later into want satisfaction. Thus, if someone wants a new boat, instead of building one, he or she will work at a job that pays enough to buy the boat. Workers on jobs that result directly in their own want satisfaction, such as building a boat, will select the most effective methods they can find and will work enthusiastically.

Profit-sharing plans are attempts at creating in employees the feeling of working for oneself. The **incentive pay** and bonus schemes are designed for the same purpose. Many managers have concluded that money in routine amounts (such as an annual raise to reflect increased productivity) is largely taken for granted, anticipated before it arrives, and viewed as a justly deserved reward for past services—not as a stimulus to new effort.

Applications of incentive wages are described in Section 10-7.

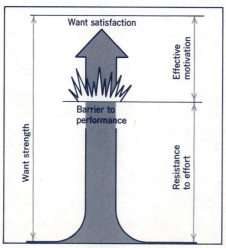

FIGURE 8.4 Want strength minus resistance to effort equals effective motivation.

Another factor affecting the reward system is the degree of immediacy. It is easier to increase effective motivation if wants are satisfied within a short time. Credit buying is a classic example of attempts to shorten the waiting period for want satisfaction. Some say it is not the employer's business how workers spend their money; but if workers waste their money and do not get their wants satisfied, job performance will suffer. Employees who have to use their wages to pay off old debts soon lose their motivation to work. Company credit unions and worker counseling help combat this type of problem.

No one has been able to pin down the all-purpose, magic motivator, if indeed it exists, but new approaches are continually being promoted. Current attention is largely devoted to behavioral approaches that attempt to make work more satisfying to the worker. A well-known example is **job enrichment,** described by its chief sponsor, Professor Herzberg, as an approach that ". . . seeks to improve both efficiency and human satisfaction by means of building into people's jobs, quite specifically, greater scope for personal achievement and recognition, more challenging and responsible work, and more opportunity for individual advancement and growth." He further stated, "The term 'job enrichment' is firmly lodged in the vocabulary of managers, behavioral scientists, and journalists. . . . The result has been that job enrichment now represents many approaches intended to increase human satisfaction and performance at work, and the differences between all the approaches are no longer clear."

> From "Job Enrichment Pays Off," by W. J. Paul, Jr., K. B. Robertson, and F. Herzberg, *Harvard Business Review*, Vol. 47, March–April 1969.

> From "The Wise Old Turk," by F. Herzberg, *Harvard Business Review*, Vol. 52, September–October 1974.

A few of the management strategies for improving worker motivation are as follows:

Job enrichment stresses opportunities for individual growth on the job by giving greater depth (responsibility) to work assignments.

> See Example 8.3.

Job enlargement concentrates on giving greater breadth (diversity), through administrative directives, to the tasks performed by a worker.

Participative management solicits employees' support by seeking their advice regarding job-related decisions.

Industrial democracy places worker representatives on all decision-making bodies in the organization.

> Developed and implemented mostly in Europe.

Organizational development attempts to change attitudes and values so as to improve communication and reduce conflicts.

> "Sensitivity training" is associated with the OD approach.

These strategies have a behavioral flavor instead of the efficiency of engineering emphasis discussed in Chapter 9. An ideal situation satisfies both the mental and physical conditions required for a motivated work force.

The choice of which motivation strategy and tactics to use depends on the particular factors present in the production setting. Job expectations differ according to the nature of the work and the work force, but safer and more pleasant working conditions are always appreciated. The opportunity to perform a variety of operations may make the work more interesting to some people, and being able to participate in managerial decisions may be valued by others. The success of any approach depends heavily on the capabilities and convictions of managers and supervisors. No approach is a guaranteed trigger for a burst of motivation, and higher motivation does not necessarily translate into production profits. That is what makes human factors intriguing.

EXAMPLE 8.3 Job simplification versus job enlargement versus job enrichment

The three philosophies of job design—simplification, enlargement, and enrichment—engender controversy. Simplification, historically the initial thrust toward work improvement, is based on the concept that jobs are best performed when fractionalized, resulting in simple, specialized, repetitive tasks. Attempts to make work more interesting by gathering several of the specialized tasks into one job assignment led to work enlargement: this horizontal job loading is a form of job enrichment. Vertical job loading, in which new responsibilities are added, is the generally accepted version of job enrichment. In a fully enriched job, employees participate in the planning, organizing, and controlling of their work.

The benefits of job enrichment have been widely proclaimed. *Work in America*, a 1973 report by a special task force to the secretary of Health, Education and Welfare, tabulates 34 job enrichment studies. Spectacular improvements are claimed: 6- to 20-percent increases in production output, 25- to 800-percent reductions in absenteeism, and 4- to 2300-percent increases in quality. Such impressive accomplishments suggest that all jobs should be enriched. Although the argument is persuasive, there are also reasons to impugn claims for automatic production improvement from enriching jobs.

Work simplification has both advantages and disadvantages when it is compared with either horizontally or vertically loaded jobs. Many jobs technically and physically conform to a chain of short operations and lend themselves to specialization: machine-paced (for example, conveyor) lines and self-paced assembly are examples. It is generally conceded that less training is required with shorter work cycles and that this reduces costs of supervision, initial training, and replacement labor for resignations, sickness, leaves, and the like. The combination of lower personnel costs and highly effective use of physical resources makes simplified jobs appealing for many production settings. However, highly repetitive jobs tend to be monotonous and may have an adverse effect on workers' mental health, or at least on their feelings of satisfaction toward their work.

The introduction of more variety (range of operations), autonomy (control of scheduling), identity (complete work on an entire item), and feedback (information on performance) enriches a job. The outcome is often a strongly motivated and satisfied jobholder. Exceptions have been noted in the case of older workers who prefer paced lines and are more compatible with assembly-line work. (R. Stagner, "Boredom on the Assembly Line: Age and Personality Variables," *Industrial Gerontology*, vol. 2, no. 1, 1975). Studies in Finland suggest that enriched jobs have a higher accident frequency rate than do preplanned, standardized jobs. (J. Saari, and J. Lahtela, "Job Enrichment Cause of Increased Accidents?," *Industrial Engineering*, October 1978.) Other studies have criticized the higher cost per worker that is associated with job enrichment and the subjectivity of reports that people prefer enriched over simplified jobs. All of which proves that there is still a lot of artistry in management, and that motivation is one of the highest forms of the art.

8-6 SAFETY

Safe working conditions contribute to effective motivation by freeing workers from physical danger. Yet safety is an enigma. Employees do not want accidents, but they often form habits or put themselves in situations in which accidents are inevitable. Employers know that accidents are expensive, but few recognize the true total cost.

We talk about a completely safe design, but there is no such thing. We cannot even agree on the worth of an arm, or a leg, or a life. Individually we would put an infinite value on the lives of our intimates, but how many lives must be endangered before we finally undertake expensive or inconvenient correctional measures? That we cannot afford to assign a lifeguard to each swimmer in a public pool or erect an automatic warning device at every intersection, crosswalk, and railroad crossing is not a sign of inhumanity; it is a realistic compromise forced on us by a lack of resources. Concern for safety has never been greater than it is today. Now the problem is to translate this concern into acts that meet our philosophical and ethical objectives within the constraints of practical economics.

The first safety problem is to admit that safety is a problem. Workers must not believe that accidents always happen to someone else. Management must believe that accident prevention is worth the cost. Furthermore, both workers and managers must remain impressed with their safety responsibilities for successful long-term programs. A particularly macabre incident such as a person being crushed, burned, or sliced to death can spur immediate dramatic safety measures, but continuous attention to more mundane accidents builds enduring safety patterns. Table 8.3 lists some benefits of safety practices that should have an obvious and prolonged appeal if those involved ponder the consequences. Each reason listed could be detailed to stress its importance. For instance, "improves reputation" could mean a wide choice of jobs and better pay to an employee; for the employer, it could build public opinion and attract better employees.

When it is accepted that safety effort is emotionally satisfying and economically logical, the next step is to determine why accidents occur. There are

Accidents are unplanned, unwanted, and improper occurrences involving the motion of a person, object, or substance that results in injury, damage, or loss. Fail-safe is a system design concept to safeguard major functions against even freak events. Only the once-in-a-million event will trigger an accident.

Accident Costs to Employer

Direct:
  Compensation
  Medical
  Legal
Indirect:
  Machine downtime
  Training and replacement
  Equipment damage
  Material waste
  Reduced production

Another responsibility of companies is consumer safety through the quality of its products.

TABLE 8.3
Reasons That Attention to Safety Is Important

To Employees	To Employers
Prevents physical suffering	Prevents waste
Prevents loss of earnings	Prevents cost of retraining
Protects future earnings	Prevents unnecessary equipment replacement
Prevents mental suffering	Reduces insurance and compensation costs
Improves reputation	Improves reputation

A conservative estimate of the ratio of hidden costs to direct costs is 4 to 1.

Human Characteristics That
Affect Safety
Recklessness
Stubbornness
Nervousness
Slowness to learn
Physical condition
Personal troubles

two basic causes: unsafe acts by workers and unsafe working conditions. Every accident has overtones of both causes. The reason behind identifying the cause is to categorize the effort that will prevent the accident.

Unsafe acts

Faulty work habits and careless worker behavior are classified as **unsafe acts.** Automobile drivers engage in unsafe acts when they exceed speed limits, disregard traffic rules, misjudge clearance, or fail to signal their intentions. Such behavioral patterns are learned over a period of time. As accident causes, unsafe acts are harder to detect and more difficult to control than are mechanical causes. They also probably account for the most accidents and injuries.

A skilled supervisor can detect unsafe acts by observing his or her work force. Too often the acts are discovered only after they cause an injury. Before the injury finally attracts corrective attention, the acts probably delay production, waste material, and cause spoilage or rejected parts. It takes a well-trained supervisor to find the acts and even more training to eradicate them. Anyone who has tried to overcome a long-standing habit knows how difficult it is to change behavior.

Unsafe Acts
Ignoring rules
Operating without authority
Teasing and horseplay
Not wearing protective
equipment
Unsafe lifting, pulling, and
pushing
Making safety devices
inoperative
Improper handling of
equipment

Special training or retraining efforts can be directed to correct specific unsafe acts. A general awareness of bad habits is possible through safety literature, posters, committee work, and meetings. Protective equipment such as steel-toed shoes or safety guards may allow a mistake or two without serious damage, but such safeguards do not develop better working habits. Protection from one hazard may even create a new hazard; protective gloves could make workers clumsy, or safety goggles might deprive them of needed vision.

Accident proneness is a convenient label to characterize some people and some jobs. The term does not really refer to the causes of accidents; instead, it describes the number and seriousness of accidents that have already occurred. It shows a human–job misfit. A reputation for accident proneness is a justifiable excuse for investigation. A job may be particularly dangerous, or a person may be a daydreamer. If so, the job can be changed to make it safer and the person can be retained, retrained, retired, or fired.

Unsafe conditions

The federally legislated
**Occupational Safety and
Health Act (OSHA)** began in
1971. Its main impact on
production has been to
heighten concern about the
condition of equipment and the
work place. It imposes fines for
uncorrected safety hazards.

Accidents attributable to physical and mechanical sources within the work environment are caused by **unsafe conditions.** Disregard for design principles in any of the environmental factors (layout, equipment arrangement, illumination, noise, atmospheric conditions, and so on) leads to unsafe conditions. Every piece of equipment, any work-space design, and all operations must be suspects for safety improvements. Both the detective work and the corrections are easier for unsafe conditions than for unsafe acts. A checklist for accident symptoms and cures is as follows:

Falling—nonslip floors, guardrails, handholds, and safety belts.

Snagging—barriers, warning signs, recessed controls, and elimination of projections that catch clothing.

Pinching—adequate room for entering and leaving, easily reached tools, correct tools, and better equipment arrangements and designs.

Bumping—increased overhead clearance, screens to catch falling objects, hard hats, warnings from moving equipment, elimination of blind corners, proper loading of material-handling equipment, and painting of objects to make them more obvious.

Burning—protective clothing; emergency showers; strict maintenance of equipment involving hot metal, glass, liquids, steam, and air; color coding and proper insulation of wiring and electrical parts; warning of repairs under way; adequate fire-fighting equipment; alarm system; and first-aid training.

Blinding—goggles, shields on abrasive equipment, and attention to flash brightness causing temporary blindness that leads to other accidents.

Slipping, tripping—painted walkways that are kept clean, well-marked risers, adequate lighting, janitor work to clean up stray objects and drippings, and treads on inclines.

The National Institute for Safety and Health (NIOSH) develops safety standards for chemicals and other toxic agents.

There are a few hazards not detectable without special equipment. Geiger counters and badges sensitive to radiation are required to detect radioactive substances. Some toxic gases give no warning to human senses. Shielding, isolation, and extra vigilance are warranted for these insidious hazards.

Many accidents are attributed to poor housekeeping. The catchall term *housekeeping* has several ramifications. Good housekeeping naturally includes litter control and janitorial cleanliness. It should also account for *consistency*. Consistent housekeeping is safer, even if it is dirty at times, because workers know where things are and can anticipate them. Work suffers when constant attention is required to react to unexpected situations, and safe habits are difficult to form. Work patterns, reliable equipment, and regularly occurring situations set the framework for safe operations.

Safety programs

From our own experiences we know that performing an unsafe act or being exposed to an unsafe condition does not result in an accident on each occasion; not every jaywalker is hit by a car, and not every faulty car gets into a wreck. But exposure to these hazards eventually causes accidents, and an occasional serious injury occurs along with minor ones.

The Travelers Insurance Company suggested the **accident pyramid** shown in Figure 8.5. Work errors are the foundation of accidents. For each 300 noninjury accidents, there are 29 minor injuries and one major injury. The major injury may result from the first accident encountered or any one of the 330 cases

The major injury can sometimes be prevented by identifying minor accidents and eliminating causes.

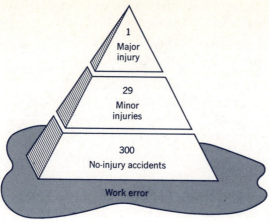

FIGURE 8.5 The accident pyramid.

Ways to Identify Unsafe Acts and Conditions

Observation of work environment

Personnel surveys

Review of past records

Victim inquiries

represented. If these figures are correct, there are 330 opportunities to correct unsafe acts and conditions for each major injury.

Most accidents are not sudden, freakish, undetectable occurrences. Every serious injury outside the truly rare "act of God" has a history. This history involves many near misses. An accident-prevention program should focus on the lower level of the accident pyramid to knock out the acts and conditions that support major injuries.

The program starts by analyzing the cause of each reported accident. Records of past accidents and current reports are examined with respect to the nature of the injury, type of accident, main cause, reason or underlying cause, and suggested remedy. From this investigation a plan is formulated to counteract the most prevalent and serious accidents. Then the responsibility for carrying out the plan is spelled out at both the supervisory and higher management levels. Finally, a follow-through routine is set to control the plan and to determine whether original objectives are being met. The success of the whole program is primarily dependent on the supervisor, the key person in determining the cause of accidents and in applying the remedy.

Control charts, as described in Chapter 15, are used to monitor safety programs. They record accidents on a time scale.

The supervisor's role is a lot easier if he or she is supported by top management. Not every plan has to be supported by a big budget for flashy signs, fancy contests, and generous awards. But top management must visibly support the program by attending safety meetings, delegating authority to make on-the-spot corrections of unsafe acts and conditions, and acting immediately when accident facts become known.

Safety programs take many forms. They may include formal classroom schooling, suggestion-box publicity, stressing maintenance or safeguarding procedures, periodic meetings and discussions, messages calling attention to particular safety features, and contests. All of these programs are designed to make work safer and more productive and to try to abort Murphy's law: "If it is possible for it to be done wrong, someone will do it sooner or later."

If awards are used in a safety program, they should be appropriate. Some human-interest areas for awards are health, recreation, achievement, culture, security, self-improvement, and mystery.

EXAMPLE 8.4 Safety program for drivers

A complete safety program is based on all the hazards that are detrimental to the performance of the tasks for which accident prevention is intended. This total system design concept can follow the planning, analysis, and control sequence. The following outline shows the principal functions of a salesperson's safe-driving program:

1. **Planning**—establishing objectives and responsibilities.

   A. Statement of policy—responsibility of drivers and managers, ground rules for reporting, goals.

   B. Problem forecast—use of personal cars makes control of unsafe conditions difficult; majority of operators are convinced they are already good drivers; importance of prompt, accurate accident reports.

   C. Appointment of program director—responsibility for direction and coordination, time and dollar budget allocations, staff selection and training plans.

2. **Analysis**—evaluating the program plan and mechanics of operation.

   A. Driver/salesperson selection—application forms, personal interviews, written test, medical (especially eye) examination.

   B. Applicant rating—for both sales work and driving.

   C. Driver testing—written and road tests.

   D. Driver training—laws, equipment care, safe practices.

   E. Vehicle inspection—periodic checks of salespersons' cars.

   F. Accident reporting—effectiveness of existing procedures.

3. **Control**—following through and checking to make the plan work.

   A. Accident investigation—accident reports, victim interviews, identification of unsafe acts and practices.

   B. Trouble spots—locating areas where more effort should be applied, recommendations for corrective action.

   C. Driver safety meetings—presentation of information from A and B to maintain awareness, extension to nonvehicular accidents.

   D. Incentive program—recognition and rewards for successful cooperation, report on standings.

8-7 SUPERVISION

As supervision is a key factor in safety, it is similarly important to all other functions of a production system. At the lowest level of the organizational pyramid is the *working supervisor*. Perhaps he or she carries no formal title, but many workers take their orders from this supervisor in his or her capacity as straw boss, crew chief, or gang leader. Higher on the supervisory ladder are managers managing managers. At any level the attitude of supervisors toward those below them can vary from mother-hen concern to dictatorial arrogance; some supervisors fluctuate between temperaments according to time and mood. Because there are individual differences among supervisors as well as among those being supervised, no strict rules fit every occasion. Therefore, two avenues to supervisory principles will be explored. One is to consider organizational structures and procedures, and the other is to give further attention to motivation resulting from human relationships.

Organizational structure

The spinal column of any business of significant size is its **organizational structure.** Information flows both ways, from president to laborer and from bottom to top. The behavior of a firm, internally and externally, is influenced by the speed and accuracy of the information flow. When a firm is small, say fewer than 30 people, the information links are short and direct; the top manager knows directly the production problems a worker is experiencing. As the firm expands, the links become tenuous, and communication noise masks the information flow. Written messages replace verbal exchanges, and several layers of authority often act as barriers to the feedback of information.

Organizational charts are an overt representation of a firm's structure. They display the formal lines of authority among divisions and show how the divisions are to interact. There are innumerable ways to slice and arrange divisions of authority. From the panoramic view, a firm can have a centralized or decentralized organization. A centralized structure takes the familiar pyramid format in which each stratum has a defined responsibility that is authorized and checked by the levels above. In a decentralized structure, usually associated with large, geographically widespread companies, the divisions are treated as somewhat autonomous or self-sufficient areas of responsibility and authority. There are advantages with either structure. Top management's tighter control and the penetrating versatility of a master staff in the centralized organization have to be weighed against the greater motivation potential of decentralization.

The web of informal lines of communication and authority in the shadow of the formal organization plan is very significant. The published plot of organizational lines may be misleading. A centralized paper layout can be run according to decentralized philosophy by making subordinate divisions essentially self-regulating. A formal supervisor–subordinate relationship may be inverted by the social standings of the two men in the positions. Staff groups can have unstructured communications through personal friendships that do not show in the chart but lead to better coordination. The closest formal recognition given to these shadow structures is the *committee.* Although the committee approach to a problem is widely criticized for its ineffectual or slow action, it does offer an opportunity to tap abilities that are hidden or distorted by organizational titles. In this manner the proposals that evolve can reflect the opinions of those affected. The shock effect of some executive decisions are thereby cushioned, and changes are often accepted with more enthusiasm because the doers helped decide what to do.

Any chain of command has some weak links. Whenever tasks are delegated to a lower link in the chain, some authority from the upper position must be allotted to the lower position. Little can be accomplished without the authority to use some resources, but the shift of authority does not relieve the upper position of the responsibility for the task. Most managers have personnel policies to handle such situations and are also guided by written or unwritten company procedures. The **management by exception** principle allows subordinates to make routine decisions, but the more important decisions must be shared with

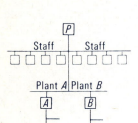

Centralized Organization

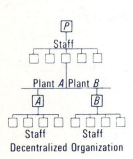

Decentralized Organization

Consultative management is a committee-oriented technique in which top managers consult with lower levels before making decisions.

From a worm's view, it usually appears that all troubles run downhill and that their speed depends on the gravity of the situation.

higher echelons. This principle encourages involvement and provides fine training unless everything unusual is treated as an exception and each exception becomes a crisis.

EXAMPLE 8.5 Span of control

Two organizational structures for supervising an 18-member work force are shown in Figure 8.6. The arrangements offer different answers to the question "How large a group can a leader effectively control?" The correct response depends on the ability of the leader, complexity of the task, importance of the objective, environmental conditions, and nature of the personnel. These elements are embodied in the concept called **span of control**. The concept provides no neat rules for optimizing the ratio of supervisors to subordinates, but it does call attention to what should be evaluated in setting the ratio.

A smaller span of control, fewer subordinates per management position, usually provides closer relationships and more strict control over performance. The significance of crew size may be overlooked if only the direct supervisor-to-worker relationships are observed. Worker-to-worker relationships also affect performance. A crew of one foreperson (F) with two subordinates ($S1$ and $S2$) has three interpersonal links: F to $S1$, F to $S2$, and $S1$ to $S2$. The multiplier effect becomes evident when we observe that a crew of three workers ($S1$, $S2$, and $S3$) with one supervisor (F) has seven interpersonal relationships: F to $S1$, F to $S2$, F to $S3$, $S1$ to $S2$, $S1$ to $S3$, $S2$ to $S3$, and $S1$ to $S2$ to $S3$. The difficulty of achieving complete cooperation, coordination, and communication within large units is obvious. Such reasoning lies behind the limited number of upper-level managers reporting to a senior manager and to the small size of research teams.

Advantages gained from smaller span sizes must be weighed against economic and intangible drawbacks. Smaller crew sizes necessarily increase the number of supervisory positions and consequently raise overhead expense. The extra supervisors cause more tiers of management because each tier is affected similarly by a given span of control philosophy. The direct results of inserting more levels of management between the policymaking head and the output-producing hands are longer lines of communication and perhaps more erratic coordination. A less measurable result of multiple management tiers is the view from below to an ambitious worker: the ladder to the top has an awful lot of rungs.

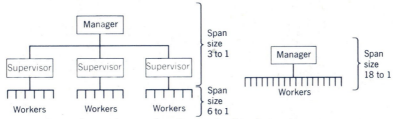

FIGURE 8.6 Span of control and organizational structure.

Human relations

Person-to-person interactions occur throughout the organizational pyramid. They are important at all levels, but where are the most people? At the bottom,

Supervisor Attributes
Not hypocritical
Confidence in higher echelons
Effective communicator
Fair disciplinarian
Respected by superiors and
subordinates (and both know it)

"Final accomplishments of an
enterprise come from the
combined work of each
organizational element and
from each of the people in
each element. Managers
assign responsibility down the
line. Each responsibility for
results that is assigned is simply
a part of a result expected
above. Managers need to
make sure that all expected
results are assigned and that
each man at the next level
down knows the result
expected of him and how it is
to blend in the total." *Goal
Setting, Work Planning, and
Review,* General Electric Co.,
Cincinnati, Ohio.

of course. This bit of geometry is about as far as we can go in quantifying **human relations.** There are few universal, timeless "truths" to rely on. Human relations is a train of episodic, variable truths. It is a people-oriented philosophy that attempts to make the human–black box opaque. That the effort is worthwhile can be documented by many work situations in which the morale of a work force has overcome obstacles flowing from poor formal organizations and physical facility handicaps.

High morale does not guarantee high production, but it does lower resistance to effort. Morale is a summation of hundreds of little things. The best directives that cascade down from top management to the operating units cannot promise high morale. It is the momentum gained from continuously healthy supervisor–worker relationships that culminates in pervasive goodwill among employees.

Working supervisors occupy a difficult, demanding, and important position. They spend most of their time in actual production and are the technical coordinators who keep their groups going. Although they seldom have the authority to hire and fire, they have to channel the directives from above to the workers. They have to speak the language of both management and "people." They get part of their authority from the organization and part from respect of the workers for their skills and personal characteristics.

It is easier to live up to supervisory responsibilities when they are well defined. Objectives and standards of performance should be clear and precise. Each managerial level has **performance expectations** of its subordinates. For summary judgments there must be standards against which actual performance can be compared. In some cases these objectives are communicated by job descriptions, specific assignments, or statements of task criteria. In other cases the employee must ferret out the implied standards from conversations with a supervisor or fellow employees. Uncommunicated expectations can hardly guide or motivate a subordinate. All workers on all jobs should know the standards by which they are judged by the boss. He or she should also know the consequences of disregarding these standards.

Discipline can be a dilemma for supervisors. They must be humane and sympathetic, yet they sometimes have to take firm action. Fewer disciplinary actions are necessary when rules are reasonable and well publicized. When punishment is deserved, it should be delivered swiftly and uniformly. For the questionable cases, Ben Franklin might have advised that "an ounce of correction is worth a pound of punishment."

EXAMPLE 8.6 How not to supervise

Sometimes it is easier to identify the things that injure than it is to prescribe measures that improve. A postmortem analysis of social troubles usually reveals where things went wrong but seldom shows how to right the situation. Avoiding the following list of *social poisons* can save a supervisor the need to find their antidotes:

1. Start each day right. Give all workers a lecture on what they did wrong the day before.

2. Be strong and stern. Make all workers toe the line; if they don't, fire them. Better still, threaten them for several days.

3. Make snap judgments. If you hesitate, workers may think you don't know your business.

4. Be calm and aloof. Think first of your own welfare, and then the workers' problems won't seem so significant. If you continually snub the workers, they will soon reach the point at which they will avoid you completely, and then all your troubles will be your own.

5. Know all the details. If you find enough little matters to occupy your time, you will nearly always overlook the important problems.

6. Gain respect. Be sure the workers know how good you are. You may have to brag a little. If this doesn't work, brag a lot.

7. Be honest and blunt. If workers look tired, tell them so. If you tell them they are idiots often enough, they may act like idiots.

8. Accept credit gracefully. If workers do a good job, take the credit for it. But don't take the blame for a wrong decision.

9. Be alert to competition. Keep your workers below you. Refuse to show them how to do anything and they will learn to do nothing.

10. Outwit your boss. Keep your workers informed of your latest exploits with the top brass. Let them know how you outwitted your own superiors and they will learn to outwit you.

8-8 CLOSE-UPS AND UPDATES

Some things in a production system do not change much: machines, materials, and processes may change swiftly, but human factors are relatively stable. The work force is human, and humans work for essentially the same reasons that they have always worked. They expect to be rewarded for their effort, want their work to be safe and pleasant, and accept the premise that jobs and discipline go together. Though their basic needs, wants, and acceptances continue unabated, managerial responses vary with the times.

Mapping the meandering motivators

It has long been realized that workers want more from their jobs than just money. The underlying drives for physical survival and security are unquestioned needs. But they tend to be taken for granted when satisfied. Higher order needs then take precedence. People seek to satisfy social needs of belonging, ego needs for respect, and self-fulfillment needs for achievement. Comparable surveys conducted in 1935 and 1982 indicate how much the relative importance of these needs has changed:

Maslow's hierarchy of needs theory was introduced in Section 2-2.

1935		1982	
Security/safety	(45%)	Ego Esteem	(30%)
Physical Needs	(35%)	Social Needs	(30%)

1935		1982	
Social Needs	(10%)	Self-fulfillment	(20%)
Ego Esteem	(7%)	Security/safety	(15%)
Self-fulfillment	(3%)	Physical Needs	(5%)

LINKING MOTIVATION TO PRODUCTIVE PERFORMANCE

Theories about motivation focus on what arouses behavior, called *content theories*, or on which behaviors lead to desired rewards, called *process theories*. The preceding lists are factors that energize behavior. Process theory presumes that workers consciously evaluate alternative behaviors and select those that will most likely yield the work-related outcomes they value.

The expected performance of individuals in specific work situations is theorized to depend on the following three factors:

1. *Capacity*—the ability to perform a task based on knowledge, skill, health, energy level, and other factors.

2. *Willingness*—the inclination to perform a task as affected by attitude, personality, task characteristics, reward expectations, and so forth.

3. *Opportunity*—factors beyond a person's control, such as tools and material availability, working conditions, coworkers, and operating policies.

These variables can be combined in a multiplicative formula to suggest the probable level of performance for a given set of conditions:

$$\text{Performance expectation} = \text{capacity} \times \text{willingness} \times \text{opportunity}$$

When each of the three factors are rated on a zero-to-one scale, the formula can be applied, for instance, to a situation in which an extremely gifted employee in an efficient, supportive work environment is discouraged by the prospects of completing the assigned task. Then,

$$\text{Performance expectation} = (1.0)(0.3)(0.9) = 0.27$$

where

capacity = 1.0 in recognition of superior personal ability

willingness = 0.3 as a result of low motivation for the assigned task

opportunity = 0.9 because the support system is nearly ideal

The relation between effort and performance was explored by V. H. Vroom in *Work and Motivation*, Wiley, New York, 1964.

Adapted from Blumberg, M., and C. D. Pringle, "The Missing Opportunity in Organizational Research: Some Implications for a Theory of Work Performance," *Academy of Management Review*, Vol 7, No 4, 1982.

See Section 5-8 for methods to develop a subjective scale.

The strength of this model is not in its exactness but in its structure. The end product is valuable only as a basis for comparison. Its application offers a reasonably objective method for evaluating a human–job fit and suggests ways to tailor a good match.

TO BUILD OR BUY BETTER BEHAVIOR

Every motivation study eventually raises the question, Which technique is most likely to improve actual work performance? The corollary question is, What practical changes are required to apply the indicated technique? An indication of the relative effectiveness of four techniques is provided in Table 8.4.

It is not surprising that money is a powerful motivator. However, noneconomic incentives can also be motivators. Titles and perquisites that go with positions, as well as such rewards as public recognition and mementos for special accomplishments, can stimulate performance improvements.

Behavior modification is a motivation technique that identifies behaviors associated with superior performance, tests the relationships, develops measures, and appropriately rewards or punishes to encourage behaviors that lead to the wanted results. The key features of behavior modification are goal setting, continued evaluation of actual behavior, and reinforcement of the consequences of both desired and undesired behavior. The intent is to develop a system in which rational people see that they cannot get what they want without helping the organization.

Those dreaded performance reviews

Human stress levels are being blamed for more and more ailments. Job **performance reviews** are generally considered by both bosses and workers to be the most stressful regular encounters. A Cornell University study

Reported by E. A. Locke and others in "The Relative Effectiveness of Four Methods of Motivating Employee Performance," Chapter 20 in *Changes in Working Life*, Wiley, New York, 1980.

Motivational incentives included in worker-involvement teams is discussed in Chapter 16.

In today's parlance, reinforcement amounts to positive strokes for positive behavior.

TABLE 8.4

Comparison of Four Motivational Methods to Improve Performance Based on the Number of Studies Shown in Parentheses

Technique and (Number of Studies)		Extent of Improvement		
		Median	Percentage = 10% Gain	Range
Monetary Incentive Systems	(10)	+30%	90%	+3% to 49%
Goal-setting Processes	(17)	+16%	94%	+2% to 57%
Job Enrichment/Design	(10)	+8.7%	50%	−1% to 61%
Employee Participation Programs	(16)	+0.5%	25%	−24% to 47%

found that the number-one cause of stress at work is employees' worrying about how supervisors rate their performance.

Trait ratings include such factors as initiative, enthusiasm, loyalty, cooperation, dependability, adaptability, and leadership. These are important but difficult to define and measure.

The intense competition of the late 1980s has put greater emphasis on the measured quality and quantity of work attained. The conventional trait-rating systems that emphasize standardized personality characteristics and behavior patterns are insufficient appraisals. More appropriate are result-oriented reviews. The foundation for such reviews are job descriptions that clearly state work responsibilities, output expectations, and the way results will be reviewed in making evaluation judgments. Some of the stress caused by work uncertainty is relieved by knowing what results are anticipated and having a well-administered review system.

Performance evaluation should not be a once-a-year event. Rather, there should be regular discussions between supervisors and subordinates. Supervisors can prepare for these meetings by keeping track of the good and bad aspects of each subordinate's job performance. When the assessments are translated into ratings, the following biases should be avoided:

Japan is one of the few countries that has little use for employee performance reviews. Because harmonious personnel relations are prized, colleague evaluation is considered to be extremely disagreeable.

Leniency error—"going easy" because the rater does not want to have to defend negative ratings.

Central tendency—rating everyone as average so as to avoid resentful feedback.

Halo effect—allowing superior or very poor performance in one category to influence other ratings.

When the person being reviewed understands the process and the reviewer conscientiously prepares the evaluation, much of the appraisal-process dread is removed.

Using new technology for high-technology training

Galloping technology is the cause, and perhaps the cure, for concerns that employees cannot be trained fast enough to operate and maintain newly developed machines and processes. Teaching technologies based on *video disc systems* and *computer-generated graphics* can deliver lively, realistic instruction to group classes or for individual learning.

New **interactive training** systems have advanced far beyond the videotape machines introduced 30 years ago. Instead of resembling books on TV screens for which students simply control their own pace and test themselves, the new devices allow students to select their own course of study, skipping material they already know or reviewing what has been forgotten, and controlling the process by critiquing, testing, retesting, and tracking time progress.

Video discs and computer-generated graphics were initially viewed as rival training technologies, but the preferred system may be a combination. A video disc offers sound and sight realism, huge storage capacity, and convenience. Computer graphics systems are less expensive and, unlike video discs, can be easily altered to update the subject matter or approach. In combination, students are exposed to the noise and pictures of actual applications, plus animation that entertains and flexible study programs. These features have led to claims that interactive training reduces learning time while increasing retention. At a minimum, on-site devices are an alternative to specially scheduled in-house courses or to sending employees away for customized schooling. Boosters claim that interactive training gives more brains for the bucks.

One side of a single disc can store 54,000 pictures or pages of text, any of which can be retrieved within one second.

A controlled study at the University of West Florida found that students taking interactive training finished faster and tested higher than did those in conventional classes.

Protecting workers from robots

ROBOT KILLS WORKER

This headline describes the first fatal accident involving an employee in a roboticized factory. The fatality occurred in 1981. Since then many thousands of new robot installations have increased workers' exposure to microelectronically-mechanically caused injuries. Although this single death is an isolated event, it is rich in overtones. It represents the dark side of a phase of production not yet fully visible, which causes apprehension. One indication of its impact is evident in the growing concern for worker protection in computer-controlled installations.

The fatal accident occurred in Japan when a repairman entered a roped-off area to adjust a malfunctioning robot. He did not stop the system's operation first. He failed to note that the robot had stopped only temporarily and was set to move the moment that electricity was supplied. When the "on" switch was accidentally nudged, the robot resumed its programmed motions, pinning the repairman against another machine.

Four differences distinguish safety considerations for industrial robots and similar automated machinery from conventional equipment:

Primary robotic safety measures are protective enclosures and sensing devices that detect the presence of humans. Both stop robot operation when a person is in the danger zone.

1. Susceptibility to stray electronic signals that can activate aberrant movements.

2. Difficulty of determining scope of movement, which is most relevant to human contacts during training and maintenance.

Runaway machinery is a danger because microprocessors are relatively sensitive to electronic interference such as lightning and electromagnetic pulse.

3. Difficulty of determining whether a machine is completely stopped or just temporarily motionless during an operation.

4. Danger of programming errors.

Mechatronics is a term coined in Japan to represent the wedding of mechanics and electronics.

> "Mechatronic" hazards are conspicuous because robots have long arms capable of independent movement at high speeds and they can handle extremely heavy objects or dangerous materials that may be dropped, thrown, spilled, or sprayed.

8-9 SUMMARY

Humans are the benefactors, components, and beneficiaries of the production systems they design, construct, and operate. They are versatile but fragile. Although abilities and physical characteristics vary widely among individuals, many guidelines are available to help people function more effectively.

Human abilities

The human ability to sense, process, and transmit data is indispensable to a production system. People can discriminate among many stimuli and then use these discriminations to make identifications. The interpretation of these identifications leads to inductive decision making—a unique and remarkable human ability. Decisions are transmitted to other people and machines by means of human motor abilities. Machines are better at certain tasks, but they cannot approach human versatility.

Employee placement

Finding the best place for each person in a production system is difficult because of the differences in particular abilities among individuals. The objective of an employment program is to detect the best employee–job fit; the criteria of success on a job are matched to a person's permanent traits associated with the success criteria. Better matches are possible if there are many applicants from which to choose. Applicants are attracted to a job if the position offers what they want. Objectivity in matching individuals to jobs is increased by use of job templates, which allow the consideration of pertinent information only.

Training programs

People cannot always do today what they can do tomorrow. Training starts with a planned orientation to a new job. Formal training for the development of special skills can take many forms. No one format is always best, but there are guidelines to follow for any training program.

Motivation methods

Effective motivation is the difference between wants and resistance to effort. People work for what they believe they want, even though these wants may seem irrational to an impartial observer. Resistance to effort is lessened by comfortable work conditions. When tasks are difficult, the want reward must be increased or the task made easier so as to maintain a level of effective motivation. Rewards are more trenchant if they are immediate. Job enrichment, participative management, and organizational development are examples of management strategies designed to increase motivation.

Safety programs

Accidents are caused by unsafe acts and unsafe conditions. Unsafe acts are more difficult to detect and cure than are unsafe mechanical conditions. Safety programs are based on investigations of accidents to determine their causes and the measures required to eliminate them. The programs may take many different forms, but success depends greatly on the supervisors involved.

Supervision is important at all levels of an organization. The formal structure of a firm can be centralized or decentralized. The centralized structure tends to have more versatility than does the decentralized format, which often encourages greater motivation. However, informal operating authority and practices may be quite different from those of the formal structure. The working supervisor at the lower levels of the organizational pyramid is the link between plans and action. The supervisor's job requires the direction and support of higher management plus the personal characteristics and experience to cultivate good human relations.

Effect of an organization's structure on supervision

Human relations

Human behavior may be modified by setting goals and rewards for goal achievement. Expected performance is theorized to be a function of an employee's capacity, willingness, and opportunity to work. A formula based on these factors gives a numerical rating for a specific situation. Performance reviews also affect behavior.

Behavior modification and expectation

Novel training devices are being developed to use evolving technologies such as video discs and computer-generated graphics. Safety training must be adapted to protect workers from mechatronic hazards.

High-tech training

8-10 REFERENCES

ANTON, T. J. *Occupational Safety and Health Management.* New York: McGraw-Hill, 1979.

BROWN, D. B. *Systems Analysis and Design for Safety.* Englewood Cliffs, N.J.: Prentice-Hall, 1976.

CLAUDE, S. G., JR. *Supervision in Action.* Reston, Va.: Reston, 1977.

DRUCKER, P. *Management: Tasks, Responsibilities, Practices.* New York: Harper & Row, 1974.

GIBSON, J. L., J. M. IVANCEVICH, and J. H. DONNELLY, JR. *Organizations: Structure, Process, Behavior,* 2nd ed. Dallas: Business Publications, 1976.

HERZBERG, F. *Work and the Nature of Man.* New York: World Publications, 1966.

KENSAI PRODUCTIVITY CENTER STUDY GROUP. *Mechatronics: The Policy Ramifications.* Tokyo: Asian Productivity Organization, 1985.

MASLOW, A. H. *Eupsychian Management.* Homewood, Ill.: Irwin, 1965.

MITCHELL, T. R. *People in Organizations: Understanding Their Behavior.* New York: McGraw-Hill, 1978.

RIGGS, J. L. *Productive Supervision.* New York: McGraw-Hill, 1985.

8-11 SELF-TEST REVIEW

Answers to the following review questions are given in Appendix A.

1. T   F   In human data sensing, the process of grouping differentiated sensations is called *discrimination.*

2. T   F   The *human ability* to make decisions based on inductive reasoning exceeds that of machines.

3. T   F   The identification of *permanent traits* associated with superior job performance is a step in the selection and placement of employees.

4. T F The screening phase of the *placement* process has the dual purpose of choosing the most promising person for a particular job and starting the orientation program.

5. T F *Effective motivation* equals want strength minus resistance to effort.

6. T F Profit shares paid frequently and regularly in modest amounts provide greater motivation than do other forms of *incentive wages*.

7. T F *Job enrichment* is accomplished by gathering several specialized tasks into one job assignment.

8. T F Accidents attributable to physical and mechanical sources are caused by *unsafe conditions*.

9. T F *OSHA* is an abbreviation for a basic rule of supervision: Only Supervisors Have Authority.

10. T F The *safety pyramid* suggests that for each 300 noninjury accidents, there are 30 minor or major injuries.

11. T F *Organizational charts* reveal formal lines of authority but may fail to show actual supervisor–subordinate relationships.

12. T F A centralized organization is generally considered to have more motivational potential than does a *decentralized organization*.

13. T F The ideal *span of command* has one supervisor for every 11 subordinates.

14. T F A numerical indicator of *performance expectation* is given by the product of capacity times willingness times opportunity.

15. T F Result-oriented *performance reviews* are more likely to improve productivity than are trait-rating systems.

8-12 DISCUSSION QUESTIONS

1. Where does advertising fit into the human-oriented production cycle of Figure 8.1? Where would the inspiration for a new product likely occur?

2. Although the assessment of humans' abilities is made somewhat easier by considering them as a black box, what dangers are inherent in such a mechanistic approach?

3. Apply the functional and physical descriptions depicted in Table 8.2 to the following operations:
a. Flavor tester in a brewery.
b. Motion picture critic.
c. Reaction to a fire alarm.

4. Name a machine that has a greater capability than humans in each of their data-sensing, processing,

and transmitting abilities. Do you think that machines will ever be built that exceed all human's abilities?

5. If the intent of a free enterprise is to make profit, why should a company consider the welfare of its employees?

6. Give an example of when each type of training method shown in the margin notes of Section 8-4 would be appropriate.

7. Relate the factors that attract applicants to a job, as described in Section 8-3, to their effectiveness as motivators. Consider their influence on want strength and resistance to effort.

8. With reference to want satisfaction, compare the employee benefits offered to build company "loyalty" with the concept of "participative management."

9. Comment by example about the historic changes in management philosophy from manipulation to motivation.

10. A report by Leland P. Deck, director of labor relations at the University of Pittsburgh, suggests that graduates are hired more on the basis of their height than their academic record. A 1967 survey revealed that starting salaries were 4-percent higher for men over six feet tall. A similar survey in 1970 upped the bonus to 10 percent. In the year of the later survey, the tallest graduate got the highest starting salary *and* had the lowest grade point average of all graduates. Deck says, "I do think I have put my finger on a truth, and that is part of the so-called sex discrimination and racial discrimination of corporation executives is in fact height discrimination."

Discuss the "hiring by height" theory in terms of performance predictors.

11. Distinguish among job simplification, job enrichment, and job enlargement. Compare them in terms of work variety, autonomy, identity, feedback, training, and management flexibility.

12. Compare the advantages and disadvantages of simplified versus enriched jobs from a production management perspective.

13. Comment on the statement "there are no accident-prone humans or jobs, just accident-prone human–job matches."

14. How does the definition of an accident fit intentionally self-inflicted wounds? How do such injuries fit into a safety program?

15. Design an accident report form to go along with the program outline in Example 8.4.

16. How can protective clothing be detrimental to the development of safe working habits?

17. What hidden costs make the ratio of indirect to direct costs of an accident so high?

18. Why are unsafe acts harder to detect and cure than unsafe mechanical conditions?

8-13 PROBLEMS AND CASES

Accident descriptions typical of everyday production operations are given in Problems 1 and 6. Analyze each according to the routine suggested in Section 8-6, page 280. The problems are adapted from material used in the Travelers Insurance Company's Safety Action course.

1. A utility man was attempting to lubricate a pump. It was necessary for him to stand near a large, unguarded pulley and reach in an unstable position to the lubrication point. While doing this, he was struck by the pulley and his right shoulder was lacerated and bruised. The tool pusher stated that the guard had been missing since he had been on the rig.

2. An operator was putting a caustic pump into operation after it had been repaired. Everything went as normal until a gasket on the pump let go, and caustic was sprayed on the operator's face and arms. Because of the pressures involved, it is not uncommon for gaskets to fail in this way. He went to a safety shower immediately but sustained skin burns and severe burns to his eyes.

3. A deckhand on his first day of work jumped from a barge to the dockside with a line to make it fast and secure the barge. He apparently misjudged the clearance and fell against the side of the landing and into the water. His heavy clothing pulled him down, and he drowned. The captain of the tow boat had instructed the deckhand. The depth of water was only 12 feet, and it was generally accepted that this was not hazardous.

4. A carpenter, using a bandsaw to cut joints, was in a hurry to finish before lunch. Another employee passing behind him tickled him, causing him to strike the blade. The tip of his finger was cut off. The carpenter was known to be "jumpy," and other employees had taken advantage of this characteristic to have a little fun.

5. An employee inhaled ammonia gas when passing the leaky discharge pipe of an ammonia compressor. There was no injury from the inhalation, but

he instinctively jumped aside, collided with the unguarded flywheel of a circulating pump, and bruised his shoulder.

6. An untrained operator overloaded a power truck, piling material too high, and a box fell off into the aisle. Another employee, who had refused to abide by instruction, came along pulling a hand truck while walking backwards. He fell over the box and was injured against the lower frame of the cart.

7. Supervisors from the human resources and manufacturing departments became engaged in a heated discussion on the allocation of money within the organization.

The manufacturing supervisor claimed that all activities and every program should be evaluated strictly on measurable results, a dollars-and-cents comparison of benefit to cost. Other departments, the manufacturing supervisor contended, were spending too much on frills and were reluctant to accept real accountability because they could not substantiate their expenses in terms of money saved or money earned.

The human resources supervisor countered that other activities were necessary to sustain manufacturing operations but simply could not be measured as easily as counting units of production. The importance of morale-building activities were cited as examples: the company newspaper, recreation program, occasional parties, and fringe benefits. The manufacturing supervisor was challenged to imagine the effect on morale if the annual company picnic were canceled, and the company-sponsored golf tournament that they had both competed in last week were abandoned.

a. Prepare a reply by the manufacturing supervisor for the next round of the conversation.

b. Prepare a rebuttal to part a. for the human resources supervisor.

Supervisor Situations: Place yourself in the position of the supervisor described in Problems 8 through 12, and respond accordingly.

8. The amount of "shrinkage" took a sudden leap in the warehousing department over the last three months. Word has leaked out that management is considering having everyone in the department take a lie detector test. A group of employees has come to see you, their supervisor, to protest. What should you say?

9. One of the men in your unit just found out that another person has received a promotion that he feels should have been his. He comes to you complaining that favoritism was shown, saying everyone knows that the person who got the promotion drinks on the job and got the job because he was a "good old boy" who belonged to the right social club. What should you do?

10. "They say they are enriched by their jobs. I say they just want me to do more for the same pay. I liked the old way when I could relax and just do things automatically." What could be done to relieve this situation?

11. A computer programmer has been assigned to your work unit for six months. She comes to work 15 to 30 minutes late. She takes more and longer breaks than do the other employees in your section. When you called her in to discuss it, she said that all programmers in the company follow the same practice. You checked and found out what she said was true. However, the rest of the people in your section are beginning to resent her. What should you do?

12. You have called Jack to your office. "O.K.," he said, "so I smoked a joint out back during the break. What harm is that? A little pot doesn't hurt anybody. Besides, I'm not the only one who smokes it. Why discriminate against me?" What should you say to Jack? If there is not a company policy on marijuana, should there be one? What should it say?

13. *The Case of Corporate Nomads*
Most major corporations, the armed services, and some government agencies have long followed the practice of rotating people to different jobs and dis-

tricts every two or three years. Advantages to the organization are that employees learn about different phases of operations, become acquainted with more people in the organization, become more versatile in their talents, and generally become "better rounded."

Some employees, particularly the latest generation of aspiring executives, have begun to criticize the transfer policy. They have expressed a desire to stay at one geographic location for several years so that they can raise their families with fewer disruptions, be able to take part in community affairs, and generally establish a permanent living style, which they believe is more rewarding and satisfying.

There are many other arguments to both sides of the question. What are your views?

14. *The Nowheresville Case*

An experimental nuclear plant is to be built on a small isolated island. The purpose of the installation is partly to produce power but mostly to investigate new methods of power generation and new techniques for using hot waste water from the nuclear reactors. The technically sound but unproved power-generating process is not considered dangerous. Innovations to dispose of unwanted cooling water include using it to heat large ocean pens for raising fish and trying other means of aquaculture. Both endeavors require scientists to conduct and evaluate the experiments plus skilled staff and managers to operate the installation.

The plant will be of the latest design, and living facilities will be constructed on the island for all staff and their families. No expense will be spared in providing comfortable homes and other conveniences, such as a small shopping center and recreation hall. Children, however, will have to be boarded at schools on the mainland.

It is apparent that the salaries will have to be considerably higher than the industry average to attract and hold qualified people. Probably additional incentives will be necessary to get and keep these kinds of people on the island.

List incentives that should contribute to higher motivation to attract employees and to keep them happy in Nowheresville.

15. *A Case of Motivation Motives*

A personnel manager and a manufacturing manager became spokespersons for two factions in an executive conference. The original question dealt with the preparation of budgets and the justification of expenditures. The production manager argued that all activities and every program should be evaluated strictly on dollars-and-cents, benefit-to-cost ratios. He contended that other departments in the organization were spending too much on frills and were reluctant to accept real accountability controls because they could not substantiate their expenses in terms of money saved or money earned.

Counterarguments came from many support departments. They claimed that their activities were necessary to sustain the manufacturing operation but that the benefits could not be measured directly, as could be done in production activities. The personnel manager mentioned the activities designed to build morale such as the company newspaper, recreation program, occasional parties, and payroll fringe benefits. She said it was not possible to measure these factors but that they assuredly contributed to better production. She challenged the manufacturing manager to get along without features such as the newspaper, the landscaping and care of the lawns and grounds, the new rug in the president's office, and the company-sponsored golf tournament that the manufacturing manager had just competed in.

a. Prepare a reply by the manufacturing manager.

b. Prepare a rebuttal to the statement made in part a.

16. *A Case of People Problems*

A year after he graduated from college and took an engineering-management post with Konrad Construction Company, Bob Conn received his first major independent assignment. With a crew of 70 men and considerable heavy equipment, his mission was to clear 42 miles of right-of-way for a four-lane superhighway through a rugged mountain pass in the West.

He felt fortunate in having a tough but respected crew chief assigned with him. The chief had been with the company for several years and was generally regarded as a spokesperson for the men, both on the job and as a representative of the union.

The crew, for the most part, had previous employment with the company and were experienced in this work. Because the right-of-way crossed uninhabited land, a company camp was set up near the work area. Wages were equal to or above union requirements and local prevailing wages.

At first work progressed smoothly, and Bob Conn believed a positive cycle of morale was in operation. The company camp was comfortable; terrain conditions were pleasant. The few gripes he heard he dismissed as a normal condition. He had often heard that a griping worker was a contented worker. It was when the worker stopped griping and became sullen that you had to be careful.

Because the men knew their jobs, Bob left the daily directions to the crew chief. Each morning the chief would get the plans for the day from him; in the evening he would report on the progress and turn in any pertinent information. Bob occupied himself with occasional checks on the workers and worked with them if they had a particular problem. The rest of the time he spent with the surveyors or in keeping the multitude of records required by the company. The men seemed to accept him, although he had no really close friends among them. He felt they respected his work. However, he remembered the training he had received at his officers' training course in the Marines and one motto in particular: familiarity breeds contempt.

Because of bad weather and unexpectedly difficult terrain features, the work began to fall behind schedule. The home office sent a message to Bob urging him to make every effort to get back on schedule because the rest of the construction outfit would soon be ready to move into his section and the contract they were working on had a time-penalty clause.

Bob called in his crew chief and showed him the communication he had received. The chief said it was not the men's fault they were behind schedule. Bob agreed but emphasized that they still had to speed up. He further suggested that they move their base camp to a new location farther along the right-of-way to cut down the daily travel time on the rough access roads. The crew chief seemed reluctant to accept the plan. He mentioned the nice streamside location they currently had and that the move would put them farther away from a small town the men found popular for off-hours.

The next day the move was made. Bob instructed the chief to tell the men that it was the only possible way to recoup the schedule losses and that they would work out the transportation problems so each man could go to the town at least one night a week. Other than being farther into the "wilds" and a bit less convenient, the new location seemed just as hospitable as the old one to the young manager. One and a half hours of travel time per day had been eliminated and he felt this would increase their output by a quarter of a mile a day. But it soon became apparent that the progress rate was just the same as before and a feeling of bitterness enveloped the crew.

One evening Bob planned to call in the crew chief and talk about the turn of events. He was worried, and it affected his work and his temper. Late in the afternoon he received a radio call for him to return to the home office to make a personal report on work progress. The trip would require at least eight hours of travel time each way. He told the dispatcher to get a vehicle ready for him and went to his tent for the necessary records. When he returned to find his transportation, the crew chief and three of the section leaders, each of the three carrying a suitcase, approached.

The crew chief said, "I heard you had to go back to the city and figured you wouldn't mind taking these boys with you. They have some personal problems they could get out of the way by the time you are ready to return."

Bob recognized the three as key men whose absence would certainly cut the capabilities of the whole crew. They would be gone at least three days. He also knew each of them was a spokesperson for a segment of the crew. Bob was undecided as to what to say or do.

What would you say and do?

This problem is well suited to the "role-playing" training method. Only two "actors" are needed: the crew chief role and the manager's role. The play can start with the last quoted words of the crew chief. Either as a player or as an observer, diagnose the interaction and record your impressions.

17. *A Case of Moonlight Motivation*

Night shifts have traditionally been less popular than regular daylight workdays. Higher wages are usually offered to make the moonlight shifts more attractive, but absenteeism and employee turnover are often a problem, even with the premium pay.

Suppose that you are the manager for a small company that is going to add a second shift to its packaging and shipping department. The additional shift is expected to be permanent and will require about a dozen people initially. About two days is required to train a person for the job. Morale has been high in the daylight shift, and working conditions are good. During the start-up phase you will supervise both shifts but will be allowed to promote from within or hire two new supervisors. Among the many questions you must answer in starting the new shift are the following:

a. What incentives should be offered to attract and maintain workers on the moonlight shift? Are higher wages enough?

b. Should some employees from the day shift be encouraged to switch shifts? Why?

c. List the inconveniences associated with working at night, and relate them to the characteristics of workers who might be interested in the job.

d. If you received no additional pay for the extra work of supervising two shifts, what motivational factors will probably provide sufficient incentives for you to do a good job and enjoy doing it?

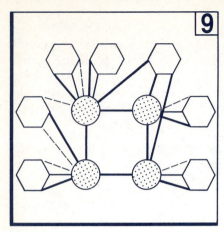

9

WORK ENVIRONMENT

LEARNING OBJECTIVES

After studying this chapter you should

- appreciate the many factors that influence plant location, layout, and working conditions.
- be able to apply a site selection model.
- understand product flow designs and plant layout principles.
- be aware of the manual and computerized methods for developing facility designs.
- have a working knowledge of plant design and control characteristics for illumination, noise, temperature, and ventilation.
- appreciate how human engineering factors are incorporated into equipment designs and arrangement.

KEY TERMS

The following words characterize subjects presented in this chapter:

work environment
site location factors
livability factor
location model
product layout
process layout
fixed-position layout
energy conservation

from-to chart
activity relationship chart
facility-design programs
programmable-path equipment
link analysis

direct and reflected glare
noise abatement and protection
effective temperature
visual display terminals (VDTs)
ergonomics

9-1 IMPORTANCE

In the previous chapter we classified basic human abilities as data sensing, processing, and transmission. In equivalent categories we can express three functions of production management as obtaining data, making decisions based on the information, and acting on the decisions. The ease with which these functions can be carried out depends strongly on the **work environment.** Thus the topics in this and the previous chapter are closely related. For example, while considering the human factors involved in attracting and placing workers, we noted the importance of the work-place location. Similarly, motivation and safety are influenced by the condition of the work environment and the layout of facilities.

Physical effects of the work environment on human performance are portrayed in Figure 9.1. Although the effects are most obvious in manufacturing, they are equally important to activities in the service sector. Performance in any production system is affected by the design of the physical environment. **Site location factors** to consider include the location and layout of facilities, arrangement of work areas, and environmental conditions such as illumination, noise, and warmth. A better work environment does not guarantee greater productivity, but it sets the stage.

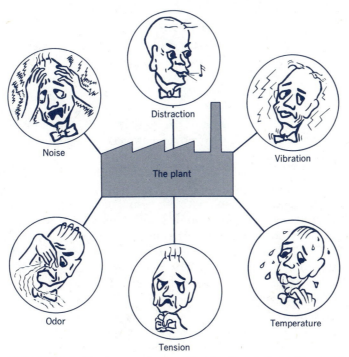

FIGURE 9.1 The human factor effect in the work environment.

9-2 SITE LOCATION

The product/service cycle in Figure 2.1 accounts for factors that lead to restarts in a fresh location.

A plant location problem is not encountered every day, but the factors that can create a problem are constantly developing. Technological improvements make existing processes noncompetitive. New products replace established lines. A requirement for different materials or a change in the source of materials alters supply costs. Power, water, or other resource needs are subject to production levels which in turn are a function of demand. Any or all of these factors can force a firm to question whether its plant should be altered at the present location or moved to another locality.

Whenever the plant location question does arise, it deserves careful attention because of the long-term consequences. Large manufacturing firms or big retail distributors often have staff groups that specialize in evaluating new sites. Consulting services can be used when a company faces an unusual relocation problem or when an objective appraisal is needed. There are also many sources of free advice. Communities here and abroad clamor for new industries to improve their tax base. To attract more businesses, they offer bait such as free land, free services, favorable financing, or tax benefits. Current location trends appear directed toward (1) industrial parks in attractive suburban settings with centralized service facilities, (2) urban renewal projects "downtown," and (3) foreign sites near the export market or natural resources. Then there are maverick owners who locate their companies near where they want to live.

Sources of Site Information

Chambers of commerce

Realtors

Periodicals such as *Factory* or *Modern Industry*

State and federal agencies

Transportation companies

Factors affecting location

At or near the top of every list of desired plant site features is labor supply. With the needed skills known, it might seem an incidental problem to locate in areas where those skills are abundant. Ideally it would be nice to have three or four times the number needed from an available labor pool to allow selectivity in hiring. On the practical side, there should be doubts when an area boasts that much available labor. Is there an unhealthy labor–management atmosphere? Are absenteeism and turnover patterns undesirable? Are labor costs and expected fringe benefits unusually high? What is the productivity expectation?

The importance of labor questions depends, of course, on the particular firm, its policies, and its products. If the firm is science oriented, it should anticipate going to an area where engineers and scientists congregate, because it is unlikely that many can be lured to remote sections. Some specialized skills, such as woodworking or art crafts, tend to be concentrated in certain regions. Semiskilled labor can often be trained from an unskilled labor pool in areas that offer compensating monetary incentives such as inexpensive land or power.

After the general geographic area has been chosen, the location still must be narrowed to a specific plot of land. The plot obviously has to be big enough for plant operations plus associated facilities such as parking and storage. Should it also allow room for expansion, landscaping, and recreational areas? Obviously the plot should have access to necessary transportation facilities and sufficient utilities. Less obvious is the cost of being distant from the homes of workers.

Site Factors

Technical skills

Cost of labor

Raw-material sources

Highway system

Railway hub

Seaport facilities

Airport

Nearness to markets

Proximity of suppliers

In-place utilities

Community facilities

Climate

Low taxes

Available financing

Land costs

Other location features that attract workers were included in Section 8-3.

If the labor force's bedroom area is far removed from the plant, workers must put in 9, 10, or 11 hours to earn 8-hour wages. And the nonpay hours spent fighting traffic are perhaps the toughest. Such travel restricts the number of workers available, lowers job performance, and takes a hefty chunk out of paychecks to meet commuting costs, especially since gasoline prices have risen so high.

A dominant physical restriction such as proximity to harbors, markets, or natural resources often overrides a location's people-oriented attributes. Then the only question is whether the "dominating" factors are truly indispensable or whether their importance is a function of historical habit. We can trace the development of industrial centers by reference to changes in physical restrictions. Years ago industrial growth began in seaports because of a reliance on inexpensive ocean traffic. As the railroad network grew, the relationship of raw materials to manufacturing to markets became more flexible. Air and trucking transportation encouraged further versatility, and industrial centers spread throughout the land. The distance in time between supply and demand is ever diminishing. Innovations will continue to tip the scale of physical resources versus human resources from its historical economic balance, and future fortunes will be built on these detected imbalances.

Service organizations frequently lease facilities rather than building them. Most of the site factors listed in the margin notes are appropriate to a lease location decision, but satisfying them is often difficult because the available space in existing structures limits the choice. A compromise between what is wanted and what can be obtained may be the only solution. For instance, public-welfare offices have to be situated convenient to the people they serve, but these locations may hinder their staffing, and a marine supply store should be near the waterfront, where rental space may be at a premium if available at all. In such situations, site location priorities decide the issue.

Evaluation of location factors

The prime criterion for a preferred location is the least total cost, the minimum delivered-to-customer cost of the product or service. A study to identify the best location typically starts with an evaluation of regional factors and progresses to particular communities within the favored region. Information of a general nature suffices to rate regions. They are compared with respect to market proximity, raw-material availability, transportation service, desirability of climate, laws, tax rates, and other characteristics of special interest to the organization seeking the site.

The factors affecting the choice of a community and a particular site within the community involve specific details. Quantitative measures are available for factors such as labor costs, tax rates, transportation expense, and building costs. Many other factors rely on subjective ratings. It is difficult to rate the community's attitude toward new industries, competition, growth potential, and the relative quality of schools or recreation. Yet these factors should be included if they influence the prosperity of the enterprise. The mixed dimension comparison

Compliance with pollution standards is a recent location constraint for heavy users of air and water resources. Reliable fuel and raw material supplies may become critical factors in the future.

Studies reveal that low-rent but inconvenient, poorly serviced locations contribute to the high mortality rate of new businesses.

Total cost for public-service organizations includes the direct cost of providing the service (for example, salaries) and indirect costs for public utilization of the service (for example, travel cost to get to a licensing bureau).

When distribution costs are the predominant concern, the transportation method described in Chapter 6 provides a location solution. The "branch and bound" technique is used when both the optimal number and location of distribution points are sought.

model described in section 5-8 incorporates both the quantitative measures and qualitative ratings needed in a location decision.

Another way to compare location factors is to establish the relative weights for the main location factors and then assign proportionate points to the contributing subfactors. Both quantitative and qualitative aspects are subjected to this scoring system. For example, an organization could use a 500-point scale for community factors, with the relative weights as follows:

See Problem 20 in Chapter 5 as an example of a plant location decision.

Labor	200 points
Livability	100
Services	75
Sites	75
Taxes	50
Total for major location factors =	500 points

Then each factor is divided into appropriate components with points assigned to evaluate the level at which each component is satisfied by each potential location. The **livability factor** could be evaluated according to the point profile given in Table 9.1. Rating categories for the other factors could include the following:

LABOR

People available in each skill, age, and sex category.

Wage rates by skill category and fringe-benefits cost.

Employment by labor-force percentage and strike records.

Degree of unionization and right-to-work laws.

SERVICES

Availability, type, and cost of personal transportation.

Reliability, speed, convenience, and cost of material transportation.

Electricity and fuel availability and cost.

SITE

Zoning, quality, and cost of land; fire and crime protection.

Adequacy of buildings to lease or buy and availability of local financing.

Availability of special features such as access, parking, railroad sidings, and storage.

TABLE 9.1
Point Ratings for Conditions That Affect the Livability of a Community (Maximum Score Is 100)

Livability factor in the site selection: 100 points

Education Subfactor: 30 points

Pupils per Teacher	$ Spent per Pupil	Higher Education	Community Support
Under 20 10	Over 1600 8	Graduate level 5	Excellent 7
20–25 6	1201–1600 4	University 4	Good 4
26–31 3	801–1200 2	Junior college 2	Poor 0
Over 31 0	Under 800 0	None 0	

Recreation and Culture Subfactor: 25 points

Parks and Programs	Live Arts and Museums	College and/or Professional Sports	Athletic Facilities
Excellent 8	Yes 7	Complete 4	Adequate 6
Good 4	No 0	Partial 3	Inadequate 0
Poor 0		None 0	

Housing Subfactor: 25 points

Population — Houses Available	Building Cost per Square Foot	Land Cost per Square Foot
Under 200 11	Under $64 7	Under $1.20 7
200–300 7	$64.00–$80.00 4	$1.20–$1.80 4
301–400 3	$80.01–$100.00 2	$1.81–$3.00 2
Over 400 0	Over $100 0	Over $3.00 0

Services Subfactor: 20 points

Doctors — 1000 Population	City Expenditures per Capita	Civic Organizations
Under 1 0	Under $150 0	Inadequate 0
1–2 4	$150–$300 3	Limited 3
Over 2 8	Over $300 6	Extensive 6

TAXES

Amount of sales, income, property, inventory, machinery, and franchise taxes.

Workmen's compensation tax and unemployment insurance tax.

City, county, and state tax trends.

All potential sites are rated according to the same point scales. The site with the most total points is preferred. The pattern of point assignments within and among categories reflects the values held by the management group directing the study. Priorities differ according to the nature of the organization and may vary from one location decision to the next by the same organization. But such variations should cause no dismay because the purpose of a systematic location study is to ensure an inclusive and thorough investigation, not to generate an unassailable score.

EXAMPLE 9.1 A location measure model for site selection

Subjective evaluations are especially critical to location studies because many important factors are not readily quantifiable. Even factors with obvious numerical measures often have to be subjectively ranked in order to allocate their degree of influence on the final decision. This example, a **location model**, explains a procedure[1] that neatly combines quantitative and qualitative evaluations.

STEP 1. Identify the factors that deserve to be included in the study, and determine which of these absolutely must be satisfied. For instance, an industry that relies on an abundant water supply would not consider a site with possible water shortages, regardless of how attractive the other factors might be. A site that fails to satisfy a critical factor is eliminated immediately.

STEP 2. Collect data for all factors that can be expressed in monetary units. Then determine an objective factor (OF) for each site (i) by multiplying that site's dollar cost (C_i) by the sum of the reciprocals of all the costs $\Sigma(1/C_i)$, and take the inverse of the product:

$$OF_i = [C_i \cdot \Sigma(1/C_i)]^{-1}$$

This calculation is illustrated by applying it to three sites for which four location factor costs are used.

[1]Adapted from P. A. Brown and D. F. Gibson, "A Quantitative Model for Facility Site Selection—Application to Multiplant Location Problem," *Technical Papers*, Twenty-second AIIE Conference, 1971.

	Annual Costs in Thousands of Dollars				Totals
Site (i)	Labor	Marketing	Utilities	Taxes	(C_i)
1	248	181	74	16	519
2	211	202	82	8	503
3	230	165	90	21	506

To obtain the objective factor (OF) rating for each site, the sum of the reciprocals $\Sigma(1/C_i)$ is calculated first:

$$\Sigma(1/C_i) = 1/519 + 1/503 + 1/506 = 0.005891$$

Then

$$OF_1 = (519 \times 0.005891)^{-1} = 0.3271$$
$$OF_2 = (503 \times 0.005891)^{-1} = 0.3374$$
$$OF_3 = (506 \times 0.005891)^{-1} = 0.3355$$

Note that ΣOF_i must equal one.

STEP 3. The intangible factors are rated according to a forced-choice procedure. This procedure is first applied to rank the importance of the factors (I_k) and is then applied to each site to rate how well that site satisfies the factors (S_{ik}). These two ratings are combined to obtain a subjective factor (SF_i) ranking for each site as

$$SF_i = \Sigma(I_k \cdot S_{ik})$$

The forced-choice procedure assigns weights by comparing each factor with all other factors, one at a time. At each pairwise comparison, one factor is selected over the other, or the two are rated equal. Using the following preference matrix the following makes the task easier. In decision 1, factor A is preferred to B, as indicated by the matrix entry of a 1 for A and a 0 for B. Next, A is compared with C, and then B is compared with C. The matrix shows that the decision maker is indifferent to B or C and that A is preferred to both of them. The number of times a factor is preferred is divided by the total number of preference marks (1's), to obtain the factor importance rating, I_k.

Factor	Comparison Decision			Sum of Preferences	Factor Rating (I_k)
	1	2	3		
A	1	1		2	$\frac{2}{4} = 0.5$
B	0		1	1	$\frac{1}{4} = 0.25$
C		0	1	1	$\frac{1}{4} = 0.25$
		Total		4	1.0

If we assume that in this matrix, factors A, B, and C represent the subjective site factors for housing, recreation, and competition, respectively, the three sites in our example are then compared in terms of how each site (1, 2, and 3) rates for each factor. Each of the following preference matrices shows the ranking of the three sites with respect to one of the subjective site factors.

The subjective factor value, SF_i, for each site is calculated as

$$SF_1 = (0.5)(0.33) + (0.25)(0)$$
$$+ (0.25)(0.25) = 0.2275$$
$$SF_2 = (0.5)(0) + (0.25)(0.67)$$
$$+ (0.25)(0.25) = 0.2300$$
$$SF_3 = (0.5)(0.67) + (0.25)(0.33)$$
$$+ (0.25)(0.50) = 0.5425$$

STEP 4. The last measure to develop is the weight (X) to be awarded the objective factors versus the subjective factors. An X value of 1 would indicate that only the objective factors are to be included in the final evaluation. At $X = 0.5$, both would exert equal influence. This decision is usually made by a committee after examining the organization's policies, past experiences, and so forth. For this example we will set $X = 0.67$, which indicates that twice as much importance is attached to the objective factors as to the subjective factors.

STEP 5. Assuming that all sites have been eliminated that failed to meet the minimum levels set for the critical factors in Step 1, the remaining sites are evaluated according to the following model:

Location measure $(LM_i) = X(OF_i) + (1 - X)(SF_i)$

Using the data generated in Steps 2, 3, and 4, we have

$$LM_{\text{Site 1}} = 0.67(0.3271) + 0.33(0.2275) = 0.29423$$
$$LM_{\text{Site 2}} = 0.67(0.3374) + 0.33(0.2300) = 0.30196$$
$$LM_{\text{Site 3}} = 0.67(0.3355) + 0.33(0.5425) = 0.40381$$

Thus, Site 3 is preferred. (A sensitivity test is suggested as an exercise, in Problem 2.)

	Factor A Housing			
Site	Decision			S_{AK}
	1	2	3	
1	1	0		0.33
2	0		0	0
3		1	1	0.67

	Factor B Recreation			
Site	Decision			S_{BK}
	1	2	3	
1	0	0		0
2	1		1	0.67
3		1	0	0.33

Factor C Competition				
	Decision			
Site	1	2	3	S_{CK}
1	1	0		0.25
2	1		0	0.25
3		1	1	0.50

Summary of Subjective Factors				
	Site Ratings			
Factor	1	2	3	Importance
A	0.33	0	0.67	0.5
B	0	0.67	0.33	0.25
C	0.25	0.25	0.50	0.25

9-3 FACILITIES LAYOUT

Plant layout is a companion problem to plant location. A decision to relocate provides an opportunity to improve total facilities and services. A decision not to relocate is often accomplished by plans to revise the current plant arrangement. The relayout may be designed to reduce increasing production costs that gradually evolve from piecemeal expansion or to introduce an entirely new process. In either situation, the relayout strives to maximize production flow and labor effectiveness.

The overall objective of facilities layout is to design a physical arrangement that most economically meets the required output quantity and quality. In satisfying this objective, it is necessary to consider the first cost in relation to the continuing production costs at both the current and foreseeable levels of demand. A fairly stable demand forecast allows the facility design to include refinements that would not be feasible for processes that are subject to rapid technological change or output sensitive to big swings in customers' tastes. Thus it is sometimes economical to provide flexibility at the expense of operating efficiency.

In this section we explore the relationship of production departments—groupings of production activities—rather than individual machines or architectural features such as supply bins, counters, and skylights. A facility layout of a hospital would include emergency rooms, operating suites, patient rooms, and even the parking lot, but it would not initially include the location of an X-ray machine or a cash register. In a few cases, a layout might not even involve a building but instead would be concerned with an outdoor sales lot, bridge construction, or a cemetery.

Product flow

The general layout for product flow follows a pattern set by the type of production anticipated:

Product layout—a line or chain of facilities and auxiliary services through which a product is progressively refined. This layout is characteristic of mass or continuous production. A logical sequence of operations reduces material handling and inventories, usually lowers the production cost per unit, and is easier to control and supervise. These advantages are achieved at the expense of flexi-

Symptoms of a Poor Layout

Lack of control

Congestion of people and materials

Excessive rehandling

Long transportation lines

Accidents

Low worker performance

Production line bottlenecks

Layout planning integrates most production subjects: forecasting, economic analysis, inventory control, material handling, and the like. Equivalently, a given layout affects all production functions.

Product layout—product moves along an "assembly line."

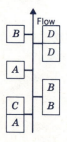

bility. The "pace" of the line is set by the slowest operation; any changes (product design, volume, and the like) in the line normally require a major investment.

Process layout—a grouping of machines and services according to common functions for the performance of distinct operations such as welding, painting, typing, or shipping. A functional arrangement is characteristic of job and batch production. It allows good flexibility and reduces the investment in machines, but it increases handling, space requirements, production time, and the need for close supervision and planning.

Fixed-position layout—an arrangement in which people and machines are brought to a product that is fixed in one position owing to its size. Shipbuilding and heavy construction of dams, bridges, and buildings are typical examples. Such operations often enjoy high worker morale and flexibility for scheduling and design changes. However, the required movement of materials and machines may be cumbersome and costly.

Each layout has relative advantages. Sometimes the advantages characteristic of one type can be imposed on another type to produce a hybrid that lowers cost for special applications. For instance, a "job shop," in which each product is produced only after securing a customer's order, could achieve the economy of an "assembly line" by mass-producing basic modules, which are later customized to specific orders. Similarly, the flexibility of a high-volume, high-investment production line is increased by planning foresight that provides space and hookups (power, waste disposal, material supply) for future modifications.

The initial planning of a layout is demanding, but the effort is justified by considering how difficult it is to alter facilities once they are in place. All sorts of product flow lines can be imagined—up or down, unidirectional or cross-flow, centralized or decentralized. The problem is to determine which design from a multitude of possibilities best fits today's needs and is amenable to tomorrow's potential.

Process layout—product moves to "departmental areas."

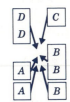

Fixed layout—facilities are centered on a static product.

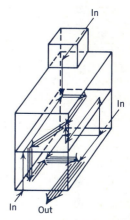

Vertical, horizontal, and convergent production flow.

EXAMPLE 9.2 Energy-efficient facility designs

Energy conservation is the newest wild card in facility design and operation. Its influence is pervasive, affecting nearly all aspects of facility planning. Energy prices exceeded the general inflation rate during the 1970s, but costs have varied since then. Because industry consumes 43 percent of the total U.S. energy demand, there are both production cost and conservation incentives to reduce consumption. The following are 10 facility design and management considerations:

ENERGY-CONSCIOUS PLANT
LOCATION STRATEGIES

1. Locate near energy supplies to avoid transmission losses. Be near good transportation services and as close to markets as possible.

2. Place plant building where it will be in summer shade and protected from winter winds, if possible. Orient the building to face south with the longer dimension running east and west.

3. Use trees and shrubs as windbreaks. Deciduous trees can provide shade in the summer and also allow winter sunlight to penetrate.

4. Consider underground structures for new facilities. Insulate well, and minimize wall perimeter areas in any structure.

5. Design for minimum glass areas, but provide openings for natural ventilation. Consider storm windows, insulating glass, awnings or roof overhangs, and tinted glass.

MANAGEMENT TACTICS FOR ENERGY CONSERVATION

6. Insulate all refrigerated or heated lines. Control thermostats: each degree set back from 72°F saves 3 to 6 percent on the heating bill and, for air conditioning, each degree set up from 72°F saves about 4 to 8 percent of cooling expense.

7. Use the most efficient light sources (see the following table). Wire lights to allow reduced lighting cutouts for specific sections. Use natural lighting as practical, and paint walls a light color so as to minimize lighting needs.

Relative Light Source Efficiency

Type	Lumen/Watt
Incandescent	17–32
Mercury	56–63
Fluorescent	67–83
Metal halide	85–100
High-pressure sodium	100–130
Low-pressure sodium	130–

8. Specify energy-efficient equipment even if the purchase price is slightly higher, because it will soon save the difference. Consider waste heat recovery systems.

9. Control energy use by metering departments or cost centers, installing computerized control systems for heating, ventilating, and air conditioning, and using timers to shut down equipment when it is unneeded and to start it when needed.

10. Prepare for energy curtailments by storing extra fuel supplies, being able to burn wood, wastes, and by-products for boiler fuel, considering renewable energy sources for supplementary and standby power (sun, wind, and waste products).

Layout analysis

A *product layout* usually allows less discretion for arrangement because it depends mostly on the technology involved. Conceptually, it is a continuous line from raw material to finished product. In practice, it is exemplified by large automobile assembly or food-processing lines and by smaller labor-intensive lines that fabricate special subassemblies or serve food in cafeterias. All of these lines are characterized by the sequence and duration of work-station activities as discussed under the title of "line balancing" in Section 11.4.

Once a production line is balanced for optimal performance, its shape is the principal layout problem. Whether to use a straight line, circle, S- or U-shape, or some other arrangement can be investigated by manipulating templates or three-dimensional scale models. The graphic and model aids assist in the visualization of a solution but lack any quantitative measures for selecting the best design.

A *process layout* challenges facility planners with a huge number of possible arrangements. Over 3 million different patterns are available for a process layout

The potential number of designs approaches infinity when dimensions of each department can vary indiscriminately; for example, a 100-sq. ft. area could be dimensioned as 100 × 1, 50 × 2, 25 × 4, and so on.

composed of 10 departments, assuming there are no constraints. Of course, there are constraints arising from architectural limitations, production requirements, safety rules, and other reasons, but these usually do not make the layout easier. Indeed, they probably complicate it.

As opposed to a product layout in which output units follow the same path, a process layout must be capable of producing a variety of outputs that follow different paths between work centers. Material-handling costs are therefore higher. Consequently, most analysis methods focus on transportation costs as the measure of effectiveness for process layouts.

Inputs to facility analysis include forecasts of future demand, which suggests the size and content of work centers. For example, specifications for a hospital in a growing community would provide medical rooms sized to accommodate additional equipment in the future and would anticipate more patient wings to be added later to the core service area. Future demands are translated into present capacity requirements by considering overtime potential, net output with existing equipment, and future new output with logical expected technological advances. These considerations, combined with limits on capital expenditures, determine the size of the work centers (departments) to be incorporated in the layout.

Several tools used by facility planners are described in the following paragraphs.

FROM-TO CHART

The flow of material between functional areas of a plant is recorded on a **from-to chart.** It resembles the mileage chart that appears on most road maps; rows and columns have identical titles in a corresponding sequence. Entries in the chart may register the distance between the centers of departments, number of material-handling trips made between departments per day, and total material movement represented by weight, quantity, or cost. Three versions are shown in Figure 9.2.

A from-to chart is a convenient way to telescope a large volume of data into a convenient form. It can also be the basis for a layout design that seeks to minimize the total material-handling costs. However, because many factors affect a layout besides interchange costs, a from-to analysis should be supplemented by additional information about location relationships.

ACTIVITY RELATIONSHIP CHART

The relative importance of having one department near another is displayed in an **activity relationship chart.** Robert Muther developed the format shown in Figure 9.3. Each diamond-shaped cell in the chart shows the relationship, if any, between two functional plant areas. Two entries appear in the cells: The top one is a letter that indicates the degree of closeness of the relationship, and the number below represents the main reason for the relationship. Definitions of the letters and typical numbered reasons are listed in Table 9.2.

3-D models often precede new construction to evaluate the vertical dimension and to sell the design.

Flow process charts as described in Section 10-2 are used to detail the passage of a product through the production system.

When facilities are completely independent, as in some warehouses, the assignment method described in Section 6-3 can identify the minimum cost layout. See Problem 3.

FIGURE 9.2 (*a*) From-to chart showing distances between departments. (*b*) From-to chart showing the number of material-handling trips per day. (*c*) From-to chart showing material-handling cost per day. These costs reflect the distance traveled, quantity moved, and transportation charge rate.

Department

Department	from \ to	#1	#2	#3	#4	#5
Shipping & receiving	#1		30	71	98	10
Assembly area A	#2	30		41*	68	80
Assembly area B	#3	82	52*		27	39
Inspection	#4	109	79	27		12
Packaging	#5	10	91	39	12	

(a)

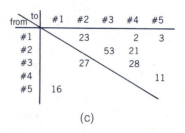

*The distance between two departments differ if there are one-way lanes or obstacles in the flow pattern.

from \ to	#1	#2	#3	#4	#5
#1		14		1	3
#2			22	6	
#3		10		18	
#4					13
#5	25				

(b)

from \ to	#1	#2	#3	#4	#5
#1		23		2	3
#2			53	21	
#3		27		28	
#4					11
#5	16				

(c)

LAYOUT EVALUATION SHEET

Alternative layouts can be compared by listing and weighing the advantages and disadvantages of each. This procedure relies heavily on the judgment of the layout evaluators. Muther suggests a "factor analysis" approach that emphasizes clear, complete definitions of the criteria and objectives of the layout:

This approach is essentially the same as the comparison model presented in Section 4-3.

1. List all significant factors in the selection.

2. Weigh the relative importance of the factors.

3. Rate each layout for each factor against each other layout.

4. Compare the sum of the weighted ratings.

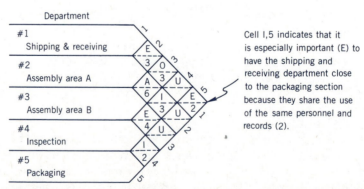

Cell I,5 indicates that it is especially important (E) to have the shipping and receiving department close to the packaging section because they share the use of the same personnel and records (2).

FIGURE 9.3 Activity relationship chart representing the same organization described in Figure 9.2. The letter in each cell shows the importance of departmental closeness, and the number represents a reason for the importance.

TABLE 9.2
Legend of Closeness Ratings and Supporting Reasons Used in an Activity Relationship Chart

Importance of Closeness	Typical Reasons for Closeness
A Absolutely necessary	1 Use same equipment or facilities
E Especially important	2 Share same personnel or records
I Important	3 Sequence of work flow
O Ordinary importance OK	4 Ease of communication
U Unimportant	5 Unsafe or unpleasant conditions
X Undesirable	6 Similar work performed

A "cost effectiveness" approach devised by Harris and Smith is based on assumptions that (1) layout design is a subproblem of the design of the total production system, (2) the layout objectives are quantifiable into a criteria set, and (3) the methods of cost effectiveness may be used to choose the best alternative. The cost-effectiveness rating (CE) for each design is the difference between the benefits (B) expected from that layout and the total cost (C) taken over the life cycle of the layout, $CE = B - C$. The broad system view urged by the CE method is commendable, but it suffers from the usual difficulties encountered in quantifying the value of the benefits.

Many of the observations made about benefit–cost analysis in Section 5-7 are applicable to cost-effectiveness studies.

EXAMPLE 9.3 Facilities layout for a sheltered workshop

The purposes of a sheltered workshop are to assist in the rehabilitation of handicapped people by providing employment, counseling, and vocational training. Hope Center is an established sheltered workshop that requires additional space in order to be able to serve more clients. A building of adequate size can be leased at an attractive price. In order to evaluate the structure, a preliminary layout conforming to the building dimensions is needed.

The proposed building will house only production departments; administrative offices and a cafeteria-classroom will be in an adjacent structure. The dimensions of the production building are 150 by 125 feet, with a railroad siding along a 150-foot side and truckloading space on a short side.

Departments in the center are named for the type of work performed. An activity relationship chart that also records the space requirements is shown in Figure 9.4.

The relationships defined in the chart are translated into a layout format by means of an activity relationship diagram, as shown in Figure 9.5. This simple diagram, constructed by trial and error, suggests the relative positions departments should take in the layout. The truckloading area (no. 2) appears to be the key department to locate because it has strong closeness links (*A-E-I*) to all other departments. It is therefore a logical choice to be placed first in the floor plan. Physical conditions decree that it has to be located on the side of the building where the truck ramps are now located. Other departments are clustered near it in an arrangement that agrees as nearly as possible with the closeness ratings.

A tentative layout is shown in Figure 9.6. De-

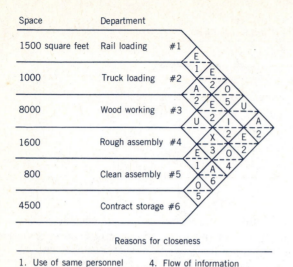

Space	Department	
1500 square feet	Rail loading	#1
1000	Truck loading	#2
8000	Wood working	#3
1600	Rough assembly	#4
800	Clean assembly	#5
4500	Contract storage	#6

Reasons for closeness

1. Use of same personnel
2. Flow of material
3. Undesirable conditions
4. Flow of information
5. Ease of supervision
6. Use of same equipment

FIGURE 9.4 Activity relationship chart for a sheltered workshop layout. The loading departments include shipping, receiving, and storage facilities. Contract storage involves receiving materials for a commercial firm and holding them until the firm wants them converted to assemblies in the rough assembly department.

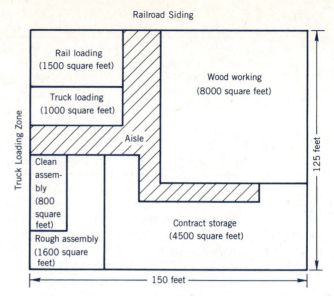

FIGURE 9.6 Preliminary workshop layout.

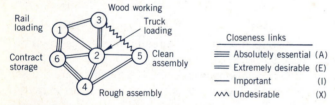

Closeness links

≡ Absolutely essential (A)
= Extremely desirable (E)
— Important (I)
∧∧∧ Undesirable (X)

FIGURE 9.5 Activity relationship diagram as a layout guide for the sheltered workshop floor plan.

partmental areas are dimensioned to conform to the required space and shaped to accommodate the departments within the building walls. The difference between the space required and that available in the building is consumed by aisles to facilitate material movement. Restrooms, utilities, and supervisors' offices can be inserted when plans are final (however, unchangeable parts of an existing structure, such as pillars and underground piping, must be inputs to a relayout). Alternative layouts could be evaluated by using material-handling costs developed by a from-to analysis; then the connections between departments in the activity relationship diagram would have a numerical rating in addition to the closeness-factor links.

Computerized facility design

The major advance in layout planning in recent years is computerized analysis. Several computer programs featuring different layout characteristics are now available. The operating details of these **facility-design programs** are too numerous to include in a short survey, but concepts of how they operate are presented in the following descriptions:

CRAFT

Computerized Relative Allocation of Facilities Technique uses material flow as the lone criterion. Input data include material flow per unit time, cost per unit per distance moved, and space requirements in the form of an initial layout. CRAFT considers exchanges between locations repeatedly until no further significant cost reductions can be found. The program output is a printed layout of facilities in a basic rectangular form that is close to, but is not guaranteed to be, the lowest cost layout. Several extensions have been made to the early versions of CRAFT to increase its realism and to make it applicable to a greater variety of situations.

CORELAP

Computerized Relationship Layout Planning uses the *A-E-I-O-U* closeness ratings, space requirements, and a maximum building length-to-width ratio to develop a layout. The program essentially asks, "Which department rates being placed next into the layout, and how is it to be entered?" Answers are obtained by considering each department in turn to see how well it satisfies the "*A*" relationships, then "*E*," then "*I*," and so forth. The CORELAP printout is an irregular arrangement that often has to be adjusted manually in order to obtain a workable layout.

ALDEP

Automated Layout Design Program requires input data for building specifications and a preference matrix of location relationships. The program starts with a randomly selected and located department. Then the relationship data are searched to find a department with a high relationship to the one just located. When no further preferences can be found, another department is randomly selected, and the procedure is repeated until all departments are processed. Completed layouts are scored on how well the wanted relationships are fulfilled. ALDEP can supply multistory layouts up to three floors, but the printouts require manual adjustments.

Many other facility-design programs have been developed. Each has special features that make it attractive for certain applications. PLANET is an accessible program that uses the relational criteria of ALDEP or the load/path considerations of CRAFT for layouts that are evaluated according to the placement of departments and the flow cost between departments. PREP (Plant Relayout and Evaluation Package) evaluates up to 99 departments at one time, analyzes mul-

Computer programs can offer valuable assistance, but they alone cannot produce a complete layout. There is more to facility planning than space arrangement. Such factors as future flexibility and cost compromises are not easily accommodated in a computerized algorithm. See L. M. Collier and R. B. Footlik, "Computerized Facilities Planning: Pro and Con," *Industrial Engineering*, March 1983.

Dozens of CAD-based applications software programs are available for facilities planning. In 1985 the list price, including training in the use of the software, ranged from $5000 to $100,000.

tilevel structures, and is based on actual footage traveled by material-handling equipment (rather than direct distances between centroids of departmental areas). COFAD II, *F*, and III are versions of the Computer Facilities Design package, which includes safety constraints on layouts (II), impact of stochastic product volume/mix on facility design (*F*), and accommodations for **programmable-path equipment** (tractor trains, robot carts, and carrier conveyors that operate over a fixed pathway on loops that connect departments served by the system).

EXAMPLE 9.4 Computer-aided design of an office

The time-saving versatility of CAD is apparent in the way a desk-top graphic computer can generate an office design. Drawing the exterior walls is the first step. This is done with control devices (cursor, thumbwheels, light pen, and the like) to define corner points on the screen. Once the walls and doors are on the plan, desks and other office equipment are brought up as a menu from the computer memory. An item, such as a bookcase, is selected from the menu and moved around the screen by the operator. It can be positioned in different locations until an ideal setting is found. It is then stopped in that location while other pieces of equipment are maneuvered. Successive replications of the equipment are positioned until the office is fully occupied. A digital plotter then makes a hard copy of the plan. A sample output is shown (Courtesy of Tektronix, Inc.).

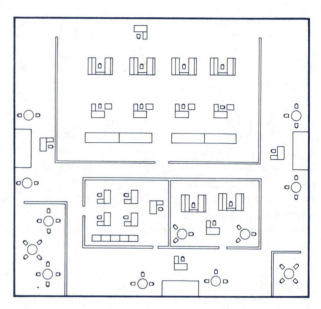

9-4 WORK-PLACE DESIGN

Parts and materials flow from one work center to another. With the exception of highly automated production processes, humans operate the work centers. Because of people's versatility, it is tempting to subordinate their welfare to the efficient arrangement of expensive equipment and machines. A long-range outlook probably would upset this precedence.

Each type of layout has human as well as product advantages and disadvantages. For example, let us consider the economically recommended mass-production assembly line. The concept is valid; it has been praised for decades. But there are still some problems. Each operator along the line has exactly the same time to accomplish his or her assigned task. Because human capabilities differ among individuals and among tasks, it is a complex problem to "balance" the work at each station and to set a pace that neither wastes nor exceeds the capabilities of individuals.

It also is difficult to establish individual incentives because work is paced by the speed of an assembly line. The open, continuous arrangement of the line hinders the control of noise, odors, heat, vibrations, and other factors detrimental to human performance. Some individuals feel insignificant in a sprawling plant. Others tire of the repetitive tasks. But some may be perfectly happy in exactly the same environment.

The importance of carefully hiring, training, and placing workers in accordance with the demands of specific situations is again evident. Special problems can usually be solved by special measures such as ear plugs for high noise levels. For each pro there is normally a con, and vice versa. System thinkers have to recognize both sides before they can weigh them.

There is no doubt that a plant should be sufficiently large to handle the operations required. This approach judges size according to technical aspects. Size may be judged also by the workers' feelings. Most workers function better when they retain some singularity and gain a feeling of security. This principle is recognized by creating optimally sized plants instead of expanding old ones to meet greater production demands. The same feelings are developed in big, integrated plants by organizing workers into administrative divisions to foster ''family'' units and, perhaps, competitive efforts. News sheets and off-work recreational teams contribute to the same effect.

The immediate work area, where workers spend their time, distinctly affects their feelings and output. To some extent, space and comfort can be gained by the choice of finishing materials. Most people feel cooler in light green or blue surroundings and warmer in dark orange or red surroundings. Rooms seem larger when the ceilings are light colored and cozier when the walls are dark. Exposed brick, concrete, and stone seem sturdier than sheet metal, glass, and smooth plaster.

Equipment arrangement

Equipment and supplies used by workers should be designed and arranged for minimum effort and maximum convenience. For equipment design the key word is *flexibility*. Adjustable equipment avoids the need for individually fitted designs as well as substandard performance from workers forced to make their work fit awkward arrangements. Supplies, materials, and tools cannot always be placed within easy reach of the worker, but they should be arranged as conveniently as possible. The extra stretch to reach a cabinet or the extra steps to get material from another floor waste the energy of a worker and, equally important, disrupt ongoing activities. Even when tools are conveniently arranged, maintaining the arrangement is still a problem. Checkout clerks at tool bins, signs reminding workers to return tools, and tool silhouettes painted on walls are methods used to avoid misplaced equipment.

More detailed analyses are warranted when a number of operators and equipment must work together to perform a task. A proper arrangement should

Minimum Office Space Guidelines

Top executives: 270 sq ft
Middle executives: 196 sq ft.
Supervisors: 120 sq ft
Specialists: 72 sq ft
Work station: 25 sq ft
Main aisle width: 5 ft
Intermediate aisles: 4 ft

In some instances, layouts are designed to force movement. Trips to obtain supplies may provide needed exercise or a break in an otherwise monotonous routine.

Occasionally a communication link uses the sense of touch, and a control link can go from worker to equipment to equipment.

be based on the communication and control relationships between workers and machines. These relationships can be represented by *links*. Communication links include visual and auditory (usually talk links) transmissions from worker to worker or from equipment to worker. Control links arise from the use of levers and switches or from direct contact with materials by operators to control machines.

A **link analysis** is conducted by determining the types of links involved, assigning a value to each link, and then evaluating different arrangements of operators and machines with respect to the links. The value of each link is determined from the *frequency* of its use and its *importance* when used. Usually a three-point scale for each factor is adequate: 3 (often)—2 (occasionally)—1 (seldom) for frequency, and 3 (vital)—2 (useful)—1 (trivial) for importance. The composite value of each link is then the product of its frequency and importance. Different arrangements of operators and equipment are evaluated according to how well the more important links are accommodated. Trial arrangements can be made to scale on paper or, for more important operations, a three-dimensional mock-up may be warranted.

EXAMPLE 9.5 Layout features for specific purposes

Specific design features are considered in conjunction with the overall facility arrangement. Sometimes the arrangement of all departments is dictated by the desire to provide a specific feature in one department. The feature could be a certain space configuration for a work area or a special-purpose machine. Whole facilities have been designed to take advantage of a specific technological advance such as a computer-controlled process line or an automated storage-retrieval system. Examples of layout considerations to serve specific purposes are provided by the following situation descriptions:

ASSEMBLY ISLANDS. The physical layout of work stations can help implement motivational concepts such as job enrichment. Under the assumption that it is more effective to delegate responsibility to groups than to individual workers, the layout should be conducive to group cohesiveness and communication. Two such designs for assembly operations are shown in Figure 9.7.

 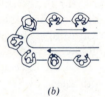

    (a)   (b)

FIGURE 9.7 The circular table (*a*) can be stationary for an assembly that is passed by hand from one operation to the next, or it can rotate when workers must abide by a definite cycle time. A conveyor belt for sequential assembly or inspection is U-shaped (*b*) to make it easier for the group to communicate. Both (*a*) and (*b*) encourage eye contact.

LOADING DOCKS. The loading dock is both the beginning and end of an industrial plant's activities. A dock is more than an appendage added to a plant layout; it is an integral part from which inefficiencies can back up to thwart all production activities. Factors that influence dock design are described in the following observations:

1. Ideally, traffic should enter the dock area counterclockwise to allow easier-to-make left-hand turns for large trucks. A truck waiting area may be needed to accommodate peak-load periods. At least 40 feet of maneuvering room is needed in front of the dock.

2. The conventional practice of placing docks at both ends of a flow-through process is being replaced by strategically located docks at several points around a structure to serve individual production departments. Sometimes the exterior appearance of a plant is waived to put the loading area along the highway frontage to provide better access.

3. Overhead cover, shallow inclines, antislip surfaces, and similar features provide safer all-weather operations. Built-in dock boards act as adjustable ramps to link the dock area with carriers of different heights. Load-activated traffic doors and dock seals between building walls and carriers can lessen the escape of heat and conditioned air.

MARKETS. Should the layout for a store be oriented to the clerk or the customer? Exclusive attention to factors that minimize handling costs may produce an efficient layout that discourages sales. Uniform, uncluttered display cases are convenient to stock and maintain, but they probably fail the retailing objective of maximizing sales per unit of display space. Customer behavioral considerations for market layouts recognize that regular, rectangular layouts provide more display space per unit of floor area, but the

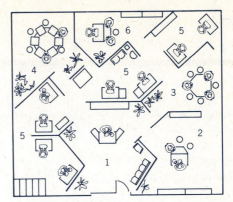

FIGURE 9.8 Office layout: 1—Receptionist/waiting area, 2—Manager's office, 3—Conference room, 4—Think tank and discussion room, 5—Secretaries, 6—Manager's office.

space may not be as valuable as are angular or freeform layouts that open more merchandise to shoppers' view. Also, aisles should be large enough to accommodate the traffic flow and allow room for lingering, especially in areas of high sales potential such as entrances to department stores and around the perimeter of supermarkets.

OFFICES. The "small-group" approach to work structuring is apparent in open designs for offices. Such layouts are characterized by generous spaces between individual work places and "think-tank" installations for undisturbed mental work or group discussions. An office layout with half open work places and separating walls is shown in Figure 9.8.

9-5 WORKING CONDITIONS

Good working conditions increase motivation by decreasing workers' resistance to effort. In extreme cases the adverse effects of bad working conditions make adequate levels of performance unattainable. For example, tasks dependent on visual acuity are obviously impossible when there is insufficient light, and the expected pace of manual labor cannot be maintained without adequate ventilation and temperature control. These extreme conditions are rare in industry today because they are foolish from both social and economic viewpoints. But many subtle defects in the working environment still can be improved.

EXAMPLE 9.6 Link analysis of operator–machine arrangements

A control tower is run by three or four controllers, depending on the amount of traffic. As depicted in Figure 9.9, position number 1 is occupied by the team leader responsible for final decisions. Operators 2 and 4 are assisted by relief help from operator 3 when traffic is heavy.

Each operator has control of a given traffic sector. Audio communication with the traffic is by radio. Other talk links are among controllers. Visual communications are composed of television screens, digital readouts, and on-off lights. Most controls are switches and buttons.

The link analysis of the existing system is conducted by having "experts" agree on a consensus rating for each observable link from operator to operator and from operator to equipment. Links between equipment are not usually included in an initial evaluation. After rating the frequency and importance of use, a composite value is obtained for each link, as represented for operator 1.

The layout is evaluated according to the way it facilitates the links. Views should not be obstructed by equipment or operators. The effect of distance and illumination on visual performance is considered. Distance is also a factor in auditory communications; shouting should never be necessary. The postural set

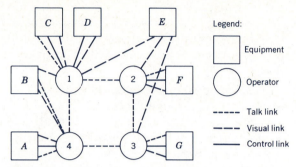

FIGURE 9.9 Present control tower arrangement.

of operators should accommodate easy visual or vocal communication. There must be sufficient room for the convenient and accurate manipulation of controls.

If it is necessary to redesign the arrangement, attention is given first to the highest valued links. A mock-up may be required. Attention is given to proper ventilation, protective safety devices, noise control, and other maintenance or convenience features that will assure smoother operations by lessening physical and mental fatigue among operators.

A possible revised layout for the control tower is shown in Figure 9.10. Link values are shown in

Operator 1	Link	Equipment or Operator	Frequency Rating	×	Importance Rating	=	Composite Rating
	- - - - - - - - -	②	2	×	3	=	6
	- - - - - - - - -	④	2	×	3	=	6
	- - - - - - - - -	C	3	×	3	=	9
	— — — — —	D	3	×	3	=	9
	— — — — —	C	2	×	2	=	4
	— — — — —	B	1	×	2	=	2
	— — — — —	E	1	×	2	=	2
	— — — — —	C	3	×	3	=	9
	— — — — —	D	3	×	2	=	6

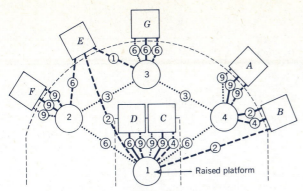

FIGURE 9.10 Revised control tower arrangement (composite link values are noted in small circles).

small circles for each link. The elevation of operator 1 is raised with respect to the other operators. A more thorough study would indicate the vertical placement of equipment; exact scaling would be used. Before finalizing a new design, thought should be devoted to unusual or emergency operating conditions. Care taken to ensure a workable design distinguishes a craftsman from a journeyman: a design that anticipates the unexpected marks a master.

Illumination

The amount of illumination desirable for a particular job depends on the individual doing the job and the nature of the task. Some individuals require more light than others do to make the same discriminations. Jobs that require exacting discriminations demand more illumination than does less detailed work. Because of these differences, it is unrealistic to follow standard illumination levels for different areas of a plant. The more useful approach is to determine the minimum illumination needed by an average person to perform certain tasks and then to alter these levels for individual preferences if necessary.

Recommended lighting levels for a multitude of situations have been published by the Illuminating Engineering Society (IES). The standard measure of light intensity used is a *footcandle*—the amount of light cast by a standard candle at a distance of one foot. Some general conditions and specific minimum task recommendations are shown in Table 9.3.

The quality as well as the quantity of light affects work performance. Women appear particularly sensitive to spectral effects of lighting. A healthy individual may appear pale or sickly under certain colored lights. Blue-tinted lights cause feelings of harshness and coldness. Amber tints provide a feeling of warmth.

Most workers prefer sunlight to artificial light. The preference is at least in part due to psychological reasons, but there are many practical disadvantages to relying on sunshine for working light. To a designer it may seem that the sun operates at only two levels, too bright and too dim. Overcast days produce as little as 50 footcandles. On cloudy days, the level varies according to the vagaries of passing clouds. Full sunshine is often distracting. By orienting a building to the position of the sun and by using transparent and translucent materials to filter and equalize the distribution of natural light, sunshine can supplement artificial light. It can also be an expensive light source when window washing, repairs, and temperature control are considered.

Glare is the most harmful effect of illumination. It can cause discomfort

Moonlight produces about 0.5 footcandle, and bright sunshine produces as much as 8000 footcandles.

There has been a long-standing controversy over reactions to fluorescent versus filament lamps. Research reports have been conflicting. In general, filament lights are considered to create a "warmer" impression.

TABLE 9.3
Recommended Levels of Illumination

Task Conditions	Footcandles	Specific Tasks or Place
Very precise work for which extreme accuracy is required	1000	Operating table for surgery, very fine assembly work
	500	Very difficult inspections
Precision tasks involving small details for prolonged time periods	300	Fine assembly work
	200	Difficult finishing and inspection work, detailed drafting, medical-dental work, fine machining
	150	Rough drafting, business machine operation
	100	Medium bench and machine work, mail sorting, filing
Prolonged tasks for which speed is not essential	70	Studying, sewing, reading clear handwriting or blackboard writing, typing
	50	Sketching, wrapping and labeling, glass grinding
Normal tasks	30	Drilling, riveting, filing, paint booths and washrooms
Casual seeing tasks that are not prolonged	20	Rough machining, stairways and corridors
	10	Shipping and receiving, auditoriums
Rough seeing tasks with low contrast	5	Storage or warehouse, theater lobby during intermission

Studies have confirmed that illumination below recommended levels leads to errors and fatigue. However, increasing the intensity well beyond the recommendations does not produce proportionate reductions in errors and fatigue.

and affect visual performance. There are two types of glare. The first type, direct glare, is caused by a light source directly in the field of vision, such as the headlights of an approaching car. The other type, reflected glare, is caused by reflectance from a bright surface. Some ways in which glare can be limited are as follows:

1. Reduce direct glare by decreasing the brightness of the light source or by increasing the brightness of the area surrounding the glare source to balance the brightness ratio. A high general level of illumination also reduces the effect of reflected glare.

2. Place light sources so that they are not in the direct field of vision and so that any reflected light is not directed toward the eye.

3. Diffuse the light at the source or with baffles, window shades, and the like.

4. Avoid glossy finishes for working-area surfaces.

5. Use light shields, hoods, or visors if glare sources cannot otherwise be reduced.

Artificial Lighting

Direct—offers maximum utilization of light at the working area; often produces shadows and glare.

Indirect—offers even lighting by directing light to ceiling or walls.

Diffused—scatters light in all directions; causes some shadows and glare.

Noise

An unwanted sound is called noise. Two aspects of noise pertain to production. One is the potential loss of hearing that can result from continued exposure to very high sound levels. The other aspect is the nuisance effect that contributes to reduced worker performance. As with illumination, there are no definitive levels that bound the regions of good to bad performance or an exact point at which hearing loss develops.

The maximum sustained noise level allowed by OSHA is 90 decibels for eight consecutive hours of exposure.

The standard unit of noise measurement is the *decibel*—a logarithmic ratio of a given sound level to a standard value that is the defined threshold of audible sound (0.0002 dyne per square centimeter). Decibel ratings of some typical sound sources are listed in Table 9.4. The ratings indicate that the sound energy of a quiet office is $10^4 = 10,000$ times greater than that of the threshold level, or that 40 decibels (db) is 100 times greater than 20 db.

There is abundant evidence that a prolonged exposure to high noise levels contributes to hearing loss, but there is little agreement as to how long is long or how high is high. Most experts state that harmful effects can be expected from noise levels above 100 db; the 100-to-90-db zone is questionable. It is believed that a greater hearing loss is experienced at higher frequencies (4000 cps range) than at lower frequencies (1000 to 2000 cps) and with noncontinuous impact noise than with continuous noise of the same intensity. Although the conditions conducive to hearing loss are not clear-cut, production managers should heed available guidelines because occupational deafness is a potentially explosive question. Compensation claims for hearing loss could be huge.

The public-relations aspect of factory noise on residential areas should be considered in plant location.

Noise well below the level of physical harm contributes to lower worker performance. Sounds of particular kinds—infrequent, irregular, high pitched, or resounding—are distracting. A siren is useful when attached to an emergency service to attract attention, but the same sound becomes a distractor of equal potency when workers have no need to be concerned. Any increase in noise level tends to increase muscular tension and consequently increase expenditure

TABLE 9.4
Noise Levels

Decibels	Noise Source
140	City warning siren, shotgun blast
130	Jet engine at 100 feet, firecrackers
120	Riveter at 4 feet, severe thunder
110	Circular saw, amplified rock music
100	Pneumatic drill at 10 feet, subway train
80	Usual factory, very heavy traffic, crowded restaurant, garbage disposal
60	Usual office, light traffic, normal conversation
40	Quiet office, household refrigerator
20	Whispered conversation, dripping faucet
0	Threshold of hearing

Sound energy becomes physically painful at about 140 decibels.

of energy. Workers are apt to become more quickly fatigued, nervous, and irritable in a noisy environment. A degree of accommodation gradually results from continued exposure, but again we have no way of extracting the exact toll on productivity.

Noise abatement and protection are the remedies for hearing loss and nuisance noise. The measures required for sound control are often complicated and costly. Acoustical engineers employ a number of techniques, which fall into three main categories:

1. Control the noise at its source by replacing noisy machines and worn parts, by modifying the mounting of equipment, and by redesigning or substituting alternatives that are inherently quieter.

2. Isolate the noise by creating barriers such as distance, baffles, or sound absorbers between the source and the listener.

3. Provide personal protection devices such as ear plugs, ear muffs, or helmets.

Insufficient noise is the modern counterpart to the traditional noise problem. When a room has a moderate background of sound, nearly all workers adjust to these sounds and do not really hear any of them. When the sound background becomes too low, each little normal sound becomes a distraction. Piped-in music is often used to break the quiet so that otherwise distracting noise blends in with the background sound.

A few years ago the possibility of using musical cadence to set the work pace initiated a frenzy of research. The extensive studies are somewhat inconclusive because of the number of uncontrollable factors in a production setting. It is generally agreed that music does facilitate the performance of repetitive tasks and that it may reduce accidents. The most beneficial effect is on the morale of workers. It relieves boredom and promotes a feeling of well-being. Polls show an almost unanimous approval by workers to music in the work place.

Atmospheric conditions

Nearly everyone in the temperate zones of the world has experienced numbing cold and exhausting heat. In such extremes, productive work is most difficult and, fortunately, is seldom expected. In most industrial settings, it is possible and economically desirable to provide comfortable atmospheric conditions for the work area. When production requirements necessitate uncomfortable working conditions in localized areas, such as extreme heat near furnaces, provision can be made for individual worker protection. For instance, jackets cooled by dry ice have been found effective for personal cooling in heat stress situations.

A healthy worker has a body temperature of approximately 98.6°F. The body is equipped to keep itself close to its normal temperature by means of blood circulation and sweat glands. When body temperature begins to drop, these and other body mechanisms raise the temperature. The resulting expe-

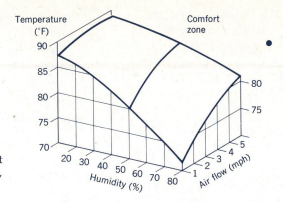

FIGURE 9.11 Approximate comfort region as a function of temperature, humidity, and air flow.

● Effective temperatures above the middle 90's impair performance

rience is one of feeling cold. A cold object feels cold because of the heat loss of the body; such heat transmission by direct contact with warmer or colder objects is called *conduction*. The other physical processes of body heat exchange are *convection*—transmission by the surrounding air, *radiation*—transmission by exchange of thermal energy between the body and objects, and *evaporation*—transmission of heat by the evaporation of perspiration and moisture from the lungs.

Feelings of warmth or cold are not dependent strictly on temperature. The interaction of temperature, humidity, and air circulation is largely responsible for comfort. These factors have been combined into a single rating scale called **effective temperature**—an index that combines into a single value the effect of temperature, humidity, and air movement on the sensation of warmth or cold felt by the human body. An approximate indication of the comfort zone is given by the curved surface in Figure 9.11.

A humidity level between 30 and 70 percent is comfortable for most people.

Gases such as carbon monoxide and natural gas are toxic. Workers cannot continue working in their presence, whether noticed or not. Dust and smoke, though not always toxic, are disturbing and will not be tolerated long without complaint. About one cubic foot of fresh air per minute per square foot of floor space is required for the average work area.

Fresh air is an ambiguous label. Although the air may have adequate oxygen content and be free of toxic gas, it still may stink. Many plants use processes that unavoidably produce unpleasant odors. Workers' adaptation to such odors is ordinarily fast. Most complaints come from new-hires and down-wind residences. Affected communities sometimes rationalize by referring to the odors as "the smell of money."

Measures to Control Air Impurities

Removing cause of impurities
Proper ventilation
Special exhaust measures
Personal protection such as facemasks

EXAMPLE 9.7 Customized working conditions

Specifications for furniture and personal accommodations are designed to fit about 90 percent of the user population. Lighting and temperature levels also are set to satisfy the bulk of the user population. The percentage served can be increased by using adjustable designs and local modifications of the work environment, but the cost of flexibility is high.

Two ways to customize working conditions are

to (1) hire only people who are suited to the conditions already established and to (2) design conditions for a specific population with known characteristics. An example of the second option is the design of a sheltered workshop, as introduced in Example 9.3. Layout considerations appropriate to the handicapped clientele include:

1. Elevated offices or observation platforms for department supervisors that allow open views of all work areas.

2. Separate aisles for pedestrians and workers in wheelchairs or on crutches. All other aisles should be extra wide.

3. Nonskid ramps to replace steps. Limited elevated storage.

4. Adjustable lighting to provide levels of illumination compatible with the type of work being done at a certain time.

5. Safety rails and painted traffic stripes on loading ramps and at congested points.

6. Abundant electrical outlets to avoid extension cords.

7. Movable screens to isolate work stations when workers tend to stare at one another and become annoyed, mainly in assembly operations.

9-6 CLOSE-UPS AND UPDATES

Easing VDT tensions with office ergonomics

Around the world more and more people are staring at **visual display terminals (VDTs)** perched on desks and tables. Nothing since the introduction of the telephone has made a greater impact on offices than VDTs and VDUs (video display units). They are becoming the dominant interface between people and machines that process data. And they are causing tensions.

The number of VDTs in use in 1986 was estimated to be 10 million, with an expected increase to 35 million by the early 1990s.

Both clerical and professional operations rely on visual displays. Traditional secretarial tasks primarily consist of data entry via a keyboard and data acquisition via the display. Professional tasks may involve word processing, CAD/CAM operations, and interactive communication. People engaged in either clerical or professional work on VDTs are subject to eye problems, aching backs, neck fatigue, and a high stress level. Because a VDT has no value without an operator and because most operating errors are caused by people, much attention has been given to the design of this important human–machine interface.

DESIGN OF TRULY USER-FRIENDLY MACHINES

The effort to shape machines to fit users is also known as human factors design, human engineering, and biotechnology.

From the time people first started using machines, they have been modifying them to make them safer and more convenient to use. This improvement process was distinguished as a separate discipline during World War II. It was called **ergonomics,** a name derived from the Greek word for work, and focused first on the design of instrument panels in aircraft. In 1940 a fighter pilot had to watch 21 instruments. By 1965 the number had more than doubled. Aircraft of the 1980s multiplied the information flow so much that several cathode ray tubes (CRTs) were needed to supplement the regular instrumentation. Rapidly changing displays on CRTs highlight the information that is most important at the moment. Such designs characterize human engineering.

The boundaries of ergonomics overlay many areas of concern to production managers. Because temperature, air movement, illumination, and other environmental factors affect human work performance, they are ergonomic considerations. However, ergonomics is most closely associated with the arrangement, shape, and operational characteristics of equipment. Making machines conform to human physical and mental limitations is the objective.

Body dimensions are particularly important to the design of office equipment. Operators are effectively glued to their equipment all day, released only by occasional breaks. Unless their machines allow comfortable movements and encourage proper posture, poor work and loud worker complaints are inevitable. Because people vary in size, habits, and postural preferences, for whom should the equipment be designed?

Table 9.5 shows body dimensions for the U.S. work force that are applicable to office equipment. The center column depicts the mythical "average person." Some designs cater to the rare individuals who have these exact dimensions, thereby forcing everyone else to adapt to less-than-ideal conditions. A preferable strategy is to design flexibility into machines that allows their convenient operation by everybody within, say, the fifth to the ninety-fifth percentile range. Similar tables are available for human capabilities and preferences. Resulting designs consider interactions among task requirements, environmental conditions, physical limitations of operators, and equipment components.

The window-level stoplight required in 1986 automobiles is an example of an ergonomically inspired safety device.

VDT DESIGNS

The explosive growth of VDTs has raised many alarms, mostly false. In the late 1970s the possibility arose that too much viewing might cause cataracts. The medical profession dismissed the threat. Then came the worry that VDTs contributed to birth defects. This one was debunked by the National Institute of Occupational Safety and Health. A 1985 report by the National Research Council, an arm of the National Academy of Sciences, confirmed that there is no scientific evidence that VDTs damage vision or cause miscarriages among pregnant operators. However, some nations have rules that limit the time that operators may spend at their machines (four hours per day in Sweden) and standards for lighting in VDT rooms (half as bright as in general offices).

The recommended dimensions and adjustments for a work sta-

Fluorescent lights in a typical office expose workers to more low-level radiation than do VDTs.

TABLE 9.5

Body Dimensions for U.S. Civilians of Age 20 to 60. Female/Male Dimensions Are Shown in Centimeters or Kilograms (1 centimeter equals approximately 0.3937 inches, and 1 kilogram is 2.2046 pounds)

	Range of Body Dimensions—Female/Male		
	Lowest 5%	*Midpoint*	*Highest 5%*
Stature (height in cm)	149.5/161.8	160.5/173.6	171.3/184.4
Eye height	138.3/151.1	148.9/162.4	159.3/172.7
Weight (in kg)	46.2/56.2	61.1/74.0	89.9/97.1
Sitting posture in cm			
Height	78.6/84.2	85.0/90.6	90.7/96.7
Eye height	67.5/72.6	73.3/78.6	78.5/84.4
Elbow rest height	18.1/19.0	23.3/24.3	28.1/29.4
Knee height	45.2/49.3	49.8/54.3	54.4/59.3
Thigh clearance height	10.5/11.4	13.7/14.4	17.5/17.7
Hip breadth	31.2/30.8	36.4/35.4	43.7/40.6
Buttock–knee length	51.8/54.0	56.9/59.4	62.5/64.2

Adapted from the article by K. H. E. Koemer and D. L. Price, "Ergonomics in the Office: Comfortable Work Stations Allow Maximum Productivity," *Industrial Engineering*, July 1982.

tion are given in Figure 9.12. They help correct the following three VDT problems:

Normal reading glasses and bifocals often cause problems during prolonged periods of work on VDTs.

1. *Eye strain* The viewing distance to a VDT screen is typically 25 inches. The ideal distance depends on the user's visual traits. Changing light sources and the angle of the screen tilt can largely eliminate glare, the most nagging vision problem. Eye strain is further eased by larger characters, better resolution, and less flickering on the screen.

2. *Aches, pains, and fatigue* A seated operator should have plenty of leg room; a seat pan contoured to support body weight; a backrest that firmly supports the trunk; supports for the arm, wrist, and hand as necessary; and a keyboard that minimizes strain. Work-station equipment obviously needs to have a wide range of adjustments to accommodate the wide range of people who use VDTs, as suggested by Table 9.5. Frequent and short rest periods can reduce much of the residual strain and fatigue.

3. *Stress* Complaints of VDT fatigue may be caused more by mental than by physical strain. The basic relationship of a human to a com-

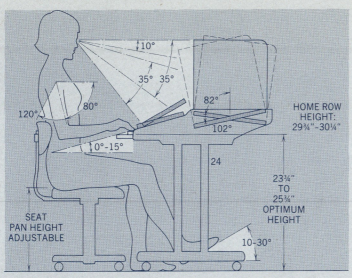

FIGURE 9.12 A well-designed terminal work station that has properly proportioned and adjustable components.

may be the root problem. Some people apparently feel overwhelming pressure to keep pace with the tireless computer and the continually changing challenge so visible on its screen. There is also the subtle threat that a supervisor can be invisibly monitoring performance from another screen keyed to the operator's work station. Though ergonomists can identify such stress factors, only management practices can lessen them.

Color me calm

Color, like music, is thought to affect performance, although solid evidence is scarce. Researchers have found that when subjects look at such warm hues as red, orange, or yellow, their blood pressure rises, brain-wave activity increases, respiration is faster, and perspiration is greater. Blue tends to have the opposite effect. But why these effects occur is still uncertain.

The human eye is sensitive to millions of colors. Cones in the back of the eye react to certain wavelengths of light. These cells send signals to the brain which, some experts believe, release hormones or trigger neurotransmitters that influence moods and activities such as heart rate.

A 1979 study led to the now-widespread use of bubble-gum-pink rooms to calm delinquents and criminals in correctional institutions. This

Color coding to indicate the purpose of wiring and machine components is used extensively.

Different color combinations for letters and backgrounds for VDT screens may ease eye strain.

pink shade tends to relax people so much that drugs and other restraining devices are often unnecessary to control inflamed tempers. Other color-behavior relationships proposed by psychologists and color consultants include the following:

- Fears are reduced in hospital rooms painted blue.
- Orange stimulates hunger in restaurants.
- IQ scores go up and disciplinary problems decline when white and orange classrooms are repainted yellow and blue.
- Earth tones make structures seem more substantial.
- Replacing traditional battleship gray with light blue or beige for industrial machinery inspires neatness and efficiency in workers.

Although the legitimacy of claims that color can modify behavior is being debated, careful and clever color choices are making offices and factories more attractive. Even if it is impossible to motivate with a paintbrush, it is still possible to paint on some pleasure.

9-7 SUMMARY

This chapter about the work environment is a companion piece to Chapter 8 about human factors. Physical and personal aspects of human performance are almost indistinguishable in effect. To say that performance will automatically improve with investments in better facilities is a rash generalization. It is safe to predict only that the better environment will relieve that particular deterrent to work performance. The complex interactions of human-oriented activities and the components of a production system are outlined in Figure 9.13.

Human factors

A plant location should satisfy both the physical needs of production (land, transportation and utility services, financial considerations) and employee wants (climate, community facilities, recreation, proximity to housing). Because many of these factors are difficult to measure, locations can be compared by assigning points to a site according to a predetermined weighting scale or by a forced-choice procedure that rates intangible factors and the degree to which each site satisfies these factors. Both methods combine intangible ratings with quantifiable measures to produce a score for each location.

Site location factors

Site selection model

Facility layout is a companion problem to plant location. There are three basic types of product flow in a plant: (1) product layout—a line or chain of stations and auxiliary services through which a product is progressively refined; (2) process layout—a grouping of machines and services according to common functions for the performance of distinct operations; and (3) fixed position layout—an arrangement in which workers and machines are brought to a stationary product. From-to charts, layout evaluation sheets, activity relationship charts,

Product flow designs

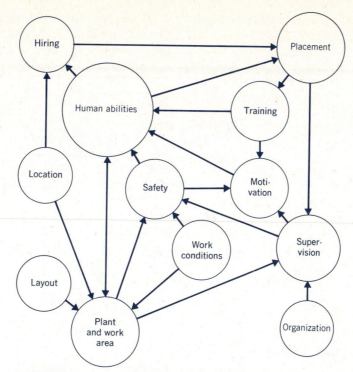

FIGURE 9.13 Human factor considerations.

and computer programs assist in evaluating existing layouts and in developing improved arrangements.

Facility-design methods

Work areas and equipment are arranged to improve human performance. Flexibility is a prime consideration in equipment design. The arrangement of equipment can be evaluated by a link analysis based on the frequency and importance of communication and control links between workers and machines. Assembly islands and open offices are examples of layout variations that can promote the implementation of management philosophies.

Equipment arrangement

Working conditions should maximize human data-sensing, processing, and transmitting abilities. Good illumination is particularly important because humans rely heavily on vision. The work area should have sufficient light intensity and quality without direct or indirect glare. Noise can cause physical damage and be detrimental to work performance. Unwanted sound can be controlled at its source, by barriers between it and workers, or by personal protective devices worn by workers. Feelings of warmth or cold are rated by effective temperature—an index that combines into a single value the effect of temperature, humidity, and air flow. Air impurities, either toxic or nontoxic, should be controlled by removal, ventilation, special exhaust, and personal protection measures.

Control of environmental factors that affect work

Ergonomic studies attempt to make machines safer and more convenient for human users. The design of VDTs has received much attention because their use is increasing so rapidly. VDTs are not inherently dangerous, but they can cause eye strain and fatigue. Proper design relieves strains.

Some experts believe that painting with appropriate colors can modify behavior. Others disagree.

9-8 REFERENCES

APPLE, J. M. *Plan Layout and Material Handling*, 3rd ed. New York: Wiley, 1977.

CAKIR, A., D. J. HART, and T. F. M. STEWART. *Visual Display Terminals.* New York: Wiley, 1981.

FRANCIS, R. L., and J. A. WHITE. *Facility Layout and Location, An Analytical Approach.* Englewood Cliffs, N.J.: Prentice-Hall, 1974.

HOLMES, W. G. *Plant Location.* New York: McGraw-Hill, 1930.

HUCHINGSON, R. D. *New Horizons for Human Factors Design.* New York: McGraw-Hill, 1981.

McCORMICK, E. J. *Human Factors in Engineering and Design,* 4th ed. New York: McGraw-Hill, 1976.

MUTHER, R., and L. HALES. *Systematic Planning of Industrial Facilities.* Kansas City, Mo.: Management & Industrial Research Publications, 1979.

TOMPKINS, J. A., and J. A. WHITE. *Facilities Planning.* New York: Wiley, 1984.

WOODSEN, W. E. *Human Factors Design Handbook.* New York: McGraw-Hill, 1981.

9-9 SELF-TEST REVIEW

Answers to the following review questions are given in Appendix A.

1. T F The most important *site location* factor is the distance from supply to demand, that is, transportation considerations.

2. T F Although such qualitative factors as *livability* cannot be measured directly, they are still included numerically in a quantitative site selection model.

3. T F Shipbuilding and dam construction are examples of a *product layout*.

4. T F In a *process layout*, a product moves from one functional area to another in which distinct operations are performed.

5. T F People, materials, and machines are brought to the product produced in a *fixed-position layout*.

6. T F A *from-to chart* displays the frequency or importance of communication links between humans and machines.

7. T F The importance of closeness factors for plant design are represented by the 0 to 10 ratings in an *activity relationship chart*.

8. T F For equipment design, the key word is *flexibility*.

9. T F Most workers prefer sunlight to *artificial light*, whereas plant designers have a reverse order of preference for reasons of practicality.

10. T F Insufficient light is the most damaging *illumination* effect on work performance.

11. T F The standard unit of noise measurement is the *decibel*—a linear ratio of a given sound level to the threshold of audible sound.

12. T F The main purposes of *noise abatement and protection* are to prevent hearing loss and nuisance noise.

13. T F *Effective temperature* is an index that indicates the combined subjective effect of temperature, humidity, and air movement.

14. T F Designs that optimize the human–machine interface are the objective of *ergonomic* studies.

15. T F A *VDT work station* should be designed to fit the average person.

9-10 DISCUSSION QUESTIONS

1. What reasons would lead to locating a plant in an urban renewal project, a domestic industrial park, and a foreign country? How might the work force view each choice?

2. Some industries require two or more dominant resources that are not located in the same geographical area. For instance, in metal-reduction industries, coal is seldom located beside the iron ore deposits, and bauxite is far from the power sources needed to produce aluminum. What location factors should be considered in choosing a plant site for these industries? Show how the location measure model (Example 9.1) could be used in the choice.

3. The effect of improved transportation facilities on the development of industrial centers is described in this chapter. Could the development of nuclear power have a similar effect on the future location of industries?

4. When a company needs more space to expand its operations, the usual method is to consider sites near the present location first. What are some advantages and disadvantages of this strategy?

5. If you were the sole owner of a small "brain-oriented" company that sold ideas as a product and wanted to locate the company in a resort area, what problems might you encounter? For the sake of discussion, say the staff is made up of two top dreamers, two half-time dreamers who also act as leg men, and two nondreaming, capable paper pushers who know the business thoroughly.

6. Compare the advantages and disadvantages of a process layout with those of a product layout with respect to material-handling cost, room required for work in process, skills of operators, machine breakdowns, use of machines, and need for supervision.

7. Compare the point system method for plant location studies with the location measure procedure. What are the advantages and disadvantages of each?

8. Why is the from-to chart considered an analysis tool for process layouts but not for product layouts?

9. How could an activity relationship chart be used to analyze the layout of a supermarket? Would this be a product or process layout?

10. Judging from the brief descriptions of computer programs for plant layouts, what are the main differences among CRAFT, CORELAP, ALDEP, and PREP?

11. Discuss the probable advantages and disadvantages of the "open" office design shown in Figure 9.8 with respect to conventional office layouts.

12. Indexes have been suggested for evaluating tentative human–machine arrangements. With reference to the link analysis technique, comment on the following suggested indexes:
a. Index of visibility (average visual links or average of individual display ratings).
b. Index of walking (walking distance of operators).
c. Index of talking (voice level needed to overcome

distance, noise, and cross-talking speech interference).

d. Index of crowding (how activities of operators interfere with one another's work).

e. Index of accessibility (adequacy of working space and ease of exit or entry).

13. Evaluate a classroom with respect to layout and working conditions.

14. What is implied by the statement "you can buy space with a paintbrush"?

15. Comment on the similarity between the measures suggested to reduce glare and those to reduce noise.

16. Compare the three forms of artificial lighting and sunlight with respect to the type of glare that can be expected from each. How can such glare be reduced or avoided?

17. Consider the noise situation in any industry with which you are familiar. Could hearing loss result from present noise levels? If so, what measures should be taken? If nuisance noise is present, how could it be reduced?

18. What measures can be used for environmental comfort control to facilitate each of the four physical processes of body heat exchange?

19. The effects of heat stress in exceptionally hot working conditions can be regulated to some extent by the management of personnel. Comment on the following possible actions:

a. Selection of workers.

b. Acclimatization.

c. Rotation of workers.

d. Work-load reductions.

e. Rules pertaining to water consumption, salt tablets, and the like.

20. Discuss the human characteristic of individual differences as it applies to design considerations for

a. Equipment arrangements.

b. Working conditions.

21. Throughout Chapters 8 and 9 there are references to economic balances between two counteralternatives such as the psychological value of sunlight versus the practical advantages of artificial light. List and comment on 10 counterbalancing alternatives.

9-11 PROBLEMS AND CASES

1. The location measure model in Example 9.1 is similar to the mixed dimension model discussed in Section 5.8. Using the factor ratings given in Example 9.1 (multiply the subjective ratings by 10 to convert them to a 10-point scale), determine the preferred site location using the mixed dimension model. Compare the results. Which model do you prefer? Why?

2. The X-value that determines the relative emphasis to be placed on the subjective and objective factors in the location measure model exerts a major influence on the final decision. One way to examine the influence is to plot the values of LM_i for different values of X. The point at which the LM_i lines cross indicates the value of X that would switch the location preference from one site to another.

Conduct a graphic sensitivity analysis for the data given in Example 9.1. Put values of X from 0 to 1 on the horizontal axis. A vertical LM scale is placed at both ends of the X scale. Thus at $X = 0$, $LM_2 = 0.2300$.

At what value of X does the site preference switch from site 3 to site 2? Comment on how a

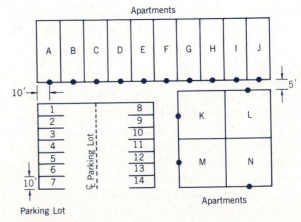

FIGURE 9.14

sensitivity analysis could assist the site selection analysts.

3. A layout for an apartment complex and its parking lot is shown in Figure 9.14.

a. Use the assignment method to assign a parking place to each apartment in a pattern that minimizes the total travel distance from the front door of each apartment (marked by a dot) to an assigned parking place.

b. What criticism from apartment dwellers could be expected from the resulting assignments?

c. What other criteria besides total travel distance could be used to evaluate the layout?

Measure distances from an apartment to a parking slot by an orthogonal route that goes from each apartment door (●), down the center line of the parking lot, and then to the center of each stall.

4. The preferred room arrangement for a new home is given by the activity relationship chart in Figure 9.15. (Baths and halls will be added to the design later.)

a. Based on the given preferences, draw an activity relationship diagram.

Area (ft²)	Room description	
480	Living room	1
196	Dining room	2
140	Kitchen	3
336	Family room	4
168	Den-study	5
192	Bedroom A	6
192	Bedroom B	7
120	Guest room	8
288	Master bedroom	9
480	Garage	10

FIGURE 9.15

b. From part a. and the areas given in the chart, construct a T-shaped (eight exterior walls) floor plan for the home. (Assume that all rooms are rectangular in shape.)

5. Assume that Figures 9.2b and 9.3 are pertinent to the design of a new facility to replace the old plant represented by Figure 9.2a.

a. Based on the given cost-data and department relationships, lay out a floor plan for a building that has 16,000 square feet and four sides. Departments 1, 4, and 5 have the same area. Departments 2 and 3 are, respectively, twice and three times as large as department 1, and all departments have a rectangular shape.

b. Develop a second design for the plant in part a that has the same total building area and the same department areas, but the building is to be six sided, and the departments can be any four- or six-sided shape.

c. Compare the two designs in parts a. and b. by determining the total travel distance for one day's operations, based on the trip frequencies given in Figure 9.2b. Assume that all travel routes are measured from the centroid of the sending department to the centroid of the receiving department along a path that parallels department walls. (For example, see Figure 9.17)

6. Seven departments are to be arranged in an L-shaped (six-sided) building. A layout designed to

Area (ft²)	Dept.	A	B	C	D	E	F	G
8100	A	–	10	100	0	200	0	10
1600	B			0	100	200	0	200
900	C				0	200	400	0
1600	D					0	10	300
3600	E						200	200
900	F							300
6400	G							–

FIGURE 9.16 Number of trips made per week between departments.

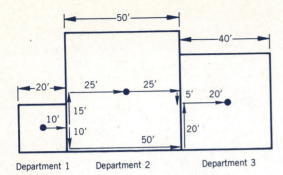

Department 1   Department 2   Department 3

FIGURE 9.17 Travel distances between the three
square-shaped department areas are
from 1 to 3 = 110 ft
from 1 to 2 = 50 ft
from 2 to 3 = 50 ft

minimize interdepartmental traffic flow is needed for
preliminary planning. Each department will have a
square shape with the area indicated in Figure 9.16.

Also shown in Figure 9.16 are the number of
trips made each week between departments. Assume
that the costs are directly proportional to the number
of trips multiplied by the distance traveled. Distances
are measured from the center of the originating department to the center of the receiving department
via the shortest path that runs along department walls.
A sample distance calculation is illustrated in Figure
9.17. Develop a layout that minimizes total transportation distance while conforming to the given restrictions.

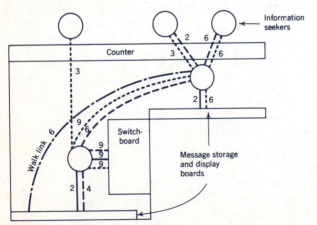

FIGURE 9.18 Message center.

a. Draw an activity relationship diagram with links
   that represent the transportation distances between departments.

b. Design and draw a layout to scale. Then determine
   weekly total transportation distance for that layout.

7. Evaluate the arrangement of the message center
shown in Figure 9.18 and suggest an alternative arrangement. The links have the same meaning as given
in Example 9.4.

8. The backseat driver has been a subject for jokes
for a long time. Assuming the backseat driver can
serve a useful purpose by helping the frontseat driver
navigate in a strange city, use a link analysis to compare the backseat arrangement with having a map-
reading helper in the frontseat.

9. Suppose you are in charge of 60 typists working
temporarily at wooden tables in one large room. They
have complained about the noise originating in the
room. New, very adequate facilities will be available
in 30 days. What would you do now? If you plan any
noise-abatement program, comment on the cost versus the expected benefits.

10. The unintentional actuation of controls has generated many accidents. They are usually caused by
unsafe conditions. Several designs aimed at alleviat-
ing the cause follow. For each, describe where you
have seen the design used (in automobiles, homes,
kitchens, or whatever), and comment on the advantages or limitations.

a. Recessed controls (nonflush mounting).

b. Raised barriers (surrounded by a guard).

c. Built-in locks (special keys or combinations required).

d. Tool actuated (screwdriver or wrench required).

e. Mechanical or electrical interlocks (prevent operating two or more controls at the same time).

f. Protective covers (completely enclosed by an identifying cover).

11. Originality and attention to human factors are key considerations in designing machines or operations to perform a new function. What designs can you dream up that perform the following functions?

a. A single gauge for an automobile that shows the direction and speed of movement plus the time of day.

b. A portable roadblock that can be used day and night to warn motorists that the road is closed to through traffic.

c. An inexpensive, self-contained, easy-to-read, and portable meter that speakers can use to warn themselves when their speaking time has elapsed.

12. The owner of a small machine shop is planning to relocate. He is moving the shop to a large midwestern city where land adjacent to transportation facilities demands a premium price. However, markets and suppliers are nearby. The shop will be all on one floor. Commuter service to the anticipated building area is adequate, and several lots are available on railroad spur tracks. The following list shows the minimum needed work areas and required personnel:

Rooms	Personnel	Purpose	Area (sq ft)
Office	2 secretaries, 1 clerk	Filing, communication, reproduction equipment	360
Owner's office	Owner	Reception room	200
Storage room	1 stockperson	Parts and material storage, catalog service, receiving area	440
Shipping dock		Storage for shipping, packing	360
Production	8 machinists	Work and equipment area	2000
Assembly	2 welders 2 fabricators	Fabrication equipment, overhead crane	1000
Shop office	1 supervisor	Administration	72
Drafting	2 draftspersons	Drafting, filing	200
Restroom		Women	40
Washroom		Men (showers, lockers)	120
Furnace room		Utilities	100

a. Make a layout showing the product flow and work divisions. Most products are custom-made. About half the parts for fabrication are produced from raw material in the shop, and the other half are catalog parts.

b. Which rooms should have special lighting, sound proofing, color, or heating and air conditioning considerations? Does the work or layout present any special safety problems?

c. What location factors should be considered in choosing the site for the machine shop?

13. *The Case of Production Frustration*
"Four Star Wood Products" is a small, but ambitious, woodworking company located in a middle-sized college town on the West Coast. Four college classmates pooled their resources six years ago to purchase a nearly bankrupt wood products outfit. Since that time they have doubled the size of the plant and increased production six times.

It came almost as a surprise to the four when they were notified that they were low bidders on a government contract to furnish 150,000 wooden packing boxes. Their plant was capable of producing the basic box with few modifications, but the specifications required certain handles, locks, and labels, which presented several problems. And the due date on the contract was only six months away.

Each of the four owners was responsible for a phase of the business: (1) office and accounting, (2) sales and buying, (3) plant operations, and (4) plant engineering. The office and sales partners handled the raw materials for the new order without difficulty. Plant engineering anticipated only minor difficulty in the cutting, hinging, and gluing of the boxes, because the work was similar to current orders. It was decided that the plant operations partner would work on the remaining problems and manage the project.

Contract specifications called for a loop-rope handle on each end of the box, a ring-in-slot lock on the front, and different decals on the boxes depending on their intended use. There was no assurance of any future contracts for the same type of product, so it was financially impractical to buy or design automatic equipment for this work. The only practical solution was to set up a hand assembly and finishing line. They were fortunate that the due date for the contract was at the end of the summer, because they could set up a temporary working shed in the company parking lot for the assembly and painting work. There was not sufficient room in the existing plant.

The first thought was to hire local women for the extra work, but the cost of additional toilet and lounge facilities was prohibitive. The alternative solution was to stockpile the raw boxes until school was out for the summer and then hire college students. When a nearby vacant building was found to be available for storage, the latter solution was chosen.

Thirty students were hired. By making the handles and locks as simple as possible, the engineer believed the workers could complete the work in 12 to 13 weeks. He allocated the jobs as follows:

10 workers putting on handles

6 workers putting on locks

8 workers painting

6 workers serving on odd jobs and acting as handlers

The six handlers could aid in the plant, transport boxes from storage, fill in for absent workers, or supplement any stage that fell behind schedule. All workers were to be paid the current local wage of $5.00 per hour for unskilled work. If they worked six days a week and finished 2000 boxes a day, the costs would be within contractual limits.

For the first three weeks the line worked at or slightly above the goal of 2000 boxes a day. The next week production averaged 1800 per day. By the middle of the following week, it was below 1700, even though the operations manager constantly checked on the line and urged them on. Two more odd-job men were hired, but the production leveled at 1700 boxes a day.

The manager questioned each worker about why he thought production was falling off. In order of frequency, he noted the following excuses:

1. We're all being paid the same, but some guys are "goofing off."
2. The outdoor shed gets too hot during the middle of the day to keep working at top efficiency.
3. Insects in the outdoor shed are annoying and distract us from work.
4. Working conditions are crowded and some operations are cramped.
5. The work is boring.

The manager also noticed the morale seemed highest in the odd-job category and that most absences from work occurred on Saturday. Another point that bothered him was the number of minor but time-consuming accidents that kept occurring.

To help improve production, several measures were immediately instituted: (1) wages were changed from an hourly to a piece-rate system, (2) an addition was put on the shed, and (3) the entire structure was screened.

In the following week, production went back up to 1800, but there were two more accidents. One worker accidentally sprayed paint into another's eyes, and another worker drove a staple into his thumb while attaching a rope handle. The crew continued to work well together and morale seemed high, but this was not reflected in increased production. At the present rate the contract could not be completed on time.

What should the manager do now?

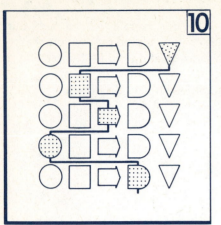

METHODS AND MEASUREMENT

LEARNING OBJECTIVES

After studying this chapter you should

- understand how work operations are analyzed, measured, and paid for.
- be able to construct flow process charts and optimize operator–machine combinations.
- be aware of the principles of motion economy and methods of motion charting.
- know how to conduct a time study using a stopwatch, synthetic times, or predetermined times.
- realize how allowances and rating factors influence the calculation of standard times.
- appreciate the statistical theory behind work sampling and how the technique can be applied to determine the actual activity pattern of a job.
- be familiar with the use and design of wage incentive plans, time-based wage payments, and management salaries.
- be aware of the challenges of the comparable worth issue and the opportunities of innovative compensation plans.

KEY TERMS

The following words characterize subjects presented in this chapter:

process analysis	motion charting
flow process charts	work elements
operator–machine process charts	stopwatch timing
motion economy	elemental selected times

k/s factor	wage incentive plans
rating factor	piece-rate wages
normal time	job evaluation
standard time	key jobs/point plan
synthetic times	salary administration
predetermined times	profit sharing
work sampling	pay-for-knowledge compensation plan
binomial distribution	
random number table	comparable worth
control limits	chronobiology

10-1 IMPORTANCE

Methods analysis and work measurement are the twin pillars supporting the design of work systems. The purpose of work design is to identify the most effective means of achieving necessary functions. In a production context, this means the analysis of present and proposed work systems to develop an optimal transformation of inputs to output.

Historically, time study as originated by Taylor was used to establish standard times for work performance; methods study as developed by the Gilbreths was aimed at improving the manner by which work is accomplished. Through the passing years the two disciplines became intertwined, the one supporting and supplementing the other. Innovations in both philosophy and techniques have provided new tools for practitioners and extended the scope of studies to even remote facets of production.

Under the heading of *methods,* we shall consider ways to improve the method of doing work. It is obviously important to know how the work is currently being done before contemplating improvements. A definitive review of the way an operation is performed results from subdividing the task into basic components. For some operations this breakdown has to provide details as fine as the movements of each finger; these microanalyses are in the province of *motion studies.* For other operations, all that is needed is the sequence of major movements such as the route of a message dispatcher; these macroevaluations are provided by a **process analysis.** Both types of studies benefit from the systematic application of well-established principles and charting techniques.

Measurement includes both the determination of time standards for work and the application of these standards to wage payments for the work done. *Time study* is the technique of setting a time to do a specific task based on the work content of that task and allowances for fatigue and delays. Another approach to work measurement is *work sampling;* a large number of observations of a process are statistically evaluated to determine the percentage of time that the process is in a given state. Results of such studies are a natural part of wage

"Work," in its occupational sense, is the effort expended to produce something useful. Both manual and mental labor fall within this definition.

Terms associated with "methods" are (1) operations analysis, (2) work simplification, and (3) methods engineering.

Frederick Taylor explained the main purpose of time study as the "transfer of skill from management to men."

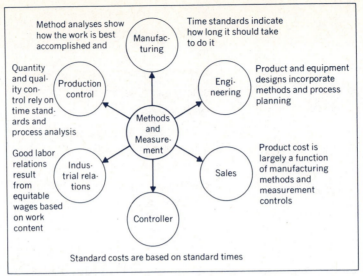

FIGURE 10.1 The influence of methods and measurement effort on other production activities.

formulation. The profuse variety of *wage payment* plans all have the target of equitable payments based on the relative worth of different work assignments.

The influence of methods analysis, time measurement, and wage payment plans permeates the entire production system (Figure 10.1). The central position of methods and measurement groups makes them a fertile training ground for new management material: neophyte managers can work with the elements of production and observe the cross-flow of ideas, ideology, and occasional idiocy. It is a prime position in which to spot suboptimization and to apply creative countermeasures. It is the wellspring of labor peace, product competitiveness, and customer satisfaction.

10-2 PROCESS ANALYSIS

The objective of a process analysis is to improve the sequence or content of operations required to complete a task. Improvement of the operations themselves is considered in Section 10-3 as motion study. The routines for both types of investigations are similar; they rely heavily on charting techniques. Graphic representations are to movement investigations what numbers are to mathematics—a language of abbreviation that aids the statement of complex relationships and makes them easier to understand.

Charting finds three major areas of usage in production system studies: survey, design, and presentation. *Survey charts* are used in the initial phase of an investigation to categorize present procedures. One difficult question to answer in this stage is the amount of detail to include. *Design charts* describe

TABLE 10.1
Types of Charts and Their Purposes

Type of Chart	Purpose of Charting	Examples of Chart Formats
Survey	To consolidate data from present operations to facilitate analysis.	Flow process charts to record current operating conditions. Link charts to evaluate layouts for communication and control. Organizational charts to show authority and responsibility.
Design	To develop changes and new concepts that will improve operations.	CPM networks for project planning. Operator human–machine charts to optimize working cycles.
Presentation	To summarize and clarify proposals in order to improve communications.	Gantt and time charts for scheduling and coordination. Break-even charts to advertise and explain the effects of different operating alternatives.

the proposed undertaking. They expose planned innovations to critical reviews which filter out the most promising designs. *Presentation charts* explain how something can be done. The purpose of the presentation is usually a mixture of clarification and selling ability. Table 10.1 shows the sequence of charting and some examples. Note that the same chart format can be used in more than one charting area. For instance, break-even charts can record the present conditions or illustrate what is expected to happen if certain changes are made.

Investigation procedures

The first step in a process analysis is to decide which process to investigate. The obvious resolution is the process whose improvement promises the greatest return, but this general objective has to be narrowed to a specific process by some preinvestigation sleuthing. Operating departments usually suggest areas of investigation when methods work is a staff function. Suggestions can come from above and below or be internally generated for line operational methods work. Background information for setting priorities for studies is obtained by reviewing reports, memoranda, and directives and by discussing the subject with conversant personnel. This preinvestigation should also indicate whose help will be needed for the study and how long it will take.

The fact-finding phase of an investigation is a delicate exercise. A brute-force search is occasionally appropriate, but diplomacy is usually more productive. Outlines of the questions to be asked and the actions to be observed should be known before anyone is contacted. Department managers or supervisors should be informed of the nature and objectives of the investigation before their

Rating factors can be used to set priorities for available study times. Rating = (potential saving/study and implementation cost) × (probability of implementation). Projects with the highest ratings receive attention first.

operating personnel are approached. More and better data result from extending similar courtesies during the investigation. Some guidelines for interviewing are

1. Be certain the workers understand the purposes and objectives of the investigation. Try to make them feel at ease.

2. Emphasize the importance of their contribution to the success of the study. Let them do most of the talking.

3. Solicit and encourage suggestions by the workers. Do not ask questions that imply their own answers.

4. Be courteous and complimentary. Do not use trick questions.

5. Do not criticize or correct the way the workers are doing anything. The mission at this stage is to find facts, not to correct faults.

Wherever possible, verbal communications should be reinforced by the investigator's own observation; an eyewitness has a better chance of getting the whole story and can only blame himself or herself if it is incorrect.

The finished chart should portray the entire operation being studied. Some common charting formats that aid completeness are described in the next section. Although neatness in charting is commendable, some analysts get a reputation for being "chart happy." A proper perspective results from recognizing that good charting improves the statement of the problem, but the ultimate value of the study depends on the creative effort expended to solve the problem.

Distinctive chart patterns often suggest handling problems such as backtracking, uneven work distribution, and duplication of work.

After a methods analyst is satisfied that there is enough information of verified quality, the analyst then attempts to synthesize a better way to do the work. He or she mentally tears the process apart with the intention of *eliminating*, *combining*, or *rearranging* operations. As a spur to the imagination, the analyst can borrow the journalist's code of Who? What? Where? When? How? And *why*?

Rudyard Kipling commented on 5W + H questions in "The Elephant's Child": "I keep six honest servingmen (They taught me all I knew); Their names are What and Why and When and How and Where and Who."

Who does it—why? Could someone else do it better or cheaper?

What is done—why? Does it need to be done?

Where is it done—why? Can it be done less expensively someplace else?

When is it done—why? Would a different sequence be better?

How is it done—why? Is there a better way of doing it?

Sometimes one question leads to another, and a chain reaction of suggestions develops. Two analysts can often generate more ideas working together than they can singly.

If a new method results from the mental gymnastics of analysis, it is committed to an "after" chart—a picture of the revised process. The initial draft of the "after" chart is reviewed with representatives of the staff and operating agencies concerned. With their concurrence, a test or full-scale application of the changes is inaugurated. The analyst usually personally supervises the first installation of the new process. He or she orients the key personnel to the changes and sets up a review schedule. The analyst should be alert to any emergencies caused by the new procedures and be prepared to make revisions promptly if the situation looks explosive.

The last step is to integrate formally the tested and reviewed process into the standard operating procedures. Minor process changes may be made by verbal agreement or insignificant adjustments. Major changes usually require modifying manuals, training programs, plant layouts, work instructions, routing slips, or other published documents. Periodic checks follow the formal installation to see that the revisions remain installed as revised.

Operation and operator process charts

The Gilbreths' early renditions of **flow process charts** had 40 symbols for plotting activities. Since that time the number of basic symbols has decreased, but the format variations for process charting are almost innumerable. And this situation is probably healthy, because customized applications can make charts much more useful for particular situations, such as computer procedures or forms flow analysis. The inherent weakness of changes is that someone unaccustomed to the unique symbols or format would have little idea of what the charts tell. We shall limit our considerations to the following standardized symbols:

F. B. and L. M. Gilbreth, "Process Charts," *Transactions of ASME*, Vol. 43, 1024–1050 (1921).

Standardized symbols are from "Operation and Flow Process Charts, ASME Standard 101," ASME, New York, 1947.

○   *Operation*—intentional changes in one or more characteristics.

⇨   *Transportation*—a movement of an object or operator that is not an integral part of an operation or inspection.

□   *Inspection*—an examination to determine quality or quantity.

◗   *Delay*—an interruption between the immediate and next planned action.

▽   *Storage*—holding an object under controlled conditions.

◎   *"Combined"*—combining two symbols indicates simultaneous activities. The symbol shown means an inspection is conducted at the same time an operation is performed.

The operation symbol is sometimes modified for paper flow analysis to
◎ Origin of papers
◓ Information added
○ All other

It is important to distinguish between an *operation* process chart that follows the routing and acts performed on an object, and an *operator* process chart that shows the sequence of activities performed by a person. The two types should not be combined. Three typical formats for process charts designed to serve different purposes are shown in Figure 10.2. The actual charts represented by the illustrations would contain entries for additional details.

Subsidiary information is often neglected during the construction of charts. Sources of data should be recorded if they add validity and emphasis to the chart or if they will later help the analyst's memory. Improper titles or confusing descriptions result from skimping on words or using personal abbreviations.

Type and Purpose	Characteristics	Illustration	
		Symbols	Description
Single column — to study the detailed steps in a relatively simple process	Charts are often drawn on printed forms; processes are shown by connecting appropriate symbols; space is provided for additional data	○ □ ⇨ D ▽	Invoice in mail room
		○ □ ⇨ D ▽	Determine addressee
		○ □ ⇨ D ▽	To addressee's secretary
		○ □ ⇨ D ▽	Placed in action tray
Multicolumn — to analyze detailed steps in the flow of work that is quite complex	Horizontal lines show operational areas; symbols are entered to show activities of the process	Operator — Activities: Mail clerk ①⇨; Secretary ②⇨ ... ④; Manager ③	
Layout diagram — to improve the layout by avoiding unnecessary steps	Lines indicate the travel and symbols show the activities on a layout often drawn to scale	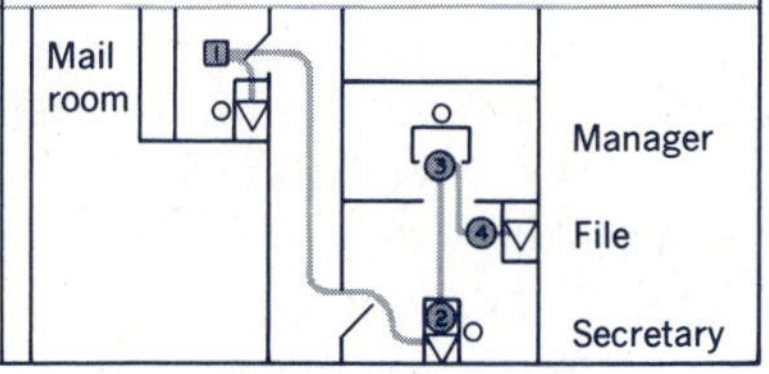 Mail room ... Manager, File, Secretary	

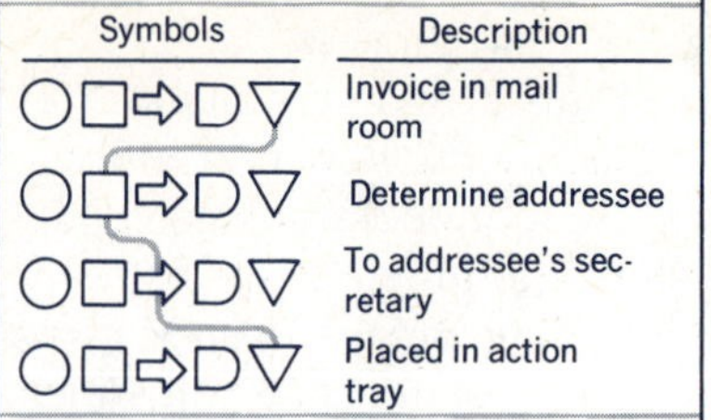

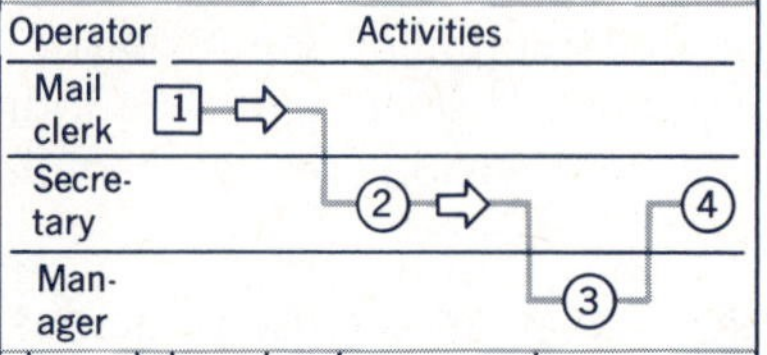

FIGURE 10.2 Illustrations of different process charts.

Legends permit a reader to understand the chart if unfamiliar conventions are used. The golden rule of charting is that everyone should get the same intended meaning from the display.

EXAMPLE 10.1 A methods study using a process chart

One department of a material-testing lab determines the compressive strength of concrete cylinders. These cylinders (one foot long and six inches in diameter) are cast at construction sites and indicate the quality of concrete being poured. They are delivered to the lab by the constructors, where they are stored in a "moist room" under controlled temperature and humidity. After the curing period, usually 7 or 28 days, the cylinders are broken to see whether they meet the specified strengths.

James, the supervisor of test section 1, reports that the price of cylinder testing will have to be increased to meet the wage hikes given to the lab technicians. The lab manager asked the methods analyst, Irma, to check the process before raising the testing price.

Irma's process chart for the present method of cylinder testing is shown in Figure 10.3. The times and distances in the chart are only approximations in the first survey. Possible corrective actions are noted in the columns on the right. From these preliminary checks, it appears that the measuring might be combined with the weighing operation (measurements taken while the cylinder is on the weighing scale), the level of the cart could be raised to allow the cylinder to be rolled from the cart to the workbench, and the entire workbench could be moved closer to the hydraulic press where the cylinders are broken. The next steps are to secure additional information about the feasibility of the indicated suggestions and to estimate the implementation costs and expected savings.

SUMMARY	UNITS	NO.	TIME	DIST.	SPACE	FLOW PROCESS CHART		Page 1 of 1
○		2	8a			☑ PRESENT	DATE	ANALYST
□		4	4a			☐ PROPOSED	11-16-76	I. C. Yu
⇨		6		60		SUBJECT & QUANTITY		
D						1 concrete test cylinder		
▽		2						

TOTALS		14	12a	60	

STEPS	PERSONS & DEPT. CONTACTED	POSSIBLE ACTION
1	James – Test Section #1	
2	Bob Wilson – Tech.	

WHY? IS IT REALLY NECESSARY? a-approx.
WHAT? WHEN? WHERE? WHO? HOW?

STEP	SYMBOL	TIME	DIST.	SPACE	PROCESS STEPS	ELIMINATE	COMBINE	SEQUE.	PLACE	PERSON	IMPROVE
1	○□⇨D▽				In moist room for 28 days						
2	○□⇨D▽	1	6		Select cylinder from rack and place in cart (holds 12)						
3	○□⇨D▽		31		Cart to test bench				✓		
4	○□⇨D▽		6		Lift cylinder to bench (up 2 ft.)						✓
5	○□⇨D▽	1½			Measure diameter & length	✓					
6	○□⇨D▽		4		Move to scales (roll)						
7	○□⇨D▽	1			Weigh	✓					
8	○□⇨D▽		5		Move to capper						
9	○□⇨D▽	6			Cap both ends and mark						✓
10	○□⇨D▽				Store under wet blanket while other cylinders are capped						
11	○□⇨D▽		8		Move to press (by cart)						
12	○□⇨D▽	½			Center in press						
13	○□⇨D▽	2			Break						
14	○□⇨D▽				Observe type of break						
	○□⇨D▽										

FIGURE 10.3 Sample flow process chart.

Operator–machine process charts

Dependent relationships of workers and machines are depicted on **operator–machine process charts.** The purpose of these charts is to analyze a process to develop an economic balance of idle time for the workers and machines. The related activities of workers and machines are often on an intermittent basis; a sequence of worker activity is required to prepare the machine for operation;

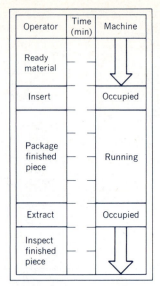

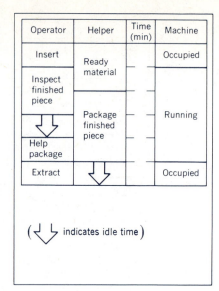

(⬇ indicates idle time)

FIGURE 10.4 (Left) Operator–machine process.

FIGURE 10.5 (Right) Modified operator–machine relationship.

the machine has a set running time to produce a finished product; and then more activities are required to extract the product and prepare for the next run. When there are long running periods, one operator can service several machines; conversely, a low ratio of running time to preparation time indicates that a "gang" of workers may be required. The crew size criterion is the contribution per piece produced: the difference between the value of the piece and the sum of machine and labor costs to produce it.

A clearer picture of the operator–machine relationship is obtained by showing the activities of each on a time scale. As illustrated in Figure 10.4, one worker operates one machine. He or she has no idle time, but the machine is idle four minutes in each cycle. The relationship could be altered by assigning two machines to the operator or by adding a helper, as shown in Figure 10.5. Assuming the following economic relationships:

$$\text{value of each piece produced} = \$ \ 4 \text{ per unit}$$

$$\text{cost of labor: operator} = \$ \ 6 \text{ per hour}$$

$$\text{helper} = \$ \ 3 \text{ per hour}$$

$$\text{burden rate of machine} = \$17 \text{ per hour}$$

the hourly contribution of each method is calculated as follows:

One Operator per One Machine		(One Operator and One Helper) per One Machine	
Output:	6 pieces per hour	Output:	10 pieces per hour
Value:	6 × $4 = $24	Value:	10 × $4 = $40
Cost:	Operator = $ 6	Cost:	Operator = $ 6
	Machine = 17		Helper = 3
			Machine = 17
	Total 23		Total 26
	Contribution per hour = $1		Contribution per hour = $14

Besides the obvious monetary attractiveness, the new method allows one minute of idle time in each cycle for unexpected delays by the operator.

EXAMPLE 10.2 Quantitative evaluation of operator–machine ratios

When several machines have the same running time and identical service requirements, a quantitative approach easily identifies the optimal operator–machine ratio. Assume the following times and costs are descriptive of a production process:

Operator:	Insert piece	0.60 min
	Remove finished piece	0.30 min
	Inspect piece	0.50 min
	File, burr, and set aside	0.20 min
	Walk to next machine	0.05 min
	Wage	$3.00 per hour
Machine:	Running time	3.95 min
	Burden rate	$4.80 per hour

The approximate number of machines per person is estimated by the ratio of the machine cycle time to the operator cycle time:

$$\frac{\text{Machine cycle time}}{\text{Operator cycle time}} = \frac{0.6 + 0.3 + 3.95}{0.6 + 0.3 + 0.5 + 0.2 + 0.05}$$

$$= \frac{4.85}{1.65}$$

$$= 2.9$$

$$= 3 \text{ machines per operator}$$

Checking the cost for an operator servicing two and three machines, we have the following calculations:

	Two Machines		Three Machines	
Labor:	$(4.85 \text{ min}) \dfrac{\$3}{60 \text{ min}}$	= $0.242	$(3)(1.65) \dfrac{3}{60}$	= $0.247
Burden:	$\dfrac{4.85 \text{ min}}{\text{machine}} \dfrac{\$4.80}{60 \text{ min}} (2 \text{ machines})$	= 0.776	$(4.95) \dfrac{\$4.80}{60} (3)$	= 1.188
Total cost per cycle		= $1.018		= $1.435
Unit cost:	$\dfrac{\$1.018}{2 \text{ pieces}}$	= $0.51	$\dfrac{\$1.435}{3 \text{ pieces}}$	= $0.48

In the two-machine case the limiting factor is the machine cycle time (4.85 minutes to complete one cycle); in the three-machine case the limiting condition is the operator cycle time ($3 \times 1.65 = 4.95$ minutes to service three machines). In the latter case the operator has no idle time. It is apparent from the cost figures that it is more economical to assign three machines to each operator.

10-3 MOTION STUDY

The purpose of motion study is to make work performance easier and more productive by improving manual motions. The most obvious difference between process and motion analysis is the scope of the study: motion studies are oriented to the body movements of one individual. Each circle on an operator process chart is a possible area for **motion economy.**

The investigation routine, techniques, and attitude suitable to a motion study are similar to those befitting a process analysis. An appropriate problem is selected for study; the current method of work is observed; and creative questioning of the operation seeks a better method. The key phase is the detection of wasteful movements.

Principles of motion economy

The question of *what* tasks workers should perform cannot be accurately answered until it is determined *how* they can perform them. This sequence of questions seems natural for machine operation but is often transposed for the performance of humans.

In Chapter 8 we considered the question: "Human or machine?" When we evaluate the motor performance of human beings in a system, we are aided by a mechanistic set of guidelines that could just as logically be applied to a machine. The principles of motion pertain to the distribution of work over different parts of the body, preferred types of motion, and the sequence of movements; the machine design equivalents of these principles are component analysis, kinematics, and programming. The simplicity and logic of the following principles are intrinsically so satisfying that one wonders how they could possibly be overlooked:

1. Minimize the number of movements—eliminate unnecessary motions and try to combine movements.

2. Minimize the lengths of movements—keep the work area within the arc of forearm motions.

3. Balance the motion pattern—use symmetrical motions in opposite directions and avoid sharp changes in direction.

4. Move hands simultaneously—start and stop hands at the same time.

5. Minimize the number of parts of the body involved in a complex motion—if necessary, change the complex motion to a single one.

6. Minimize the muscular force required for movements—slide objects instead of lifting them.

7. Minimize the muscular effort required for control—use prepositioned stops and mechanical guides.

8. Minimize the number of eye fixations and the distance of eye movements required—preposition tools and materials to allow a touch system of operations.

9. Minimize the eye fixation time required for perception when this is a controlling factor—use color, shape, and size coding; provide qualitative indicators if possible.

10. Distribute actions among the members of the body in accordance with their natural capacities—relieve hands of simple repetitive jobs by using foot controls.

11. Provide holding devices—free one or both hands for more useful motions.

12. Provide for the intermittent use of different muscles—try to alternate sitting/standing or similar arrangements to relieve muscle tension by varying motion patterns.

13. Take advantage of gravity—use gravity feed for materials and dispose of objects by drop delivery.

14. Take advantage of natural rhythm—arrange work to allow easy, continuous, and repeated movements that develop good work habits.

Motion analysis

A motion study microscopically examines an operation. The power of magnification depends on the nature and importance of the operation. Visual observations of work in progress are sufficient for gross movements such as loading trucks or felling trees. When an operation is frequently repeated and requires many small movements, a more detailed analysis may be warranted. The rapid micromovements found in some assembly operations are difficult to follow with unaided vision. Motion pictures can be used to assist such studies.

A process analysis is a macromovement study, but it also benefits from attention to the principles of motion economy.

CAMERA STUDIES

A *micromotion* study uses a high-speed (around 960 frames per minute) camera to photograph short-cycle (two minutes or less) operations. After filming the operation, an analyst can retire to a more secluded spot and replay the film at a slower speed to detect improper motions. This approach has advantages of the thoroughness resulting from unlimited playback opportunities and instructional applications, because the films document both good and bad practices. It has the disadvantage of being expensive when compared with the cost of direct observation.

The analysis potential of slow-motion replays is well publicized in televised sporting events.

A *memomotion* study uses a slow-speed (60 frames per minute or less) automatic camera to record a long-long cycle operation or several long cycles of the same operation. The resulting films are run at higher speeds to give continuity to motion analyses. It is a relatively inexpensive way to obtain the general operational characteristics without an on-site analyst. The recorded work patterns serve as training guides and are useful in labor disputes.

Time-lapse cameras also are used as "electronic watchdogs" in banks and for important control records.

Video cameras enable the immediate analysis of taped operations.

Camera studies are often resented by workers. Even cooperative employees may become suspicious or overreact when their performance is photographed unless analysts explain their missions. Part of the analysts' reassurance to workers should be a caution to act naturally. Although hamming antics provide darkroom chuckles for the analysts, they defeat the fact-finding mission.

The first elements of work divisions were called *therbligs* after their originators, the Gilbreths, spelled backwards, almost.

MOTION CHARTING

Whether the motion patterns are observed directly or recorded on film, they are often committed to paper **(motion charting)** using shorthand symbols. Some of the relatively standardized symbols are shown in Figure 10.6. Two degrees of refinement are indicated. The top set, called *get-and-place* elements, are appropriate to longer cycles involving arm and hand movements. The other set, for *microscopic* analyses, are associated with very short cycles detailing hand and finger movements.

The flavor of motion charting is well displayed by diagrams of concurrent hand movements. Two charts of this type are called *Left-hand Right-hand* and *SIMO* (*SImultaneous MOtions*). Both charts detail the simultaneous movements

Symbol	Motion	Description
	(Larger elements of an operation for "get and place" analysis)	
G	Get	Reaching and securing control of an object
P	Place	Moving an object to an intended position
U	Use	Employing a tool or instrument to accomplish a purpose
A	Assemble	Joining two objects in an intended manner
H	Hold	Supporting an object with one hand while the other hand performs work on the object
	(Smaller elements of an operation for "microscopic" analysis)	
R	Reach	Moving an unloaded hand or finger to within one inch of the intended position
G	Grasp	Securing control of an object with the fingers
H	Hold	Supporting an object with one hand while work is being performed on the object
M	Move	Moving a loaded hand or finger to within one inch of the intended target
P	Position	Aligning, orienting, or engaging one object with another
T	Turn	Rotating the forearm about its long axis
D	Delay	Hesitating while awaiting termination of an act or event
RL	Release	Relinquishing control of an object previously grasped in fingers

FIGURE 10.6 Work element descriptions and symbols.

of an operator's right and left hands. SIMO charts are usually applied to more minute motions which are keyed to a time scale. The times are determined from a frame count or a film clock from camera studies and by stopwatch studies. Symbols like those shown in Figure 10.6 are used to describe the movements.

The ideal area for microcharting is where a large number of workers are doing identical jobs on identical machines. Under such conditions, the considerable expense needed to analyze and optimize the motion pattern of an operation is paid for by the quantity of work benefiting from the study. The area appropriate to microanalysis is the very same area where mechanical processes are most profitably applied. From a system viewpoint, a study in this area should not have the pure objective of improving manual methods; it should also be a feasibility study for new and different methods to perform the work.

An assumption that "improving the parts will improve the whole" can be misleading unless system effectiveness is considered concurrently with operation efficiency.

EXAMPLE 10.3 A macroview of microanalysis

In an attempt to stem an ever-increasing burden of paperwork, a Methods Department investigation is requested. A preliminary survey indicates that the existing equipment and staff are sufficient to handle anticipated demands if paper handling and form design are improved. Analysts interview the clerks and

Part:	Form 3291			Layout or part sketch	
Operation:	Requisition				
Method:	Present			I A	
Date:	9–23			B Work space	
Analyst:	SC yu				

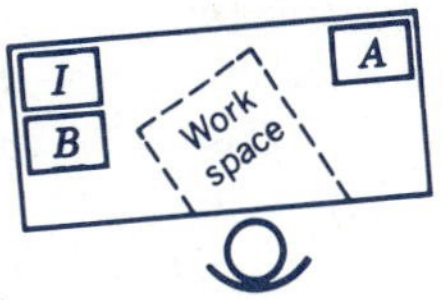

Left hand		Right hand	
Description	Symbol	Symbol	Description
Form 3291 from Box I	G	H	Pen
Center of desk	P		
Expose page 2	G		
	H	U	Check and sign
Expose page 4	G	H	Pen
	H	U	Mark routing
Expose page 5	G	H	Pen
	H	U	Assign account number and endorse
		G	Tear off last page (5)
		P	Last page to Box A
Pages 1–4 to Box B	P	G	Pen

FIGURE 10.7

supervisors to get their suggestions. Attention is focused on the requisition and billing system. Procedures for certain key operations are charted. A get-and-place analysis is deemed appropriate because there are too few identical, high-volume operations to warrant the application of micro-analysis techniques. An example of the charted procedure for checking and endorsing one requisition form is shown by the Left-hand Right-hand chart in Figure 10.7.

The movements shown in the chart appear unnecessarily repetitive. Are three entries necessary? Does each entry have to be on a different page? Could a stamp or initial be used? Could carbon copying be incorporated? Can the in-box be designed to make the forms easier to grasp? Should several forms be piled on top of one another to allow continuous marking? Can the completed forms be removed by a drop disposal?

Answers to these questions depend on considerations beyond the immediate problem of improving one operator's motions. For instance, all entries might be consolidated on one page to improve this operation, but inconvenience or misinterpretation could result in other operations. Such a solution could be a suboptimization of the system objective.

10-4 TIME STUDY

Time study is the handmaiden of motion study. The proof of an improved method is confirmed by a time reduction. A valid time standard for an operation depends on first identifying the best method to perform that operation. This relationship establishes the need and justification for time study: it measures the labor needed to produce a product and thereby is the basis of wage payments.

The objective of time study is to determine the *standard* time for an operation—the time required by a qualified and fully trained operator to perform the operation by a specified method while working at a normal tempo. This definition sets the routine for time studies. An operation is evaluated, and redesigned if necessary, by methods analysis. The work elements thus obtained are clocked. These clock readings are normalized by applying a rating factor to account for the operator's tempo. Then allowances are included to compensate for production interruptions. The end product is a realistic evaluation of labor accomplishment.

A *work count* is an informal method of measuring work by counting accomplishments (units produced, postings made, inspections completed, and so on) during a period of time.

Study preparation

The initial phase of a time study is to acquire sufficient familiarity and knowledge of the operation, equipment, and working conditions. An ideal situation is one in which a highly skilled, cooperative operator is performing a task by approved methods under typical working conditions. When the situation is not so perfect, recorded data must be amended later to make it representative of conceived performance. It is important to record a thorough description of the actual situation at the time of the study: a sketch of the layout, tools and materials being used, unusual environmental conditions, and the like.

Standard data recorded for each time study include (1) operator's name, (2) observer's name, (3) operation, (4) speed and feed of machines, (5) times the study starts and ends, and (6) date.

A job is described by dividing an operator's activities into **work elements,** or groupings of basic movements. Occasionally a work element may be as short as one of the micromotion elements in Figure 10.6, but more commonly the divisions are larger. The following guidelines provide a framework by which to identify distinct work elements that simplify the timing and the evaluation of observed times.

1. *Irregular* elements, those not repeated in every operation cycle, should be separated from *regular* elements. If a machine is cleaned off after every five parts produced, "cleaning" is an irregular element. Its time should be prorated over five cycles when one cycle is required to produce one part.

2. *Constant* elements should be separated from *variable* elements. When the time to perform a work element depends on the size, weight, length, or shape of the piece produced, it is termed variable. "Sew edges" is a variable element if the perimeter of the piece changes from one cycle to the next.

3. *Machine-paced* elements should be separated from *operator-controlled* elements. This division helps recognize delays.

4. *Foreign* or *accidental* elements should be identified. Such activities would be considered expected occurrences if labeled as irregular elements. Therefore, a separate list of foreign elements is kept during each timing session.

5. The *end points* of each element should be easily recognized. A distinct sound or a change in contact forms a readily identifiable break-off point.

6. It is desirable to have *shorter* elements when a choice must be made, but durations of less than 0.03 to 0.04 of a minute are difficult to time accurately.

7. *Unnecessary* movements and activities should be separated from those deemed essential. A skilled operator, by definition, has fewer unnecessary actions than does an untrained operator. Familiarity with motion principles is a definite aid in recognizing improper movements and in designing better motion patterns.

Because work elements are essentially microactivities, they form an activity list similar to those for CPM (Section 7-3) and share many of the same considerations.

After all the work elements in a cycle have been identified and any needed motion corrections have been incorporated, a complete description of each element is entered on a time-study record form in the expected sequence of occurrence.

Data collection

Element times are taken directly at the work place by clock readings or remotely by motion-picture analysis. For either course, it is advisable to seek the active cooperation of the operator being observed. The timer should stand behind and out of the way of the operator. **Stopwatch-timing** studies are far more numerous than are camera studies owing to their greater versatility and lower costs.

There are differently calibrated stopwatches. The two most commonly used are (1) a minute decimal watch calibrated in increments of $\frac{1}{100}$ of a minute and (2) an hour decimal watch showing increments of $\frac{1}{100}$ of an hour. Both have large sweep hands that make one revolution per 100 increments, and both have smaller dials for a cumulative count of the revolutions.

There are two widely used techniques for recording elemental times:

Dual and triple watch arrangements linked together on time-study boards allow direct readings while retaining features of continuous timing.

1. *Snap-back* timing directly determines the time to complete each element. The sweep hand is at zero when each work element begins, and it turns while the work element is being performed. When the element is completed, the time consumed for its performance is read from the watch, and the sweep hand is snapped back to zero, ready to begin the next element.

2. *Continuous* timing allows the stopwatch to run without interruption through the entire duration of the study. Accumulated elemental times are recorded in sequence while the watch is running. Then each reading must be subtracted from the preceding reading to show the elapsed time of the work elements.

The continuous method is the most commonly used timing technique.

Both methods have their adherents. The snap-back method requires less clerical work, and delays in work procedures are somewhat easier to handle. But errors may be introduced if snap-back reflexes are slow. Continuous timing allows better readings for very short elements but requires strict concentration to note the passing times. It forces a full accounting of all the time involved in the study.

Selected time

The times recorded for each element will vary among observations. Variations are natural because operators do not always move at a uniform pace; motion patterns are not always identical; and tools and materials are not always replaced in exactly the same spot. A small degree of variation can even be expected from the timer as he or she reads the watch. However, variations traceable to unusual occurrences, such as a dropped tool or a coughing fit by the operator, are accounted for by deleting the affected observation. The remaining observations, now considered "representative," are averaged to determine the **elemental selected times:**

Selected time to complete an element $= ST_e$

$$ST_e = \frac{\Sigma \text{ (elemental times of representative observations)}}{\text{number of representative observations}}$$

By summing the selected elemental times, we have our best estimate of the *task selected time*, ST_t. Thus

$$ST_t = \Sigma ST_e$$

where the summation includes all the work elements required to complete the task. ST_t is then the average time the operator needs to perform the task being studied. At this stage we are not prepared to say that ST_t is the standard of performance we should expect from all operators. In fact, we are not even sure that it is the true time for the observed task. What we have is a sample that should be statistically checked to determine how much confidence we can have in it.

One method of deciding on the number of timing cycles is to use a formula based on the standard error of the average for an element:

$$N' = \left(\frac{k/s \sqrt{N \Sigma X^2 - (\Sigma X)^2}}{\Sigma X}\right)^2$$

where

N' = number of cycles needed to produce desired precision and confidence level

k/s = confidence-precision factor

X = representative element times

N = number of representative element times

Quicker methods for estimating the number of cycles required include alignment charts and tables relating the average cycle time and the range of cycle times to the number of observations. Each chart applies to only one accuracy and confidence level.

Two decisions must be made in applying the formula. The first is to decide which element in the task to use for the X values. A conservative solution is ensured by choosing the element with the greatest variation: the larger range of readings produces a larger variance, which in turn necessitates a greater sample size for a given confidence level.

The second decision is the selection of a **k/s factor**—the confidence level that the desired accuracy is attained. A k/s of $2/0.05 = 40$ indicates a 95-percent confidence level and a ±5-percent precision. The numerator, 2, is derived from the relationship that 95 percent of the sample observations fall within the range of ±2 standard errors of the mean. The denominator, 0.05, sets the allowable error at ±0.05 of the true element time. Similarly, the insertion in the formula of

A k/s factor of 20 is generally accepted as reasonable with respect to the overall accuracy expected in a time study.

$$k/s = 3/0.03 = 100$$

allows the calculation of the number of cycles that must be timed so that 997 times out of 1000 the ST_t will be within ±3 percent of the true task time.

EXAMPLE 10.4 Stopwatch timing

The concrete-cylinder-breaking operation of the material-testing lab introduced in Example 10.1 is still under investigation. The macroanalysis conducted by process charting implied that further attention should be given to the capping operation. This task is to put a hot liquid chemical compound in a mold on the end of a cylinder. The liquid quickly dries into a very hard cap. The purpose of the caps is to square off the cylinder ends and to provide a smooth surface for the uniform application of force to break the con-

crete. A time study will define the labor content of the capping cost for cylinder testing.

A breakdown of the capping operation reveals the nine basic work elements shown in the observation sheet (Figure 10.8). The capping operator, James Randle, is well trained and qualified to perform the task. The procedure he follows is efficient, and the working conditions are favorable. A sketch of the workbench arrangement is shown at the top of the observation sheet.

Observation Sheet

Sheet _1_ of _1_ sheet
Operation _cylinder capping_
Operator _James Randle_
Conditions _experienced_
Machine _Capper model 3_
Speed _____
Material _capping compound_
Date _11-21-76_   Begin _1021_   End _1053_   Observer _I.C.Yu_

Sketches and notes

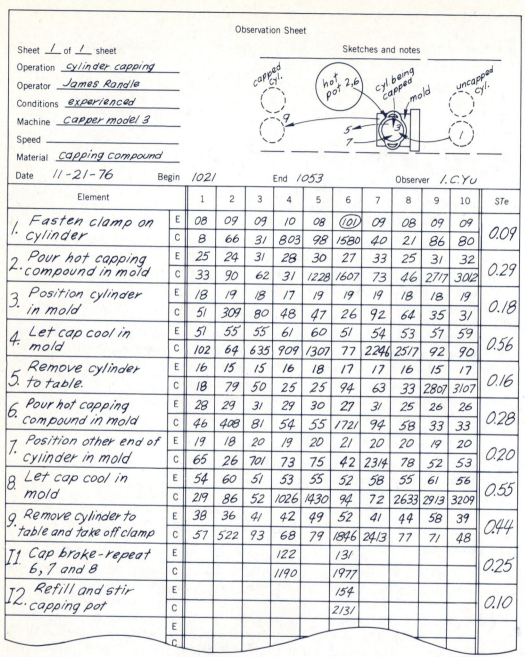

Element		1	2	3	4	5	6	7	8	9	10	STe
1. Fasten clamp on cylinder	E	08	09	09	10	08	(101)	09	08	09	09	0.09
	C	8	66	31	803	98	1580	40	21	86	80	
2. Pour hot capping compound in mold	E	25	24	31	28	30	27	33	25	31	32	0.29
	C	33	90	62	31	1228	1607	73	46	2717	3012	
3. Position cylinder in mold	E	18	19	18	17	19	19	19	18	18	19	0.18
	C	51	309	80	48	47	26	92	64	35	31	
4. Let cap cool in mold	E	51	55	55	61	60	51	54	53	57	59	0.56
	C	102	64	635	909	1307	77	2246	2517	92	90	
5. Remove cylinder to table.	E	16	15	15	16	18	17	17	16	15	17	0.16
	C	18	79	50	25	25	94	63	33	2807	3107	
6. Pour hot capping compound in mold	E	28	29	31	29	30	27	31	25	26	26	0.28
	C	46	408	81	54	55	1721	94	58	33	33	
7. Position other end of cylinder in mold	E	19	18	20	19	20	21	20	20	19	20	0.20
	C	65	26	701	73	75	42	2314	78	52	53	
8. Let cap cool in mold	E	54	60	51	53	55	52	58	55	61	56	0.55
	C	219	86	52	1026	1430	94	72	2633	2913	3209	
9. Remove cylinder to table and take off clamp	E	38	36	41	42	49	52	41	44	58	39	0.44
	C	57	522	93	68	79	1846	2413	77	71	48	
11. Cap broke-repeat 6, 7 and 8	E				122		131					0.25
	C				1190		1977					
12. Refill and stir capping pot	E						154					0.10
	C						2131					
	E											
	C											

FIGURE 10.8

The observer uses the continuous timing technique; consecutive readings from the stopwatch are entered in the rows marked "C." All readings are in hundredths of a minute. During the fourth and sixth cycles an irregular element (I1) occurs when Randle breaks a cap as he removes the cylinder from the mold. A replacement cap is formed by repeating elements 6, 7, and 8. The need to add more capping compound to the heating pot is another irregular element (I2) which Randle reports occurs about once every 30 caps or 15 cylinder cycles. Irregular element descriptions are added as necessary below the regular element sequence.

Ten cycles are timed in about 32 minutes. Sometime later, the work elemental times are calculated as the differences between the continuous time readings. The differences are entered in the "E" rows. The average value of each row establishes the selected time for the work elements, ST_e. During the sixth observation for the first element, the operator caught his finger in the clamp while watching a girl pass by. He explained that this was the first time he had managed to pinch a finger, and he doubts that it will ever happen again. Therefore, the observation is circled and discarded before calculating ST_t. The selected time for I2 is based on the assumption that the element will occur once every 15 cycles.

The number of cycles timed will be considered adequate if the precision is within ±10 percent at a 95-percent confidence level. Work element 9 has the greatest variation and is therefore used in the formula check.

$$
\begin{aligned}
N' &= \left(\frac{2/0.10\sqrt{10 \times 19{,}792 - (440)^2}}{440} \right)^2 \\[2mm]
&= \left(\frac{20\sqrt{197{,}920 - 193{,}600}}{440} \right)^2 \\[2mm]
&= \left(\frac{20\sqrt{4320}}{440} \right)^2 = \left(\frac{66}{22} \right)^2 \\[2mm]
&= 9 \text{ cycles}
\end{aligned}
$$

where

$$
\begin{aligned}
X &= \text{observation of element 9} \\
\Sigma X &= 38 + 36 + \cdots + 39 = 440 \\
\text{and } \Sigma X^2 &= (38)^2 + (36)^2 + \cdots + (39)^2 = 19{,}792
\end{aligned}
$$

It thus appears that enough observations have been taken and that the selected time for the capping operation is equal to

$$
\begin{aligned}
ST_t &= 0.09 + 0.29 + 0.18 + 0.56 + 0.16 + 0.28 + 0.20 + 0.55 + 0.44 + 0.25 + 0.10 \\
&= 3.10 \text{ minutes}
\end{aligned}
$$

Normal time

At the conclusion of a timing session, the observer faces the disagreeable prospect of assigning a leveling or **rating factor** to the operation just observed. The rating function compares the tempo of an operator's performance with the observer's conception of a **normal time** or pace. This comparison is difficult because there are very few solid supports on which to base the concept of normal, and there is a shortage of aids by which to maintain a lasting conception once it is conceived. But regardless of the difficulties, some factor must be applied to ST's to account for individual differences among operators before performance times can be standardized.

Rating starts with a definition of normal. A generally accepted launching point is an agreement that *a normal pace is one that can be attained and maintained by an average worker during a typical working day without undue fatigue.* Trouble comes when the adjectives in the definition are put to practice. What is an "average" worker, a "typical" workday, or "undue" fatigue? Jobs are so different that opinions are bound to vary. Thus, some time-study practitioners acquire reputations for "tight" or "loose" ratings, the opinions of the workers about the opinions of the timers. Even the most skilled and experienced timing practitioners can expect disagreements over their interpretation of pace because rating factors are inherently subjective.

A measure of objectivity is introduced by recognizing certain benchmarks of normal performance. For instance, there is general agreement that the normal time to deal a deck of 52 cards into four equal piles is 0.50 minute and a normal pace for walking on a smooth, level surface is 3 miles per hour. Motion pictures of operators performing different tasks at consensus percentages of normal are available. Some films simultaneously show multiple shots of one operation as it is performed at levels above, at, and below normal tempo. Films taken for micromotion studies can be shown at different projector speeds to qualify or retrain raters.

There will probably never be a yardstick of normality that is universally applicable, because different companies will use contrasting methods of measuring their versions of normal. The important aspect is the uniform maintenance of whatever level is chosen as normal.

Numerous methods of setting rating factors have been proposed. A strict reliance on speed or tempo as appraised by an observer is one of the oldest and most used methods; if the rater judges the operator to be working at a speed of 90 percent of normal, the rater normalizes the selected time by multiplying it by 0.90. Another method, called *objective rating*, considers both the speed and the relative difficulty of the task being performed, and so both a speed factor and a difficulty factor are applied to the selected time to obtain a normal time. More elaborate methods attempt to consider skill, effort, conditions, and consistency through the use of tables or tabulated time expectations. Because there is no industrywide preference for one method, each company has an option as to technique. But once the option is exercised, it should be followed consistently if rating is to be understood and respected by the work force.

Rating factors are applied to elemental selected times (ST_e) or to the task selected time (ST_t) when the operator performs at the same level throughout the task. Ratings are formulated during the timing sessions by noting the effort and skill of the operator as well as the conditions and nature of the task in relation to the observer's image of normal. The normal time is the product of this rating and the selected time:

$$\text{Normal time} = \text{selected time} \times \text{rating factor}$$

Assume that the selected time for a particular task in which the operator gave a consistent performance for all the elements is 0.75 minute. If the observer rated the performance at 12 percent of normal, then

$$\text{Normal time} = 0.75 \times 1.20 = 0.90 \text{ minute}$$

The time thus stated for the task represents the time in which a qualified and well-trained operator working at normal tempo is expected to perform one cycle of the task.

Standard time

Securing selected and normal times are intermediate steps in the development of **standard time**—the time an operation will normally require when allowances are made for interruptions. These interruptions are caused by factors external to the job itself and are therefore not accounted for by the observed elements, corrections to observations, or the rating factors.

Personal allowances are granted for the workers' physical needs. Rest periods for light work average about 5 percent of the typical 8-hour work day. More exhausting work or unfavorable working conditions are reflected by allowing more personal time.

Fatigue allowances are intended to compensate for below-normal performance resulting from fatigue effects. By setting a 5-percent allowance, for example, the time within which a task is expected to be completed is extended by 5 percent, giving the worker a fair opportunity to meet normal standards. Controversy over such reasoning stems from a contention that performance deterioration does not set in until fatigue reaches serious proportions. Therefore, the normal pace should be defined in terms of the effort required to meet normal output under deleterious or strenuous work conditions. The question is further complicated by (1) the lack of any satisfactory way of measuring fatigue and (2) modern working conditions which have almost eliminated actual muscle exhaustion. It is not surprising, then, to find the value of fatigue allowances subject to considerable debate and the percentages a topic of arbitration.

Delay allowances compensate for unavoidable work delays. These are inescapable interruptions to productivity caused by external forces such as power outages, defective materials, waiting lines, late deliveries, and other events over which a worker has no control. Unintentional but avoidable delays by the operators are not included. The irregularity of occurrence and duration of unavoidable delays make this allowance another suitable target for collective bargaining.

Special allowances are sometimes added to give redress for unusual and frequently temporary conditions that curtail workers' output through no fault of their own. Exceptional maintenance requirements, temporary machine interference, and short production runs during which workers are always in an initial learning period are examples. Such cases are individually treated and seldom cause much debate.

Allowances are frequently a topic for collective bargaining, in which increased allowances tend to be treated as new fringe benefits.

Delay allowances are often developed by "ratio delay" studies using techniques described in Section 10-5.

Extra allowances are often provided when a large portion of cycle time is machine controlled and operators are paid incentive wages. They may be as high as 30 percent.

A composite total allowance is determined by adding individual allowance percentages applicable to a certain job. The standard times for tasks included in the job are calculated by adding to each normal time the product of its total allowance percentage and its duration:

Standard time = normal time + (normal time)(total allowance)

or

Standard time = normal time × allowance factor

where

$$\text{Allowance factor} = 1 + \frac{\text{total allowance \%}}{100}$$

EXAMPLE 10.5 Standard time calculation

Let us return again to the cylinder-breaking process encountered in Examples 10.1 and 10.4. The evaluation of the capping operation can now be completed. Because the operator, Randle, is a competent cylinder capper, the main concern of the observer is to rate Randle's tempo with respect to a normal pace. Each capping task element is rated separately as shown; rating factors are applicable to an entire task only when all the elements in the task are subject to and performed at a uniform pace. In the cylinder-capping operation, the times for two elements, 4 and 8, are not controlled by the operator's pace; they depend on the time it takes the hot liquid capping compound to solidify in the mold. Such material-, product-, or machine-controlled elements are usually assigned a rating of 1.00.

Element	ST_e (Minutes)	Rating Factor	Normal Element Time (Minutes)
6. Pour	0.28	1.10	0.31
7. Position	0.20	1.00	0.20
8. Cool	0.55	1.00	0.55
9. Remove	0.44	1.00	0.44
I1. Repeat 6 to 8	0.25	1.30	0.32
I2. Refill pot	0.10	1.00	0.10
		Normal task time =	3.25 minutes

It is apparent from the ratings that the observer thought Randle was working at a pace that an average worker could not maintain. This is not an unusual situation when an operator is impressed because he or she has been selected for a time study. An observer should be sensitive to carelessness or overexertion, as was apparently indicated when Randle had to remake the broken caps in element I1. Sometimes an uncooperative operator may play guessing games with the observer by varying the pace, a game neither can win completely.

The applicable allowance factor is more a function of the entire job, in this case the complete cylinder-testing process, than of the operation of capping the cylinders:

Element	ST_e (Minutes)	Rating Factor	Normal Element Time (Minutes)
1. Clamp	0.09	1.20	0.11
2. Pour	0.29	1.10	0.32
3. Position	0.18	1.00	0.18
4. Cool	0.56	1.00	0.56
5. Remove	0.16	1.00	0.16

Customary personal allowance for all lab employees = 5%

Required handling of 30-pound cylinders and hot capping material is considered in the fatigue allowance = 8%

Interruptions primarily due to impromptu visits by the concrete suppliers are accounted for in the delay allowance = 7%

Total allowance = 20%

Adding 20 percent to the normal time does not mean that each cycle is expected to take $0.20 \times 3.25 = 0.65$ minute longer to complete. Instead, the 0.65-minute increments build up during normal cycles to afford time for rest breaks and other work interruptions. With an allowance factor of 1.20, the cylinder capping task has a standard time of

$$3.25 \times 1.20 = 3.9 \text{ minutes}$$

Synthetic times

Many operations in a plant require the same types of movements. The data collected from previous time studies are valuable for estimating times for similar work in the future. **Synthetic timing** is the name given to the use of standard data to synthesize task durations. Using already-developed normal times in this fashion overcomes two prominent deficiencies of stopwatch timing: (1) time standards can be determined before a process actually is in operation, and (2) after appropriate data have been developed, the time and cost of setting new time standards are significantly lowered.

The applicability of synthetic timing to an individual firm depends on the relative slopes of the two lines shown in Figure 10.9. The solid line depicts the cost of stopwatch studies as directly related to the number of time standards established. The flatter slope of the dashed line, representing synthetic timing, indicates that the variable cost per time standard set by this method is less. Where the two lines intersect depends substantially on the initial cost of developing the synthetic times. This cost is largely a function of the time-study information already available and the degree of job similarity for which times must be established. Whether this extra development expenditure should be made hinges on its cost and the number of time standards the organization expects to set.

The procedures and some of the techniques used to develop standard data are close to those we encountered for forecasting. In estimating sales we looked for some relationship between past sales data and independent variables such

Synthetic timing is best suited to production involving a few basic operations on products that vary in size or material rather than design.

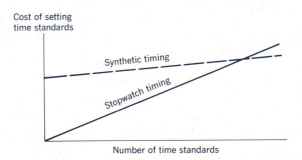

FIGURE 10.9 Cost relationship of setting time standards.

Figure 10.9 is a good illustration of possible temporal suboptimization—not looking far enough into the future to substantiate a current expenditure.

as time. For task times we hypothesize the independent variables that will significantly affect the normal time for an operation. In both cases the next steps are to collect pertinent data and to test the validity of the hypothesized relationships. Line fitting and correlation techniques are useful tools for processing the data.

The most common form of standard data is based on elemental times and takes the form

$$\text{Normal time for task} = NT_1 + NT_2 + \cdots + NT_n$$

where NT_n = normal time for an element characteristic of the task. The equation is merely a summation of elemental times that correspond to the characteristics of the job. For instance, if a drilling operation is frequently encountered, a table of normal times indexed by different drill sizes versus the depth of the hole required could be developed. Whenever the element "drill hole" is obliged while synthesizing a task time, the appropriate normal time is extracted from the table according to the task characteristic of hole size and depth. This approach, using predetermined times for micromotions, is discussed in detail in the next section.

Another method of obtaining synthetic times is to relate by formula the elemental normal times to a characteristic of the task:

$$\text{Normal time for task} = fV_1 + fV_2 + \cdots + fV_n$$

where fV_n = normal time expressed as a function of a characteristic task variable, V_n.

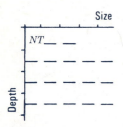

Size

NT____ __

Depth

Element: "drill hole."

Normal times must be developed as a function of frequently encountered task characteristics before this method can be used. These times may already be available from earlier studies and need only be related to the desired variable. For instance, the element "paint" could be formulated as a function of the area to be painted. Least squares or another regression technique described in Sections 3-6 and 3-7 could be used to formulate the relationship. Supplementary time studies may be necessary if available records do not correspond to the desired range of the variable or if innovations are anticipated. Once the formulas are ready, normal times are calculated by using only the magnitude of the task characteristic—pounds to lift, inches to drill, square feet to paint, number to load, miles to travel, and the like.

Work elements of different classifications such as regular and irregular may bear different relationships to the same task characteristic.

Appropriate allowances are applied after the normal time computations to ascertain standard times. The synthesized times should be tested against the actual task times to confirm the hypothesized relationships. They should also be checked periodically or whenever design specifications are changed, to reaffirm the validity of the parameters in the formula.

EXAMPLE 10.6 Synthetic timing by task formula

Synthetic times are being developed for a warehouse facility. One operation is transporting palleted loads from the receiving yard to the appropriate storage building. A single worker operating a lift truck is normally assigned to each shipment. A synthetic time formula is the preferred method to establish standard times for incentive pay because the number of pallets and the distance they must be moved are different for each shipment.

It is initially assumed that the time to transport a shipment will be a function of the size of the shipment and the distance from the receiving point to the storage point. The work elements composing the task are identified and classified as follows:

task and is therefore a constant in the formula: constant = 12.30 minutes. Elements 2, 3, 5, and 6 depend on the number of pallets to be moved, N; $f(\text{number}) = (0.88 + 0.24 + 1.06 + 0.33)N = 2.51N$. The relationship between normal travel time and distance is shown in Figure 10.10. Each dot represents a previous stopwatch time for a certain distance traveled. The lines are least square fits relating travel time, loaded and empty, to the distance moved, D; $f(\text{distance}) = (0.0016 + 0.0021)D$. Because this dis-

Element	Classification	Task Characteristic	Normal Time
1. Check in shipment	Irregular-constant		12.30
2. Position truck	Regular-constant	Number of pallets	0.88N
3. Pick up load	Regular-constant	Number of pallets	0.24N
4. Transport load	Regular-variable	Travel distance	(chart)
5. Position truck	Regular-constant	Number of pallets	1.06N
6. Deposit load	Regular-constant	Number of pallets	0.33N
7. Return empty	Regular-variable	Travel distance	(chart)

The natural division of elements according to task characteristics takes the form

$$\text{Normal time} = f(\text{number of pallets})$$
$$+ f(\text{travel distance}) + \text{constant}$$

Element 1, the storage assignment and check-in list for the shipment, occurs only once during the

tance must be traversed for each pallet, total travel time is also a function of the number of pallets: $f(\text{total travel}) = 0.0037DN$, where D is the one-way distance and N is the number of pallet trips. Combining the two functional relationships involving N and adding the constant term provide the complete formula:

$$\text{Normal time} = (2.51 + 0.0037D)N + 12.30$$

The standard time for the job is calculated by applying an appropriate allowance to the normal time. If a 17-percent allowance is provided for personal time and unavoidable delays in loading and unloading, the task of transporting to storage a shipment of 10 pallets received at a dock 600 feet from the assigned warehouse will have

$$\text{Standard time} = (\text{normal time})(\text{allowance factor})$$
$$= [(2.51 + 0.0037 \times 600)10$$
$$+ 12.30](1.17)$$

$$\text{Standard time} = 59.60 \times 1.17 = 69.73 \text{ minutes}$$

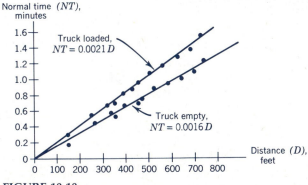

Normal time (NT), minutes

Truck loaded, NT = 0.0021 D

Truck empty, NT = 0.0016 D

Distance (D), feet

FIGURE 10.10

Predetermined times

Predetermined times are the tabulated values of the normal time required to perform individual movements, such as moving an arm from one position to another. An entire operation is described by a series of these basic motions. The total time to perform the operation is the sum of the times needed for the basic motions. By arranging the basic motions and aggregating associated times, an existing task can be analyzed or a proposed operation can be timed without actually performing it.

The broad acceptance of predetermined timing is due in part to the advantages previously mentioned for synthetic times: convenience and the opportunity to estimate the time of an operation before it takes place. In addition, predetermined times have the following attributes:

Camera studies of factory operations were used extensively in developing predetermined times.

1. They eliminate the need for an observer to clock the manual movements of an operator, thereby avoiding distortions possible from observer bias.

2. They bestow credence to task times by virtue of the very large samples from which the component times were developed.

3. They bypass the troublesome performance rating factor by including the pace ratings in the tabulated values.

4. They yield standard times for operations that are reliable enough to be accepted by most trade unions.

5. They are adaptable to computerization, allowing electronic assistance for the design of motion patterns and the collection of operation times.

The application of predetermined times still requires judgment of task content and procedures, but it is usually less obvious and thereby less objectionable to time-study critics.

Recall Section 1-2.

HISTORY OF PREDETERMINED TIMES

An article in 1945, "Work Factor System," was the first published announcement of a procedure for predetermining motion times.

Early in this century both Taylor and the Gilbreths, the pioneers in work design, suggested the establishment of predetermined time standards for each element in an operation. The first practical version of elementary time standards resulted from research done by A. B. Segur during the 1920s. His system, called *Motion Time Analysis*, was made available only to his clients. In 1934, a group of industrial engineers headed by Joseph H. Quick originated a system called *Work Factor*. During the 1940s, several companies developed systems of predetermined motion times: General Electric, Honeywell, and Westinghouse. The first book on the subject was published in 1948. It described *Methods–Time Measurement* (MTM), the first and only system of predetermined times whose complete data are publicly available.

H. B. Maynard, G. J. Stegmerten, and J. L. Schwad, *Methods-Time Measurement*, McGraw-Hill, New York, 1944.

The MTM family of work measurement data systems is international in scope and application. The basic system is known as MTM-1. The MTM-2 and MTM-3 systems are versions that shorten analysis at the expense of some precision. Other versions (for example, MTM-GPD, 4M, MTM-V, MTM-M, MTM-C) are adaptations applicable to particular activities such as clerical, tool use, and microassembly. The following example describes the base system, MTM-1.

EXAMPLE 10.7 Predetermined times according to methods–time measurement (MTM-1)

The official definition of MTM has remained unchanged since its origin:

Methods–Time Measurement is a procedure that analyzes any manual operation or method into the basic motions required to perform it and assigns to each motion a predetermined time standard that is determined by the nature of the motion and the conditions under which it is made.

In MTM-1, nine basic motions are delineated on data cards, and a tenth card shows the ease of difficulty of performing motions simultaneously with both hands. Durations of motions are given in *Time Measurement Units (TMU)*, with each equal to 0.0006 minutes, or 0.036 seconds. Sample data cards for the basic motions are displayed in Table 10.2.

A task is recorded or synthesized by listing the constituent movements, extracting the appropriate table values, and summing the values to determine the normal time of the operation. Using the "reach" table as an example, the motion has three parameters:

DISTANCE. True path in inches traveled by the hand between two points (first column on the left of the "Reach" card in Table 10.2).

TARGET. Nature and position of the objective sought (right-hand side of the card—Cases A to E).

MOTION STATE. Whether the hand is already in motion at the beginning of the reach or continues moving at the end (columns A and B in the middle of the card). A continuous motion requires less time because neither acceleration nor deceleration is included.

Motions are described by shorthand notations. An $R12E$ symbol indicates a 12-inch hand movement from a rest position stopping at an indefinite spot; the required time is 11.8 TMU, or 0.0708 minutes, which is 0.425 seconds. An $mR12A$ symbol indicates the hand is already in motion as the reach to a fixed object begins (8.1 TMU); an m at the right of the "reach" symbol means the hand is in motion at the end of the movement. An $R12E$ reach followed by a select grasp ($4B$) of a small object that is moved to an exact location 9 inches away ($M9C$) is a series of movements that totals $11.8 + 9.1 + 12.7 = 33.6$ TMU, or 1.21 seconds. A continuation of this type of buildup produces a complete description of a task, which can be used for training purposes, and the leveled time by which it should be completed.

PREDETERMINED TIMES PRECAUTIONS

A statement is posted on MTM cards that warns against using the data without thoroughly understanding its proper application. The warning is equally applicable to all predetermined times systems. The process for determining task times may appear so simple that a beginner can become overeager, overconfident, and overwhelmed. Blatant errors, such as the failure to note that the limiting time for an operation requiring both hands is set by the hand used longest, can lead to loose standards. More subtle errors, from overlooking or not recognizing the difficulty of motions, produce tight standards. Quality training and thorough familiarity with the tasks will reduce errors of judgment and

Training tends to erode when one generation of time analysts trains the next generation. A professional predetermined-times course takes from 12 hours (simplified system) to 120 hours (detailed system).

TABLE 10.2
MTM Data Cards.

EFFECTIVE NET WEIGHT

Effective Net Weight (ENW)	No. of Hands	Spatial	Sliding
	1	W	W × F_c
	2	W/2	W/2 × F_c

W = Weight in pounds
F_c = Coefficient of Friction

TABLE IX — BODY, LEG, AND FOOT MOTIONS

TYPE		SYMBOL	TMU	DISTANCE	DESCRIPTION
LEG–FOOT MOTION		FM	8.5	To 4"	Hinged at ankle.
		FMP	19.1	To 4"	With heavy pressure.
		LM___	7.1	To 6"	Hinged at knee or hip in any direction.
			1.2	Ea. add'l inch	
HORIZONTAL MOTION	SIDE STEP	.SS___C1	*	<12"	Use Reach or Move time when less than 12". Complete when leading leg contacts floor.
			17.0	12"	
			0.6	Ea. add'l inch	
		SS___C2	34.1	12"	Lagging leg must contact floor before next motion can be made.
			1.1	Ea. add'l inch	
	TURN BODY	TBC1	18.6	———	Complete when leading leg contacts floor.
		TBC2	37.2	———	Lagging leg must contact floor before next motion can be made
	WALK	W___FT	5.3	Per Foot	Unobstructed.
		W___P	15.0	Per Pace	Unobstructed.
		W___PO	17.0	Per Pace	When obstructed or with weight.
VERTICAL MOTION	SIT	SIT	34.7	———	From standing position.
	STD	STD	43.4	———	From sitting position.
		B,S,KOK	29.0	———	Bend, Stoop, Kneel on One Knee.
		AB,AS,AKOK	31.9	———	Arise from Bend, Stoop, Kneel on One Knee
		KBK	69.4	———	Kneel on Both Knees.
		AKBK	76.7	———	Arise from Kneel on Both Knees.

TABLE X — SIMULTANEOUS MOTIONS

(simultaneous motions chart)

□ EASY to perform simultaneously.

▨ Can be performed simultaneously with PRACTICE.

■ DIFFICULT to perform simultaneously even after long practice. Allow both times.

MOTIONS NOT INCLUDED IN ABOVE TABLE

TURN—Normally EASY with all motions except when TURN is controlled or with DISENGAGE.
APPLY PRESSURE—May be EASY, PRACTICE, or DIFFICULT. Each case must be analyzed.
POSITION—Class 3—Always DIFFICULT.
DISENGAGE—Class 3—Normally DIFFICULT.
RELEASE—Always EASY.
DISENGAGE—Any class may be DIFFICULT if care must be exercised to avoid injury or damage to object.

*W = Within the area of normal vision.
O = Outside the area of normal vision.
**E = EASY to Handle.
D = DIFFICULT to Handle.

SUPPLEMENTARY MTM DATA
TABLE 1 — POSITION — P

Class of Fit and Clearance	Case of † Symmetry	Align Only	Depth of Insertion (per ¼")		
			0 >0≤1/8"	2 >1/8≤¾	4 >¾≤1¼	6 >1¼≤1¾
21 .150" – .350"	S	3.0	3.4	6.6	7.7	8.8
	SS	3.0	10.3	13.5	14.6	15.7
	NS	4.8	15.5	18.7	19.8	20.9
22 .025" – .149"	S	7.2	7.2	11.9	13.0	14.2
	SS	8.0	14.9	19.6	20.7	21.9
	NS	9.5	20.2	24.9	26.0	27.2
23* .005" – .024"	S	9.5	9.5	16.3	18.7	21.0
	SS	10.4	17.3	24.1	26.5	28.8
	NS	12.2	22.9	29.7	32.1	34.4

*BINDING—Add observed number of Apply Pressures.
DIFFICULT HANDLING—Add observed number of G2's.

†Determine symmetry by geometric properties, except use S case when object is oriented prior to preceding Move.

TABLE 1A — SECONDARY ENGAGE — E2

CLASS OF FIT	DEPTH OF INSERTION (PER 1/4")		
	2	4	6
21	3.2	4.3	5.4
22	4.7	5.8	7.0
23	6.8	9.2	11.5

TABLE 2 — CRANK (LIGHT RESISTANCE) — C

DIAMETER OF CRANKING (INCHES)	TMU (T) PER REVOLUTION	DIAMETER OF CRANKING (INCHES)	TMU (T) PER REVOLUTION
1	8.5	9	14.0
2	9.7	10	14.4
3	10.6	11	14.7
4	11.4	12	15.0
5	12.1	14	15.5
6	12.7	16	16.0
7	13.2	18	16.4
8	13.6	20	16.7

FORMULAS:

A. CONTINUOUS CRANKING (Start at beginning and stop at end of cycle only)
$$TMU = [(N \times T) + 5.2] \cdot F + C$$

B. INTERMITTENT CRANKING (Start at beginning and stop at end of each revolution)
$$TMU = [(T + 5.2) F + C] \cdot N$$

C	=	Static component TMU weight allowance constant from move table
F	=	Dynamic component weight allowance factor from move table
N	=	Number of revolutions
T	=	TMU per revolution (Type III Motion)
5.2	=	TMU for start and stop

METHODS-TIME MEASUREMENT
MTM-I APPLICATION DATA

1 TMU	=	.00001	hour		1 hour	=	100,000.0 TMU
	=	.0006	minute		1 minute	=	1,666.7 TMU
	=	.036	seconds		1 second	=	27.8 TMU

Do not attempt to use this chart or apply Methods-Time Measurement in any way unless you understand the proper application of the data. This statement is included as a word of caution to prevent difficulties resulting from mis-application of the data.

**MTM ASSOCIATION
FOR STANDARDS
AND RESEARCH**

16-01 Broadway
Fair Lawn, N.J. 07410

TABLE 10.2
continued

TABLE I – REACH – R

Distance Moved Inches	Time TMU A	B	C or D	E	Hand In Motion A	B	CASE AND DESCRIPTION	
3/4 or less	2.0	2.0	2.0	2.0	1.6	1.6	**A** Reach to object in fixed location, or to object in other hand or on which other hand rests.	
1	2.5	2.5	3.6	2.4	2.3	2.3		
2	4.0	4.0	5.9	3.8	3.5	2.7		
3	5.3	5.3	7.3	5.3	4.5	3.6	**B** Reach to single object in location which may vary slightly from cycle to cycle.	
4	6.1	6.4	8.4	6.8	4.9	4.3		
5	6.5	7.8	9.4	7.4	5.3	5.0		
6	7.0	8.6	10.1	8.0	5.7	5.7		
7	7.4	9.3	10.8	8.7	6.1	6.5		
8	7.9	10.1	11.5	9.3	6.5	7.2	**C** Reach to object jumbled with other objects in a group so that search and select occur.	
9	8.3	10.8	12.2	9.9	6.9	7.9		
10	8.7	11.5	12.9	10.5	7.3	8.6		
12	9.6	12.9	14.2	11.8	8.1	10.1		
14	10.5	14.4	15.6	13.0	8.9	11.5	**D** Reach to a very small object or where accurate grasp is required.	
16	11.4	15.8	17.0	14.2	9.7	12.9		
18	12.3	17.2	18.4	15.5	10.5	14.4		
20	13.1	18.6	19.8	16.7	11.3	15.8		
22	14.0	20.1	21.2	18.0	12.1	17.3	**E** Reach to indefinite location to get hand in position for body balance or next motion or out of way.	
24	14.9	21.5	22.5	19.2	12.9	18.8		
26	15.8	22.9	23.9	20.4	13.7	20.2		
28	16.7	24.4	25.3	21.7	14.5	21.7		
30	17.5	25.8	26.7	22.9	15.3	23.2		
Additional	0.4	0.7	0.7	0.6			TMU per inch over 30 inches	

TABLE II – MOVE – M

Distance Moved Inches	Time TMU A	B	C	Hand In Motion B	Wt. Allowance Wt. (lb.) Up to	Dynamic Factor	Static Constant TMU	CASE AND DESCRIPTION
3/4 or less	2.0	2.0	2.0	1.7				
1	2.5	2.9	3.4	2.3	2.5	1.00	0	
2	3.6	4.6	5.2	2.9				
3	4.9	5.7	6.7	3.6	7.5	1.06	2.2	**A** Move object to other hand or against stop.
4	6.1	6.9	8.0	4.3				
5	7.3	8.0	9.2	5.0	12.5	1.11	3.9	
6	8.1	8.9	10.3	5.7				
7	8.9	9.7	11.1	6.5	17.5	1.17	5.6	
8	9.7	10.6	11.8	7.2				
9	10.5	11.5	12.7	7.9	22.5	1.22	7.4	**B** Move object to approximate or indefinite location.
10	11.3	12.2	13.5	8.6				
12	12.9	13.4	15.2	10.0	27.5	1.28	9.1	
14	14.4	14.6	16.9	11.4				
16	16.0	15.8	18.7	12.8	32.5	1.33	10.8	
18	17.6	17.0	20.4	14.2				
20	19.2	18.2	22.1	15.6	37.5	1.39	12.5	
22	20.8	19.4	23.8	17.0				
24	22.4	20.6	25.5	18.4	42.5	1.44	14.3	**C** Move object to exact location.
26	24.0	21.8	27.3	19.8				
28	25.5	23.1	29.0	21.2	47.5	1.50	16.0	
30	27.1	24.3	30.7	22.7				
Additional	0.8	0.6	0.85			TMU per inch over 30 inches		

TABLE III A – TURN – T

Weight	Time TMU for Degrees Turned 30°	45°	60°	75°	90°	105°	120°	135°	150°	165°	180°
Small – 0 to 2 Pounds	2.8	3.5	4.1	4.8	5.4	6.1	6.8	7.4	8.1	8.7	9.4
Medium – 2.1 to 10 Pounds	4.4	5.5	6.5	7.5	8.5	9.6	10.6	11.6	12.7	13.7	14.8
Large – 10.1 to 35 Pounds	8.4	10.5	12.3	14.4	16.2	18.3	20.4	22.2	24.3	26.1	28.2

TABLE III B – APPLY PRESSURE – AP

FULL CYCLE SYMBOL	TMU	DESCRIPTION	COMPONENTS SYMBOL	TMU	DESCRIPTION
APA	10.6	AF + DM + RLF	AF	3.4	Apply Force
			DM	4.2	Dwell, Minimum
APB	16.2	APA + G2	RLF	3.0	Release Force

TABLE IV – GRASP – G

TYPE OF GRASP	Case	Time TMU	DESCRIPTION		
PICK-UP	1A	2.0	Any size object by itself, easily grasped		
	1B	3.5	Object very small or lying close against a flat surface		
	1C1	7.3	Diameter larger than 1/2″	Interference with Grasp on bottom and one side of nearly cylindrical object.	
	1C2	8.7	Diameter 1/4″ to 1/2″		
	1C3	10.8	Diameter less than 1/4″		
REGRASP	2	5.6	Change grasp without relinquishing control		
TRANSFER	3	5.6	Control transferred from one hand to the other.		
SELECT	4A	7.3	Larger than 1″ x 1″ x 1″	Object jumbled with other objects so that search and select occur.	
	4B	9.1	1/4″ x 1/4″ x 1/8″ to 1″ x 1″ x 1″		
	4C	12.9	Smaller than 1/4″ x 1/4″ x 1/8″		
CONTACT	5	0	Contact, Sliding, or Hook Grasp.		

TABLE V – POSITION* – P

CLASS OF FIT		Symmetry	Easy To Handle	Difficult To Handle
1—Loose	No pressure required	S	5.6	11.2
		SS	9.1	14.7
		NS	10.4	16.0
2—Close	Light pressure required	S	16.2	21.8
		SS	19.7	25.3
		NS	21.0	26.6
3—Exact	Heavy pressure required.	S	43.0	48.6
		SS	46.5	52.1
		NS	47.8	53.4

SUPPLEMENTARY RULE FOR SURFACE ALIGNMENT	
P1SE per alignment: > 1/16 ≤ 1/4″	P2SE per alignment: ≤ 1/16″

*Distance moved to engage—1″ or less.

TABLE VI – RELEASE – RL

Case	Time TMU	DESCRIPTION
1	2.0	Normal release performed by opening fingers as independent motion.
2	0	Contact Release

TABLE VII – DISENGAGE – D

CLASS OF FIT	HEIGHT OF RECOIL	EASY TO HANDLE	DIFFICULT TO HANDLE
1—LOOSE—Very slight effort, blends with subsequent move.	Up to 1″	4.0	5.7
2—CLOSE—Normal effort, slight recoil.	Over 1″ to 5″	7.5	11.8
3—TIGHT—Considerable effort, hand recoils markedly.	Over 5″ to 12″	22.9	34.7

SUPPLEMENTARY

CLASS OF FIT	CARE IN HANDLING	BINDING
1— LOOSE	Allow Class 2	———
2—CLOSE	Allow Class 3	One G2 per Bind
3— TIGHT	Change Method	One APB per Bind

TABLE VIII – EYE TRAVEL AND EYE FOCUS – ET AND EF

Eye Travel Time = 15.2 x $\frac{T}{D}$ TMU, with a maximum value of 20 TMU.

where T = the distance between points from and to which the eye travels.
D = the perpendicular distance from the eye to the line of travel T.

Eye Focus Time = 7.3 TMU.

SUPPLEMENTARY INFORMATION

– Area of Normal Vision = Circle 4″ in Diameter 16″ from Eyes

– Reading Formula = 5.05 N Where N = The Number of Words.

smooth the way toward worker confidence that is vital to the acceptance of any method for setting standards.

10-5 WORK SAMPLING

L. H. C. Tippet pioneered the use of work sampling in the British textile industry in the 1930s.

Work sampling is a statistically based technique for analyzing work performance and machine utilization by direct observation but without a stopwatch. The analyst takes a relatively large number of observations of a process at random intervals. Each observation is categorized as to the state of the process at the instant it is observed. In a machine utilization study, the categories could be "idle" and "working." Then the ratio of the times the machine was observed working to the total number of observations indicates the utilization of the machine.

The growing usage of work sampling stems from its advantages over conventional time-study techniques for certain types of studies. It is particularly suitable for estimating unavoidable delays to establish delay allowances, for investigating the use of high investment assets, and for estimating the distribution of time spent by workers on different job activities. Such information is obtained by work sampling without continuous observation by an analyst. The observer arrives, determines the state of the process at a glance, checkmarks the state on a record sheet, and leaves. The process is not disturbed by long scrutiny periods, and the simple checking procedure diminishes the clerical time needed.

OPTIONAL MATERIAL _____

Sampling theory

In a multiactivity study, each observation is in a binary state for every activity taken separately.

The whole concept of work sampling is based on the fundamental laws of probability. If we are dealing with an operation that can be in only two states (on or off, yes or no, working or idle), we know that the percentage of time the two states can occur must total 100 percent. In terms of probabilities, we have the relationship expressed as

$$p + q = 1$$

where

p = probability of a single observation in one state, say W for working
$q = (1 - p)$ = probability of no observation in state W

The relationship is extended to include n observations in the form

$$(p + q)^n = 1$$

where n = number of observations in the sample. This expression is expanded according to the binomial theorem, to give the probability that a certain number of observations will be in state W out of a total of n observations.

The distribution of the probabilities resulting from the binomial expansion follows the **binomial distribution.** The mean of this distribution is equal to np, and the standard deviation is equal to $\sqrt{npq}$. As n becomes large, the binomial distribution takes on the properties of the normal distribution. To use the normal approximation of the binomial distribution, we have to divide both the mean and standard deviation by the sample size:

$$\text{Sample mean} = \frac{np}{n} = p$$

$$\text{Sample standard deviation} = \frac{\sqrt{npq}}{n} = \sqrt{\frac{pq}{n}}$$

$$= \sqrt{\frac{p(1 - p)}{n}}$$

Relying on the normal-distribution approximation, we can answer the common question plaguing time studies: How many observations should be made? The approach is identical to that employed for estimating the number of stopwatch observations needed, except that we now use the preceding sample standard deviation. The selected k/s factor again sets the confidence measure that the true value of the mean falls within the range $p \pm sp$. Where N' is the number of observations taken to provide a sufficient sample for the management-defined k/s factor, we have

Shortcut methods for estimating N' include alignment charts, graphs, and tables.

$$N' = \left(\frac{k}{sp} \sqrt{p(1 - p)}\right)^2 = \left(\frac{k}{s}\right)^2 \left(\frac{1 - p}{p}\right)$$

To apply the equation for sample size, an estimate of p is required. This estimate can be derived from previous experience with the work being sampled; or a small preliminary sample, say 50 observations, can be taken for an initial indication. As a more accurate estimation of p develops from additional observations, the equation can be solved again in terms of the new p to reaffirm the required N'. When p measures the percentage of time spent among several activities, the activity requiring the least time is used in the equation. Using the activity with the smallest p ensures the most conservative (largest) value of N'.

Using the smallest value of p is in the same vein as using the element with the greatest range of times to calculate N'.

Work-sampling procedures

In addition to the number of observations taken, the accuracy of a work-sampling study depends on the randomness of the sampling and how well each state of the work is defined. Sampling is effective only if every moment of time has an equal chance of being selected as the occasion for an observation. An easy method of ensuring randomness is to use a **random number table**—a listing of numbers that follows no pattern or discernible order. The period available for

A table of random numbers is in Appendix C.

A faster method of scheduling observations is to decide how many will be taken and then to enter a random number table, letting each three-digit sequence set an hour (first digit) and minutes (two other digits) for each observation. Impossible times are ignored.

the study is divided into numbered increments such as the number of minutes in a working day, 1 through 480. Each increment can be associated with a two-digit random number. The range of the random numbers that indicates a time increment in which an observation is to take place corresponds to the percentage of increments that will be used. For instance, if we wish to sample 400 out of 4000 minutes available (10 percent of the one-minute increments), the range of random numbers will be 10. Letting numbers 90 through 99 represent the range, we look through 4000 random numbers corresponding to 4000 successive increments and schedule an observation whenever the digits 90 to 99 appear.

Descriptions of each state of interest for the work must be definitive enough to allow the observer to make a classification. Better descriptions require less judgment. A study of whether a machine is idle or working is relatively simple because each state is quite apparent without judgment. However, a similar study of whether an individual is idle or working is far from simple. Is the individual working if he or she is talking to the supervisor? Is a worker idle when he or she is leaning back in apparent thought? Such questions must be faced and overcome by ground rules.

EXAMPLE 10.8 Work sampling to review unavoidable delay allowances

The present delay allowance of 9 percent for assemblers in the capacitor plant is being reviewed. Management believes that the allowance is too high because it is based on conditions four years ago. Many improvements have been made since then. The review is to be made by work sampling. About 300 women work on capacitors. Those working on soldering and winding are paid on an incentive plan.

The work is divided into three states: (1) idle owing to unavoidable delays, (2) idle for any reason other than unavoidable delay, and (3) working. For the study, an unavoidable delay is identified by the following conditions:

1. Conversation with a supervisor at the work station.
2. Lack of material, as apparent from empty in-trays and a signal for service.
3. Equipment maintenance or repair, indicated by a signal for service or the presence of repair personnel.
4. General disruption of service, as evidenced by all workers being idle.

A "working" state occurs whenever an assembler is at her work station, even if she is apparently relaxing or gossiping with a coworker. Any activity not falling into the "working" or "unavoidable delay" states is classified as "unaccounted for" time. This category includes personal, fatigue, and special allowances.

A tally sheet is developed for recording observations. Then the observers have several practice rounds with a supervisor to become thoroughly familiar with the work-state descriptions in relation to the workers' postures. Observations are taken from a balcony on one side of the work area. The specific women to be observed are chosen randomly.

Management wants the study completed in two weeks. The standard study design has a 10-percent precision at a confidence level of 95 percent. The current allowance, 9 percent, is used to determine the approximate number of observations needed:

$$N' = \left(\frac{2}{0.10}\right)^2 \left(\frac{1 - 0.09}{0.09}\right) = (20)^2 \left(\frac{0.91}{0.09}\right)$$

$$= 4000+ \text{ observations}$$

Based on this estimate, 45 observations of 10 workers per observation are taken each day. A different schedule for the observations is made every day, by reference to a random number table. The chance that workers will anticipate the observations is further reduced by randomizing the order in which individuals are checked during each observation trip.

After 4500 observations, the totals for the three states show the following proportions:

States	Observations	Percentages
Unavoidable delay	395	8.8
Unaccounted for time	316	7.0
Working time	3789	84.2
Totals	4500	100.0

The number of observations is again checked for the key statistic, unavoidable delay:

$$N' = (20)^2 \left(\frac{1 - 0.088}{0.088} \right) = 4145 \text{ observations}$$

With the adequacy of the total established for the main state, it is of interest to know whether the other states meet the study standards. The "working" state obviously meets the standard, but the smaller "unaccounted for" category is questionable. It is checked by rearranging the N' equation:

$$s = k \sqrt{\frac{1 - p}{p \times n}}$$

where n = actual number of observations taken of the state with percentage p

$$s = 2 \sqrt{\frac{1 - 0.07}{0.07 \times 4500}} = 2 \sqrt{\frac{0.93}{315}}$$

$$s = 2 \times 0.055 = 0.11 \text{ or } 11\%$$

Thus the present unavoidable delay allowance appears about right, and so management is in a position to question the adequacy of other allowances.

Work-sampling applications

Work-sampling studies are highly susceptible to the "get it done yesterday" attitude. The sense of urgency that often accompanies a decision finally to undertake a study might tempt the study designer to meet the N' criterion by compressing all the observations into a single day. If enough observers were available, the study could be completed in a day, but the results would probably be misleading unless the chosen day were truly representative. The advantage of spreading observations over several days is that the effects of unusual or intermittent conditions are included without monopolizing results. A study of absenteeism would show alarming dimensions if it were conducted only on days preceding holidays; such days are only part of the overall pattern and should contribute just a proportionate influence.

It is necessary to consider the route the observers must take when their daily rounds are scheduled. A single observation is a "snap reading," but it may take considerable time to reach the point at which it can be made. Time should also be allotted for observers to judge activities, if there is a choice of several states, and to record various entries when crews are being sampled.

Control charts are useful aids when work sampling is applied to check continually an activity of particular concern. The charts portray the successive percentages obtained from sampling as a function of time. The pattern formed by the plotted points gives an indication of the trend. An "out of control"

Control charting is discussed in Chapter 15.

condition is conventionally indicated by values outside a three-standard-deviation **control limit.** The limit is calculated as

$$\text{Control limits} = \bar{p} \pm 3 \text{ standard deviations}$$

where $\bar{p}$ = average percentage obtained from periodic samples

$$\text{Control limits} = \bar{p} \pm 3 \sqrt{\frac{\bar{p}(1 - \bar{p})}{n}}$$

where n = number of observations in each periodic sample.

A typical control-chart format is shown in Figure 10.11. The value of $\bar{p}$ is based on the first 10 samples shown in Table 10.3. To set the table in perspective, assume that 50 observations are taken daily of the time that the maintenance crews spend in Department A. The purpose of the study is to settle a dispute over the cost allocation of maintenance expense.

Based on $\bar{p} = 0.153$, the control limits are calculated as

$$\text{Upper control limits} = 0.153 + 3 \sqrt{\frac{0.153(1 - 0.153)}{50}}$$

$$= 0.153 + 3 \times 0.0509$$

$$= 0.153 + 0.153 = 0.306$$

$$\text{Lower control limit} = 0.153 - 0.153 = 0$$

The limits so calculated should contain 997 out of 1000 sample proportions as

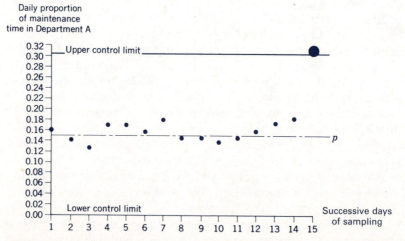

FIGURE 10.11 Control chart for successive samples, 50 observations each, taken of the presence of maintenance crews in Department A.

TABLE 10.3
p Values for Samples of 50 Observations Each

		Proportion of Time Maintenance Crews Spend in Department A			
Day	p	Day	p	Day	p
1	0.16	6	0.16	11	0.15
2	0.14	7	0.18	12	0.16
3	0.13	8	0.15	13	0.18
4	0.17	9	0.15	14	0.19
5	0.17	10	0.14	15	0.32

*0.153 average sample for days 1 to 10.

long as the mean of the samples remains as first computed. The first 14 samples are well within the limits. The last sample, above the upper control limit, can be attributed to one of three causes: (1) it is 1 of the 3 cases per 1000 that a point falls outside the limits by chance alone; (2) there is an assignable reason for its magnitude; or (3) it represents a new condition in which $\bar{p} = 0.153$ no longer represents the mean.

The immediate consequence of a point outside the control limits is an investigation. If an assignable cause is found for the maverick point, it should be excluded from the other data, and an additional sample should be taken if necessary for the desired accuracy. The trend of the plotted points serves as a warning of potential change in $\bar{p}$.

Aside from the psychological aspects of observing and being observed, the application of work sampling to time studies is a matter of money. Observers for work sampling do not need the costly training required for a good stopwatch or predetermined time analyst. However, any rating factors applied by semi-trained observers should be suspect. It is difficult and expensive to apply these other timing techniques to nonrepetitive work or when behavior over a long period is sought. The same cost and convenience factors go against using work sampling for a single worker or for micromotion studies.

Each time-study technique has its own advantages and disadvantages. The choice of one over the others depends on the familiar strategy of fitting the techniques to the occasion. It is easy to become enamored with a "progressive" management tool. The temptation is to fit the occasion to the tool and hope that some progressiveness rubs off on the user. Such hopes of secondhand prestige appear petty when compared with the possible system cost of myopic malfeasance.

END OPTIONAL MATERIAL

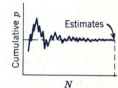

A chart of the cumulative proportions similar to Figure 10.11 can be used to estimate $\bar{p}$ and N'. The estimates are taken when the line connecting the cumulative p values stabilizes as shown above.

10-6 WAGE PAYMENTS

The study of work measurement inevitably leads to wage payments. In theory at least, the wages paid to a worker are a measurement of the amount of work he or she does. In the relatively rare instances on which pay is directly proportional to the pieces produced, money is a linear measure of work. For most industries in which wages are a function of an employment period, the relationship is obscure. Then why go to all the bother of measuring work if the wages paid for its accomplishment are so nebulously related? Although it has "cart before the horse" aspects, one reason is to find out how much is accomplished by a traditional pay scale. This information is necessary for cost accounting and provides a yardstick of job performance for wage adjustments. Another reason is that the attention devoted to element and task analysis tends to define responsibilities and to make better use of workers' time.

Wages are supposed to increase effective motivation. We observed in Section 8-5 that effective motivation is the difference between want strength and resistance to effort. Wages do not directly affect either want strength or effort resistance. They contribute to want satisfaction as the conversion factor between effort expended and want rewards. The equity of this conversion is a personal judgment by individual workers. It is taken in context of the standard of living they aspire to, the overall impression of their job, and their position in relation to their coworkers.

Employers also have a complicated framework by which to judge wages. Their basic objectives are to produce a competitive product and to secure a reasonable return. Higher wages paid for production either detract from competitiveness or decrease return. Both reactions have serious long-range consequences. Loss of competitiveness means cutbacks with a subsequent worsening of the wage–output ratio. A lower return has the chain reaction of blocking expansion, loss of investor confidence, and higher interest rates for loans. Lower wages make it difficult for employers to secure skilled craftspersons and competent managers vital to a competitive position. Decreasing wages obviously discourages motivation, increases turnover which raises training costs, and may even raise unit labor costs.

Somewhere between the two unattractive extremes is a wage level neither too high nor too low. To discover this point once is difficult. To maintain it requires an ear sensitive to labor demands, an eye to the future, and a delicate touch on the pulse of changing economic conditions.

Wage policies are burdened by accepted wage variances among types of industries, geographical areas, and even companies within the same industry and region. Some degree of consistency is introduced by "industry averages," "geographical averages," and standard classification systems, such as "grade definitions" of the U.S. Civil Service. National yardsticks such as the "cost of living" index help relate geographical wage differences to the common denominator of buying power. Internal company wage frictions are considerably reduced by carefully designed incentive wage plans and by objectivity in job evaluations.

Improving the method by which work is done increases effective motivation without changing wages. It lowers the resistance to effort by making a task easier to perform.

"Labor is prior to, and independent of, capital. Capital is only the fruit of labor, and could never have existed if labor had not first existed." Abraham Lincoln, 1861.

Minimum wage laws create a floor from which all wage policies build.

Wage incentive plans

Wage incentive plans are intended to increase workers' motivation by letting them earn proportionately higher returns from greater effort. Historically, incentive plans were part of the scientific management movement at the turn of this century. Attention was focused on questions of profit sharing, work divisions, and financial incentives for factory operations. A major accomplishment of the movement was the recognition given to the establishment of a standard rate of work for a job as a critical component of all incentive plans. The importance of realistic and acceptable standards developed by work measurement techniques becomes apparent when the different plans are considered.

STRAIGHT PIECE RATE

The most direct incentive plan is a **straight piece-rate wages** scale—payment at a constant amount per unit of output. Payments are calculated from an equation based on hourly rates as

$$E = R_h \left(\frac{A}{T} \right)$$

where

E = earnings in dollars per hour
R_h = hourly rate in dollars per hour
T = time taken per piece produced
A = time allowed or standard time per piece

or from the product of output times payment per piece as

$$E = R_p O$$

where

R_p = piece rate in dollars per unit
O = output in pieces per hour

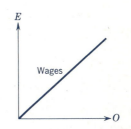

Straight piece-rate wages.

Harvest workers and many types of contract labor are paid by this method. It is easy to understand and apply. If R_h or R_p are high enough to ensure average workers an equitable wage for their normal output, it can inspire greater effort. Conversely, it can discourage workers if they feel they can never achieve a level compatible with their wants.

PIECE RATE WITH A GUARANTEED BASE

A piece-rate plan is made more socially and motivationally acceptable by placing a minimum level on the amount earned. The manner in which incentives are added to the minimum is the subject of numerous proposals.

The basic "one-for-one" plan has a set wage for production up to a level designated as 100 percent—a normal day's work determined from production standards set by work measurement. Beyond the 100-percent level, workers are

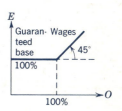

One-for-one incentive plan for output above 100 percent.

paid an additional 1 percent of their base pay for each 1 percent they exceed the normal standard. For example, if the standard rate of output is 10 (100-percent level) pieces per hour and an operator produces 96 pieces during an eight-hour day, his or her base pay for the day will be increased by the fraction

$$\frac{\text{Actual production} = 96 \text{ pieces}}{\text{Standard production} = 10 \times 8 = 80 \text{ pieces}} = \frac{6}{5} = 120\%$$

Assuming an hourly base pay of $3.00, the result of bettering the standard by 20 percent will increase the pay per hour to $3.00 × ⅗ = $3.60. If the same operator produces only 75 pieces the following day, his or her hourly rate for the day will be $3.00, owing to the guaranteed minimum. That this payment plan is just a delayed-action straight piece-rate scale is apparent from the equations for calculating earnings:

$$E = R_h \qquad \text{when } T \geq A$$

$$E = R_h \left(\frac{A}{T}\right) \text{ when } T \leq A$$

Modifications of this method vary the point at which the incentives start and vary the percentage received by the worker for production over the output corresponding to the guaranteed base. Three representative versions are described next and are illustrated in the margin notes.

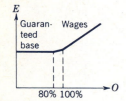

Two-stage incentive rates starting at 80 percent.

1. *Shortened-base incentive.* By starting incentives at less than the 100-percent standard, some low-production workers may be encouraged to reach the output level at which incentives begin. This condition is especially applicable during the installation of new pay plans and during the training or retraining periods for new workers. The incentives that apply between the shortened base and 100 percent are usually less than the rates for output above 100 percent.

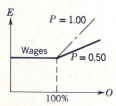

Profit sharing at $P = 0.50$ compared with a one-for-one plan.

2. *Profit-sharing incentive.* This plan is similar to the basic one-for-one plan, except that the reward for output above 100 percent is shared with the employer. The basic equation for earnings is

$$E = R_h \left[1 + P\left(\frac{A}{T} - 1\right)\right]$$

where P = proportion of reward above 100-percent output paid to the worker. A "50-50" plan ($P = 0.50$) would increase the base rate by 0.50 percent for every 1-percent increase in output over the standard. Other versions of this basic plan vary P as a function of output increments above 100 percent. For instance, P could be 0.50 for output from 100 to 120 percent and 0.75 above 120 percent. Note that when $P = 1$, the one-for-one plan is in effect: when $P = 0$, the incentive reverts to a straight hourly wage.

3. *Step incentive.* The use of a single bonus for meeting an output level set at or above the 100-percent standard dates back to the earliest incentive proposals. Although it is not too common today, it presents an interesting concept. The guaranteed base is pegged below the prevailing average wage. When output reaches a certain level, say 110 percent, the rate jumps to a level well above the average wage. The net effect of this plan is to reward exceptional workers and discourage others. In line with the mechanistic thinking prevalent during the period wherein the plan was developed, workers that did not meet the bonus criterion were supposed to be shamed into seeking employment elsewhere. Then only the "best" workers would be left. The modern version tends toward the usual guaranteed base with provisions for an immediate reward for making the 100-percent standard and a profit-sharing incentive beyond.

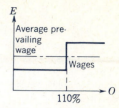

Original step incentive plan.

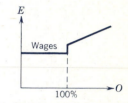

Modern step incentive plan for meeting 100 percent.

OTHER INCENTIVE PLANS

Most incentive plans are linked to individual units of output. When output is difficult to measure or is an inappropriate measure, other bases are used. Incentives can be based on the percent of rejects "saved" below a nominal standard, the amount of scrap reduction, the percentage utilization of machines, or combinations of similar factors. Attempts to include supervisors in incentive programs usually offer a percentage bonus based on improvements in crew or departmental ratings, as judged by measurement of scrap, spoilage, delays, and the like.

EVALUATION OF INCENTIVES

Most trade unions have opposed incentive plans for wage payment at some point in their history. Today, the majority opinion is still negative, but there is wider acceptance that workers can be adequately protected by collective bargaining while incentives are in effect. The hourly base, rather than a per-piece base, and the guaranteed minimum make incentives more palatable to organized labor.

The following advantages and disadvantages apply generally to all incentive plans, as compared with time-based wages:

1. Labor costs per unit can be estimated more accurately.

2. Less supervision is needed to keep output up to a reasonable level.

3. Work studies associated with incentives stimulate method improvements.

4. Quality and safety may suffer when workers work too fast to achieve bonus rates.

5. Payment by results may lead to opposition to or restriction of output when new machines, materials, or methods are proposed.

6. Clerical work is increased.

Wages set by job evaluation

Motivationally, time wages are at a disadvantage because rewards for better performances are cumulative and delayed. The annual raise is seldom inspirational.

Wage payments based on time are far more common than are payments based on output. Time scales may be short-term hourly rates or long-term salaries. From quality, convenience, record-keeping, and labor relations viewpoints, a time-based system is preferable to an output system. The most serious drawback of a time wage is the difficulty of determining wage plans and levels that accurately reward workers for work performed. Because there are no precise standards of performance, a superior level of execution is largely a subjective judgment over which the worker has little control.

The purpose of **job evaluation** is to establish base rates for incentive plans and hourly rates for time-based wages. The general guidelines are easy to identify: scarcer and more skillful personnel holding more difficult and responsible jobs should be paid more. Getting these generalities down to dollars-and-cents realities is not so easy. Some evaluation methods are described in the following paragraphs.

Figure 10.12 illustrates key-job interpolation.

Job ranking of occupations within a company is the simplest method of evaluation. It is based on the logical assumption that wages should be proportionate to the importance of the job to the company's welfare. An evaluation is conducted by having a committee rank jobs according to their duties and responsibilities. Then rates for the **key jobs** are decided on; other job rates are determined from interpolation between wages for the key jobs. The simplicity of the system is its undoing. It presupposes that the rankers have unlimited familiarity with all the jobs and insert no bias in their appraisals.

The U.S. Civil Service modifies its classification system by using points to assign grade numbers to certain occupations.

The *classification* method improves on job ranking by establishing predetermined labor classifications. The widest application is found in federal classifications for civil service and military ranks. The hitch in application comes when jobs are fit into the broad classifications. The questionable attribute of committee agreement on ranking is often used. On the positive side, the classification system allows a range of pay within each category to reflect the employees' merit ratings. This range also provides flexibility in recruiting and reranking when a particularly fine performance or an initial placement error is recognized.

The *factor comparison* method is similar to point plans. Dollar values replace points to rate basic job requirements such as (1) mental, (2) skill, (3) physical, (4) responsibility, and (5) working conditions.

A **point plan** injects objectivity into job evaluation by assigning points to the qualifications needed to perform a job. It is the most complex and widely used of all evaluation methods. Factors common to most jobs in an organization are determined first. A typical number is from 8 to 12. The 4 large categories and 11 subfactors used by the National Metal Trades Association are shown in Table 10.4.

After the appropriate factors are selected and completely described, preferably with examples, the points corresponding to each degree of the factors are assigned. Then the points associated with key jobs in the company are developed by reference to appropriate degree points for each factor. A plot of the key-job totals versus the prevailing wages for the respective jobs gives a framework for pegging wages to other job point totals. Figure 10.12 shows a

TABLE 10.4
Points Assigned to Factor Degrees for Job Evaluation

Factors	Degrees				
	1st	*2nd*	*3rd*	*4th*	*5th*
Skill:					
1. Education	14	28	42	56	70
2. Experience	22	44	66	88	110
3. Initiative and ingenuity	14	28	42	56	70
Effort:					
4. Physical demand	10	20	30	40	50
5. Mental and visual demand	5	10	15	20	25
Responsibility:					
6. Equipment or process	5	10	15	20	25
7. Material or product	5	10	15	20	25
8. Safety of others	5	10	15	20	25
9. Work of others	5	10	15	20	25
Job conditions:					
10. Working conditions	10	20	30	40	50
11. Hazards	5	10	15	20	25

line fitted to the scattergram of key-job points versus wage data and an interpolated job wage.

The administration of wage plans should be considered during the job-evaluation stage. For instance, when a point plan is being initiated, there are bound to be some workers that are currently underpaid or overpaid as indexed by the point totals for their jobs. This condition is normally anticipated by

Details of many different wage plans and their administration are available in several references at the end of this chapter.

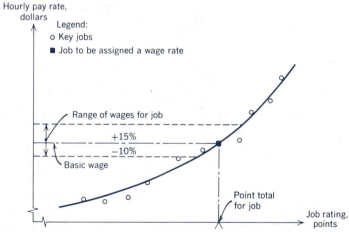

FIGURE 10.12 Key-job concept for assigning wage rates by interpolation.

allowing rates to vary in a range of perhaps 15 percent above the indicated wage and 10 percent below (Figure 10.12). But what about the workers still outside the range? Adjustments upward to the lower edge of the range present no problem. But the instigation of wage cuts, however well justified by theory, creates hostility. A compromise to avoid undermining the entire program allows exceptional wage rates to exist until their bearers leave the jobs by turnover or retraining. The need for this action should be anticipated when wage plans are developed.

Recent inflation fears have fostered demands for built-in "cost of living" wage increases.

Periodic, broad-scale wage hikes frequently result from negotiated labor contracts. Over a period of years these hikes can seriously distort a wage payment plan unless delicate adjustments are made. A "flat increase," the same amount given to all workers regardless of the jobs they hold, reduces the percentage of differential among the wages of different job classifications. Then a promotion tends to take the form of a new title instead of more money. When an "across the board" percentage increase is applied, the wage differential is exaggerated in the other direction. A wage settlement that grants wage increases commensurate with point totals would be more difficult to negotiate but would maintain the character of the established wage policy.

The personal nature of any wage plan reinforces the need for objectivity in evaluation and openness in application. In the early stages of incentive plans, it was considered an attribute to have complicated formulas for wages. Still, workers found ways to decipher the confusion factors. The sociological structure today has no room for such chicanery. There is enough dissatisfaction about the fairness of individual wages, whether justified or not, without hiding the standards of evaluation. Because wages are increasingly settled in a political arena, the best advice could be President Lincoln's:

> If you once forfeit the confidence of your fellow citizens, you can never regain their respect and esteem. It is true that you may fool all the people some of the time; you can even fool some of the people all the time, but you can't fool all of the people all the time.

EXAMPLE 10.9 A point system for management salaries

A salary structure for managerial positions should have *internal equity*—an equitable alignment of positions within the organization based on each person's worth and performance—and *external equality*—a compensation scale that is competitive with other organizations that draw from the same labor supply. It may be difficult to satisfy simultaneously these objectives because of the dynamics of the labor market. An equitable salary ladder could become dis-jointed when there is a shortage of certain skills, for instance, management information specialists or manufacturing engineers, that inflates salary offers for people qualified in the scarce specialties. If compensation levels for other positions remain essentially static while specialists' salaries soar, the formerly equitable salary ladder will be shattered. On the other side of the ledger, maintaining a salary scale that is not responsive to market fluctuations almost guarantees employee defections to higher paying jobs. A mission of **salary administration** is to balance internal equity with external equality.

Most large firms use a factor-point evaluation system to establish internal alignment. The points are assigned to weight the five factors described as follows:

1. *Personal knowledge.* What must be known to understand the nature of the required duties and have familiarity with the information needed to perform the duties.

2. *Managerial skills.* Aptitude and dexterity required to coordinate activities and communication skills to relate well with superiors and subordinates inside and outside the organization.

3. *Mental demands.* Latitude permitted for independent judgment and the extent of required analysis and problem solving.

4. *Accountability.* Extent of supervision or direction in the nature of the position.

5. *Position's impact.* Degree of effect the position has on the organization's overall performance: direct action that controls and achieves results, directly supporting services that influence results, and incidental activities that impinge indirectly on results.

Different organizations obviously score the factors differently, depending on the nature, size, and thrust of their activities. However, the different factor-point scales can be normalized to achieve com-

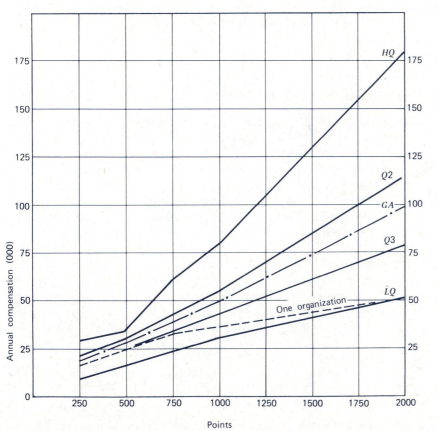

HQ Represents the highest company practice for salaries, but not necessarily the highest individual salary nor is the same company represented at all of the point-scale intervals.

Q2 and Interquartile range of
Q3 salaries encompassing middle 50% of salaries from reporting companies.

GA Grand average of all company compensation practices.

LQ Lowest company compensation practice but not necessarily the same organization at each point level.

Dashed One company's com-
Line pensation profile.

FIGURE 10.13 A graph of salary compensation with respect to a point scale representing managerial positions. Industry averages are compared with the salary practices of one organization to evaluate its compensation policy.

parability among organizations. This is done in surveys conducted by government agencies, trade associations, and consulting companies. These surveys provide data that can be used to check both the internal alignment and the external equality of compensation.

The initial step is to convert an organization's compensation posture to the point scaling used in available surveys. The result is a profile similar to Figure 10.12, in which the horizontal scale is the survey's point scale. Then this profile is superimposed on a chart of point-compensation values for comparable companies in the survey. An example is shown in Figure 10.13.

Point evaluations for positions are numbered on the chart's horizontal axis. Annual cash compensation in thousands of dollars is registered vertically. Survey data are divided into quartiles representing the salary practices of the organizations participating in the survey.

Analysis of the plotted data reveals how the internal compensation profile compares with those of other organizations, in terms of both its steepness and its relative position. A profile that departs significantly from the trend of the grand average indicates a compensation policy that is either unnecessarily lucrative or conservatively risky. Such departures could arise from a high proportion of employees who are in limited supply, collective bargaining agreements that have pushed nonexempt wages up and forced salaries to rise correspondingly, and special arrangements for outstanding producers. Sharp breaks in a compensation profile suggest different point ranges; then, either the scaling procedures or the pay policy should be examined to avoid inequities in internal compensation.

Wages are disincentives for productivity when slow performance is rewarded by overtime pay or when inept workers receive higher pay than productive workers do.

Rather low regular wages combined with substantial semiannual bonuses are credited in Japan with stabilizing employment, by allowing firms to avoid layoffs during slack business periods.

10-7 CLOSE-UPS AND UPDATES

Employers paying wages to employees to perform certain tasks is taken for granted, but the way those wages are paid and the nature of tasks vary over time. Once-popular bonuses were losing favor until they regained stature as part of the Japanese productivity success story. Concern for pay equality increased as more women entered the work force. Efforts to get workers to try harder resulted in offering ownership shares and innovative incentives. Changing compensation patterns reflect the employers' drive to improve performance and the employees' desire for greater security.

Bonuses attract workers' attention

Pure inventive wages based on units of output are time honored, but their application is limited by the necessity of having readily countable outputs. In the expanding service sector, the most readily available measure of output is total sales or profit. Bonuses based on profit are widely used for both executive compensation packages and as supplements to regular hourly wages or salaries.

Profit sharing typically is an annual or semiannual payment to employees of a percentage of the profit earned by the firm during the previous accounting period. The advantages to the firm of this plan are that personnel costs are lower when the firm is less profitable and that employees are more concerned with the organization's total performance. However, bonuses lose their effectiveness when all workers get the same amount, irrespective of their individual performance. Bonuses are the most effective when they reward specific accomplishments, are proportional to the size of an accomplishment, and occur soon after the accomplishment.

Formula bonus plans mathematically define the size of the accomplishment and proportionate shares. Although the formulas vary considerably, their similarities include provisions for frequent bonus payments, production-based rather than sales-based measurement of gains, and an emphasis on employee involvement. The three most frequently cited plans are briefly described as follows:

Executive bonuses are criticized for both their size and worth. Suggested changes include eliminating them, delaying them to see the long-term effect of the decisions, and linking salary to a formula bonus plan with specified objectives.

SCANLON PLAN

Since Joseph Scanlon conceived his plan in the 1930s, it has undergone numerous modifications, but its basic concepts of employee involvement and recognition have remained intact. Employees participate through a formalized suggestion system that operates at two levels: (1) monthly meetings of a production committee composed of elected employee representatives to review suggestions and (2) a higher-level screening committee composed of representatives from the production committee and management, meeting monthly to discuss company operations.

Recognition for increased productivity comes from a monthly bonus based on a single ratio:

$$\text{Base ratio} = \frac{\text{Payroll costs}}{\text{Value of production}}$$

Assuming that the historical base ratio is 0.20 and the employees' share is 75 percent, a production value (sales plus or minus inventory) of $1 million leads to an allowed labor cost of $1,000,000 \times 0.20 = \$200,000$. If the actual labor cost during the month were $160,000, the employee bonus pool would be 75 percent of the difference between allowed and actual cost:

$$(\$200,000 - \$160,000) \times 0.75 = \$30,000$$

A portion of the bonus is usually withheld to cover any deficit months, and the accumulated leftovers are paid as a year-end jackpot.

RUCKER PLAN

The Rucker plan operates similarly through a suggestion system and Rucker committees to improve on a historical relationship between labor cost and value added. For example, if the Rucker standard is 0.60 and

$$\text{Net sales} - \text{Purchases} = \$1,000,000 - \$400,000$$
$$= \$600,000$$

then the allowable payroll will be $600,000 \times 0.60 = \$360,000$. Any month

Built-in employee involvement that stimulates performance improvement is the main reason that the Scanlon and Rucker plans have had lasting popularity.

in which the actual labor cost is less than 60 percent of the production value, a bonus is earned. The formula encourages employees to save on materials and supplies because they share the gains.

IMPROSHARE PLAN

Mitchell Fein is the architect of Improshare: Improving Productivity through Sharing. The plan compares the number of work hours needed to produce a certain quantity of units in the current period with the number of work hours needed to produce the same quantity of units during a base period. Both direct and indirect work hours are included to allow both production and support workers to share the gains achieved together. For example, assume that 100,000 work hours were consumed in producing 10,000 units in the base period. If it took only 95,000 hours to produce the same 10,000 units in the current period, the productivity gain would be 5,000 hours. This gain is split 50-50 between the company and all the workers in the plant. When the gain is facilitated by capital investment, the sharing proportion is changed to reflect the new equipment's contribution, perhaps 80-20 in favor of the company.

Pay-for-knowledge and pay-in-ownership plans can pay off

Companies and some government units are experimenting with innovative compensation plans. Various forms of merit pay are being used in education and public agencies. The better plans focus on the quality of service and combine bonuses with conventional step increases in salary. Such recognition can be personalized and is clearly a reward for superior performance.

Pay-for-knowledge compensation plans—also known as skill-based pay and multiskill compensation—are based on workers' knowledge and mastery of specified jobs in an organization. These plans typically start with a wage lower than the normal entry level. Employees then earn pay increases by learning the skills required to perform different jobs or for acquiring additional expertise in a single job that has several levels of competency. Both types of pay-for-knowledge may exist in the same organization.

A company gains by having employees who understand the entire operation, can do a variety of jobs, and have high skill levels. Because each worker knows several jobs, management has more flexibility in meeting product demand changes without hiring and firing. With a stable work force, there is less need to overstaff in anticipation of absenteeism and supply–demand fluctuations. The stronger links forged between the company and its employees by the pay-for-knowledge system tend to increase output quality and to decrease personnel problems.

In place of money, a bonus may be a free trip, extra vacation time, savings bonds, a bike or car, and so forth.

The concept of rewarding workers for competency in a range of jobs was well publicized in the early 1970s by its application at the Gaines Pet Food plant in Topeka, Kansas.

Studies indicate that employees who have knowledge-based pay tend to be more satisfied, committed, and motivated.

But the advantages are gained at a steep price. Hourly labor costs are higher after the start-up phase than under traditional compensation systems. Training costs are much higher, and overhead costs are increased by the added bookkeeping required for job rotation, pay grade documentation, and human resource planning. Other pay-for-knowledge problems include "maxing out" (the lid on top pay), knowledge-versus-performance questions (knowing the theory but not being able to apply it), and pay inequity (those not in the system may be senior but are paid less). Although skill-based plans are organizationally complex, paying for knowledge can pay off under the right conditions.

Promoting a sense of ownership among workers through employee compensation is another effective way to secure cooperation and commitment. An employee stock-ownership plan is one method. An employee has the option of buying company stock for a tax-sheltered investment plan, and the company contributes an amount ranging from 10 to 50 percent of the employee's investment. By contributing, the company not only saves cash and possibly obtains tax advantages but also gives its employees a vested interest in the company's success.

Some companies tie the size of their contributions to a specific performance criterion such as absenteeism.

Another version of stock ownership was instituted at People Express Airline. From its founding in 1980, it rose to a billion-dollar business by 1986. Its unusually flat organizational structure had only three management levels, and these managers collectively owned one-third of the company. Every new employee was required to buy and hold 100 shares of stock, offered at a 70-percent discount. Base wages were lower than the industry average, but profit sharing could add up to 27 percent to employees' paychecks. The sense of ownership engendered by the physical ownership of stock and mental affiliation promoted by profit sharing stimulates esprit de corps when things go well, but problems arise when stock prices and profits tumble.

Complex questions of comparable worth

Should an administrative assistant, electrician, or nurse receive higher pay? The answer is fraught with emotional biases. The **comparable worth** issue became inflammatory in the 1980s and will continue to elicit strong feelings until a solution is developed that is acceptable to the many affected parties.

The main purpose of the movement is to identify "undervalued" job classifications—those predominately held by women that pay significantly less than do the classifications predominately filled by men that require the same education, experience, and skills. The difficulty comes in defining job characteristics in terms of equivalent qualifications that substantiate pay differentials, rather than relying on historical market values to set wage

Reclassifications in government agencies usually raise the pay of undervalued jobs instead of lowering the pay of overvalued jobs.

scales. Studies to determine actual job values examine the following categories:

- Position information—the purpose of the position in meeting the objectives of a program or an organization.
- Description of duties—assignments that fulfill the purpose of the position.
- Working conditions—specific hostilities, unpleasant conditions, and risks of injury associated with duties.
- Work contacts and job-related decisions—people with whom contacts must be maintained and types of decisions involved in the work.
- Supervisory duties—managerial and supervisory activities required to meet work responsibilities.

Information of this nature forms the baseline for ascertaining comparability. The final determination is expected to merge technological requirements, human capabilities, work responsibilities, worker competencies, and special conditions into a consistent pattern of compensation.

Relieving the stress of rotating shifts

One-quarter of the manufacturing workers in America are on rotating shifts. Capital-intensive industries have to operate around the clock in order to earn an adequate return on investment. This necessitates shift work which causes high personnel turnover, high rates of accidents and illness, low morale, and low productivity. According to a Harvard medical study, the problem stems from the human biological clock, which operates on a 25-hour cycle to govern waking and sleep, instead of the familiar 24-hour cycle on which shift work is based. **Chronobiology** is the burgeoning field of research that examines our biological clocks.

Most companies rotate their workers from days to nights (graveyard shifts) to evenings (swing shifts), and the shifts change weekly. Chronobiologists say that the order should be reversed—days to evenings to nights—and that the time between shift changes should be extended to two or three weeks. The advised rotation allows workers to sleep longer on new shifts, which appeases their biological clocks, and the extended period on each shift permits a more complete adjustment.

Other simple changes, such as sleeping at the same time each day while on the same shift, can overcome many of the rotating-shift blues. Estimates that rotating shifts can reduce productivity by a third and may cut employees' life spans by a fifth make the effort well worthwhile.

Longer stretches on a night shift cause workers to miss the company of their friends or to revert to normal hours on weekends, which leads to exhaustion.

10-8 SUMMARY

The objective of a process analysis is to improve the sequence or content of the operations required to complete a task. A preinvestigation indicates which areas will benefit most from the analysis. Fact finding is done mostly by observation and interviewing. The analyst improves a process by seeking ways of eliminating, combining, and rearranging operations.

Method improvement information is often displayed on charts. Survey charting records present procedures; design charting develops innovations; and presentation charts explain proposals to help them be accepted. Process charts employ symbols for operation ($\bigcirc$), transportation ($\Rightarrow$), inspection ($\square$), delay (D), and storage ($\triangledown$) to portray the flow of activities concerning an operator or a product. Operator–machine charts aid the visualization of dependent relationships between operators and machines; an economic balance is sought for the idle time of workers versus idle machines.

Motion studies attempt to make work performance easier and more productive by improving manual motions. Principles of better motions serve as reminders for evaluating an operation. Evaluations can be conducted by camera studies that use high camera speeds for micromotion and slow speeds for memomotion studies. A cycle-graph study records on a single time exposure the movements of an operator as evidenced by lights attached to his or her limbs. Motions are charted using get-and-place elements for longer cycles or more detailed elements for microscopic analyses of very short cycles. The elements are displayed in Left-hand Right-hand and SIMO charts.

A time study often follows a method change. The work elements identified in the preliminary methods study are categorized according to their cyclic characteristics of regular or irregular, constant or variable, machine or operator controlled, and foreign. Selected times for work elements are the averages of representative observations. The number of observations required for a given confidence-precision level (k/s factor) is determined from the equation

Flow process charting

Motion economy

Motion charting

Time study

$$N' = \left(\frac{k/s\sqrt{N\,\Sigma X^2 - (\Sigma X)^2}}{\Sigma X} \right)^2$$

where

X = representative element times
N = number of representative element times

A rating factor applied to a selected element time yields normal time. Standard times are secured by applying an allowance factor to the normal times. Personal, fatigue, delay, and special allowances are totaled to set the allowance factor. Different sequences leading to standard times are as follows:

Standard times

Selected time $\times$ rating factor

$$\left.\begin{array}{l} \text{Synthetic time} \\ \text{Predetermined time} \end{array}\right\rangle \times \begin{array}{c} \text{Allowance} \\ \text{factor} \end{array} = \begin{array}{c} \text{Standard} \\ \text{time} \end{array}$$

Synthetic times   Synthetic timing incorporates the characteristics of an operation or product in an equation to provide the normal time for a task. It is appropriate to frequently encountered tasks and allows times to be calculated before a process is actually in operation.

Predetermined times   Predetermined times use tabulated values of normal times for very small motion elements. Task times are synthesized from the individual times of component micromotions. The technique can be used in place of stopwatch timing and in conjunction with synthetic timing.

Work sampling   Work sampling, based on the fundamental laws of probability, employs a large number of observations to determine the percentage of time that a process is in a certain state. The number of samples needed to meet a selected k/s level is calculated from

$$N' = \left(\frac{k}{s}\right)^2\left(\frac{1-p}{p}\right)$$

where p = sample mean of the probability that the process is in a given state. A schedule of observations is developed by reference to a random number table to help ensure the representativeness of a sample. Control charts based on

$$\text{Control limits} = \bar{p} \pm 3\sqrt{\frac{\bar{p}(1-\bar{p})}{n}}$$

where

$\bar{p}$ = average probability obtained from periodic samples
n = number of observations in each sample

provide a warning for developing state changes in a process. A point outside the limits is an alarm for investigation.

Wage incentive plans are designed to increase workers' motivation by letting them earn proportionately higher returns from greater effort. On a straight piece-rate plan, earnings are directly proportional to output. Piece rates are often **Incentive plans** modified to provide a guaranteed minimum wage with a fractional bonus per piece added for production beyond a certain level. A step incentive introduces a one-shot bonus for reaching a specified output. Other incentive plans combine these features or are linked to a desired function other than output. Attempts to spur productivity have led to executive bonuses, formula-based bonus plans for the entire work force, profit sharing, stock-ownership plans, and increased compensation for learning several jobs.

Job evaluation methods are required to establish base rates for both incentive plans and hourly wages of time-based plans. Job ranking is a simple approach that relates wages to the importance of the jobs in a company. An industrywide classification method improves on job ranking by predetermining labor classifications with associated wage ranges by which jobs can be categorized. A point plan is most widely used by industry. It identifies the main factors that contribute to job performance and assigns points to degrees of quality for

Time-based wages

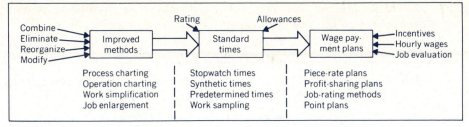

FIGURE 10.14 Tools and techniques of work design.

each factor. Points for key jobs are related to wages. Then the base rates for other point totals are set by interpolation from key-job versus wage relationships. Questions concerning equal pay for jobs of comparable worth are difficult to answer but are being examined.

The kinship of methods analyses, time-measurement objectives and techniques, and wage payment is depicted in Figure 10.14.

10-9 REFERENCES

BACKMAN, J. *Wage Determination: An Analysis of Wage Criteria.* Princeton, N.J.: Van Nostrand, 1960.

BARNES, R. M. *Motion and Time Study*, 7th ed. New York: Wiley, 1980.

CARROLL, P. *Better Wage Incentives.* New York: McGraw-Hill, 1957.

DOYLE, R. J. *Gainsharing and Productivity.* New York: American Management Associations, 1983.

ELIOT, J. *Equitable Payment.* New York: Wiley, 1961.

GANTT, H. L. *Work, Wages and Profits.* New York: Engineering Management, 1913.

GILBRETH, F. B., and L. M. GILBRETH. *Applied Motion Study.* New York: Sturgis and Walton, 1917.

KARGER, D. W., and W. HANCOCK. *Advanced Work Measurement*, 3rd ed. New York: Industrial Press, 1982.

KONZ, S. *Work Design.* Columbus, Ohio: Grid Publishing, 1979.

LAWLER, E. E., III. *Pay and Organization Development.* Reading, Mass.: Addison-Wesley, 1981.

MUNDEL, M. E. *Motion and Time Study—Improving Productivity*, 5th ed. Englewood Cliffs, N.J.: Prentice-Hall, 1978.

POLK, E. J. *Methods Analysis and Work Measurement.* New York: McGraw-Hill, 1984.

SLOANE, A. A. *Personnel: Managing Human Resources.* Englewood Cliffs, N.J.: Prentice-Hall, 1983.

VROOM, V. H. *Work and Motivation.* New York: Wiley, 1964.

WHISLER, T. L., and S. F. HARPER. *Performance Appraisal.* New York: Holt, Rinehart and Winston, 1962.

10-10 SELF-TEST REVIEW

Answers to the following review questions are given in Appendix A.

1. T   F   *Charting* has three major areas of use in production system studies: survey, design, and presentation.

2. T   F   A *process* is analyzed by questioning whether its operations can be eliminated, duplicated, or rearranged.

3. T F A *flow process chart* symbol of an arrow in a circle indicates an operation is being performed during transportation.

4. T F Contribution in an *operator–machine process* is the sum of the output value, labor cost, and machine expense.

5. T F The principles of *motion economy* determine the hourly compensation rate for a job.

6. T F *Motion charting* employs symbols that represent work elements.

7. T F Two techniques for *stopwatch studies* are snap-back and continuous timing.

8. T F The *normal time* for an operation equals the sum of the elemental selected times.

9. T F A *k/s factor* indicates the confidence that the desired accuracy is attained.

10. T F Personal, fatigue, delay, and/or special allowances are included in *standard times.*

11. T F MTM data is a form of *synthetic times.*

12. T F Observations in a *work-sampling study* are expected to follow a binomial distribution.

13. T F The most direct *incentive wage* is a straight piece-rate scale.

14. T F *Time-based wages* require more clerical work to administer than do incentive wages, but they are likely to beget greater quality and safety.

15. T F Attributing Japanese productivity gains to bonus wages spurred the creation of *formula bonus plans* such as the Scanlon and Rucker plans.

16. T F The *comparable worth* of a job is calculated for the correction of sex bias by contrasting it with the point score of a key job.

17. T F Biological clocks are studies in *chronobiology* research.

10-11 DISCUSSION QUESTIONS

1. Why would early training in methods analysis and work measurement benefit someone interested in pursuing the following careers?
a. Sales.
b. Accounting.
c. Industrial engineering.
d. Industrial relations.
e. Production manager.
f. Research.
g. Advertising.
h. Purchasing.

2. Substantiate or refute the statement: "Methods and measurement are the wellspring of labor peace, product competitiveness, and customer satisfaction."

3. Explain how each of the following diagrams could be used to fulfill two or more of the basic purposes of charting:
a. Architectural drawing.
b. Pie chart.
c. Wiring diagram.

d. Bar chart.

e. Histogram.

f. Road map.

4. What factors affect a decision to make a macro-motion or a micromotion analysis?

5. How does the principle of "diminishing returns" apply to motion studies?

6. What general categories of operations are appropriate for camera studies made at 1000 frames per minute and at 1 frame per minute?

7. Some unions require their permission before camera studies can be conducted in a plant. What do you think might have caused this requirement? Why does it still exist?

8. How could micromotion studies be useful for training new workers?

9. Why are delays (D) sometimes subdivided into avoidable and unavoidable delays? Give an example of each as it might apply to an assembly operation.

10. When representatives of the Westinghouse Corporation approached Commodore Vanderbilt about an air brake, he stated: "I have no time to listen to fools who want to blow air on wheels to stop trains." How would this attitude at the top, middle, and lower levels of management affect a work simplification program?

11. Why is continuous timing generally preferred to snap-back timing?

12. Name and describe seven general classifications of work elements. How does recognizing the classifications aid a time analyst in breaking an operation into its component parts? Classify each work element in Example 10.5.

13. Defend or rebut the following statements:

a. *"We will not bandy words. Time study permits a meaningful rough fix to be made on the system's manpower output and costs. However, because leveling must be used, real precision can never be obtained."*

M. K. STARR, *Production Management: Systems and Synthesis* (Englewood Cliffs, N.J.: Prentice-Hall, 1964.)

b. *"A time study is often conducted in such a manner as to substantiate an answer that the time-study man ar-*

rived at before starting to time the operator in question, as the result of experience on his part or of his taking desired take-home pay into account."

(E. V. KRICK, *Methods Engineering*.)

c. *"Other engineers may have their problems, but nothing like these. If the civil engineer wants to understand what it's like to be a time-study engineer, let him visualize a dark, eerie Halloween night on which spirits, animate and inanimate, are abroad. Along comes the bridge, which is his pride and joy, spanning a majestic river and it addresses him in these accents, 'Hey, jerk, do you know that I could have remained standing and carried just as big a load if you used one-quarter the tonnage of steel that my poor piles must hold up?' This is just an everyday experience for the time-study man."*

(W. GOMBERG, *A Trade Union Analysis of Time Study*.)

14. Why is it so difficult for all industries to agree on a universal conception of "normal performance"?

15. How does the practice of including allowances as part of the standard time for an operation promote "effective motivation"?

16. What is the reasoning behind providing special allowances for operators paid on incentive plans when their operations are process or machine controlled?

17. What are the principal advantages and disadvantages of synthetic timing?

18. What measures can be taken to ensure representative work samples?

19. Binomial theory allows only one event either to occur or not occur. Is this premise violated by using work sampling to determine the percentage of time a worker spends on each of several activities required in his or her job? Why?

20. Why should an "out of control" sample be eliminated from the calculation of $\bar{p}$ after an assignable cause for its occurrence has been found?

21. Describe the effect on (1) wages received by the worker and (2) the direct labor cost for the units being produced when output increases above the 100-percent standard using each of the following wage plans:

a. Straight piece rate.

b. One-for-one plan with a guaranteed 100-percent base.

c. Profit-sharing plan where $P = 0.40$ after 100-percent output.

d. Step incentive of 5 percent followed by a "50-50" plan at and above the 100-percent standard.

e. Hourly wage.

22. Some jobs defy time measurements. Tasks requiring mostly mental effort are normally paid on a time system such as a salary per month. What would be the advantages and disadvantages associated with an incentive plan for such jobs? Is there any reasonable way that such jobs could be put on an incentive basis?

23. Why are time-based wages often paid weekly when it would be more economical for a company to issue paychecks only once a month?

24. A "rate buster" is an individual who works at an exceptional pace. When incentive pay is involved, this individual receives very high wages compared with those of other workers because his or her output is far above the average on which the incentives are based. This individual is frequently unpopular with the other workers, who feel that standards will be raised by management when it is observed how much the rate buster earns.

a. Comment on the validity of the workers' concern about standards being raised.

b. Should management encourage the rate buster as a good example, or should he or she be discouraged in order to reduce conflict among the other workers?

25. Do you think the original version of the step incentive wage plan would work today? Why?

26. Within the context of motivation, discuss the relationship of work measurement, methods analysis, incentive plans, and hourly wages.

10-12 PROBLEMS AND CASES

1. Construct an operator process chart by using the format shown in Example 10.1 to depict the sequence of operations required to change the front tire on an automobile. First indicate the way you would presently change a tire. Then analyze the chart to see whether you can discover a better way to do it.

2. Develop a process chart to

a. Show how a page should be inserted in a typewriter when three carbon copies are required. Start the task by removing the paper and carbons from a desk drawer.

b. Water a lawn when the hose and sprinkler must be removed from a garden house before being attached to a water outlet. Use a layout diagram.

3. The following three projects have been submitted to the methods staff as possible study areas:

Study	Estimated Annual Savings	Estimated Study and Implementation Cost	Probability of Implementation
A	$3300	$600	0.80
B	$6500	$1000	0.70
C	$9000	$1900	0.90

According to the rating factor formula of Section 9-2, which project should receive first attention? What other considerations should influence the choice?

4. Draw an operator–machine chart to check the solution to the problem posed in Example 10.2.

5. A large number of semiautomatic machines produces identical products. Time studies reveal the following time in minutes for one worker to service one machine:

Load machine	3.1
Remove finished product	0.6
Inspect finished product	2.4
Pack finished product	1.9
Walk to next machine	0.4

The machine takes 41.3 minutes to produce a finished product. Machine operators are paid $4.90 per hour, and the burden rate for the machine is $18.00 per hour. What is the lowest cost per unit to produce the product with the optimal ratio of operators to machines?

6. Develop a Left-hand Right-hand chart for

a. Opening a pop bottle with a hand-operated bottle opener.

b. Opening a checkbook, writing a check, and handing it to a cashier.

7. For a 95-percent confidence level, what value was used for the required precision in the following formula?

$$N' = \left(\frac{10\sqrt{N\,\Sigma X^2 - (\Sigma X)^2}}{\Sigma X}\right)^2$$

8. The following times were recorded for one element: 7, 8, 6, 9, 7, 7, 6, and 6. How many cycles are necessary for the time study if the error is not to exceed 5 percent as often as 997 times out of 1000?

9. A portion of a time-study observation sheet is as follows. The continuous method of stopwatch timing was used. For this operation it is estimated that an operator has 420 minutes out of a 480-minute workday to apply to production. Determine the standard time for the task and the number of pieces produced per standard hour.

Four 10-minute coffee breaks.

Thirty minutes for personal time.

Twenty minutes for unavoidable delay.

Standard time for an operation is six minutes.

a. How many complete operations can be performed in a day if an operator has a speed rating of 120 percent?

b. What is the personal time percentage allowance?

11. The normal time for performing a certain task depends on the weight and surface area of the part being handled, as shown in Figure 10.15. Determine the synthetic time formula for the normal time of the operation.

12. You are given the following conditions in which the rating for the observed times is 1.0:

a. A pail is in the shape of a rectangle. The bottom has a 144-square inch area. The sides are calibrated in 1-inch increments from 6 to 12. The pail weighs 6 pounds.

b. A faucet delivers 4 cubic inches of water per second at a constant rate. Water weighs 62.4 pounds per cubic foot.

Element		1	2	3	4	5	6	7	8	9	10	ST_e	Rating Factor
1	E												0.95
	C	11	45	81	13	48	83	20	55	93	30		
2	E												1.10
	C	28	63	97	30	65	200	36	75	310	48		
3	E												1.05
	C	35	70	104	37	73	08	44	82	18	55		

10. In an eight-hour day the following conditions occur:

Four setups of 20 minutes each.

c. An empty pail can be carried 200 feet in 1 minute. A full pail (12-inch level) can be carried 200 feet in 4 minutes. (Assume linear distribution between points.)

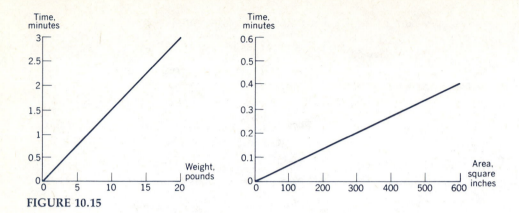

FIGURE 10.15

(1) Determine the simplest possible formula for the synthetic time to fill and transport a pail of water.

(2) What is the standard time for filling a pail to 3 inches and carrying it 100 feet when the allowance factor is 1.15?

13. Beginning with the relationship $sp = k\sqrt{(pq)/n}$, develop the equation for determining the number of work-sampling observations required to conform to a given k/s factor.

14. If one state occurred four times in 100 observations while using the work-sampling technique, determine the precision of the study using a 95-percent confidence level.

If the task from which the observation described above takes 20 minutes to complete, within what range in minutes should you expect the observations to fall 95 times out of 100?

15. Figure 10.16 indicates the distribution of three activity states: (1) productive time, (2) necessary delays, and (3) avoidable delays for a lift-truck operator during an eight-day period.

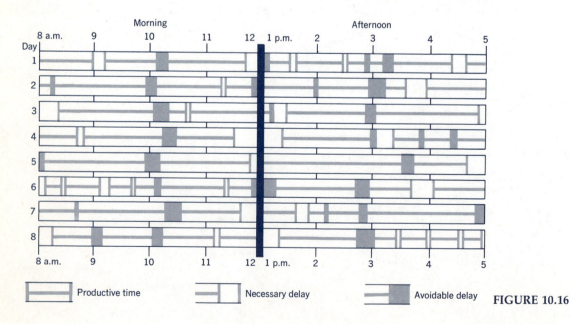

FIGURE 10.16

a. Develop a tally sheet, determine a random observation schedule for 80 observations during the eight-day period, and make a work sampling of the lift-truck operator's activities.

b. How many observations would be required for a k/s factor of 20 based on the limiting state?

c. Comment on the study with respect to the following: (1) the difficulties that might be encountered in making the study, (2) explaining to the lift-truck operator what you are doing and how it is supposed to be used by management, (3) definitions of each state (both for an actual study and as the definition could apply to Figure 10.16) and the possibility that too many observations were taken in the morning or afternoon.

16. A nine-day study of work operation has been completed. During each eight-hour day, 100 random observations were made of workers engaged in the operation. The number of "operator idle" observations found for each day are given as follows:

Day	1	2	3	4	5	6	7	8	9
Operator idle	7	9	16	18	9	27	9	12	13

a. Before this study was conducted, the total number of observations to be taken was determined by selecting a desired relative accuracy of ±5 percent at a 95-percent confidence level. What anticipated idle-time percentage was used in the preliminary work to set $N' = 900$?

b. What was the actual precision obtained for the above study at a 95-percent confidence level?

c. From part b. you could say you are 95 percent sure that the idle time will be in the range from ——— to ——— minutes.

d. If a control chart were made for this operation using a three-standard-deviation confidence limit, which points in the study would be considered out of control?

17. A company hopes to increase machine utilization by means of an incentive wage plan. Each machine has a clock on it that records running time; 330 minutes of use in an eight-hour day is considered 100-percent utilization. Operators receive a base pay of $4.00 per hour. The incentive plan calls for a 2-percent increase in base pay for each 5-percent machine utilization between 70 percent and the 100-percent standard; the 40–60 arrangement is linear between 231 and 330 minutes running time. A step bonus of 5 percent is offered for meeting the 100-percent standard. Above this point the incentives are 50–50 profit sharing.

a. Graphically portray the basic relationship of machine utilization and earnings.

b. How much will an operator earn if his or her machine is running 400 minutes a day?

18. Key-job points and associated wages are shown in Figure 10.17.

a. What would an across-the-board increase of 10 percent do to the slope of the line from which wages for other point totals are interpolated?

b. What would a "flat increase" of 20 cents per hour do to the slope?

c. Under conditions a, b, and the original rates, what wages should be paid to a worker performing a job with a point total of 340?

19. *A Case of Unsteady Standards*
The construction manager for Roadbuilders, Inc., George Kelly, called a meeting of the six project managers. The "call order" gave the purpose as establishing performance standards for engineers and construction crews. Although George was relatively new in his position, he wanted his people to develop an understanding of their responsibilities in terms of the new standards the company was implementing.

All six were experienced people and equally balanced between three recent college graduates and three up-from-the-ranks construction foremen. During the meeting it was apparent that one old-timer, Ed Johns, was not cooperating in the project. He did not raise objections, however, and a few realistic

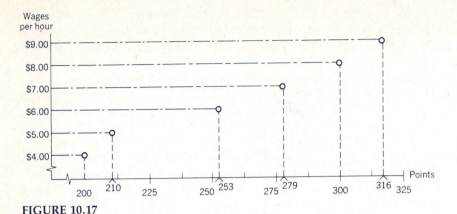

FIGURE 10.17

standards were mutually agreed on. The other five managers seemed to be participating sincerely. The new boss was so pleased with their participation that he assumed that Ed Johns was just quiet by nature. All in all, George Kelly felt it was a successful standards-setting meeting; realistic performance expectations that had been developed by consultants were introduced, and the project managers seemed to accept the need for them.

When the construction job was well under way, George Kelly sat down with each of the project managers to review the achievements that had been experienced with the new standards. All went well until Kelly talked to Johns. Johns was impatient and told Kelly: "I sat around while you wasted almost all day in fooling around with all that standards stuff, but I didn't know you really meant it for me and my crew. We got a highway to build. That's my job. I know it. I do it. Right now I'd be doing more good by planning tomorrow's schedule than wasting my time hashing over last week's stuff."

Kelly replied, "Ed, you sat in our meeting two months ago and didn't object when we set up these performance standards. They apply to you as well as the others. I noticed then you didn't say much. Now I take it you think we wasted a lot of time."

"As far as I'm concerned, we wasted a day," said Ed. "I've been around, George. I know what I'm doing. If I don't do it to suit you, say so. If you can't tell how well it's done without those fancy time

limits, that's your problem. I've got eight foremen under me. I don't have to waste time on standards with them. They know what I want because I've told them. If they goof off, I tell 'em quick. If they do O.K., they know it."

"I appreciate that, but maybe you might find ways to do better if you regularly reviewed these standards with them," countered George.

"Could be, but I don't get the point in going at it this way. Now don't take this personal. I've gotten a fair shake so far. But believe me, this standards business is a waste of time. It makes no sense to gab about what we're going to do, fussing with all those job times you bought from somewhere, and then hashing it all over again if we didn't get it done like the book says. Every job is different. No book can tell you how to do it. After all, I've been doing pretty good for the past five years managing my road building without it," said Ed.

George paused to squarely meet John's eyes, lowered his voice, and gravely asked, "How do you know? Can you prove it, Ed?"

a. In your opinion, was this a good answer? If Kelly made a mistake, where was it? If Ed was wrong, how so?

b. What advice do you suggest for implementing a performance appraisal system in an organization in which such systems are unknown?

MACHINES AND MAINTENANCE

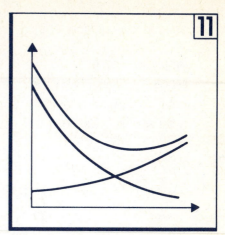

LEARNING OBJECTIVES

After studying this chapter you should

- realize the importance of obtaining the maximum utilization of machines and be aware of analytical methods for evaluating the costs of different options.
- be aware of the types of production equipment used in factories, both conventional and automated.
- be able to apply analysis methods that maximize assembly-line efficiency and minimize maintenance and replacement costs.
- understand the basic concepts of queuing theory and how waiting lines can be quantitatively analyzed.
- be aware of the use of simulation for studying production systems and how the Monte Carlo technique is applied.
- appreciate the impact of computerization on manufacturing, realizing both the limitations and the advantages.

KEY WORDS

The following words characterize subjects presented in this chapter:

general- or single-purpose machines	life cycle evaluation
tool engineering	standby machines
sequencing	preventive maintenance
assembly-line balancing	queuing theory
cycle time	queue discipline
group replacement policy	Poisson (arrival) distribution

exponential (service time) distribution

simulation model

Monte Carlo simulation

random number generation

manufacturing automation protocol (MAP)

group technology (GT)

cellular manufacturing systems (CMS)

flexible manufacturing systems (FMS)

hard automation

robots

11-1 IMPORTANCE

Machines are employed in a production sense to generate or facilitate output. They are directly involved in the process or provide auxiliary services.

"Made by hand" is a banner proclaiming an oddity that merits a higher price. Not too many years ago, similar pride was proclaimed by signs like "machine tested." Today we live in a machine age. Many say we are in a second industrial revolution that is characterized by automated machines, just as the steam engine characterized the first industrial revolution. The question of "human or machine" is again raised, but a more appropriate issue is the coordination of humans *and* machines.

There are no completely independent machine production systems; there is always a human–machine interface, as shown in Figure 11.1. A highly automated plant with its scarcity of operating personnel gives an impression of near-complete machine control, but the observation is deceptive. People planned the facility, maintain it, feed it raw materials, and distribute its output. In the usual situation in which operators directly control machines, activities at the interface are evaluated by method and measurement studies. The objective of the studies is to improve the coordinated performance of operators and machines.

Focusing on the machine side of the interface, we encounter two variables present in most evaluations: time and money. That duration and dollars interact was established in Chapter 5 by time-value-of-money relationships. A machine is a capital asset. Its loss in capital value over time is theoretically recovered

Amortization is the maintenance of a capital fund over a period of time.

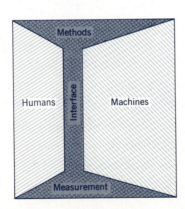

FIGURE 11.1 Human–machine interface.

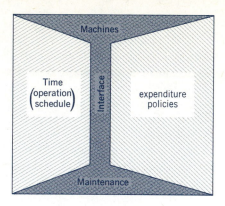

FIGURE 11.2 Machine-oriented interface of time and money.

from the worth it contributes to production; the investment is amortized. How fast it loses value is, in part, a function of the maintenance program. In turn, the maintenance program is predicated on the value of the machine as part of the total production system (its efficiency and the amount of disruption caused by its breakdown), the production schedule (degree of machine utilization), and the policy of the organization toward the retention of assets (whether the product line warrants the continued upgrading of equipment). These relationships are expressed in Figure 11.2 as the machine-maintenance interface of *time* (utilization and need for the machine in the schedule of operations) and *money* (capital expenditure policy of the organization).

In the following sections, after an overview of machines in manufacturing, we begin to look at the time side of the interface for machine scheduling. Then we switch our perspective to the money side to observe the balance between maintenance cost and the cost of production delays that are caused by machine outages. Finally, both money and time are considered in waiting-line situations in which the cost of waiting in a queue is compared with the cost of providing facilities that reduce the wait.

Computer simulation is introduced as a method for analyzing situations that are awkward to formulate.

11-2 MACHINES IN MANUFACTURING

It has been suggested that we are about to enter the third industrial revolution. During the first, muscle power was replaced by generated power, whereas the second wedded humans to machines in automated production. The third industrial revolution would extend the integration of humans and machines, with machines contributing more to production operations by means of versatile, programmable robots and process control through computerized "thinking machines."

One of the early decisions in the selection of equipment concerns the degree of flexibility or adaptability desired. Although designs are seldom "frozen," some products are subject only to infrequent design changes; others are more

Design standardization, the use of common parts for several versions of a product, is a way to customize products and still benefit from economies of scale.

unstable, depending on fickle consumer acceptance or technological innovations. The more likely it is that frequent changes in design will occur, the more necessary it becomes to build flexibility into the productive equipment.

Machine tools can be classified as either **general purpose** or **single purpose.** General-purpose machines are more flexible and constitute the bulk of machine tools in use today. Included in this category are equipment such as engine and turret lathes, bar machines, universal grinders, and welding machines. Each of these types of machines is designed to perform one or more operations on a variety of sizes and items. A change in product design simply means changing tools, gears, or fixtures, and the same piece of equipment is again ready to start production.

By contrast, special-purpose machines are designed to do one job and that job alone, but to do it very efficiently. Such machines generally have the advantage of performing specific operations more rapidly than do general-purpose machines—an important factor in volume manufacture. However, they are likely to be inflexible, and a change in product design may require scrapping or a complete conversion. Usually the choice between general- and single-purpose equipment is a matter of economics: (1) initial cost that must be charged off during the anticipated useful life of the equipment, (2) direct labor cost, and (3) preparation cost including tooling and setup.

Machine tooling refers to the selection and design of cutting tools, jigs, fixtures, and dies required to perform a specific operation. This field, which is generally called **tool engineering,** has much to do with the operating efficiency of machines. A decision as to whether an air chuck or a mechanical chuck should be used; whether a cemented carbide tool or one made from high-speed steel will perform better; what speeds and feeds produce the most satisfactory results; whether water, oil, or air will serve best as a coolant—these are typical problems encountered by the tool engineer.

Many manufacturing industries have adopted automation principles because quite often they lead to more output per labor dollar and per dollar invested in facilities. Quality is more uniform; work is safer; and production scheduling, once the planning of the system has been completed, is virtually automatic. In most instances in-process inventories are substantially reduced because there are few, if any, stopping points in the process.

But automation also has disadvantages. There is considerable investment in facilities, requiring a long period of heavy utilization to recapture the investment. Furthermore, there may be greater manufacturing inflexibility. Product designs must be frozen for long time intervals. In industry, in which change is rapid or unpredictable, this loss of fluidity may be serious. During a business recession or other times when volume decreases, management cannot lay off an automated line or readily employ it in other work. Finally, the downtime of the equipment is apt to be devilishly expensive. There are more mechanical elements to go wrong, and the interdependency of equipment tends to make the system only as reliable as its weakest element. Downtime is cumulative. One failure may shut down the entire line. Tool and tool-change costs tend to

Key features in the design or selection of any machine are that (1) it be easy to set up, operate, service, and repair and (2) that it be provided with safety devices to prevent costly difficulties as a result of improper operation.

Social considerations associated with automation were discussed in Chapter 9. A 1969 report by Professor Eva Mueller of the University of Wisconsin's Institute for Social Research concluded: "For the most part, Americans value the equipment they work with, and believe that automation is a good thing for people in their line of work." (See Case 19 at the end of this chapter.)

rise, as all tools are removed simultaneously for inspection and sharpening at regular intervals, whether dull or not.

11-3 SEQUENCING

The order in which jobs pass through machines or work stations is a **sequencing** problem. When there are very few different types of jobs or machines, the problem is solved informally by sketching the flow mentally or on a time chart. Consider the simple 2×2 sequencing problem represented by the tableau in Figure 11.3: two jobs require work by two machines, $M1$ and $M2$, in that order. There are only two possible sequences, job 1 first and job 2 second, or $J2 < J1$. The schedule resulting from each sequence is shown by the time charts. Based on a preference for the shortest total elapsed time to complete the work, the $J1 < J2$ sequence is selected.

Time chart drawing and application details are in Section 14-4.

Sequencing problems quickly become more tedious as the number of jobs and machines increases. With n jobs passing only from $M1$ to $M2$, there are $n!$ alternatives; that is, it would take almost 40 million time charts to show all the sequence patterns possible for just 11 jobs. Thus, charting is not a very practical solution tool for larger exercises, and unfortunately, we have exact analytical methods for only the smaller problems. More complex problems are treated by simulation techniques, but the cost and time required to produce a satisfactory solution are still very large. Much research remains to be done in this area.

Simulation is discussed in Section 11-7.

Sequencing *n* jobs through two machines

A quick, simple technique provides the least-elapsed-time solution to the problem of sequencing any number of jobs through the same two-step process. Two-station sequences are frequently found in job shops where the "process layout" has machines and services functionally grouped. The solution procedure uses the following steps:

1. List job times to pass from $M1$ to $M2$.

2. Select the shortest job listed.

3. If the shortest time is by $M1$, that job is placed as early as possible in the job sequence. If the shortest job is by $M2$, it is placed as late as possible in the sequence. A tie between shortest job times is broken arbitrarily because it cannot affect the minimum elapsed time to complete all the jobs.

The $n \times 2$ solution technique is sometimes referred to as "Johnson's rule" after S. M. Johnson, "Optional Two- and Three-Stage Production Schedules with Setup Times Included," *Naval Research Logistics Quarterly* 1 (March 1954).

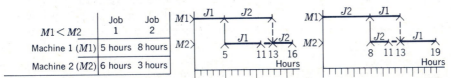

FIGURE 11.3 Total elapsed time to pass two jobs through a two-machine process.

4. Delete the job selected in Step 2 from the listing in Step 1. Then repeat Steps 2, 3, and 4 until a complete sequence is obtained.

The technique applies to single-unit or single-type batch jobs in which there is no priority of completion. It is assumed that there is sufficient in-process storage space and that the cost of in-process inventory is the same or varies insignificantly for all units. For short processes, these assumptions are usually valid. Extended processes subject to closer inventory cost controls and expediting priorities may have an optimality criterion other than minimizing total elapsed time. Additional complicating factors include variable transportation time between machines, rework of defective units, machine breakdowns, and variable processing times caused by operator proficiencies or working conditions.

EXAMPLE 11.1 Sequencing six jobs through two machines

The time to wash and cook batches of produce depends on their condition, type, quality, quantity, and intended end product. The time in hours to process six known batches, $J1$–$J6$, through the washer and cooker is as follows:

Batches	$J1$	$J2$	$J3$	$J4$	$J5$	$J6$
Washer ($M1$)	0.4	0.7	0.3	1.2	1.1	0.9
Cooker ($M2$)	1.1	0.7	1.0	0.8	1.0	1.3

The smallest number in the listing is 0.3 for $J3$. Because it occurs for $M1$, it is placed first in the job sequence and is then deleted from the tableau:

Job Sequence

$J3$					

	$J1$	$J2$	$J4$	$J5$	$J6$
$M1$	0.4	0.7	1.2	1.1	0.9
$M2$	1.1	0.7	0.8	1.0	1.3

The next smallest job time is $J1$ by $M1$. Therefore, $J1$ is placed as near the start of the sequence as possible, the second position, and is then dropped from the tableau:

	$J2$	$J4$	$J5$	$J6$
$M1$	0.7	1.2	1.1	0.9
$M2$	0.7	0.8	1.0	1.3

Job Sequence

$J3$	$J1$				

Now there is a tie between the job times for $J2$ by $M1$ and $M2$. A decision to put $J2$ in the third or sixth position, as early or late as possible in the sequence, will not affect the total elapsed time. The choice can be resolved by other priority considerations. By letting the time in $M2$ be the deciding factor, $J2$ is placed last in the job sequence:

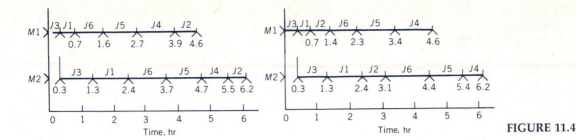

FIGURE 11.4

	J4	J5	J6
M1	1.2	1.1	0.9
M2	0.8	1.0	1.3

Job Sequence

J3	J1				J2

J3	J1	J6	J5	J4	J2

An alternative sequence for J2 placed early instead of late is

J3	J1	J2	J6	J5	J4

The remaining jobs are sequenced by the same rules. In order of attention, J4 is placed next to J2, J6 goes after J1, and J5 fits in the last open position. The completed sequence is

That the breaking of a tie does not affect the minimum elapsed time is demonstrated by the time charts in Figure 11.4.

Sequencing two jobs through *n* machines

A reverse twist to the "$n \times 2$" sequence exposes the problem of routing two jobs through n facilities, $2 \times n$. This time a graphical approach is used. The axes of the graph represent the two jobs that must be processed through the same n facilities in whatever order the facilities are needed. The areas in the chart where both jobs could simultaneously be assigned to the same facility are blocked out. These areas must be avoided in determining the schedule of facility usage that minimizes the time to complete the jobs.

Any line drawn on the chart that avoids the blocked off areas while connecting the origin to the point representing the completion of both jobs is a feasible schedule. When a line runs horizontal, it represents exclusive work on the job scaled to the horizontal axis. Conversely, a vertical line represents work on the other job. A line 45° to the base indicates concurrent work on both jobs. Consequently, the optimal sequence is identified by the origin-to-completion-point line with the greatest amount of 45° travel.

EXAMPLE 11.2 Sequencing two jobs through six machines

Custom products pass through six machine centers during fabrication. The time spent in each center depends on the design. For instance, one design requires a large amount of welding and very little drilling, whereas another design has reversed proportions. The order of work also varies for different jobs as a function of product requirements and the state of completion. The expected times obliged by two jobs in six machine centers are as follows:

		Machine Center Order and Work Times					
J1:	Sequence	$C \rightarrow A \rightarrow E \rightarrow F \rightarrow D \rightarrow B$					
	Time, hours	1	3	3	5	4	2
J2:	Sequence	$B \rightarrow A \rightarrow E \rightarrow F \rightarrow C \rightarrow D$					
	Time, hours	3	2	4	4	1	2

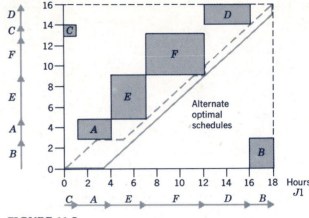

FIGURE 11.5

This information is translated into a chart by scaling each sequence on one axis of the graph. Rectangular blocks representing simultaneous work on both jobs in one machine center are formed by overlapping projections of each machine center's time.

Two work-order paths are shown on Figure 11.5. Both paths are alternative optimal solutions. The solid line indicates that no work is done on J2 for the first 3 hours; then both jobs are in process until J1 is completed, and one more hour is spent to complete J2. The other path has the same amount of 45° travel, but idle times for the jobs are spaced irregularly. Time charts for the two sequences, showing that the minimum total time required for both jobs is 19 hours by either path, are in Figure 11.6.

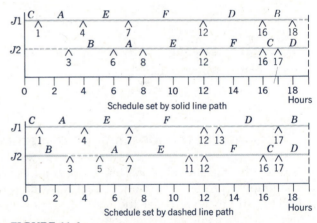

FIGURE 11.6

11-4 LINE BALANCING

Assembly-line balancing is associated with a "product layout" in which products are refined as they pass through a line of work centers. A designated number of work elements is performed at each center. The times to perform work elements are derived from work measurement studies. The period allowed to

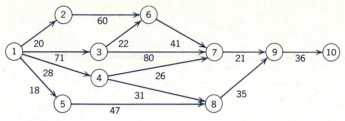

FIGURE 11.7 Sequence of operations required to complete a product.

complete operations at each station is determined by the speed of the assembly line; all work stations share the same allowed **cycle time.** Idle or float time is created for a station when the work assigned to it takes less time than does the set cycle time. The objective of line balancing is to minimize idle time while assigning operations to work stations according to a predetermined technological sequence. A perfect balance, from a theoretical viewpoint, is obtained when the assignments provide no idle time.

Line-balancing problems have received a great deal of attention, perhaps more than the prevalence of assembly lines warrants. Some techniques yield exact solutions for the given assumptions. Others are designed to yield approximate solutions based on practical considerations. The exclusive emphasis should not be to get a perfect balance but to obtain the optimal layout and flow in relation to the rest of the production operations.

An arrow network representing a product's manufacturing sequence is shown in Figure 11.7. Each arrow denotes an operation required on the product. The duration of each operation, a sum of the work elements for that operation, is shown below its arrow. The complete product network maps the metamorphic sequence from raw materials and parts to finished work.

With the times and order of operations known, the remaining point to decide before making the assignments is a desirable cycle time. The sum of the durations for all 14 operations in the network is 536 minutes. The largest individual duration establishes the minimum cycle time: 80 minutes for operation 3,7. In turn, the minimum cycle time sets a lower limit on the number of work stations needed, $536/80 = 6.7$; at least seven stations are required. There may also be other commanding considerations, such as a limit on machines available or workers with the skills to perform certain operations. For example, assume that operations (1,5), (3,6), and (6,7) require time on a sophisticated machine. If only one machine of this type is available, the limiting cycle time will be increased to $18 + 22 + 41 = 81$ minutes.

The assignment of operations to work stations is conducted in a straightforward manner. A preliminary assignment sheet allocates operations to the desired number of work stations in conformance with the network's sequence restrictions. The most time-consuming operations with the earliest start times are assigned first. Special restrictions, such as the assignment of certain operations to one station, noted by double squares in the network, require extra

Assembly lines are most common in continuous manufacturing industries such as automotive and electronics. Line operations may be manual, by machines, or combinations.

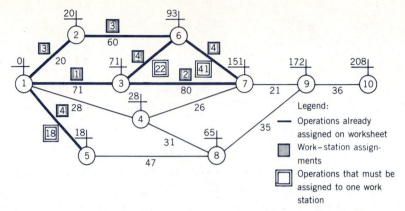

FIGURE 11.8 Product network showing special constraints, earliest start times, and partial work station allocations.

vigilance. Record keeping is improved by marking network operation arrows as they are entered in the assignment sheet.

Figure 11.8 shows the line-balancing network after four work stations have been allocated their operations. Number 3 work station has to finish operations 1,2 and 2,6 before the special machine station, 4, can perform operation 6,7. An initial list is completed, as shown in Table 11.1, by carefully adhering to the prescribed sequence restrictions and by keeping the total time for the operations assigned to one station near the limiting cycle time, 81 minutes.

Improvements on the initial effort are initiated by noting the work-station times that exceed the minimum, 81. By restructuring the assignments, an improved schedule is developed, as shown in Table 11.2. A check of the assignments shows that all the restriction requirements have been satisfied and that no times exceed the limiting sequence in station 4. Further juggling might be

TABLE 11.1
Assignment Sheet for the Initial Allocation of Operations to Work Stations

Work Station	Assigned Operations			Total Time
1	1,3			71
2	3,7			80
3	1,2	2,6		20 + 60 = 80
4	1,5	3,6	6,7	18 + 22 + 41 = 81
5	1,4	4,8	4,7	28 + 31 + 26 = 85*
6	5,8	8,9		47 + 35 = 82
7	7,9	9,10		21 + 36 = 57

*Cycle time is 85 minutes.

TABLE 11.2
Improved Work-Station Assignments That Decrease Cycle Time

Work Station	Assigned Operations			Total Time
1	1,3			71
2	3,7			80
3	1,2	2,6		20 + 60 = 80
4	1,5	3,6	6,7	18 + 22 + 41 = 81*
5	1,4	4,7	7,9	28 + 26 + 21 = 75
6	5,8	4,8		47 + 31 = 78
7	8,9	9,10		35 + 36 = 71

*Cycle time is 81 minutes.

done to alter specific assignments that would benefit work performance at certain stations. For instance, the quality of output might be improved by assigning a certain operation to a work station run by an operator known to be particularly proficient at that operation.

The approach to line balancing just considered is mathematically unsophisticated. It relies more on resourcefulness and judgment than on analysis techniques. The legitimacy of such an approach is realized when the dynamic nature of a production line is considered. The sequence of operations seldom varies, but the operation times are far from constant. Performance by human operators is continually modulated by enthusiasm, health, and social conditions. Fluctuating flow between stations means the operators must cooperate to balance internally one another's work output with the pace of the assembly line. Human behavior mocks exact formulation, but it remains a major factor in effective line balancing. Perhaps minimizing idle time is an adjunct to maximizing conditions for individual and collective human effectiveness.

> Operation times can occasionally be adjusted to reduce idleness by redesigning equipment, changing machines, and applying motion economy principles.

> Impromptu line balancing occurs whenever one operator is absent and others "take up the slack."

EXAMPLE 11.3 Balancing a miniproduction line

An auction is being planned for a charitable organization. A thousand announcements are to be sent out to a select mailing list. The announcements contain a list of items to be auctioned in a specially designed folder, with slits in it to hold the formal invitation, and a ribboned citation. Expenses are minimized by using volunteer workers who will assemble the announcements as well as personalize them by adding color highlights and signing the invitations.

A few preliminary runs are made to organize the assembly operations by identifying distinct activities and determining activity durations. The results of the trials indicate the activity sequence and times, as shown in Figure 11.9.

Although the labor is free, the organizers feel a responsibility to make the miniassembly operation as efficient as possible. Analysis starts with the usual assumptions for a line-balancing study:

1. The activities are indivisible (cannot be broken down into finer elements).

2. The work method is fixed and the activity durations are constant.

3. There is no operator learning or fatigue (adequate allowances are built into the activity durations).

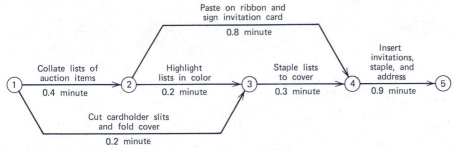

FIGURE 11.9 Sequence of assembly operations for the announcements.

4. The cycle time must be greater than, or equal to, the maximum activity duration, and the activities performed at one work station cannot exceed the cycle time.

If just one person did all of the activities to complete an announcement, it would take

$$0.4 + 0.2 + 0.2 + 0.8 + 0.3 + 0.9 = 2.8 \text{ minutes}$$

This is the minimum duration, but it is not necessarily the most effective one. Whereas job enlargement, which allows a worker to produce a finished product, is often a motivating work design, there are occasions when it is not feasible because of facility limitations or skill requirements. In this volunteer activity, both practical and motivational considerations are present: the "togetherness" generated by a team project helps morale, and the pace set by faster volunteers keeps the operation moving better. Therefore, a team approach is selected. Because the largest activity time is 0.9 minutes,

$$\text{Minimum number of work stations} = \frac{2.8}{0.9} \doteq 3$$

which means that at least three operators are required.

Tasks for the work stations are easily grouped:

Work Station A

Activity 1,2 Collate	0.4 minute
Activity 3,4 Staple	0.3 minute
Activity 1,3 Highlight or Activity 2,3 Cover	0.2 minute
Total time required =	0.9 minute

Work Station B

Activity 2,4 Invitation	0.8 minute
Activity 1,3 Highlight or Activity 2,3 Cover	0.2 minute
Total time required =	1.0 minute

Work Station C

Activity 4,5 Assemble and address	0.9 minute
Total time required =	0.9 minute

No matter which work-station activities 1,3 and 2,3 are assigned to, the cycle time is 1.0 minute. Efficiency of this arrangement is

$$\text{Efficiency} = \frac{\text{sum of activity times}}{(\text{cycle time})(\text{number of work stations})}$$

$$= \frac{2.8}{(1.0)(3)}$$

$$= 93.3\%$$

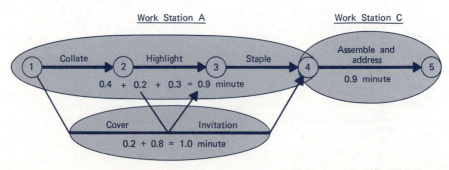

FIGURE 11.10 Work-station assignments for the auction-announcement assembly line.

This level of efficiency cannot be improved on with the given composition and sequence of activities. Because several work elements are included in most activities, it might be feasible to redefine activities by trading elements among them (for example, switch the insertion of the invitation cards into the folder slits from activity 4,5 to 2,4). This type of analysis is appropriate to expensive industrial assembly lines but is hardly worthwhile for the small service project that takes only about 50 hours of assembly. Therefore, activity 1,3 is arbitrarily coupled with activity 2,3 in Work Station B to establish the assembly configuration shown in Figure 11.10.

11-5 MAINTENANCE

Maintenance programs are closely linked to replacement policies. All manufacturing industries follow some maintenance routine because the cost of lost production from unexpected breakdowns is significant and the capital cost of owning an asset is usually lower when the asset receives proper care. The quality of production may also be higher with better maintenance. The economic balance for a maintenance policy takes the familiar form shown in Figure 11.11.

An alternative for any decision is to do nothing. Some machines are operated without servicing until they expire. Occasionally this is the least expensive maintenance policy.

As indicated in this figure, maintenance costs are lower when a machine is new. They increase with age because more work is needed to maintain a given level of performance. Capital costs are normally high in the first part of a machine's life and level out with age, but the cost of repairs is often more than offset by lower capital and outage costs (Figure 11.11). The better policy is the one that provides the lowest total cost.

There are many versions of maintenance programs. In our personal lives we informally practice different programs for different items. Hand tools, small electrical appliances, and lightbulbs are normally used until they break down and are then replaced. The frequency of replacement is primarily a function of the quality purchased. When an asset serves a particularly important purpose, such as tires on an automobile, the policy is to carry a spare. The maintenance program is to check periodically the condition of the standby asset, the spare

Government manuals define *maintainability* as a characteristic of design and installation expressed as a probability that an item will conform to specified conditions for a given time period when prescribed maintenance procedures are followed.

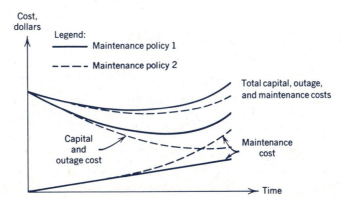

FIGURE 11.11 Relationship of maintenance policy to total cost.

tire. More automobile owners follow a policy of periodic maintenance (oil changes, lubrication jobs, and so on) to get less expensive and more reliable transportation. The policy is to prevent the inconvenience and high cost of a breakdown by keeping parts in a near-new condition through care and replacement.

Policies of industry are similar to personal maintenance routines, but the scale is magnified. Smaller items are often replaced before they fail. The question is how long to keep them in service. Standby machines are frequently held to reduce the impact of a breakdown of key machines. The question is how many standbys to have. Preventive maintenance is used to reduce the frequency and magnitude of major repairs. The question is whether preventive maintenance is more economical than are repairs made as needed, and if it is more economical, how often the preventive maintenance checks should be made. All these questions can be treated by probability models that are similar to the ones described in the following sections.

The size of repair facilities is another important consideration in a maintenance program. An economic balance is sought for the cost of idle repair facilities versus the cost of machines waiting to be repaired when facilities are crowded. This topic will be considered in the discussion of queuing theory in Section 11-6.

Preventive maintenance is so well accepted that the concept is simply nicknamed PM.

Group replacement policy

A **group replacement policy** is often feasible when a large number of identical low-cost items are increasingly prone to fail as they age. The classic example is the replacement of streetlights at set intervals. Each lightbulb is so inexpensive that keeping individual service records is not warranted. The main cost of replacement is incurred by the truck and crew that removes a burned-out light and puts in a new one. Once the crew reaches the location where a renewal is required, little additional expense is involved.

We now consider an individual versus group replacement model based on the following simplifying assumptions:

1. Only one type of asset with a known failure distribution is considered at a time.

2. Costs of individual and group replacements can be accurately estimated.

3. When items fail during an interval, they are replaced individually at the end of the interval. For instance, each item that fails during a week is replaced on the last day of that week.

4. When a group replacement is made, all items in the group are replaced at one time, regardless of age. Thus an individual replacement made the day before a scheduled group replacement is still renewed, although it has only one day of service.

TABLE 11.3
Data Required to Determine a Group Replacement Policy

	Failure Distribution							Service Cost for Replacement	
Age, n, in weeks	1	2	3	4	5	6	7	Individual	$3.00
Probability of failure by age n	0.10	0.05	0.05	0.10	0.20	0.20	0.30	Group	$0.50

To illustrate the comparison method for the two replacement plans, we use the data in Table 11.3. It presents the failure pattern of 1000 high-intensity drying lamps used in a manufacturing process. The lamps cost $1.75 each. It is difficult to change a lamp during a working day without disrupting production. A much lower per-lamp service cost results from group replacements scheduled on weekends when the production line is idle.

The comparison is made by calculating the first minimum of a curve derived from the cost of group replacements at different intervals. If this minimum, as shown in Figure 11.12, is less than the average individual replacement cost, the indicated policy is to replace all items at the interval shown by the first minimum.

According to the data in Table 11.3, each time the entire bank of lamps is replaced, the cost will be $500 = 1000 lamps $\times$ $0.50 per lamp. A group replacement after one week would cost $500 plus the cost of individual replacements required during the week; 10 percent of the 1000 lamps will expire and need renewal at $3 per lamp. Then a one-week group replacement plan will have a total cost of

Individual replacement + group replacement = cost per week
$$1000 \times \$3 \times 0.10 + \$500 \qquad = \$800$$

A two-week plan has the same group cost of $500, but individual replacements must be made for failures during two weeks for the original 1000 lamps plus renewals for the burned-out lamps from the 100 replaced at the end of week 1:

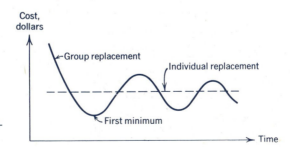

FIGURE 11.12 Cost pattern of different group replacement intervals.

Total cost of two-week group replacement policy	Replacements for lamps failing = in week 1 = $3000 × 0.10 = $300	Replacements for lamps failing + in week 2 + $3000 × 0.05 + $150	Replacements in week 2 for lamps replaced + in week 1 + $3000 × 0.10 × 0.10 + $30	Group replacement cost at end + of week 2 + $500 + $500 = $980

Then, the cost per week = $980/2 weeks = $490.

Each additional interval forms a similar but extended pattern of replacements for replacements for replacements, and on and on. A tabular form (Figure 11.13) simplifies the bookkeeping for the calculations. Each row corresponds to an alternative group replacement period. The entries in the body of the tableau are developed successively from the periodic expected costs: the cost for the previous period becomes the first entry for the next period. In the last columns the group replacement costs are added to the cumulative weekly costs to obtain the total cost for every interval. The weekly cost of each replacement period is its total cost divided by the number of weeks in the period.

The tabulated calculations from Figure 11.13 disclose a minimum expected weekly cost of $376 for a four-week replacement cycle. This value is now compared with the average weekly cost of strictly individual replacements. The expected life of a lamp is the sum of the products of each weekly interval multiplied by its associated failure probability:

$$1 \text{ week} \times 0.10 + 2 \text{ weeks} \times 0.05$$
$$+ \ 3 \text{ weeks} \times 0.05 + 4 \text{ weeks} \times 0.10$$
$$+ \ 5 \text{ weeks} \times 0.20 + 6 \text{ weeks} \times 0.20$$
$$+ \ 7 \text{ weeks} \times 0.30 = 5.05 \text{ weeks}$$

With the cost of individual replacements at $3 per lamp, the average weekly cost is

$$\frac{1000 \text{ lamps} \times \$3 \text{ per lamp}}{5.05 \text{ weeks}} = \$594$$

Replacement Period, Week	Individual Failure Cost during Week							Expected Cost		Group Cost	Total Cost	Weekly Cost
Week: Probability:	1 0.10	2 0.05	3 0.05	4 0.10	5 0.20	6 0.20	7 0.30	Week	Cumulative			
1	3000	–	–	–	–	–	–	300	300	500	800	800
2	300	3000	–	–	–	–	–	180	480	500	980	490
3	180	300	3000	–	–	–	–	183	663	500	1163	388
4	183	180	300	3000	–	–	–	342	1005	500	1505	376
5	342	183	180	300	3000	–	–	682	1687	500	2187	437
6	682	342	183	180	300	3000	–	772	2459	500	2959	493
7	772	682	342	183	180	300	3000	1142	3601	500	4101	586

FIGURE 11.13 Expected cost of group replacement periods.

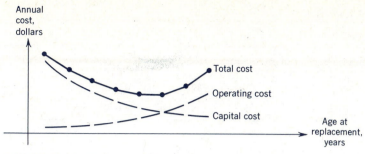

FIGURE 11.14 Cost pattern for increasing asset age.

Thus a group replacement policy for the given conditions is far superior to replacing each lamp as it expires.

Replacement cycle evaluation

A cyclic replacement of a machine by another machine of the same type is a special case of the general economic comparison method discussed in Section 5-3. It is of interest because it establishes the economic life by combined capital and operating cost considerations. The relationship is displayed in Figure 11.14, in which capital costs vary inversely with time and operating costs increase with time. This pattern is typical of most production equipment.

The objective of a **life cycle evaluation** is to find the replacement age that minimizes an asset's annual cost. The calculations follow an iterative routing in which the capital cost—the decrease in salvage value to a certain age—is added to the accumulated operating cost for the same period. Each sum is divided by the respective age to give an average cost for that replacement interval. Interest can be included by determining the present worth for all the figures. For shorter intervals, say 10 or fewer periods, and relatively low interest rates, discounting seldom makes a difference in the optimal replacement age.

EXAMPLE 11.4 Replacement study based on a life cycle evaluation

Small personnel vans carry work crews to remote facilities. The remote stations are expected to be run indefinitely by the same transportation arrangement.

The present replacement cycle for the vans is based on the following annual cost pattern:

Table 11.4 indicates that the optimal replace-

Year	0	1	2	3	4	5	6	7	8
Salvage value	$4000	$2900	$2200	$1600	$1200	$900	$600	$400	$300
Operating cost	—	800	900	1100	1300	1700	2200	2700	3300

TABLE 11.4

End of Year	1	2	3	4	5	6	7	8
Total depreciation	$1100	$1800	$2400	$2800	$3100	$3400	$3600	$3700
Cumulative operating cost	800	1700	2800	4100	5800	8000	10700	14000
Total cost	1900	3500	5200	6900	8900	11400	14300	17700
Cost per year	$1900	$1750	$1733	$1725	$1780	$1900	$2043	$2213

ment interval is four years with an average total cost, (depreciation + operating cost)/number of years, of $1725 per year. The small difference in cost for years 2, 3, and 4 throws the final replacement age decision open to less tangible considerations.

An equipment supplier has offered to buy three 1-year-old vans for $2900 each as trade-ins on two vans that have a capacity equal to the three smaller carriers. The large vans have a three-year optimal replacement cycle with an average per year cost of $2500. The dealer points out that the switch will serve the routes just as well, and the total average cost per year for the two vans is 2 × $2500 = $5000, com-pared with the three-van average yearly cost of 3 × $1725 = $5175.

The offer is refused because the incremental cost increase for the three present vans during the next year (end of year 1 to the end of year 2) is 3($3500 − $1900) = $4800. This increment, the cost for next year, is less than the average cost of the larger vans, $5000 per year. If the dealer accepted three 2-year-old vans at $2200 apiece, the deal would be ad-vantageous because the incremental cost during year 2 is 3($5200 − $3500) = $5100. Any deviation from a cyclic replacement schedule should be evaluated on an incremental basis.

Standby machines

A provision for activating a **standby machine** when an on-line machine fails is another way to maintain service. This alternative does not take the place of routine maintenance but serves as an insurance policy for conditions that could seriously hurt production. A regular premium must be paid for the insurance. A standby machine takes up space, must be periodically checked, and depre-ciates in value whether or not it is used. On the other side of the ledger, the availability of a substitute to keep a production line functioning can avert a slowdown or a shutdown of a whole series of dependent operations.

An evaluation of standby costs compared with benefits is conducted in about the same manner as is a group replacement study. The failure pattern and cost of failures must be known or estimated. It is assumed that a standby machine will always operate when placed in service. An adequate inspection program for standbys is postulated to justify the assumption, and the cost of the program is included with the other expenses of keeping a spare.

A tableau similar to Figure 11.13 is used for the calculations. Each row represents the expected cost of a different number of standbys. The cost for the first alternative, no standby machine, is composed entirely of lost production expense. As the number of standbys increases, lost production costs decrease, whereas the holding costs for the standbys rise. The alternative providing the minimum total cost is preferred.

EXAMPLE 11.5 Selection of the optimal number of standby machines

Forty identical machines are operated at one work station. The cost in lost production and disruption to subsequent operations is estimated to be $100 per day per machine out of operation. The exact distribution of failures is not known, but it is believed to closely follow a Poisson distribution with a mean of two failures per day. Conditions become serious when six or more machines are out of operation at once. Then it is necessary to curtail activity for the rest of the line, causing an additional loss of $1000 per day. Standby machines can provide a hedge against lost production. The cost of maintaining a standby is estimated at $20 per day.

A tableau is set up for the calculations. The likelihood of machine failures is computed from the formula of Poisson probabilities:

$$P_n = \frac{e^{-\lambda} \lambda^n}{n!}$$

where

n = number out of order
λ = mean number of failures

$e = 2.7183$
P_n = probability of n failures

For instance, the probability that three machines are out of order at one time is

$$P_3 = \frac{2.7183^{-2} \times 2^3}{3!} = \frac{8}{6(2.7183)^2} = 0.180$$

The entries in the body of the tableau are derived from the cost accrued by the number of machines out of order at one time. The costs increase linearly from zero for no failures to $500 for five machines concurrently out of service. Beyond five, $1000 are added to account for curtailed operations. As is apparent from the completed tableau of Figure 11.15, the failure costs form a recurring pattern. The cost expectation for each standby alternative is added to the expense of maintaining that number of standbys. The resulting total cost figures indicate that three standby machines should be used.

Number of Standby Machines n:0 P_n:0.135	Cost of Lost Production for Machine Failures								Cost of Lost Production	Standby Cost	Total Cost
	1 0.270	2 0.270	3 0.180	4 0.090	5 0.036	6 0.012	7 0.003	8 0.001			
0 0	100	200	300	400	500	1600	1700	1800	215	0	215
1 —	0	100	200	300	400	500	1600	1700	117	20	137
2 —	—	0	100	200	300	400	500	1600	55	40	95
3 —	—	—	0	100	200	300	400	500	22	60	82
4 —	—	—	—	0	100	200	300	400	7	80	87
5 —	—	—	—	—	0	100	200	300	2	100	102
6 —	—	—	—	—	—	0	100	200	1	120	121
7 —	—	—	—	—	—	—	0	100	0	140	140
8 —	—	—	—	—	—	—	—	0	0	160	160

FIGURE 11.15

Preventive maintenance

Routine **preventive maintenance** is about the least glamorous job in production but one of the most important. As processes continually become more mechanized, maintenance correspondingly becomes more complex, and the damage potential of malfunctions soars. The backbone of a healthy preventive maintenance program is good planning and capable maintenance personnel backed by a supporting management policy.

We have already observed mathematical evaluations of maintenance alternatives involving replacement cycles and standby machines. The same tableau format can be used to determine whether preventive maintenance (replacement of parts, overhauls, and the like) is less expensive than repairing on call. It also sets the most economical period for preventive maintenance checks by the same routine used for replacement intervals.

There are many more elaborate models for special maintenance problems, such as an inspection policy for equipment that deteriorates with age (a guided missile or a fire hose), a renewal policy for equipment that can be restored to an operating condition (recapping old tires), and replacement policies for equipment renewed after a certain length of service (replacing all tubes after they have been in service one-half their expected life). Such sophisticated analyses are appropriate when the dollar volume is large or service reliability is critical.

Typical preventive maintenance planning is more mundane than mathematical modeling. The guiding rule is that the time spent on preventive maintenance should be less than the time required for repairs and that the value imparted to machines by preventive maintenance should exceed the program cost. The logic of the rule is unassailable, but collecting the cost data to put it into effect creates a paradox: You have to try the program first to find out whether it is good. The following list offers some practical considerations that should be weighed in maintenance planning.

1. Machines can be "overdesigned" to improve reliability. Redundant circuits and extra bracing increase initial costs but may produce greater-than-proportionate savings in maintenance costs.

2. Increasing in-process inventories can provide a buffer against the effect of machine failure. Extra stock increases carrying costs, but it insulates the rest of a production line until breakdown repairs can be completed.

3. Inspections should dovetail with periods of cleaning, adjusting, and other maintenance work to reduce cost and inconvenience.

4. Training programs and disciplinary policies have a distinct effect on the amount of maintenance needed. Operators should be responsible for preventive maintenance work on their machines whenever possible. Periodic checks can be made by maintenance personnel to ensure the level of care.

5. Friction, vibrations, corrosion, and erosion are physical conditions that should be detected and controlled before they develop into major problems.

All mathematical models rely on information collected on the job. The value of keeping complete, accurate, and detailed records of repairs and servicing can hardly be exaggerated.

Objectives of preventive maintenance are
1. To attain longer life.
2. To increase useful time.
3. To maintain the design level of performance.

An adjunct but difficult part of planning is the distribution of PM costs among operating departments.

A move to put more responsibility on operators for their machines is called *productive maintenance*. Operators do all the routine maintenance so as to free maintenance workers for actual preventive work.

Few employees receive less recognition than do maintenance personnel. Their good work is taken for granted, but an occasional failure, often not their fault, may create a minor panic with hasty accusations. The ample responsibility in maintenance work is quite obvious; the difficulty of carrying out the work is not so apparent. Careful records of repairs and servicing are necessary, and often the data can be acquired only through the cooperation of operating departments. The coordination of an extensive maintenance program needs the backing of higher management. Besides the special skills required to diagnose failures, persistent attention to details is compulsory for inspection duties. The watchword should be a cross between the Coast Guard's motto and Murphy's law: Be prepared; if something can go wrong, it will.

EXAMPLE 11.6 Preventive maintenance emphasizes uptime instead of downtime

Productivity rises as the use of production equipment increases. A machine is a drag on productivity during its downtime, and it still fails to make its full contribution if it is not operating efficiently during uptime. Therefore, preventive maintenance is aimed at increasing effective uptime, and success is becoming more crucial as production systems rely more and more on electronic equipment. Measurable areas by which to judge success are

1. Percentage of uptime.
2. Quality of output.
3. Energy conservation.
4. Cost of maintenance labor and material.
5. Actual replacement pattern versus the expected life cycle.

A successful preventive maintenance program has the following characteristics:

• An inventory of equipment that categorizes items according to their return-to-service urgency: immediate, 24 hours, or as convenient.

• A checklist for each major piece of equipment that includes the manufacturer's service manual, schedule of routine maintenance activities, and the on-hand supply of critical spare parts.

• A *maintenance operation center*, endorsed by top management, that has the resources to meet emergency repair demands and the authority to schedule corrective maintenance and planned overhauls to best use total resources, instead of fitting maintenance into production gaps.

11-6 WAITING LINES

Everyone has experienced the frustration of waiting in lines to obtain service. It usually seems like an unnecessary waste of time. In our private lives we have the option of seeking service elsewhere or going without the service. Such defections have direct economic consequences for the organization providing the service. When a customer leaves a waiting line, he or she becomes an opportunity cost; the opportunity to make a profit by providing the service is lost. An important aspect of system design is to balance this cost against the expense of additional capacity.

The delayed effect of ill will from irate defectors is a serious penalty created by waiting lines.

A machine needing repair is a customer for a service facility. When the facility is busy, the machine waits. Production capacity is lost during the waiting period. The waiting time could be reduced by cutting down the average time to repair a machine, by providing more repair areas, and by instituting special procedures to give priority to repair work when it is seriously behind demand. The net effect on waiting time from the implementation of any of these alternatives is not at all obvious. Doubling the number of service facilities does not necessarily reduce the waiting time by half because breakdowns do not occur at regular, predictable intervals and repair time varies with the extent of damage. Attempts to analyze such situations have led to the development of **queuing theory**.

The first recognized effort to analyze queues was made by a Danish engineer, A. K. Erlang, in his attempts to eliminate bottlenecks created by telephone calls on switching circuits.

There are an amazing number of waiting-time situations in industry. At almost every stage of production, something is in temporary storage: papers wait for an executive to sign them; parts wait to be assembled; orders wait to be processed; and materials wait to be inspected or transported. In most instances such storage is justified on the grounds that the cost of waiting is less than the cost of providing service to eliminate the wait. But in a few situations, waiting lines cause significant congestion and a corresponding increase in operating costs. For example, ships wait to be unloaded at docks; projects await attention by the engineering staff; aircraft wait to land at an airport; and breakdowns await repair by maintenance crews. Only those conditions in which substantial cost is incurred from waiting warrant analysis by queuing techniques. The cost of a thorough study may be considerable, assuming that there are analysts available and capable of conducting the study.

Queuing theory rests on some formidable mathematics. It is primarily concerned with the properties of the waiting lines—the distribution of arrivals and service times, service policy, and similar considerations—not with the cost evaluations. Once a suitable queuing model is determined or developed, the cost comparisons are relatively straightforward. Here we consider some elementary applications to gain an appreciation of queuing fundamentals.

Most managerial decision makers do not have the time, inclination, or training to develop a sophisticated queuing study. Yet they should have enough background to decide whether a study is justified and to question the reasonableness of a solution developed by specialists. It is important to recognize that theoretically correct equations will produce erroneous solutions when based on unfounded assumptions.

Queuing concepts

To simplify the mathematical formulations, it is generally assumed that the population of customers is infinite.

The language of queuing theory is refreshingly descriptive. A *customer* is a person, machine, or object that requires *servicing*—an action performed for the customer. The customer is serviced by a *service facility*. When several customers can be serviced at one time, the facility is said to have several *channels*. A *waiting line* or *queue* is formed whenever there are waiting customers.

Customers arrive at a service facility according to an *arrival distribution*—constant intervals, random rate, or other patterns. The time taken to provide the desired service follows a *service time distribution*. If every customer required exactly the same service and it were provided by automatic equipment, all the service times would be constant. A more typical case is a machine breakdown service in which the service time to repair it depends on the type of machine, its condition, and the seriousness of the damage; that is, the repair time is irregular.

At first glance it might seem reasonable that a queue would never form when the average arrival rate is less than the average service rate. The reasoning is valid when both patterns are constant but is fallacious for irregular patterns. Customers always have to wait when they arrive in groups exceeding in size the number of channels. Similarly, the occasional service time that is far larger than the average rate probably causes a waiting line to form. An ever-increasing queue is caused by the explosive condition of an average arrival rate greater than the average service rate. In general, the number of customers that has to wait increases proportionately as the arrival rate approaches the service rate.

The manner in which customers are served is called **queue discipline**. The most common assumption is first in–first out with no defectors: customers are not discouraged from joining by the length of the queue and patiently wait their turn for service. Other orders include (1) last in–first out as encountered in some inventory policies, (2) a priority arrangement that allows a rush or emergency case to be serviced out of turn, and (3) a random selection that gives every waiting customer an equal chance of being the next to receive service.

An automatic production line provides a rare example in which products (customers) ride a constant-speed conveyor (uniform arrival rate) and have the same cycle time at each work station (constant service rate).

"Bulk" arrivals occur when a large shipment arrives or a carrier unloads.

Unused service time, like sleep, cannot be stored for use in a busy period.

OPTIONAL MATERIAL

Queuing formulas

Four basic queue structures are illustrated in Figure 11.16. When one crew does all the work to repair a machine, we have a single-channel, single-phase case.

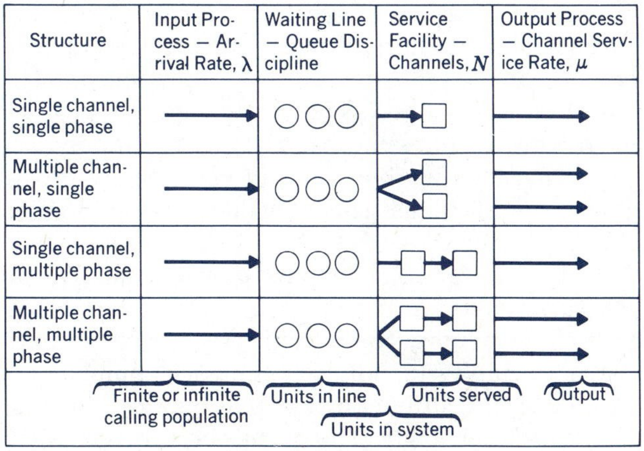

FIGURE 11.16 Waiting-line structures.

Increasing the number of complete crews converts it to a multiple-channel, single-phase structure. By dividing the one crew into specialist teams that sequentially perform operations to repair a single machine, the structure changes to single channel, multiple phase. Adding another crew of team specialists that progressively completes the servicing of one machine transforms the previous structure to multiple channel, multiple phase. A first come–first served queue discipline is diagrammed in Figure 11.16, but other disciplines could as easily apply. There are also other possible combinations, such as a multiple-channel, first phase followed by a single-channel, second phase.

Different formulas are associated with each queue structure. The formulas also differ as a function of the applicable arrival and service time distribution. The answers usually sought through queuing formulas are

P_n = probability of n customers being serviced or waiting to be serviced
P_0 = probability that the service facility is idle (no customers in the system, $n = 0$)
L_q = mean number of customers in the waiting line
L = mean number of customers in the system (number in the queue plus number being serviced)
W_q = mean waiting time for a customer before being serviced
W = mean time in the system (customer waiting time plus servicing time)

We shall consider only single-phase cases. The applicable formulas for the desired queue characteristics are shown in Table 11.5. The Greek symbols traditionally associated with queuing theory are

λ(lambda) = average number of arrivals per unit time
μ(mu) = average number of servicings that each channel can perform per unit time

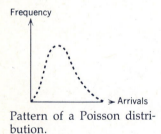

Frequency

Pattern of a Poisson distribution.

$P(>t)$

Negative exponential distribution.

The formulas in Table 11.5 are based on a definitive set of assumptions. Because calculations quickly become overwhelming when different assumptions are incorporated, it is tempting to apply the simpler expressions to all waiting-line studies. The effect is similar to a quartermaster's assuming that all persons in an army are the same size in order to simplify logistics; the policy makes the quartermaster's work easier and occasionally provides the right fit, but some results are ridiculous. The following assumptions pertain to the given formulas:

1. Arrivals follow a **Poisson distribution**. We previously encountered this distribution in Example 11.5, in which it described the probability of machine failures. The formula given in the example would be modified to fit a queuing situation by letting n = number of customers. In general, the Poisson distribution says that on a continuous time scale, there is a very small probability of an event's occurring at any particular instant, and there

TABLE 11.5
Single- and Multiple-Channel, Single-Phase Queuing Formulas

Single Channel	Multiple Channel
$P_n = \left(1 - \dfrac{\lambda}{\mu}\right)\left(\dfrac{\lambda}{\mu}\right)^n$	$P_0 = \dfrac{1}{\left[\displaystyle\sum_{n=0}^{N-1} \dfrac{(\lambda/\mu)_n}{n!}\right] + \left[\dfrac{(\lambda/\mu)_N}{N!\,(1 - \lambda/\mu N)}\right]}$
$P_0 = 1 - \dfrac{\lambda}{\mu}$	
$L_q = \dfrac{\lambda^2}{\mu(\mu - \lambda)}$	$L_q = \dfrac{(\lambda/\mu)^{N+1}}{(N-1)!(N - \lambda/\mu)^2} \times P_0$
$L = \dfrac{\lambda}{\mu - \lambda} = L_q + \dfrac{\lambda}{\mu}$	$L = L_q + \dfrac{\lambda}{\mu}$
$W_q = \dfrac{\lambda}{\mu(\mu - \lambda)} = \dfrac{L_q}{\lambda}$	$W_q = \dfrac{(\lambda/\mu)^{N+1}}{\lambda(N-1)!(N - \lambda/\mu)^2} \times P_0 = \dfrac{L_q}{\lambda}$
$W = \dfrac{1}{\mu - \lambda} = W_q + \dfrac{1}{\mu} = \dfrac{L}{\lambda}$	$W = W_q + \dfrac{1}{\mu} = \dfrac{L}{\lambda}$

The single-channel formulas are modified for constant service times to

$$L_q = \frac{\lambda^2}{2\mu(\mu - \lambda)}$$

and

$$W_q = \frac{\lambda}{2\mu(\mu - \lambda)}$$

is a very large number of times an event could occur. Translating these implications into a maintenance situation, we would say a breakdown requiring attention could occur at any instant during a day and the probability for any one instant is the same regardless of what has happened before or the length of the waiting line.

2. Servicings follow an **exponential distribution**. The equation for the distribution,

$$P(>t) = e^{-\mu t} \quad \text{where } t = \text{duration of a servicing}$$

gives a reasonable approximation of many industrial servicing situations. A graph of the density function slopes downward to the right to show that the deviation from μ is greater for longer servicings than for shorter servicings. For maintenance work, this condition is interpreted as a greater probability of shorter repair times, but occasionally a task will far exceed the average repair time.

The quotient of λ/μ, called the utilization factor, is expressed in units of "Erlangs" in honor of the queuing pioneer.

3. The average service rate is greater than the average arrival rate, $\mu > \lambda$.

4. Queue discipline is first in–first out with no defections from the waiting line.

5. The number of customers is infinite, and the waiting-line size is unlimited. Few real problems truly satisfy this assumption, but the theoretical effect is not seriously violated so long as the number could grow relatively large through returns to the queue after previous servicing.

EXAMPLE 11.7 Comparison of waiting time for different maintenance facility structures

Plans are being made for a plant enlargement. Repair facilities for machine breakdowns are barely adequate in the existing plant and will certainly not provide acceptable service when more machines are added. Records of recent repair activities show an average of four breakdowns per eight-hour shift. The pattern of breakdowns closely follows a Poisson distribution. When the new additions to the plant are completed, an average of six breakdowns per shift following the present distribution pattern is expected.

An exponentially distributed service rate of six repairs per shift is the capacity of the present repair facility. Without enlarging the capacity when the new addition is made, the utilization factor would be $\lambda/\mu = 6/6 = 1$, which looks commendable until the impossibly large waiting time is considered:

$$W_q = \frac{\lambda}{\mu(\mu - \lambda)} = \frac{6}{6(6 - 6)} = \infty$$

Two alternatives with equivalent annual cost are available. New equipment and a larger crew for the existing station would increase the average servicing rate to 11 repairs per shift, or a second servicing station could be built in the new addition. In the latter alternative, the capacity of the two service stations would be five servicings per shift in each. Repair times would still be exponentially distributed.

The single-channel alternative would have the following characteristics when $\lambda = 6$ and $\mu = 11$:

Probability of being idle
$$= P_0 = 1 - \frac{6}{11} = 0.45$$

Mean number of machines waiting for service
$$= L_q = \frac{6^2}{11(11 - 6)} = 0.66$$

Mean time before a machine is repaired (waiting time plus repair time)
$$= W = \frac{1}{11 - 6} = 0.2 \text{ shift}$$

$$= 0.2 \times 8 \text{ hr/shift} = 1.6 \text{ hr}$$

The values of the same characteristics for the second alternative providing two channels, $N = 2$, each with $\mu = 5$, are

$$P_0 = \frac{1}{\left[\displaystyle\sum_{n=0}^{2-1} \frac{(6/5)^n}{n!}\right] + \left[\dfrac{(6/5)^2}{2! \, (1 - [6/(5 \times 2)])}\right]}$$

$$= \frac{1}{\left[\dfrac{(6/5)^0}{1} + \dfrac{(6/5)^1}{1}\right] + \left[\dfrac{(6/5)^2}{2(1 - [6/10])}\right]}$$

$$= \frac{1}{\left[1 + \dfrac{6}{5}\right] + \left[\dfrac{36/25}{2(4/10)}\right]} = \frac{1}{2.2 + 1.8} = 0.25$$

$$L_q = \frac{(6/5)^{2+1}}{(2 - 1)!(2 - 6/5)^2} \times P_0$$

$$= \frac{(1.2)^3}{(1)(0.8)^2} \times 0.25 = 0.68 \text{ machine}$$

$$W = \frac{L_q}{\lambda} + \frac{1}{\mu} = \frac{0.68}{6} + \frac{1}{5} = 0.31 \text{ shift}$$

$$= 0.31 \times 8 \text{ hr/shift} = 2.5 \text{ hr}$$

By using the mean time in a repair facility (waiting time plus servicing time) as the comparison criterion, the single enlarged facility where $\mu = 11$ is preferred. Quicker return to production means less loss of output and thereby lower repair cost for the single-channel station, if the equivalent annual capital costs plus operating costs for the two alternatives are indeed the same. The single-channel alternative would be preferred, with $\mu = 11$, even if the servicing time of the two-channel alternative could be increased to $\mu = 6$:

$$P_0 = \frac{1}{\left[\displaystyle\sum_{n=0}^{1} \frac{(6/6)^n}{n!}\right] + \left[\dfrac{(6/6)^2}{2!(1 - [6/(6 \times 2)])}\right]}$$

$$= \frac{1}{1 + 1 + 1} = 0.33$$

$$W = \left[\frac{(6/6)^{2+1}}{6(2-1)!(2-6/6)^2} \times P_0 \right] + \frac{1}{6}$$

$$= \left[\frac{1^3}{6(1)\,(1)^2} \times \frac{1}{3} \right] + \frac{1}{6} = 0.22 \text{ shift}$$

$$= 1.76 \text{ hr}$$

One large facility clearly gives better theoretical service than does an equivalent number of smaller facilities. A complete evaluation should check the time advantage by assigning values to hours saved. The study should also include other factors such as the logistic conveniences of a centralized location versus the greater average transportation distance between breakdowns and a repair facility when there is only one station.

END OPTIONAL MATERIAL

Queuing applications

Queuing theory occupies an interesting position among management tools. It is applicable to an omnipresent problem, idleness versus congestion. It is well publicized and an "in" topic for management scientists. But according to polls conducted by universities and professional journals, it is not applied as widely as are other, less glamorous tools. Perhaps practitioners shy away from the facade created by the more sophisticated formulations. Possibly the barrier is a lack of data or the feeling that analysis effort would not provide an answer significantly better than that found in less formal procedures. Whatever the cause, there is an unfilled potential for waiting-line studies, but an effective analyst must have more than a casual familiarity with the subject.

The distribution of actual arrivals and servicing times may not fit one of the theoretical distributions—Poisson, normal, exponential, or binomial. If formulas are available for a reasonably close fit, it is a matter of judgment whether the extra effort required to develop custom formulas or to use simulation methods is compensated by improved accuracy. It should be remembered that distributions are forecasts of future patterns, and so it therefore makes sense to apply the precautionary measures discussed for forecasting in Chapter 3.

The cost of servicing customers is fairly easy to determine. Past performances give a good indication of operating times and material requirements. Capital costs are given by the depression policy. The cost of idle facilities is largely a function of depreciation expense and wages.

Predicting the cost of waiting customers is usually uncertain and always difficult. When customers arrive from internal sources, such as machines in a factory sent to a maintenance facility, the expense of waiting in line is at least definable—lost production from idle machines, wages wasted for idle operators, inefficiencies in a production line caused by a vacant station, extra cost of standbys or inventory stock to take up the slack for a "down" machine, congestion at a repair facility caused by a rash of breakdowns, and similar observable disruptions.

Estimating waiting costs for external customers is more difficult because of their uncertain behavior. Some customers may enter a queue and later defect to a competitor when they tire of waiting. Seeing a long queue may discourage

The advantages of well-behaved theoretical functions stem from the tables and analytical solutions already developed for these distributions.

"Truncated" queues are waiting lines shortened by defectors or the lack of space in which to wait. Formulas have been developed to cope with truncation.

Customers at a restaurant can
be observed leaving a line or
passing by when a line exists.
Whether they discover a more
attractive eating place after
leaving is hard to ascertain.

potential customers from joining the waiting line. Former customers may permanently take their business elsewhere after enduring a long waiting period. Direct observations can indicate the short-range effects of customers not joining or defecting from a queue. Long-range reactions can only be guessed at in most situations. Some form of experimentation aimed at identifying the customer population and its behavior is a guide to guessing.

EXAMPLE 11.8 Cost comparison of queuing alternatives

A company plans to redesign its maintenance facilities. The line supervisors complain that the existing service is too slow. The cost controller claims that the five workers in the facility are idle one-third of the time and that the only reason repairs appear slow is that they sometimes occur in bunches, causing delay. All agree that a priority system or any other queue discipline refinement would not be feasible. A compromise solution appears to be the installation of more automatic equipment to reduce the size of the maintenance crew.

A complete redesign of the present facility will cost $60,000 and allow repairs at the rate of 40 per week with two maintenance workers. Adding $10,000 of new equipment will increase the repair rate from the present level of 15 per week to 20 per week with a crew of four. Both alternatives will have

an economic life of six years, no salvage value, and the same repair costs. A 10-percent return on invested capital is required.

An average of 10 breakdowns occurs each week. The probability of a breakdown at any time is approximately uniform, so an assumption of a Poisson distribution of arrivals is appropriate. The pattern and mean number of breakdowns are expected to remain about the same in the future. Wages for maintenance personnel average $18,000 per year, and the opportunity cost of having a breakdown is estimated at $30 for each day of lost production time. The variation of repair times is quite close to an exponential distribution.

The annual costs of the two equipment change alternatives are compared to determine the most favorable course of action.

Complete redesign ($\mu = 40$).

Capital recovery cost: $60,000 (a/p, 10,8) = $60,000(0.22961)$ $13,776

Annual operating cost: 2 workers × $18,000 per work-year   36,000

Annual lost production cost:

$$\frac{\text{Time in a repair facility}}{\text{machine breakdown}} = W = \frac{1}{40 - 10} = 0.033 \text{ week}$$

$$\frac{\text{Total lost time}}{\text{Year}} = \frac{10 \text{ breakdowns}}{\text{week}} \times \frac{52 \text{ weeks}}{\text{year}} \times \frac{0.033 \text{ wk}}{\text{breakdown}} = \frac{17.3 \text{ weeks}}{\text{year}}$$

$$\frac{\text{Annual cost of waiting}}{\text{plus repair time}} = \frac{17.3 \text{ weeks}}{\text{year}} \times \frac{5 \text{ days}}{\text{week}} \times \frac{\$30}{\text{day}} \qquad 2595$$

Total annual cost: $52,371

Adding equipment (μ = 20).

Capital recovery cost: $10,000(a/p, 10,8) = $10,000(0.22961) $ 2296

Annual operating cost: 4 workers × $18,000 per work-year   72,000

Annual lost production cost:

$$W = \frac{1}{20 - 10} = 0.10 \text{ week}$$

Total lost time per year = 10 × 52 × 0.10 = 52 weeks per year

$$\frac{\text{Annual cost of waiting}}{\text{plus repair time}} = \frac{52 \text{ weeks}}{\text{year}} \times \frac{5 \text{ days}}{\text{week}} \times \frac{\$30}{\text{day}} \qquad 7800$$

Total annual cost: $82,096

The complete redesign of the present facility at a cost of $60,000 is clearly preferable. Now this challenger is compared with the existing facility, the defender. Assuming the defender has no capital cost, a conservative assumption, we have the following:

Present facility (μ = 15).

Annual operating cost: 5 workers × $18,000 per work-year $ 90,000

Annual lost production cost:

$$W = \frac{1}{15 - 10} = 0.20$$

Total lost time per year = 10 × 52 × 0.20 = 104 weeks per year

$$\frac{\text{Annual cost of waiting}}{\text{plus repair time}} = \frac{104 \text{ weeks}}{\text{year}} \times \frac{5 \text{ days}}{\text{week}} \times \frac{\$30}{\text{day}} \qquad 15,600$$

Total annual cost: $105,600

With the redesign alternative still preferred, the cost controller should be happier because the idle work-hours are reduced from

$$5 \text{ workers} \times 40 \text{ hr per week} \left(1 - \frac{10}{15}\right) = 67 \text{ per week to } 2 \times 40 \left(1 - \frac{10}{40}\right) = 60 \text{ work-hours per week}$$

The line supervisors should also be pleased that the time a machine will be lost from production should average only

$$\frac{17.3 \text{ weeks per year}}{104 \text{ weeks per year}} = 0.166$$

or about 17 percent of the former level.

11-7 SIMULATION

To simulate, in a general sense, means to feign or to assume the appearance of something without being the real thing. In a management sense, we use simulation to feign a real system in order to observe and learn from the behavior of the replica. This departure from reality has several advantages over observing the real system. It is usually easier and less expensive and may be more illuminating by confining attention to characteristics of particular interest.

The use of simulation techniques is increasing mainly because of the wider availability of computers, along with the greater sophistication of today's managers and systems analysts. A computer simulation is an effective way for managers to treat complex business relationships without suffering the penalties of real trial-and-error experiences. Some problems that managers face do not yield to the convenient solution methods encountered in previous sections, and some managers cannot cope with the solution methods. In such instances the problems can be "run" to determine the numerical effect of different alternatives, instead of solved by analysis. Both the learning and the problem-solving simulations rely on a model, a program that directs the responses of the simulated system.

Computerized simulation models

When systems analysts talk of simulation, they usually mean **simulation model** experimentation conducted on a digital computer. As is described in Section 1-4, a problem from the real world is abstracted in a form that can be assimilated by computers. The problem may be too large and complex to tackle with standard optimization techniques, or it may be an attempt to evaluate the internal workings and environmental responses of a new design before it is physically implemented.

The methodology of simulation is portrayed in Figure 11.17. A problem is defined by identifying the variables and parameters to be included in the model. Two classes of variables are normally present: (1) variables controllable by the decision maker (for example, which customer will be served next?), and (2) uncontrolled variables associated with the environment (for example, how many customers will arrive?). Parameters are properties of the system being studied that remain constant during the study period (for example, the distribution of the customer population).

A simulation model is constructed by setting the starting conditions for the study, specifying the time increments (fixed or variable) at which the system will be observed, determining the decision rules that affect the behavior of the system, and specifying the probability distributions for variables. As a childish example of model design, consider the problem of where and when to operate a lemonade stand. The initial conditions are a full pitcher of lemonade and the stand placed at a certain intersection. Conditions at the stand will be observed every five minutes, in which the operating rules are that lemonade will be sold to anyone who will pay a quarter and that the stand will be open from midafternoon until dinner time (or until the pitcher is empty). The distributions of

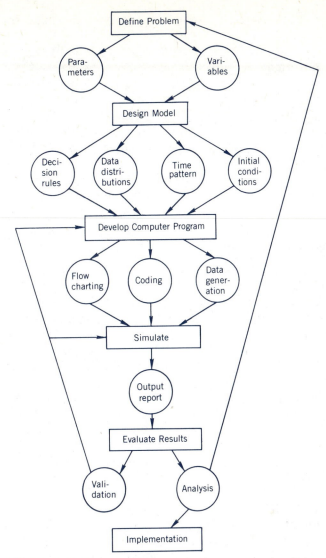

FIGURE 11.17 Descriptive network of computer simulation methodology. Results of a validated simulation exercise can lead to further experimentation, in which the problem definition is changed slightly or different decision rules are employed, or it can be used for actual operations.

interest are the number of people who will pass the stand at a given location and the probability that they will buy when they pass.

A computer program is developed by *flowcharting* the problem to display how the system dynamically responds to the conditions of interest, translating the flowchart into computer language, and obtaining a random number generator.

The output of a simulation exercise will, of course, depend on the program governing the run. This output must be checked to determine whether it adequately represents actual conditions. Errors occur from mistakes in flowchart logic and coding. Basic data and assumptions can also be in error. Program calculations can be checked independently to verify accuracy, and the conditions of the modeled system can be compared with an existing system to validate its authenticity. However, the real validation depends on how closely the simulated conditions resemble actual performance after the modeled system is implemented. Then the analyst can either rejoice if the advice was wise or be reminded of decision-making perils inherent under conditions of risk.

EXAMPLE 11.9 Simulation applications

Many industrial and service applications have been reported in professional journals. Most frequently cited are inventory and queuing experiences, but a wide array of other problems have been examined through computer simulation.

Investment policies are investigated by constructing simulation models that consider new markets and growth in existing businesses, development of new products, methods of financing and reinvestment of earnings, effects of mergers and acquisitions, and additional factors that are important to strategic planning. Similar studies assist hospitals in planning for future patient care.

A simulation model of the manufacture of telephone cables revealed that costs could be reduced by *not* repairing defective wires before they were bundled into cable units. Simulation of where and how defects occur in a production process helps locate the best place for inspection stations.

Simulation of anticipated operations reveals the expected utilization rate for the involved equipment and thereby assists in the design of new facilities.

For instance, the utilization of an overhead crane depends on the level of activity of the operations it serves, downtime of the crane, idle time in the operations, timing of demands for crane service, and so forth. The resulting performance expectations indicate the design and type of crane that should provide satisfactory service.

The passage of trains in a railroad system is simulated to determine the most economical use of engines, preferred times for maintenance, where to improve tracks, effects of delays on schedules and support facilities, and so on. Similar studies are conducted by air, ocean, and trucking lines to coordinate loading, transshipments, and unloading.

Very ambitious, computerized, management competition games have been developed. The pedagogical intent differs among games: practicing decision making, gaining analytic ability, learning from interpersonal experiences, and appreciating particular problems. Most games are programmed to emphasize a variety of problems in a relatively short time rather than to achieve a penetrating analysis.

OPTIONAL MATERIAL

Monte Carlo technique

Monte Carlo technique is the colorful name of a technique by which random numbers are generated to select events from a probability distribution of occurrences. The name is derived from possible **random number generators:** a flipped coin, a tossed die, a cut from a deck of cards, a draw from a hat, or even a roulette wheel. However, the most used generator is a random number table, shown in Appendix C.

The Monte Carlo technique is applied to problem solving; it does not include human decision making, other than possible activities represented by a known distribution of outcomes. The approach is more of a controlled experiment than a direct assault on problem variables. Instead of deriving a formula to solve or a heuristic process to follow, the conditions of the problem are experienced in trial runs to see what would happen as a result of different decision alternatives. The concepts of simulation by sampling are so simple that it may take restraint to keep from applying the method to every problem. It undeniably produces an answer, but there are some reservations:

A heuristic solution method directs the way to a preferred solution by comparison with previous solutions, but it claims no optimality for the indicated preference. It shares with simulation and trial-and-error methods the goal of *discovering* a solution, not *deriving* it.

1. Many rounds of simulated trials are needed to produce a solution in which confidence can be placed.

2. The trials may be more expensive in computer time or clerical effort than would be required for an analytic solution.

3. There is no exact measure of the precision of the solution. Because digital simulation is in numbers, the outflow of figures may be more precise than is really justified by experimental conditions.

4. A slight change in conditions usually means that the entire simulation process must be repeated to incorporate the revised data. A similar change in an analytically derived model can be handled by a less demanding change in a legendary "variable constant."

The routine for simulated sampling can be quickly described by a simple illustration. Suppose the arrival rate of workers calling on a supply crib to pick up equipment and expendable supplies has the following pattern:

Arrivals per minute, A	0	1	2	3
Probability of A	$\frac{1}{3}$	$\frac{1}{3}$	$\frac{1}{6}$	$\frac{1}{6}$

When a worker wants to check out a piece of equipment, it takes two minutes, on the average, for an attendant to serve him. Expendable supply requests average one minute to fulfill. There is an equal likelihood for equipment and supply requests. Service is on a first come–first served basis, and arrivals are assumed to appear at the beginning of each one-minute increment.

The described situation obviously leads to a queuing problem in which the cost of making workers wait is compared with the cost of idle crib attendants.

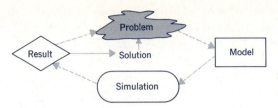

FIGURE 11.18 Process for Monte Carlo simulation.

It is also apparent that the distributions of arrivals and service times do not follow the familiar Poisson-exponential assumptions used in the developed queuing formulas. We, therefore, have a candidate for Monte Carlo simulation.

The general approach is represented in Figure 11.18. From the problem we draw the information to establish a model and an alternative to test by simulation. The end result is evaluated as a possible solution to the problem and may suggest the advisability of testing other alternatives.

For the problem under consideration, we have enough information to set up a simulation process. However, we need to know costs in order to select a sensible testing alternative. The appropriate costs are crib attendant, $0.12 per minute or $7.20 per hour; and waiting time of workers, $0.20 per minute. Based on the costs, service times, and arrival rate, a crib operated by two attendants will be the test model.

The cumulative frequency distributions for arrivals and service times are shown in Figure 11.19. Also noted in the figure are the random number generators: one die is thrown to designate the number of arrivals in each trial, and a coin is flipped to see how long it takes to serve customers after their arrival. The purpose of these maneuvers is to provide unbiased entries for the probability distributions. Thus a throw of a die revealing six spots would mean three arrivals during the minute represented by the throw. Similarly, because there is an equal likelihood of a one- or two-minute service time and a corresponding likelihood

> It is easy to eliminate the alternative of one attendant because the arrival rate averages more than one per minute and the service time is 1.5 minutes per arrival—an explosive queue.

> It is customarily assumed that a die, deck of cards, or a coin used in simulation is "unloaded" and manipulated honestly.

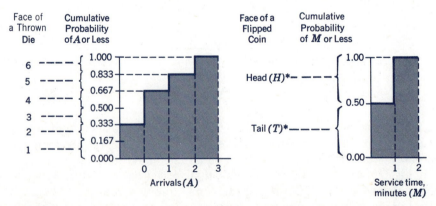

FIGURE 11.19 Problem data and associated random number–generating assignments.

of a head or tail from the flip of a coin, one side of the coin represents each servicing.

A record-keeping table (Table 11.6) is convenient when simulations are conducted manually; the record shows the data generated by die and coin throws. The simulation starts at cumulative time 0 with a thrown die showing a 2. This means that there are no arrivals during the first minute. The second die toss is a 5, to represent two arrivals at the beginning of the second minute. Two coin flips, each representing one of the arrivals, are a head (H_1) and a tail (T_1). Because there are two operators available, operator A spends her next 2 minutes satisfying the H_1 flip, and operator B serves 1 minute in accordance to the T_1 flip. The next die cast, at the beginning of the third minute, is a 3 to indicate one arrival. Because operator A is still working to service the previous arrival, operator B performs the next 2-minute servicing in conformance to the H_2 flip for the arrival at the end of minute 2. The tabulation for 50 samples, 50 minutes, is contained in Table 11.6.

First admitting that there are too few samples to obtain a definitive pattern, we at least can make some observations from the available results. The waiting time for arrivals is easily determined by comparing the time that each servicing actually starts with the earliest time that it could have started. The circled numbers to the right of the service arrows show the waiting periods. The total of all the circled numbers is 37, indicating that the workers were required to wait 37

TABLE 11.6
Record of Monte Carlo Simulation for the Solution of a Queuing Problem

Cumulative Time	$RN_{(A)}$*	Arrivals	$RN_{(S)}$*	Servicing* O_A		O_B	
0	2	0					
1	5	2	H_1T_1	H_1 ↓		T_1 ↓	
2	3	1	H_2	↓		H_2 ↓	
3	1	0				↓	
4	2	0				↓	
5	6	3	$H_5T_5T_5$	H_5 ↓		T_5 ↓	
6	1	0		↓		T_5 ↓	①
7	5	2	T_7T_7	T_7 ↓		T_7 ↓	
8	1	0				↓	
9	1	0					
10	2	0					
11	6	3	$H_{11}H_{11}T_{11}$	H_{11} ↓		H_{11} ↓	
12	5	2	$T_{12}H_{12}$	↓		↓	
13	3	1	H_{13}	T_{11} ↓	②	T_{12} ↓	①
14	4	1	T_{14}	H_{12} ↓	②	H_{13} ↓	①
15	2	0		↓		↓	

TABLE 11.6
Record of Monte Carlo Simulation for the Solution of a Queuing Problem

Cumulative Time	$RN_{(A)}$*	Arrivals	$RN_{(S)}$*	Servicing* O_A		Servicing* O_B	
16	1	0		T_{14} ↓	②		
17	6	3	$T_{17}H_{17}T_{17}$	T_{17} ↓		H_{17} ↓	
18	1	0		T_{17} ↓	①	↓	
19	6	3	$T_{19}H_{19}H_{19}$	T_{19} ↓		H_{19} ↓	
20	6	3	$H_{20}T_{20}T_{20}$	H_{19} ↓	①	↓	
21	3	1	T_{21}	↓		H_{20} ↓	①
22	5	2	$H_{22}H_{22}$	T_{20} ↓	②	↓	
23	2	0		T_{20} ↓	③	T_{21} ↓	②
24	1	0		H_{22} ↓	②	H_{22} ↓	②
25	4	1	H_{25}	↓		↓	
26	2	0		H_{25} ↓	①	↓	
27	4	1	T_{27}	↓		T_{27} ↓	
28	1	0				↓	
29	5	2	$T_{29}T_{29}$	T_{29} ↓		T_{29} ↓	
30	4	1	H_{30}	H_{30} ↓		↓	
31	6	3	$H_{31}H_{31}T_{31}$	↓		H_{31} ↓	
32	1	0		H_{31} ↓	①	↓	
33	3	1	H_{33}	↓		T_{31} ↓	②
34	2	0				H_{33} ↓	①
35	5	2	$H_{35}H_{35}$	H_{35} ↓		↓	
36	3	1	T_{36}	↓		H_{35} ↓	①
37	4	1	H_{37}	T_{36} ↓	①	↓	
38	5	2	$T_{38}T_{38}$	H_{37} ↓	①	T_{38} ↓	
39	5	2	$H_{39}T_{39}$	↓		T_{38} ↓	①
40	4	1	T_{40}	H_{39} ↓	①	T_{39} ↓	①
41	5	2	$T_{41}H_{41}$			T_{40} ↓	①
42	2	0		T_{41} ↓	①	H_{41} ↓	①
43	1	0				↓	
44	5	2	$T_{44}H_{44}$	T_{44} ↓		H_{44} ↓	
45	4	1	H_{45}	H_{45} ↓		↓	
46	4	1	H_{46}	↓		H_{46} ↓	
47	1	0				↓	
48	2	0					
49	1	0					

*$RN_{(A)}$ = .random for arrivals; $RN_{(S)}$ = random number for servicings, $O_{(a)}$ = operator A in tool crib, O_B = operator B in tool crib.

minutes during the 50-minute simulation period. With a cost of $0.20 per minute for idle workers, the apparent average cost of congestion is

$$\frac{37 \text{ min} \times \$0.20 \text{ min}}{^{50}/_{60} \text{ hr}} = \$8.88/\text{hr}$$

Adding a third attendant to the crib would cost $7.20 per hour and would reduce, but probably not eliminate, the workers' waiting time. More samples for the two-attendant model and a comparable experiment with a three-attendant model would decide the issue.

The value of a digital computer to carry out the tedious record keeping is apparent even from a model as limited as the one just considered. It is difficult to say when enough samples have been taken. Practical considerations allow more samples when the correctness of a decision is more important. It is also logical not to stop a program when the mean of the generated data differs appreciably from the empirical mean; for instance, if the coin flipped for service times showed heads 45 out of 50 times, one would conclude that the highly irregular streak should not be left as the representative pattern of service requirements.

END OPTIONAL MATERIAL _____

11-8 CLOSE-UPS AND UPDATES

Many things have been done physically in recent years to upgrade working conditions and to uplift attitudes in the work force through more meaningful involvement in operations, but these changes pale against the wondrous advances made in machines. So rapid have been technological improvements that it is difficult to keep up with the vocabulary, and even harder to comprehend what the advances mean to production systems. Although the details can be shunted to specialists, production managers should at least be aware of the relationship of such developments as CIM, CMS, GT, MAP, LAN, FMS, and so on—the alphabet of technological progress.

CIM and group technology (plus MAP)

Computer-integrated manufacturing—CIM, as introduced in Chapter 1— is the central nervous system of a factory organization in which computers assist information flow in both directions, incorporating feedback from real-time operations. The blessing expected from CAM is a level of factory coordination that instantly detects overproduction or defects in any part

General Motors spearheaded the development of MAP in 1980.

MAP is not patented.

"Scientific Principles of Group Technology" by S. P. Mitrofanov is the original text in Russian published in 1959.

Groups and families can be identified by Production Flow Analysis (PFA) which is championed by J. L. Burbidge in "The Simplification of Material Systems," *International Journal of Production Research*, 20(3) (1982).

Characteristics of different facility layouts are given in Section 9-3.

and then reprograms production to switch machines or to reroute subassemblies. Lack of communication among machines was limiting the promise of CIM until MAP came about.

Manufacturing automation protocol (MAP) is a standardized communication system that lets machines converse electronically in a common language. The purpose of MAP is to establish a multivendor *local area network* (LAN) to facilitate the data flow among all intelligent devices in a factory. That is, equipment makers who want to sell to MAP-oriented manufacturers will agree to conform to MAP specifications. By overcoming machine incompatibility, companies save time and money through elimination of transmission delays and easier correction of errors.

Production efficiencies spawned by CIM often lead users to **group technology (GT).** The concept of arranging machines that produce similar but not necessarily identical parts in a pattern that minimizes setup time came from the Soviet Union. It has been applied much more widely in Europe than in the United States.

The underlying premise of group technology is that similar parts can be grouped into families. Family similarities are defined by attributes such as the surface finish, shape, strength, or function of a part and its manufacturing method. Parts are classified into clusters and coded for identification. A part numbering system may use as many as 36 digits, which permits very precise resolution. A designer can retrieve a part design by referencing a chain of digits even when the function and exact design are unknown. Potential economies include floor space and data-file savings, reduced inventory, fewer jigs and fixtures, and other savings that can be realized from taking advantage of similarities.

GT and flexible manufacturing systems (plus CMS)

Detection of a family of parts that are flowing, or could be reasonably made to flow, through the same cluster of machines identifies a manufacturing cell. This distinct area of the factory can be upgraded to produce, optimally, the detected part family. As a result, output quality and worker morale may go up simply from working with an entire assembly and being able to build a finished product rather than performing repetitive tasks forever.

Cellular manufacturing system (CMS) is the name given to applied group technology. It collects into groups the people, processes, and machines needed to produce a part family, typically a complete component or subcomponent. The difficult transition from a functional layout, such as a job shop, to a CMS layout apparently stalled the movement until recently. Enormous reductions in queuing times and work-in-process inventories enabled by CMS are now hastening its acceptance.

Simple forms of cellular manufacturing have existed for many years, especially in small shops where machines were grouped together for convenience in making the dominant product. Little thought was given to the versatility of the arrangement. A premeditated design for cellular manufacturing features flexibility. A cluster or collection of carefully selected machines is arranged to facilitate the production of a specific group of component parts. When the cluster is capable of handling a variety of family members and the cells are linked together, a **flexible manufacturing system (FMS)** has been created.

The greatest advantage of FMS is its capacity to manufacture goods cheaply in small quantities. The old fixed-position machines and rigid transfer lines of **hard automation** could stamp out look-alike parts at low cost in huge volume, but economies of scale disappear when small batches of different parts are required. Because 75 percent of all machined parts in the United States are now produced in batches of 50 or fewer, flexibility has become a necessity. A flexible manufacturing system can turn out a small batch almost as efficiently as can an inflexible production line producing a large batch. This capability allows a manufacturer to change products rapidly in response to market gyrations. It also permits the construction of smaller plants that can be located closer to markets.

Hard automation is typified by fixed-purpose machinery. *Soft automation* is represented by computers and their associated apparatus.

Enthusiasts call it "economy of scope."

FMS and robots (plus people)

A robotic cell is the crown jewel of a cellular manufacturing system. When **robots** can communicate better and acquire mobility, depopulated factories will achieve huge productivity gains. Until then, piecemeal robotization provides degrees of consistency and precision that are breaking all records for product quality. A possible sequence of operations in a robotic cell is shown in Figure 11.20. It displays the versatility of robots and suggests their expanding role in manufacturing.

The word **robot** was coined by playwright Karel Capek in 1920 from the Czechoslovak work *robota*, meaning "forced work." A more practical definition from the Robot Institute of America says that a robot is "a reprogrammable multifunctional manipulator designed to move material, parts, tools, or specialized devices through variable programmed motions for the performance of a variety of tasks." These tasks for metal working include welding, painting, cutting, deburring, gluing, assembling, polishing, riveting, and positioning materials for other machines. Some robots can learn a new task by being led through the motions just once. All robots are essentially indefatigable, passionless, precise, and insensitive to toxic chemicals, radioactivity, and dust.

Robot applications outside the factory include sentry duty, bomb squads, fire fighting, traffic control, and so forth.

Robots are also nearly senseless and stupid, compared with people, and they are costly when compared with simpler operator-controlled machines. The relationship among investment in factory automation, employment, and flexibility of manufacturing operations is shown in Figure

Robots do only what they are told to do, no more and no less, and they never offer innovative ideas.

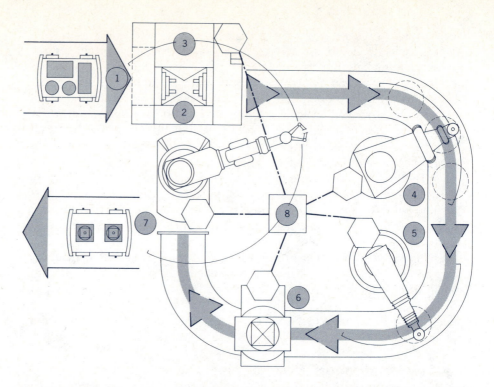

FIGURE 11.20 A robot cell in which parts are machined, assembled, welded, and inspected. Individual machines and operations are indicated by circled numbers. ① A computer-directed carrier brings raw materials. It has been loaded automatically in the storage area and is guided by low-frequency radio signals from wires buried in the floor. ② A robot unloads material and places it in the lathe. This "pick and place" robot also handles finished parts. ③ A lathe automatically selects the proper cutting tool and performs the prescribed operations. ④ An assembly robot puts the parts together, a more difficult task than machining. ⑤ A welding robot joins the parts, making as many welds as necessary on all sides. ⑥ The completed part is measured for deviations and inspected for imperfections. ⑦ Items that pass inspection are placed in a precise position on the automatic cart for transportation to a shipping area where they are automatically sorted and tagged for shipment. ⑧ A programmable controller directs the work through the microprocessors that control each machine. It determines the number and types of parts being made. Should something go wrong, red lights alert human "tenders."

11.21. At the left of the graph is a factory full of robots, highly flexible in terms of machine movement but inflexible compared with labor-intensive operations. The maximum use of machines to cut employment drastically is extremely expensive and severely reduces the flexibility of operations,

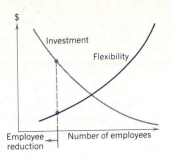

FIGURE 11.21 Relationships among investment, employment, and degree of flexibility in an automated factory. The high level of investment in automated equipment indicated by the dotted line greatly reduces the number of workers in the factory and correspondingly diminishes the flexibility of the system to respond quickly to unexpected events.

because human workers can do so many things impossible to duplicate with machines, including the most advanced robots. But the long-range effect on society as robots become more sophisticated could be enormous. Automated factories that can operate without humans and that can reproduce themselves could lead to an entirely new era in production systems—and in the history of civilization.

The social implications of robotization are discussed in Section 3-9.

11-9 SUMMARY

Time and money are the parameters by which machine performance and maintenance policies are measured. Time is the criterion for utilization, and money is the yardstick for investment comparisons.

Sequencing procedures establish the minimum time route for processing units through work stations. Convenient methods are available for two (products) $\times$ n (stations) and n $\times$ two patterns of process layouts. The sequences can be displayed on time charts. Line-balancing procedures are used to sequence n operations for one product in product layouts. A network is drawn to describe the problem. A minimum cycle time and number of work stations are set by the largest operation in the line. Then operations are assigned to work stations in conformance to the network's sequence restrictions, with the objective of minimizing idle time and maximizing proficiency.

Sequencing

Line balancing

Four methods of maintaining service are cyclic replacement of individual machines, group replacement of worn parts, standby machines, and preventive maintenance. The most economic machine replacement cycle is the age that minimizes the sum of life cycle operating and capital costs. Cost evaluations for the other three methods use a tableau relating the cost-to-failure pattern to alternatives—replacement period, number of standbys, or time between preventive maintenance checks. The preferred policy is the one with the lowest

Machine replacement policies

cost based on the expected value of expense associated with the failure probability distribution of each alternative.

Preventive maintenance

Other practical considerations for determining a maintenance policy include overdesign, larger in-process inventories, coordinated inspection schedules, training programs, disciplinary measures, detection of harmful physical conditions, and recognition for maintenance personnel.

Queuing theory

Waiting-line evaluations compare the cost of congestion with the cost of providing facilities to relieve congestion. Queuing theory is concerned with the properties and behavior of waiting lines. The language associated with queuing theory includes the following terms:

1. *Customer*—a person, machine, or object requiring service.

2. *Service*—action desired by the customer.

3. *Channels*—number of servicing facilities open to customers.

4. *Queue*—the line formed when customers are idle.

5. *Queue discipline*—manner in which customers are served.

Arrival and service time distributions

6. *Arrival distribution*—pattern of customer arrivals for servicing (λ = average number of arrivals per unit time).

7. *Service time distribution*—pattern of servicing times (μ = average number of servicings each channel can perform per unit time).

Queue discipline

The queuing formulas for the single-phase case illustrated in the chapter are based on the assumptions of a Poisson arrival distribution, exponential service-time distribution, an arrival rate less than the servicing rate, first in–first out queue discipline, and an infinite calling population with no defectors or deserters from the queue. Based on these assumptions, formulas are available for the probability of a certain number of customers being serviced or waiting to be served, the mean number of customers in the system, and the mean time required by a customer in the system.

Monte Carlo simulation

Simulation has become a practical management tool as a result of computers' data-generating and data-digesting capabilities. Monte Carlo simulation is a problem-solving technique that uses randomly generated numbers to experience the conditions of a problem. The method is particularly useful for problems that are difficult to formulate for analytic solutions. From the circumstances of the problem, a certain alternative is selected for the simulation experiment. Samples are taken from a relevant distribution by converting random numbers to influencing factors from the distribution. The samples are then inserted in the model being tested to determine their effect on the system. The process is repeated many times to increase the reality of and confidence in the results.

Random number generation

Factory automation is increasing, and the direction is toward computer-integrated systems. The information flow among machines is improved by the use of a standard communication system such as MAP—manufacturing automation protocol. A cellular manufacturing system groups the people, processes, and machines needed to produce a family of parts; this concept is embodied in group technology. When cells capable of producing a variety of family members are linked together, a flexible manufacturing system (FMS) is formed. FMS permits the inexpensive production of small batches, which accounts for its growing use. A robotic cell in an FMS provides maximum flexibility, a highly prized attribute, and consistent performance even under adverse conditions. But the wider use of robots awaits improvements in their mobility and sensing capabilities.

Group technology

Flexible manufacturing system

Robots

11-10 REFERENCES

ASHFAHL, C. R. *Robots and Manufacturing Automation*. New York: Wiley, 1985.

BOOTHROYD, G. C., C. POLI, and L. E. MURCH. *Automatic Assembly*. New York: Marcel Dekker, 1982.

GIFFIN, W. C. *Queueing: Basic Theory and Applications*. Columbus, Ohio: Grid Publishing, 1978.

GRADON, F. *Maintenance Engineering: Organization and Management*. New York: Wiley, 1973.

GRAYBEAL, W., and U. W. POOCH. *Simulation: Principles and Methods*. Cambridge, Mass.: Winthrop, 1980.

GROOVER, M. P. *Automation, Production Systems, and Computer-aided Manufacturing*. Englewood Cliffs, N.J.: Prentice-Hall, 1980.

HIGGINS, L. R., and L. C. MORROW. *Maintenance Engineering Handbook*. New York: McGraw-Hill, 1977.

MENIPAZ, E. *Essentials of Production and Operations Management*. Englewood Cliffs, N.J.: Prentice-Hall, 1984.

RIGGS, J. L., L. L. BETHEL, F. S. ATWATER, G. H. E. SMITH, and H. A. STACKMAN, JR. *Industrial Organization and Management*, 6th ed. New York: McGraw-Hill, 1979.

SKINNER, W. *Manufacturing in the Corporate Strategy*. New York: Wiley, 1978.

ULLRICH, R. A. *The Robotics Primer*. Englewood Cliffs, N.J.: Prentice-Hall, 1983.

WAGNER, H. M. *Principles of Operations Research with Applications to Managerial Decisions*, 2nd ed. Englewood Cliffs, N.J.: Prentice-Hall, 1975.

WHITE, J. A., J. W. SCHMIDT, and G. K. BENNET. *Analysis of Queueing Systems*. New York: Academic Press, 1975.

11-11 SELF-TEST REVIEW

Answers to the following review questions are given in Appendix A.

1. T F Most *machine tools* in use today are single- or special-purpose machines.

2. T F Inflexibility requires plants with *hard automation* to have large production runs in order to be economical.

3. T F Johnson's rule reveals the minimum *sequencing* time for any number of jobs to pass through any number of machines.

4. T F *Line balancing* is associated with a job-shop layout.

5. T F A *group replacement* plan is most commonly applied to identical high-cost items that lose value very slowly.

6. T F A *life cycle replacement* evaluation determines the age at which the average cost of ownership plus operation is minimized.

7. T F   *Preventive maintenance* is not a very glamorous job, but it is important and easy to do well.

8. T F   According to *queuing theory,* doubling the number of service facilities will reduce the waiting by half or more.

9. T F   The statistical distributions assumed for arrivals and service times are *Poisson* and *negative exponential,* respectively.

10. T F   The most commonly assumed *queue discipline* is first come–first served.

11. T F   *Computer simulation* is a technique by which a computer simulates actual problem conditions so as to determine the ideal solution.

12. T F   *Monte Carlo simulation* uses a random number generator.

13. T F   *CIM* is a standardized communication system that lets machines converse electronically in a common language.

14. T F   Arranging machines to produce a family of parts is a central feature of *group technology.*

15. T F   The greatest advantage of a *flexible manufacturing system* is its capacity to produce goods cheaply in small quantities.

16. T F   Highly *automated factories* employ fewer workers but are less flexible than are factories that use mostly operator-controlled machinery.

11-12 DISCUSSION QUESTIONS

1. Under what conditions are special-purpose machines preferred to general-purpose machines? In which category do industrial robots fall?

2. How do standardization of product parts and economy of scale (magnitude of production) contribute to lower production costs?

3. One observer has defined *man* as a carefully engineered control mechanism weighing about 160 pounds, with five senses, capable of producing circular as well as linear motions, completely self-contained, self-lubricated, and self-powered (except for refueling three times a day). Furthermore, no other control device yet invented is so readily produced by inexperienced labor.

In light of this definition, how does a human worker compare with automation labor?

4. Why is it necessary to assume sufficient in-process storage space and insignificant in-process inventory costs to apply "Johnson's rule"?

5. Why is a perfect line balance (no idle time) the-oretically ideal but less than perfect in practice? How do the human factors mentioned in Chapter 8 relate to assembly-line balancing?

6. Is the assumption justified that fatigue and learning need not be considered in assembly-line balancing? Why?

7. Why is it necessary to assume insignificant setup costs and uniform fixed costs when applying the transportation method to machine loading?

8. What effect, if any, would you expect from applying discounted cash flow analysis to the values in Table 11.4 to determine the most economical cycle time for an asset?

9. Why is energy conservation, as mentioned in Example 11.6, a criterion for the success of a preventive maintenance program?

10. Nearly everyone follows some variety of preventive maintenance for his or her personal automobile. Comment on why the practice is so common when few owners actually know whether it saves time or money, as compared with making repairs as needed. How does a preventive maintenance policy differ for a single machine versus ownership of a number of the same machines?

11. How can maintenance personnel be recognized for good work and protected from abuse when they are not at fault? Base comments on the motivational and supervisory considerations of Chapter 8. Could practical standards be developed for maintenance personnel?

12. How can a disciplinary policy affect a preventive maintenance program?

13. State a production example for each of the waiting-line structures described in Figure 11.16. For instance, a barbershop could be a multiple-channel, single-phase case.

14. Why is μ assumed to be greater than λ in queuing formulas?

15. Plan and describe a mechanical device that could simulate a single-channel, single-phase queue structure. Let marbles be the customers and their arrival and servicing rates be set by tapes with holes punched at prearranged intervals. A marble falling through a hole in the arrival tape represents a customer for the servicing station. Similarly, a marble passing through the servicing tape represents a customer served.

16. What are the practical limitations to solving waiting-line problems by mathematical formulas?

17. Why are tables of random numbers more widely used for Monte Carlo simulation than for gambling devices such as dice, cards, or coins?

18. How does the number of trials in simulated sampling affect the accuracy of results?

19. Your company daily averages $12,000 in cash receipts and $10,000 in cash disbursements. Both are normally distributed with a standard deviation of $2000 for receipts and $1000 for disbursements. How could simulation be used to help predict your average cash balance at the end of the day, maximum cash on hand, and the duration of bank loans to allow prompt payment of disbursements? Could the problem be treated as an inventory evaluation in which cash is the material? Discuss.

11-13 PROBLEMS AND CASES

1. A dump truck with a first cost of $30,000 has the following depreciation and service-cost pattern:

a. Assuming that no interest charges are necessary for the evaluation, how many years should the truck be kept in service before being replaced?

b. Assume that the truck presently owned is two years old. It is known that a truck will be needed for only six more years. When should another truck having the same cost pattern be purchased in order to minimize ownership and operating costs during the six years?

End of Year	1	2	3	4	5	6
Depreciation during year	$10,800	$7200	$4800	$1200	$1200	$1200
Service cost during year	$4800	$6600	$9000	$11,400	$14,400	$18,000

2. Find the sequence (or sequences) that minimizes the total elapsed time for the 10 jobs through the following two-machine process:

Job	A	B	C	D	E	F	G	H	I	J
Time on M1	7	3	10	8	13	9	5	11	7	10
Time on M2	6	5	15	7	12	12	2	8	5	11

3. Show the optimal schedule or schedules from Problem 2 on a time-chart format.

4. Use the graphical method to determine the minimum time necessary to process the two jobs through the five machines with the following time relationships:

J1:	Sequence	B	C	A	D	E
	Time	3	2	4	4	2
J2:	Sequence	D	C	B	A	E
	Time	4	5	3	3	3

Week	Probability of Failure during Week
1	0.3
2	0.1
3	0.1
4	0.2
5	0.3

5. On a time chart, show the optimal schedule from Problem 4 that delays the start of *J*2 as long as possible without increasing the minimum total processing time.

6. A line-balancing problem is represented by the network in Figure 11.22. Determine the assignment of operations to work stations that provides the minimum possible cycle time. Arrange the assignments to prorate the idle time as evenly as possible among the work stations.

7. Assume that the sequential relationships of the network in Problem 6 still apply but that operations 1,4; 4,7; and 5,8 must be assigned to the same station. Complete the remaining assignments of operations to work stations to minimize the cycle time.

8. A data-processing firm is considering a policy of replacing certain key electrical components on a group replacement basis instead of making repairs as needed. There are approximately 100 parts of one type that have the mortality distribution shown below. The cost of replacing the parts on an individual basis is estimated to be $1 per part; the cost of group replacement averages $0.30 per part. Compare the average weekly cost of the two replacement alternatives.

9. Operators working in a "clean room" environment use magnifying equipment that has a breakdown pattern following an arithmetic progression where on 50 percent of the shifts no equipment fails, 25 percent one fails, 12.5 percent two fail, 6.25 percent three fail, and so on. The 30 operators now have five standby machines. The cost of each standby is $40 per shift; the cost of lost production and servicing for a down machine averages $300 per shift. There is also a lost time cost of $30 to get the replacement machine in position when a breakdown occurs. Show the calculations that prove whether the present number of standby machines is the optimal number.

10. A plastic extrusion plant has 30 machines, each capable of producing any of the plant's product mix. An average of three machines undergo repairs, with a loss in production and service amounting to $400 per machine per day. The plant has sufficient space to mount standby equipment. The cost of each standby machine is $40 per day. The number of breakdowns has a Poisson distribution. How many standbys should the plant have to minimize total cost?

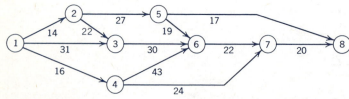

FIGURE 11.22

Number of Machines	Number of Hours between Major Breakdowns
5	3001–4000
20	4001–5000
90	5001–6000
120	6001–7000
135	7001–8000
70	8001–9000
50	9001–10,000
10	10,001–11,000

11. A maintenance manager has developed the following data for 500 machines that exhibit similar breakdown characteristics. The present policy is to give a complete overhaul whenever a major breakdown occurs. The cost of the overhaul and repairs is $350 per machine. If the preventive maintenance work were done on a scheduled basis, the overhaul cost would be reduced to $100 per machine. Individual repairs made between preventive maintenance periods are expected to average $150 per machine.

a. If the policy were to overhaul each machine after 8000 hours of service, how many would be expected to break down before the scheduled overhaul?

b. What is the average number of hours between breakdowns? Assume all breakdowns occur at the end of each interval.

c. What is the maintenance cost per 1000 hours of service for a machine when overhauls are made as breakdowns occur?

d. What is the maintenance cost per 1000 hours of machine service if a preventive maintenance policy on a scheduled basis is used? What interval between preventive maintenance checks provides the minimum cost?

12. Repairs in a large printing shop are handled by outside facilities. The service provided is not satis-

factory; therefore, an internal repair facility is planned. Major repairs average four per week and follow a Poisson distribution. The cost in lost production is estimated at $100 per week during each repair, including waiting and servicing. Which of the following three alternatives will provide the lowest total cost for repairs?

a. One repairman, paid $175 a week, who can complete an average of eight repairs per week.

b. One repairman, paid $120 a week, who can complete an average of five repairs per week.

c. Two repairmen with the wages and output given in alternative b.

Assume that the servicing times follow an exponential distribution.

13. Arrivals at an unloading dock follow a Poisson distribution with an average of three per hour. Trucks are unloaded at an exponential rate averaging four per hour. The cost of an idle truck is $20 per hour. The facility operates 2000 hours a year. If improvements to the dock could increase the servicing rate to six per hour, what would be the maximum investment that could be made for the improvements, in order not to increase the cost of unloading? The economic life of the improvements will be six years with no salvage value. A rate of return of 10 percent is required for all investments.

14. An average of 50 calls (following a Poisson distribution) is received per hour at the dispatching office of a taxi service. There are 10 taxicabs in the fleet. The average length of a trip is eight minutes, including travel to the customer, waiting, loading, and delivery time. Trip times follow an exponential distribution.

a. What is the probability that a taxi is idle?

b. It is estimated that the profit from each trip is $0.20 per minute of trip time and that the fixed cost of each taxi is $5 per hour. If the opportunity cost for a waiting customer is $0.10 per minute, would it be advisable to retire one taxi from service?

15. A service department has a constant service time of two minutes for all servicings. The following data on arrival rates have been observed:

Time Between Arrivals (Minutes)	Number of Occurrences
0–0.99	150
1–1.99	250
2–2.99	550
3–3.99	300
4–4.99	150
5–5.99	50
6 or more	50

a. Simulate 60 minutes to estimate the average length of the queue. Assume arrivals come at the beginning of each one-minute increment.

b. Compare the answer in part a. to a solution obtained by using a queuing formula selected from Section 11-6. How do you account for the difference?

16. Use Monte Carlo sampling to develop a solution to part a. of Problem 14.

17. Treat Question 19 quantitatively. Use 40 trials to determine a cash flow pattern when no money is borrowed or deposited outside the firm. Comment on the results with reference to risk and cash payment of bills.

18. An unhealthy taxi service has eight cabs. The company tries to keep six of the dilapidated cabs operating all the time, with the other two on standby or being repaired. When a cab is started in the morning, there is one chance in 10 that it will not operate (if it starts, it is assumed it will operate all day). If a cab does not start, two-thirds of the time it will be repaired to rejoin the fleet by the next day; there is a one-third chance that the repair will take two days (even after repair there is a 0.1 probability of the cab's not starting). From a simulation of 100 days, what is the probability of having six, five, four, or three cabs in operating order on any day?

19. *A Case of Work and Worker Satisfaction*
A 1977 study, conducted for the Labor Department by the University of Michigan's Survey Research Center, attempted to ascertain how satisfied Americans are with their employment. Previous surveys in 1969 and 1973 investigated the same subject. In a direct question that asked how satisfied people are with their jobs, 88 percent said they were at least somewhat satisfied, about the same percentage as in 1969 and 1973. However, study analysts reported that the responses to that question contradicted responses to indirect questions about job satisfaction. The survey showed a decline since 1973 in the percentage of people expressing satisfaction about their working environment, job security, interest in their work, financial compensation, and relations with co-workers. The lowest overall levels of job satisfaction were reported by people under 30, the semiskilled, and those employed in manufacturing.

Automation often bears the brunt of criticism regarding a decreasing quality of work life in manufacturing. However, some studies report that life on an assembly line is not so bad, and when alienated young workers proclaim their intention to shun tedious, unpleasant tasks, they blithely say, "Let the machines do them."

Considering the motivational considerations of job enrichment (Chapter 9), work design and pay considerations (Chapter 10), and the widely heralded need to improve productivity, present your views regarding automation, automatons, and work satisfaction in today's society.

20. *The Case of the Maintainability Mission*
A preponderance of production interruptions manifest themselves in mechanical problems. Five aspects of mechanical problems are as follows:

a. *"Fair wear and tear"* is the colloquialism for normal aging of equipment; old machines are increasingly likely to break down.

b. *Operator-avoidable damage* is estimated to cause 20 to 50 percent of all production interruptions.

c. *Weak original design* is a result of design compromises, inadequate design data, or lack of expertise.

d. *Operating beyond design limits* occurs when production schedules are pushed or operators become overeager and careless.

e. *"Bathtub mortality"* is a phrase that recognizes the familiar bath-tub reliability curve in which the life cycle of an item shows a very high failure rate at birth, followed by a long period of stability with few failures, and ends with an upsurging failure rate. This means that for some products there is a significant probability that an item will not work at all or will fail quickly, but if it survives the initial phase, it will likely have a long life.

(1) Consider the ownership of a motor vehicle in terms of the five mechanical-problem characteristics. Briefly discuss each characteristic with references to personal experiences with automobiles or what you have read about motor vehicle ownership (recalls, guarantees, and the like).

(2) What maintenance activities can be conducted to control mechanical problems? Briefly discuss some of the options available for each of the five categories.

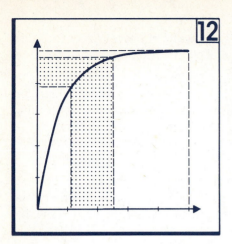

MATERIAL AND INVENTORY MANAGEMENT

LEARNING OBJECTIVES

After studying this chapter you should

- understand the flow of material through a production system from purchasing to finished goods inventory.
- be familiar with purchasing functions and procedures, including value engineering.
- appreciate the nature of material costs and the purpose of holding inventory when demand is independent.
- be able to calculate economic lot sizes using EOQ, EPQ, and risk models.
- be aware of expedient inventory methods, ABC analysis, and perpetual and periodic inventory management systems.
- realize the significance of material handling and control in production systems and be aware of the types of equipment available to move, store, and retrieve material.

KEY TERMS

The following words characterize subjects presented in this chapter:

purchasing function

centralized purchasing

lead time

value analysis, value engineering

supplier rating system

material management

capital/order/holding/opportunity cost

independent/dependent demand

economic order quantity (EOQ)

quantity discounts

economic production quantity (EPQ)

safety stock

ABC analysis

perpetual/periodic
inventory systems

material-handling
principles

automatic guided vehicle
systems (AGVS)

conveyor systems

work-in-process (WIP) inventory

automated storage and retrieval
systems (AS/RS)

12-1 IMPORTANCE

The interrelationships and cross-flow of objectives in production are nowhere more visible than in the measures taken to procure, store, and distribute materials. The system objective is to have the right materials in the right amount at the right place at the right time. Implementation problems arise in deciding which the proper materials are, how much of them is needed, how to get them there, and what the best time to act is. It is indeed a formidable undertaking to make such decisions for each of the multitude of items required to sustain production facilities and to provide the inputs that are transformed into product outputs.

"Like men with sore eyes: they find the light painful, while the darkness, which permits them to see nothing, is restful and agreeable."
Dio Chrysostom.

The functions of material management are grossly represented in Figure 12.1. The first three functions may be handled within one department responsible for the entire material system. In other organizations, the functions are fragmented, but the need for a comprehensive policy still remains. Guidelines are mandatory to resolve opposing dogmas, such as a desire by the production department to replenish stock by many small orders, as against purchasing's claim that the stock will be less expensive if purchased in larger quantities, and inventory control's disagreement with both.

Purchasing has the responsibility of getting the most value from supply expenditures. To do so, as indicated by the combined operation and inspection symbol in Figure 12.1, the merits of internal purchase requests are reviewed, and external trends are closely watched to determine the direction of price, service, and quality. In addition to its surveillance duties, it coordinates the administrative routines for issuing purchasing orders, following the progress of delivery, and paying for the material received.

Stored material acts as a buffer between the demands of production activities and purchasing or between stages of production. Supplies cannot always

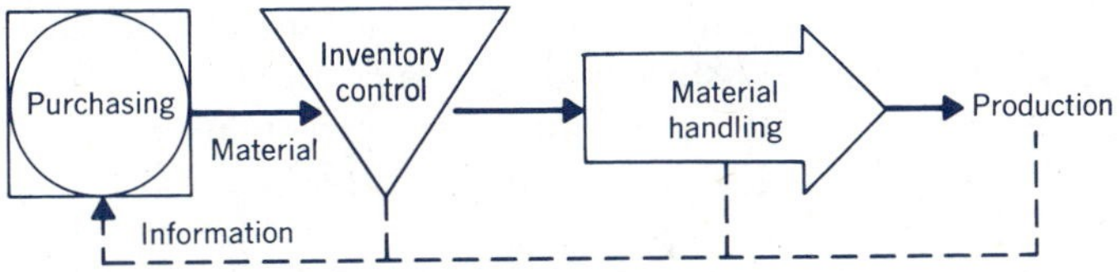

FIGURE 12.1 Flow and functions involved in material management.

be purchased or produced as needed; lead time is a normal delay in receiving purchased goods, and setup times are frequently required to get facilities ready to produce a desired item. Although the sense of maintaining an inventory of supplies is apparent, it is not obvious how much it will cost if the supplies are not available when needed, or which procedures to follow for recording, checking, and issuing material.

Determining the most effective means of transporting raw materials, parts, subassemblies, and finished products falls within the province of material handling. Many of the topics and techniques that we already have encountered are applicable to material handling: linear programming, equipment arrangement, methods and motion study, plant layout, and queuing theory. The attention devoted to material-handling problems is understandable from estimates that up to 80 percent of indirect labor cost in a plant, or 20 percent to 50 percent of total production cost, is consumed in transporting items from one place to the next.

The following sections treat the three material management functions in the order shown in Figure 12.1. This separation is more pedagogical than natural. To appreciate the function of a whole system, it is first necessary to comprehend the functioning parts. Each stage of material management is irrevocably linked to the other stages and to the aggregate production process it serves.

The discussion of material management and its role in the production system is continued in Chapter 13.

12-2 PURCHASING

The **purchasing function** is the interface between a company and its suppliers. From the supplier side of the interface, the company is viewed as a customer. Accordingly, the company is catered to by the sales forces of the vendors and is susceptible to their marketing strategies. On the other side of the interface, purchasing functions as a monitor, clearinghouse, and pipeline to supply materials needed to maintain production. The operating units feed in their requisitions; the requests are reviewed and converted to orders; and the filled orders flow back to restock production supplies.

Purchases are roughly divided into two categories, maintenance supplies and raw materials. Spare parts, replacement tools, new machines, office supplies, and housekeeping provisions are routine but inevitable purchases. The main question is how much to keep on hand: a temporary shortage of cleaning compounds has far less impact on production than does the absence of a critical replacement part for a production machine. "Raw" materials may be truly unrefined substances such as ore, oil, or plants, or they may take the form of subassemblies, manufactured components, or even complete products that are retailed by a mail-order company. For such items the buyers rely heavily on forecasts. They juggle forecasted demands for material against the expected price and delivery time for supplies.

Purchasing function

Purchasing is a service function that supports the activities of other operations.

In turn, it receives assistance from other operating units. Effective functioning requires a steady and reliable flow of information between concerned departments. The relationship of the purchasing structure to other parts of the production system is displayed in Figure 12.2. The dotted lines represent the interchange of information, and the solid lines show the movement of cash and material.

Marketing information provides an indication of anticipated production output. Reliable forecasts are needed because it takes considerable time to review and process requisitions, select suppliers, issue purchase orders, and obtain delivery. Advance clues allow purchasing to shop for the best price. By watching price trends, inventories of items likely to increase in cost can be built up in advance. The speculative practice of "hedging" is possible for materials carried on an organized commodity exchange. Purchasing and marketing also work together in reciprocity agreements, arrangements made to purchase materials from a vendor who reciprocates by buying products produced by the purchaser. These "you scratch my back while I scratch yours" deals can work out nicely, but care should be given to the danger of being at the mercy of an exclusive, complacent supplier.

Production is the terminal point for most material flow and is the initiating point for most material requests. Two age-old customs color the dealings between purchasing and production. The first is a "squirrel complex" which leads production supervisors and managers to hoard supplies. This protective policy certainly limits the chance of work delays owing to a shortage of parts or materials, but it builds a big inventory that is subject to damage, loss, and obsolescence. The second custom is a "brand X complex," a preference for a particular brand that has previously provided good service. Again, there are legitimate bases for these feelings because past performance is an indicator of future satisfaction. The troublesome aspect is that new products are continually being developed, and a loyalty to one brand eliminates the chance to recognize equal

> "Hedging" is buying and selling commodity "futures." Fearing a commodity price increase, a firm can buy at a given price, usually higher than the present price but lower than the anticipated future price. The quantity purchased is to be delivered at a *future* date.

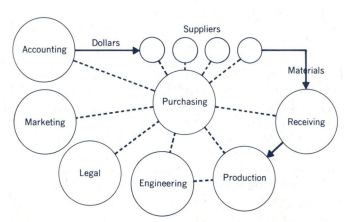

FIGURE 12.2 Information and material flow in the purchasing function.

or even higher quality at a lower price. Though it is easy to scoff at such attitudes in other people, another guilty person is visible in most mirrors.

Engineering personnel also are influenced by a brand X complex. Their preferences stem from the reputation of a supplier and are reinforced by the services rendered by the preferred vendor. When these options and actions lead to competent product design and competitive prices, they are eminently acceptable. But when they represent a path of least resistance for writing specifications, they are questionable. Purchasing agents, as contacts for the firm with sales personnel from prospective suppliers, often are aware of new, lower-cost materials before engineering departments are. The agents usually do not have the technical knowledge to evaluate new material developments, but they should have them appraised by those who do. Distribution of appropriate literature or meetings arranged between engineers and suppliers can produce new ideas and lower procurement costs.

Legal aspects require attention whenever binding agreements are made between the company and supplier. Standard forms, already checked for legality, are available for most routine purchases. The larger, longer contracts, in which special conditions are incorporated, rate a thorough legal review to ensure that both sides understand their responsibilities. Company lawyers can also assist purchasing by interpreting new legislation or existing laws applicable to fair pricing, misrepresentation, rebates, freight rates, and similar subjects.

Receiving personnel report the quantity and quality of supplies received. Purchasing uses this information to appraise the service of suppliers. When shipments are late, purchasing initiates contacts with vendors to determine the state of progress. Reimbursements or allowances are negotiated for shipments damaged in transit. After a shipment is accepted by receiving, it is uncrated and moved to production facilities or to inventory storage areas.

Accounting pays for the shipments after notification of their acceptance. Prompt notification is necessary to take advantage of cash discounts awarded for quick payments. Actions taken to speed deliveries and to secure reimbursement for damaged shipments should be reported because accounting also handles the internal paperwork of inventory records, invoice checking, and other financial details for material transactions.

Purchasing functions are usually coordinated by a purchasing department under the auspices of finance or marketing. There appears to be a trend toward grouping purchasing along with other material-oriented functions within a single material management department with a top-level standing. The intricate network of information channels and the high dollar volume of activity make purchasing a prime candidate for system synthesis. By combining material procurement with control, many communications lines would be shortened, and purchasing policies would likely achieve greater strategic effectiveness.

Centralized purchasing, as practiced by most companies, is a step toward systemization. Passing all orders through one office allows consolidated purchasing. Buying in greater quantities leads to cash discounts. Purchasing agents can specialize in certain accounts and negotiate more effectively with suppliers.

However, centralization tends to reduce flexibility and may be less sensitive to the local needs of geographically separated operations. Frequently, these disadvantages are reduced by central purchasing of high-cost, high-volume supplies while allowing local control of small or rush orders. The decentralization of authority for some purchases does not dilute the system principle if the wielders of authority follow system objectives. In fact, placing controlled authority at the point of responsibility can make believers out of system skeptics.

Purchasing procedures

A purchasing cycle begins with a decision to buy material and ends when the material is accepted by the unit that instigated the order. Purchasing responsibilities extend from one end point to the other and include many procedures in between. Some of the more important operations are described by the following sequence of purchasing activities:

1. *Receive requisitions.* Purchase requisitions are made out by personnel from all functional areas of the firm. The requisition forms include information as to what and how many items are wanted, when they should be available, and who is making the request. A column for "quantity on hand" is frequently included to force the requisitioner to check whether the items are truly needed.

   The elapsed time between placing an order and receiving it is known as **lead time.** It plays a significant role in purchasing and inventory control. Most purchasing departments urge requisitioners to anticipate material demands well ahead of actual need. Early requests act as buffers for unexpected delivery delays. Allowing a long lead time also means larger inventories must be kept to carry operations over the longer wait between ordering and receiving. The practical consequence of allowing longer times for delivery seems to be that present lead times just grow to take up whatever slack is allowed. Perhaps this is due to the "squeaky wheel principle": buyers who expect the shortest lead times complain the loudest when deliveries are late and thereby receive the most attention from suppliers. Requisitioners certainly should be aware of a minimum lead time but should attempt to correct suppliers' delivery delays instead of automatically increasing allowed lead times.

2. *Review requisitions.* **Value analysis** and **value engineering** are generic terms used to describe a study of the functions materials are supposed to accommodate. From a purchasing viewpoint, value analysis represents a relatively recent change from concentrating on finding the best price for a certain item to finding the lowest cost for any item that will satisfy an intended function. The analysis answers such questions as Could a less expensive material serve the same function? Is the function necessary? Could it be eliminated? Could other items serve the same function? Can

Creative substitutes such as new alloys and synthetics and unusual designs such as spiral nails and nylon lock fasteners have recorded huge savings.

they be simplified? Could the supplier reduce the price by a cooperative redesign or revised specifications?

Purchasing usually does not have the authority to substitute or modify materials designated by operating units. It does have the responsibilities to question requisitions and to suggest alternatives that would lead to better prices. When the originators know their requisitions are going to be scrutinized, more attention is normally devoted to their preparation.

A value engineering approach identifies three types of value, each having a distinct relationship to cost and function:

Use value—a monetary measure of the qualities of an item that contribute to its performance.

Esteem value—a monetary measure of the properties of an item that contribute to its salability or desirability of ownership.

Exchange value—a monetary measure of the qualities and properties of an item that enable it to be exchanged for something else.

Incentive clauses to promote the use of value engineering are often included in government contracts. Savings from the value program are split at a given rate such as 90 percent to the government and 10 percent to the contractor.

The equation that guides purchasing is

$$\text{Use value} + \text{esteem value} = \text{exchange value}$$

As applied to the purchase of stationery, the equation would require assigning a dollar value first for any writing material and then placing a price on the worth of accessories such as fancy letterheads or colored paper. The exchange value is the amount that must be paid to satisfy the other functions; it is a value established by comparison and no other means.

The most attention is devoted to the use value. One way to pinpoint the use function is to force a description by two words, a verb and a noun. For instance, the primary function of a shipping container is to "protect contents." A secondary function could be to "create impression" or to "explain contents." The purpose of functional definitions of use is to focus study on frivolous or secondary functions that may unnecessarily increase cost.

Six "whats of value engineering":
1. What is it?
2. What does it do?
3. What does it cost?
4. What is it worth?
5. What else will do the job?
6. What does that cost?

The intent underlying the surface techniques of value engineering is to foster a creative questioning attitude. Purchasing has the knowledge of the suppliers' offerings and competitive prices. Engineering has the know-how for technical comparisons. Operating personnel know what services they need and the practical limitations for substitutions. It takes a cooperative effort to identify new ways to do something, to evaluate them, and then to get them accepted by the users. Some practices that discourage the recognition and acceptance of change are (1) the inability or refusal to gather all the facts, (2) the failure to explore all possible ways to perform a function, (3) decisions made on what is believed to be true rather than what is known to be true, and (4) habits that were formed in the past and attitudes that keep them from changing.

3. *Select suppliers.* Sources of supplies originate from sales personnel contacts, ads in trade journals, descriptions of products in buyer's guides, correspondence, inspection tours of plants, and experience with a supplier's products. From such sources an *approved supplier list* is developed. The list results from rating vendors as to the quality of their products, prices, services, and delivery reliability. An approved list is usually drawn up for each class of supplies. Then a purchasing agent has only to contact a few acceptable suppliers to obtain quotes on price and delivery.

A complete, up-to-date list of suppliers is instrumental in obtaining better prices and services. New names should be continually sought, and the performance of existing suppliers should be continually monitored because poor suppliers can improve and better suppliers can become careless. Some suppliers offer exceptional technical assistance, training programs, equipment-borrowing privileges, and other inducements. When services of this kind will be helpful, the appropriate vendor should be known. Freight rates for some purchases may swing the total cost advantage from a distant supplier with low prices to a closer supplier with slightly higher prices. Most companies distribute large orders among two or more suppliers to provide a competitive check on the major supplier (one supplier often receives 50 to 70 percent of the orders) and to protect against delivery defaults stemming from mismanagement, strikes, fire, or natural catastrophes.

The possibility of a firm's being its own supplier occurs in a "make or buy" decision. Purchasing represents the "buy" side of the question. The "make" side was considered in Section 4-3.

4. *Place orders.* The normal purchasing routine of processing individual item requisitions is modified for very large, very small, and continuous purchases. Major acquisitions of unique machines and custom production facilities are *one-of-a-kind contracts.* Bids are often solicited for the entire investment—design, construction, and installation. Negotiations are conducted with several potential suppliers and usually extended over a considerable period because specifications for the asset are seldom confirmed until the supplier's resources are explored. The development of new designs is frequently a cooperative effort between the staffs of the buyer and the seller.

The purchase price of low-cost, infrequently needed items may be less than the cost to process a purchase order. Processing costs are typically $8 to $40 per order or even higher. The absurdity of incurring processing costs greater than the amount of the order is obvious. Small organizations often have a petty cash fund for petty purchases. More commonly, an open account is established with a supplier who inventories many minor items occasionally required. The supplier keeps track of direct orders and periodically bills the buyer. Purchasing negotiates the original *open contract* and monitors payments to keep the practice from getting out of control.

Individual purchasing orders are avoided for items in continuous demand by means of a *blanket purchase order.* This contract differs from an open contract in that the orders are generally predictable and for homogeneous items. A price for the items may be negotiated annually, with deliveries made on request during the year. The list price is charged for each delivery, but a discount based on the total annual quantity is usually obtained at the end of the year. Both the buyer and seller receive benefits from the agreement. The buyer can negotiate at one time for a substantial portion of annual supplies, and the order-processing cost is reduced. The seller, assured of a market for his or her products, can reduce advertising and other selling expenses.

5. *Monitor orders.* Important or lengthy orders are routinely checked by purchasing to determine whether anticipated deliveries are on schedule. Production difficulties of the supplier and change orders from within his or her own firm can occasionally put the purchasing agent in an uncomfortable position. Production-control colleagues blame late delivery on the agent's lack of follow-up. The supplier blames order changes concocted by the agent's production-control colleagues for the delays. The purchasing agent replies to both that there is not enough time to monitor every order with every supplier, especially when so many are marked "CHANGE–RUSH."

Such sensitive situations are not uncommon and merely underscore the coordination problems inherent in material management. Rescheduling is occasionally unavoidable. Disturbances are allayed by a give-and-take attitude. Internal production schedules can sometimes be altered; at times the supplier can fall behind on some deliveries so as to concentrate on more important demands. The key to coordination is to keep open channels of communication, both within the firm and to the supplier. Purchasing acts as the switchboard.

Acceptance sampling applicable to the inspection of shipments is discussed in Section 15-5.

6. *Receive orders.* Receipt of a contracted quantity in an acceptable condition is a signal to complete the purchase transaction. Records of the purchase are consolidated, and payment is made. The final price is possibly subject to a discount—trade, quantity, or cash.

The boast "I can get it for you wholesale" is an example of a *trade discount,* a price level determined by the classification of the buyer such as manufacturer, wholesaler, or retailer. *Quantity discounts* are based on the number of items ordered. The justification for price reductions on larger orders results from the decreased cost of selling, shipping, handling, and record keeping. *Cash discounts* are awarded for the prompt payment of bills. A policy such as "2/10 net 30" means the purchaser can deduct 2 percent of the list price if the order is paid for within the first 10 days of the month following receipt of the bill. Payment is expected within 30 days, even if the discount is not used. Offering cash discounts is justified by the time value of money considerations. Failure to take advantage of such discounts is difficult to justify; discount calculations are the pleasant aftermath of prudent purchasing.

EXAMPLE 12.1 Two sides of the purchaser's supplier rating system

The National Association of Purchasing Agents outlined a cost-ratio plan for numerically rating suppliers in their pamphlet "Evaluation of Supplier Performance" (New York, 1963). This **supplier rating system** attempts to attach a dollar value to each of four major procurement factors: price, quality, delivery, and service. The supplier who can consistently provide the required material at the lowest net value cost is, in theory, the most frequently selected. The procedure for determining the net value cost is summarized as follows:

Step 1. Net delivery price
$$= \text{list price} - \text{discounts}$$
$$+ \text{ freight cost} + \text{insurance, taxes, etc.}$$

Step 2. Quality cost ratio
$$= \frac{\text{material quality costs}}{\text{total value of purchases}}$$

The material quality costs are taken from past quality reports on purchases made from each supplier. These expenses include the cost for laboratory tests, incoming inspections, processing inspection reports, handling and packaging rejects, spoilage and waste, and manufacturing losses. Most of these costs are prepared by production and quality control departments. The yearly trend of the ratio indicates whether quality levels are being maintained or improved by the supplier.

Step 3. Acquisition cost ratio =
$$\frac{\text{acquisition and continuity costs}}{\text{total value of purchases}}$$

The denominator of the equation is the same as in Step 2. The numerator is derived by the purchasing department from the cost of sale negotiations, communication tolls, surveys, premium transportation, monitoring, and progress reporting.

Step 4. Delivery cost ratio =
$$\text{acquisition cost ratio} + \text{promises-kept penalty}$$

The cost of deliveries later than promised is expressed as a percentage of the total value of purchases delivered.

Step 5. Service cost ratio =
$$\frac{\text{maximum possible rating} - \text{supplier rating}}{\text{maximum possible rating}}$$

(A supplier rating below a given level, such as 60, automatically makes the ratio = 1.0.) Service costs are determined from absolute ratings of special considerations that suppliers offer with products and services. These ratings are converted to a penalty percentage and charged against the supplier lacking the considerations. The following list illustrates how a supplier service cost ratio of $(100 - 70)/100 = 0.3$ could be obtained:

Maximum Points	Category	Supplier Rating
	Competence and Ability:	
15	Product development and advancement	11
15	Product leadership and reputation	9
10	Technical ability of staff	9
10	Capacity for volume production	8
10	Financial solvency and profitability	8
	Attitudes and Special Considerations:	
5	Labor relations record	2
10	Business approach	8
5	Field service and adaptability to changes	2
10	Warranty conditions	6
10	Communication of progress data	7
100	Total Points	70

Step 6. Net value cost = net delivery price
$$+ \text{ (net delivery price}$$
$$\times \text{ sum of ratios from Steps 2, 4, and 5)}$$

The comparison of suppliers is based on their present net delivery price modified by additional costs expected from the history of their past performance. For instance, a net value cost for one supplier could result in the following price and penalty pattern:

Step 1. Price quote ($114,300) − discount (10% × $114,300) + freight ($600)

= net delivery price ($103,470)

Steps 2–5. Sum of ratios = quality cost (2.2%) + delivery cost (1.2%) + service cost (0.3%) = 3.7%

Step 6. Net value = $103,470 + ($103,470 × 0.037)

= $107,300

Such ratings are then compared for each order so as to identify the preferred supplier.

An interesting viewpoint is developed by reviewing the rating system as it appears to the supplier. The most pertinent source of information by which suppliers can evaluate their products and services is the response of the purchaser. A purchasing policy is dedicated to maximizing returns from supply expenditures—at least that is what suppliers must believe if they are to use their sales as a criterion for improving their competitive position.

Suppliers recognize that few purchasing agents will give 100 percent of their orders to one vendor, but there is some maximum percentage they will give. The difference between this maximum percentage and the percentage being supplied is a measure of the opportunity cost for a supplier. Assuming that a purchaser will reveal the maximum percentage, the supplier can see what potential sales are available from each buyer. The potential is an indication of how much the supplier can afford to spend to make his or her product and services more attractive to the purchaser. Then the problem narrows to determining what must be done to improve performance enough to obtain more orders.

The supplier rating scale pinpoints exactly where the purchaser believes the supplier is lacking. However, this information is typically considered proprietary. There are sound competitive reasons for keeping current ratings confidential. There are also sound financial reasons for revealing at least enough information to show a supplier his or her relative standing in each area considered important: the information can guide the supplier in the effort to improve performance, and the result of the effort will give purchasers more of the qualities they desire. The health of the supplier can contribute to the wealth of the user—and vice versa.

12-3 INVENTORY CONCEPTS

Hiring more people than needed for current operations in anticipation of receiving a major contract creates a labor inventory. The shortage of some skills, such as engineering, has been blamed on stockpiling.

Inventory, in a production context, is an idle resource. The resource can be animate or inanimate. Most commonly it is production material: tools, purchased parts, raw materials, office supplies, products in-process, and so forth. That the resource is idle does not mean it is serving no purpose. It is available when needed. It serves as an insurance policy against the unexpected breakdowns, delays, and other disturbances that could disrupt ongoing production. Insurance is not free. The idle resource can be damaged or become obsolete before it serves any purpose. The task is again to secure an economic balance between the cost of loss and the cost of preventing it.

By the early 1900s, formulas were developed to analyze inventory problems, but it was not until the 1940s that the theories were widely put into practice. Inventory problems are natural candidates for formal analysis. The problem area is common to all industries, its costs warrant detailed attention, and few intangible considerations are embraced. These factors all lead to neat mathematical formulations for general situations. When the situations are more specific, and consequently more realistic, more sophisticated formulations are needed. We shall investigate the basic formulas and some of their more frequently encountered refinements for inventory control, or **material management,** when the demand for items is *independent* of the production process.

Dependent demand conditions for materials are discussed in Chapter 13.

Inventory functions

It would be nice if we could completely exclude the human factor from inventory considerations. We cannot. Inventories serve many functions. The people associated with each function would prefer an inventory policy that first satisfied their pet function. The opposing nature of preferences is shown in Table 12.1.

The most important function of inventories is insulation. A reserve of material can be tapped whenever a delay in a preceding stage threatens to curtail operations in the following stage. The stages stretch the length of the production cycle, from initial inputs to delivery of final output. Material buffers are used to cushion the production process from the uncertainty of material deliveries, to decouple progressive stages of product development from disruptions in earlier stages, and to provide a steady supply of finished output for the unsteady demands of customers.

The apparent functions of an inventory policy obscure the many subtle ways in which it affects operations throughout a production system. Daily and seasonal work loads are stabilized by inventory. Stable work requirements allow workers to establish consistent work patterns: incentive wages are ineffective, and supervising is more difficult when work loads fluctuate from day to day. By smoothing out the peaks and troughs of customer demand through inventory

The importance of reliable sales forecasting is conspicuous in deciding inventory size.

TABLE 12.1
Conflicting Objectives for Material Management

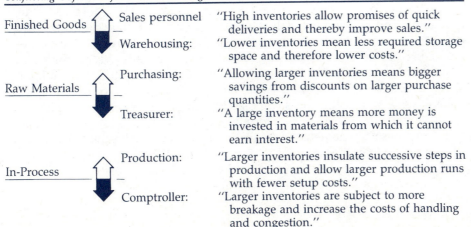

Finished Goods	Sales personnel	"High inventories allow promises of quick deliveries and thereby improve sales."
	Warehousing:	"Lower inventories mean less required storage space and therefore lower costs."
Raw Materials	Purchasing:	"Allowing larger inventories means bigger savings from discounts on larger purchase quantities."
	Treasurer:	"A large inventory means more money is invested in materials from which it cannot earn interest."
In-Process	Production:	"Larger inventories insulate successive steps in production and allow larger production runs with fewer setup costs."
	Comptroller:	"Larger inventories are subject to more breakage and increase the costs of handling and congestion."

FIGURE 12.3 Effect of buying down and speculative purchases on inventory levels.

Queuing theory is directly applicable to many inventory problems. For instance, an inventory of cars in the showroom for "demonstrators" could be the number of servicing channels, and the arrivals could be potential buyers.

The model for determining the optimal number of standby machines also fits the queuing–inventory pattern.

Inventory accumulation in anticipation of a strike is an example of stockpiling. The rationale followed is that the stockpile will allow continued production during supplier's labor–management strife and that the prices before negotiations will probably be lower than afterward.

buildups during slack sales periods, a relatively constant work force can be maintained. This avoids frantic hiring during rush periods and damaged worker morale or community relations during layoff periods.

The less obvious effects of finished-product inventories are similar to the considerations for servicing facilities in queuing theory. The "idle resource" definition for inventory could easily accommodate the policy of providing extra servicing stations for emergency service. The behavior of arrivals to or in a waiting line is about the same as that of customers in a retail store. If the items they want from the store are not on hand, they may go elsewhere to make purchases instead of waiting for the retailer to order a supply. Items on display may create a surface "want" when the deeper "need" is still embryonic. This impulse buying is akin to a customer's deserting a waiting line to try the service at a competing station with no waiting line. Such considerations were excluded from queuing formulas by assumption, and they are treated in the same way by most inventory formulas. But the potential cost of vindictive human behavior is still very real.

Inventory policy is closely related to purchasing policy. Most prices fluctuate above and below a general trend line. A policy of "buying down" is an attempt to make purchases whenever the price dips below the trend line. Because purchasing has no control over the general price structure, its purchases will accumulate stock at uneven intervals and in unequal quantities, depending on market conditions (Figure 12.3). When it appears that a source of supply will be temporarily shut off or prices will rise sharply, large purchases at a favorable price will markedly increase inventory levels. As shown in Figure 12.3, a speculative purchase can sometimes facilitate impressive purchasing bargains and will always create impressive purchasing stockpiles. The apparent savings from these purchasing policies must be weighed against the inventory costs of storage, handling, depreciation, and interest charges.

Inventory costs

Costs must be assigned to the diverse inventory considerations to evaluate properly the merits of opposing functions. The more relevant costs and the symbols by which they are denoted in formulas are itemized in the following paragraphs.

PRICE (P)

The value of an item is its unit purchase price if it is obtained from an outside supplier, or its unit production cost if it is produced internally. The amount invested in an item being manufactured is a function of its degree of refinement. The value of a product during its initial stage of development is little more than the aggregate cost of collecting raw materials. As it progresses through the production cycle, it accumulates a share of the fixed cost of production facilities, direct and indirect labor costs for refining operations, and direct cost of material additions. The price per piece of outside purchases also can vary as a function of quantity discounts.

Normal production costs can be increased by overtime labor required to meet special rush orders.

CAPITAL COST (iP)

The amount invested in an item, or the **capital cost,** is an amount of capital not available for other purposes. If the money were invested elsewhere, a return on the investment would be expected. A charge to inventory expense is made to account for this unreceived return. The amount of the charge reflects the percentage return expected from other investments. The interest charged, i, is applied against the price P, to support a claim for the annual capital cost.

ORDER COST (O)

Procurement costs originate from the expense of issuing an order to an outside supplier or from internal production setup costs. **Order costs** include the fixed cost of maintaining an order department and the variable costs of preparing and executing purchase requisitions. Even when orders are delivered from other parts of the same company, order costs still apply. The same purchasing routine of checking inventory levels, issuing orders, follow-up, inspection, and updating inventory records pertains to internal procurement.

Setup costs account for the physical work incurred in preparing for a production run (setting up equipment and adjusting machines) and include the clerical costs of shop orders, scheduling, and expediting. External orders, internal procurement, and setup costs remain relatively constant regardless of the order size.

HOLDING COST (H)

Costs originating from many sources are consolidated under the heading of **holding cost.** One percentage or dollar value is usually placed on the conglomerate total to account for all the sources itemized as follows. In general, holding costs remain fixed to a certain inventory capacity and then vary with the additional quantity stored.

1. *Storage facilities.* Buildings have to be owned or leased to store the inventory. The expense includes the equivalent annual cost of the investment

if the facilities are owned or the rent, if leased, heat, lights, and property taxes.

2. *Handling.* The cost of moving items to, from, and within storage includes damages, wages, and equipment expense.

3. *Depreciation.* The change in value of an item during storage is caused by physical deterioration, mutilation and pilferage not covered by insurance, and obsolescence.

4. *Insurance.* A conservative policy is to insure goods during storage. The protection is usually based on the average dollar value of the inventory.

5. *Taxes.* Some states levy an inventory tax periodically during a year on the amount in storage at the time. Particularly in retail outlets, such as automobile dealerships, it is possible to manipulate inventory levels to have dips coincide with assessment dates.

OPPORTUNITY COST (OC)

Two types of costs are associated with running out of stock when there is still a demand for the product. The first is the cost of emergency measures to expedite a rush delivery. This cost is easily identified as the difference between the usual cost of procurement and the extra cost for accelerated service. The other cost, the **opportunity cost,** is much more difficult to divine because it involves people. When emergency procedures cannot provide a wanted item, the customer is left unsatisfied. The only apparent cost is the profit lost from the potential retail sale or the production lost. The reaction of a dissatisfied customer in terms of future business is a cost estimate of the roughest nature.

Some firms stress customer satisfaction to the point that they give greater value as a substitute for items that should have been available.

12-4 NATURE OF INVENTORY DEMAND

Answers to the two cardinal questions of inventory policy—*how much* and *when* to order—depend on the nature of inventory demand. As is implied in the preceding discussion of inventory functions and costs, *how much* to order is mainly a matter of acquisition and holding expense, whereas *when* to order depends on the purpose and usage rate of the material. The interplay of these factors is determined by the practical situation in which the material management occurs: whether material demand is independent or dependent.

An **independent demand** for an item is experienced when its demand is not significantly influenced by other items. There is always some degree of dependence among stocked items because they all compete for shelf or storage space and the attention of inventory managers, but there is no physical relationship among items when the demand is independent. Conversely, a **dependent demand** exists when an item is an integral part of another item and the integration ensues according to a production plan. Dependence is most obvious in a manufacturing setting in which components are combined to pro-

TABLE 12.2

Percentage Distribution of Inventories in Three U.S. Industries, As Averaged from Several Estimates

	Raw Material	In-process Material	Manufacturer's Finished Goods	Distributor's Finished Goods
Capital goods	60%	20%	20%	0
Garment industry	30%	55%	5%	10%
Consumer goods	5%	10%	30%	55%

duce a finished product and the requirements for components with respect to other components are fixed by design.

It is estimated that 50 percent of the value of all inventories is accounted for in raw materials and work-in-process in manufacturing. The other one half is in finished goods held by manufacturers, wholesalers, and retailers. A rough indication of inventory distribution in selected U.S. industries is shown in Table 12.2. It is apparent from these percentages that inventory managers in different industries are more concerned with certain types of inventories than with others. Manufacturers of capital goods focus on raw material because they produce primarily to fill specific orders. In the garment industry the in-process inventories are large, presumably to allow a quick response to changes in market preferences. Consumer goods are stocked heavily at distribution points to provide immediate delivery of finished goods to customers.

Models have been developed to analyze inventory practices for both independent and dependent demands. Order-quantity and order-point models are most closely associated with independent demand. Several of these models are described in the following sections. *Material Requirements Planning* (MRP) is a relatively recent methodology developed for high-volume, dependent demands. It is more of a coordination and scheduling technique than an economic ordering model and is consequently a computerized procedure. Because MRP is a synthesis of analysis and control procedures abetted by electronic assistance, it is described in Chapter 13 along with other new improvements for production systems.

"Pipeline" inventories include goods in transit, on trucks or in pipelines, and goods between work stations or distribution levels within a production system.

The models discussed in Sections 12-5 and 12-6 are appropriate when demand is independent and predictable.

12-5 INVENTORY MODELS ASSUMING CERTAINTY

Analyses of inventory costs recognize just two patterns: costs that vary directly with the size of an order and costs that vary inversely with order quantity. All of the costs we have considered fit into these two categories. Capital and holding costs increase as the order size increases because larger orders mean higher inventory levels. These "carrying costs" are decreased by ordering smaller amounts. For a given demand, ordering smaller amounts means that more orders must be placed. Placing more orders increases the total annual ordering cost. Because stock levels are allowed to dip more often when more orders are placed, there

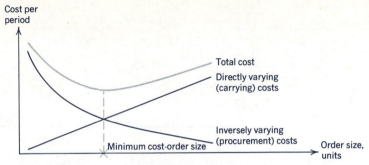

FIGURE 12.4 Cost-order size relationship.

are more chances to run out of stock, and opportunity costs consequently increase. The total inventory cost is then the sum of the carrying and procurement costs. The relationship of directly varying, inversely varying, and total costs is shown in Figure 12.4. Inventory models quantify the relationship to identify the order size that minimizes total cost.

The inventory models considered under an assumption of certainty are based on premises that greatly simplify their structure but diminish their reality. The following assumptions and their effects are applicable to all formulas in this section:

1. The total number of units required for one year is known exactly:

   Annual demand $= D =$ yearly usage of items

2. Demand is constant. The exact number of items needed during any time period is known when the usage rate is steady.

3. Orders are received instantaneously. This condition is not as absurd or restrictive as it first appears. It means an ordered quantity will be available when expected, and the corollary is that the lead time is known and constant. This assumption erases the possibility of opportunity costs' being incurred: if orders can be received instantaneously, there can never be an unfilled demand.

4. Ordering costs are the same regardless of order size. Similarly, setup costs are constant, and the rate at which products are produced is known:

   Manufacturing rate $= M =$ annual rate at which items can be produced

5. The purchase or production price does not fluctuate during the period considered, but the price can vary as a function of order quantity.

6. There is sufficient space, handling capacity, and money to allow the procurement of any quantity desired. The limiting condition is to order only once a year:

   Procurement quantity $= Q$

   $=$ number of items ordered each replenishment period

Then, $D = Q$ when all the material required for a year's operations is secured at one time.

Economic order quantity

The size of an order that minimizes the total inventory cost is known as the **economic order quantity, EOQ.** The usage and replenishment pattern for the EOQ based on the given assumptions is shown in Figure 12.5. The vertical lines indicate instantaneous receipt of an order size Q. A constant usage rate, represented by the sloping lines, takes the inventory level down to zero during the interval between orders, t. The average number of items in storage is $Q/2$.

The unknowns in the triangular pattern are the peaks—amount to order, Q, and the bases—time between orders, t. Both the procurement and carrying costs graphed in Figure 12.4 are a function of Q. The annual procurement cost is the number of orders submitted per year times the cost per order:

$$\text{Annual procurement cost} = O\frac{D}{Q}$$

$$= \text{cost per order}$$

$$\times \frac{\text{number of units demanded per year}}{\text{number of units per order}}$$

Assuming the handling cost and capital cost are based on the average inventory level, we have

Holding and interest costs are often combined into one percentage charge or a single monetary charge per item.

$$\text{Annual carrying cost} = (H + iP)\frac{Q}{2}$$

$$= (\text{holding} + \text{interest charge per unit per year}) \times \text{average inventory}$$

Combining the expressions, we get the formula

$$\text{Total annual inventory cost} = O\frac{D}{Q} + (H + iP)\frac{Q}{2}$$

When holding costs are based on maximum inventory, $2H$ replaces H in the carrying cost expression.

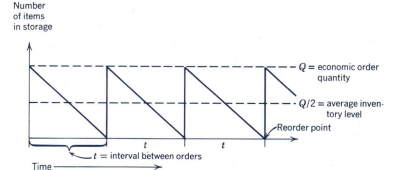

Number
of items
in storage

Q = economic order quantity

$Q/2$ = average inventory level

Reorder point

t = interval between orders

Time

FIGURE 12.5 Inventory pattern of instantaneous replenishment and constant demand.

Recalling the shape of the total cost curve from Figure 12.4, we know the slope is zero at the minimum point. Thus, one way to determine the value of Q that minimizes total annual cost is to differentiate the expression with respect to Q and set the differential equal to zero:

$$\frac{d}{dQ} \text{ (total annual cost)} = -\frac{OD}{Q_2} + \frac{H + iP}{2} = 0$$

Then solving the equation for Q, we get the *EOQ* formula

$$Q = \sqrt{\frac{2OD}{H + iP}}$$

Another method to obtain the same formula is to recall from Section 4-3 that a minimum (or maximum) occurs when the rates of change of two opposing trends in one function are equal. The marginal balance of direct and inversely varying costs for inventory occurs when procurement costs equal carrying costs. From this equality,

$$\frac{OD}{Q} = \frac{(H + iP)Q}{2}$$

we get

$$\frac{Q^2}{2} = \frac{OD}{H + iP} \quad \text{and} \quad Q = \sqrt{\frac{2OD}{H + iP}}$$

Several other statistics of interest can be calculated once Q is known. The number of orders to place in one year is given by the quotient of D/Q. If we assume there are 200 working days in a year, the number of working days between orders will be

$$\text{Order interval} = t = \frac{200}{D/Q}$$

The complete cost of stocking an item is calculated by adding the purchase price for a year's supply to the total annual inventory cost:

$$\text{Total annual stocking cost} = \frac{OD}{Q} + \frac{(H + iP)Q}{2} + PD$$

EXAMPLE 12.2 Calculation of an economic order quantity with no purchase discounts

The Moore-Fun Novelty Company buys 80,000 shipping containers each year. The following costs are applicable:

P = $0.40 per container

O = $80.00 per order

H = $0.10 per container per year

i = 15 percent, including a charge for taxes and insurance as well as interest

One warehouse is used exclusively to hold paper products. Because the space is not used for other storage when inventory supplies are low, holding costs are based on the maximum rather than average inventory level. The *EOQ* formula thus takes the form

$$Q = \sqrt{\frac{2OD}{2H + iP}} = \sqrt{\frac{2 \times 80 \times 80,000}{(2 \times 0.10) + (0.15 \times 0.40)}}$$

$$= \sqrt{49,230,769} = 7016$$

The number of orders to place in one year is

$$\frac{D}{Q} = \frac{80,000}{7016} = 11$$

and the time between orders, based on 220 working days per year, is

$$t = \frac{220}{11} = 20$$

The total annual stocking cost for shipping containers then becomes

$$\frac{OD}{Q} + \frac{(2H + iP)Q}{2} + PD = \frac{(80)(80,000)}{7016}$$

$$+ \frac{(2 \times \$0.10 + 0.15 \times \$0.40)7016}{2}$$

$$+ \$0.40 \times 80,000$$

which is

$$\$912 + \$912 + \$32,000 = \$33,824$$

Note the equality between procurement and carrying costs.

Quantity discounts

Price discounts are often offered by suppliers to encourage larger orders. Benefits for the purchaser from bigger orders include the reduction in unit price, lower shipping and handling costs, and a reduction in ordering costs owing to fewer orders. These benefits have to be measured against the incremental increase in carrying costs. As the order size increases, more space must be provided for storage, and the costs of holding the larger inventory level correspondingly increase. Another pertinent, though difficult to quantify, consideration is the risk of obsolescence or functional depreciation. Larger inventories magnify the loss that would result if design or demand changes made the stored supplies less valuable.

The lowest-cost ordering policy when price breaks are in effect is determined by calculating the total annual stocking cost for each *feasible* economic order quantity and the minimum quantity at which the **quantity discount** is allowed. A feasible order quantity is found when the calculated Q is within the price-break range of the P used in the calculation.

Fewer computations are needed for common quantity discount patterns when the procedures on page 467 are used. However, all likely order quantities should be checked whenever an unusual pattern is encountered.

1. Compute Q using the lowest unit price. If Q is feasible (Q is large enough to qualify for the lowest price), it is the optimal order quantity. Stop.

2. If Q is not feasible, calculate the total annual stocking cost for the minimum order quantity allowed at that price.

3. Compute Q using the next highest unit price.
   a. If Q is feasible, calculate the total annual stocking cost using this *EOQ*,

Typical quantity discount pattern.

EOQ calculations are expedited by tables or nomographs of Q as a function of frequently encountered carrying and procurement costs.

and compare it with the cost obtained in Step 2. The lot size producing the lower total cost is the optimal order quantity. Stop.

b. If Q is not feasible, repeat Steps 2 and 3a until the minimum cost solution is identified.

Any savings allowed by ordering larger quantities should be evaluated against the risks incurred from maintaining higher stock levels. Risks are gauged by the stability of past demand, resale value of stock, and market trends.

EXAMPLE 12.3 Calculation of order size when quantity discounts are available

A supplier from which the Moore-Funn Company (see Example 12.2) could obtain shipping containers has offered the following quantity discount schedule:

Price	Quantity
$P1$ = $0.40 per container	All orders up to 9999 containers
$P2$ = $0.36 per container	Orders from 10,000 to 19,999
$P3$ = $0.35 per container	Any order above 19,999 containers

Starting with the lowest price per container ($P3$ = $0.35), Q is computed as

$$Q3 = \sqrt{\frac{2 \times 80 \times 80,000}{(2 \times 0.10) + (0.15 \times 0.35)}} = 7120$$

Because the order quantity is below the number needed to qualify for $P3$ = $0.35 per container, the annual stocking cost from ordering the minimum lot size (20,000 containers) that qualifies for the lowest price break is computed as

Total annual stocking cost $= \dfrac{(\$80)(80,000)}{20,000} +$

$$\frac{(2 \times \$0.10 + 0.15 \times \$0.35)20,000}{2}$$

$$+ \$0.35 \times 80,000$$

$$= \$320 + \$2525 + \$28,000$$

$$= \$30,845$$

Step 3 is to calculate Q using the next highest price break ($P2$ = $0.36):

$$Q2 = \sqrt{\frac{2 \times 80 \times 80,000}{(2 \times 0.10) + (0.15 \times 0.36)}} = 7099$$

Again, Q does not fit within the range of the quantities required for the price used in its computation. Therefore, the total annual stocking cost for the minimum lot size (10,000 containers) required to use the $P2$ = $0.36 is calculated:

$$\frac{(\$80)(80,000)}{10,000} + \frac{(2 \times \$0.10 + 0.15 \times \$0.36)10,000}{2}$$

$$+ \$0.36 \times 80,000$$

which is

$$\$640 + \$1270 + \$28,800 = \$30,710$$

(Note that the reduction in carrying costs for $Q2$ = 10,000 compared with $Q3$ = 20,000 outweighs the added costs for purchasing and procurement.)

The total cost for using a lot size of 10,000 versus 20,000 containers is compared and discloses a savings of $30,845 − $30,710 = $135.

Finally, the highest purchase price alternative must be checked because a feasible Q was not calculated previously. In Example 12.2, the total annual stocking cost was found to be $33,824 for an order quantity of 7016 when $P1$ = $0.40 was the price assigned per container. The policy of ordering 10,000 containers at a time is considerably less expensive, $33,824 − $30,710 = $3114 savings. However, the preference for an order size of 10,000 containers should receive a final check with marketing and production to ascertain whether any changes are anticipated in container specifications, and with warehousing to be sure that the storage facilities can handle the larger level with no increase in per-unit carrying charges.

Economic production quantity

The conditions for the instantaneous replenishment of supplies are modified slightly when the supplies are manufactured on order rather than shipped from a stockpile of already manufactured items. The difference is that supplies are shipped as soon as they are manufactured. This means that the items are used during the replenishment period, as represented by the sloping lines rising from each reorder point in Figure 12.6.

The principal expense of procurement is the setup cost when a firm produces its own supplies. The inventory pattern in Figure 12.6 shows production beginning the moment the supplies on hand are exhausted. In practice, the reorder point would be set at some inventory level above zero to notify production that supplies soon would be needed. This lead time should allow sufficient leeway for scheduling the setup procedures.

The replenishment period, t', is the length of time required to produce the economic production quantity, *EPQ*:

$$t' = \frac{Q}{M} = \frac{\text{quantity ordered}}{\text{production output per day}}$$

When D and M are stated in daily rates, the inventory level increases each day during the replenishment period by the amount $M - D$. The stock on hand reaches its peak at the end of the replenishment period when

$$\text{Maximum inventory level} = (M - D)t' = (M - D)\frac{Q}{M}$$

$$= \left(1 - \frac{D}{M}\right)Q$$

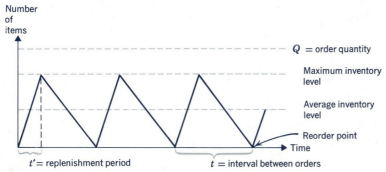

FIGURE 12.6 Inventory pattern for usage during a replenishment period.

Then,

$$\text{Average inventory level} = \left(1 - \frac{D}{M}\right)\frac{Q}{2}$$

which makes

$$\text{Total annual } EPQ \text{ cost} = \frac{OD}{Q} + \frac{(H + iP)(1 - D/M)Q}{2}$$

where Q includes setup costs and P is the production cost, and it leads to

$$Q = \sqrt{\frac{2OD}{(H + iP)(1 - D/M)}}$$

EXAMPLE 12.4 Calculation of an economic production quantity

One product produced by Moore-Funn Novelties is a voodoo doll. It has a fairly constant demand of 40,000 per year. The soft plastic body is the same for all the dolls, but the clothing is changed periodically to conform to fad hysterics. Production runs for different products require changing the molds and settings of plastic-forming machines, new patterns for the cutters and sewers, and some adjustments in the assembly area. The production rate of previous runs has averaged 2000 dolls per day. Setup costs are estimated at $350 per production run.

A doll that sells for $2.50 at a retail outlet is valued at $0.90 when it comes off the production line. Complete carrying costs for production items are set at 20 percent of the production cost and are based on the average inventory level. From these cost figures, the economic production quantity is calculated as

$$Q = \sqrt{\frac{2OD}{iP(1 - D/M)}}$$

$$= \sqrt{\frac{2 \times \$350 \times 40{,}000}{(0.20 \times \$0.90)(1 - 40{,}000/400{,}000)}}$$

$$= \sqrt{172{,}840{,}000} = 13{,}146$$

where

$$M = \frac{2000 \text{ dolls}}{\text{day}} \times \frac{200 \text{ days}}{\text{year}} = \frac{400{,}000 \text{ dolls}}{\text{year}}$$

Using the calculated Q value, production can anticipate

Number of production runs per year

$$= \frac{D}{Q} = \frac{40{,}000}{13{,}146} = 3$$

Length of production run, t'

$$= \frac{Q}{M} = \frac{13{,}146}{2000} = 6.6 \text{ days}$$

and warehousing can expect

Maximum inventory level

$$= \left(1 - \frac{D}{M}\right)Q$$

$$= (1 - 0.1)(13{,}146) = 11{,}831$$

12-6 INVENTORY MODELS RECOGNIZING RISK

The *EOQ* and *EPQ* analyses represent idealized versions of material flow. The idealization contributes to quick calculations and serves as a convenient reference condition. The basic formulas clarify the sensitivity of order size to estimating errors for the interacting variables; for instance, errors in demand forecasts are less significant than might be supposed because the order quantity has a square root relationship to demand, and furthermore, the total cost is relatively unresponsive to small changes in Q.

Additional insights into material flow are gained from a familiarity with inventory models that recognize risk. Some of the restrictions imposed for conditions of certainty are relaxed to make the risk models more realistic. Many different models are available to fit specific situations. We limit our attention to one model for treating single inventory orders and to another for a continuous inventory policy. Each solution illustrates considerations common to many related inventory problems.

Single-order inventory policy

A grocer stocking perishable fresh produce, a clothing buyer selecting seasonal merchandise for a ready-to-wear department, and a production planner placing parts orders for an untried product all face essentially the same risk: How large should the unique, one-of-a-kind order be when the demand is unknown? A large order protects against the opportunity costs of running out of stock, and a small order minimizes the loss for products that cannot be sold.

The first step in analyzing the problem is to estimate the probable demand. In a one-only situation, there are usually few data available for estimating purposes. However, the original decision to act in the unique manner must have been made on some assumptions of success. A typical recourse is to forecast the probability of discrete blocks of demand, such as the increments and probabilities shown in Table 12.3.

TABLE 12.3
Distribution of Demand

Demand, y Units	Probability of Selling Fewer Than y Units
100	0.00
200	0.20
300	0.50
400	0.90

TABLE 12.4
Inventory Pattern for a Single-Period Sale

	Possible Demand			
Order Size	100	200	300	400
100	0	− 100	− 200	− 300
200	100	0	− 100	− 200
300	200	100	0	− 100
400	300	200	100	0

By letting the increments of demand be the order size alternatives, the effect on inventory from following each alternative can be itemized for each possible future, as shown in Table 12.4. The symmetrical tableau has a diagonal of zeros that represent the ideal order size for each demand. To the right of the zeros are negative quantities showing the amount that could have been sold if units were available—stockout quantities. To the left of the zero diagonal are the quantities left over from each order size when demand fails to meet expectations—depreciated quantities. The economic effect of these quantities is evaluated by assigning stockout and depreciation costs associated with shortages and oversupply.

By allotting a stockout cost of $0.50 for each unit (or $50 per 100 units) and a $1 depreciation charge for each unit purchased but not sold, the tableau of Table 12.4 is converted to the expected value format shown in Table 12.5. Thus, an oversupply of 300 units (lower left entry) is represented by a loss of $300, and a 300-unit shortage (upper right entry) is a $150 loss. The incremental probability for each demand level is taken from the cumulative totals in Table 12.3. The expected costs are calculated as the sums of the probability of each demand level multiplied by the cost resulting from that demand for each alternative order size. The alternative with the lowest expected cost is the preferred inventory policy. From Table 12.5, it is an order size of 200 with an expected cost of ($100 × 0.20) + (0 × 0.30) + ($50 × 0.40) + ($100 × 0.10) = $50.

A much shorter though less descriptive method of solving a single-order model is to use the ratio of stockout cost to the sum of stockout plus depreciation

TABLE 12.5
Expected Value of Stockout and Depreciation Costs for Order Size Alternatives

Order Size	Demand: Probability	100 0.20	200 0.30	300 0.40	400 0.10	Expected Cost
100		$0	$50	$100	$150	$70
200		100	0	50	100	$50
300		200	100	0	50	$75
400		300	200	100	0	$160

costs to indicate the demand probability associated with the preferred order size. Specifically,

$$P(y) \leq \frac{\text{stockout cost}}{\text{stockout cost} + \text{depreciation cost}}$$

where $P(y)$ is the cumulative probability of demand *less than or equal* to a level that will produce the minimum-cost order size. From the cumulative probabilities in Table 12.3 and the cost data used to develop Table 12.5, the order size is directed by

$$P(y) \leq \frac{\$0.50}{\$1.00 + \$0.50} \leq 0.34$$

to the largest demand with a cumulative probability less than or equal to $0.34:200$ units. This demand level is then equated to a preferred order size and, fortunately, agrees with the order size previously determined in Table 12.5.

A more general version of the same technique is

$$\sum^{y^*} P(y) \leq \frac{P - C}{P - S}$$

where

P = purchase price
C = wholesale or production cost
S = salvage value
y^* = demand level just below the preferred order increment

EXAMPLE 12.5 Order size for a single inventory buildup

Most single-order inventory situations are associated with perishable products. A decision as to the number of Christmas trees to stock during the short selling season could produce a situation in which the salvage value would be negative. A retailer buys trees at a delivered cost of $2 each and sells them at an average price of $5. Any trees left over after the selling season cost $0.50 each for removal. The expected demand based on previous years and salted with optimistic guesswork is

Then,

$$\sum^{y^*} P(y) \leq \frac{P - C}{P - S} = \frac{\$5.00 - \$2.00}{\$5.00 - (-\$0.50)} \leq 0.55$$

The order size is 400 trees, the largest demand level with a probability of sales less than or equal to the cost ratio ($0.10 + 0.15 = 0.25$). If the retailer could get free disposal of unsold trees, he should raise his order size to 500 trees.

Sales of N trees	200	300	400	500	600	700	800
Probability of N sales	0.10	0.15	0.35	0.20	0.10	0.05	0.05
Probability of selling fewer than N	0.00	0.10	0.25	0.60	0.80	0.90	0.95

Continuing inventory policy

The risk of running out of supplies continuously in demand is created by variations in the usage rate and the replenishment lead time. As displayed in Figure 12.7, three conditions contribute to stockouts after a replenishment order has been placed: accelerated demand, extended lead time, or a spurt in demand coupled with a delivery delay. The way to avoid running out of stock is to hold a buffer supply beyond the amount consumed by average usage during an average lead time. This **safety stock** obviously increases the holding cost. The problem thus centers on determining a safety-stock level that balances the opportunity costs of stockouts against the carrying costs for the extra stock in storage.

We now consider the case in which demand is relatively stable but lead times vary. The techniques used for the analyses can also be used for the case of constant lead time but variable demand. The most practical method for evaluating the case in which both demand and lead times vary is by simulation.

The approach for determining a lead-time safety stock for continuing inventory is very similar to that developed for a single-order policy. The expected patterns of lead-time probabilities, unit opportunity costs, and carrying costs for stored items are needed. The following data will be used to illustrate the calculations:

A continuous distribution for demand and lead times could be used as shown in Figures 12.8 and 12.9.

$$\text{Ordering cost} = O = \$60 \text{ per order}$$

$$\text{Holding cost} = H = \$8 \text{ per unit per year, based on the average inventory level}$$

$$\text{Opportunity cost} = OC = \$5 \text{ per day per item demanded but not available}$$

$$\text{Average demand} = D = 10 \text{ units per day or 2000 annually}$$

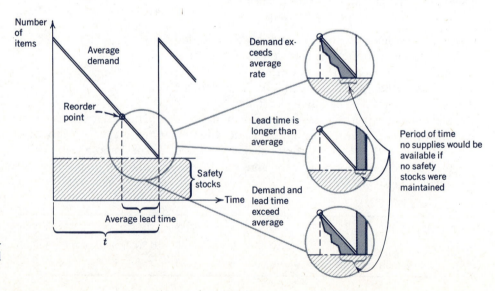

FIGURE 12.7 Causes of stockouts and the buffer provided by safety stock.

The distribution of lead times shows there is a 10-percent chance for the shortest delivery time of 5 days, but it can take up to 11 days to receive delivery after an order is placed. The cost incurred from running out of supplies is $5 per unit × 10 units per day = $50 per day. Conversely, the cost of holding an extra day's supply in a safety stock that is not used increases the holding cost by $8 × 10 = $80 per year. The expectation of incurring each of these costs is the basis for selecting the optimal allowed lead time. This selection in turn establishes the reorder level, which is another way of stating the safety-stock level.

Shortages and overages resulting from different lead-time alternatives could be shown in a tabular form analogous to Table 12.4. For each day the actual lead time exceeds the lead time provided, a shortage of 10 units is incurred, and 10 units accumulate for each day the lead time is shorter than anticipated. We bypass this tableau to enter the costs directly for each "lead time provided"–"lead time required" match in an expected value format, as shown in Table 12.6.

The expected value for each lead-time alternative is calculated in the usual manner as the average of costs weighted by their probability of occurrence. For instance, the seven-day lead-time alternative has an expected opportunity cost of

$$E(OC) = (\$50 \times 0.20) + (\$100 \times 0.15) + (\$150 \times 0.10)$$
$$+ (\$200 \times 0.05) = \$50$$

and the carrying cost expectation is

$$E(CC) = (\$160 \times 0.10) + (\$80 \times 0.15) = \$28$$

The opportunity costs to the right of the zero diagonal are isolated in a separate column from the carrying costs to the left of the zeros, because each affects total inventory cost differently. The opportunity costs occur only at the end of an order period, whereas carrying costs go on through the entire year. Therefore,

Lead Time, Days	Relative Frequency
5	0.10
6	0.15
7	0.25
8	0.20
9	0.15
10	0.10
11	0.05

Opportunity costs are difficult to estimate. Fortunately, the total inventory cost is not very sensitive to OC estimates. For instance, doubling OC estimates does not change the preferred lead time alternative from the 10-day preference shown in Table 12.7.

TABLE 12.6
Expected Costs for Lead-Time Alternatives in a Continuous Inventory System

Lead Time Provided	Lead Time Required							Expected Value	
	5 0.10	6 0.15	7 0.25	8 0.20	9 0.15	10 0.10	11 0.05	Carrying Cost(CC)	Opportunity Cost(OC)
5	0	50	100	150	200	250	300	0	$132.50
6	80	0	50	100	150	200	250	$8	87.50
7	160	80	0	50	100	150	200	28	50.00
8	240	160	80	0	50	100	150	68	25.00
9	320	240	160	80	0	50	100	124	10.00
10	400	320	240	160	80	0	50	192	2.50
11	480	400	320	240	160	80	0	268	0.00

TABLE 12.7
Total Cost of Lead-Time Alternatives

Lead-Time Alternative (LT)	Order Size (Q)	Total Inventory Cost
5	310	$2482
6	272	2181
7	235	1904
8	206	1717
9	187	1621
10	177	1606*
11	173	1654

*Minimum cost policy.

opportunity costs are treated as an addition to order costs in the *EOQ* formula, and carrying costs are considered a safety-stock expense that is added as an increment to the total annual inventory cost formula. These conditions dictate the procedure for determining the lowest-cost lead-time policy:

1. Calculate Q using the formula

$$Q_{LT} = \sqrt{\frac{2(O + OC)D}{H}}$$

where

Q_{LT} = order size for a given lead-time alternative
OC = expected value of opportunity costs for the lead-time alternative

2. Calculate the total annual inventory cost for each lead-time alternative from

$$\frac{\text{Total}}{\text{inventory cost}} = \frac{(O + OC)D}{Q_{LT}} + \frac{HQ_{LT}}{2} + CC$$

where CC = expected value of carrying costs for the lead-time alternative. Using the lead-time alternative $LT = 7$ as an example,

$$Q_7 = \sqrt{\frac{2(60 + 50)2000}{8}} = 235 \text{ units}$$

$$\frac{\text{Total}}{\text{inventory cost}} = \frac{(60 + 50)2000}{235} + \frac{8(235)}{2} + 28$$

$$= 936 + 940 + 28 = \$1904$$

Steps 1 and 2 are repeated for all the lead-time alternatives to complete the array of costs shown in Table 12.7. The minimum at the 10-day alternative means an order should be placed when the inventory level drops to 10 days × 10 units per day = 100 units. Comparison of this level to a policy of providing just enough units to meet the average lead time, 7 days, shows that a safety stock of 30 units will save $1904 − $1606 = $298 a year.

END OPTIONAL MATERIAL _____

Expedient inventory policies

Selection of a preferred single-order inventory policy requires considerable computation. The techniques illustrated for these computations are rather crude compared with the more complete mathematical formulations developed for some problems. For instance, visualize the complexity that would be introduced for the lead-time alternatives if we had not assumed that demand was steady during lead time. Also consider how many different items may be included in one order. For example, ordering and carrying costs for the single housekeeping item "screws" could be determined without too much difficulty. But what is the individual demand for wood screws, brass screws, steel screws, short screws, long screws, and so on? Estimating the huge number of items that could be analyzed with the complexity of thorough analyses is such a formidable task that even the most ardent disciples of quantitative solutions recognize the need for abridged methods.

The items in an inventory that deserve more attention are classified in Section 12-7.

It is not surprising that a wide variety of expedient measures are recommended by practitioners. Several routines are described next; many more versions are available from the references at the end of the chapter. Although the versions vary, all recognize the risk imposed by usage variations during lead time.

One alternative is to ignore risk and to use the conditions of assumed certainty. Then the reorder point equals the product of the average demand and the average lead time.

1. *Ultraconservative method.* Multiply the largest daily usage ever incurred for an item by the longest delivery time ever subjected by the supplier. The result is a huge reorder level that is as close as possible to a foolproof guarantee of never running out of stock. An item would have to be almost indispensable to the operation to afford the disproportionate carrying costs.

2. *Safety-stock percentage method.* Carry a safety stock equal to average demand times average lead-time times a percentage factor. A 25- to 40-percent safety factor is typically applied. If the average daily demand were 10 units, the average lead time were nine days, and a 30-percent factor were used, the reorder point would be the average usage during lead time plus the safety stock:

$$\text{Reorder point} = (10 \times 9) + (10 \times 9)0.30 = 117 \text{ units}$$

3. *Square root of lead-time usage method.* Experience indicates that the lead time seldom varies from its normal length by more than the square root of that length. This relationship can be used directly to set a safety-stock level when the demand is fairly constant: safety stock = square root of

Daily Demand (D units)	Number of Days Demand Occurred
6 or less	3
7	15
8	57
9	75
10	66
11	54
12	18
13	9
14 or more	3

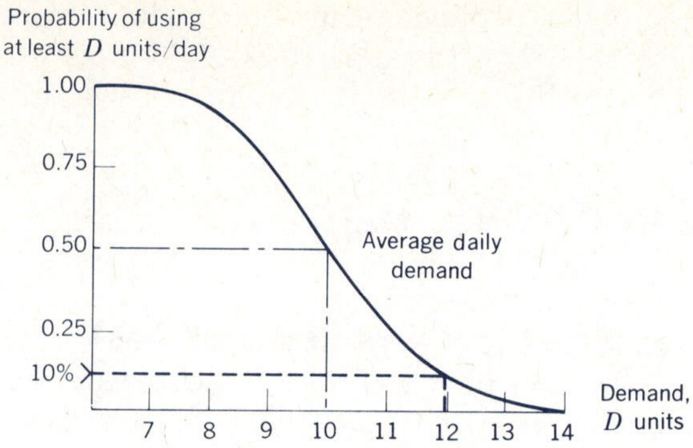

FIGURE 12.8 Cumulative distribution for the probability of daily demand.

average usage during lead time. For the conditions introduced in method 2 ($D = 10$ per day and $LT = 9$),

$$\text{Reorder point} = (10 \times 9) + \sqrt{10 \times 9} = 100 \text{ units}$$

4. ***Demand percentage method.*** Plot past records of daily demand on a cumulative distribution graph, as shown in Figure 12.8. Decide what percentage of time stockouts can be incurred without seriously damaging operations. Spot this percentage on the vertical axis, read across to the curve, and pass directly down to the horizontal axis to determine the demand associated with the acceptable stock percentage. Then multiply this demand by the average lead time to set the reorder level. For the pattern marked in Figure 12.8, a demand of more than 12 units will occur only 10 percent of the time. Allowing a 10-percent risk of stockouts for a fairly constant lead time of nine days will produce

$$\text{Reorder point} = 12 \times 9 = 108 \text{ units}$$

Because the mean of the distribution is 10 units per day, the calculated point is equivalent to a safety stock of

$$(12 \times 9) - (10 \times 9) = 18 \text{ units}$$

The Z factor used in PERT calculations (Section 7-5) serves a purpose equivalent to its use in safety-stock calculations.

The standard deviation is
$$\sqrt{\frac{\Sigma(x - \bar{x})^2}{n - 1}}$$
where

x = daily demand
$\bar{x}$ = mean demand
n = number of days demand occurred

5. ***Combination method.*** Methods 2 and 3 are combined by calculating a safety-stock quantity as the product of a factor giving the desired probability of being out of stock, the standard deviation of demand variation, and the square root of the average lead time. The probability factor (Z) is the standard normal deviate corresponding to the percentage of times per year a stockout is allowable. For instance, if annual demand is 2000 units and $Q = 200$, there will be 10 order periods per year. A customer service policy of only one stockout per year would allow 10 percent of the replenishment periods to reach a zero stock level before delivery. From the table in Appendix B of probabilities for the normal distribution, a probability of 0.10 yields a Z factor of 1.28. Then assuming that the standard deviation is 1.1

units, as developed from Figure 12.8, and the average lead time is nine days,

$$\text{Safety stock} = 1.28 \times 1.1 \times \sqrt{9} = 5 \text{ units}$$

and the reorder point is $(10 \times 9) + 5 = 95$ units

6. *Simulation.* Although simulation is a more demanding exercise than are the other five methods for considering risk, it also provides a closer approximation of actual conditions. More factors can be included in a simulation model, such as costs that increase geometrically with late deliveries. Example 12.6 describes the use of random numbers to sample different demands during lead time, and Table 12.7 shows how the simulated trials suggest a preferred reorder level.

Which method to use to calculate the reorder point is, of course, a management prerogative. The goal is to obtain a "reasonable" procedure for implementing an inventory policy: a reasonable method takes a reasonable effort to achieve a reasonable balance between risks and the related carrying cost burden.

EXAMPLE 12.6 Application of Monte Carlo simulation to determination of an inventory safety-stock level

A new product is being stocked. Experience suggests that the daily demand rate will follow a normal distribution, with a mean of 200 units and a standard deviation of 20 units. Lead times have a narrow range of from 7 to 11 days distributed as shown:

Lead time, LT	7	8	9	10	11
Probability of LT	0.15	0.20	0.40	0.15	0.10

Monte Carlo sampling is used to investigate a safety-stock level for a fixed-order inventory policy.

Two cumulative distribution patterns are developed as shown in Figure 12.9. The cumulative normal distribution comprises three standard deviations on either side of the mean. The values are adapted from the areas under a normal curve given in Appendix B. Thus a $-3Z$ value is associated with a demand of $200 - (3 \times 20) = 140$ and a probability of occurrence of 0.001.

A random number table is used to enter the distributions. From each block of five numbers, the first two are used to select a lead time, and the last two represent the demand. A separate block could be used for each entry, or four consecutive numbers in one block could suffice. The important concern is not to impose a pattern by repeatedly using the same set of five numbers or skipping around a table to select blocks that "seem" more random. A safe approach is to extract numbers in a consistent, noncyclic manner—by row or column, diagonally, up or down, and the like.

Each simulated lead time and demand are multiplied together to set a demand during lead time, as shown by column 3 in Table 12.8. Column 4 is the average reorder level (1800) resulting from the product of the average lead time (nine days) and the average demand rate (200 per day). The difference obtained by subtracting column 3 from column 4 is the net effect on the inventory level after each replenishment period.

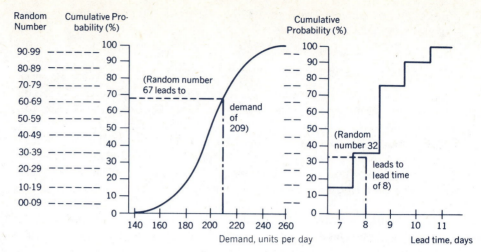

Random Number		Cumulative Probability (%)
90-99	----	
80-89	----	
70-79	----	
60-69	----	
50-59	----	
40-49	----	
30-39	----	
20-29	----	
10-19	----	
00-09	----	

FIGURE 12.9

Given an initial random number block of 32867, the lead time obtained from entering the distribution at 32 is eight days. The last two digits of the block, 67, lead to a demand of 209. These two values are shown on the cumulative distribution charts and are entered in the first row of the record-keeping table. Then the entry for column 3 is 209 × 8 = 1672, which leads to the value for column 5 of 1800 − 1672 = 128.

The 10 trials show inventory levels varying from +617 to −763 before replenishments are received. For this limited sample, increasing the reorder point to 1800 + 763 = 2563 would have eliminated all stockouts, at the added expense of carrying a safety stock of 763 units. A much larger sample would provide sound bases for setting a reorder level that limits stockouts to a figure deemed acceptable with respect to customer service and holding cost.

TABLE 12.8
Record of Monte Carlo Simulation Trials

Random Number LT D	Lead Time (1)	Daily Demand (2)	Demand during Lead Time (1) × (2) = (3)	Reorder Level (4)	(4) − (3) (5)
32 8 67	8	209	1672	1800	128
43 1 11	9	176	1584	1800	116
38 9 47	9	198	1782	1800	18
71 6 84	9	220	1980	1800	−180
14 6 06	7	169	1183	1800	617
82 2 44	10	197	1970	1800	−170
55 8 47	9	198	1782	1800	18
94 0 95	11	233	2563	1800	−763
11 7 49	7	200	1400	1800	400
69 9 02	9	159	1431	1800	369

12-7 INVENTORY MANAGEMENT

Lot size and reorder point calculations are the most spectacular aspects of inventory management. Once the calculations are complete, the continuing routine commences for checking deliveries and physically keeping count of the amount on hand. It is easy to cast these problems off as paper-shuffling procedures, but a well-designed record-keeping system can contribute as much value as can elaborate quantity specifications. More pointedly, the quantity policies are worthless exercises of logic without the physical controls to implement them.

Priorities

It is obviously uneconomical to devote the same amount of time and attention to inconsequential items and to vital supplies. This widely applicable concept has become famous as "the Pareto principle," named after the Italian economist Vilfredo Pareto. In simple terms, it says that a few activities in a group of activities, or a few items in a group of items made, purchased, sold, or stored, account for the larger part of the resources used or gained. Its application to inventory policy recognizes that a small number of production supplies accounts for the bulk of the total value used.

The division of inventory into three classes according to dollar usage is known as **ABC analysis.** The usage rating for each item is the product of its annual usage and its unit purchase or production cost. The typical pattern of dollar usage is depicted in Figure 12.10. The *A* class, on which attention is concentrated, includes high-value items whose dollar volume typically accounts for 75 to 80 percent of the material expenditures but represents only 15 to 20 percent of the quantity volume. The proportions are reversed in passing from *A* to *C* items.

A selective treatment of inventory items directs formal analyses to areas where they will do the most good. The information required to develop a value-quantity distribution curve is usually easy to obtain. The plotted values guide progressive analyses: the most important items are treated first, and successive evaluations are conducted as time allows. The general effect is to "buy" analysis

The Pareto principle translated into general management functions concentrates on a few important tasks that should receive the most skillful treatment because those functions produce the most good in the organization.

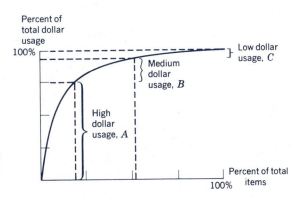

FIGURE 12.10 Distribution of supply expenditures with respect to quantity supplied.

time for the high-volume items by overstocking the low-value items. More specific procedures are noted next.

A ITEMS

Order quantities and order points are carefully determined. Procurement costs and usage rates are reviewed each time an order is placed. Tight controls are applied to stock records and to lead-time developments.

B ITEMS

EOQ and reorder-level calculations are conducted, and the variables are reviewed quarterly or semiannually. Normal controls and good records are expected to detect any major changes in usage.

C ITEMS

No formal calculations are made. The reorder quantity is usually a one- or two-year supply. Simple notations record when replenishment stocks are received, but no attempt is made to keep a running account of the stock level. A periodic review, perhaps once a year, physically checks the amount in storage.

The supply of *A* and *B* items is controlled by subjecting them to a perpetual or periodic inventory-control system.

Perpetual inventory system

A **perpetual inventory system** keeps a running record of the amount in storage and replenishes the stock when it drops to a certain level, by ordering a *fixed quantity*. Each time a withdrawal is made, the amount is subtracted from the previous level on a stockcard to portray accurately the quantity still on hand. In some larger, more modern establishments, stock records are kept by a *real-time* computer system; the withdrawal amount is immediately fed to a computer in which the current status of *A* and *B* items is maintained. The computer signals when a reorder point is reached and may even be programmed to change the *EOQ* when the demand pattern appears permanently altered.

A "two-bin" system is a version of the fixed-order plan in which one bin holds a quantity equal to the reorder level, and the second bin holds the difference between *Q* and the reorder point. Items are used from the second bin until it is empty. This signals the need for a replenishment order. During the lead time for resupply, withdrawals are made from the other bin holding the expected lead-time usage quantity plus a safety stock. The cyclic procedure eliminates the need for stockcard entries, but it requires control to be sure that withdrawals are not made from the reserve bin until the other is empty and that purchasing is notified when the reserve bin is first used.

An actual second bin is not always required. Especially when the material is protected by a controlled environment, the reorder point quantity is separated from normal stock by a simple attention-getting partition such as paper.

Periodic inventory system

In a **periodic inventory system,** the number of items in storage is reviewed at a fixed interval: weekly, monthly, and so on. The intervals essentially follow the *ABC* concept: items with high dollar usage are checked more frequently than are C items. After each review, an order is placed. The size of the order depends on the rate of usage during the period between checks. This variable order size is designed to bring the stock level near a maximum desired number, such as the *EOQ* plus a safety stock:

Order quantity = *EOQ* − present inventory
+ usage during lead time + safety stock

Thus the order size is larger when the demand between reviews is high.

The fixed-order interval system is especially suitable for inventory situations in which there are many small stock withdrawals and the order costs are low. A department store is a good example: the inventory on the shelves and racks is checked visually at frequent intervals. Large orders made up of many different items may be placed at one time with the warehouse. The visual checks are much more practical than is recording individual withdrawals, and the periodic large orders reduce trucking charges. A combined order may also allow quantity discounts. The weaknesses of the system are caused by human fallibilities of not making the periodic checks on time and not finding all the stock present because it has been mislaid or stored in more than one location.

Backorders, when present, would also be subtracted in calculating the order quantity.

Expecting customers to mark withdrawals in a department store is, of course, absurd. It is about as bad to expect workers to record withdrawals of routinely used items such as typing paper, nails, and punch cards.

Comparison of perpetual and periodic systems

The operations of the two systems are best illustrated by example. Assume that the annual demand for a class *A* item is 20,000 units. Procurement cost is $20 per order, and carrying costs are $0.05 per unit per year. Then

$$Q = \sqrt{\frac{2 \times 20 \times 20,000}{0.05}} = 4000 \text{ units}$$

Further assuming that there are 200 working days per year and the average lead time is 10 days,

$$\text{Average usage during lead time} = \frac{D}{200} \times LT$$

$$= \frac{20,000 \text{ units}}{200 \text{ days}} \times 10 \text{ days}$$

$$= 1000 \text{ units}$$

and

$$\text{Reorder interval} = \frac{200}{D/Q} = \frac{200}{5} = 40 \text{ days}$$

Based on this information, we can follow the ups and downs of inventory levels subject to each ordering system as they react to variable demands.

The inventory patterns for both systems during a 120-day working period are shown in Figure 12.11. For the moment, we shall let the safety stock, abbreviated *SS*, be some indefinite quantity that fully protects against stockouts.

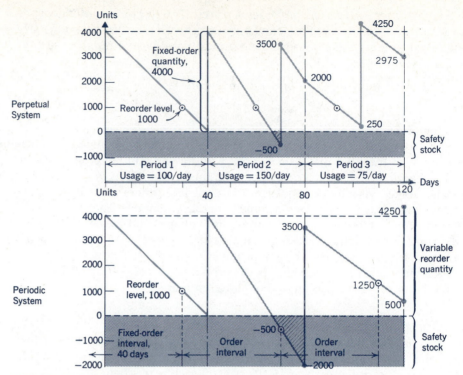

FIGURE 12.11 Stock-level patterns for identical demands placed on perpetual and periodic inventory systems.

The span of time includes three replenishment periods of 40 days each. The effects of three different usage rates during the period are discussed next.

REPLENISHMENT PERIOD 1—USAGE RATE OF 100 UNITS PER DAY

Both the perpetual and periodic systems exhibit the same inventory pattern when the demand follows the expected average. The fixed-order quantity is 4000 units ordered when the stock level reaches 1000 units plus safety stock. The fixed-order interval is 40 days measured from one reorder point to the next. The desired maximum inventory level for both systems is 4000 units plus safety stock.

REPLENISHMENT PERIOD 2—USAGE RATE OF 150 UNITS PER DAY

The fixed-order quantity is placed after $(4000 - 1000)/150 = 20$ days. During the lead time, the stock level drops to $1000 - (150 \times 10) = -500$ units, or equivalently, 500 units of safety stock are withdrawn. An order of 4000 units brings the stock level up to 3500 units plus safety stock on day 70. Then, during the last 10 days of period 2, the stock level continues to decline at the rate of 150 per day to reach 2000 units on day 80.

In the first 10 days of the fixed 40-day order interval (days 30 to 40 on the chart), the stock level drops to zero at the rate of 100 units per day. During the remaining 30 days of the fixed interval, the stock level declines at the greater-than-average rate of 150 units per day to reach a -500 level when the order is placed. The order size is then $4000 - (SS - 500) + 1000 + SS = 5500$. Because the order was placed when the stock level was at -500 and the actual usage during the lead time is $150 + 10 = 1500$ units, the stock level for the beginning of period 3 is $(5500 + SS) - 500 - 1500 = 3500 + SS$.

The first 10 days of the first fixed-order interval are not shown in Figure 12.11.

Usage rate during lead time is assumed to follow the historical average of 100 per day, or 1000 during the 10-day lead time.

REPLENISHMENT PERIOD 3—USAGE RATE OF 75 UNITS PER DAY

Measured from day 80 when the stock level is 2000 units, the fixed-order quantity reorder level is reached in $(2000 - 1000)/75 = 13$ days. The continuous usage of 75 units per day puts the inventory level at 250 plus safety stock when the 4000-unit order is received on day 103. Usage during the remaining days of period 3 takes the final inventory level to 2975 units at day 120.

With the stock level at 3500 units plus safety stock at day 80, 10 days into the 40-day fixed-order interval, an order is placed at day 110 when the stock level is at $(3500 + SS) - (75 \times 30) = 1250 + SS$. The order quantity is then $(4000 + SS) - (1250 + SS) + 1000 = 3750$ units. During the lead time, withdrawals reduce the stock level by $10 \times 75 = 750$ units. On the final day of period 3, day 120, the stock level goes from a low of 500 to a high, just after delivery of the order, of $3750 + SS + 500 = 4250 + SS$.

The patterns in Figure 12.11 reveal that stock levels tend to fluctuate more widely in the fixed-order interval system. The deeper wedges into the safety-stock region indicate that the reserve should be greater for the periodic than for the perpetual system. The reason is that the safety stock for the perpetual system protects only from usage variation during the lead time, but the safety stock must protect the periodic system from usage variations during the entire fixed interval between orders. The magnitude of the added protection is implied by the changes required in some of the practical inventory methods listed in Section 12-6. The formulas in methods 3 and 5, based on a fixed-order size system, must be changed to recognize the greater risk of stockouts in a fixed-interval system by changing the $\sqrt{\text{lead time}}$ in the equations to $\sqrt{\text{order interval}}$.

The greater cost of safety stock in a periodic system is usually compensated for by lower clerical and data-processing costs. Loose or tight control can be enforced under either system. The management decision as to which system to use must be based on the unique characteristics of each situation. The profuse variables that could influence this decision range from a close appraisal of data-processing capabilities and cost to reliance on worn tenets such as "you can't sell what you ain't got" or "the more you carry, the faster you fall."

EXAMPLE 12.7 A "supermarket" inventory policy

Establishing a semiautonomous market within a firm to supply operating departments is a recent inventory wrinkle. It has been effective for housekeeping supplies and repair work.

A quantity of expendable items such as office and janitorial supplies is traditionally issued by departments several times a year. Between issues the departments run out of some items and accumulate a surplus of others. A stockroom operates as a supermarket by displaying items on shelves with the prices marked. Departmental representatives call on the market and pick out needed supplies. The purchases are charged against the department budget. The system eliminates filling out requisitions and associated processing, reduces delivery time, lowers departmental hoarding of excess items, and still controls the dollar usage of supplies.

A similar procedure can be followed for simple maintenance and repair. Each department has a "charge-a-plate." Items requiring repair are delivered to a maintenance depot accompanied by a charge-a-plate. The time and material required to fix the item are billed on a job order to the plate address. On-the-spot comparison of repair cost versus replacement cost can often be made by representatives of the department involved.

In both cases the departments are considered customers to an independent service facility. The advantages for such inventory management are (1) the centralization of supplies and record-keeping functions and (2) the improved sensitivity of the stocking system to current demand.

12-8 MATERIAL HANDLING AND CONTROL

The transportation of materials can be called the curse of production because it adds little value to the product but consumes a major portion of the manufacturing budget. The entire production cycle relies on material handling to link phases of product development, as is shown in Figure 12.12.

Material flow in a factory is analogous to a river system. Water is the material of the river. A breakdown or slowdown of a transportation link is similar to an obstruction in a river. Activities upstream from the blockade are inundated by the stagnant surplus; downstream activities are restricted to leakage or supply from other tributaries. When a number of dams are purposely built along the river, the water flow can be controlled because the inventory behind each dam can be fed into the system as needed. This is the rationale for in-process inventory.

Just as a river system is influenced by the terrain and vegetation on the terrain, the material-handling system is dependent on the building design and the equipment in the buildings. The ideal way to treat material-handling problems is to anticipate them before they occur and to provide facilities to overcome them. In new plant designs this approach is feasible; material-handling plans using the latest model equipment are developed in conjunction with other production requirements. A more common situation is the introduction of a new production machine that makes existing material flow inadequate. The danger in exclusively treating one segment of the production line is that the solution to one problem may create another one just as severe. Again using the river-

Characteristics of Ineffective Material Flow

Idle machines

Buildup of unplanned inventory

Damaged or lost parts

Injuries to workers

Late shipments

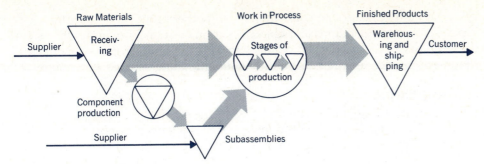

FIGURE 12.12 Relationship of material handling and inventory in a production process.

system analogy, clearing one section of a riverbed to smooth flow does not improve the volume flowing into the improved section.

Principles

Material-handling principles are macroextensions of the micromovement canons presented as the principles of motion economy in Section 10-3. The principles for material handling are less definitive than are those for hand movements because material flow is so closely related to other production considerations such as inventory, purchasing, scheduling, and similar product-oriented policies. Therefore, the principles should be treated as guidelines or thought provokers that may lead to improved performance. Complete obedience to the principle that "the shortest distance between two points is a straight line" is thwarted when an effectively immovable object sits between the two points, but its contemplation might inspire a better route that goes up and over instead of through.

1. *Eliminate.* If not, make the transportation distances as short as possible.

2. *Keep moving.* If not, reduce the time spent at the terminal points of a route to as short as possible.

3. *Use simple patterns.* If not, reduce backtracking, crossovers, and other congestion-producing patterns as much as facilities allow.

4. *Carry payloads both ways.* If not, minimize the time spent in "transport empty" by means of speed changes and route relocations.

5. *Carry full loads.* If not, consider increasing the size of unit loads, decreasing carrying capacity, lowering speed, or acquiring more versatile equipment.

6. *Use gravity.* If not, try to find another source of power that is as reliable and inexpensive.

In the transportation industry, the loading and unloading functions have not kept pace with the ever-increasing travel speed. Getting passengers on and off airplanes is a classic example.

Although these principles might mask it, there are important aspects of material handling besides geometry and hardware. Material considerations should include the movement of people, machines, tools, and information. The flow system must support the objectives of receiving, sorting, inspecting, inventorying, accounting, packaging, assembling, and other production functions. And typically, the considerations and objectives conflict. It takes an in-depth system decision followed by delicate diplomacy to establish a material-movement plan that meets service requirements without subordinating safety and economy.

Applications

Because material is required in all phases of the production process, there are innumerable facets to concern the material-handling specialists. Here we consider briefly three categories common to most processes: mass transit, tool control, and information flow.

Equipment for the horizontal and vertical transportation of massed material may be classified in the following three categories:

See Section 12-9 for close-ups of computer-assisted transportation and storage equipment.

1. *Cranes* handle material in the air above ground level to free the floor for other handling devices. Heavy, bulky, and awkward objects are logical candidates for airborne movement. Clearances, communication between floor workers and crane operators, and safety are problems associated with overhead transportation.

2. *Conveyors* may take the form of moving belts; gravity- or power-operated rollers; screws for shoveling bulk material; chains from which carriers are suspended; pipes for the pressurized flow of liquids, gases, or powdered material; and gravity chutes. The gravity-powered devices are convenient for between-floor movements and are easily adaptable to layout changes.

3. *Trucks* include both hand- and power-operated vehicles. Hand-operated wagons, platforms, and leverage trucks are suited to light loads, short hauls, and cramped quarters. Lift trucks with all their multiduty attachments and tractors with or without a chain of trailers can move heavy objects and palletized or bulk containers to various locations. Safety, visibility, and maneuvering space are the main limitations.

Three carbons of a tool receipt are usually made. One goes to the worker, one receipts the tool in the crib, and the last one is a permanent record.

Tool control is a limited but pesky material-handling problem. One of F. W. Taylor's first applications of scientific management was concerned with tool utilization. The first requirement is to procure the correct tools for the job. Once procured, the handling problem begins. Tools are usually issued to workers from centralized tool cribs. A "brass ring" control has the worker exchange his or her personalized brass ring for a tool, and the ring is returned when the tool is returned. A more acceptable method replaces the brass ring with a charge-a-plate or receipt slip made out by the crib attendant. The advantage of the paper record is that it indicates the amount of tool usage.

Reporting by exception: Reporting only the unusual; no report means no change. *Tickle file*: Reminders of future events filed by the anticipated date of occurrence. *Coding*: Replacing word messages by symbols such as numbers for a postal zip code.

No one denies the importance of *information flow*, and everyone decries the amount of *paper flow*; the two flows can seldom be equated. Unending efforts appear to be directed at stemming the mounting flow of paper while retaining the information content—reporting by exception, tickle files, coding, and so on. Campaigns often dent the flow temporarily; computerized systems cut out many reports when first initiated, but the "explanatory memo" soon filled the void. Some office innovations, such as automatic copying machines, make it easier than ever to flood operations with questionably needed data.

Information handling can benefit from concepts drawn from material-handling principles. The objective is to carry information from one point to another. The first question is whether the information needs to be moved. If so, how can it be done most economically? Voice communication, the oldest form of information transmission, is still effective, and its use is extended by mechanical aids such as message recorders and answering services for telephones. Radios and intercom systems are being augmented by closed-circuit television systems. This greater dissemination ability proves effective only when it is combined with a policy that limits proliferation to a "need to know."

" 'The horror of that moment,' the King went on, 'I shall never, *never* forget!' 'You will, though,' the Queen said, 'if you don't make a memorandum of it.' " Lewis Carroll, *Alice's Adventures in Wonderland.*

12-9 CLOSE-UPS AND UPDATES

The saying goes, "Moving is 90 percent of making." If this is even remotely accurate, the way to improve making is to improve moving, which means *material handling.* An improvement may be as simple as a new attachment for a lift truck or as advanced as a robot stockpicker.

Material handling has been progressing for 50 years. Ram-equipped lift trucks to shift metal coils and trackless trains to supply assembly lines were introduced in the 1930s. By the 1950s live-storage and order-picking devices were being built. Soon came air flotation equipment, which allowed a worker to push 15 tons of coiled steel manually, as well as driverless factory trains. The 1970s ushered in computer-controlled warehouses which were refined in the 1980s. The parade continues.

A 1979 study found that a part moving through a metal-cutting operation was being worked on only 5 percent of the time. The other 95 percent was spent in transit or in queues.

From here to there faster and cheaper

The ways to move material are as numerous as the types of materials to move. Some methods have been in use for decades and are essentially unchanged. For instance, it is difficult to improve on a gravity slide. Two of the most common transportation methods in production systems are automatic guided vehicles and conveyors.

Automatic guided vehicle systems (AGVS) have been in use for over 30 years, but the newer ones are smarter. An automatic guided vehicle is battery powered, driverless, and equipped to follow a flexible path while either pulling or carrying a load. Its guide path consists of a visible line or a buried wire laid out to optimize the material flow pattern. Computer-controlled AGVS can shunt vehicles with low battery readings to the recharge area, route vehicles to multiple destinations or reroute them around blockages, and be reprogrammed to meet new material flow requirements.

Today's short manufacturing cycles put a premium on adaptability. New AGVS paths are relatively inexpensive to reconfigure.

The basic vehicle types and their applications include the following:

- *Towing*—the most common type of AGVS whose vehicles pull up to six times what they can carry at speeds up to 3 mph.
- *Unit load*—frequently used to connect conveyors with storage/retrieval systems and on specific mission assignments for individual pallet movement, often involving automatic pickup and delivery; maximum capacity is about 6000 pounds.
- *Light load*—designed to operate in areas with limited space at a speed of 1 mph with loads of 100 to 500 pounds.

Conveyor systems encompass several broad classes of equipment, ranging from well-known roller belts to new overhead monorails. Some basic systems include

- *Package handling*—a wide class of conveyors for package weights up to 80 pounds that use slider beds or powered rollers for horizontal transportation, rough belts for inclines, and spiral lifts for vertical movements.
- *Unit load*—a heavy drive chain mounted below floor level that tows carts capable of carrying heavy loads. Towline layouts are expensive to change, and cart storage requires considerable space.
- *Automated electrified monorails*—overhead systems that share many of the advantages of AGVS. Motorized carriers that hold 500 to 7000 pounds run on overhead tracks, following either programmed paths or receiving commands while in transit.
- *Carousels*—transporter systems that consist of horizontal revolving bins that are microprocessor controlled to bring a requested bin to a work station where items are taken out or put into storage. They are frequently used in electronic assembly operations and occasionally integrated with storage and retriever robots.

Putting it away and getting it back again

Products and parts are typically put in storage by a producer, moved to another storage by a wholesaler or retailer, and perhaps stored again by the purchaser before final disposition. An increasing percentage of stock is found in automated storage systems, especially when response times for getting into or out of storage are important.

Warehousing has borne the brunt of many jokes: To purchasers a warehouse is a boundless area waiting to be filled with endless bargains, but to a salesperson it is an overstocked cavern that never contains what customers order. Both views contain some truth. Warehousing can reduce

As used in assembly lines in Europe, a light-load vehicle carries a car body through progressive work stages.

Power and control signals are picked up from bus bars—electrical conductor bars typically located above or alongside the track.

Manufactured parts are also stored as **WIP—work-in-process Inventory.**

A dictionary says a warehouse is a "structure or room for storage of merchandise or commodities."

material costs by astutely timed quantity purchases and should allow customers' needs to be met promptly. To serve both functions, warehouses should provide secure storage at low cost with rapid access and accurate accounting. The modern answer to these requirements is the spreading use of **automated storage and retrieval systems, AS/RS.**

An AS/R system is composed of four elements: *S/R machines* that move loads into and out of the *racks* or shelves, *conveyors* that feed or discharge the system, and *controls* that coordinate everything. Early versions in the 1960s had racks 60 to 70 feet tall mounted inside a building with punch-card controls for computerized S/R machines. By the 1970s, some rack heights exceeded 80 feet, with the racks supporting the rest of the building and capable of holding unit loads of 20,000 pounds. The more sophisticated systems are completely computer controlled, but others still have manual S/R machines.

Different types of load extractors require special types of racks. Rack design must also consider fire, seismic, and local building code criteria.

Storage/retrieval machines, or stacker cranes, traverse the aisles between racks guided by rails at the bottom and top. Automatic S/R machines are computer controlled to travel up and down aisles collecting or depositing pallet loads with motor-driven shuttle assemblies. Manual S/R machines have a person in the cab who positions the machine for handling unit loads with shuttle forks or for manual picking of stock.

Manual S/R machines have the advantages of lower initial cost and greater versatility, especially for low-volume and less-than-unit-load picking.

AS/RS are not limited to manufacturing operations. Figure 12.13 shows a system for storing forms used by an insurance company. Each rack is 32 feet tall, or seven loads high, 490 feet long, and has a maximum load weight of 2000 pounds. A peripheral conveyor system transports loads to and from S/R machines. In this installation, the AS/R system supplies picker stations where orders are filled for customers. When an order picker gets low on material, an AS/R system operator informs the computer of the shortage, and replenishments arrive automatically.

Benefits of AS/RS include reduced labor costs, improved throughput, reduced damage and theft, space and energy savings, and better inventory records.

Two factors halted the bigger-is-better design philosophy for AS/RS. The unprecedented jump in interest rates in the early 1980s not only made the price of all new investments difficult to justify, but it also caused the whole concept of safety stocks to be questioned. Efforts to reduce the amount of capital tied up in inventory in order to reduce interest charges, and the growing fame of just-in-time production, in which all material storage is minimized, were the instigators. As shown in Figure 12.14, work-in-process stores have been largely eliminated in automated manufacturing operations.

JIT systems are examined in Chapter 13.

Future AS/RS likely will be smaller and have more sophisticated controls. The smaller size reflects the decline in in-process storage and the use of multiple minisystems to feed parts to and from automated assembly lines and work stations in cellular manufacturing systems. The versatility of the material flow system must match the flexibility of the manufacturing system.

Micromine AS/RS are integral components of new production lines. They are typically 20 feet high and handle loads up to 150 pounds in 18-x-24-inch tote sizes.

FIGURE 12.13 An automated storage and retrieval system to supply picking stations that fill orders for forms from insurance offices and clients. As shown, the AS/RS vehicles have emerged from the racks to face the integrated conveyor system. (Courtesy of Allis-Chalmers Corporation.)

FIGURE 12.14 Comparison of material flow in a conventional manufacturing operation with one that has been automated. Elimination of in-process storage is the significant difference.

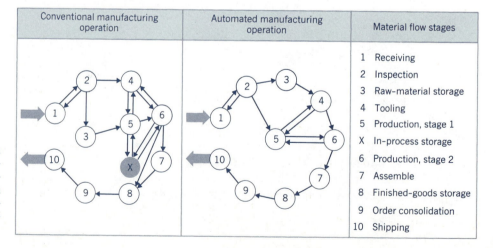

Conventional manufacturing operation	Automated manufacturing operation	Material flow stages	
		1	Receiving
		2	Inspection
		3	Raw-material storage
		4	Tooling
		5	Production, stage 1
		X	In-process storage
		6	Production, stage 2
		7	Assemble
		8	Finished-goods storage
		9	Order consolidation
		10	Shipping

12-10 SUMMARY

The purchasing function is the interface between a company and its suppliers. A purchasing department is treated as a customer by suppliers and as a support function by other operating units. In turn, it receives assistance from both suppliers and the operating units.

A more systematic purchasing policy usually results from centralization. The flexibility of centralized purchasing is increased by the local control of small or rush orders.

Purchasing

Purchasing activities follow this sequence:

1. *Receive requisition.* Requisitioners should recognize the need for lead time, the elapsed time between placing an order and its receipt.

2. *Review requisition.* Value analysis is used to review the true need for requisitioned supplies. A value engineering approach evaluates the worth of an item by partitioning its cost according to

   Value engineering

   Use value + esteem value = exchange value

   The use value receives the most attention by forcing a verb-and-noun description for each primary and secondary function.

3. *Select suppliers.* An "approved suppliers list" is developed by purchasing to expedite ordering. Orders are usually spread over several suppliers to avoid relying completely on one vendor.

4. *Place orders.* One-of-a-kind contracts are usually granted on the basis of competitive bids. Open contracts are open accounts with one supplier for the direct ordering of low-cost, infrequently used supplies. Items with a continuous demand are purchased under blanket purchase orders. Open blanket contracts are designed to avoid the cost of numerous individual purchase orders.

5. *Monitor orders.* Purchasing keeps tab on the supplier's delivery schedule.

6. *Receive orders.* Purchasing or payment policies may allow trade, quantity, or cash discounts.

Inventory is an idle resource available when needed but subject to storage costs. Its main purpose is to insulate production from delays caused by lack of material. It also affects work stability, customer satisfaction, material-handling requirements, and purchasing policy.

Purpose of inventory

Inventory costs are composed of expenses that vary directly and inversely with the size of an order. Order costs (O) and opportunity costs (OC) increase with smaller order sizes. Holding costs (H) and capital costs (iP) decrease with smaller order sizes. The order size (Q) that balances these two cost categories under the assumptions of constant demand and a known lead time is calculated by the *EOQ* formula:

Economic order quantity

$$Q = \sqrt{\frac{2OD}{H + iP}}$$

where D = annual demand. When the items are expended as they are produced, the *EPQ* formula is

Economic production quantity

$$Q = \sqrt{\frac{2OD}{(H + iP)(1 - D/M)}}$$

where M = annual manufacturing rate. The effect of quantity discounts for larger orders is evaluated by calculating the total annual cost of each level of discount by

$$\text{Total annual stocking cost} = O\frac{D}{Q} + (H + iP)\frac{Q}{2} + PD$$

where the first two terms are the annual inventory cost and the last term is the purchase or production cost for a year's stock.

A continuing order policy, assuming that demand is relatively stable but lead time varies, uses a tableau format to aid calculation of an expected value for the carrying costs (*CC*) and opportunity costs (*OC*) of each lead-time alter-

Inventory risk models

native. Then,

$$\text{Total inventory cost} = \frac{(O + OC)D}{Q_{LT}} + \frac{HQ_{LT}}{2} + CC$$

where the *EOQ* for each lead-time alternative (*LT*) is

$$Q_{LT} = \sqrt{\frac{2(O + OC)D}{H}}$$

Expedient inventory models

Less formal methods of including risk in an inventory policy are provided by expedient measures that use different assumptions about demand and lead time to identify a reorder point. Simulation is also used to determine a preferred ordering policy.

ABC analysis

The inventory items that deserve the most attention are determined by an *ABC* analysis. Class *A* items, on which attention is concentrated, typically account for 75 to 80 percent of the total dollar usage but only 15 to 20 percent of the quantity volume. The effect of following an *ABC* policy is to "buy" analysis time for the high-value items by overstocking the low-value items.

Inventory management systems

A perpetual inventory system replenishes stock when it drops to a certain level by ordering a fixed quantity. A two-bin version of the system simplifies record keeping.

A periodic inventory system has fixed review periods when a variable order quantity is placed to bring the stock level near a desired maximum. The stock levels vary more widely in a fixed-order interval than in a fixed-order quantity system. The greater safety stock required by the periodic review system is compensated for by lower clerical and data-processing costs.

Material handling is a necessary evil that consumes a major portion of total production cost while contributing little value to the actual refinement of a product. Guidelines for handling are extensions of the principles of motion economy.

Tool control is a limited but pesky material-handling problem. It is best treated by multiple copy receipts by the tool crib attendant.

Information flow is linked to the paper flow problem. The extended use of voice communications and strict attention to the "need to know" principle can help stem the increase in paper flow while maintaining information flow.

Equipment to move and store material is steadily evolving in response to the changing practices of production. The basic designs for transporting and warehousing materials have been in use for many years. Computer controls largely distinguish the newer designs for automated guided vehicle and storage/retrieval systems. The trend toward reducing inventory levels probably will lead to smaller, more flexible material-handling and storage equipment.

Material handling

Equipment for moving, storing, and retrieving material

12-11 REFERENCES

ASTME. *Value Engineering in Manufacturing.* Englewood Cliffs, N.J.: Prentice-Hall, 1963.

BAILY, P. J. H. *Purchasing and Supply Management*, 4th ed. New York: Wiley, 1978.

BROWN, R. G. *Material Management Systems.* New York: Wiley, 1977.

BUCHAN, J., and E. KOENIGSBERG. *Scientific Inventory Management.* Englewood Cliffs, N.J.: Prentice-Hall, 1963.

BUFFA, E. S., and R. G. MILLER. *Production Inventory Systems: Planning and Control*, 3rd ed. Homewood, Ill.: Irwin, 1979.

HADLEY, G. M., and T. M. WHITIN. *Analysis of Inventory Systems.* Englewood Cliffs, N.J.: Prentice-Hall, 1963.

HARRIS, F. W. *Operations and Costs.* Chicago: A. W. Shaw Company, 1915.

HAX, A. C., and D. CANDEA. *Production and Inventory Management.* Englewood Cliffs, N.J.: Prentice-Hall, 1984.

LOVE, S. F. *Inventory Control.* New York: McGraw-Hill, 1979.

McCLAIN, J. O., and L. J. THOMAS. *Operations Management: Production of Goods and Services.* Englewood Cliffs, N.J.: Prentice-Hall, 1980.

SILVER, E. A., and R. PETERSON. *Decision Systems for Inventory Management and Production Planning.* New York: Wiley, 1984.

WAGNER, H. M. *Statistical Management of Inventory Systems.* New York: Wiley, 1962.

WRIGHT, O. W. *Production and Inventory Management in the Computer Age.* Boston: Cahners Books, 1975.

12-12 SELF-TEST REVIEW

Answers to the following review questions are given in Appendix A.

1. T F It is estimated that up to 80 percent of the indirect cost in a plant, or 20 to 50 percent of the total production cost, is consumed in *transporting materials.*

2. T F "Hedging" is a *purchasing* practice that limits the amount and type of supplies that production managers can order.

3. T F The elapsed time between placing an order and receiving the ordered materials is known as *float time*.

4. T F The *value engineering* equation that guides purchasing is use value + esteem value = exchange value.

5. T F A *trade discount* is awarded for the prompt payment of bills.

6. T F *Inventory*, in a production sense, is an idle resource.

7. T F *Holding cost* is the expense associated with running out of stock.

8. T F *Economic order quantities* can be estimated for demand that is dependent and predictable.

9. T F The maximum number of items in storage at any time is lower than the *economic production quantity*.

10. T F Maintaining a *safety stock* increases the holding cost.

11. T F *ABC analysis* is based on the Pareto principle.

12. T F In a *periodic inventory system*, a fixed-order quantity is ordered at fixed intervals.

13. T F *Work-in-process* inventory smooths the flow of material between stages of production.

14. T F Information flow can benefit from the application of *material-handling principles*.

15. T F The main elements of an *automated storage and retrieval system* are racks, electrified monorails, carousels, and controls.

16. T F Successful efforts to reduce *inventory levels* in manufacturing are leading to smaller-scale AS/RS.

12-13 DISCUSSION QUESTIONS

1. When a purchasing agent assumes the role of a customer, he or she is susceptible to many ethical questions of conduct. Comment on the following situations:

a. A purchasing agent owns stock in a company from which the firm she represents could buy material. Should she deal with this company?

b. Some firms provide expensive Christmas presents for purchasing agents with whom they do business. Should an agent accept all gifts offered, just modest gifts, or no gifts at all?

2. What type of requisition could be expected for the two rough divisions of purchases—maintenance supplies and raw materials?

3. Information channels between purchasing and other parts of the production system are shown in Figure 12.2. Which of these channels are more important for each of the six operations performed by purchasing (refer to Section 12-2)?

4. How could "squirrel" and "brand X" complexes affect the need for requisition reviews? Would you say these complexes are natural reactions to past experiences? What arguments would you use as a purchasing director in rebuttal to statements defending each complex?

5. Should an engineer feel insulted when a purchasing director questions particularly tight tolerances set for a part that is to be purchased? Why?

6. Is reciprocity an ethical practice? What problems can result from its extensive use?

7. Discuss the advantages and disadvantages of centralized versus decentralized purchasing as it would apply to each of the following situations:

a. A firm operating three "99¢ burger" shops in the same city.

b. A firm operating three "99¢ burger" shops in three cities, each 200 miles apart.

c. A large discount house located in one sprawling building subdivided into 18 operating departments: household supplies, drugs, hardware, sporting goods, and so on.

8. A tie clasp issued to all new recruits in a branch of the armed forces is chrome plated, has a pivoted clasp, and contains a replica of the insignia of the service branch. The description of the working function of the clasp could be "hold parts" whose parts are the tie and shirt. A secondary (or perhaps the basic) function is "creates impression." Show how the creation of substitute means and the evaluation depend on the item's function description. If the present tie clasp costs $0.93, determine a less expensive alternative based on the "holds parts" function. Does this alternative satisfy the function "creates impression?"

9. Comment on the following quotation from ASTME's publication, *Value Engineering in Manufacturing:*

It must be repeated that value engineering cannot be practiced where there are restrictions to freedom to change. If a product can only be refined superficially (and there can be many perfectly valid reasons for this), value engineering is not an appropriate cost-reduction tool. For instance, it was found that wide-mouth catsup bottles could not be sold under market conditions in the late Fifties, although the wider-mouth jar was less expensive, took less storage space, and would pour at a more uniform rate. If there is substantial tooling involved for a product of doubtful sales life, it would be wiser to concentrate on substitution of materials, changes in tolerances, or methods improvements, since it is likely that these would be the types of proposals developed after an exhaustive value engineering study was completed on a project with such restrictions.

But if there is freedom to change, value engineering will bring forth the full benefits of change. Where normal cost improvement effort refines by chipping away at the subject, value engineering blasts out great chunks of cost. Many firms with a product line facing extinction have in desperation turned to value engineering when normal cost improvement has proved inadequate. But many more firms are unknowingly allowing available profit to slip by them through clinging to traditional methods of product development and refinement.

10. Should a supplier be told why he or she is not on an approved supplier list?

11. What are the advantages of splitting an order among suppliers? Can competitive pricing be maintained with an exclusive supplier?

12. Would it be wise to borrow money from a bank for one month at 6-percent simple interest in order to take advantage of a cash discount on terms "$3/10$ net 30"?

13. The engineering staffs of a buyer and a seller often cooperate in developing new equipment wanted by the buyer. Such activities are common for one-of-a-kind contracts and are not uncommon for blanket contracts. Why might purchasing encourage this cooperation, and why might a vendor agree to it? Why might a vendor refuse it?

14. Give an example of the type or class of items for which each of the three ordering contracts would be appropriate.

15. If you were a purchasing director, would you follow the policy suggested in Example 12.1 of informing your suppliers about their ratings? What advantages and drawbacks should influence your decision?

16. Using the "idle resource" definition of inventory, comment on the following physical, political, and sociological examples:

a. An untapped oil reserve that would lower the price of gasoline if developed.

b. A standing army maintained during periods of peace.

c. The "land bank" plan to keep land from producing crops.

d. Hiring extra engineers and scientists to substantiate a claim for research and development potential when bidding for major contracts.

17. Comment on the effect of inventory as it influences:
a. Hiring and placement departments.
b. Public relations.
c. A customer of a wholesaler.
d. A customer of a retailer.

18. Distinguish between "buying down" and "speculative buying." How does each practice influence an inventory policy?

19. Why are opportunity costs not included in basic EOQ formulas?

20. Give examples of production situations in which $D = M$ and $D > M$.

21. Comment on the application of the Pareto principle to the following production situations:
a. A few departments perform the bulk of production work that contributes to the firm's profit.
b. A small number of suppliers causes most of the delays in procuring purchased production supplies.
c. A few operators produce most of the scrap and rejects.
d. Most orders from purchasing come from a few vendors.

22. Would the holding costs for a two-bin inventory system more likely be based on the maximum or the average inventory level? How does the two-bin method reduce clerical costs in a perpetual inventory system?

23. Would you prefer a perpetual or a periodic inventory system for class C items? Why?

24. Use the value engineering approach to describe the use function for each class in the ABC inventory divisions.

25. Which categories of equipment are more likely to be associated with a product layout and a process layout?

26. Can material handling be evaluated independently from purchasing and inventory policies?

27. What ways can you suggest for reducing terminal times between material movements?

28. What ways can you suggest for reducing "travel empty" time?

29. How does the size of unit loads affect material handling time?

30. Is the "technological information explosion" affecting paper flow? Should it?

31. What is the advantage of knowing tool usage for a tool control plan?

32. How can a "need to know" limitation reduce paper flow? Could an overzealous application lead to legitimate accusations of favoritism and censorship?

12-14 PROBLEMS AND CASES

1. A wholesaler forecasts annual sales of 200,000 units for one product. Order costs are $75 per order. Holding costs of $0.04 per unit per year and interest charges of 12 percent are based on the average inventory level. The cost to the wholesaler is $0.80 per unit acquired from the factory. What is the EOQ? What is the total annual stocking cost? What is the time between orders based on 250 working days per year?

2. Parts used in assembly work are purchased from a supplier who has a remarkable record for prompt delivery. The inventory history closely follows an instantaneous replenishment pattern. However, to be on the safe side, a policy is followed of never planning an inventory level below 500 parts. The following costs are applicable:

Procurement cost = $60 per order

Carrying cost = $0.20 per unit per year
based on average inventory

If the demand is 40,000 parts per year used at a steady rate, what will be the total annual inventory cost?

3. Determine the economic order size for the following conditions, where n refers to the number of units and carrying costs (H) are based on average inventory size:

$O = \$10 + \$0.01n$

$H = \$2 + 0.01\ P/n^2$

$P = \$40$ per unit

$D = 800$ units per year

4. The annual amount ordered from one raw-materials supplier is $260,000. Annual order costs are 1 percent of the raw material cost, and carrying costs are 18 percent of the average inventory level. How many weeks of supply should be ordered at one time?

5. A company currently buys 20,000 parts per year from a supplier. Each part costs $1. Ordering costs are $50 per order, and carrying costs are 20 percent of the purchase price based on maximum inventory. It has been suggested that the company buy a machine that can produce the parts for half the price now being paid, but setup costs will be double the order cost. The machine can produce at an annual rate of 200,000 parts and has an economic life of 10 years, at which time the salvage value will be zero. If the savings from producing this part are expected to pay half the equivalent annual cost of the machine, what will be the maximum amount that can be paid for it? Capital recovery is based on an 8-percent interest charge.

6. Four thousand tons of raw materials are used each year. Order costs are $20 per order; carrying costs are $8 per ton based on the maximum storage requirement. The supplier has offered to reduce the $40 per-ton price by 5 percent if the minimum order size is 500 tons and by another 5 percent if the minimum order is 1000 tons. The capacity of the present storage facility is 250 tons. Any increase in capacity will increase the carrying costs in a direct ratio. Show your calculations and assumptions to prove whether the quantity discounts should be used.

7. A knitting mill is closing its swimming suit division to devote full production to other sportswear. It has offered its stock of 400 daring, net, see-through suits at a closeout price of $4 each. Similar models sold last year for $25 per suit, but some bathers were barred from beaches when they wore them. By spending $1600 on advertising, a retailer estimates the probability of sales this year as

Suit sales, S	150	200	250	300	350	400
Probability of S	0.10	0.10	0.20	0.25	0.20	0.15

If the suits not sold at the regular price can be sold at special sales for $2 each, how many of the 400 suits should the retailer take?

8. A shipbuilding yard has an annual demand for 1150 bottles of acetylene. Each bottle costs $40; order and inspection costs are $55 per order. Carrying costs are 25 percent of the value of average inventory in storage. The yard operates 230 days per year. Interruptions to production cause opportunity costs of $40 per bottle for each day it is unavailable.

a. What will be the order interval if the lead time is instantaneous?

b. What will be the total inventory cost and the optimal order size if the lead time follows the pattern shown in the table?

9. All differential gears used by a manufacturer are furnished by one supplier. The inventory system operates on a fixed-order quantity basis. Procurement costs are $20 per order, and the annual usage is 3200 gears. Carrying costs are $0.75 per unit per year. The average lead time is four days, and daily demand follows the pattern shown in the table. (Note the Poisson distribution with a mean of 16.)

What is the order size, reorder point, and expected minimum inventory level when risk is accounted for by the

a. Ultraconservative method? (Assume that the maximum demand is 25 gears and the maximum lead time is seven days.)

b. Safety-stock percentage method? (Assume a safety factor of 30 percent.)

c. Square root of lead-time usage method?

d. Demand percentage method?

e. Combination method?

costs associated with wood-chip inventory are holding cost, $60 per year per unit for maximum order; order cost, $300 per order including rental of unloading equipment; and opportunity cost, $80 per unit per week for disruption of production schedules.

Data for Problem 8

Lead time	6	7	8	9	10	11	12	13	14
Relative frequency	0.00	0.04	0.08	0.38	0.24	0.12	0.09	0.03	0.02

Data for Problem 9

Demand	13 or less	14	15	16	17	18	19	20 or more
Occurrence	5	15	225	225	17	11	6	3

10. Continue the simulation begun in Example 12.5. After 50 trials, determine the size of a safety stock that allows stockouts on no more than two consecutive days only 5 percent of the time.

11. A small plant uses wood chips as a raw material for one of its products. The chips are delivered in units of railway cars. The average demand is four cars per month, but the usage varies according to the following pattern:

Demand	Probability
6	0.10
5	0.20
4	0.50
3	0.20

Lead Time (Weeks)	Probability
1	0.10
2	0.10
3	0.30
4	0.50

Because special unloading equipment is rented when a delivery arrives, it is desirable to have large orders. Space is available to store any size order, but stored inventory is subject to weather damage. The

The lead time varies from one to four weeks according to the distribution shown in the margin area.

Determine by simulation the preferred reorder level for wood chips.

12. Throw-away filters for production machines cost $1 each and are used at an average daily rate of 25. The *EOQ* for filters is 500. Average delivery time is 4 working days. At the start of a year, the inventory level is 3000 filters. During the first 30 days, production drops, and the usage of filters decreases to 15 per day. In the next 30 days, usage is at the average level, 25 per day. For the following 30 days (cumulative days 61 to 90), usage climbs to an average of 50 per day and stays at this level for the last 30 days of the 120-day span.

a. Graph the pattern of inventory on hand when a perpetual system is used. Numerically label all the break points on the plotted line.

b. Graph the pattern of inventory on hand when a periodic system is used. Numerically label all breaks in the plotted line.

c. If the square root of lead-time usage method is used to calculate the safety stock, what would the safety stock be for (1) the fixed-order quantity system and (2) the fixed-order interval system? Compare the indicated safety-stock level with need for a reserve supply in the patterns from parts a. and b.

13. *The Case of Inventory Ups and Downs*
A new, very eager scheduler has been hired to coordinate production and inventory control. It is a new position created by the need to increase the production of a relatively small firm to meet a large contract. After a work-simplification study, the assembly department has a surge in output. The corresponding increase in demand for parts to assemble overtaxes the capacity of the production department. Consequently, the inventory of parts on hand falls well below the normal replenishment level. To avoid an increase in lead time, the eager scheduler raises the expected manufacturing rate. Then additional workers are hired by production to meet the greater demands.

After a while, the pace of the assembly workers drifts back to a rate above the former output. But the rate is not as high as the enthusiastic surge that so often accompanies the installation of a new program. Now inventory levels are above original levels and are still growing larger owing to the higher production rates made possible by the new workers. The scheduler again revises the reorder level and manufacturing rate, this time downward. This action means that no order will go to production while the excess inventory is being used in assembly work.

With no orders coming in to use the greater manufacturing rate, production must lay off workers.

What errors caused this overly responsive situation?

How was the concept of safety stock violated? Under what condition should reorder points be changed?

14. *A Case of Material Mismanagement—Maybe.*
"I propose that we apply *EOQ* calculations to all our inventory procurements. At the present time we have one all-purpose purchasing agent who operates in a very informal manner, so informal that we don't even find out how much it costs us to place an order. We know he gets paid $25,000 a year, occupies an office that we figure costs $3500 a year, has a telephone bill averaging $900 per year, and annual incidental expenses of about $600. Because our small company has no major material requirements, he just circulates around the building to see what supplies are needed. Then he phones the orders to local merchants and writes out the checks himself to make all the payments. Apparently he places about 2500 orders each year, ranging in size from a few dollars to $800 or $900. That comes out to about $12 ordering cost per order. We can probably cut the number of orders down to 1500, which means a savings of (2500 − 1500 orders) × $12 per order = $12,000 per year."

Name three attributes of the present system that should be considered before changes are made to formalize the inventory policy. Comment on the savings expectations.

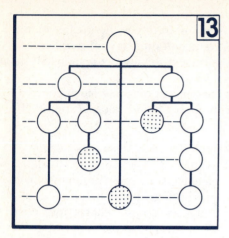

SYSTEMS SYNTHESIS: MRP, MRP II, AND JIT

LEARNING OBJECTIVES

After studying this chapter you should

- realize the importance of systems thinking and how material management interacts with the rest of the production system.
- understand lot sizing for time-varying demand and be able to apply the Wagner–Whitlin algorithm and the Silver–Meal heuristic.
- be familiar with the concepts and terminology of MRP and know how an MRP program operates.
- comprehend the intent and characteristics of MRP II.
- understand the philosophy of JIT and the nature of a kanban system. Realize the significant impact that JIT applications have had on production management.
- be able to distinguish the features of pull- and push-manufacturing systems.
- be aware of emerging systems synthesis programs such as OPT.

KEY TERMS

The following words characterize subjects presented in this chapter:

distribution system

corporate crime

time-varying demand

heuristic

material requirements planning (MRP)

time phasing

capacity planning

bill of material

level coding

time buckets

master production scheduling

manufacturing resources planning (MRP II)

500

just-in-time (JIT) production
kanban
move/production cards

pull/push manufacturing systems
optimized production
technology (OPT)

13-1 IMPORTANCE

If asked which development has had the greatest impact on manufacturing in
the 1980s, most production managers would likely answer, "Material flow control." They would be referring to newly adopted techniques and machines that
have slashed inventory levels throughout the production process. Indirectly they
would be referring to the modern information networks that monitor material
flows and support decision making based on system requirements rather than
local needs. The coordinated flow of material through automated processes
epitomizes system synthesis.

Systems synthesis is effectively pictured by its core position in the emblem
of this text. As shown in Figure 13.1, systems synthesis incorporates planning,
analysis, and control throughout the production function. It is concerned with
people, machines, and materials from a perspective that emphasizes the integration of production activities for the benefit of the total system. Fulfilling this
objective is as difficult as the goal is worthwhile.

A broad perspective demands that allegiance to pet projects and concerns
be waived to avoid favoritism and narrowness of attention. The appropriate
breadth depends on the level of management involved. At the supervisory level,
the relationship of hourly activities and immediate production objectives are
prime considerations, and synthesis efforts are consequently directed toward
smaller systems. Higher in the management hierarchy, the boundaries of systems synthesis expand to encompass the smaller systems considered at lower
levels and the adjacent environment.

> To synthesize is to form a whole
> by bringing together separate
> parts. Systems synthesis is a
> cause-to-effect process that
> integrates the elements of
> production.

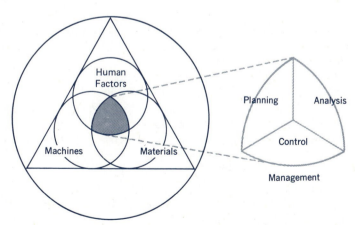

FIGURE 13.1 The central position of systems synthesis.

Analysis and synthesis are lockstep stages in the development of a more effective system. The elements of a system are analyzed individually to improve their productivity. Then the improved elements are reunited, causing occasional trade-offs to make functions compatible enough to form a coordinated body. The coordination is often assisted by computer technology. In this chapter we examine a **distribution system** as an example of systems thinking, consider an extension to economic lot sizing that accommodates time-varying demand, study material requirement planning for dependent demand, observe how this approach is expanding to incorporate influencing factors from the rest of the production system, and survey the remarkable impact of just-in-time production, a movement that penetrates the entire production system.

13-2 SYSTEMS THINKING

The gambling addict has his or her pet system. An automobile has a cooling system. Animals have circulatory systems. Sales personnel no longer sell computers; they sell complete information systems. There are military systems, judicial systems, educational systems, and production systems.

There have always been systems, but the modern ones are more impressive. They are bigger, tighter, and more complex because civilization has need for them and technology is available to support them. Standards of quality and reliability taken for granted today were inconceivable just a few decades ago. To raise or just maintain these standards while increasing the quantity and diversification of products is a continuous challenge. Old techniques merge with the new as managers race to meet the demands of the systems they manage.

One significant trend has been in the way management thinks about production problems. It is easy to say effectiveness of the whole is more important than efficiency of the parts or that profit optimization should take the place of cost reduction. But only skillful managers can put into practice what they know is theoretically correct. They have to be familiar with the idiosyncrasies of the system to recognize the interplay of modifications, and they must have the knowledge to evaluate the effects. If either the familiarity or the knowledge is lacking, systems thinking will suffer.

Let us explore just one example to solidify system concepts. We could make an example out of an "ice cream man" selling his products from a pedal-operated cart. He has to consider how much to order, his best route, competition, and other business factors. Or we could discuss a huge "general" system—General Motors, General Foods, or General Electric. Certainly these multiproduct, multiservice giants are guided by a set of objectives that directs continuity of purpose in their diverse activities. Instead of either extreme, we shall survey a system within a system.

Again there are many choices. Advertising, packaging, or research and development are broad but distinct areas that require coordination of effort among many operating units. A good selling job, new container design, or a

A short, inclusive definition says that a system is a set of related elements. The implied task is to define the set and identify the relationships.

Production systems can be technically classified as functional–purposive–mechanical as opposed to, say, an electromagnetic field system that is structural–nonpurposive–organismic.

successful research program does not spring from the isolated efforts of one department. They are a composite of ideas, demands, reservations, and enterprise from many directions. The objective, also a function of cooperative effort, is accomplished by coordinated action of components from other systems. The descriptive nouns that commonly designate operating departments, such as marketing, sales, and purchasing, are functional adjectives for subsystem objectives. The verb that unites systems within a system to implement strategic objectives is *coordinate*.

Our example is a distribution system. It is a subsystem within the confines of the larger corporate system. Although a separate department may not be devoted strictly to distribution problems, every industry gives attention to the delivery of products and services to consumers. The ratio of physical distribution costs to sales is prominent in some industries: about 25 percent for the food industry and 22 percent for the chemical, rubber, petroleum, and primary metal industries.

The objective of the distribution system, as depicted in Figure 13.2, is to serve the customer; it starts with a sale secured from the customer and ends with the delivery of the ordered product. The top halves of the circles read like

Distribution is the last stage in the material flow process, in which the customer receives the product or service that represents the main objective of the entire production system, and the system is judged accordingly.

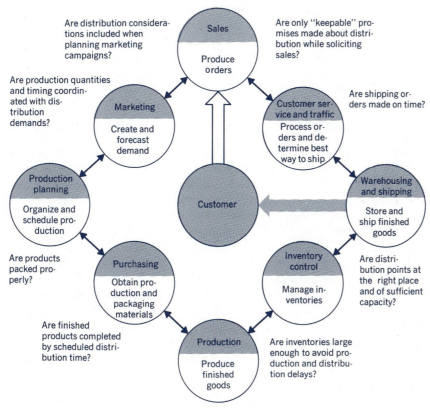

FIGURE 13.2 Interfaces of functional areas in a distribution system.

an organizational chart. The lower half of each circle relates representative functions to physical distribution. A change in function originating in any organizational division will subsequently affect the overall distribution policy.

Consider a successful new marketing development. A greater customer demand increases sales and requires new production schedules to meet expected deliveries. Purchasing works with production planners to develop a procurement program. The new production schedules are converted to increased capacity at the manufacturing plant by higher machine utilization, overtime, altered material-handling routines, more workers, or other measures requiring assistance from operating divisions such as payroll, recruiting, training, engineering, and maintenance. Meanwhile, customers' orders are being processed. The orders form the basis for inventory planning to facilitate storage and for traffic planning to determine shipping. Warehousing and shipping personnel arrange the physical facilities and administrative details to distribute the increased flow of products.

Each interface of specialty interests is an area of potential conflict and compromise. *Marketing* may disagree with *sales* as to the timing of campaigns. *Purchasing* could press for larger ordering quantities to take advantage of quantity discounts, whereas *inventory management* warns against the increase in holding costs. Longer production runs with fewer setups are opposed by the need to find storage space for the larger output before distribution. The greater stock in warehouses pleases *customer service* because the lead time before shipment is lower. *Sales* pushes for a greater range of styles against the caution of *manufacturing* that diversification complicates production. A new packaging design could be favored by *marketing*, *sales*, and *purchasing* while being opposed by *production planning*, *warehousing*, and *inventory management*.

These opposing views arise from honest and legitimate opinions about the best tactics for *each operational area*. The disappointing aspect is that a strong conviction, even though based on valid facts and a conscientious evaluation, can be detrimental to overall performance if the tactics are confined to local considerations. Leaders are expected to be resolute, but stubborn concern coupled with a myopic outlook destroys their effectiveness.

One way to resolve differences of opinion is by executive decree. The top person can dictate policy. If he or she is unusually astute and diplomatically gifted, the executive will select the best course of action and persuade participants to cooperate. But a dictatorial regime often attains its efficiency at the expense of destroyed initiative.

Another way to achieve unity is through management orientation and an environment for systems thinking. The orientation includes formal training for evaluating interrelated functions and familiarity with strategic objectives. The environment includes a fast, accurate, communication network and a spirit of cooperation. Most managers are naturally inclined to devote their attention to activities immediately surrounding them; they protect subordinates and promote policies with proprietary interest. These practices are commendable so long as they do not interfere with the welfare of strategic objectives. Interference is

Sensitivity analysis is a natural prelude to synthesis when there are strongly held, opposing options. The analysis reveals how much give-and-take is available before a major change must be considered.

Suboptimization, as introduced in Chapter 4, is a major threat to effective systems synthesis.

recognized through the inability to integrate activities into a coherent program. Unfortunately, it is much easier to preach systems thinking than to practice it.

EXAMPLE 13.1 The negative side of systems thinking—Corporate and computerized crime

Petty theft has always been a business problem. Both customers and employees steal goods or engage in illegal activities that can wipe out profits, causing some small businesses to fail. Large corporations can be seriously hurt, too, by grand theft in which whole shipments are diverted or intercepted. Theft control has influenced the design of warehouses and material-handling equipment through the use of containerized shipments and sealed unit loads. But the direct pilferage of physical goods is only one aspect of **corporate crime,** though the most obvious, albeit not necessarily the most damaging.

Corporations may themselves be the offenders. Several defense contractors have been found guilty of fraudulent billing on contracts. One financial institution pleaded guilty to over 2000 counts of mail and wire fraud in a scheme to overdraw checking accounts. A U.S. congressional study found that criminal acts by insiders were major contributors to roughly half of the failures of savings and loan associations between 1980 and 1984. Although there may be no lessening of business ethics, in general, currently committed corporate crimes are likely to be larger and have much wider impact now that far-flung operations are electronically connected.

The crux of systems thievery is data security. As computer applications have grown and management information systems have proliferated, many benefits have accrued to users. However, the opportunities to manipulate data for personal gain have also grown. Greater usage has increased the number of people with access to the computerized environment, and has made it more vulnerable to attack. Consider some of these publicized computer crimes:

• A bank teller in New York embezzled $1.4 million over three years to pay gambling debts; he escaped detection by the bank's auditors until he was implicated by a police raid on his bookmaker.

• A 15-year-old high school student gained access to the University of California's computer with $60 worth of used terminals. He stole over 200 hours of computer time and printouts of many university reports.

• In a classic swindle, an insurance company's assets were bloated for a year by recording the sale of 97,000 policies in the computer, when only 33,000 were actually sold.

• Organized burglars used a computer in Chicago to compile lists of promising robbery targets and then stole over $1 million in negotiable securities from private homes.

These examples just bare the tip of the proverbial iceberg of computer fraud. It is estimated that more than 85 percent of computer crimes go undetected or unreported. In addition to outright crime, crippling damage to computer-based data can result from accidental or intentional tampering. Both industrial sabotage and the theft of confidential information pose serious threats to production systems; a slight modification to a program could destroy a design and would be very difficult to track down, as would the theft of company secrets, because no one would know if or when it had occurred.

The majority of reported losses, either through error or intent, are caused by an organization's own employees. This makes security difficult because most of the employees have access to computers as part of their normal work assignments. An effective countermeasure is a procedure that authorizes jobs, rather than users, to access protected data. Another approach is the use of passwords to gain access. Access logs and tight scheduling of computer times are other security measures. But even as computer technology and its associated security systems advance, so does the sophistication of intruders who try to match their wits against the systems.

13-3 LOT SIZES FOR TIME-VARYING DEMAND

In a 1939 paper entitled "Mathematical Methods in the Organization and Planning of Production," the Russian mathematician L. V. Kantorovich observed:

> *There are two ways of increasing the efficiency of the work in a shop and enterprise, or a whole branch of industry. One way is by various improvements of technology—that is—new attachments for individual machines, changes in technological processes, and the discovery of new and better kinds of raw materials. The other way, thus far much less used, is by improvement in the organization of planning and production.*

Applying Kantorovich's observation narrowly to material management, new technology in the form of faster, more versatile, and smarter machines has indeed increased efficiency, and mathematical techniques for organizing material flow have had comparable effect. The evolution of inventory practices is an apt example.

Early inventory models dealt with finished goods and salable items whose future disposition was predictable. Calculations were simplified by assuming that demand was level and known. It was also recognized that there was another class of material usage in which the need for certain items depended on the demand for other items and that this demand was not level. Little attention was given to **time-varying demand** until computers made feasible the collection and processing of real-time data for inventory control.

In this section, the assumption of level demand that was in effect in Chapter 12 is relaxed to consider models that specify different replenishment quantities for successive order periods. This type of replenishment pattern is associated with **material requirements planning (MRP),** the topic of the following section, and is representative of the following situations:

Other deviations for which customized *EOQ*-type models have been developed include

- Nonzero constant lead time known with certainty.
- Specified minimum-order quantity.
- One-time opportunity to procure material at a very attractive price.

- Demand for components or an assembly process is relatively deterministic but varies appreciably according to the production schedule for finished assemblies.

- Demand follows a seasonal pattern.

- Demand is determined by production contracts.

- Demand meets given maintenance, replacement, or availability requirements.

Three approaches can be used to cope with varying demand. When the variation is minor, it might be ignored by applying the *EOQ* formula based on the average demand within the time horizon being considered, but the solution could be costly. The minimum-cost solution can be found with the Wagner–Whitin algorithm, but computations are tedious. A compromise is the use of an approximate or heuristic method that promises near-minimal costs with moderate computational effort.

Wagner–Whitin algorithm

Both the exact and heuristic methods are based on known periodic demands over a given planning horizon, block replenishments that occur only at the beginning of a period and holding costs charged only to the inventory carried over from one period to the next. The *Wagner–Whitin algorithm* uses these assumptions with rules that replenishment takes place only when the inventory is zero and that the replenishment period is timed to occur when the carrying costs become higher than the cost of replenishment. Thus the algorithm works backward in time from the planning horizon, considering the demand experienced since the last replenishment and asking this question at the beginning of each time unit: If a replenishment is needed now, what size is most economical? The procedure is demonstrated in Example 13.2.

H. M. Wagner, and T. M. Whitin, "Dynamic Version of the Economic Lot Size Model," *Management Science*, October 1958.

EXAMPLE 13.2 Wagner–Whitin method applied to a time-varying demand over a nine-month planning horizon

The demand for an item during the next nine months is scheduled to follow the pattern shown:

Month	1	2	3	4	5	6	7	8	9	Total
Demand	31	14	7	0	87	44	10	51	8	252

There is no beginning inventory, and the setup cost is $100 per replenishment. Holding costs are $4 per item per month.

A quick way to obtain a replenishment policy is to calculate the *EOQ* by using the average monthly demand, which is 252/9 = 28 units, and

$$EOQ = \sqrt{\frac{2 \times 100 \times 28}{4}} = 53 \text{ units (rounded to the nearest integer)}$$

Then a practical replenishment policy that takes advantage of the given demand pattern is to accumulate monthly requirements until the closest total to the *EOQ* is found and then to replenish that amount. For instance, in month 1 the demand is 31 units, which is below the *EOQ* of 53 units. The sum of the demands for the first two months, 31 + 14 = 45 units, is still below the *EOQ*. The accumulated demand for three months (31 + 14 + 7 = 52 units) is closer to the *EOQ*. Because there is no demand in month 4, the cumulative demand for five months is calculated next (31 + 14 + 7 + 0 + 87 = 139 units) and found to be far above the *EOQ*. Therefore, a replenishment of 52 units is selected as being closest to the *EOQ*, and this amount is assigned to the setup at time zero. The procedure is continued, starting next from a setup at the beginning of month 5, and eventually yields the complete replenishment pattern and associated costs shown in Table 13.1.

The Wagner–Whitin algorithm operates in a somewhat similar fashion but starts with the most distant demand in the planning horizon and works back to time zero. A decision is made at each time period by comparing the cost to replenish at that point with the cost to replenish at the previous time period. When the cost at an earlier period ($T - 1$) exceeds the previous one, a preference is indicated for replenishment at period T.

Applying the algorithm to the given data, the cost for a replenishment at period 9 is the setup charge of $100 to furnish 8 units at the beginning of the period; there is obviously no holding cost.

TABLE 13.1

Replenishment Pattern and Costs for a Policy of Replenishing Whenever the Cumulative Demand Since the Last Replenishment is Closest to the EOQ (53 Units); the Closest Amount of Demand is Then the Replenishment Quantity

Month	1	2	3	4	5	6	7	8	9	Total
Demand	31	14	7	0	87	44	10	51	8	252
Starting inventory	0	21	7	0	0	0	10	0	0	
Replenishment	52	—	—	—	87	54	—	51	8	252
Ending inventory	21	7	0	0	0	10	0	0	0	
Setup cost	100	0	0	0	100	100	0	100	100	$500
Holding cost	84	28	0	0	0	40	0	0	0	152

Total setup and holding cost = $652

Each replenishment has a setup cost of $100. Holding costs are the number of items held over to the next month at a charge of $4 per item per month.

Moving to month 8 ($T - 1$), there are two options:

OPTION 1. Replenish just enough to cover month 8, and accept the cost calculated for the later period, month 9. Cost = $100 + $100 = $200 (two setups and no holding costs).

OPTION 2. Replenish enough for two months at the beginning of period 8. Cost equals setup cost at month 8 plus holding charges for 8 units during month 8 = $100 + 8($4) = $132.

Option 2 is preferred. It is still necessary to see whether there is yet a better option in the next earlier period.

At the beginning of month 7, there are three options:

OPTION 1. Replenish just enough to cover month 7 (10 units), and accept the lowest cost for the later periods, months 8 and 9. Cost equals setup cost in month 7 plus option 2 from the previous comparison = $100 + $132 = $232. Thus this replenishment plan is for a setup in month 7 for 10 units followed

TABLE 13.2

Minimum-Cost Replenishment Pattern Determined from the Wagner–Whitin Algorithm

Month	1	2	3	4	5	6	7	8	9	Total
Demand	31	14	7	0	87	44	10	51	8	252
Starting inventory	0	21	7	0	0	0	10	0	8	
Replenishment	53	—	—	—	87	54	—	59	—	252
Ending inventory	21	7	0	0	0	10	0	8	0	
Setup cost	100	0	0	0	100	100	0	100	0	$400
Holding cost	84	28	0	0	0	40	0	32	0	184

Total setup and holding cost = $584

by a setup in month 8 for 59 units; the option to replenish in both months 8 and 9 is ignored because it was shown to be less economical in the previous round of calculations.

OPTION 2. Replenish enough for months 7 and 8 (10 + 51 = 61 units), and accept the cost calculated for the last period, month 9. Cost equals setup costs for months 7 and 9 plus holding charges for 51 units during month 7 = $100 + 51($4) + $100 = $404.

OPTION 3. Replenish enough for months 7, 8, and 9 (10 + 51 + 8 = 69 units). Cost equals setup cost at month 7 plus holding charges for 51 + 8 =

59 units during month 7 and 8 units in month 8 = $100 + (59 + 8)($4) = $368.

Option 1 provides the lowest cost for the examination of the last three months in the planning period. However, there may yet be a better schedule for replenishments. And indeed there is. The algorithm calls for a continuation of the comparisons, backward in time, month by month, until the complete set of replenishment options is revealed. Table 13.2 shows the resulting minimum-cost pattern. Note that the replenishment option at month 8, but not at month 7, was retained.

Silver–Meal heuristic

Heuristic decision rules provide a simpler approach. Several are available, but only one will be examined. The *Silver–Meal heuristic* is well regarded, easy to use, and yields admirable replenishment patterns as compared with other heuristics. Its operation is similar to the *EOQ* heuristic described in Example 13.2, except that the demand used is based on variable replenishment periods rather than the total demand over the entire planning horizon.

Letting T equal the number of time units in a replenishment period,

E. A. Silver, and H. C. Meal, "A Heuristic for Selecting Lot Size Requirements for the Case of Deterministic Time-Varying Demand Rate and Discrete Opportunities for Replenishment," *Production and Inventory Management*, Vol. 14, No. 2, 1973.

$$\text{Average cost per time unit} = \frac{(\text{setup cost}) + \left(\begin{array}{c}\text{total holding cost} \\ \text{to end of period } T\end{array}\right)}{T}$$

or

$$\frac{AC}{TU} = \frac{O + [(1 - 1)D_1 + (2 - 1)D_2 + (3 - 1)D_3 + \cdots + (T - 1)D_T](H)}{T}$$

where

O = setup cost per replenishment
D_T = demand during the last time unit in replenishment period T
H = holding cost per item per time unit

D_1 drops from the equation because there are no holding charges for inventory used during the time unit in which the replenishment is received. The decision rule is to calculate AC/TU for increasingly longer replenishment periods until the AC/TU no longer decreases; the lowest-cost AC/TU is the replenishment period (T), and the lot size is the sum of the demands during the period, $Q_T = D_1 + D_2 + \cdots + D_T$.

As a sample application of the Silver–Meal heuristic, let consecutive de-

A heuristic helps solve a problem but does not provide a justified solution.

mands for a nine-month planning horizon be 31, 14, 7, 0, 87, 44, 10, 51, 8 (the same demand pattern in Example 12.5). Then the heuristic is applied with $H = \$4$ per item per month and $O = \$100$ per setup.

$T = 1$

$$\frac{AC}{TU} = \frac{\$100}{1} = \$100$$

When just one month is in the replenishment period, only the setup cost is incurred because demand D_1 is never held in inventory.

$T = 2$

$$\frac{AC}{TU} = \frac{O + (T - 1)D_T(H)}{T} = \frac{\$100 + 14(\$4)}{2} = \$78$$

One setup provides 45 items to meet the demand for two months. The 14 items in the second demand unit are charged the holding cost for one month.

$T = 3$

$$\frac{AC}{TU} = \frac{O + [(2 - 1)D_2 + (T - 1)D_T](H)}{T}$$

$$= \frac{\$100 + [14 + (3 - 1)7]\,\$4}{3} = \$70.67$$

The 7 units required during the third month must be held in inventory for two months. The total holding cost results from carrying 14 units for one month and 7 units for two months.

$T = 4$

$$\frac{AC}{TU} = \frac{\$100 + [14 + (2)7 + (4 - 1)0]\,\$4}{4} = \$53$$

Because no items are required during the fourth month, costs remain constant from the three-month replenishment period but are spread over four months to reduce further the average cost per month.

$T = 5$

$$\frac{AC}{TU} = \frac{O + [D_2 + 2D_3 + 3D_4 + (T - 1)D_T](H)}{T}$$

$$= \frac{\$100 + [14 + (2)7 + (3)0 + (5 - 1)87]\,\$4}{5}$$

$$= \$320.80$$

The Silver–Meal heuristic is simpler to use than is the Wagner–Whitin algorithm and gives approximately equivalent results; comparisons indicate less than a 1-percent penalty.

The sharp rise in the AC/TU when the replenishment period is extended from four to five months indicates a preference for a four-month period. A setup at time zero produces a lot size of $Q = 31 + 14 + 7 + 0 = 52$ items. The next replenishment will occur at the beginning of month 5. Applying the heuristic to the remaining months produces the same replenishment pattern shown in Table 13.2 for the Wagner–Whitin solution. In this case the heuristic yields the minimum-cost replenishment plan.

13-4 MATERIAL REQUIREMENTS PLANNING (MRP)

The underlying concepts for the techniques collected and unified under the name **material requirements planning (MRP)** have been known for many years, but they could not be exploited fully without the data-processing power of modern computers. Record-keeping time and costs were formerly prohibitive because MRP combines inventory control with production planning; the time required manually to modify production schedules, as unpredictable demands and delays occurred, was so long that adequate inventory adjustments could not be made fast enough to satisfy the material requirements of manufacturing. After computers eliminated the time constraints, it became possible to unify forecasting, order points, lot sizes, master scheduling, lead times, and inventory record keeping under an umbrella approach. Thus, material requirements planning is an ideal example of an MIS application and systems synthesis. Its role as a computerized "systems balancer" is portrayed in Figure 13.3.

MRP synthesizes inventory concepts from Chapter 12 with control considerations presented in Chapter 14.

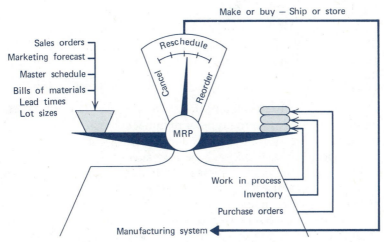

FIGURE 13.3 The MRP "balance" weighs material requirements for product demand against the supply on hand and work in process. In accordance with imbalances, MRP coordinates product characteristics with manufacturing capabilities to schedule production.

Advantages and reservations

The growth of MRP has paralleled developments in computer technology. Its origin, in about 1960, was in line with the movement toward acceptance of quantitative management tools that embraced the data-digesting power of computers (for example, critical path scheduling and linear programming). Early spokesmen for MRP included G. W. Plossl, J. A. Orlicky, and O. W. Wright. It was nurtured by the American Production and Inventory Control Society. Several organizations now offer software packages and consulting services for MRP implementation.

"To Order Point or Not to Order Point," by O. W. Wright, *Production and Inventory Management*, Vol. 9, No. 3, 1968.

Dr. Orlicky suggested that successful MRP users enjoy manufacturing inventory investment levels reduced by 20 to 35 percent. Other claimed benefits are a reduction in production and purchasing costs, and improved delivery service. However, not all implementations are successful.

See "MRP: Is It A Myth Or Panacea," by Don Swam, *Industrial Engineering*, June, 1983.

A switchover from an existing inventory control system to an MRP system is a demanding exercise, one which is fraught with human-relations and technical difficulties. Successful installations of MRP systems are estimated to range from a high of 50 percent to lows of 20 percent, or less. Many reasons have been offered for this low success ratio, most of which center on an eager client's jumping into an MRP installation unprepared and expecting too much too soon. Nearly all new management tools have suffered as a result of woes during their introduction. The cure is to ascertain whether production operations are truly suitable for conversion to MRP; what can be gained in comparison with existing practices; how MRP fits the rest of the production system; and whether the level of competence of existing personnel in inventory management and computer programming is adequate or can be developed.

Overview of MRP

MRP is best applied to manufacturing industries. As we observed in Chapter 12, *EOQ* formulas are associated with independent demand. In manufacturing, the end product is subject to the independent demand of the marketplace, but given that demand, requirements for raw materials and components depend on the manufacturing schedule. This dependent demand condition is served by MRP.

A decision to install an MRP system, and which features to include in an installation, hinges on the type of manufacturing and delivery time required by the marketplace. More complex manufacturing (job lot, manufacturers of capital goods, electrical systems, high performance valves, and the like) and manufacturers with shorter delivery times (finished products available from stock) benefit most from successful MRP implementations. Even in the logical MRP application areas, it may be economical to use order point techniques for high-volume, low-value items that would cost more to control through MRP than simply to carry enough inventory to prevent frequent stockouts.

Process industries (for example, chemical, oil) have relatively simple inventory loads compared with job and flow shops.

MRP is most appropriate for class *A* inventory items.

How MRP meets the needs of manufacturing can be observed by a brief overview of its operation. An MRP system links the planned production schedule with the bill of materials needed to make the product and examines the manufacturing inventory to see which parts and raw materials have to be ordered. By considering when various components of the end product are scheduled to be produced and the necessary lead times for supply, it **time phases** replenishment orders so that parts and materials are available when they are needed at work stations. Ideally, replenishments would arrive at the stations exactly when they are required for use, thus minimizing the amount of work-in-process inventory. But because ideals are rarely attainable, the system reexamines the production schedule, continuously or periodically, to recognize schedule disruptions and irregularities as soon as possible; this reduces stockpiles of inventory in the production area. Success of the adjustments relies on the accuracy of the forecast, bills of materials, processing times, routings, and inventory records.

Perhaps the most distinguishing feature of MRP is that in response to stock level a *time-phased order point* triggers the *arrival* of an order, instead of the *release* of an order.

Another way of viewing MRP is shown in Figure 13.4. Manufacturing control results from forecasts that guide capacity planning and inventory control. Master scheduling considers available capacity, stock status, forecast demand, and customer orders to develop a production schedule. The MRP system issues planned replenishment orders based on information about job status and work-in-process inventory levels. Production results from releasing the planned orders

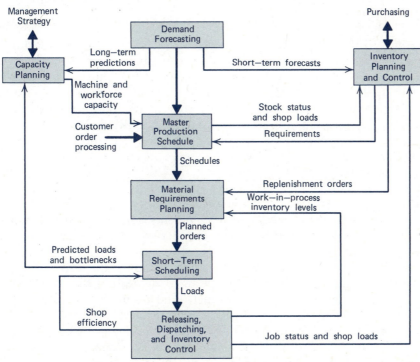

FIGURE 13.4 Position of MRP among the conventional elements of manufacturing control.

in terms of work loads. Data needed to control operations are obtained by measuring work-center output, work-in-process inventory, and job status.

Tools and terminology

To understand the workings of MRP, it is first necessary to become familiar with its tools and terminology. The vocabulary of MRP is composed mostly of terms from MIS, inventory management, and production control. A computer powers MRP, but the data come from conventional tools of production planning. The following definitions and descriptions are pertinent.

MASTER PRODUCTION SCHEDULE

Capacity planning is a management function that allocates capital to production functions in accordance with long-term objectives of the organization (see Chapter 4).

The backbone of an MRP system is the master production schedule. It is built from inputs from marketing and sales that forecast product demand, **capacity planning** that allocates resources to production, and manufacturing operations that reveal status of machines, facilities, operators, and inventories. The resulting schedule shows planned production quantities by time periods, called **time buckets** in MRP parlance. Time buckets are usually a week in duration but may be mixed with longer durations. For instance, a *planning horizon* (the future period to which plans can legitimately be extended) of one year could be composed of 32 time buckets: 26 week-long periods followed by six month-long periods. Shorter periods are typically associated with firm demands supported by customer orders.

A master production schedule is *not* a forecast of *demand*. It is a plan for *production*.

Objectives and constraints of aggregate planning, as discussed in Chapter 6, are included in the preparation of a master schedule.

The development and use of a master schedule is still more skill than science. The complicated, often unpredictable nature of a manufacturing environment precludes strict optimization to minimize cost. Instead, the master schedule is laid out to keep costs within reasonable limits while *satisficing* other objectives. Trial schedules can be developed from *load profiles*—the load on production resources required to produce one unit of a product (for example, standard hours required at each work center). The MRP system can produce load profiles to simulate the use of resources according to different versions of the master schedule. Managers can then observe the effects of various product mixes and the ways to employ resources.

BILL OF MATERIALS

Parent/component relationships of modular bills of materials are similar to prerequisite–postrequisite restrictions found in critical path scheduling. A restriction list is equivalent to the list of named and numbered components of a parent.

The dependent nature of material requirements is shown by the **bill of materials,** also known as a *product structure* or *assembly parts list*. It describes how a product is made from its component parts and assemblies. In the more common *modular form*, a bill of materials for an inventory item (termed the *parent*) shows just the *immediate* components required to produce one unit of the parent (see Figure 13.5). For instance, a finished product could result from the assembly of one unit of component *A*, two units of *B*, and three units of *C*; the finished product is the parent in this case. At the next level, component *A* would be the parent, and its modular bill of material would indicate the components needed to produce one unit of *A*. (In Figure 13.5, component *A* requires one unit of subcomponent *D* and two units of subcomponent *E*.)

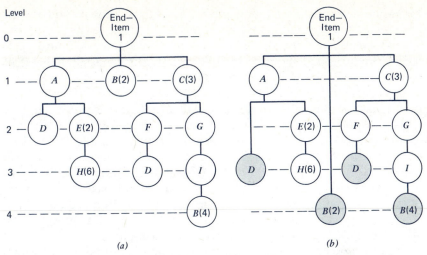

Level

(a)

(b)

FIGURE 13.5 Explosion of a bill of materials according to physical structure and in conformance with low-level coding. Levels suggest the modular form of a bill of materials. An item is the parent for all components linked to it from the next higher-numbered level. (a) Physical structure levels. (b) Low-level coding of components.

LEVEL CODING

Each bill of materials is assigned a **level code** according to lineage from the end product:

Level 0. A finished product, not used as a component in any other product.

Level 1. A component of a level-0 parent. If a component is at once a finished product sold to customers *and* a component of another finished product, it is classified at the lower level. This practice is called *low-level coding* and is used to increase the efficiency of data processing.

Level n. An item at level n is a component of a parent at level $n - 1$. When the item exists at two levels, it is classified at the lower level in conformance to the low-level code.

In Figure 13.5, component *B* is classified as a level-4 item, even though it is a component of end-item 1, because it is also required for subcomponent *I*. Component *D* exists at two levels in the physical structure (levels 2 and 3) and is classified as level 3 in accordance with low-level coding.

LEAD TIMES

The time between issuing a purchase order and receiving the material from a vendor is the lead time. For purchased parts and raw materials, this lead time is assumed to be known, based on negotiations and past experience. Lead times for in-house production are similarly assumed to be known and have the added advantage of being controllable.

Assume that 100 units of component A in Figure 13.5 must be available for the production of end-item 1 on August 22. Further assume that it takes two weeks to produce 200 units of component A, the usual lot size that balances setup costs with demand. This means that subcomponents D and E must arrive by August 8 at the work station where their parent A is produced; that is, 200 units of D and 400 units of E should be available before setting up to make A. (Actually, work could begin on A before the full amount of D and E arrive.) This cascade effect continues with the time that 12 units of H must be available to produce each couple of E units needed in each A unit to make each unit of end-item 1. The whole concept of time phasing is based on scheduling order releases (time to start production) for components in a sequence that allows adequate time for production but cuts storage as much as possible while still allowing some slack or stockpiling to meet uncertainties.

A construct of MRP is that manufacturing lead time is not an automatic "given." Lead time is considered to be a controllable resource and should be managed in the same way as are other resources so as to produce maximum return on investment. When the time between order release and order availability increases by a day, it means that the work-in-process, or "backlog," is behind by one full day of output. Controlling the queues behind critical machines and work centers is necessary to avoid "passing along the backlog" until it delays end-item output.

Information about potential or actual delivery delays is secured by *transaction processing*. Receipts are recorded for each movement of material through the production system. The tracking of material through the stages shown in Figure 13.6 yields the inventory tickets needed to modify order schedules when the planned lead times are jeopardized.

OPERATIONS ROUTING AND WORK CENTERS

The **master** routing for each component is maintained by the industrial engineering department. It lists the operations to produce the component and the sequence in which they are usually performed. Also specified are the work center in which each operation is performed, required setup and processing time, tooling, and materials. Routing records provide data to calculate capacity requirements and to control work-in-process.

Work-center records define manufacturing resources that have the same or similar functions. Data include the number of machines and operators, standard capacity, number of shifts, and work center efficiency. Work-in-process details may also be included: run and setup times, move and queue times, buffer remaining, quantity completed, and the like.

All of the information from operation routings and work centers is included in the data base from which the MRP system guides short-term scheduling or work loads.

Time phasing is the same concept as activity sequencing in critical path scheduling.

In theory, no safety stock should exist in an MRP system, but in practice many organizations build in stock buffers.

Transaction history provides an *audit trail* that allows irregularities to be checked.

Figure 13.3 should be reviewed again to observe how MRP functions in the total control of manufacturing.

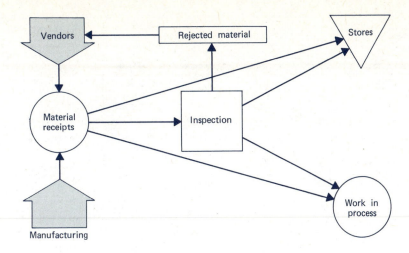

EXAMPLE 13.3 What MRP did for one company, and how

This is the story of a successful MRP installation. (Adapted from "Production Control: MRP Ends Guessing at Southwire," by C. H. Boyer, *Industrial Engineering*, March 1977.) According to Stan Morgan, manager of Corporate Production Control at Southwire Company, pre-MRP inventory practices were wanting: "Our production planners were running the show by the seat of their pants. Reordering for manufacturing plants was being done by a mish-mash of safety-stock levels and gut feelings. And some of the production data, such as manufacturing specifications, had to be transferred onto production orders from hard copy. The system was slow and cumbersome."

It took a year and a half, from the time a decision was made to go with MRP, to start up and debug the system. Controls had to be established over 4900 unique stock numbers from 700 different end-item products with approximately seven components on each product's bill of materials.

Morgan tells what happened: "Our first output from the system was garbage. We soon found out why. It's the computer analyst's axiom: garbage in–garbage out. The information we were feeding the system was not accurate. One major problem was our work-in-process tagging system. The tags were not taken seriously by the employees. Consequently, the number and types of tags turned in from the shop floor did not match the inventory consumed. We implemented an improvement program, with management emphasis, to educate everyone about the importance of the tag system. Tags released on the shop floor are now accounted for at release, at turn in, and once during every shift."

The accuracy of work-in-process inventory went from 69 to 96 percent after the improved tag control became effective. Since the MRP start-up, the number of unique stock items in the work-in process inventory dropped by 50 percent and the total for all items is expected to be reduced by 30 percent.

Figure 13.7 shows how the system operates. The stages of the operation are described next.

① Finished goods demand is generated by customers or by Southwire's 48 distribution warehouses.

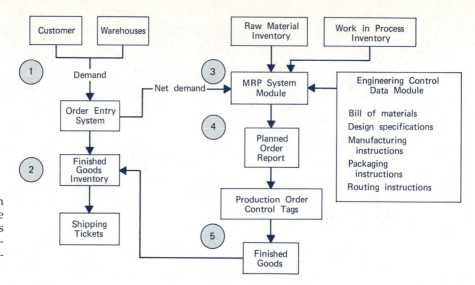

FIGURE 13.7 MRP system employed at the Southwire Company. Circled numbers indicate the stages of MRP operations discussed in the example.

This demand, in the form of sales orders and replenishment orders, is entered in an on-line order entry system controlled by a computer.

② The computer checks demand against a finished goods inventory stored on a real-time access disk file. Finished goods in stock are allocated to the demand, and a shipping ticket is automatically printed. Demand that is not covered by finished goods available for shipment becomes "net demand," the source demand for the production system.

③ Net demand is accumulated daily, entered on magnetic tape in batch form, and directed to the MRP module. The initial output from the module is the Planned Order Report which acts as a master production schedule. It shows product description, source (manufacture or purchase) order number, amount, release and due date, manufacturing time,

and lead time. The MRP module then compares work-in-process inventory with the net demand requirements and determines what materials must be ordered, and when, to meet the production requirements.

④ Production control personnel use the Planned Order Report to initiate and release the actual production orders and to control tags that accompany the products' components through the manufacturing process. Production orders contain manufacturing and packaging instructions, bills of materials, routing instructions, and design specifications.

⑤ As the finished goods are produced, on-line terminals are used to record the information and enter it in the real-time finished goods inventory disk file.

13-5 MECHANICS OF MRP

There are no elaborate formulas involved in MRP. As will be apparent shortly, the reaction of an MRP system to inputs from the inventory records and the master schedule is intuitively logical; the system simply considers the question: What materials have to be available to get the product completed on time? To answer this question, it is necessary to know the gross requirements, how much is on hand, the lead time, and any special conditions such as scrap allowances during production.

To get a feel for what the computer accomplishes and how to read the printouts, sample applications will be examined. Although several conditions are discussed, the sample is an oversimplified version of how the MRP system operates. It must be remembered that the computer may be handling thousands of components in each run, and production/inventory/demand conditions frequently fluctuate between runs.

A composite bill of materials for Product A is shown in Figure 13.8. In a modular explosion of components, each parent would have its own bill of materials, and the components would be classified by levels. Here these conditions are pictorially consolidated for convenience.

The master schedule drives the MRP system by establishing the demand. The projected *independent* demand is derived from external orders already received or from quantities expected to be ordered. Demands are shown in Table 13.3 for 10 periods for end-item A and component D; the external demand for component D results from sales as a subassembly to another manufacturer.

The time buckets could represent any duration; in this case they are assumed to be weeks. A question could arise as to when the demand actually takes place *within* a week. Conventions such as the first day or midpoint of the period could be followed; a common convention we shall adopt is that transactions may occur any time during the period but no later than its last day. Note that the requirements for component D do not include its dependent demand as a component required to produce product A. This will be incorporated later.

Assume that product A has a one-week lead time and can be produced in lot sizes equal to its demand. Then components B, C, and D have a dependent demand equal to the demand for A, but occurring one week earlier. Because one unit of component B is required for each unit of the end item, as prescribed in the bill of materials for product A, it will have projected requirements of 200 units in the first week and 300 in the second. This one-week offset of demand and the special conditions for the production of B (order quantity, lead time, and amount available at time zero) are shown in Table 13.4.

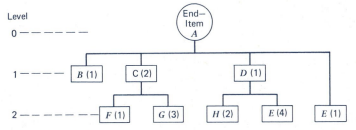

FIGURE 13.8 The bill of materials for end-item A is laid out according to low-level coding. The numbers in each box indicate the quantity of components required to produce one parent.

TABLE 13.3

Projected External Requirements for Product A and Component D.

Time Bucket	1	2	3	4	5	6	7	8	9	10
Product A requirements		200	300		500		400			600
Component D requirements	80	80	80	80	60	60	60	40	40	40

The replenishment pattern in Table 13.4 is an ideal and, naturally, one that is seldom possible: a replenishment of any size is instantly available. Thus the schedule receipts equal the planned order release because there is no time lag between them. For this situation, a planner would wait until the demand is certain and then order that exact quantity. *The quantity available in period* n *is equal to the scheduled receipts in that period plus the carry-over amount available from period n-1.* It is zero throughout the shown pattern because scheduled receipts are coordinated with the demand. Because scheduled receipts are not normally shown on a plan until an order has actually been authorized, entries beyond the first week are shaded to indicate their hypothetical nature. They would appear as shown (without shading) in an audit if the projected requirements turned out to be the actual demand.

A more realistic situation for component B is shown in Table 13.5. Components are ordered in lot sizes of 450, and the lead time is two weeks. Apparently an order was placed one week before the first week in the table, and its receipt is scheduled in week 2. Note that 200 units are available in inventory from a previous receipt.

Again, as in Table 13.4, the shaded cells indicate hypothetical transactions over the 10-week planning horizon. Orders are released in weeks 2, 4, and 7;

TABLE 13.4

Material Requirements Plan for Component B When Any Quantity Can Be Ordered and Delivery Is Immediate. Orders Are Time Phased

Component B
Order quantity: variable
Lead time = 0

	1	2	3	4	5	6	7	8	9	10
Projected requirements	200	300		500		400			600	
Available 0										
Scheduled receipts	200	300		500		400			600	
Planned order release	200	300		500		400			600	

Available indicates the number of units on hand.

Scheduled receipts are the quantities to be delivered from orders already placed.

Planned order release is the premeditated size of a production order.

TABLE 13.5

Material Requirements Plan for Component B When Its Order Quantity Is 450 and Lead Time Is Two Weeks

Component B
Order quantity = 450
Lead time = 2 weeks

		1	2	3	4	5	6	7	8	9	10
Projected requirements		200	300		500		400			600	
Available	200		150	150	100	100	150	150	150		
Scheduled receipts			450		450		450			450	
Planned order release			450		450			450			

each arrives as a scheduled receipt two weeks after release. The quantity available in week 2 is

$$450 \text{ (amount received)} - 300 \text{ (demand requirement)} = 150$$

This amount is carried over to week 3. Then the quantity available in week 4 is

$$450 + 150 \text{ (available carry-over)} - 500 = 100$$

Recall that component D is required as a subassembly by another manufacturer in addition to being part of end-item A (Table 13.3). Therefore, the two demands are combined to obtain its total projected requirements, as is shown in Table 13.6. Special conditions for component D include an order quantity of 500 units with a lead time of one week, 160 units on hand at the beginning of the planning period with receipt of 500 units scheduled during week 1, and a policy of requiring a safety stock of 40 units.

Holding a safety stock for components is contrary to the principles of MRP. It is recognized that safety stock is appropriate to end-item products that are

A negative quantity available would indicate a shortage. Shortages are never purposely planned in an MRP system.

MRP philosophy recognizes the need for a safety stock for components when *supply* is uncertain. Therefore, safety stocks are usually limited to purchased items.

TABLE 13.6

Material Requirements Plan for Component D In Which Projected Requirements Are Obtained From Combining Internal and External Demands

Component D
Order quantity = 500
Lead time = 1 week
Safety stock = 40

		1	2	3	4	5	6	7	8	9	10
Projected requirements		280	380	80	580	60	460	60	40	640	40
Available SS 40 120		340	460	380	300	240	280	220	180	40	
Scheduled receipts		500	500		500		500			500	
Planned order release		500		500		500			500		

A safety stock of 40 units is "buried" in the quantity available, 120 + 40 = 160.

subject to independent demand, but the inclusion of safety stocks for components overstates the requirements and creates "dead" inventory that is carried along but is never used. This reasoning is supported by the realization that the demand for an individual component is not being forecast: the demand is a function of its end-item production schedule, which has already been adjusted to account for market fluctuations.

Because component D is subject to independent demand from sales to another manufacturer, it is both an end-item and a component. The safety stock is applied in recognition of its independent demand. As is shown in Table 13.6, the safety stock is buried in the quantity available. That is, the safety stock is subtracted from the supply on hand, and the reduced quantity is carried forward in the material requirements plan. The same effect could be obtained by adding the safety stock to the projected requirements in all time buckets. The available quantities would then be actual measures, and the projected requirements would be artificially enlarged.

Safety stock applied to projected requirements would alter the first row of Table 13.4 to Projected requirements: 320, 420, 120, 620, 100, and so on.

In the requirements plans for components B and D, the order quantities are held constant throughout the planning horizon. In Section 13-2, models were introduced to calculate variable lot sizes for time-varying demand. Most MRP applications use such models to reduce the quantity available during periods of low demand. The Silver–Meal heuristic is applied to the requirements plan for component E.

Low-level coding forces the collection of both demands for component E into one set of requirements.

The bill of materials in Figure 13.8 stipulates that one unit of component E is required for the assembly of each end-item A and that four units of E are used in each unit of component D. It is assumed that the previously used order releases for A and D are applicable. That is, the requirements for product A from Table 13.3 are offset one week to account for the lead time, and the order releases for component D from Table 13.6 are each multiplied by 4 to represent the number of units of component E per unit of D. Every lot size of 500 units of D demands 2000 units of E. These requirements are summarized as follows:

Week	1	2	3	4	5	6	7	8	9	10
Requirements of component E for assembly of end-item A	200	300		500		400			600	
Requirements of four units of E in each unit of component D	2000		2000		2000			2000		
Total projected requirements for component E	2200	300	2000	500	2000	400		2000	600	

Other lot size heuristics are called Least Unit Cost, Least Total Cost, and Part-Period Balancing. All resemble the Silver–Meal approach.

Special conditions for component E show no units on hand at time zero, 2500 scheduled to be received during week 1, holding cost of $0.07 per unit per week, and setup cost of $119. Based on these conditions and application of the Silver–Meal heuristic, the material requirements plan evolves as is shown in Table 13.7.

TABLE 13.7
Material Requirements Plan for Component E to Satisfy Demands for End Item A and Component D

Component E
Lead time = 1 week
Setup cost = $119
H = $0.07 per unit per week

	1	2	3	4	5	6	7	8	9	10
Projected requirements	2200	300	2000	500	2000	400		2000	600	
Available 0	300		500		400			600		
Scheduled receipts	2500		2500		2400			2600		
Planned order release		2500		2400			2600			

Order sizes are determined from the Silver–Meal heuristic. The average cost per week for satisfying demands during weeks 3 and 4 with a single order released in the second week is [$119 + 500($0.07)]/2 = $77 per week, as opposed to $119 per week if separate orders were placed.

As a final illustration of MRP concepts, assume that G, a subcomponent of C, has a production history of 5-percent defectives. This means that roughly 105 units must be ordered for every 100 units needed (100/0.95 = 105.26). The 5 defective units per 100 manufactured are called the *reject allowance*. A reject allowance, rounded to the nearest whole unit, must be applied to each order release for subcomponent G to provide enough good units to meet the requirements for component C.

To find out when to produce subcomponent G, it is first necessary to determine the order release dates for component C. A combination material requirements plan for end-item A, component C, and subcomponent G is shown in Table 13.8. The combination illustrates the "explosion" of requirements from the bill of materials. The cascade effect of lead-time offsets is graphically apparent. The expansion of projected requirements is caused by the need for two units of C in each product A (2C/A) and four units of G in each unit of C (4G/C).

An exception to the MRP rule against holding a safety stock for a component is allowed when the manufacturing process cannot avoid producing defective items. An order size larger than the scheduled receipts compensates for the *average* number of defectives in each lot. However, the actual number rejected in a lot may deviate considerably from the mean. To protect against the rare situation when a lot is spoiled by a very high rejection rate, a safety stock can be held. This condition is shown in the material requirements plan for component G in Table 13.8: 36 units are kept available to make up deficiencies in a spoiled lot. The size of the safety stock can be calculated according to the methods suggested in Chapter 12.

Assuming a binomial distribution and a mean lot size of 3000, the safety stock for component G could be calculated as
$3\sqrt{3000(.05)(1 - .05)} =$
36 units when a three-standard-deviation protection is sufficient.

Pegging
The demand for some items results from several sources, as is demonstrated for components D and E in the sample application, but the source identity is

TABLE 13.8
Material Requirements Plans for End Term A, Component C, and Subcomponent G

End-Item A
Lead time = 1 week

	1	2	3	4	5	6	7	8	9	10
Projected requirements		200	300		500		400			600
Available 0										
Scheduled receipts		200	300		500		400			600
Planned order release	200	300		500		400			600	

(2C/A) (2C/A) (2C/A) (2C/A)

Component C
Lead time = 2 weeks

	1	2	3	4	5	6	7	8	9	10
Projected requirements	400	600		1000		800			1200	
Available 0	600									
Scheduled receipts	1000			1000		800			1200	
Planned order release		1000		800			1200			

(4G/C) (4G/C) (4G/C)

Subcomponent G
Lead time = 2 weeks
5-percent reject allowance
Safety stock = 25

	1	2	3	4	5	6	7	8	9	10
Projected requirements		4000		3200			4800			
Available $|SS = 36|0$										
Scheduled receipts		4000		3200			4800			
Planned order release		3368			5053					

Explosion of the bill of materials and lead-time units of product A in week 10 causes an order release for 5053 units f subcomponent G in week 5 (dashed lines). The order side for G includes a 5-percent reject allowance (in week 5: 4800/0.95 = 5053). Lot sizes for planned order releases are determined heuristically.

lost when the total requirements are collected for an item's material requirements plan. For some occasions it may be important to know the status of the items intended for certain destinations. For instance, if production delays are likely to create an imminent shortage of a particular item, it is helpful to know which assemblies, finished products, and, ultimately, customer orders may be adversely affected. This is accomplished by *pegging*: an item is "pegged" with information to identify its parent in a given time bucket. Because pegging consumes considerable file space and data-processing time, it should be applied only when the information it supplies is of special consequence to management.

Accommodation of changes

Production plans are subject to frequent changes both from external supply or demand requirements and from internal factors that contribute to or detract from manufacturing capabilities. An MRP system accommodates these changes through one of the following methods:

Schedule regeneration. Material requirements are updated by literally throwing away the previous plan and periodically reexploding all end-item requirements, usually each week. In doing so, every bill of material must be retrieved, and the inventory status of every item must be recomputed. All changes that have taken place since the last regeneration are incorporated in the new run.

Net change. Replanning continuously takes place in a net change system. Whereas the regeneration system is set off with the submission of a master schedule, the net change system is initiated by inventory transactions. Each transaction is fed into the system as it occurs. Because a transaction affects only predecessor requirements, just those items affected by the change are updated.

Which of the two updating methods to use depends on the nature of the production setting. Most users start with the regeneration system. As they become more familiar with MRP, they may elect to switch to the net change system to gain the benefits of reduced computer time and more specific information. The benefits are bought at the price of stricter discipline for system inputs and more "nervousness" in the system: indications that frequent actions should be initiated to comply with the latest fluctuations in scheduling information (for example, moves to remedy temporary shortages or surpluses at various points in the system in reaction to the most recent computer printout).

13-6 MANUFACTURING RESOURCES PLANNING (MRP II)

As originally conceived and applied, material requirements planning was a technique that drew on computers' data-processing power to print detailed plans for producing dependent items just when they were needed. Although the savings attributable to MRP did not always meet expectations, it gained respect as a planning tool. Improved computer hardware and software made MRP implementations less arduous. An MRP methodology continued to evolve.

The expansion of MRP beyond the planning function to include control was a natural progression. It was readily apparent that master production scheduling would be improved by harnessing material needs to other dependent

Typical changes: customer orders, bill of materials, bill of production orders, lead times, and discrepancies in inventory records.

Only affected parts of the master schedule must be exploded.

A compromise between the two updating extremes is to process batches of changes as needed, perhaps two or three times a week.

requirements such as machine hours, labor hours, and capital. This development tied material requirements to capacity requirements. Then shop-floor progress and vendor relations were added to control output quantity and time. When feedback from operations is connected to management planning activities, the whole manufacturing process becomes a closed-loop system. The name given

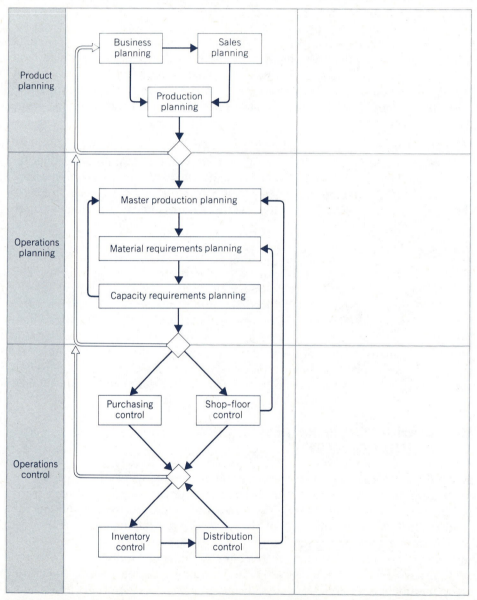

FIGURE 13.9 Functions and feedback channels in a closed-loop MRP II system.

to this concept is **manufacturing resources planning (MRP II),** also known as closed loop MRP.

Under MRP II, complete product cycles are analyzed, from corporate plans about which products to emphasize to control over finished goods distribution. MRP II thus provides an integrated database for information sharing throughout the organization. By knowing the current status of orders, inventories, output, and other operating levels, questions can be routinely answered that were not even asked before because of data difficulties.

A manufacturing resources planning system is modeled in Figure 13.9. It is divided into three parts composed of *product-planning* functions which take place mostly at the upper-management level, *operations planning* handled by staff units, and *operations control* functions conducted by manufacturing line and staff supervisors. Checkpoints among the three divisions provide feedback, in descending order, about the adequacy of overall resources, completeness of resource commitments, and the quality of performance in carrying out the plans. Feedback based on these checks permits a quick response to changing conditions using the latest operating data.

Strategic considerations translate in the top section of Figure 13.9 into corporate business objectives, into sales objectives by product lines, and into product forecasts by part number. The resulting production plan becomes a commitment for *finance* to provide enough financial resources, *production* to manufacture the agreed-upon quantity, and *sales* to sell the amount produced.

The middle section of the MRP II model is basically the MRP function with added attention to both long- and short-term physical constraints. The master production scheduler compares what is needed with what is on hand and makes adjustments considering lead-time and work-center utilization factors. The traditional MRP then explodes end-item schedules to organize orders for lower-level parts.

Orders become products in the execution section where parts are purchased, stock levels are tracked, work orders are carried out, parts movements are monitored, labor performance is checked, and distribution activities are reviewed. Evaluation of the data from these sources influence the implementation of the master schedule and provide input for future business planning.

Few changes in manufacturing practices are promoted by MRP II. Its value is derived from information that gives all departments access to relevant data and the application of its database for system improvements. The methodology of manufacturing resource planning is a strong advance toward total systems synthesis.

MRP II software packages are bifurcated. One part describes manufacturing (bills of material, part routings, and so forth), and the other portion controls the production process.

Top MRP II systems are called class *A*. Among other accomplishments they must have the following:

- inventory records 95-percent accurate.
- bills of material 99-percent accurate.
- single-digit set-ups established.
- weekly or shorter deliveries to point of usage for 80 percent of all material.
- individual operator responsible for quality.

The MRP II database is ideal for simulation studies that ask "what-if" questions about manufacturing.

13-7 JUST-IN-TIME (JIT) PRODUCTION

If inventory is thought of as water in a lake covering a bed of boulders, in which the boulders represent problems, it is logical that lowering the water level (inventory) exposes the boulders (problems), making them easier to find and remove. By likening the lake bed to a factory floor, the analogy conveys a powerful message: The way to detect what is holding back production is to reduce stock levels enough to expose operating inefficiencies that are normally masked by buffers of stockpiled parts. In essence, this happens when parts are produced or received *just* when they are needed in the production process, thereby eliminating buffer stocks. The name given to this manufacturing philosophy is **just-in-time (JIT) production.**

Advocates of JIT claim it is a revolutionary concept that all manufacturers will have to adopt in order to remain competitive. Also known as kanban, zero inventory production (ZIP), demand scheduling, stockless production, and minimum inventory production system (MIPS), the basic approach is continually to reduce product costs by stressing the elimination of waste: no rejects, no delays, no stockpiles, no queues, no idleness, and no useless motion.

To achieve low-cost, high-quality, on-time production, the JIT system removes stock accumulations between successive operations. It does so by organizing around a production quantity of "1," which means the ideal lot size for each part is 1. Because no safety stock is allowed, no parts can be faulty. The responsibility for eradicating defective work and equipment failure is placed on individual operators. Output quotas are inviolable, and fluctuations in daily schedules are minimized to maintain a nearly uniform flow rate. Results from applying these principles, along with a concerted effort to improve productivity, have frequently been spectacular.

Kanban system: Pull instead of push

The phrase "just-in-time" originated in Japan, and its most famous application is at the Toyota Motor Company. The Toyota system, known as **kanban** after the Japanese word meaning "visible record," uses only two types of cards (kanbans) to signal the amount and timing of material flow:

A **move card** authorizes the transfer of one standard container of a specified part from the work station where the part was produced to the station where it will be used.

A **production card** authorizes the production of one standard container of a specified part at the work station from which a container has been transferred.

A card travels with a container and typically is marked with an identification number, a part number, a part description, a place of issue, and the number of items in the standard container. Cards thus substitute for a computer in tracking and controlling material flow.

Kanban cards constitute a simple, flexible scheduling system that promotes close coordination among work centers in repetitive manufacturing. The amount of material in the system is controlled by having a prescribed number of containers in circulation at any time. A *user* work centers "pulls" containers from

MRP II emphasizes planning to anticipate problems. JIT's emphasis is on execution to identify problems.

Typical claims three years after implementing JIT are a 30-percent increase in labor productivity, a 60-percent reduction in inventory, a 90-percent reduction in reject rates, and a 15-percent reduction in used plant space.

Special kanbans may be used for subcontractors and emergency situations. Automatic machines equipped with limit switches that conform to the as-needed principle are called *electric kanban.*

Bar-coded plastic tags and overhead signal lights have replaced conventional kanbans in some installations.

a *supplier* work center with a *move* card. Equivalently, a supplier cannot "push" a container out to a user because no movement can occur until the user is ready, as indicated by the arrival of the move card. Moreover, the supplier cannot produce until it receives a go-ahead in the form of a *production* card.

The difference between a **pull manufacturing system** and a **push manufacturing system** is the difference between *producing to order* and *producing to schedule*. In a pull arrangement, upstream activities are geared to match final assembly needs. When all component parts and materials are pulled through production in exact correspondence to end-item demands, the theoretical ideal of *stockless production* is achieved. However, a pure pull system is susceptible to almost-instant stoppage by a breakdown in any upstream activity.

Because a kanban moves with a batch of parts, specifying what is to be done with the parts, any change in flow is automatically signaled by a comparable change in the movement of kanbans.

EXAMPLE 13.4 Toyota's kanban system*

The idea for just-in-time production is said to have been conceived by the president of Toyota Motor Company in 1940, but it was the mid-1950s before the idea became a practice. Toyota started producing trucks after World War II. As output increased from 1000 vehicles per month to 5000, the plant was gradually automated. Automation (*jidohka*) was the first reform made to strengthen Toyota. The second was just-in-time production. Both were spurred by competitive pressure. The initial effort at JIT was to change the flow of production, transportation, and delivery.

The kanban system was introduced in 1953 as a production tool. Cards served as production orders in the line-manufacturing departments and as withdrawal indicators for the subsequent departments. To make the system work, the production control department is responsible for overcoming manufacturing troubles, and so a competent department is essential. If the manufacturing organization is compared to a human body, production control is the brain, and kanban is the nervous system.

The touted *Toyota Production Method* is more than production tracking by kanbans. The following features support the kanban concept and are themselves major programs:

*Adapted from H. Hayashida and T. Kondo, "Kanban System in Practice," *Proceedings* (Tokyo: 1982). International Congress on Productivity and Quality Improvement.

DEFECT-FREE PRODUCTION. The kanban method requires that production be stopped when defects are found. Stops are minimized by eliminating the major causes of defects: operator carelessness, excessive force, irregular procedures, and waste. It has been proved that the rate of defective production is consistently below 1 percent when these four causes are removed.

SINGLE-UNIT PRODUCTION. The implementation of kanban usually reveals imbalances in production that can be corrected by smoothing the flow when material accumulates, unnecessary reloading occurs, and inefficient deliveries take place.

Several different products can be made on the same assembly line, as each product, such as one model of an automobile, is a lot size of 1. To achieve single-unit production, workers must have several skills; cycle times must be leveled using very short tasks; and facilities must allow rapid changeovers.

INTEGRATED PRODUCTION. When an end-item manufacturer uses the kanban method, the suppliers of that manufacturer must also be willing to adopt the method. Complete production information must flow between the user and the suppliers, even when they are separated by great distances. The resulting coordination should increase profits for both.

Comparison of pull and push system characteristics

The logic behind just-in-time production is unassailable. Its emphasis on personal dedication to higher quality, job enrichment from the application of multiple skills, elimination of idle resources, trust in suppliers, and a continuous effort to increase productivity are practices endorsed by nearly everyone in

TABLE 13.9

Comparison of Manufacturing Pull and Push Systems Representing, Respectively, Kanban/ JIT and MRP II/Batch Production

System Characteristics	Manufacturing Push System (MRP II and Batch Production)	Manufacturing Pull System (Kanban—JIT Production)
General Approach	Balanced, nonstop production to meet a predetermined schedule.	Simple, flexible production that responds quickly to demand changes.
	Extensive computerization to handle complex procedures.	Minimized record keeping and simplified methods.
	Task-oriented workers with specialized skills.	Thinking workers with multiple skills.
Machines	Specialized machines.	General-purpose machines.
	A few large "super" machines.	Many small machines.
	Single-purpose tooling.	Flexible tooling and simplified setups.
	Sophisticated material-handling devices to move large quantities of material.	Material handling limited by placing work centers near one another.
	One operator per machine.	Multiple machine responsibility.
Materials	Multiple suppliers to avoid disruption of deliveries caused by single sourcing.	Limited number of suppliers to establish close coordination with each one.
	Large, infrequent deliveries.	Small, frequent deliveries.
	Obtain stock from storage.	Obtain stock from production.
Production Practices	Fixed model production runs.	Mixed-model production runs.
	Keep labor busy.	Keep material moving.
	Identify manufacturing defects.	Prevent manufacturing defects.
	Performance measured by output and individual accomplishments.	Performance measures such as individual quality and team productivity improvements.
	Extensive planning by the engineering department to correct problems before they occur.	Joint problem solving by workers, engineering staff, and management to correct problems as they occur.
	Quality control department monitors quality.	Individual workers are responsible for quality.
	Inspect quality in.	Build quality in.

production management. Why, then, does not everyone implement JIT? The reasons are expense, difficulty, and tradition.

A JIT system is not easy to install. Not only does it require special training, new changeover tooling, reorganization of policies and procedures, revised supplier relations, and reindoctrination of employees, but management commitment and patience also are necessary. Everyone involved must have great faith in the process to get through the turmoil of implementation before the expected results are realized. And the results are not guaranteed, but potentially large returns make the gamble attractive.

The characteristics of a manufacturing pull system, as represented by kanban, are compared in Table 13.9 with traditional push systems, such as MRP II and batch manufacturing. The contrasts are vivid. They suggest that in U.S. manufacturing, a blend of push and pull systems will emerge. Kanban procedures may be modified to accommodate distant suppliers and more frequent revisions to production plans. Equivalently, conventional practices will likely put more emphasis on quality at all levels, integrated supply systems, lean inventories, and increased concern for productivity. The blend is a recipe for the must-prophesized reindustrialization of the United States.

Design changes that assist the introduction of JIT include
- lenient tolerancing.
- specifications that allow the use of simple, flexible equipment.
- producibility with common tooling.

13-8 CLOSE-UPS AND UPDATES

Production doctors that make factory calls

Riding the crest of the electronic tidal wave is a raft of computer consultants with programmed answers to production problems. Their programs range from spreadsheet solutions for small business problems that run on personal computers to elaborate packages that include training and myriad managerial services, along with computerized assistance. **Optimized production technology (OPT)** fits the second category.

OPT is the brainchild of M. E. Goldratt and is marketed as a proprietary package. Since its U.S. introduction in 1979 it has generated considerable interest as a production-planning and -scheduling tool. OPT has characteristics of both MRP II and JIT, using product information similar to that required for MRP and attempting to immerse managers in a complete philosophy similar to the commitment needed to apply JIT effectively.

Throughput maximization is the OPT objective, and *bottleneck elimination* is the action plan, in which any resource that must be used at 100-percent capacity to maintain maximum output is considered to be a potential bottleneck. By breaking orders into a number of small batches and scheduling these minibatches to be run consecutively, setup times are reduced, and queues in front of bottleneck machines or work centers ensure continuous work for limiting resources. By considering such priorities and capacities, OPT calculates a nearly optimal schedule.

OPT uses a set of "management coefficients" that weight factors that control production efficiency such as product mix, plant capacity, work in progress, due dates, setup times, subcontracts, and safety stocks.

A set of rules, called *thoughtware* by OPT developers, describes the program's premises:

- *Balance flow, not capacity.* Do not be concerned whether work is distributed evenly. Concentrate instead on throughput.
- *Realize that losses from bottlenecks cannot be recaptured.* Get 100-percent utilization from bottleneck resources. Do not worry if less important resources are occasionally idle.
- *Vary lot sizes.* Different lot sizes can be used for each operation, and they can change in size as a function of the production schedule. Lead times are not fixed.

There are many algorithmic programs for production scheduling besides OPT and MRP, of course, and more will be available in the future. All will have advocates and detractors. The challenge to production managers is to decide which, if any, are worth using. Trenchant observations by Sumer C. Aggarwal in *Harvard Business Review*, September–October, 1985, in his article "MRP, JIT, OPT, FMS?", focused on implementation considerations for these celebrated techniques:

If a common obstacle exists to successful implementation of computerized production systems, it's getting employees to perform appropriately. . . .

Indirectly, the kanban system addresses these problems, which is probably why most of its users are reporting successful results. Kanban is a simple and transparent system. Employees are responsible for making it work, and results indicate that they are willing to accept the challenge.

MRP offers no challenge to employees but requires that they be extremely disciplined and committed at all levels—which helps explain why 90% of users are unhappy with results.

OPT tolerates minor disturbances and requires moderate discipline and limited data accuracy. The contract rules its consultants impose force top executives to make procedural, cost-accounting, and work-method changes, which may explain why problems with employees get resolved indirectly and the limited number (to date) of OPT users seems to be reasonably happy with the system.

13-9 SUMMARY

Systems synthesis

Systems synthesis is a cause-to-effect process that integrates the elements of production. A distribution system, for example, interfaces with all of the production functions. Systems thinking treats distribution as a means to unite systems within systems. Suboptimization within the total system weakens the coordination potential of the distribution subsystem.

Replenishment quantities and scheduling for time-varying demands can be determined by the Wagner–Whitin model for the least-cost solution or by heuristic decision rules, such as the Silver–Meal heuristic, which are quicker but do not guarantee cost minimization. These methods are applicable to deterministic demand patterns, which occur in manufacturing systems that use MRP.

Lot sizing for time-varying demands

Material requirements planning synthesizes inventory management and production planning. It is associated with the dependent inventory demands found in manufacturing. An MRP system is driven by the master production schedule. It compares projected material requirements with the amount in inventory and the scheduled work-in-process to determine when an order should be released for the manufacture of the next lot.

MRP concepts

The master schedule states the quantity of each item needed in each scheduling period, called a time bucket. A modular bill of materials shows the components required to produce the next level of an item, called the parent. Low-level coding classifies a component according to its lowest level of use in the structure of a product. Lead times are believed to be controllable and therefore of fixed duration. Time phasing offsets the release of a component's production order from its scheduled receipt: the offset in the lead time between order release and the time the parent needs the component. Holding a safety stock for components is contrary to the principles of MRP, except for items with independent demand or reject allowances.

The mechanics of an MRP system can be illustrated by a tabular format that relates quantities required, available, on order, and planned. The quantity available in period n is equal to the scheduled receipts in that period plus the carry-over amount available in period $n - 1$ minus the requirements for period n. Lot sizes are typically determined by a heuristic method that recognizes time-varying demand. An order release may be increased by a reject allowance to account for defective items from the production process.

MRP procedures

Material requirements are updated by periodic *regeneration* of the entire schedule or by continuous *net-change* replanning that involves only items affected by a change. The critical areas to control for successful operation of an MRP system are data base input, engineering changes, and transactions reporting.

Closed-loop MRP, known as manufacturing resources planning (MRP II), combines planning functions with operational controls to form an information network that promotes systems synthesis.

MRP II concepts

An extensive set of management–worker practices and work-place innovations are included in the just-in-time (JIT) approach to production. The Japanese kanban method that uses cards to control material movement and emphasizes defect-free, single-unit manufacturing has attracted wide attention. Kanban is a pull system in which part production matches assembly output, rather than a push system, such as batch manufacturing, in which production levels are based on the forecast demand.

JIT concepts

Kanban

Pull/push systems

Optimized production technology (OPT) is an example of computerized attempts to maximize efficiency by including total-system considerations in production scheduling. OPT focuses on eliminating bottlenecks.

OPT

ADAM, E. E., JR., and R. J. EBERT. *Production and Operations Management*, 3rd ed. Englewood Cliffs, N.J.: Prentice-Hall, 1986.

BEER, S. *Cybernetics and Management.* New York: Wiley, 1964.

HALL, R. W. *Zero Inventories.* Homewood, Ill.: Dow Jones–Irwin, 1983.

HAX, A. C. and D. CANDEA. *Production and Inventory Management.* Englewood Cliffs, N.J.: Prentice-Hall, 1984.

KRAUSS, L. I., and A. MACGAHAN. *Computer Fraud and Counter-measures.* Englewood Cliffs, N.J.: Prentice-Hall, 1979.

ORLICKY, J. *Material Requirements Planning.* New York: McGraw-Hill, 1975.

SCHONBERGER, R. J. *Japanese Manufacturing Techniques: Nine Hidden Lessons in Simplicity.* New York: Free Press, 1982.

STEVENSON, W. J. *Production/Operations Management.* Homewood, Ill.: Irwin, 1982.

13-11 SELF-TEST REVIEW

Answers to the following review questions are given in Appendix A.

1. T F *Distribution* costs can account for as much as 25 percent of the price of some products.

2. T F The majority of reported losses from *computer fraud* are caused by the organization's own employees.

3. T F The *Wagner–Whitin* algorithm is simpler to use by hand than is the *Silver–Meal* heuristic.

4. T F *MRP* concepts could not be fully exploited without the data-processing power of computers.

5. T F The majority of organizations that attempt to implement *MRP* have highly successful results.

6. T F A *bill of material* shows planned production quantities by time periods, called *time buckets.*

7. T F Estimates in an MRP system of the *dependent demand* for an end item are made from the external orders received and expected.

8. T F In the *level coding* of a product structure, level zero represents the finished product.

9. T F *Schedule regeneration* and *net change* are two methods for updating in MRP.

10. T F *MRP II* includes MRP in its planning section.

11. T F In the analogy in which water is likened to *inventory*, lowering the water level exposes boulders that are likened to marketing problems.

12. T F *JIT* provides an effective database for simulation exercises.

13. T F The Japanese *kanban* system involves computerized tracking of *move* and *production cards.*

14. T F A manufacturing *pull* system, such as kanban, matches production to final assembly requirements.

15. T F *Just-in-time* production is associated with task-oriented workers with specialized skills.

16. T F According to *OPT* principles, eliminating bottlenecks is far more important than eliminating idle time throughout the production system.

13-12 DISCUSSION QUESTIONS

1. A margin note in Section 13-2 uses three paired words to characterize a system: (1) structural (static) or functional (dynamic), (2) purposive (goal oriented) or nonpurposive, and (3) mechanical or organismic (if an element of the system can be removed without affecting the remaining elements in the system, it is organismic). Classify the following systems according to these "bipolar dimensions":

a. The solar system.
b. A road network.
c. A standard operating procedure (SOP).
d. The human nervous system.

2. Another way to classify systems (besides the categories in Question 1 is according to the following divisions: (1) living or nonliving, (2) abstract or concrete, and (3) open or closed.

a. Identify two systems with exactly opposing classifications.
b. What is the purpose of classifying the characteristics of a system according to a set of dimensions?

3. From a systems perspective, what is meant by the advice that "profit maximization should take precedence over cost reduction"?

4. Compare Figures 13.2 and 12.2. Both represent functions within a larger system. What differences in perspective are evident in the two characterizations of material flow? Are they compatible?

5. Two of the difficulties encountered in attempting to use the potential of computers (and, correspondingly, systems analysis) are captured by the descriptions of "Maslow's maxim": "If the only tool you have is a hammer, you tend to treat everything as though it were a nail," and the "Cargo cult": a custom supposedly practiced by Melanesian natives who, having seen "silver birds" come out of the sky to land on cleared jungle runways during the war, still periodically cut down the brush and wait for the silver birds to return. Discuss computer application problems analogous to Maslow's maxim and the Cargo cult.

6. It is known that a very small proportion of computer frauds are ever recorded. Suggest reasons that they often go undetected and, if they are detected, that they are still often unreported.

7. Data security is achieved by preventing the unauthorized use and accidental loss of data. Both events can be expensive. The three main trouble sources, with examples of each, are listed in order of likelihood:

a. *Error and omission*—transactions routed to the wrong place and duplicate entries that result in double ordering that causes inventory imbalances.
b. *Disgruntled and dishonest employees*—clever operators who feel unappreciated and find that by moving around accounts, shipments can be made to a nonexistent warehouse.
c. *Damage from fire, water, and external attack*—ruined equipment or theft by competitors.

   Discuss the three trouble areas in terms of ethics and ways to avoid or minimize loss.

8. A property of the Wagner–Whitin algorithm and all of the heuristics for determining the replenishment period for time-varying demand is that the replenishment never occurs during a period when the demand is zero. Why?

9. Why are class *C* inventory items often more economical to handle by using order point techniques, even when the rest of the manufacturing system uses MRP?

10. Compare the *EOQ* and *MRP* responses to stock-level changes that indicate the need for inventory replenishment (Distinguish between an order release and an order arrival in each situation).

11. How does an MRP system epitomize the concepts of systems synthesis? Mention the different parts of a production system that are affected by implementing MRP.

12. Why is MRP associated with manufacturing industries rather than retail organizations?

13. What is the primary output of a master production schedule as it pertains to MRP? Where does this information come from?

14. How can MRP in a simulation mode help construct a master production schedule?

15. How does low-level coding increase the efficiency of data processing in an MRP system?

16. How does the concept of time phasing in MRP differ from a perpetual ordering system for lot size inventory applications? Draw a chart to show the accumulation and consumption of stock under both systems. What savings should result from efficient time phasing?

17. What is the purpose of "exploding" requirements from end-items backward through the predecessors until raw materials or parts having no predecessors are reached?

18. *Netting* is the name given to the calculation that determines how many units are required during a period to meet projected requirements in an MRP system. The net requirement in time bucket n is the projected requirement that cannot be satisfied by available units or by scheduled receipts. Net requirements are thus the dependent demand pattern for a component.
a. Calculate the net requirements for the time buckets in Table 13.4 (Assume that the shaded values in the table have not been entered).
b. To what use could the netting pattern be put?

19. Why is the practice of holding safety stocks for inventory items contrary to the intent of MRP? What exceptions are allowed in MRP systems?

20. Dr. Orlicky said, "The net change MRP is a more advanced, more powerful version of regeneration MRP. It is a successor to the regeneration."

Distinguish between the two updating systems, and indicate the advantages and disadvantages of each.

21. Discuss how lack of control over the three important areas of MRP implementation might contribute to the number of MRP installations that are disappointing.

22. Distinguish between MRP II and MRP.

23. To carry out manufacturing operations, three conditions have to be satisfied. These are
a. Material has to be available.
b. Tooling has to be available (cutting tools, dies, jigs, and fixtures).
c. A production center has to be available (machinery and an appropriate crew).
It has been said that because MRP II evolved from MRP, the first condition is met, but the other two are not as well satisfied. Discuss the criticism in terms of how information flow should avoid the difficulties.

24. The success of kanban in Japan is attributed to a synergistic relation of setup times, quality, and worker discipline. Explain the interaction in terms of lot sizes.

25. What does the word *pull* imply in describing material flow in a kanban system?

26. Select the factor you feel is most significant in each of the system characteristics in Table 13.9 for the manufacturing push and pull systems. State why you selected each factor.

27. Explain why eliminating bottlenecks is a logical way for OPT to maximize throughput.

13-13 PROBLEMS AND CASES

1. An item with a time-varying demand has a setup cost of $55 and holding cost of $0.40 per unit per month. At time zero the inventory level has dropped to zero. There is negligible lead time. The demand for the next 12 months is as follows:

Data for Problem 1											
Month	1	2	3	4	5	6	7	8	9	10	11	12
Demand	10	60	15	130	155	130	90	50	125	160	235	40

All of the demand requirements must be met at the beginning of the month. No shortages are allowed. Replenishments should be scheduled to have a zero ending inventory for the 12-month plan. Using the methods listed, determine the replenishment periods, order sizes, and associated costs.

a. One replenishment at the start of the first month to meet all requirements for 12 months.

b. A replenishment policy of a setup each month to meet the demand for that month.

2. The demand for a product averages 100 per month, but the actual monthly pattern for one year is as follows:

Month	1	2	3	4	5	6	7	8	9	10	11	12
Demand	10	62	12	130	154	129	88	52	124	160	238	41

The ordering cost is $54, and the holding cost is 40 cents per item per month. The carry-over inventory level is zero at the beginning of month 1, and the items needed each month must be available at the beginning of the month. No stockouts are permitted. Develop tables similar to Table 13.1, and calculate the total order and holding costs for 12 months using the following techniques:

a. A fixed economic order quantity.

b. The Wagner–Whitin algorithm.

c. The Silver–Meal heuristic.

3. Assembly X is the parent of subassembly Y and part Z is used in the fabrication of Y. Complete a material requirements plan for each component based on the following conditions:

a. *Assembly X.* Ninety units are on hand, and no order receipts are already scheduled. The demand is as follows:

Order quantity = 90 Lead time = 1 week	Data for Problem 3											
	1	2	3	4	5	6	7	8	9	10	11	12
Projected requirements		60	10	90	80		40	80	20	70		90

b. *Subassembly Y*. Three units of subassembly Y are needed to produce one unit of Assembly X. The lead time is two weeks, and the lot size is based on the *EOQ* model in which the setup cost is $100, and the holding cost is $0.02 per unit per week. There are no units on hand, but an order will be received during the first week.

c. *Part Z*. Two units of part Z per unit of subassembly Y are required. Also, part Z has a constant independent demand of 300 units per week for sales to another manufacturer. Use the Silver–Meal heuristic to determine the lot size (Setup cost =

$150 and holding cost = $0.03/unit/week.) Lead time is one week. There are currently 140 units available in stock, and an order of 700 units is due during the first week. The company does not allow a planned shortage (demand must always be satisfied by scheduled receipts or available units). In addition to determining the material requirements plan for part Z, calculate its total cost of setups and carrying charges over the planning horizon.

4. Information about end-item I is as follows:

Item	Parent	Item Units Parent Unit	On Hand	Lead Time	Scheduled Receipts in Week (n)	Special Conditions
I	—	—	90	1 week	300 in (1)	Order quantity = 300
II	I	1	300	2 weeks	—	Order quantity = 300
III	II	2	600	2 weeks	—	Reject allowance = 10% Minimum constant lot size
IV	I, II	1/I, 2/II	300	1 week	600 in (1)	One unit is included as a spare part in I and two units are needed per II. Order quantity = 600
V	1	2	600	1 week	—	Has independent demand (see below). Order size is calculated from the *EOQ* formula where O = $392 and H = $0.03 per unit per week
VI	III, V	2/III, 1/V	500	1 week	1500 in (1)	One-time-only sales order for delivery of 1000 units in week (8). Setup cost = $500 Holding cost = $0.01 per unit per week

Independent demands for end-item I and component V are shown in the following table:

Week	1	2	3	4	5	6	7	8	9	10	11	12
End-item I	180	310	40	250		300	240	90	150	220	80	330
Component V		500		750			500	750		500		750

a. Determine the material requirements plan for item II.

b. Determine the material requirements plan for item III.

c. Determine the material requirements plan for item IV.

d. Determine the material requirements plan for item V.

e. Determine the material requirements plan for item VI using the Silver–Meal heuristic.

f. Which items would likely carry a safety stock? Why?

g. Compare the sum of setup and holding costs for item IV (as determined in part c.) with the total that results from the replenishment pattern suggested by applying the Silver–Meal heuristic when setup cost = $250 and holding cost = $0.02 per unit per week.

5. A company produces two models, A and T, for which the product structures are as follows. The lead time in weeks for each procurement or production stage is indicated in parentheses on each diagram, and the number of items that go into the next higher component level is given before the code letter (for example, $4C(2)$ means 4 units of component C go into 1 unit of component A).

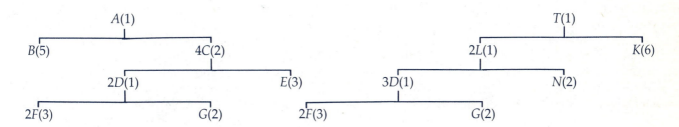

The company has on-hand balances, scheduled receipts, and rejection allowances as indicated in the following table:

Item	On-hand Balance	Scheduled Receipts	Reject Allowance	Ordering Policy
B	0			Each period
C	50	300 in week 2	10%	$EOQ = 300$
D	2,000		5%	$EOQ = 500$
F	1,000	5,000 in week 2		Silver–Meal algorithm
G	500		3%	Each period
L	125			Each period

The master schedule indicating when the final assembly of models A and T will begin is as follows.

Construct material requirement plans for components C, L, D, and F.

Master Schedule

Week	1	2	3	4	5	6	7	8	9	10	11
Model *A*								100	200	150	100
Model *T*								225	180	210	250

6. *The Case of Good Intentions Gone Awry*

In late 1983, the managers of a manufacturing company in Virginia heard about JIT. They liked what they heard and decided to go all out to make it work in their organization. The decision was part desperation. The company had been operating at a loss for over a year as a result of foreign competition.

The manager of the plant was not completely convinced that JIT could do all the things his staff promised, but he pledged his full support—initially. The 155 employees also had reservations after they heard what was expected of them, but they knew that something had to be done, because their jobs were in jeopardy.

New equipment was purchased, and manufacturing practices were changed in many ways, often at the employees' suggestion. At first the changes created a lot of tension between the workers and the managers. But even the managers became anxious when productivity appeared to be declining during the first six months. But by the end of one year, spirits rose along with productivity. Aided by a strong national economy, sales doubled, but the work force increased by only 10 percent. The time to produce a finished product was cut by half, and the reduced in-process inventory freed 25 percent of the floor space in the factory.

Some of the most notable savings came from reduced raw-material inventory. Before JIT, the company held a six-month supply, receiving a replenishment order each month. This buffer stock covered surges in manufacturing and allowed material that failed acceptance inspections to be returned without delaying production. But it was expensive insurance. Then, by working with the vendors to improve quality and to make weekly shipments of smaller amounts, the holding costs were cut by 60 percent.

Then the plant manager reverted to an old habit. A vendor made an offer he could not resist: He bought a seven-month supply of material at a bargain rate for bulk purchases. The workers were shocked when the largest order they had ever seen filled all available storage space. Other managers were dismayed at the reversal of policy, even after the comptroller confirmed that the price for the big order was indeed favorable. The other vendors that had been cooperating with the company's new ordering policies became skeptical. The board of directors questioned why the plant manager was undercutting a program that had been producing excellent results.

a. How might the plant manager have justified to himself that the large purchase would be a wise decision?

b. What arguments would you give against the purchase if you were one of the managers who had observed the savings being generated by the JIT implementation?

c. Because this case is true, what do you think was the result of the plant manager's action?

QUANTITY CONTROL

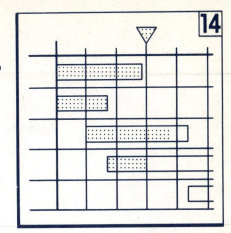

LEARNING OBJECTIVES

After studying this chapter you should

- appreciate the many factors in controlling a production system.
- understand the features of different control designs and the relations among pre-production planning, dispatching, and expediting.
- know how to use critical ratios to set priorities, Gantt and CPM time charts for scheduling and monitoring, and the line-of-balance method for progress control.
- be aware of the effect of learning rates on production schedules and be able to calculate the time to produce successive units based on a learning curve.
- realize the significance of energy costs and how energy can be conserved.
- be familiar with the various types of automatic identification systems and their use in production control.

KEY TERMS

The following words characterize subjects presented in this chapter:

flow/order/special project control

backward/forward scheduling

centralized/decentralized dispatching

expediting

Gantt charts

perpetual/periodic/order schedules

CPM time chart

resource grid

line of balance

objective chart

program plan

progress chart

learning curves

learning/forgetting rates

energy audit
cogeneration
energy load
control

automatic identification
systems (AIS)
bar coding
voice recognition

14-1 IMPORTANCE

Quality is a contemporary consideration with quantity and will be treated in the next chapter.

Production control serves the dual purpose of directing the implementation of previously planned activities and monitoring their progress to discover and correct irregularities. Quantity control concentrates on delivering the desired output within the expected delivery date. In this respect, the control function is the action phase of production. Plans are converted into action notices that spell out exactly which workers and machines will operate, what the operations will be, and when they must be done. Then the actions are compared with planned performance to provide the feedback for replanning or initiating corrective actions.

The dynamic nature of ongoing control activities makes quantity control difficult to program; it benefits from well-designed procedures, good training, and pictorial management tools, but it is not adaptable to rigid formulations.

It is said, "An army travels on its stomach." Equivalently, it can be said that production rolls on material and energy supplies.

There is an interesting resemblance between production control and a military operation. Before a battle (production process), logistic planning puts the troops and supplies in a preferred strategic position (allocation and scheduling of workers, machines, and materials). Tactical plans are developed for battlefield maneuvers (sequencing operations, inventory policies, machine loadings, and so on). But even the best military plans cannot fully predict the actions of the enemy (forecasting business conditions, competitors' actions, delays and breakdowns, and so on). Therefore, performance in the actual battle depends largely on individual training, equipment, supervision, and tactics. A communication net is supposed to feed back current battle reports (production operations) to the command post (production supervisors) to allow adjustments in the tactical plans. Calls for reinforcements or coordinated actions are relayed farther back to headquarters (production control department). There the adjustments are evaluated and, if accepted, translated into orders. Prompt and appropriate actions are necessary at all levels, from the front lines (production line) to the general staff (manufacturing staff), for a sensitive and reliable control system.

14-2 CONTROL DESIGNS

There are many production control designs. We shall consider three basic types, but innumerable mutations have been made to fit specific situations—an eminently sound practice. A control system designed for one plant might not work in another and might not even remain effective for the original plant as production requirements change. Three divisions of production, corresponding to the plant layouts introduced in Section 9-3 and associated with the three types of control designs, are described here.

1. *Continuous production* (uses a product layout).
   Standardized end product and manufacturing routine.
   High volume of output produced by specialized equipment.
   Low in-process inventory and long production runs.
   Low worker-skill levels.
   Limited flexibility of process.

2. *Intermittent production* (uses a process layout).
   Nonstandard end product requiring extensive production controls.
   Medium volume of output produced by general-purpose equipment.
   High in-process inventory and shorter production runs.
   Medium to high worker-skill levels.
   More flexible process owing to versatile material-handling equipment.

3. *Special projects* (frequently uses fixed-position layout).
   Unique end product requiring extreme production controls.
   Low volume of output often requiring the cooperation of several subcontractors.
   High in-process inventory with a single production run.
   High worker-skill levels.
   High flexibility of process.

Flow control

Flow, or serialized, **control** applies to the control of continuous production as found in oil refineries, bottling works, cigarette-making factories, papermaking mills, and other mass-manufacturing plants. The standardization of products, equipment, and work assignments allows the controls to be standardized also. The main concern is to maintain a continuous, ample supply of materials.

In continuous production the lines are balanced, and the sequence of operations is seldom changed. Economy results from operating near maximum capacity.

The high-volume production means that huge quantities of raw materials must be accumulated and stored until needed. The usual balance of storage cost to opportunity cost is biased by the seriousness of stockouts. Because of the inflexibility of the process, the entire operation is curtailed by a shortage of material in any part of the sequence. The volume of output requires strict attention to the finished-goods inventory and a smoothly operating distribution system.

Work-order dispatching is largely unnecessary because of the long production runs. The workers know their repetitive work assignments without special instructions. The closest equivalent to work orders is the issuance of production releases, which state the output level expected during a certain time interval. These are usually issued to the production managers and supervisors who control the process flow instead of to the workers on the production line.

"Batch" or "block" control is an offshoot of flow control, in which the same process produces products modified as to taste, style, color, or size. Control is exercised over individual blocks or batches as they pass through the process.

Order control

A bill of material for an order includes

1. Name and model of product.
2. Order number and quantity.
3. Raw materials.
4. Parts by name and number.
5. Appropriate specifications, drawings, and other references.

Order control, associated with intermittent production, is far more complex than is flow control. The job-shop nature of the work means that production orders may come from different sources and for different quantities and designs; the time allowed for production also may vary as a result of salespeople's delivery promises. These conditions make prior planning difficult and necessitate a high degree of control over each order.

The receipt of a "job" or "shop order" initiates action to determine what raw materials and parts are required and which production operations should be scheduled. The bill of material, routing information as to the sequence of operations, and desired delivery dates are considered in making the work schedule.

The two principal methods of scheduling are **backward scheduling** to meet a deadline and **forward scheduling** to produce as soon as possible. The former method starts with the required delivery date and calculates backward to determine the release date for the order. When several subassemblies with different lead times are involved, the scheduler must work backward along each subassembly line to set the times for component work orders (see Figure 14.1).

More detailed scheduling formats are described in Section 14-4.

Forward scheduling is used most frequently for products whose components do not require assembly. In a metal mill where a certain thickness of a certain alloy is ordered, customers usually want the material as soon as they can get it or at least will take it as soon as it is available. Under these conditions the scheduler issues orders to begin production as soon as machine time is available. When there is a backlog of demand, he or she checks requested delivery dates to set a priority; in effect, the scheduler is then combining backward and forward scheduling.

When there is a distinct limiting factor in the production flow, it receives special attention and is referred to as *load control*. For instance, if there is a very expensive or one-of-a-kind machine required for a number of different orders, all activities concerning these orders are geared to the time available on the critical machine. The purpose of load control is to ease as much as possible the effects of the bottleneck by ensuring maximum utilization of the decisive asset.

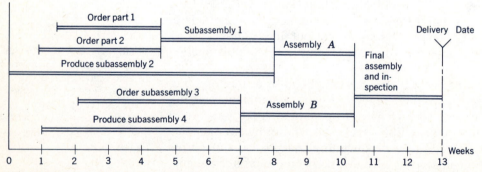

FIGURE 14.1 Assembly chart based on lead times required to meet a promised delivery date.

Special project control

Special project control is reserved for distinctive or particularly important undertakings with unusual features. The most common examples are construction projects—dams, factory modifications, buildings, bridges, and the like. The distinguishing characteristic is personal contact. Most orders are issued by managers and supervisors directly to the workers responsible for performance. The same supervisory personnel monitor the progress and initiate corrective actions when the work falls behind expectation. These less formal and more expensive procedures are warranted by the coordination problems entailed in unifying the efforts of diverse trades to meet rigid time limitations.

The "trades" in contracted projects include specialized skills represented by different unions and subcontracting firms.

Because each phase of a special project tends to rely on the completion of a previous phase, expediting measures often take the form of crash scheduling for specific segments. Big dosages of extra resources applied intermittently rapidly increase direct costs. It is difficult to control the application of added resources because the supervisory force is seldom increased during periods of accelerated activity. The use of short-range activity charting, as described in Section 14-4, by capable controllers gives protection from soaring costs arising from congested working conditions and inefficient operations.

Crash scheduling is not necessarily a reflection of poor project control. It can result from uncontrollable conditions such as weather. Extra resources applied to "buy" time include overtime and employing more workers and materials than originally planned.

14-3 CONTROL ACTIONS

The multitude of considerations that go into preproduction planning have filled the previous chapters; some of the more direct tactical activities are shown in Figure 14.2. The result of these activities is a master production schedule—the foundation of all production actions. It shows how many and when products will be ready for distribution. This information is used for material control and the development of operator and machine assignments for individual orders.

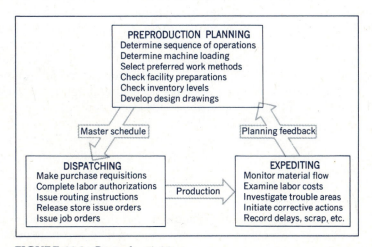

FIGURE 14.2 Control activities.

Dispatching

The dispatching function marks the "go" state of production. The action that triggers manufacturing is the release of job and material orders. These orders implement operations from the master schedule. When several parts are involved, production orders are broken down into job cards for successive tasks and the methods to do them. Material issues and information about the next destination of the component are also recorded.

The exact duties of dispatching and the manner of carrying out the duties vary among companies. **Centralized dispatching** has the production control department issue detailed work orders directly to the operators performing the tasks. **Decentralized dispatching** sends general work orders to department supervisors who decide which machines to use, who should use them, and when to perform the work. After the orders are released, the dispatching function sometimes continues with the responsibility for recording actual operation times, reasons for idleness, causes of breakdowns, and similar relevant information about the schedule. At other times, these responsibilities fall into the province of expediting.

Expediting

Whereas dispatching primarily consists of a flow of information, **expediting** is concerned with the flow of materials and components. The two functions tend to overlap both chronologically and in responsibility. The events recorded under the dispatching function are adjusted by the expediting activities.

An expeditor follows the development of an order from the raw-material stage to the finished product. He or she often is given the authority and facilities to move materials or semifinished products to relieve congestion in production flow. For instance, the failure of components to reach a certain work station is detected at the dispatching office from progress charts or data-processing output. The expeditor investigates to determine the cause of the delay. If it is caused by misinterpretation of work orders, the expeditor can clarify the points and possibly use equipment under his or her command to relieve a material jam. If flow circulation problems extend beyond the expeditor's authority, he or she will refer them to the production-planning office for revised assignments or reallocations of resources. The resulting alterations are generally returned to the expeditor with the responsibility for implementation.

Expediting and dispatching are frequently performed under the same agency, particularly in special project control. Under a flow control design, combining the functions is also logical: dispatching is relatively straightforward because of the need for few detailed instructions, whereas expediting is more important owing to the seriousness of temporary flow interruptions. In order control design the two functions complement each other in the close control needed to chase orders through the plant. Whether or not the functions are combined, both must contribute to postoperation evaluations. The recommendations derived from correcting past mistakes lead to control designs that reduce repeated mistakes.

Critical ratios

The priority by which jobs are assigned to people or machines can be determined by many routes. Several quantitative techniques for job ordering were presented in preceding chapters (for example, sequencing methods, line balancing, assignment and transportation models). Less formal procedures are to assign jobs by simple priority rules, such as

- First come, first served (schedule work as it arrives from customers).

- Preferred customer, priority service (give priority to orders from the "best" customers).

- Strongest complaint, next served (customers who howl the loudest about poor service have their delayed jobs moved ahead).

- Any time, any order (a random selection of a job from work that needs doing to any open work station).

Each of the rules more or less assists a scheduler in deciding what work to assign next, but none of them provides an "alert" to indicate the relative urgency of all jobs waiting to be completed. Better control over the entire menu of work orders is attained by calculating the *critical ratio* of each unscheduled work commitment—an index of the relative priority of jobs.

Priorities must be established for orders and operations. *Order priority* ranks customer orders. *Operation priority* ranks the production operations required to complete an order.

The general form of a critical ratio (*CR*) is

$$CR = \frac{\text{time remaining before a job should be done}}{\text{usual time required to complete the job}}$$

For example, if

$$CR = \frac{\text{product is promised in 15 days}}{\text{process time for the product is 20 days}} = \frac{3}{4}$$

it is apparent that the product will not be delivered on time unless it is expedited to reduce the normal processing time. When several jobs are stacked up waiting for a slot on the work schedule, critical ratios are measures of urgency: the lower the ratio is, the greater the urgency will be.

Critical ratios are effective for both advance scheduling and current reviews of existing schedules. When generated as part of a computerized work schedule, ratios can be updated daily to show which jobs most urgently need work to meet shipping schedules. By adjusting schedules in accordance to critical ratios, expediting efforts are partly curtailed and the effects of production delays on customers' orders are reduced.

EXAMPLE 14.1 Priority ranking of customers' orders according to critical ratios

A backlog of customers' orders, as of working day 15, is as follows:

Customer order:	A	B	C	D	E
Promised delivery date:	14	15	17	20	25

All five customer orders have an estimated shipping date of working day 20. This date is obtained from correcting the normal time it takes to complete work on an order to account for other orders already in queues behind critical work stations and any work previously done on the order. Critical ratios are calculated from the following formula:

$$CR = \frac{\text{estimated shipping date} - \text{today's date}}{\text{estimated shipping date} - \text{promised delivery date}}$$

Rankings based on critical ratios for the four orders are as follows:

Critical ratios with positive values imply probable failures in meeting promised delivery times: the lower the ratio is, the larger will be the expediting effort required to catch up to a scheduled shipping date. An on-time order has an infinite ratio, and orders ahead of schedule (float time is available) have negative ratios. Priority rankings correspond to the degree of urgency shown by the ratios.

Customer Order	Critical Ratio	Priority	Interpretation
A	$\dfrac{20-15}{20-14} = \dfrac{5}{6} = 0.83$	1	Delivery date has already been missed. Expedite to avoid further delay.
B	$\dfrac{20-15}{20-15} = \dfrac{5}{5} = 1.0$	2	Order will be late. Control closely to avoid additional schedule slippage.
C	$\dfrac{20-15}{20-17} = \dfrac{5}{3} = 1.67$	3	Delivery date will be missed unless a major overtime schedule is implemented.
D	$\dfrac{20-15}{20-20} = \dfrac{5}{0} = \infty$	4	Order is currently on schedule, but it must be watched carefully for slippage.
E	$\dfrac{20-15}{20-25} = \dfrac{5}{-5} = -1.0$	5	Order should meet the delivery date unless other orders postpone its production.

First introduced by A. O. Putnam, "How to Prevent Stockouts," *American Machinist*, February 1964.

Another version of the critical-ratio technique focuses on the material requirements to complete customer orders.

$$CR = \frac{\text{stock ratio}}{\text{manufacturing ratio}}$$

$$= \frac{\text{stock on hand/order point quantity}}{\text{remaining manufacturing lead time/total planned lead time}}$$

The stock ratio is sometimes defined as (stock on hand minus safety stock) divided by average daily usage. Then the critical ratio is the stock ratio divided by the remaining lead time.

This version is applicable when an order point quantity has been established for inventory control and stock depletion is gradual and steady. The numerator is a measure of the availability of materials needed to maintain supply; it is assumed that the material will be produced to have it available close to the time the supply is exhausted. The denominator relates the amount of work that still needs to be done to the total time originally allotted to the manufacturing operation.

Table 14.1 shows the status of four jobs. No work has been done on operation *A*, yet only 50 percent of the expected stock level is on hand; the operation is awarded a high priority to indicate a behind-schedule condition. Operation

TABLE 14.1
Manufacturing Priorities Determined by Critical Ratios

	Job A	Job B	Job C	Job D
Stock on hand	50	50	20	20
Order point	100	100	100	100
Stock ratio	50/100 = 0.5	50/100 = 0.5	20/100 = 0.2	20/100 = 0.2
Remaining lead time	20	10	10	2
Planned lead time	20	20	20	20
Manufacturing ratio	20/20 = 1.0	10/20 = 0.5	10/20 = 0.5	2/20 = 0.1
Critical ratio	0.5/1.0 = 0.5	0.5/0.5 = 1	0.2/0.5 = 0.4	0.2/0.1 = 2
Priority	2	3	1	4

A *CR* lower than 1.0 indicates work completion lags the stock demand, or correspondingly, stock depletion has exceeded the rate of work completion. A critical ratio greater than 1.0 indicates the reverse. The lower the value of the critical ratio is, the higher the priority of the job will be.

B is on schedule; the stock is 50-percent depleted, and the work is 50-percent completed to provide the replenishment quantity. When only 20 percent of the stock is on hand when there is still 50 percent of the work yet to be done, as in operation *C*, the critical ratio warns that the operation deserves a high-priority ranking. A low priority is given to operation *D* because apparently enough material is on hand to satisfy demand until the operation is completed.

14-4 CONTROL TOOLS AND TECHNIQUES

Each company and even controllers within the same company use control devices uniquely adapted to their own requirements. The printed forms for master schedules, load analyses, factory orders, stores requisitions, stores credit memos, scrap tickets, material move orders, inspection tickets, and other commonly used control records vary widely in design, size, and detail.

Visual charting is also customized. Charts vary from bar graph notes tacked to walls in a message center to massive, permanently fixed boards with chrome-plated slots for colored plastic time bars. Although the form and symbols differ, the uniform intent is to transmit information and follow the progress of actions triggered by the information. Here we consider several common management tools and techniques designed to accomplish the control mission.

Perhaps it would be more accurate to call the formats in this section "Gantt-type" charts because many modifications have been made since the original version.

Gantt charts

When Henry L. Gantt worked at the Frankford Arsenal in 1917, he recognized the need for a formal device to cope with scheduling problems. The device he developed was a graph of output activities plotted as bars on a time scale. The same types of chart are still used for essentially the same purpose.

The principal virtue of a **Gantt chart** is its simplicity. No attempt is made to recognize risk or alternative actions. Activities are committed to dates ac-

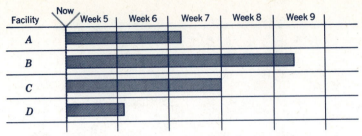

FIGURE 14.3 Gantt load chart for perpetual scheduling.

cording to the preferred schedule. Deviations from the anticipated calendar are recorded to show current conditions. Through this routine, operators are given their assignments, the pattern of delays is revealed, and the changing distribution of production loads is forcefully exposed.

Three versions of Gantt charts are illustrated in Figures 14.3, 14.4, and 14.5. The charts are applied to three methods of scheduling: perpetual, periodic, and order. The different conventions and symbols illustrated could be interchanged among the types of schedules, and entirely different descriptive renditions could supplant or complement those shown. Some restraint in elaboration is advisable so as to retain the virtue of simplicity; cluttering a chart with too much information may destroy its data-recording and transmitting capacity.

A **perpetual schedule** is developed by reviewing the status of all jobs in an open order file. The amount of time required for all the jobs from each department, machine, or facility can be posted in a form similar to Figure 14.3. Postings are typically made once a week. Horizontal bars show the time "reserved" in each facility to accomplish work already on order. The relative loads of the facilities and the overall work load of the plant are clearly displayed; facilities with smaller loads may be used to relieve overloaded departments. The weekly updating also provides a graphic record of the changing work loads. The character of loading patterns is useful in allotting method-improvement studies, selecting capital investments, and forecasting personnel and maintenance needs.

At the beginning of a period, the supervisor of a facility is given a work list of the expected output. The list may contain assignments above 100-percent capacity on a standby basis: if all goes well, there may be time to do the extra orders.

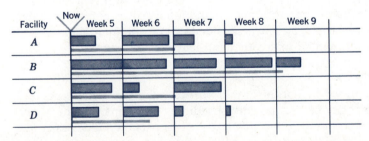

FIGURE 14.4 Gantt load chart showing accumulated loads for periodic scheduling.

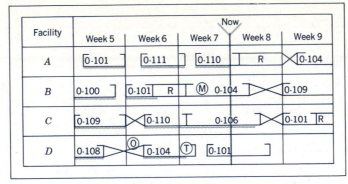

FIGURE 14.5 Gantt chart for order scheduling and progress reporting.

In **periodic scheduling,** the work to be accomplished within a set period, typically a week, is loaded to appropriate facilities, as shown in Figure 14.4. The jobs required to complete individual orders are allotted to the facilities with no directive other than that they must be completed within the period. The length of the bars in the chart represents the amount of time scheduled in each facility during the period. The line below the bar indicates the cumulative work load already scheduled ahead. Again, the backlog of work accumulated at each facility conveys to the scheduler at a glance the changing pattern of work loads and calls attention to overloads. In Figure 14.4, facility B has an accumulated work load of about four weeks, which could portend trouble in meeting future commitments.

A Gantt chart is also useful for specific **order scheduling.** The simplest form uses bars whose lengths represent the time to complete work on the order. Bars are positioned in the appropriate facility row, tagged with numbers identifying the order, and located along the time scale in accordance with the completion schedule. Regard is given to facility availability and to the sequence of operations that characterizes each order; the completion schedule corresponds to the lead-time considerations of an assembly chart, as displayed in Figure 14.1.

Additional value is gained from the Gantt order chart by recording what actually occurs as work progresses. After starting with a schedule laid out by bars on a time scale, marks and symbols are imposed on the chart at each updating to show the current status of the work and the cause of deviations from planned performance. Some useful symbols for updating are given in Table 14.2.

The data-recording and transmitting capacity of Gantt charts is evident in Figure 14.5. The present date is the end of week 7, Now. Immediate work assignments for each facility are apparent. Facility A is on schedule for order 110. Facilities B and D are behind schedule owing to tool trouble for order 104 in facility D and lack of material in facility B. If the work on order 104 in facility

TABLE 14.2
Typical Symbols Used in Updating Gantt Charts

Symbol	Meaning
	Marks present date on time scale.
0-101	Bracketed line shows time allowed for completion of the designated order; in the symbol shown the order number is 101
0-101	Double or colored line below bracket denotes the portion of work completed by present date
⋈	X between brackets reserves time to make up for delays and to increase the schedule flexibility
R	Shows repair time for machine breakdowns or maintenance
M	Shows delay caused by missing or improper materials
T	Shows delay caused by tool trouble
P	Shows delay caused by power failure or poor operator performance (inexperience, slowness, lack of instructions, etc.)
O	Shows delay caused by operator's error

Interrelationships of orders are more easily shown by the CPM time chart discussed in the next section.

D must be completed before the work on the same order can be initiated in facility *B*, special expediting measures are indicated. Order 106 in facility *C* is the only operation ahead of schedule. The updating strongly suggests that the jobs should be rescheduled; perhaps overtime, multishift operations, or additional equipment are advisable.

The given Gantt chart examples only hint at the uses that have been made of the bar-chart format. The record could be of individual workers instead of facilities, and the time scale could be hours or days in place of weeks. Other notations could be used to depict information succinctly. The bars are often color coded to show distinguishing characteristics. Large and elaborate boards based on Gantt charting principles are commercially available. They use a pegboard design, revolving disks, colored strings, plastic slide inserts, and similar devices to provide a visual picture of current and planned production.

Some commercial mechanical scheduling aids are called "Productrol Boards," "Schedugraph," and "Boardmaster."

It is difficult to extend the use of bar charts to analysis. Their fine communicative attributes work against the evaluation of opposing alternatives. Adding to the bars the costs and dependency relationships required for an evaluation seriously clutters the charts. The bar format suffices for simple problems such as planning assignments for small machine centers. For more complex problems, we have CPM techniques available.

CPM time charts

Critical path scheduling is most commonly applied to special projects. The graphic version used for control purposes is called a **time chart**—an arrow network drawn to a time scale. All the information available from an arrow network is incorporated in a time chart, but slightly different conventions make the time chart easier to understand and more useful for control purposes.

Converting an arrow network to a time chart amounts to making the network conform to a rectangular coordinate system. The horizontal axis indicates time; the vertical axis represents activity chains that may be associated with a division of the project or facility assignments, as shown in the previously discussed bar charts. The conversion process is largely a mechanical effort, as the basic data are already tabulated in the network and the boundary timetable.

Each symbol used in an arrow network has its counterpart in a time chart, as shown in Table 14.3. Solid vertical lines indicate restrictions in both directions; all activities connected to the left of a vertical line are prerequisite to all activities starting to the right of that line. Events are distinct points in time marked by carets—time indicators that display key moments in the left-to-right flow of time. Horizontal lines between carets represent activities. There is no need to mark activity durations because they are implied by the time carets; the caret at the start of an activity is the earliest start, and the one at the end is the earliest finish, making the activity duration $= EF - ES$. Activity descriptions are still

A time chart is the end product of a CPM application. Previous steps were presented in Sections 7-3 and 7-4.

Event numbers can be included as part of the time carets to facilitate schedule updating by computers.

TABLE 14.3
Comparison of Arrow Network and Time Chart Symbols

Arrow Network	Symbol	Time Chart	Description
A / 10 (arrow)	Activity	A / 0—10 / Caret time indicators	No arrow heads are needed in the time chart because time is known to flow from left to right.
A / 10 (open arrow)	Critical activity	A / 0—10 / ES_A EF_A	Double lines mark the critical path. Activity durations are the difference between earliest start and finish times.
A (2) → B, C	Event	Event / 2→B / A C / $ES_{B\ and\ C}$	Event 2 is the vertical line at the end of activity A. Separate time carets are not needed at an activity merge or burst.
A → (3) → C; B → (2) → D	Dummy	A C / B----→ D	Dummies are shown by dotted arrows drawn vertically or sloping to the right. Activities should have a word description above the line.

'entered above the activity line. Other information, such as the cost or resources required to complete an activity, can be inserted below the activity line.

The pictorial representation of total float is one of the most revealing features of a time chart. Total float is shown as a horizontal dotted line following the last activity in a chain. All the activities in the chain share the indicated float. As shown in Figure 14.6, activities B, C, and D form a chain. The total float for each of these activities, as indicated by $TF = LS - ES$ from the arrow network, is three time units. Therefore, if B or C is delayed by three time units, no float will be left for leeway in scheduling D; D thus becomes a critical activity.

All the float is shown after the last activity in a chain only for convenience in the initial chart construction: it shows that leeway exists but in no way infers that it should be so assigned in the final schedule. Float lines point out areas of schedule flexibility, and the critical path sets limits to schedule juggling. Additional restrictions result from resource constraints associated with particular activities. A working schedule evolves when float time is used to improve resource utilization.

An illustration of a resource grid is shown in Example 14.2.

The versatility of time charts is further enhanced by adding a **resource grid** with the same time scale as the chart. The purpose of the grid is to record individual resource assignments. In a project's planning stage the grid can be used to experiment with different resource assignments. In the control stage it can serve as a record for comparing actual with planned resource expenditures. For smaller projects a grid is seldom necessary, because the information can be shown by entries directly on the chart without cluttering it too much.

When a project has fallen behind the anticipated pace, three alternatives are open to the project manager:

1. ***Do nothing.*** The duration of the project will likely be longer than planned, but the lost time might be recovered by shorter-than-expected completion times for remaining activities.

Applying measures that shorten one activity can place a previously noncritical activity on the critical path.

2. ***Apply additional resources.*** Certain activities can benefit from crash procedures, but such expense should be limited to activities with little or no float.

3. ***Change methods.*** A substitute resource, such as an automatic machine replacing a hand-operated model, or subcontracting the entire job, can sometimes alter the method of accomplishing an activity to reduce its duration.

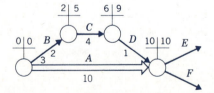
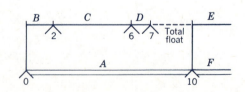

FIGURE 14.6 Arrow network segment and corresponding time-chart segment.

The time chart helps select these alternatives by displaying the total effect of possible alterations.

The appropriateness of time charts versus Gantt bar charts can be judged by comparing the two formats according to the charts' uses.

Use	Comparison
Survey	A chronicle of information about present conditions is easier to compile with bars. If the activities have important interrelationships, a more complete representation can be made via time chart conventions.
Design	A new or revised solution is usually shown better on a time chart because more data are displayed, especially when a resource grid is used in conjunction with the time diagram. The full effect of outcomes is easier to observe when float and dependencies are known.
Presentation	A Gantt chart is admirably suited for presenting a summary solution and action instructions. Its simplicity eliminates most misinterpretations. When resources of several agencies interact to accomplish a project, the more complete descriptions in a time chart more clearly reveal schedule flexibility and responsibilities for cooperative operations.

EXAMPLE 14.2 Construction and use of a CPM time chart

In Examples 7.1 and 7.3, a network was developed to construct a prototype product. The initial arrow network was refined for better control by subdividing key activities. For the final version, repeated in Figure 14.7, a time chart is to be constructed for control purposes.

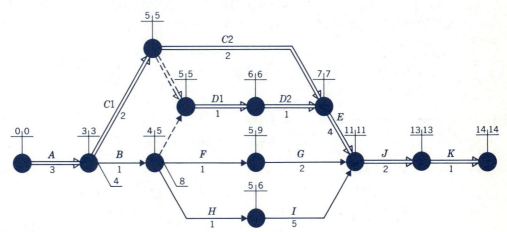

FIGURE 14.7

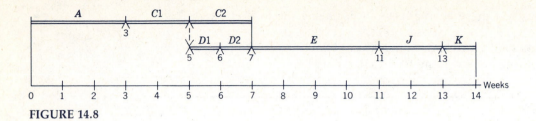

FIGURE 14.8

The first step in time-chart construction is to set a scale large enough to present the symbols and activity descriptions clearly. In Figure 14.8, letters are used to represent activity descriptions to allow the chart to conform to the book-page size; actual applications use charts up to 60 feet or longer. It simplifies the drawing to lay out the most important activities first.

With the critical path in place, remaining activities are entered according to network restraints. Activities are shown starting at their earliest possible times in the preliminary chart. Float is inserted to complete noncritical activity chains. Float can later be distributed to schedule noncritical activities at their most economical and convenient times (see Figure 14.9).

As the project progresses, the time chart is used for updating. Because the project is relatively small, an auxiliary resource accounting grid can be used for progress reporting. For larger projects, several copies of the basic chart are reproduced to facilitate more entries at each updating. Entries on these periodic progress reports then serve as records of performance.

A chart updated at the end of week 9 is shown in Figure 14.10. The critical activity E is one week

behind schedule. As noted in the comments, the engineering analysis (activity B) took one week longer than scheduled. This delay postponed the start of all postrequisite activities. Besides activity B, the work on the subcontracted parts (activity I), as determined by liaison with the subcontractor by the purchasing and engineering departments, is also a week behind schedule. Because a week of float is available for activity I, the condition is not too serious; the activity is considered on schedule but critical because the float no longer exists. The main decision facing the project manager is whether to accept an extension of the total project duration to 15 weeks or to apply extra resources to make up the week's delay.

The following chart and the resource accounting grid provide information helpful in deciding what to do. Department responsibilities are noted below the activity lines in the chart. The grid shows the cumulative variance of actual work-hours (above the diagonal in each cell of the grid) and costs (below the diagonal in units of $100) from the amounts budgeted. It is apparent that engineering expenditures considerably exceed the amount expected. The negative variance in manufacturing might be deceiving because the budgeted amounts are for an on-schedule project. Part of the low expenditures in manu-

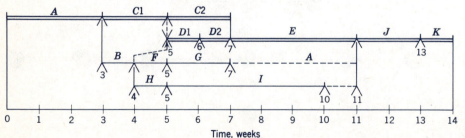

Time, weeks

FIGURE 14.9

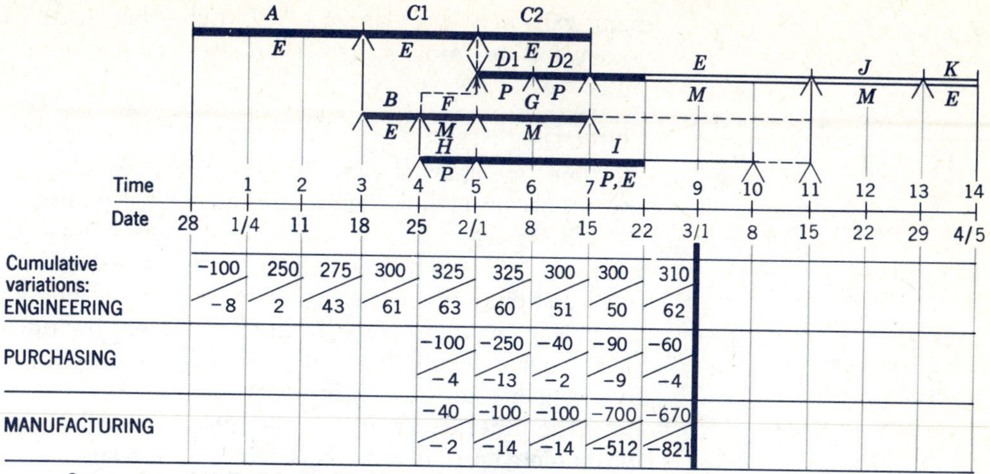

Chart with activities A, C1, C2, B, F, D1, D2, G, P, E, M, H, I, J, K and Time/Date axes.

	1	2	3	4	5	6	7		9	10	11	12	13	14
Time														
Date	28	1/4	11	18	25	2/1	8	15	22	3/1	8	15	22	29	4/5

Cumulative variations:										
ENGINEERING	−100	250	275	300	325	325	300	300	310	
	−8	2	43	61	63	60	51	50	62	
PURCHASING					−100	−250	−40	−90	−60	
					−4	−13	−2	−9	−4	
MANUFACTURING					−40	−100	−100	−700	−670	
					−2	−14	−14	−512	−821	

Comments: 1/4–Final design (A) delayed by material change; 3 workers added
1/25–Analysis (B) extended 1 week to check stress concentrations
2/1–Analysis (B) completed; original specifications changed

FIGURE 14.10

facturing is due to the late start for activity *E*. A detailed budget should be consulted to determine actual resource utilization. The grid thus serves as a warning device to signal the need for remedial action or close surveillance for current and future activities.

Line of balance

The **line-of-balance** method for production control combines features from a Gantt bar chart and a CPM time chart with graphs of material requirements. The method was developed by the navy during World War II. It is most appropriate for assembly operations involving a number of distinct components. In essence, it uses the principle of management-by-exception through a comparison of progress on individual components with the time schedule for completed assemblies. Regular progress checks reveal the future effect of any current delays and indicate the degree of urgency for corrective action.

The application of the line-of-balance technique consists of four main stages, all using graphic aids:

1. A graphic representation of the delivery objective.

2. A chart of the production program showing the sequence and duration of all activities required to produce a product.

3. A progress chart of the current status of component completion.

4. A line of balance drawn to show the relationship of component progress to the output needed to meet the delivery schedule.

Diagrams for the first three stages are displayed on one composite chart. The fourth stage relates data from all the diagrams. The complexity of the tech-

Line of balance, like CPM, was forcefully imposed for government contracts. Many companies were initially skeptical but later recognized the value of the technique and adopted it for their own use. Again, like CPM, some individuals felt it cost more to implement than it was worth for managerial control. Any management tool must be applied properly and appropriately or it will become an expensive frill.

nique falls between Gantt and CPM charting. Most of the effort is demanded to develop the initial stages. Thereafter, succeeding updatings require less skill and time.

OBJECTIVE CHART

An **objective chart** shows the expected completion schedule of products and the actual completion rate. It is designated as an **objective chart** because it shows the output objective and how well the objective is being met. As displayed in Figure 14.11, the schedule delivery curve begins at the origin and slopes upward as a function of time. The curved shape is associated with a new product design when learning takes place during production. In anticipation of more proficient performance during the later phases of the program, the rate of promised deliveries is greater than in the initial phase.

Learning effects are discussed in the next section.

The other curve in the objective chart is derived from the actual number of completed products delivered by a certain date. At each updating, the number of products delivered since the last progress check is entered as an extension of the "actual deliveries" line. A dip in this line below the "scheduled deliveries" line is an obvious cause for alarm. Because an established trend toward an inadequate production pace is more difficult to reverse than is an incipient trend, the main intent of the line-of-balance technique is to identify areas that might cause difficulties in the future.

PROGRAM PLAN

A chart of the operations required to complete one unit of the finished product is called the **program plan.** It closely resembles a time chart in which all activities are scheduled at their latest start times. Each major row of activities is associated with one component of the final assembly. The nodes (circles) along the rows mark the events that must be completed by the date indicated on the time scale. Thus the chart is event oriented. In Figure 14.12 the point at which the final assembly can begin is marked by the completion of components A and B.

A convention that locates activities by latest start and finish times eliminates float. Scheduling flexibility is possible only by starting earlier than shown.

Other symbols are sometimes used in place of circled events to distinguish the source of activities.

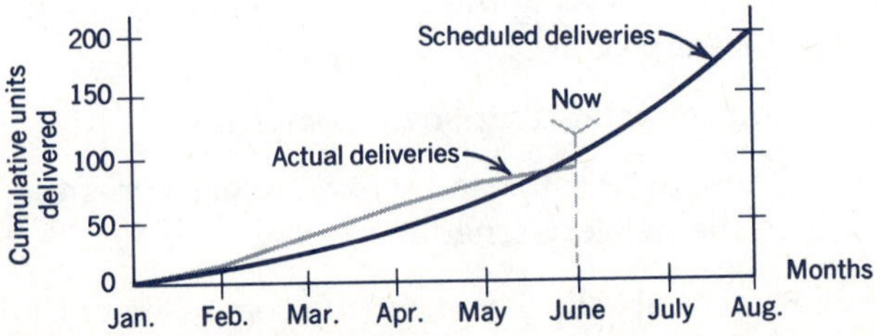

The completed chart serves as a reference to the amount of lead time by which each event must precede final completion. The time scale is reversed from the normal left-to-right sequence to show this lead-time requirement. The raw material for component A must be received nine weeks before the product completion date. The other events must also be accomplished by their respective lead times to maintain anticipated output.

FIGURE 14.11 Objective chart.

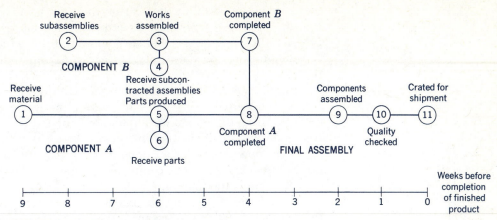

FIGURE 14.12 Program plan.

PROGRESS CHART

The *current* status of each operation designated by an event in the program plan is shown on the **progress chart.** Vertical bars show the physical inventory of the parts, components, and assemblies associated with the events. At each updating, the heights of the bars are increased by the number of units passing each event since the last progress check. The vertical scale of the progress chart is set adjacent and equal to the vertical axis of the objective chart, as shown in Figure 14.13.

Bars are often color coded to define areas of responsibility.

LINE OF BALANCE

A line from the objective chart coursing along the bars of the progress chart to show the *expected* number of completed items forms the line of balance. The line starts from the scheduled delivery curve of the objective chart at a point associated with the time of the progress updating, Now. The number of units on the vertical scale corresponds to the number of units that *should have* passed the last

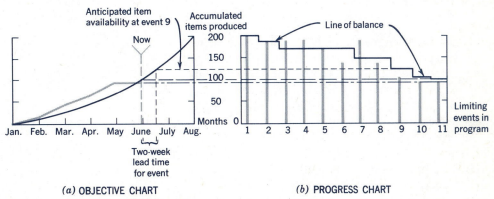

FIGURE 14.13 Line of balance: (*a*) objective chart and (*b*) progress chart.

event, represented by the last bar on the right in the progress chart. A similar reading of what the physical inventory should be for each event to keep the program on schedule is obtained by adding the lead time for the event under consideration to the date Now. The lead time thus added sets a point on the horizontal scale from which a line drawn vertically to the curve will establish the accumulated number of items expected through the event.

As illustrated in Figure 14.13, adding the two-week lead time for event 9 (obtained from the program plan) to the Now date establishes a point on the curve corresponding to 130 items. This means that 130 completed assemblies (event 9) should now be available to maintain the scheduled deliveries—the number of products expected to be completed by the middle of June.

Starting at the right end of the progress chart, the failure to meet expected deliveries is clearly revealed by bar 11: 90 units have been delivered instead of the contracted 100 units. Passing to the left, it is also clear there is no immediate prospect for catching up, because events 9 and 10 are even farther below expectation. The main deficiency appears to be the subcontracted work of event 6. Until this shortage is overcome, the output of component A will remain behind schedule.

A closer analysis of the line of balance shows that the events for component B (3, 4, and 7) are out of balance on the positive side. This condition could result from an overapplication of resources to component B: a too-rapid completion record can cause production problems by siphoning off resources that could be used more advantageously elsewhere and by accumulating stockpiles of components that must be handled and stored until they are needed for subsequent assemblies. From a more optimistic viewpoint, the accelerated completion of some events may allow resources to be diverted to lagging events. Although the decision for a preferred way to remedy production difficulties awaits a thorough cause-and-effect investigation, the line of balance directs attention to likely solution alternatives.

> For evaluation purposes, bars can be grouped by categories, such as subcontracted components, purchased parts, and internally produced assemblies.

EXAMPLE 14.3 Production control by a line of balance

An order has been placed for 300 prefabricated cabins. The delivery schedule is set at 30 cabins per week. The prime contractor is a lumber dealer with subsidiary prefabricating facilities. The electrical and plumbing components will be subcontracted. Hinges, plates, metal window casings, and other hardware components will be purchased. The major portion of the work is in cutting, assembling, and packaging wood components. For this work, the company has well-trained personnel and excellent facilities. The program plan is shown in Figure 14.14, in which

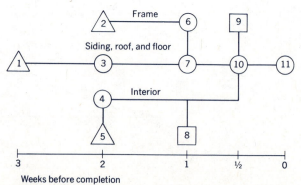

FIGURE 14.14

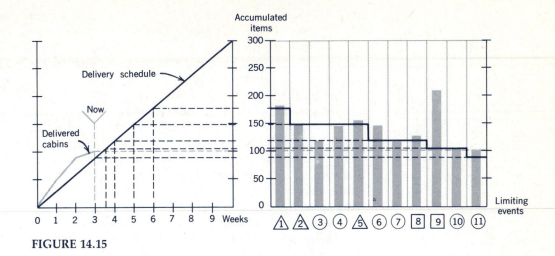

FIGURE 14.15

event symbols are circles for company-made assemblies, triangles for purchased parts, and squares for subcontracted components.

The first updating is made two weeks after the delivery of the first 30 cabin assemblies. The result of the study is shown in Figure 14.15. From the line of balance, it appears that the present favorable delivery status will be jeopardized by difficulties in the "siding, roof, and floor" assembly, event 3. This event is almost a week behind the production schedule, and the shortage will soon show in the final output count, event 11, unless remedial action is taken. An examination of the other company-made components, circled events, does not suggest excessive resource utilization. No blame can be put on the suppliers of purchased parts because all the triangled events have bars well in balance. Therefore, the chart suggests the application of additional resources for the activities preceding event 3. An investigation of these activities will indicate the most advantageous use of resources—overtime, equipment, more supervision, more working space, improved material handling, and the like.

It also appears advisable for the company to look into the delivery contract with the subcontractor responsible for event 9. By accepting early delivery, the company assumes responsibility for holding costs and possible damage before use. Prompt deliveries are nice, but too much of a good thing can be just as disruptive as a shortage is.

OVERVIEW

An appraisal of the previous examples suggests the reasons that the line of balance has not been more widely used for control. Initially there is the usual skepticism that accompanies the introduction of any new tool, particularly one that appears to require considerable graphing time. Familiarity and computer assistance reduce these objections. A more serious objection is that production is not for specified orders but, rather, for the general market. A market-demand forecast has to be substituted for a known demand as the objective curve. A similar forecasting difficulty exists when the lead time for purchased or subcontracted parts is questionable.

In progress reporting, it is normally not too difficult to obtain a physical

Line-of-balance concepts are very similar to MRP. Both use requirement explosions and time phasing. Both provide close control over dependent demands and rely on computers for large applications.

count of inventory. But estimating a percentage completion for major items still being produced may be very difficult. Forecasts and estimations are limelighted because they are probabilistic inputs to an essentially deterministic control procedure. The opposing assumptions are somewhat reconciled by treating uncertainties as though they were certainties, but meaningful control then depends on the reliability of the prognostications.

An acceptable range is particularly appropriate for a computerized line of balance. The computer can be programmed to sort out items requiring special attention without committing the updating to the artwork of a progress chart.

A practical variant to mollify conditions of uncertainty uses a critical range in conjunction with the line of balance. The range is established from the minimum and maximum allowable variations for events in the program plan. These extremes provide two entry points for each event in the objective chart. For instance, if an event has a lead time of 8 weeks with a minimum of 7 weeks and a maximum of 10 weeks, the expected accumulated item count would be read from the curve at points corresponding to entry time Now and 7 and 10 weeks. When these limits are transposed to the line of balance, they establish a range of acceptable performance. In tune with the management-by-exception principle, attention is called only to items outside the acceptable range.

In research and development work, the objective chart may show budgeted funds versus actual expenditures.

The similarity between the line-of-balance and CPM methods suggests a marriage of the two techniques. The planning merits of CPM, integrated with the control features of a line of balance, offer a routine for project management from the preliminary planning stages through the final delivery. In the initial stages, critical activities of parallel critical paths are candidates for line-of-balance control. During the production stage, both operations and components can be followed in progress reporting. The purpose of graphically portraying the typically computerized CPM data is to summarize the project's progress.

OPTIONAL MATERIAL

Learning curves

As stated earlier, performance today is not necessarily a guarantee of what it will be tomorrow. Individuals can learn new working skills. Groups can learn cooperative skills. The increased productivity resulting from learning can radically influence the completion time of an order. The production schedule for special projects in which learning is a governing factor should take into account the change of output with experience, or the **learning curve.**

An 80-percent learning rate has been found to be descriptive of certain operations in various industries, such as aircraft instruments and frame assemblies, electronic data-processing equipment, ship construction, and automatic machine production. The meaning of an 80-percent learning rate is that the accumulated average time to double the number of units produced is 80 percent of the time required to produce the previous increment.

No learning takes place at a 100-percent learning rate; 100 percent is already known. The other limit approaches 50 percent when the second unit takes almost no time to produce. Picking up bad habits that must be "unlearned" could produce a theoretical learning rate greater than the 100-percent limit.

Assume that it took 100 hours to produce the first unit and that the rate of learning is 80 percent. The accumulated average rate to double the present output, from one to two units, is 100 hours $\times$ 0.80 = 80 hours. Because this is the *average* time per unit, the total time to produce two units is 80 hours × 2 = 160 hours. To again double production from two to four units, the average per unit time decreases to 80 percent of the previous average, 80 hours × 0.80 = 64 hours. This makes the total time to produce four units equal to 64 hours × 4 = 256 hours. Continuing this progression, we have the values tabulated in Table 14.4, in which the hours are generalized by converting 100 hours to 100 percent—the time to produce the first unit expressed as a percentage.

Figure 14.16 shows the total production hours with and without learning when the first unit took 100 hours to produce. The bottom curve corresponds to the upper curve multiplied by the factor in column 2 of Table 14.4. The middle curve is based on a 90-percent learning rate.

Column 2 in Table 14.4 is simply the learning rate raised to a power corresponding to the number of times that production is doubled. The relationship is convenient for illustrative purposes but is unwieldy for practical applications. A mathematical expression of the relationship between production times and quantities is more useful.

The nonlinear relationship of column 1 to column 2 suggests an exponential function. Furthermore, the inverse relationship of the two columns indicates a negative exponent. Letting

Y = accumulated average time per unit
N = number of units produced
a = time required to produce the first unit
b = exponent associated with the learning rate

the equation takes the form

$$Y = aN^{-b} \quad \text{or} \quad Y = a/N^b$$

When production times are known for two levels of production bearing a binary

The learning-curve equation is an application of exponential line fitting discussed in Section 3-6.

TABLE 14.4
Production Time Percentages Based on an 80-Percent Rate of Learning

Number of Units Produced (1)	Accumulated Average Time Percentage per Unit (2)	Total Percentage of Initial Production Time (1) × (2) = (3)
1	100.00	100.00
2	80.00	160.00
4	64.00	256.00
8	51.20	409.60
16	40.96	655.36
32	32.77	1048.64
64	26.21	1677.44
128	20.97	2684.16
256	16.78	4295.68

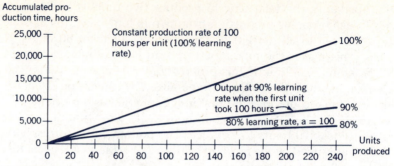

FIGURE 14.16 Production times at 80-percent, 90-percent, and 100-percent learning rates.

relationship, $Y_1 = aN_1^{-b}$ and $Y_2 = aN_2^{-b}$, we can determine the exponent b that defines the proportionality. Using the data from Table 14.4 for $N = 2$ and $N = 4$ in the ratio

$$\frac{Y_1}{Y_2} = \frac{a/N_1^b}{a/N_2^b} = \frac{N_2^b}{N_1^b} = \left(\frac{2N_1}{N_1}\right)^b$$

where $N_2 = 2N_1$, we get

$$\frac{80}{64} = \left(\frac{4}{2}\right)^b = 1.25 = 2^b$$

which yields

A table of logarithms is provided in Appendix D.

$$b = \frac{\log 1.25}{\log 2} = \frac{0.0969}{0.3010} = 0.322$$

By knowing that the first unit took 100 percent (or 100 hours as in the illustration), we can write the expression for the accumulated average time as

A learning curve plots as a straight line on log-log paper. Two widely separated points suffice to set the line required to obtain readings for other points.

$$Y = 100N^{-0.322}$$

Checking another point from Table 14.4 for corroboration, say $N = 64$, we have

$$Y = \frac{a}{N^b} = \frac{100}{64^{0.322}}$$

or

$$\log Y = \log 100 - (\log 64)(0.322)$$
$$= 2.000 - 1.8062(0.322)$$
$$= 2.000 - 0.5816 = 1.4184$$
$$Y = \text{antilog of } 1.4184 = 26.21$$

which, as advertised, agrees with the tabulated value.

The major influence that learning has on the completion time of special

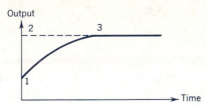

FIGURE 14.17 Learning curve for an individual worker.

projects is apparent, but the mechanics of applying a learning factor can easily lead to deceptive results. By definition, special projects comprise unique products, often entailing untried designs and unfamiliar production techniques. For such projects it cannot be known with certainty the exact rate of learning or even to which portions of the project the learning factor should be applied. The most conspicuous learning area is in manual performance, which determines direct labor cost. But correlative costs of direct materials and supervision overhead can be expected to decline with improved manual performance, owing to less waste, fewer rejects, and reduced supervisory demands. During the initial phase of a project, it is generally wise to calculate learning curves from several bases to discover where learning *actually* affects operations, instead of where it is *supposed* to be influential.

An investigation of the rate of learning for individual operators may attract attention to instructional costs and hiring policies. Figure 14.17 shows a smoothed learning curve. The area of triangle 1-2-3 represents lost production during the learning period. It accounts for extra labor costs to maintain a desired level of production, higher material cost because of inefficient usage, training expense, and fixed costs of overhead. In applying a learning curve, it should be recognized that not all operators start at the same point on the curve; the curve may follow a step pattern in which improvements are made from one plateau to the next.

A learning curve, like other control tools, can engender a false sense of security when operations follow an anticipated pattern for a while but then veer away because the true cause for performance deviations has not been identified. Similarly, other factors such as method changes or model redesigns can obscure an anticipated, actual trend. The difficult search for elusive cause-and-effect data is unending. The decision to adopt a probable relationship for production control rests with the manager—the terminal screen for operating information and the final weighing station for operational alternatives.

Learning curves for intermittent production are often well defined from previous experience with similar operations. Continuous production is seldom influenced by group learning, but individuals entering the system can be expected to pass through a learning period.

Operations controlled more by machine speeds and capacity than by operator skills are less subject to the learning process.

EXAMPLE 14.4 Calculation of individual unit times when production is subject to learning

A government contract for the construction of five prototype missile assemblies is based on an 87-percent **learning rate.** The first unit produced took 190,000 work hours. To compare the actual performance with the contracted production rate, the time to complete each additional unit is calculated. The total production time is determined from the expression

$$Y_i = aN^c$$

where $c = 1 - b$ and i = cumulative unit of output.

Assuming 87 percent is indeed the applicable learning rate, the total production time for the first two units must be 2×87 percent = 174 percent of the time required for the first unit. Then, using 100 percent as equivalent to the 190,000 work hours taken for the first unit, the ingredients for the total production time equation are $Y_2 = 174$ percent, $a = 100$ percent, and $N_2 = 2$, which produce

$$174 = 100(2)^c$$

and

$$2^c = \frac{174}{100} = 1.74 \quad \text{or} \quad c(\log 2) = \log 1.74$$

to yield

$$c = \frac{\log 1.74}{\log 2} = \frac{0.2406}{0.310} = 0.799$$

The time to produce each individual unit is then calculated from

$$y_i = a \, [N_i^{0.799} - (N_i - 1)^{0.799}]$$

By using this equation, the expected time to produce the third unit is

$$y_3 = 190,000 \, [3^{0.799} - (3 - 1)^{0.799}]$$

Then, $(3 - 1)^{0.799} = (2)^{0.799} = 1.74$ (from previous calculations) and

$$3^{0.799} = \text{antilog } 0.799(\log 3)$$
$$= \text{antilog } 0.799(0.4771) = \text{antilog } 0.3812$$
$$= 2.41$$

so

$$Y_3 = 190,000(2.41 - 1.74) = 127,300 \text{ work hours}$$

The individual unit times and total production time based on an 87-percent learning rate for all five units are as follows:

Units	Individual (y_i)	Total (Y_i)
1	190,000	190,000
2	140,600	330,600
3	127,300	457,900
4	117,800	575,700
5	112,100	687,800

EXAMPLE 14.5 Forgetting curves

Studies of performance improvement as a function of time have revealed various patterns for learning by individuals. Sometimes learning by operators spurts ahead quickly, levels off, and then climbs again at a lower rate than the initial surge. At other times the learning is at a steady but decreasing rate. More often, learning is slow during an incipient phase when the operator is getting acquainted with the work, accelerates with familiarity with work conditions, and then levels off as there are fewer chances to reduce errors and improve movements. This pattern is called an *S-curve*. Regardless of the shape of the learning curve, there is always a **forgetting rate** that begins when an operator leaves the previously learned work.

J. G. Carlson and A. J. Rowe ("How Much Does Forgetting Cost?," *Industrial Engineering*, September 1976) suggested that an S-curve is the most representative learning model and that it is affected by forgetting in the following ways:

1. Some forgetting is always to be expected, but total forgetting does not occur during short periods of interruption.

2. Forgetting curves show rapid initial decreases in performance followed by a gradual leveling off as a function of the interruption interval.

3. The rate and amount of forgetting decreases as an

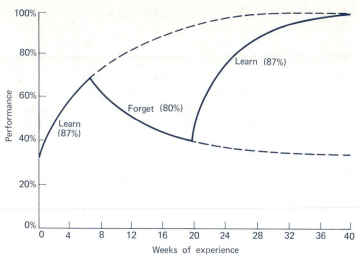

FIGURE 14.18 Individual performance during a learn-forget-learn cycle for an interruption of 13 weeks with the given learning and forgetting rates.

increased number of units are completed before an interruption occurs.

According to Carlson and Rowe, a "learn–forget–learn" function could appear as shown in Figure 14.18. An operator is interrupted after seven weeks of production involving a complicated task. A learning rate of 87 percent had almost doubled performance during that period. A 13-week interruption at a forgetting rate of 80 percent dropped performance from 70 to 52 percent, the level previously attained after 2.5 weeks of task performance. At this point, the operator returns to the task, and learning again takes place at the 87-percent rate. The implications of interruptions on production costs are obvious.

END OPTIONAL MATERIAL _____

14-5 ENERGY CONTROL

Energy conservation has become of paramount importance since the mid-1970s. The rising costs and dangers of supply curtailments worry industrial and private users alike. The United States consumes far more energy than it produces, accounting for one-third of the world's consumption while producing about the same proportion of the gross world product. It took 50 years (1900 to 1950) for the total annual U.S. energy consumption to go from 4 million barrels of oil equivalent per day to 16 million, but it took only 20 years (1950 to 1970) to go from 16 to 32. The rate of increase has slowed recently, largely as a result of conservation efforts.

Coal produces only 18 percent of U.S. energy needs yet accounts for 90 percent of its proven reserves.

Energy audits and load control

Energy management starts with an **energy audit** to determine an organization's

Energy sources are converted to a standard measurement unit in consumption studies, usually the British Thermal Unit (BTU) (for example, 1 gal gasoline = 130,000 BTU, 1 kWh electricity = 3413 BTU)

current energy utilization. A *gross energy audit* measures total consumption and identifies the types of energy used. This flow information reveals the seriousness of the energy situation and suggests the most profitable areas for conservation. A *detailed energy audit* determines where each type of energy is being consumed—which equipment, process, and operations rely on what type of energy, and how much. It documents the energy consumption from the gross audit by collecting policies, procedures, usage rates, and costs associated with each subsystem in the total energy system. For instance, a lighting audit would include data on the number of bulbs, power requirement per bulb, location, hours of

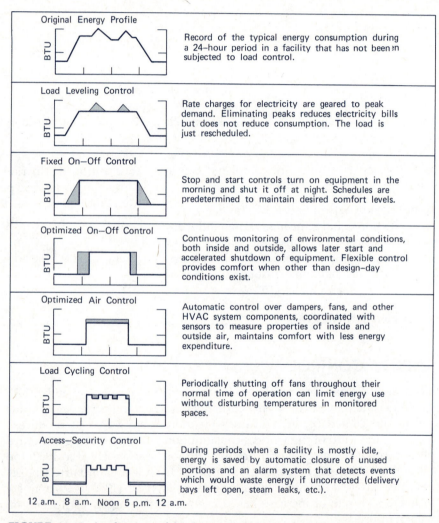

FIGURE 14.19 Application of load-control techniques. (Adapted from D. Foley, "Reduce Waste Energy with Load Controls," *Industrial Engineering*, July 1979.)

operation, light meter readings, and illumination policies. Forms for making audits are readily available from government, educational, and professional groups.

Large buildings erected in the 1950s and 1960s have seldom been energy efficient, particularly the glass-box skyscrapers that sprouted in big cities. New energy conscious designs feature sunlit atriums, black finish on north and east sides to absorb heat, silver finish on the other sides to reflect heat, and reflecting surfaces to bounce natural light into the interior. Industrial plant design is receiving equivalent attention (see Example 9.2). Old and new buildings and plants are being outfitted with computer-controlled systems that sense changing conditions, alter energy usage to conform to current needs, and sound alarms when an unnecessary energy expenditure is detected.

New designs can reduce energy needs by 60 percent in similar glass and steel buildings.

Underground construction is increasing to take advantage of the insulating properties of earth.

In most buildings, lighting and heating-ventilating-air conditioning (HVAC) systems account for 80 percent of energy consumption. **Energy load control** techniques may reduce energy costs by 15 to 30 percent. Several of the techniques are shown in Figure 14.19. Centralized controls can be provided by a minicomputer for large installations. Less expensive programmable controllers in dispersed locations are suitable for smaller facilities.

Energy conservation

Significant energy savings can be accumulated from many small but conspicuous actions to conserve energy: turning off lights, insulating buildings, driving less, using more energy-efficient machines, and the like. A more institutionalized way to reduce energy consumption is to recapture waste heat. Process steam and direct process heat together account for about two-thirds of the current industrial use of energy. By the year 2000 it is forecast that industry's electric power requirements will exceed its thermal power requirements. This represents a major change, as today's thermal energy requirements are much larger than electrical energy usage.

ERDA Report TID-27384/1-3, January 1977.

Waste heat is found in condenser cooling water, contaminated process water, and boiler or furnace exhaust. It exists in various stages of product refinement, for example, in the multiple stages of heating and cooling of iron as it moves from the blast furnace to a piece of sheet steel. The thermal wastes from industrial processes are estimated to be as large as the fossil fuel input for electric power generation.

Rising industrial demand for electric power suggests more attention will be given to *energy cascade* systems—those designed to produce both heat and electricity. Energy cascading attempts to match the quality (temperature) of available energy to the needs of the task. For instance, leftover heat from a steel rolling mill furnace (3500°–2000°F) would be an input to a gas turbine (2000°–1000°F), from which the output would flow to a steam turbine (1000°–400°F) where the exhaust would supply process steam. This cycle is a form of **cogeneration**—the production of electricity and process steam in one system. The cycle could begin at a lower temperature when the generation of electricity is

District heating is more common in Europe than in the United States. A municipally owned utility may supply electricity plus process steam to industries and electricity plus heat to residences.

the initial stage and the waste heat is used for industrial or commercial heating; this pattern is most evident in *district heating* in which electricity and hot water/steam are produced for residential and commercial users.

Efforts to control pollution frequently lead to energy generation. Energy recovered from burning industrial, agricultural, and municipal wastes partially finances investments to clean up the environment and occasionally more than pays for the pollution abatement. As an example, a paper mill was told in 1973 to shut down or install a new recovery system to meet Florida's particulate emission standards. A new system was installed in 1976 that met the standards and enabled the company to increase production by 15 percent while saving the equivalent of 150,000 barrels of oil a year in energy usage. With a continued escalation of energy prices, pollution control and energy control are likely to become entwined, and both energy consumers and the environment should benefit.

14-6 CLOSE-UPS AND UPDATES

Increased information flow is both a bane and a boon for production managers. In theory it is impossible to have too much information about production operations. In practice, however, managers may not be able to use all the information available for decision making. Data are also cursed by inaccuracy, collection expense, and tardiness. many of the baneful aspects of information are avoided by **automatic identification systems (AIS).** As prices for automatic identification devices have declined, AIS have expanded rapidly in manufacturing, distribution, and retailing industries. They are installed for one purpose: better control.

AIS can reduce the amount of paperwork, time, and labor needed to generate, acquire, and use information. Devices read source data or codes and translate them into digital form that can be computer processed for controlling operations, generating reports, and conducting transactions. The systems are fast. They process hundreds of characters per second, as compared with maximum rates of 5 to 7 by hand and 10 to 15 via a keyboard. They are also accurate. The error rates for AIS are about one in 3 million entries. Handwritten records have about 1 error in 30 entries, and the keyboard input error rate is about 1 in 300.

Among the AIS technologies now available are optical, radio frequency, magnetic, and voice equipment. The following factors influence the selection of which technology to adopt:

- *Amount of information required*—most bar codes are limited to 32 characters.

- *How the information is attached to the product or process*—directly on the product, on a transport device, or at a fixed position such as a work station.

• *Operating environment*—temperature extremes, high humidity, corrosive liquids or gases, dirt or grime, and exposure to weather or radiation.

Although AIS are usually associated with material flow control, they can collect factory-floor data for payroll purposes, product/process costing, defect analysis, and machine utilization.

Devices for AIS that use low-level radiation must carry precautionary labels.

Bar-code technology:
Reading between the lines

Bar codes have become an international language for encoding data that can be captured by automatic and semiautomatic devices. In-place laser scanners can extract bar-coded data from products whipping by at 500 feet per minute. Hand-held wands can read employee identification badges and work tickets. **Bar coding** is the most widely used automatic identification method of collecting manufacturing information.

Manual light pens must touch the bar code to "make a read." Read rates depend on an operator's dexterity.

Some bar-code scanners can read codes through protective coverings and holographic versions, like those used in supermarket checkouts, and can read codes on cylindrical and soft objects. The major components of a scanner are an illumination source, photodetector, and microcomputer. A focus light is moved across the barred symbol. Reflected light is received by the photodetector which generates a voltage proportional to the light received. The signal is preprocessed and sent to a microcomputer, where it is decoded and translated for printing, display, or transmission.

Bar-code applications are best suited for the repetitive identification of large numbers of items.

To "bar code" an item means to print a *code* (encodation of information to make it shorter) in the form of a *symbol* (a block of bars of varying widths) so that a computerized scanner can read it. Symbols are produced by mechanical, laser, and photographic means directly on an item or as labels that can be placed on items. Reading problems arise when bars cannot be clearly distinguished from the background color: reds, oranges, and browns cause the most difficulty. Quality assurance procedures have been established to verify that bar-code printing conforms to specifications.

Other types of identifiers:
Machines that look and listen

The capabilities of machines to recognize images and signals are increasing and becoming more reliable. *Optical character readers* have a limited ability to extract information from "human-readable" text. *Magnetic strips* found on credit cards and *magnetic ink characters* used on bank checks can be read easily and rapidly by machines. *Radio frequency* and *acoustical wave* systems used in automated production processes consist of a control unit, antenna, and code tag. The tag is mounted on carriers such as a railroad car or an automobile frame. When the tag enters the antenna's capture window, the tag is excited by radio or radar pulses and returns a coded signal. Such signals can reveal location information or activate machines in a production line or transport area.

The idea of giving oral commands to a machine has long been fascinating. Within limits it is now practical. **Voice recognition** by machines is ideal for operators whose work fully occupies their hands and eyes; they can control other machines by talking to them. Most voice systems require the receiving machine to be "trained" to understand a particular speaker, who speaks one word or short phrases at a time. The current technology limits the vocabulary to a few hundred words, which is adequate for many data-entry, control-activation, and security applications. Future developments should yield speaker-independent systems with large vocabularies that "hear" normal speech.

Machines that see also use video technology to present a picture to a processor that digitizes the image. Numbers are assigned to different shades of gray to define a pattern. The pattern is then matched to stored images for interpretative purposes. The process is expensive because huge amounts of data must be handled, but it holds great promise.

14-7 SUMMARY

Quantity control concentrates on delivering the desired output within the expected delivery date. Production plans are converted into action notices that spell out exactly which workers and machines will operate, what the operations will be, and when they must be done. Then the actions are compared with planned performance to provide feedback for replanning or initiating corrective action. The choice of control designs and the control techniques depends on the type of production to be controlled.

Flow control is applied to continuous production when products, equipment, and work assignments are standardized. The emphasis on continuous flow for high-volume output requires high raw-material inventory levels but a relatively low in-process inventory level.

Order control is associated with intermittent production. The work schedules must be closely controlled because shop orders are sequenced through different production patterns. The two principal methods of scheduling are (1) *backward* to meet a deadline and (2) *forward* to produce as soon as possible. Load control is applied when there is a distinct limiting factor in the production flow, and particular attention is given to funneling orders through the bottleneck.

Special project control is characterized by personal contact. Managers and

Control factors

Control designs

supervisors issue orders directly to workers, monitor progress, and initiate corrective action.

A master production schedule consolidates information from preproduction planning activities. It shows how and when products will be ready for distribution. The dispatching function implements the operations from the master schedule. Expediting is often combined with the dispatching function. Expediting is concerned with the flow of material and components, whereas dispatching pertains more to the flow of information. An expeditor follows the development of an order from the raw-material stage to finished products.

Critical ratios are calculated to assist the scheduling of work. Job orders are given priority according to the urgency of completing them to meet promised delivery times or to maintain a desired level of supply. Lower ratios indicate greater urgency.

Gantt charts display production activities as bars on a horizontal time scale. Their principal virtue is simplicity, which makes them commendable communication aids. They are used to show assignments derived from one of the three scheduling methods: perpetual, periodic, and order. Through the use of special marks, symbols, and colors, a Gantt-chart format can also be used to record deviations from the time schedule and the causes of these irregularities.

CPM time charts are generally more appropriate than are bar charts for analysis and design. A time chart is essentially an arrow network drawn to a time scale. Activity relationships and the amount of float time available are shown in a rectangular coordinate system. The versatility of the chart is enhanced by adding a resource grid that conforms to the chart's time scale.

Line of balance uses the principle of management by exception for production control by a comparison of progress on individual components with the time schedule for their completed assemblies. An objective chart shows the scheduled delivery of finished products and is updated to show actual deliveries. A program plan, an event-oriented chart similar to a time chart, shows the lead time by which each event must precede the final completion. The current status of each event is shown by a bar on the progress chart. The vertical scale of the progress chart is set adjacent and equal to the vertical axis of the objective chart. A line of balance relates the two charts to show the expected and actual number of completed items associated with each event.

Learning curves help anticipate output levels when operating times decrease with experience. An 80-percent learning rate, typical of several industries, means that the accumulated average time to double the number of units produced is 80 percent of the time required to produce the previous increment. The nonlinear relationship of accumulated average time, Y, to the number of units produced, N, is given by the expression

$$Y = aN^{-b}$$

where a equals the time required to produce the first unit and b equals the exponent associated with the learning rate.

Group learning curves should be calculated for a number of production activities to discover how learning actually affects a project. Individual learning curves highlight costs incurred before an operator reaches an acceptable level of performance. Forgetting must be considered too.

Energy control

Control over energy consumption becomes more important as the costs of energy rise and the supply of energy becomes less reliable. Energy audits determine the amount, location, and type of energy usage. Improved building design can reduce requirements, and load controls can decrease energy consumption by programming the operation of equipment to minimize waste without adversely affecting the work environment. Waste heat can be recovered by energy cascade systems designed to produce heat, electricity, and steam in successive stages of certain kinds of production. Using waste products for fuel may cut energy costs and simultaneously reduce pollution.

Machine identification

Automatic identification systems are improving the accuracy, timeliness, and cost of data collection. Bar coding is the most widely used method. Other devices recognize radio signals, spoken words, and visual images.

14-8 REFERENCES

Buffa, E. S. *Modern Production/Operations Management*, 6th ed. New York: Wiley, 1980.

Cochran, I. B. *Planning Production Costs*. New York: Chandler, 1968.

Department of Commerce. *Waste Heat Management Guidelines, 121.* Washington, D.C.: U.S. Government Printing Office, 1977.

Department of the Navy. *Line of Balance Technology*. Washington, D.C.: Office of Naval Materiel, U.S. Government Printing Office, 1962.

Doolittle, J. S. *Energy*. Champaign, Ill.: Matrix Publishers, 1977.

Eilon, S. *Elements of Production Planning and Control*. New York: Macmillan, 1962.

McLeavy, D. W., and S. L. Narasimhan. *Production Planning and Inventory Control*. Boston: Allyn & Bacon, 1985.

Riggs, J. L., L. L. Bethel, F. S. Atwater, G. H. E. Smith, and H. A. Stackman, Jr. *Industrial Organization and Management*, 6th ed. New York: McGraw-Hill, 1979.

Vollman, T. E., W. L. Berry, and D. C. Whybark. *Manufacturing Planning and Control Systems*. Homewood, Ill.: Irwin, 1984.

14-9 SELF-TEST REVIEW

Answers to the following review questions are given in Appendix A.

1. T F *Flow control* is associated with continuous production.

2. T F *Order control* is simpler to apply than flow control is because it is used for intermittent or job-shop production.

3. T F When a required delivery date is known, *forward scheduling* is applied.

4. T F *Load control* is the name for both a scheduling method used when a limiting factor affects production flow and a technique to reduce energy consumption.

5. T F *Expediting* is associated with issuing work orders.

6. T F A *critical ratio* is the percentage of a project's activities that are on the critical path.

7. T F A common form of a *Gantt chart* uses bars to represent jobs scheduled for different production facilities.

8. T F Activities, dummies, and float are indicated by solid or dotted lines on both CPM networks and *CPM time charts*.

9. T F A *resource grid* displays the allocation of resources for a project scheduled on a time chart.

10. T F An LOB chart of the operations required to complete one unit of the finished product is called an *objective chart*.

11. T F The line of balance can be drawn on a *progress chart*.

12. T F A 100-percent *learning rate* means that no learning occurs.

13. T F In the *learning curve* equation, *a* indicates the time required to produce the Nth unit.

14. T F An *energy audit* determines an organization's energy utilization.

15. T F *Cogeneration* is the production of electricity and process steam in one system.

16. T F *Bar coding* is the most widely used automatic identification method.

14-10 DISCUSSION QUESTIONS

1. Select a production plant or product characterized by each of the three divisions of production discussed in Section 14-2. Compare the selections according to the characteristics listed for each division.

2. Could dispatching be considered the interface between production planning and production control? Why?

3. Compare the duties of a police officer with those of an expeditor.

4. Two methods of expediting are
a. Exception method—devote attention only to orders or situations known to be in trouble.
b. Fathering method—watch over and attend to all problems for a group of orders.
Comment on both methods, and suggest production situations for which each would be well suited.

5. What is the danger in trying to compensate for too little production capacity by increasing the number and authority of expeditors?

6. What relationship does the type of production (continuous, intermittent, and special) have to the type of scheduling used (perpetual, periodic, and order)?

7. Why is it important to distinguish between customer order priority and operation priority? Which priority rules are appropriate to each?

8. Compare the advantages and disadvantages of Gantt charts and CPM time charts.

9. Why bother to update control charts? Would lists be an adequate substitute for charts?

10. The outstanding virtue of Gantt charts was said to be simplicity. What is the value of simplicity? For what uses is simplicity most important?

11. Why do dummies in a time chart never slope to the left?

12. Assume that the construction plans for a small vacation home have been committed to a time chart.

The owner-builder plans to use the cabin in August. At the end of May he finds he is 5 working days behind schedule with only 20 possible working days left before August. Give and discuss an example of how he could apply each of the three scheduling alternatives now open to him.

13. A time chart is based on activity descriptions, whereas a program plan is based on event descriptions. Why? How could a time chart be converted to a program plan?

14. How does the line-of-balance method use the principle of management by exception?

15. Describe a computer printout based on an acceptable range of item inventory levels that could avoid the requirement for committing the information to a progress chart.

16. What can be learned from comparing the vertical bars in the progress chart with the line of balance at each updating? What can be learned from a succession of the same comparisons made over several updating periods?

17. How would you explain the meaning of a 90-percent rate of learning to a potential supplier? Refer to the upper and lower limits for the learning rate.

18. What support activities would be affected by the discovery that production on a contract was actually following a 90-percent learning curve instead of the expected 80-percent learning rate?

19. Explain the relationship between the c and b exponents of learning-rate expressions.

20. Comment on the importance of individual learning rates in deciding whether to carry larger inventories that level out the production schedule or to carry a smaller safety stock and alter production output by laying off or hiring workers according to market demands.

21. In what ways are efforts directed toward pollution control and energy conservation compatible? In what ways do they oppose each other? Provide an example of each situation.

22. What production scheduling tactics are suggested by "forgetting curves"?

23. Discuss how an energy audit corresponds to the cause side of a C&E diagram for energy management and how load control corresponds to the effect side of the diagram.

24. Which institutional, economic, and technological constraints limit the practial application of
a. Energy cascade of waste heat.
b. Cogeneration
c. District heating.
d. Waste products burned as fuel.

25. Dr. Orlicky uses an example of "lumpy" stock depletion to show a weakness of critical ratio priorities. ("Requirements Planning Systems," *Proceedings*, APICS 13th International Conference, 1970). A situation is described in which the beginning inventory level is 95 units, the order point is 60, and the order quantity is 100. As shown, 65 units are required during the first period. This drops the amount on hand to $95 - 65 = 30$, which is half the order point level, causing a critical ratio of $0.5/1.0 = 0.5$. A CR less than 1.0 awards a high priority to the job; consequently work begins on it, and 50 percent of the job is completed during the second period. Then the critical ratio becomes $0.5/0.5 = 1$, which means "on schedule," and erases the job's high priority status. Nothing is done during the third period; the job is still low on the priority list. In period 4, 35 units are needed, but only 30 are on hand—a stockout. At this point the critical ratio drops to $0/0.5 = 0$, which is the highest possible priority. The emergency condition came about, even though there were two periods in which work could have been done, but the job was untouched because its $CR = 1$ priority raised no warning of the potential stockout.

Period	1	2	3	4
Material requirement	65	0	0	35
Remaining work to be completed	100%	50%	50%	50%
Critical ratio	0.5	1.0	1.0	0.0

What basic assumption of the critical ratio technique has been thwarted by the demand pattern? Comment on the possibility of this type of demand with respect to manufacturing schedules discussed in Chapter 13.

14-11 PROBLEMS

1. The order quantity for a job is 200 units of material, and its planned completion duration (lead time) is 12 days. Complete the following table, which marks the progress of the job through the production process.

Days in process	0	3	6	9	12
Stock on hand (units)	210		100	40	
Work completed (%)	0	10		75	80
Critical ratio		1.11	1.0		0.5

2. The following orders have been received by a job shop. The date today is working day 100.

Job Number	Date Received	Promised Delivery Date	Production Days Required	Estimated Shipping Date
711	91	111	20	112
712	92	105	12	110
713	94	99	5	102
714	96	111	15	110

a. In what sequence would the jobs be scheduled under a policy of "first come, first served"?

b. Determine the critical ratio for each job and the corresponding priority.

3. Use a Gantt-chart format to show the major activities you plan to accomplish during a day. Update the chart at noon and at 6 P.M. Note the causes of deviations from the schedule. Comment on the process and possible use of the record.

4. A box factory has received an order for 40,000 fancy fruit boxes to be delivered in 90 working days. The four pieces for the sides and ends are cut and grooved in saw line 1, where the capacity is 1300 units (4 pieces each) a day. The thinner slats for the identical tops and bottoms are cut in saw line 2, where the maximum daily output is enough slats to fabricate 6000 tops *or* bottoms. The slats are stapled to form the top and bottom assemblies at the rate of 3500 assemblies (top or bottom for a box) per day. The top assemblies are to be spray painted. The paint shop can spray and dry a maximum of 5000 lids each day. The last operation is to stack the side and end pieces between the top and bottom assemblies, collect parts for 50 boxes in a bundle, and band the bundle together: 300 bundles can be stacked and banded in one day. One week (5 working days) must be allowed for shipment from the factory to the buyer.

The policy is not to commence work on a new operation until the material has completely passed through the prerequisite operation. One day is allowed for leeway in sequencing. Because of the high work load, backward scheduling is used. No overtime should be included.

Develop a schedule for processing the order. Display the developed schedule in a Gantt chart.

5. Convert the arrow network shown in Figure 7.8 to a time chart. Show the float at the end of appropriate activity chains.

6. The project plan for reconstructing a portion of the manufacturing facilities is as follows. Develop a time chart for the program in which total float is shown by the conventions described in Section 14-4 and illustrated in Example 14.2.

Duration (weeks)	Symbol	Activities Description	Restrictions	(Data for Problem 6)
8	A	Prepare drawings and specifications	$A < B,C$	
4	B	Secure bids and award contract	$B < D,H,I$	
2	C	Remove existing equipment	$C < E$	
1	D	Reschedule production work loads	$D < E$	
1	E	Form and pour foundations	$E < F,G$	
2	F	Electrical modifications and renovations	$F < J$	
1	G	Paint	$G < J$	
1	H	Purchase and delivery of auxiliary equipment	$H < K$	
6	I	Construct new production machine	$I < J$	
1	J	Install new production machine	$J < K$	
2	K	Test and debug new machine and equipment	$K < L$	
1	L	Reschedule and start production in reconstructed facility		

7. Assume you are the chairperson of a committee to plan and conduct a regional convention. The date of the convention is now 20 weeks away. Most of the activities that must be accomplished are as follows:

a. Outline plans and check with the committee.

b. Select and secure speakers.

c. Reserve meeting place.

d. Arrange catering for banquet.

e. Design and print programs.

f. Distribute programs.

g. Select and purchase awards.

h. Arrange publicity.

i. Reaffirm arrangements and make last-minute adjustments.

j. Conduct banquet.

k. Write acknowledgments.

l. File report and statements.

Complete the list and allocate times you deem appropriate for each activity. Prepare a time chart that can be used to explain your plans to the rest of the committee.

8. Three-hundred swamp buggies have been ordered by the Forest Service for timber cruising and forest development of inaccessible lands. The supplier has selected the line-of-balance method for production control. The program naturally divides into five divisions: securing traction subassemblies from a subcontractor, fabricating frames and cabs, obtaining motors and developing controls, building accessories, and assembling all components. The program plan is shown in Figure 14.20.

The delivery schedule calls for six buggies to be finished each week during the contract period of 50 weeks. The tenth-week progress report reveals the following physical count of items associated with each event:

1–141	5–112	9–103	13–75
2–135	6–100	10–96	14–76
3–124	7–146	11–89	15–71
4–106	8–121	12–88	16–60

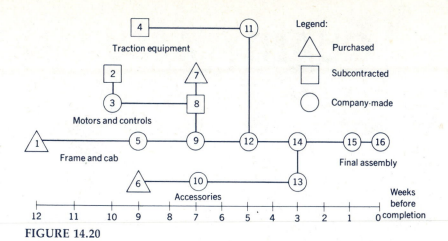

Legend:

△ Purchased

□ Subcontracted

○ Company-made

Traction equipment

Motors and controls

Frame and cab

Final assembly

Accessories

Weeks before completion

| 12 | 11 | 10 | 9 | 8 | 7 | 6 | 5 | 4 | 3 | 2 | 1 | 0 |

FIGURE 14.20

a. Construct a line of balance for the current status of the program.

b. Comment on items out of balance. Which event appears most critical? Compare the progress of purchased, subcontracted, and company-made components.

9. A cost-plus contract has been awarded for 90 technologically new components. From similar experiences with new designs, the company has found an 85-percent learning rate descriptive of the process, exclusive of material costs. The first unit costs $2400, of which $600 was for material. If the contract specifies a profit of 18 percent on the total cost of each unit, what will be the expected cost of the entire contract (total price paid for 90 units)?

10. Workers being trained to wire a complex control panel have generally followed a learning pattern where the accumulated average time is given by $Y = aN^{-0.15}$.

a. What learning rate corresponds to the given $b = 0.15$?

b. If a new worker completed wiring on his or her first panel in six hours, how much time can the worker be expected to take for the second, third, and fourth panels?

c. How many panels will the worker likely complete in his or her first 40-hour week?

11. Given a learning rate of 70 percent and assuming that it took 20 hours to produce the first 25 units, how long should it take to produce 100 units?

12. From the production records of a recent contract, it is found that the accumulated direct-labor hours expended in producing the first 20 units were 250; 30 units later the total hours were up to 450 (time to produce 50 units). What is the learning rate?

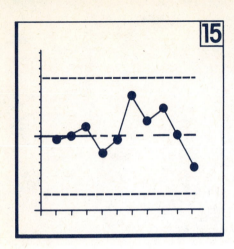

15 QUALITY ASSURANCE

LEARNING OBJECTIVES

After studying this chapter you should

- realize that superior quality is a crucial attribute of both service and manufacturing systems and that there are many ways to pursue it.
- understand the relationships of value to cost, vigilance to error, and precision to accuracy.
- be familiar with the organizational aspects of quality assurance and zero defect programs.
- know when, where, and how to inspect.
- be familiar with the nature and methods of acceptance sampling, including the variety of sampling plans that can be used.
- understand the statistical theory that supports quality control charting and be able to construct $\bar{x}$, R, p, and c charts, knowing when each is appropriate.
- recognize the advantages and limitations of applying Japanese practices and techniques such as total quality control (TQC), CEDAC, and quality control circles.

KEY TERMS

The following words characterize subjects presented in this chapter:

precision/accuracy

vigilance versus error

quality assurance

zero defect (ZD) approach

variables/attribute inspection

acceptance sampling

operating characteristic (OC) curves

acceptable quality level (AQL)

lot tolerance percent defectives (LTPD)

producer's/consumer's risk

sample mean

average outgoing quality (AOQ)

p- and c-charts

sampling plans

total quality control (TQC)

upper and lower control limits

quality circles

$\bar{x}$ and R control charts

CEDAC

15-1 IMPORTANCE

Within a production system, quality can take many meanings and impart different considerations. All of them are important to the managers of the system.

To retail customers, quality is a characteristic of the product they may buy. They can accurately measure quality in only a few cases because they lack the ability, equipment, or inclination. Instead, they rely on brand names, reputation, previous experience, and general appearance. After a product is purchased, its performance is measured against competitors and its own advertised image.

Wholesale buyers or industrial customers are better prepared to measure quality. They know that the purchased inputs to their system will affect the quality of their output and, thereby, their reputation. Their large volume supports a staff and technology sufficient to verify the quality and quantity of purchases. Yet they generally cannot afford, nor do they prefer, to measure the quality of every item they buy. The problem is to select the important criteria of quality and then to develop a sampling plan that ensures conformance to the criteria with reasonable inspection costs.

Product designers stand astride the quality demands of customers and the quality capabilities of producers. Their first responsibility is to design a product wanted by consumers. They are assisted by market research and other staff efforts. Then the product design specifications must be set within the production capabilities of the producer and the requisites of the buyer.

The quality theme underscoring production activities is continuous control. Its manifestations take many forms and provoke diverse attitudes. To the company statisticians, it is a challenge to develop statistical formulas compatible with the production process and quality directives. Inspectors make measurements and observations to effectuate the statistical design. Supervisors are at the interface between the quality goals set by top management and the execution of programs to attain the goals, an important position in all quality control efforts. Finally, the workers who are the basic source of quality are subjected to exhortations from quality-conscious supervisors, to motivational propaganda of quality programs, and to performance ratings from inspection procedures.

All pieces of the quality picture have to be fitted into a functional entity, or it can degenerate into sporadic campaigns of hectic activity founded on a pile of inspection reports. As represented in Figure 15.1, the quality subsystem is closely linked with the governing production system. A quality product is the result of careful design specifications, conformance to specifications, and feedback about product performance. The crucial position of customers is evident

Consumer protection is the goal of the Consumer Federation of America and the National Consumer Congress. Tighter controls on advertising, safety, and pricing have resulted from legislation and rulings by federal regulating agencies.

Product specifications are delineated by *tolerances*—permissible variations in size—and by *allowances*—differences in size between parts that fit together. For instance, a shaft with a tolerance of 1 inch ± 0.002 inch could be designed to fit into a hole with a minimum diameter of 1.003 inches, an allowance of 0.003 − 0.002 = 0.001 inch.

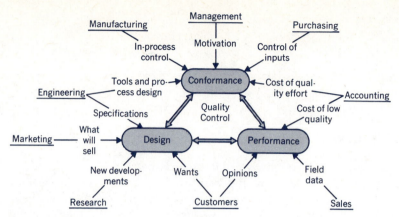

FIGURE 15.1 Functional responsibility for quality within the total production system.

from the design that seeks to satisfy their wants and the performance feedback that checks whether they are indeed satisfied.

Government and service organizations also strive for higher quality in operations whose product is service to the consuming public. Police, waiters, bankers, repairpersons, teachers, nurses, and a multitude of other government and service-industry employees do work that can be enhanced by quality assurance procedures. Their output is usually more difficult to measure than are manufactured products, and the effects of poor quality are often masked. Instead of faulty products returned for a refund, poor service gradually builds resentment that may eventually flare into irate complaints, calls for dismissals, or switched patronage.

In this chapter we emphasize the procedures and tools available to improve conformance to quality specifications. This phase of quality control has received dramatic documentary endorsements and has a well-developed body of technology associated with it. But it cannot operate in isolation. The best statistical tools and the most clever quality programs can be effective only if they are backed by all levels of management and are honed by feedback from the production system.

15-2 ECONOMICS OF QUALITY ASSURANCE

Precision describes the refinement of the product, and *accuracy* pertains to conformance to design specifications.

Production quality starts with a process capable of producing to the design specifications and continues with an inspection program that ascertains whether the standards are being met. The initial decision concerning specifications is based on the **precision** sought by customers and the **accuracy** attainable by production facilities. Given a process capable of obtaining the required precision with desired accuracy, unacceptable variations may, and usually do, still occur. Blunt tools, misalignments caused by wear and tear on machinery, and workers' carelessness contribute to inferior output from a process inherently capable of acceptable quality. The preferable means to achieve a quality output is revealed by an economic evaluation of the cost of quality assurance compared with the increase in product value financed by the cost.

Cost of vigilance versus cost of error

The interlocking nature of design decisions and continuous control is implied by the cost–value relationship shown in Figure 15.2. An initial evaluation relates the returns gained from greater product refinement to the cost of providing the refinements. The value of providing additional product refinements is determined in the marketplace, perhaps with some help from advertising. Production cost includes both the capital investment to attain the necessary accuracy and the quality assurance program to sustain it.

Tactical implementation of the strategic cost–value relationships is complicated by the shades of accuracy possible for each level of refinement. Would it be more economical to accept a lower price for items that have to be held within stated tolerances 90 percent of the time, or to ask a higher price and promise that 95 percent will meet the same specifications? Should a higher-priced machine be purchased to produce items consistently within narrow tolerances, or will a lower-priced machine, coupled with intensive sorting, produce equivalent output quality? Answers to such questions are determined mainly by referring to the cost pattern shown in Figure 15.3.

In most production situations, the costs of **vigilance** and **error** vary inversely. Greater vigilance may take the form of extra time taken by individual crafts people, closer supervision, additional tests for products, and inspection of all or a portion of output. The cost of errors includes rework, rejects, and customer dissatisfaction. The tip-up at the lower end of the error-cost curve represents a condition in which too much attention to errors tends to provoke mistakes; a common illustration is in typing (typical of many worker–machine operations), in which concentration on perfection leads to more errors than usual. Somewhere between the extremes of no vigilance and ultravigilance is a point at which control over the magnitude of errors produces a minimum total cost.

Try to think about how you write your signature as you sign your name, and it will probably look like a forgery.

When the cost of vigilance is primarily constituted by inspection, we may encounter a seemingly contradictory inspection cost–value relationship. Higher inspection costs, if the money is wisely spent, will detect a greater proportion

FIGURE 15.2 Product value and cost as a function of refinement.

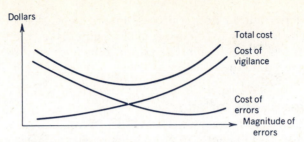

Dollars

Total cost

Cost of
vigilance

Cost of
errors

Magnitude of
errors

FIGURE 15.3 Relationship of vigilance to errors.

Inspection as described in this section serves to rectify the process—bad items are sorted from the output. The rectification can upgrade output from a process otherwise not capable of meeting specifications.

Acceptance sampling examines a portion of the items being purchased to determine whether the overall quality meets the standards expected from the supplier.

of defects and thereby provide a greater degree of protection from the effects of defective output. As the number of undetected defects decreases, a point is reached at which further improvement is unnecessary because the higher quality does not proportionately increase the value of a product. For instance, if 98-percent viability of grass seeds is considered acceptable, decreasing the undetected nonviable seeds below 2 percent would be uneconomical unless the higher quality were rewarded with a higher price. Thus, it is not uncommon to find the minimum total cost obtained by an inspection plan that purposely passes some defectives.

In some processes the purpose of inspection is to examine all the output to discover any defective workmanship. For special projects such as construction work, a single item may be inspected several times during its development. In other processes (inevitably, those in which the tests destroy the product), inspection means examining samples from the total output. The statistical design of the sampling plan attempts to minimize the inspection cost while maintaining a desired level of confidence that the process is within established control limits.

EXAMPLE 15.1 Balancing public costs and services

A state department of motor vehicles issues registration papers for all motor vehicles operated in the state. The original information is taken from handwritten application forms and entered into computer files. Errors occur in the original data and in the transfer from the application to the computer. Some errors are minor and some are major, such as entering the wrong serial numbers. It would be possible to check and double-check every document, but the cost would be high, and some errors would still slip through owing to human oversights. Therefore, a program was needed that balanced the cost of errors

to the public and the cost of extra vigilance which is also a charge to the public.

Analysts in the registration department decided that 98-percent accuracy was a realistic goal: only two registrations per 100 could have any type of error. In the rare instance when an error was serious enough to cause financial loss to the owner of a vehicle, the department assumed full liability for the loss. Experience proved that such losses were lower than the expense involved even to approach 100-percent accuracy.

To ensure that 98-percent accuracy was maintained, the work of each operator involved with registration was randomly inspected, and a chart of the results was posted at each work station. When the samples indicated that an operator was allowing too many errors, *all* of that operator's printouts were inspected until the operator was consistently attaining the desired quality level. This inspection routine contributed to correct documents, job motivation, and employee training.

History of quality control

Formal control of quality was unnecessary when production was the province of individual crafts people; then the producer's personal reputation was at stake with each unit of output. With mass production, division of labor, and interchangeable parts, the individual pride of performance had to be buttressed by more formal controls.

The course of quality control was set in 1924 by the work of Walter A. Shewhart of Bell Telephone Laboratories. He first applied a statistical control chart to manufactured products and later suggested statistical refinements for process control. Two other men from the Bell system, H. F. Dodge and H. C. Romig, applied statistical theory to sampling inspection to produce their widely used *Sampling Inspection Tables*.

The advent of World War II awakened an otherwise lethargic interest in statistical techniques for quality control. The armed forces adopted scientifically designed sampling inspection plans that eventually culminated in the publication of Military Standard 105 for acceptance sampling of attributes. This action put pressure on suppliers to adopt equivalent inspection procedures for their output to keep it from being rejected by the military services. The training and research that accompanied the original and subsequent governmental applications spawned an enthusiastic following and aroused interest in related statistical control techniques. Today many organizations internationally promote quality control, and numerous books and journals report new developments.

Shewhart's pioneering efforts produced the book *Economic Control of Quality of Manufactured Product.*

The American Society for Quality Control (ASQC) was founded in 1946.

In the 1980s, the premium placed on defect-free production in MRP and JIT has raised the concern for quality to unprecedented levels.

15-3 ORGANIZATION FOR QUALITY

Management for quality assurance varies from an informal arrangement in which operators perform quality checks and supervisors have the final say on acceptability, to an extensive organization that designs and conducts inspection checks for incoming, in-process, and outgoing material. An autonomous quality department may even have the authority to shut down production until the source of defective output has been corrected. In most firms, the authority and responsibilities of the quality group lie somewhere between the extremes. The quality staff inspects where requested and *recommends* additional controls or corrective action. When the heads of production, purchasing, or other departments disagree with the quality recommendations, the disputes are pushed upstairs to a higher management level.

The company's interests may best be served by viewing the quality organization as representatives of the customers and as consultants hired to suggest means for quality improvement. The customers represented may be consumers, distributors, or another operating division in the same company. The consulting effort may be research into new techniques, development of revised inspection plans and devices, or investigations of the source of defects. By treating the quality staff as a pseudoindependent organization, personality conflicts and pressure tactics by affected parties can be considerably reduced.

Because **quality assurance** enters into so many linkages within the production system (between receiving and inventory, between operations in production, between production and shipping, and so on), more support is needed from all levels of management than for most other functions. No single department or staff can ensure quality by itself; it takes the cooperation of on-line workers, their supervisors, and related staff organizations. Training, informational seminars, publicity, and special programs cross organizational lines. The crossovers are easier and more rewarding when they have active backing on both sides and from above.

Maintenance of a quality organization is justified by the return from expenditures for its support. When the quality control staff has slowly increased with the company's growth, it may be top-heavy with "empire" supervisors and relegated to outmoded functions. An excess in the other direction has the quality group faddishly pursuing short-range objectives by overexploiting gimmicky quality promotions. The overall value of the quality organization should be judged by the ratio of costs incurred to costs saved, not by the glamour of its own advertisements.

Quality assurance is a skill. Like other skills, it takes a while to develop, and if it is not continually exercised, it will deteriorate.

It has been said that "quality is everybody's concern," but a job that belongs to everybody can easily become a job that nobody does.

EXAMPLE 15.2 Quality improvement through a zero defect program

From its introduction in aerospace-defense industries in 1962, the **zero defect (ZD) approach** to quality improvement has spread dramatically. The reported successes have been impressive. The program aim is aptly described by its name; ZD means *no* defects. It emphasizes prevention instead of cure, to get a job done correctly the first time, every time.

In essence, ZD is a motivational program sparked by a feedback routine. It recognizes that the people best prepared to eliminate errors are those who create them. It receives its impetus from the enthusiastic support of all tiers of management and depends on the pride of workers to identify error-prone situations.

Two main causes for mistakes are lack of knowledge and lack of attention. The easier of the two to correct is knowledge. Errors caused by improper or insufficient training can be pointed out by more experienced workers. Revisions of training programs, with subsequent follow-ups to check their adequacy, would satisfy quality demands for knowledge.

Lack of attention is more serious and more difficult to combat. Sometimes the environment can be altered to compensate for a lack of attention, for example, better communications, easier material identification, simpler movements, more complete specifications and procedures, and the like. Of at least equal importance to the physical conditions is the mental indoctrination to convince workers they *can* produce zero defects. The "people are human and humans make errors" attitude must be replaced by a belief that mistakes are not normal; they do not *have* to happen.

"Error cause removal" is a key feature of a ZD campaign. All workers are encouraged to identify and point attention to all causes for errors in their jobs. They fill out "error cause identification" forms to describe the error situation and submit them to their supervisors. If they know a solution, it too is requested. As in any suggestion plan, rapid response is vital. Potential causes for errors should be investigated, and suggested solutions should be evaluated promptly. Individuals cannot be expected to apply themselves unless they are convinced the organization is serious and recognizes their efforts.

Several features typically included in the planning, initiation, and follow-through of a ZD program are noted as follows:

1. *Planning.* An administrator is selected and a ZD committee is organized. Representatives from all departments are included in the horizontal makeup of the committee. After planning the steps necessary to launch the program, the committee acts as a liaison team.

2. *Management.* Upper and middle management are sold on the program and, even better, are personally committed to participating in and sustaining it. With active support from above, the orientation and indoctrination of supervisors are completed before the program kickoff. Communication between workers and supervisors makes or breaks the program. Support also is sought from union officials.

3. *Promotion.* Interest in the program is stirred by

publicity. A "teaser" approach uses posters, bulletin board notices, and announcements in company papers or over the address system to build up enthusiasm for the climactic kickoff. The inaugural should be impressive. Mass meetings followed by individual department meetings are common. Top managers speak at the mass meeting and supervisors explain details at group get-togethers. Banners, posters, pennants, and wandering visitations remind employees throughout the kickoff day that the program is not a one-shot promotion.

4. *Continuance.* Each employee is encouraged to sign a pledge card signifying his or her intent to reduce mistakes. The employee's contribution is further emphasized by the "error cause removal" campaign. Individual achievements are recognized by pins, plaques, dinners, and announcements. Departments also receive recognition for meeting goals they set earlier. "Days without errors" scorecards are posted. Particularly good ideas are exchanged among departments, with credit given to the originators.

Whether the ZD approach to quality improvement and maintenance is a guide or a gimmick, momentary or momentous, will be decided by experience. Its heavy reliance on motivation may make it difficult to sustain, but this same emphasis on personal pride makes the program applicable throughout the production system. Practitioners report that ZD programs contribute to safety, suggestion systems, good worker–management relations, and quality control. However, a good quality assurance program must be already operating before the motivational impetus can be incorporated.

15-4 INSPECTION

The single act most closely associated with quality control is inspection. A casual acquaintance with quality inspection may invoke the image of a person in a white coat peering through a magnifying glass at passing products. Such situations do exist, but inspections predominately take other forms. For special projects the inspectors may be engineers measuring the grade of a roadbed in foul weather, or technicians testing concrete samples in the comfort of a lab. Food inspectors examine the sanitation of bakeries, dairies, and restaurants. Workers producing parts use calipers and gauges to determine their acceptability. Finished products such as lumber are graded for quality by inspectors. In some situations, inspections can be completely mechanized. Though the means and purposes of inspections vary widely, they share the goal of distinguishing a level of accomplishment.

When and where to inspect

Where to inspect depends largely on *when* the inspection is scheduled. The location of most inspection stations is at the site of production—the receiving dock for incoming shipments, the assembly area, the construction site, distribution points, and so forth. In a fixed-position layout, inspectors must come to the product to check quality at various stages of development. In product layouts, particularly mechanized production lines, products come to the inspectors at special stations built into the line. Roving *floor inspectors* examine output from the individual work stations associated with a process layout.

Some inspections requiring special equipment and skills can be performed only in testing laboratories. On other occasions it may be more economical to transport products to a central laboratory than to take the specialists into the field or to the factory floor. It is not uncommon for products to be subjected to on-site inspection during early stages of their development and then to be sent to a laboratory for special tests such as chemical composition, compressive or tensile strength, structural qualities, or length of life.

Laboratories may allow the use of less skilled inspectors accompanied by the use of standardized procedures, more automated equipment, and closer supervision. These same conditions contribute to more consistent and therefore more reliable tests. When there is a choice between laboratory and on-site inspections, the listed advantages of lab examinations may be offset by the added costs of handling and congestion. Floor operators can often conveniently make on-site inspections, checked occasionally by senior floor inspectors. The same roving inspectors can check machine setups and material inputs to avoid potential errors. Mistakes caught in a lab have to be traced back to their origin for correction.

Deciding when to inspect during a production process is simply a matter of common sense—when it will do the most good. Logical choices are the beginning and end of the production process. Raw-material and component inputs should be inspected to see whether they meet expected standards. Acceptance of substandard inputs obviously jeopardizes outgoing quality and may damage equipment or disturb process continuity. Outgoing products are examined to protect the producer from customer discontent or buyer rejection.

During the production process, inspections are scheduled in front of operations that are costly, irreversible, or masking. Considerable expense is avoided by eliminating defective units before they undergo a costly phase of their development or before they pass through a process that cannot be undone, such as welding, pouring concrete, or mixing chemicals. Operations such as painting and encapsulating may hide defects easily detected before the masking operation.

From the foregoing it may appear that products are continually under inspection. This view is valid in the way artists inspect pictures they are painting: they are aware of their pictures' quality as they develop, but only at the end or at critical points do they really study the quality. Similarly, workers continually check the quality of their own or a machine's output, but there are just a few distinct inspection stations. Constant formal surveillance would not only increase cost, but it would also create an uncomfortable, "big brother is watching you" atmosphere. The timing and location of inspection points are key features in the design of any testing program.

Even if *all* finished items are inspected, some output will still likely be defective because of inspection errors. Equivalently, good products may be rejected.

How to inspect

When precise measurements are made of dimensions, weight, or other critical characteristics capable of expression on a continuous scale, the products are being subjected to **variables inspection.** The alternative to exact measurements is to set limits within which the product is judged acceptable or defective. A binary yes–no rating results from an **attribute inspection.** Because a good or bad grading normally requires less time and skill to make and uses lower-cost equipment than do exact measurements, attribute inspection is usually less expensive than is variables inspection.

It is generally assumed that the variables measured have a normal distribution.

Precise measurements require closely calibrated devices such as rulers, micrometers, scales, and meters capable of measuring the product's fineness standard. Devices to check attributes are designed to provide a quick verdict of acceptability—go–no go gauges, snap gauges, templates, balances, and the like. Although it is preferable to have all verdicts made objectively, some attribute ratings rely on subjective judgments. A decision whether a batch of brew tastes "bad enough" or a crack is "serious enough" to cause rejection is certainly subject to adjudication. Training, experience, supervision, and well-defined guidelines help guard against inconsistent judgments.

New devices are continually being developed to improve inspecting. Hidden flaws are revealed by magnetic, X-ray, and ultrasonic tests. Radio waves detect metal impurity. Television cameras allow vision where humans cannot venture. Computers hooked to mechanical scanning devices can inspect a rapidly moving process during which humans would not have time to react.

Statistical sampling techniques frequently reduce inspection costs. The use of samples to replace 100-percent inspection is usually appropriate for machine output whose units are not so likely to vary as much as are hand-crafted products. High production quantities and expensive inspections also suggest sampling. Then there is destructive testing (the performance test destroys the unit tested) which absolutely rules out 100-percent inspection. The characteristics and designs of sampling plans are the subjects of the following section.

EXAMPLE 15.3 Nursing quality

Quality assurance programs in service industries follow the same theory developed for product quality, but practices differ because the output frequently has more characteristics to evaluate than does a manufactured product. Attribute inspections are commonly used. The most important characteristics required of a service are carefully defined, and inspections reveal whether or not each characteristic is being satisfied. The quality rating for the service is a composite of weighted scores from inspections.

Nursing activities in hospitals illustrate most of the considerations required in quality control of services. Design of the program starts with recognition of nursing specialties (medical, pediatrics, intensive care, and so on). Specific characteristics are identified within each specialty area and are collected under categories of nursing responsibilities such as patient care, patient environment, nursing care plan, and record keeping.

A check sheet is drawn up that contains questions to be answered *yes* or *no* (or *not applicable*) that describe all the characteristics to be satisfied by good performance. Typical questions include

Have adequate measures been taken to make the patient with no need for immediate attention as comfortable as possible?

Is the patient clean and dry?

Is the bed neatly made and in the proper position?

Are the bedside stand and personal effects within easy reach?

Is the appropriate armband identification being worn?

Have discontinued drugs been returned to pharmacy?

Are progress notes complete and current?

About 10 to 15 questions are typically included in four to eight categories, depending on the objectives of the quality assurance program.

A nursing supervisor, accompanied by a head nurse, makes unannounced observation visits to each unit on each shift. He or she marks answers on the check sheet and writes comments to accompany each "no" answer. All personnel remain anonymous on the examiner's report, to maintain the objectivity of the inspections. Feedback is given when the reports are returned to the nursing units.

Composite ratings are obtained by using a conversion table to give different weights to the scores in each category. For instance, an inspection of a nursing unit could reveal 90 yes and 12 no (88 percent yes) answers for "patient care" and 58 yes, 15 no (79 percent yes) for "patient environment," which converts to scores of 35 (40 is perfect) for patient care and 16 (20 is perfect) for patient environment. This conversion obviously gives more importance to patient care than to patient environment. The highest possible *quality index* is 100, the total of the maximum conversion scales for all categories of nursing responsibilities. The value of the quality index obtained from each inspection is recorded on a chart to show performance trends.

15-5 ACCEPTANCE SAMPLING

The purpose of **acceptance sampling** is to recommend a specific action; it is not an attempt to estimate quality or to control quality directly. The basic action recommendation is to accept or reject the items represented by the sample. The rejection alternative can be temporized by resorting to 100-percent inspection to confirm the suspicion of inferior quality or by seeking a price concession to accept apparently inferior goods from a supplier.

When a buyer uses acceptance sampling at the supplier's plant, a rejection cancels shipment on the lot sampled.

When acceptance sampling is applied to incoming shipments, the usual practice is to accept shipments accredited by the sampling plan and to return rejected shipments to the supplier for rectification or to charge the cost of 100-percent screening to the vendor. A high rate of rejected samples may lead the buyer to seek a new supplier. The high rate is also a clear warning to the supplier: either lower quality claims or produce up to them.

Acceptance sampling can be used at inspection stations within the production process. When the inspection procedure is particularly fatiguing or monotonous, sampling may produce as good or better results than does 100-percent inspection. More careful examinations of fewer units may actually decrease the percentage of defective items that escape detection. A lot rejected by the sampling plan is typically returned to the responsible department for 100-percent inspection and rework of the items found defective.

The psychological effect of 100-percent inspection may lead workers to rely on the inspection process to catch all their mistakes. A "quantity in place of quality" attitude can be costly in scrap and rework.

Of the several types of acceptance sampling plans in use, we shall pay particular attention to lot-by-lot acceptance sampling by attributes, using a single sample from each lot. The sample is examined by attribute-type measuring devices. When more defective items are found in the sample than are allowed by the statistical plan, the entire lot is considered of inferior quality. This version of acceptance sampling is currently the most widely used because it is easy to understand and apply.

Other versions of acceptance sampling include double, sequential, and multiple sampling plans.

OPTIONAL MATERIAL

Operating characteristic curves

A sampling plan specifies the sample size (n) and the associated number of defectives (c) that cannot be exceeded without rejecting the lot from which the sample was taken. The capability of the plan to discriminate between acceptable and unacceptable lots is revealed by its **operating characteristic (OC) curve.**

As always, samples are assumed to be randomly drawn.

An OC curve for a single-sample, percentage-defective plan is shown in Figure 15.4. The horizontal axis of the graph indicates the percentage defective

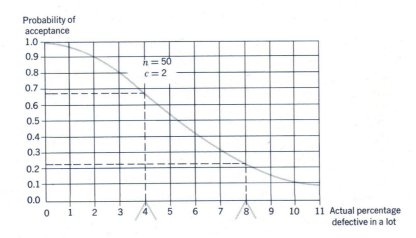

FIGURE 15.4 OC curve for sample size of 50 and acceptance number of 2.

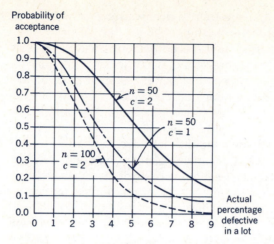

FIGURE 15.5 Relative discrimination of *OC* curves.

An assumption that the lot is large in comparison with the sample suggests that the probabilities for an *OC* curve will follow a binomial distribution. When the sample is also large, calculations are simplified by using the Poisson or normal distribution to construct *OC* curves.

in the lot being sampled; the vertical axis shows the probability that the lot will be accepted. A condition common to all *OC* curves is that a lot with no defectives will always be accepted. As the percentage of defectives increases, the probability that it will be rejected also increases. The curve shown is based on a sample size of 50 ($n = 50$) and an acceptance number of 2 ($c = 2$). For this curve, a lot with 4 percent defectives has a probability of acceptance of about 0.68. Expressed another way, in a random sample of 50 items, there is about a 68-percent chance that two or fewer defects will be found when 4 percent of the lot is defective. When the lot has 8 percent defectives, the probability of its acceptance decreases to 0.24.

Two ways to make *OC* curves more discriminating are to increase *n* while maintaining *c* and to decrease *c* while maintaining *n*. These recourses are intuitively logical; larger samples tend to represent the lot more accurately, and

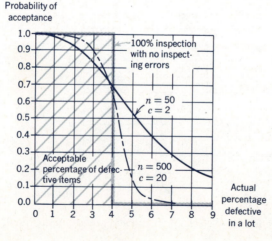

FIGURE 15.6 The ideal *OC* curve and approaches to it.

a decrease in the acceptance number tightens the restrictions. The shift toward the origin for these conditions is displayed in Figure 15.5.

The ideal *OC* curve can be obtained only by a 100-percent inspection of the entire lot made without inspection errors. It takes the form shown by the bold lines in Figure 15.6. If the total number of defectives in the lot is more than the indicated acceptable 4 percent (shaded area), the lot will be rejected without question. The two curves in the figure show that increasing the sample size while maintaining the same acceptance proportion ($^{20}/_{500} = 0.04 = {}^{2}/_{50}$) tends to make the *OC* curve come closer to the ideal Z-shaped curve. The best sample size to apply for each set of inspecting conditions is a compromise between the value of greater precision resulting from larger samples and the inspecting cost of collecting the larger sample.

Development of a single-sample, percentage-defective plan

Sampling plans protect against accepting defective lots and ensure that good lots will be accepted. A plan is designed around these two characteristics to safeguard the interests of both the receiver and supplier.

A grade of material considered adequate by the receiver is called the **acceptable quality level, AQL.** This is the level the receiver prefers, but he or she recognizes that a sampling plan will accept some lots with a higher percentage of defectives. The level at which the receiver can tolerate no further increase in defectives is called the **lot tolerance percent defectives, LTPD.** The range of quality levels between "good" lots, above *AQL*, and "bad" lots, below *LTPD*, is the "indifference" range. Ideally, lots above the *AQL* will seldom be rejected and lots below the *LTPD* will seldom be accepted.

Both the buyer and the seller accept some risk that the acceptance sampling will pass lots outside the *AQL-LTPD* range. The chance that lots will be rejected while actually having fewer defectives than specified by the *AQL* is called the **producer's risk, α.** The probability of a lot's being accepted with a greater percentage of defectives than set by the *LTPD* is termed **consumer's risk, β.** Typical values for the sampling risks are α = 5 percent and β = 10 percent. As an example, assume that a contract calls for an *AQL* of 2 percent and a *LTPD* of 12 percent. Thus the buyer seeks only 2 percent defectives and will rebel if more than 12 percent of the items are defective. For a perfect sampling plan designed to conform to these *AQL-LTPD* values with α = 0.05 and β = 0.10, no more than 5 percent of the lots rejected will have less than 2 percent defectives and no more than 10 percent of the lots accepted will have greater than 12 percent defectives. The *OC* curve for these theoretical relationships is shown in Figure 15.7.

As is apparent from the figure, two points on the *OC* curve are established by specifying the *AQL*, α, and the *LTPD*, β. The shape of the curve that passes through these points is a function of the sample size, *n*, and the acceptance number, *c*. Without tabular assistance, the procedure for designing a plan to conform to given conditions is essentially trial and error. Values for *n* and *c* are

"Indifference" in a decision-making situation is an area where risk is so delicately balanced by return that the decision maker can claim no distinct preference.

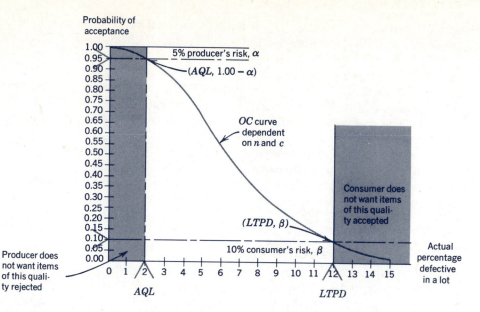

Probability of
acceptance

5% producer's risk, α

(AQL, 1.00 − α)

OC curve
dependent
on n and c

Consumer does
not want items
of this quali-
ty accepted

(LTPD, β)

10% consumer's risk, β

Actual
percentage
defective
in a lot

Producer does
not want items
of this quali-
ty rejected

AQL

LTPD

FIGURE 15.7 Relationship of producer's and consumer's risks to the OC curve.

arbitrarily selected, and the probabilities for accepting lots with the given *AQL* and *LTPD* are calculated. The calculated values are then compared with the desired (1.00 − α) and β values. Repeated trials eventually reveal a set of n and c values that produce a curve that comes very close to the (*AQL*: 1.00 − α) and (*LTPD*: β) points: an exact fit is unlikely because n and c must be integers.

Extensive tables for designing sampling plans are available in H. F. Dodge and H. G. Romig, *Sampling Inspection Tables*.

Fortunately, tables and charts are available that relieve the tedious search for n and c values and speed up calculations. Table 15.1 suggests n and c values under the uniform conditions of $\alpha = 0.05$ and $\beta = 0.10$. A sample size and acceptance number are indicated by the row corresponding to the chosen *AQL* and the column associated with the *LTPD*. For instance, a 2-percent *AQL* and a 12-percent *LTPD* indicates $n = 40$ and $c = 2$. Therefore, a lot from which a sample of 40 items is taken will be rejected when over 2 items are found defective, and approximately 5 percent of the time the rejected lot will have fewer than 2 percent defectives. Similarly, there is a probability of about 0.10 that fewer than two items in the sample of 40 will be found defective when the lot actually contains more than 12 percent defective items.

EXAMPLE 15.4 Design of a single-sample, percentage-defective sampling plan

An acceptance sampling plan is to be designed to protect the consumer from more than 7 percent defectives with a risk of 10 percent. The producer is expected to provide lots with about 1 percent defectives; if this is done, the plan should accept such material at least 95 percent of the time. The two de-

TABLE 15.1

Suggested n and c Values for Single Sampling Plans in Which $\alpha = 0.05$ and $\beta = 0.10$, Given AQL and LTPD

AQL \ LTPD	4.51 to 5.60	5.61 to 7.10	7.11 to 9.00	9.01 to 11.2	11.3 to 14.0	14.1 to 18.0	18.1 to 22.4
0.451 to 0.560	80	60	60	50	15	15	10
	1	1	1	1	0	0	0
0.561 to 0.710	100	80	50	50	40	10	10
	2	1	1	1	1	0	0
0.711 to 0.900	100	80	50	40	40	30	7
	2	2	1	1	1	1	0
0.901 to 1.12	120	80	60	40	30	30	25
	3	2	2	1	1	1	1
1.13 to 1.40	150	100	60	50	30	25	25
	4	3	2	2	1	1	1
1.41 to 1.80	200	120	80	50	40	25	20
	6	4	3	2	1	1	1
1.81 to 2.24	300	150	100	60	40	30	20
	10	6	4	3	2	2	1
2.25 to 2.80	n	250	120	70	50	30	25
	c	10	6	4	3	2	2
2.81 to 3.55	n	n	200	100	60	40	25
	c	c	10	6	4	3	2
3.56 to 4.50	n	n	n	150	80	50	30
	c	c	c	10	6	4	3
4.51 to 5.60	n	n	n	n	120	60	40
	c	c	c	c	10	6	4

sign points for the plan are thus designated to be

$$AQL = 0.01 \quad LTPD = 0.07$$
$$\alpha = 0.05 \quad\quad \beta = 0.10$$

Each lot contains 1000 items of the same type. The sample will also be large and thus justifies the use of the Poisson distribution to calculate acceptance probabilities:

$$P_{(x)} = \frac{(np)^x}{x!} e^{-np}$$

where

n = sample size taken from the lot
p = percentage defectives in the lot
x = number of defectives in the sample

A starting point for the calculations is available from Table 15.1. For the given AQL and $LTPD$, the apparent sample size is $n = 80$, and the acceptance number is $c = 2$. The adequacy of the indicated OC curve is determined by summing the probabilities for 0, 1, and 2 defectives occurring in a sample of 80 from lots having 1 percent and 7 percent defectives. Calculating the probability of acceptance (P_a) for lots containing 1 percent defectives, we have

$$P_{(0)} = \frac{(80 \times 0.01)^0}{0!} e^{-80 \times 0.01}$$

$$= \frac{(0.8)^0}{1} (0.4493) \quad = 0.4493$$

$$P_{(1)} = \frac{(0.8)^1}{1!} e^{-0.8}$$

$$= \frac{0.8}{1} (0.4493) \quad = 0.3594$$

$$P_{(2)} = \frac{(0.8)^2}{2!} e^{-0.8}$$

$$= \frac{0.64}{2} (0.4493) \quad = \underline{0.1438}$$

$$P_a = P_{(0)} + P_{(1)} + P_{(2)} = 0.9525$$

Then the probability of rejecting a lot of 1 percent quality is $1.000 - 0.9525 = 0.0475$. This value compares favorably with the anticipated producer's risk of $\alpha = 5$ percent.

The same calculations with p changed to 0.07 reveal the probability of accepting a lot with more than 7 percent defectives:

$$P_{(0)} = \frac{(80 \times 0.07)^2}{0!} e^{-80 \times 0.07}$$

$$= \frac{(5.6)^0}{1} (0.0037) \quad = 0.0037$$

$$P_{(1)} = \frac{(5.6)^1}{1!} e^{-5.6}$$

$$= \frac{5.6}{1} (0.037) \quad = 0.0207$$

$$P_{(2)} = \frac{(5.6)^2}{2!} e^{-5.6}$$

$$= \frac{31.36}{2} (0.0037) \quad = \underline{0.0580}$$

$$P_a = P_{(0)} + P_{(1)} + P_{(2)} = 0.0824$$

For this plan there is only an 8.24 percent consumer's risk. Because both the producer's and the consumer's risks are less than the design goals, a smaller sample size may still provide adequate protection while reducing inspection cost. Using the same acceptance number, 2, with a sample size of 75 changes the acceptance probabilities to 0.041 and 0.106 for lot-defective percentages of 1 and 7, respectively.

Average outgoing quality

Average outgoing quality (AOQ) is a measure of the quality level resulting from an acceptance sampling plan when the rejected lots are subjected to 100-percent rectifying inspection and defectives found in the sample are replaced by ac-

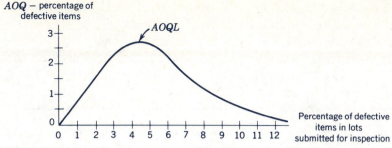

FIGURE 15.8 *AOQ* curve for a rectifying inspection process.

ceptable items. The quality is thus a function of the *n*, *c*, and lot fraction defective percentage, *p*. When a sample of size *n* has *c* or fewer defectives, the lot is passed, and only the bad items in the sample are rectified by replacement with good items. When the sample contains more than *c* defectives, it undergoes 100-percent inspection, in which all bad items found are replaced with good ones. The rejected lots then theoretically have no defectives. Under this method of rectifying inspection, the extreme condition of perfect quality occurs only when the lot has 0 defectives or 100 percent defectives. In the former case, the lot is never rejected; in the latter case, it is always rejected to force a complete rectification.

Lots between the all-good and all-bad bounds contain a certain percentage of defectives. This percentage (*AOQ*) is calculated from the equation

$$AOQ = \frac{P_a \times p(N - n)}{N} \qquad \text{where } N = \text{lot size}$$

Values of P_a for different values of p are read from the *OC* curve or tables.

The rationale behind the equation is that there will be no defectives in the lots rejected and in the sample taken from accepted lots. However, in the accepted lots there are $(N - n)$ items that contain the lot percentage defective, *p*. The product of $p(N - n)$ gives the number of defectives in the accepted lot. Multiplying this product by the probability of accepting a lot with *p* defectives, P_a, indicates the average number of defective items in sampled lots. This number, divided by the lot size *N*, is the average proportion of defectives remaining in the inspected lots, *AOQ*.

When *N* is large compared with *n*, a close approximation is $AOQ = pP_a$.

A plot of *AOQ* values for all single-sample, rectifying sampling plans follows the pattern shown in Figure 15.8. The maximum point in the *AOQ* curve is known as the *average outgoing quality limit*, AOQL. This limiting value is the worst *AOQ* possible from the rectifying inspection process, regardless of incoming quality.

EXAMPLE 15.5 Calculation of *AOQ* and *AOQL*

Consumers using the sampling plan described in Example 15.4 want to know the *AOQ* they can expect when the plan includes rectifying inspection. If the producer is successful in providing lots with 1 percent (p) defective items and the preferred sampling plan uses a sample of 80 (n) from a lot of 1000 (N) with an acceptance number of 2 (c), the probability of accepting such lots (P_a) will be 0.9525. Then the AOQ is calculated as

$$AOQ = \frac{P_a \times p(N - n)}{N}$$

$$= \frac{0.9525 \times 0.01(1000 - 80)}{1000} = 0.0088$$

which means that the actual quality received is expected to contain even fewer than 1 percent defectives. The improved quality results from replacing good for bad items discovered in the sampling and the complete rectification of lots rejected as producer's risk.

As a further check, consumers want to know the worst average quality they can receive from the inspection process. Because n is much smaller than N, the $AOQ = pP_a$ approximation can be used. Note that this approximation would have given $AOQ = 0.01 \times 0.9525 = 0.0095$ in place of 0.0088 in the previous calculations for $p = 0.01$. The values of P_a for each p can be calculated by the routine shown in Example 15.4, or it can be taken from OC curves or tables when available.

From the adjacent table it appears that the AOQL is about 1.68 percent when the incoming lot percentage defective is between 3 and 3.5:

p	n	np	P_a	$pP_a = AOQ$
0.010	80	0.8	0.95	0.0095
0.020	80	1.6	0.78	0.0156
0.030	80	2.4	0.56	0.0168 } AOQL
0.035	80	2.8	0.48	0.0168 }
0.040	80	3.2	0.34	0.0136

Because the AOQ for a single-sample, percentage-defective plan with rectifying inspection always follows the shape shown in Figure 15.8, there is no need to carry the calculations beyond $p = 0.04$. The curve has crested and will continue down as more lots receive 100-percent inspection owing to their greater rejection rate as p increases—a rare illustration of ever-worsening inputs generating ever-better output.

END OPTIONAL MATERIAL

Other acceptance sampling plans

The type of acceptance **sampling plan** to employ depends on the characteristics of the product and the economics of the inspection process. When exact dimensions of a product are critical, the important variables must be measured. Just saying that an item is defective because one variable violates a specified limit does not provide the needed information. A single sample, as discussed previously, may not be the most economical plan design. The amount of inspection is often reduced by taking two or more smaller samples and applying statistical decision rules for accepting or rejecting the lot.

VARIABLES SAMPLING

A single-sample, variables acceptance plan has basically the same considerations as described for attributes sampling. The main difference is the inspection process. Actual records of the measurements taken during the examination of an item are used. These measurements are expected to follow a normal distribution, not the binomial or Poisson distributions recognized for attributes. This change alters the relationships of n, c, α, β, AQL, and $LTPD$. The appropriate OC curve still shows the discriminating power of a specific plan, but it may be based on upper and lower tolerance limits instead of one tolerance grade defined by α and β.

An inspection requiring actual measurement is typically more expensive than is a pass-fail evaluation. To some extent this higher cost of variables inspection is counterbalanced by the smaller sample size required to provide the same sampling discrimination as attribute inspection does. The fewer items sampled per lot makes variables sampling especially attractive when the items are destroyed by the inspection process.

Another feature of variables sampling is that information about a measured characteristic, mean and variance, is more useful for process control than are reports of items being simply good or bad. Such cost and value considerations determine the preference when there is a choice between attribute and variables sampling.

MULTIPLE SAMPLING

A decision to use attribute sampling is accompanied by a decision about plan design. In some instances a double sampling plan is more economical than is a single sample taken from each lot. The mechanics of double sampling are pictured in Figure 15.9. The cost advantage of double sampling emerges when lots of very good or bad quality are accepted or rejected on the basis of the first sample, n_1. In a double sampling plan of a given discriminating power, n_1 is always smaller than n for a single sampling plan of corresponding power. Lots of mediocre quality require a second sample (n_2) in the double sampling scheme.

Two tolerance limits occur when an item is rejected because its measurements fall on either side of an acceptance zone, for example, too large or too small. The other approach is to maintain an average quality through α and β risks of extreme measurements.

When several characteristics of an item are measured, each must have its own sampling plan because the means and variances of the characteristics differ.

Double sampling has the psychological appeal of providing a "second chance" for a rejected lot. Actually the reprieve is illusory, and the appeal is naive because the plan can be just as discriminating as can a single-sample inspection plan.

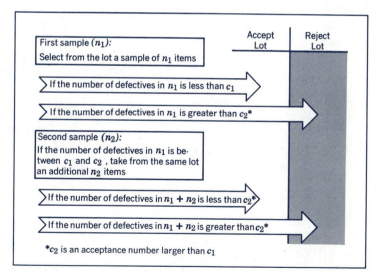

FIGURE 15.9 Procedure for a percentage-defective double sampling plan.

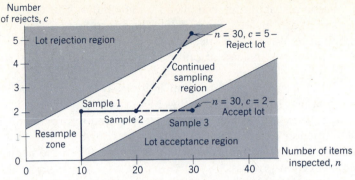

FIGURE 15.10 Sequential sampling plan illustrated by three samples of 10 items each.

Because $n_1 + n_2$ is greater than n of a single sample, inspection loads can be expected to increase for lots of intermediate quality.

SEQUENTIAL SAMPLING

The size of a sample can be pared even further by a sequential sampling plan. As displayed in Figure 15.10, sequential sampling is characterized by acceptance, rejection, and resampling zones. The initial sample is relatively small: in the pictured plan the minimum sample size is 10. If all of the first 10 items inspected pass, the lot will be accepted. Conversely, the lot will be rejected if over two items are defective in the first 10. When the number of defectives places the sample in the intermediate zone, additional items are inspected until the sample passes into the reject-or-accept category. The continued inspection can be by group samples or individual items.

The sequence of samples recorded in Figure 15.10 starts with two defectives found in the first sample of 10. A second sample of 10 adds no more defectives, but the total sample size of 20 with two defectives is still inside the resampling zone. A third successive sample of 10 with no defectives would put the lot in the acceptance zone ($n = 30$, $c = 2$), or a third sample revealing three additional defectives would reject the lot ($n = 30$, $c = 5$). Further sampling would have been indicated if one or two defectives had been found in the third sample.

Both double and sequential sampling plans are described by *OC* curves with relevant α, β, *AQL*, and *LTPD* characteristics. Charts and tables aid in the development of appropriate plans. In general, multiple and sequential plans require less inspection than does a single-sample plan for equivalent protection, but continued sampling means that inspection loads usually vary considerably between lots.

15-6 THEORY OF CONTROL CHARTS

A control chart is a graphic aid to detect quality variations in output from a production process. As opposed to the aim of acceptance sampling (to accept

Sequential sampling can be based on an item-by-item inspection plan instead of on successive samples containing several items.

Many sequential sampling plans are truncated to force a lot rejection or acceptance decision after a specified number of items have been inspected.

Details for different sampling plan designs are well described in A. J. Duncan, *Quality Control and Industrial Statistics.*

or reject products already produced), control charts help produce a better product. The charts have three main applications: (1) to determine the actual capability of a production process, (2) to guide modifications for improving the output quality of the process, and (3) to monitor the output. The monitoring function shows the current status of output quality and provides an early warning of deviations from quality goals.

Variations

Almost every production process is subject to some degree of *natural variability*. Innumerable small causes contribute to overall chance variations in the quality of output. The individual causes are so slight that no major portion of the variation can be traced to a single cause. These deviations are a function of the accuracy of the process, should be expected, and largely determine whether a process can deliver the precision stated for output specifications.

A record of natural variations reveals the capability of a process to meet specifications. It is generally uneconomical to attempt to improve by inspection the output quality of a process lacking inherent capabilities.

Another type of process variation is produced by *assignable causes*. As opposed to natural variations, these causes produce a relatively large variation traceable to a specific reason. Most commonly, these causes are owed to differences among machines, differences among operators, differences in materials, and differences caused by the interaction of workers, materials, and/or machines.

The control problem is to distinguish between natural variations and variations due to assignable causes. A statistical knowledge of the behavior of chance variations is the foundation on which control charting rests. A process is said to be "under control" when deviations in output are the result of chance variations. When the pattern of output deviations does not follow the distribution expected from chance causes, the process is considered "out of control" and the cause is probably assignable.

Control limits

Variations produced by chance follow statistical laws. From experience with past process variations, the distributions of future variations can be anticipated. After the distribution has been identified by the mean and variance, the dispersion of samples from the described process indicates the state of the process. For instance, a sample reading far from the mean reading is a clue that the distribution may no longer be descriptive of the process, that the process is out of control. The risk associated with a mistaken "out of control" assumption determines the control limits.

The frequent use of the normal distribution stems from its appropriateness to many variable sampling situations and to the central limit theorem: the means of small samples tend to be distributed according to the properties of the normal distribution, regardless of the distribution from which the samples were taken.

We encountered control limits as applied to forecasting in Chapter 3. Figure 15.11 shows commonly used control limits for the normal distribution. When limits are set three standard deviations ($\pm 3\sigma$) away from the mean, a sample from a normally distributed population has only about three chances in a thousand of appearing outside the limits when the process is under control. Stated another way, the odds of a wrong conclusion that the process is out of control when a sample reading falls outside the $\pm 3\sigma$ limit are only 26 in 10,000.

In the following discussions and examples, we confine our applications to

Because the true mean (μ) and standard deviation (σ) are seldom known, control limits are based on the mean of sample means ($\bar{x}$) and the standard error of the estimate (s).

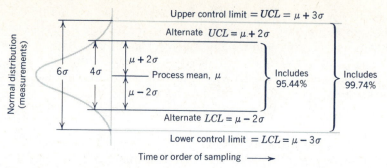

FIGURE 15.11 Theoretical statistical basis for control charts.

control limits based on ±3 standard deviations from the mean, the limits most used in this country.

EXAMPLE 15.6 Police control charts

Control charts have been adapted to help police departments identify where and when to assign their limited resources. Demands for police resources vary during a day, week, and year according to areas served and type of service required. Monitored activities indicate trends or new demands to be met by reassignment of personnel and equipment.

If a state of perfect quality is defined as a condition in which no police services are needed, each incident of activity will represent a defect. Deviations from the perfect state are caused by criminal acts (homicides, robberies, assaults, and the like) and public assistance (funeral escorts, traffic directions, and the like). The number of incidents in each category per time period is analogous to the number of defects per product unit in industrial quality control.

Separate control charts are constructed for different areas and categories of incidents classified according to a uniform code. These charts are used in the same way as are industrial control charts. Deviations beyond control limits are investigated for assignable causes. Trends are spotted to anticipate future needs and to suggest immediate reassignments to reverse threatening patterns. Because the size of the police force is considered constant, each increased allocation to an area must be matched by equivalent decreases in other areas. This limited adjustment capability makes the early-warning features of control charting especially valuable to law enforcement agencies.

15-7 TYPES OF CONTROL CHARTS

Control charts have the same principal divisions as does acceptance sampling: attributes and variables. Attribute control can be further divided into charts for percentage defectives and charts for the number of defects per unit. The main interest in variables is control over changes in the average and the range of measurements. Control charts for all these considerations follow the same basic format of a mean value bounded by **upper and lower control limits.** It is the calculation of the control limits that distinguishes the types of charts.

Control charts for variables

The best-known control charts for variables record the process average, $\bar{x}$, and the range, R. The $\bar{x}$ **charts** show how individual measurements or the means of samples compare with the overall mean or the desired average. The R **charts** record the variability of individual readings within a sample. These two charts complement each other because a sample must have both an acceptable average and a reasonable range of measurements before the process it represents can be considered under control.

As shown in Figure 15.12, output from a process can fail to meet tolerance limits by a change in the mean of the process or an increase in the variation of output measurements. The condition is compounded when both the mean and variation are experiencing changes from the expected values. Information about both trends is necessary to make reasonably sound decisions about process capabilities and in-process control.

One method of setting control limits for the process mean is to use the average of the **sample means** for the central value and the range to estimate the variance of the process. With the sample mean calculated as

$$\bar{x} = \frac{x_1 + x_2 + \ldots x_n}{n}$$

where x's are measurements and n = number of measurements in the sample,

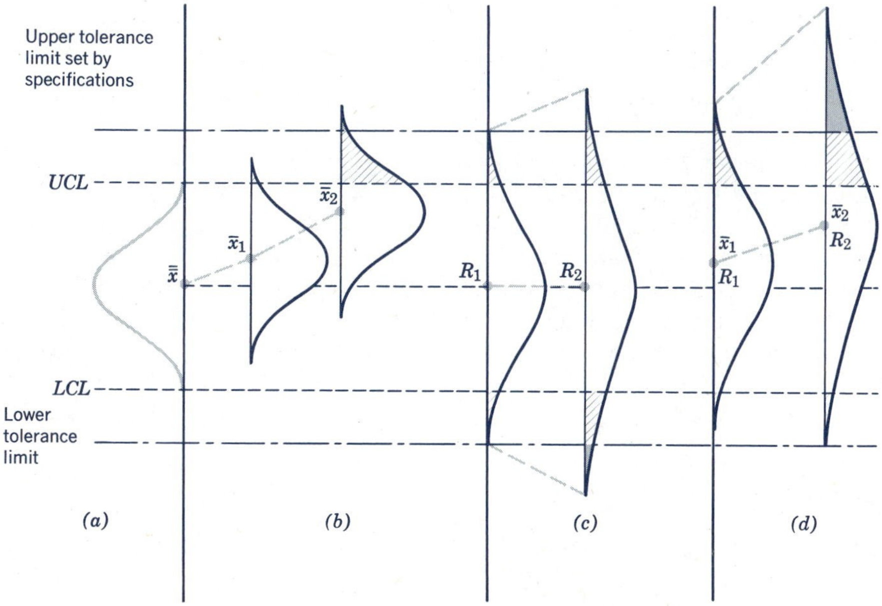

The procedure described is for control limits based on observed values. Often the limits are set by standard values based on what a process is expected to accomplish. In this case the center line, $\bar{\bar{x}}$, and the standard deviation, σ, are given. Then the control limits are $\bar{\bar{x}} \pm 3\,\sigma/\sqrt{n}$.

FIGURE 15.12 Effect of changes in the means and variability on the process output quality: (a) expected distribution, (b) drift of sample means away from expected mean while range remains constant, (c) increase in range while sample mean remains constant, and (d) drift of sample mean while range increases.

Control Limit Factors for $\bar{x}$ Charts, A

n	A
2	1.880
3	1.023
4	0.729
5	0.577
6	0.483
7	0.419
8	0.373
9	0.337
10	0.308
12	0.266
14	0.235
16	0.212

Control Limit Factors for R Charts, B and C

n	B	C
2	3.268	0.000
3	2.574	0.000
4	2.282	0.000
5	2.114	0.000
6	2.004	0.000
7	1.924	0.076
8	1.864	0.136
9	1.816	0.184
10	1.777	0.223
12	1.716	0.284
14	1.671	0.329
16	1.636	0.364

If it is necessary to control more than one quality characteristic of a product, individual $\bar{x}$ and R charts must be developed for each variable.

the estimate of the process mean is

$$\bar{\bar{x}} = \frac{\bar{x}_1 + \bar{x}_2 + \ldots + \bar{x}_n}{n}$$

where N = number of samples. The range for each sample, R, is the difference between the highest and the lowest measurements in the sample. The mean range is then

$$\bar{R} = \frac{R_1 + R_2 + \ldots + R_N}{N}$$

where R = maximum difference in measurements for each sample.

The standard deviation of the sample means is estimated from the mean range by applying a factor A. This factor is based on the sample size. The product of $A \times \bar{R}$ sets the three standard deviation boundaries around the center line as

Upper control limit $= \bar{\bar{x}} + A\bar{R}$

Center line $= \bar{\bar{x}}$

Lower control limit $= \bar{\bar{x}} - A\bar{R}$

The associated range control chart is developed in a similar fashion: $\bar{R}$ is the center line. The upper and lower control limits use two factors, B and C, respectively. These factors are also a function of sample size and set three standard deviation bounds. The range control chart is thus defined by

Upper control limit $= B\bar{R}$

Center line $= \bar{R}$

Lower control limit $= C\bar{R}$

The question of sample size and frequency of sampling is usually a judgment based on the inspectors' experience. The considerations include facilities and inspectors available, inspection costs, sensitivity of the process to abrupt trend changes, and cost incurred from producing rejects. In general, sample sizes of four or five taken more often are preferred to larger samples. It is important to keep the samples the same size because control limits are applicable only to a distinct value of n; when n varies, different control limits are associated with each sample.

EXAMPLE 15.7 $\bar{x}$ and R control charts

A seafood processor uses his facilities to process crab, shrimp, or fish, according to the season of the year. Most operations are manual, owing to the nature of the work, need for versatility, and lack of investment capital. Complaints from buyers about discrepancies in the quoted weights of packed shrimp have led the processor to initiate a control program.

The shrimp are packed in 1- and 5-pound containers. A worker scoops the small shrimp into a container and places it on a balance scale. The final weight adjustment is made by hand, adding or removing a few small shrimp from the container. It is necessary to perform the weighing quickly because microbial growth can spoil the shrimp if they are left too long without refrigeration.

The results of weighing 10 samples of five 1-pound containers are as follows. Two samples were taken each day by a random selection of containers packed during the day and stored in the cold room.

Sample	Weight Measurements (pounds)					Sum	$\bar{x}$	R
1	1.04	1.01	0.98	1.02	1.00	5.05	1.010	0.06
2	1.02	0.97	0.96	1.01	1.02	4.98	0.996	0.06
3	1.01	1.07	0.99	1.03	1.00	5.10	1.020	0.08
4	0.98	0.97	1.02	0.98	0.98	4.93	0.986	0.05
5	0.99	1.03	0.98	1.02	1.01	5.03	1.006	0.04
6	1.02	0.95	1.04	1.02	0.95	4.98	0.996	0.09
7	1.00	0.99	1.01	1.02	1.01	5.03	1.006	0.03
8	0.99	1.02	1.00	1.04	1.09	5.14	1.028	0.10
9	1.03	1.04	0.99	1.02	0.94	5.02	1.004	0.10
10	1.02	0.98	1.00	0.99	1.02	5.01	1.002	0.04
					Totals	10.054	0.65

The average weight of "1-pound" containers is

$$\bar{\bar{x}} = \frac{\Sigma \bar{x}}{N} = \frac{10.054}{10} = 1.0054$$

and

$$\bar{R} = \frac{\Sigma R}{N} = \frac{0.65}{10} = 0.065$$

These measures are the bases for the $\bar{x}$ control chart. The upper and lower control limits, which should enclose 99.73 percent of the measurements from the present process, are calculated as

$$UCL = \bar{\bar{x}} + A\bar{R}$$
$$= 1.0054 + 0.577\,(0.065)$$
$$= 1.0054 + 0.0375 = 1.0429$$
$$LCL = \bar{\bar{x}} - A\bar{R}$$
$$= 1.0054 - 0.0375 = 0.9679$$

where A is based on the sample size of 5.

The control limits bounding the mean range, $\bar{R} = 0.065$, are determined as

$$UCL = B\bar{R} = 2.114(0.065) = 0.137$$
$$LCL = C\bar{R} = 0.0(0.065) = 0.0$$

where B for $n = 5$ is 2.114 and C for $n = 5$ is 0.0. From this information, $\bar{x}$ and R control charts are developed (see Figures 15.13 and 15.14).

The plot of individual sample means and ranges indicates that the weighing operation is under control. That is, the variation among the samples ($\bar{x}$) and within the samples (R) can be explained by random influences. Continued sampling may reveal a sample reading outside the control limits. Such a point should be investigated to confirm whether there is an assignable cause for the deviation.

The apparently satisfactory process variation should still be questioned with reference to customer complaints about "short weights." Meeting process variation limits has little worth if the limits do not satisfy output standards. Perhaps the scales should be set to produce an average weight greater than $\bar{x} = 1.0054$ so as to reduce the number of containers weighing less than 1 pound. Another approach to the same objective would be to mechanize the weighing operation to reduce the weight variations. Which, if either, course of action to undertake depends on the opportunity cost of dissatisfied customers in relation to the cost of the cure. Quality control charts for variables contribute data for the decision and subsequently promote the process consistency needed to carry out the decision.

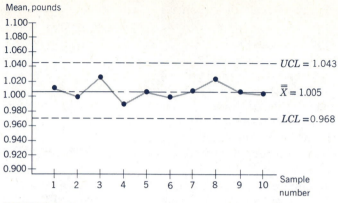

Mean, pounds

UCL = 1.043

$\bar{\bar{X}} = 1.005$

LCL = 0.968

Sample number

FIGURE 15.13

Control charts for attributes

Process quality control charts for attributes take one of two forms depending on the type of output. A **p-chart** is used when individual units are judged acceptable or defective. A **c-chart** is appropriate when the quality is best measured by the number of defects in a constant unit of output such as lineal feet of wire or square feet of cloth. Accordingly, a p-chart shows the variation in percentage defective and the c-chart shows the pattern of defects per unit of output.

EXAMPLE 15.8 Development of a *p*-chart for percentage-defective control

A random sample of 50 wiring boards used in electrical assemblies is taken from the wiring assembly line each day. The boards are activated in a testing machine, in which a light glows if the wiring is acceptable.

Whenever a new wiring board design is re-

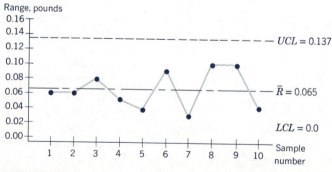

Range, pounds

UCL = 0.137

$\bar{R} = 0.065$

LCL = 0.0

Sample number

FIGURE 15.14

quired, a new control chart must be developed to measure process characteristics for that particular design. A preliminary chart is based on data recorded during the initial phase of production. The number and percentage rejected during 20 consecutive days of production are as follows:

Date Sampled	Number Sampled	Number Rejected	Percentage Defective
Sept. 8	50	4	8
9	50	3	6
10	50	2	4
11	50	6	12
12	50	3	6
15	50	1	2
16	50	3	6
17	50	2	4
18	50	9	18
19	50	5	10
22	50	3	6
23	50	2	4
24	50	5	10
25	50	2	4
26	50	2	4
29	50	1	2
30	50	3	6
Oct. 1	50	2	4
2	50	1	2
3	50	3	6
Total	1000	62	124

The preliminary estimate of the average percentage defective is

$$\bar{p} = \frac{62}{20 \times 50}(100\%) = 6.2\%$$

or, equivalently,

$$\bar{p} = \frac{124\%}{20} = 6.2\%$$

Assuming that the process is already under control, $\bar{p}$ is the best available estimate of the mean and is used to estimate the standard deviation of the process as

$$\sqrt{\frac{\bar{p}(100 - \bar{p})}{n}} = \sqrt{\frac{6.2(100 - 6.2)}{50}} = 3.4\%$$

Then the control limits based on three standard deviations from the mean are

$$UCL = 6.2\% + 3(3.4\%) = 16.4\%$$

$$LCL = 6.2\% - 3(3.4\%) = 0 \quad \text{(because negative defectives are impossible)}$$

Comparing the recorded data with the calculated control limits shows that the sample taken on September 18 (18 percent rejected) is the only observation outside the control lines. An investigation of the production on this date reveals that two inexperienced workers were assigned to the wiring line, and some of their output was included in the sample before the floor supervisor had a chance to inspect and rectify the work. Because the occasion is not deemed typical and the assigned cause was remedied, the reading should be omitted from the data, to produce the following revised chart lines:

$$\bar{p} = \frac{62 - 9}{19 \times 50}(100\%) = 5.6\%$$

$$UCL = 5.6 + 3\sqrt{\frac{5.6(100 - 5.6)}{50}} = 5.6 + 3(2.5)$$

$$= 14.3\%$$

$$LCL = 5.6 - 3(2.9) = 0$$

The original and recalculated control limits are shown in Figure 15.15. The plotted points indicate that the process is under control for the revised limits (dashed lines); the point with an assignable cause is still shown but is no longer considered a random fluctuation. The past data thus analyzed are adequate for immediate production control.

The next stage of process control is to see whether samples of the same size continue to fall within the designated limits. As workers gain pro-

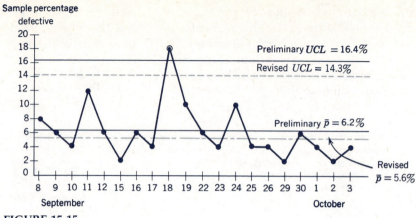

FIGURE 15.15

ficiency with new wiring requirements, the average percentage of defectives may decrease. If future points on the chart randomly fluctuate around a lower center line, the improved process capability should be recognized by lower $\bar{p}$ and UCL values.

The object is to set the center line as close as possible to the level at which the process actually operates. The control limits then provide a more sensitive signal to deviations outside normal variations.

p-CHARTS

By definition, the output population classified by percentage defective is divided into two factions, acceptable or not acceptable. This two-way classification naturally leads to a binomial description of the standard error of the mean. With the central line determined by the average percentage defective during a trial period, $\bar{p}$, the typical boundaries are developed from

$$\text{Control limits} = \bar{p} \pm 3\sqrt{\frac{\bar{p}(1 - \bar{p})}{n}}$$

where $\bar{p} = \Sigma p/n$ and n = sample size when p is expressed as a decimal or

$$\text{Control limits} = \bar{p} \pm 3\sqrt{\frac{\bar{p}(100 - \bar{p})}{n}}$$

when p is expressed as a percentage.

The p-chart format and its interpretation are essentially the same as for the control charts for variables. It is assumed that process variations due to chance are largely accounted for by the calculated control limits and that points falling outside the limits are owed to extraneous factors affecting the process. Points thus isolated are investigated for an assignable cause, as described in Example 15.5.

c-CHARTS

A defect is a flaw in a product. When a single defect is significant enough to cause the product's rejection, the percentage defective (p-chart) is the measure of control. When the purpose of inspection is to measure the *number* of defects per unit of production, a c-chart is appropriate. The inspection unit may be a single item like an airplane wing, a set of items such as a certain number of TV sets, or a block of output such as a length of belting, an area painted, or a weight produced. The number of defects (c) in the unit inspected is recorded on a c-chart.

The size of the "inspection unit," like the number sampled for $\bar{p}$ or $\bar{x}$ charts, must remain constant for the continued relevance of control limits.

For output that has many opportunities for defects to occur and the probability of a defect in any particular spot is small, the Poisson distribution nicely describes the sample fluctuations. With $\bar{c}$ as the average number of defects per unit, the standard deviation is $\sqrt{\bar{c}}$, and the working control limits are

$$UCL = \bar{c} + 3\sqrt{\bar{c}}$$

$$LCL = \bar{c} - 3\sqrt{\bar{c}}$$

Again, like $\bar{p}$ and $\bar{\bar{x}}$ values, $\bar{c}$ is usually estimated from past data. The given control limits will encompass practically all the values if the process is in control.

EXAMPLE 15.9 Application of a c-chart for control of defects

A c-chart is used to evaluate the performance of an automated process for producing special impregnated material designed for use in extreme cold. The "inspection unit" is a continuous 10-yard roll of the cloth. Both sides of the material are inspected under high-intensity lights and magnifying screens. Defects occur from imperfections in the weave or inadequate coating. The blemishes are typically very small, 2 square centimeters or less. Checks for major imperfections due to machine malfunctions are conducted visually at the point of production.

In previous runs of the impregnated cloth, the number of defects per 10 yards has been 40, $\bar{c} = 40$. From this center line the control limits were set at

$$UCL = 40 + 3\sqrt{40} = 40 + 3(6.3) = 59$$

$$LCL = 40 - 3\sqrt{40} = 40 - 19 = 21$$

The present production run is represented by the following data for the successive samples 81 through 100. The number of defects in each sample is plotted on the accompanying control chart (Figure 15.16).

Sample Number	Number of Defects per 10 Yards	Sample Number	Number of Defects per 10 Yards
81	33	91	35
82	16	92	28
83	19	93	24
84	26	94	31
85	36	95	34
86	38	96	40
87	37	97	30
88	41	98	31
89	32	99	22
90	30	100	28

Special attention was given to the new run because a modified impregnating method was introduced. The initial samples, with only 33, 16, 19, and 26 defects, immediately raised hopes that the new method would radically improve the quality of the cloth. In conformance to control chart principles, the

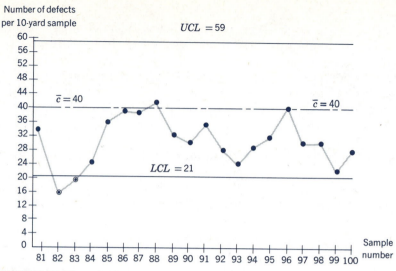

Number of defects per 10-yard sample

UCL = 59

$\bar{c} = 40$

$\bar{c} = 40$

LCL = 21

Sample number

FIGURE 15.16

values outside the historical control limits were investigated for assignable causes. Some of the incipient optimism was drained by the revelation that an inspector unfamiliar with the characteristics of the special cloth had made the first inspections. She had missed blemishes that should have been counted as defects. Observations 82 and 83 were thus discounted.

The pattern of subsequent inspection reports substantiates a beneficial, but not really dramatic, decrease in defects attributable to the refined process. Most of the recent points are below the former average number of defects, $\bar{c} = 40$. Because the process average appears to have changed as a result of the new methods employed, a revision of control limits appears advisable. Samples 82 and 83 were previously eliminated for assignable causes. During the review of these samples, it was determined that the inspector's inexperience may also have affected the results for samples 81 and 84. Therefore, the revised chart is based on the sequence of samples 85 through 100 to provide

$\bar{c}_r$ = revised process average

$$= \frac{\Sigma(c_{85} + c_{86} + \ldots + c_{99})}{N} = \frac{517}{16} = 32.3$$

$$UCL_{(revised)} = \bar{c} + 3\sqrt{\bar{c}_r} = 32.3 + 3\sqrt{32.3}$$

$$= 32.3 + 17.1 \doteq 49$$

$$LCL_{(revised)} = \bar{c}_r - 3\sqrt{\bar{c}_r} = 32.2 - 3(5.7)$$

$$= 32.3 - 17.1 \doteq 15$$

All of the points recorded on the chart fall within the revised limits, indicating that the process is apparently in control for the new process average. The revised limits are then applicable to future output using the improved impregnating technique.

15-8 APPLICATION OF QUALITY CONTROL

Quality control is receiving greater attention than ever before. Some critics say that quality is currently dropping, and they place the blame on a lack of pride and craftsmanship by today's workers. Others place the blame for questionable quality on high-speed assembly lines in which it is difficult to catch mistakes before the product is out of the factory.

Another argument states that there has been no decrease at all in quality; instead, modern customers simply expect more than ever before. Elaborate and saturating advertising campaigns stress the perfection of products, and this emphasis is likely to produce expectations exceeding production capabilities. Consumer-minded marketers with their pledges and warranties set a scrambling pace for responsible production-minded managers. Few would care to argue against the worthiness of the trend, but many are wondering how to support it.

Both sampling plans and control charts are powerful allies in the pursuit of quality. They are grounded in solid mathematics and have been refined to a level at which they are simple to apply. They are working evidence of the principle of management by exception. They embody more depth and facets than are immediately apparent.

We can return to control charts to illustrate some of the quality considerations recognized by experienced practitioners. The considerations evolve from a blend of theoretical knowledge and a working familiarity with the process subject to control.

A primary objective of control charting is to obtain the fastest possible warning of a deviating condition. This objective may be applied to the operation of the process itself or to the acceptance of products from the process. Sampling designs are different for the two divisions. A sample intended for process control should be selected to yield small variations *within* each sample and large variations *among* samples. When the purpose is to control outgoing quality, a sample should be *representative* of the output period, a random sample from a dated block of products.

We have observed the importance and logic of carefully examining all points falling outside the control limits. Much can also be learned from observing the pattern of points still inside the limits. The purpose is to identify a trend that can be investigated for an underlying cause before the possible cause damages production. The *theory of runs* is a probability-based routine for determining the chance occurrence of a sequence of points consistently crowding one side of the process mean. Rules-of-thumb are applied by some practitioners to serve the same purpose. The *rule of seven* indicates that the process should be investigated for nonrandom influences whenever seven consecutive points occur above or below the mean. Similar "rules" suggest scrutiny for two consecutive points "very" close to the control limit and a run of five points successively higher or lower.

The manner in which a control chart is maintained and acted upon can influence its value. When charts are outdated or their signals are disregarded, workers soon become cynical about the quality program. It would be better to hide a neglected chart because its presence advertises the lack of concern by management; the cynicism it engenders could be contagious. Conversely, a

The concept that quality should be *built in*, not *inspected in*, does not diminish the value of control charts. As described in Example 15.1, operators can chart their own performance to detect irregularities and trends.

Runs are often classed as "runs above" or "runs below" (sequence consistently above or below the mean) and "runs up" or "runs down" (continuous sequence of increasing or decreasing values).

Unrealistic quality goals, like any subjectively unobtainable standards, may discourage or frustrate the worker's concern for quality.

properly treated chart can encourage better performance. A prominently displayed control chart with achievable standards advances a strong psychological incentive.

Charts are conspicuous and reputable parts of a quality control program, but they are not the whole program. Education, supervision, and training are necessary complements. In turn, a quality control program is only one area in the total production program. A concern for quality depends on other functions of the production system, and reciprocally, the concern supports the other functions. Exclusive attention to the engineering implications of quality control forfeits the quality program's potential for administrative and managerial benefits; its motivational potential could be its most significant contribution.

EXAMPLE 15.10 Total quality control

The three letters—*T–Q–C*—for **total quality control,** are associated with Japanese production successes. Although the term may have been borrowed from A. V. Feigenbaum's 1961 book *Total Quality Control,* TQC is now an apt description of the way that Japanese manufacturers obtain total involvement and total commitment of the total company in a total quest for quality.

Because quality is a companywide concern, no single department is responsible for the results. The whole factory is accountable. This means that everyone has to be convinced of the importance of quality, that each person is given an opportunity to contribute, and that their contributions are recognized.

Several of the more evident practices that promote quality production follow. Many of them are also used in companies other than those in Japan. What distinguishes the Japanese effort is the combined effect gained from applying all the practices together.

• *No defectives.* Japanese quality goals are extremely high. According to R. J. Schonberger,[1] "Today the term AQL is likely to draw looks of wry amusement from knowledgeable Japanese. The AQL . . . has long since fallen out of favor. In its place is a true long-term target, which is, simply, *no defectives.*" Unlike

the United States where zero defects is an overused and largely unrealized idea, the Japanese consciously and continually seek quality improvements that closely approach no defects.

• *Good housekeeping.* Plant tidiness and cleanliness set the basic conditions for safety which, in turn, influences the quality of performance.

• *Buffer time.* Production quotas are set below maximum output levels so as to allow time for workers to deal immediately with quality problems whenever they arise and to complete rework when needed.

• *Pictures and analyses.* Charts, diagrams, electric signs, and all kinds of display boards illustrate quality accomplishments in a factory. Visible signs of attention to quality remind everyone of the quality mission, provoke discussion, and prove to visitors that the company's commitment to quality is total. Japanese workers construct their own control charts, C&E diagrams, and other visual tools to analyze performance.

• *Total preventive maintenance.* A companion to TQC is TPM—an all-out effort to keep machines functioning perfectly. In addition to giving the prime responsibility for maintenance to machine operators, additional preventive maintenance work is squeezed in between shifts, creating a two-shift pattern when the production shifts are separated by four hours of machine servicing.

[1]"Production Workers Bear Major Quality Responsibility in Japanese Industry," *Industrial Engineering,* December 1982.

- *Sample size of 2.* The Japanese $N = 2$ concept refutes random-sampling principles by inspecting only the first and last items in a batch. If the first one is good, the machine is assumed to be adjusted correctly. If the last one is also good, the adjustment is probably still correct, and the intervening items are likely to be good.

- *Rework responsibility.* Instead of collecting defective items during inspections for later rework, Japanese workers are responsible for the inspections *and* the performance of the rework. If rework cannot be completed during the regular shift, workers are expected to do it on their own time. Workers' authority to stop the production line whenever something appears to be out of conformance tends to limit the amount of rework needed.

- *Top-to-bottom participation.* It is generally conceded that top-management commitment is a prerequisite to quality production. Once the necessary worker/supervisor skills are developed and high-performance machines and processes are in place, further improvements rely heavily on employee input. This input often takes the form of *Jishu Kanri* activities—small groups of employees who meet voluntarily, select their own leaders, decide quality improvement goals, and work together to attain these goals. (The QC circles described in Section 15-9 are JK activities.) The whole concept of involvement was nicely stated by Y. Kobayashi, president of Fuji Xerox, "If you lack either the QC circles or similar activities at the bottom, or if you have someone who feels that these kinds of activities can be delegated to some of the key managers rather than top management, you should forget about the whole concept of TQC."[2]

[2]"Reshaping a Company through TQC," *Proceedings*, International Conference on Productivity and Quality Improvement, Tokyo, October 1982.

15-9 CLOSE-UPS AND UPDATES

Both producers and consumers are now seeking quality more fervently than in the past. Frustrating experiences with machines that refuse to run, utensils that break, and clothes that lack stitching have made consumers wary. A National Family Opinion survey in 1983 revealed that 49 percent of the public felt that the quality of American goods had slipped in the previous five years, and 53 percent expected it to remain at the same level or decline further during the next five years.

A survey by the American Society of Quality Control (ASQC) found that two-thirds of the respondents doubted that U.S. workers were really concerned about quality. Experts defend U.S. quality by noting that new products are more complex and that consumers forget the extra attributes of today's goods compared with those in "the good old days." They also note that the public is more aware of product failures, as a result of nationwide publicity, and that a shortage of qualified repair workers causes minor repairs to become major annoyances when they are not fixed on time—the first time.

Two aspects of quality are apparent: *objective quality*—the measurable properties of products—is stressed by producers to impart a feeling of value

Quality has become a heavily advertised virtue, with companies making direct comparisons with competing products.

In the ASQC survey, 73 percent felt that U.S. industry is more concerned with profit than with quality.

for their products to consumers—*subjective quality*. Workers who have the necessary training, tools, and pride can provide objective quality. Customers with faith in a product (subjective quality) consistently choose that product over competitors' products. Higher sales then motivate higher quality consciousness and performance. Quality breeds quality.

The voyages of Dr. Deming: United States to Japan to United States

The modern guru of quality is Dr. W. Edwards Deming, an American expert with an important message that was first heard across the Pacific. Dr. Deming took his expertise on statistical analysis of quality to Japan in 1950, gained fame as the father of the Deming prize for quality control, and returned to the United States as an acclaimed consultant. His sobering message for his homeland is: "American management has no idea what quality control is and how to achieve it."

Deming aims his sharpest barbs at top management for not paying attention to quality, especially statistical methods for quality assurance. His recipe for quality control is deceptively simple: tally defects, examine them to find their source, correct the cause, and then keep a record of what happens afterward. He preaches that statistical analysis, not investments in equipment and automation, is the way for America to garner the gains the Japanese have enjoyed, offering the following advice to do so:

- Rely on statistical evidence of quality *during* the process, not at the end of the process. The earlier an error is caught, the less it costs to correct it.
- Rely on suppliers that have historically provided quality, not on sampling inspections to determine the quality of each delivery. Instead of a number of vendors, select and stick with a few sources that furnish consistently satisfactory quality.
- Rely on training and retraining to give employees the skills to use statistical methods in their jobs, not on slogans to improve quality. Employees should feel free to report any conditions that detract from quality.
- Rely on supervision guided by statistical methods to help people do their work better, not on production work standards. Statistical techniques detect the sources of waste and teams of designers, supervisors, and workers eliminate the sources.
- Rely on the doctrine that poor quality is flatly unacceptable. Defective materials, workmanship, products, and service will not be tolerated.[3]

[3]See W. Edwards Deming, *Quality, Productivity, and Competitive Position* (Cambridge, Mass.: MIT University Press, 1982).

The Deming prize was created in 1951 to reward innovation in quality control annually. It has become so famous in Japan that the ceremony is telecast live nationally.

See Deming, W. Edwards, Quality, *Productivity, and Competitive Position,* MIT Press, Cambridge, MA, 1982.

According to the Deming philosophy, there is little use in exhorting hourly workers to improve quality, because they do not control the resources needed to do it, such as tools, materials, scheduling, and facilities. Management controls the resources, and 85 percent of all quality problems originate in the system itself, not from the workers. The other 15 percent result from special causes, such as defective tools and negligent acts. Deming placed the responsibility for improvement squarely on management, believing that when managers remove the barriers that stand between hourly workers and their right to pride of workmanship, quality soon surfaces.

Quality control circles encircle the globe

One of the most celebrated programs to involve workers in quality improvement is the **quality circle (QC)** movement. Based on glowing accounts of how circles inspire greater concern for quality and productivity in Japan, the movement quickly girdled the globe, with circles springing up in most industrialized nations. Although many applications have been disappointing, enough have harvested good results to enthuse continued interest.

The academic origin of the quality circle concept is in doubt, but the fame assuredly came from Japan. QC principles were postulated in the United States long before taking hold in Japan, as were the statistical quality control techniques emphasized in circle activities. Nonetheless, it was the work of Kaoru Ishikawa and the Union of Japanese Scientists and Engineers that sparked the growth of quality control circles from their inception in 1962. The worldwide surge of interest is a direct result of Japan's enormous economic success since then.

According to the *QC Circle Koryo*, the biannual 1971 publication that lays out the principles of the QC movement, "the QC Circle is a small group to voluntarily perform quality control activities within the workshop to which they belong. This small group with every member participating to the full carries on continuously, as a part of companywide control activities, self-development and mutual-development within the workshop utilizing quality control techniques."[4] Observers of Japanese productivity attribute much of its growth to the morale-, skill-, and quality-building benefits of QC circles.

Circles were first introduced in the United States at the Lockheed works in Sunnyvale, California, in 1974. From there, they spread across the continent, picking up momentum in the early 1980s. With growth came change. The Japanese prototype concentrated on quality, and early U.S. implementations were similar. Later applications were restructured to fit American management styles and to stress particular production functions.

The number of Japanese workers engaged in quality control circle activities in 1982 was estimated to be over 5 million participants in some 56,000 circles, according to T. Sugimoto.

From Sugimoto, T., "Present Status and Result of QC Circle Activity in Japan," *Proceedings, International Conference on Productivity and Quality Improvement*, Tokyo, 1982.

[4]T. Sugimoto, "Present Status and Result of QC Circle Activity in Japan," *Proceedings,* International Conference on Productivity and Quality Improvement, Tokyo, 1982.

Some of the most acclaimed QC applications are in city, state, and federal government units.

The more significant changes have been *increased training* for circle members, *greater staff support* for circle operations, and *expanded productivity measurement* of circle activities.

Membership in a quality circle is typically voluntary. Each circle is composed of six to 10 employees engaged in similar work. The group is normally led by a supervisor, as the QC philosophy is to maintain the existing organizational structure while creating an opportunity for all those in the work force to express themselves and become more involved in operations. A circle usually meets for about an hour once a week during regular work hours in a meeting room near the work area. Agendas for the meetings include open discussions, problem-oriented discussions, training, and progress reports. Circles can be formed at any level of the organization, although most are at the lower levels.

Most circles operate in an informal manner. Members are encouraged to exchange ideas freely but to keep their focus on quality and productivity. As circles mature, they are expected not only to discover operating troubles but also to design ways to overcome the difficulties. Additional benefits realized from a flourishing QC program include better employee communications and work attitudes, development of employees' abilities and confidence, lower turnover and absenteeism rates, fewer grievances, and greater job satisfaction.

Not all of these expectations are met in every application, of course, and some implementations have failed miserably. There are innumerable reasons that any employee participation program can fail. The most common are lack of preparation before starting one, selection of weak or undedicated leaders, insufficient commitment by management, lack of cooperation from unions, and expecting great results too soon.

A QC program's chances of success are improved by extensive planning before it is launched. In a unionized plant, union officials must be included in the planning, as should representatives from the work force in nonunion organizations. Because supervisors carry so much responsibility for success, they must be convinced of the value of circles and assured that their authority will not be undermined. Failure is almost guaranteed if union leaders oppose the program or if supervisors feel threatened by it. Circle members themselves may lose interest if activities become routine and ritualistic. Spontaneity can be generated by varying the emphasis, having joint meetings, and offering different forms of recognition. Genuine enthusiasm is more likely when participants realize that their involvement is appreciated and they can see the results.

Improving quality by pinning tags on a fishbone: CEDAC

The second generation of cause and effect diagramming is **CEDAC**—**Cause**

and Effect Diagram with the Addition of Cards. As reported by Ryuji Fukuda, in *Managerial Engineering* (English edition published by Productivity Inc., Stamford, CT, in 1983), "CEDAC grew out of the diverse approaches of the individual group members. It is a modification of the cause-and-effect diagram, which was already well known and widely used by QC circles in our plants when I started our quality improvement project. . . . CEDAC grew out of this diagram. It emphasizes the importance of both the engineering knowledge and workers' practical experience that lie behind simple words such as 'temperature' or 'dryness' found on the cause-and-effect diagram."[5]

A CEDAC is constructed as follows:

C&E diagrams are described in Section 2-4.

1. Select a quality problem to investigate that has measurable dimensions.

2. Construct a large conventional cause diagram for each dimension, leaving room to place the cards on the fishbone.

3. Provide a control chart on which performance data can be entered as the *effect* portion of the diagram.

4. Hang the completed diagram in a conspicuous location that is within easy reach to add cards.

All employees are encouraged to pin cards on the diagram, but the appearance of points near or outside the control limits of the chart is a visual call for cards that suggest ways to remedy the situation. A quality circle may take the responsibility for updating the control chart and following up on cards attached to the CEDAC. The format of a CEDAC is shown in Figure 15.17.

Quality is quality, be it in products or services, public or private

Service is a commodity. Just as a product's quality can be evaluated by its reliability, finish, and conformance to specifications, so too can a service be measured by timeliness, accuracy, and completeness. Some dimensions are obvious. A new roof should not leak, a cashier should give the correct change, and deliveries should be made to the right address. Other aspects of service are less distinct. How thoroughly should the roofer clean up after the job? Should a cashier be a lookout for shoplifters? What is the value of special effort made to deliver, safely, an incorrectly addressed parcel?

[5]*Managerial Engineering*, English ed. (Stamford, Conn.: Productivity, Inc.: 1983).

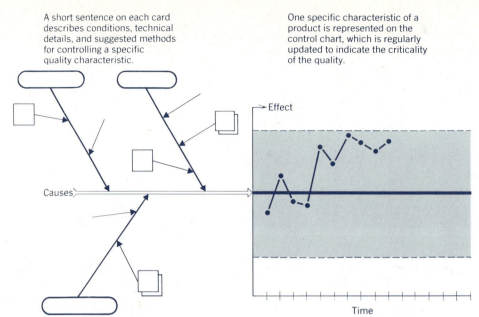

A short sentence on each card describes conditions, technical details, and suggested methods for controlling a specific quality characteristic.

One specific characteristic of a product is represented on the control chart, which is regularly updated to indicate the criticality of the quality.

Causes

Effect

Time

FIGURE 15.17 Basic structure of the cause-and-effect portions of a CEDAC. A single quality characteristic is displayed on the diagram.

The objectives matrix described in Chapter 16 is an effective method of measuring the performance of service workers.

Essentially the same statistical techniques used to inspect manufactured products are applicable to services. Only the setting and the gauges differ. In a service setting an inspector counts the number of errors on a page, asks customers whether they are satisfied, and collects performance data that indicate the consistency and level of output. As in manufacturing, the service dispensers themselves are often the best data collectors. Counting errors and tallying complaints are sobering motivators for service improvement. Statistical sampling methods and how to interpret the results should be part of all employees' training if they are going to police their own performance. Employees who know what is expected of them, how to provide it, how to measure results, and why, are the vanguard for quality assurance.

Not only do service providers now outnumber product manufacturers two to one, which means service performance is everywhere visible and criticized, but inferior service quality also is castigated just as vehemently, or perhaps more so, than are defective products. Clues to improving quality are often embodied in the nature of the service being criticized. For example:

• Nearly half the buyer complaints reported to Better Business Bureaus concern mail-order companies. One complaint told of sending

$10 to a mail-order house for a "valuable engraved picture of Abraham Lincoln." A Lincoln penny was sent by return mail. This near fraud still beats no reply at all to paid-for orders, as often reported. **Objective:** Have a legitimate product before worrying about quality.

• A dress was returned to a department store for a larger size. Bills sent for both dresses were excused due to "computer error." **Objective:** Build in quality checks for automated procedures (and people procedures, too).

• Taxpayers complain about the cost of schools. During the past decade the student population in Boston schools declined by 33 percent, but the teaching staff increased by 13 percent and the nonteaching staff by 150 percent. **Objective:** If quality is up, be sure customers know it (and taxpayers, too).

With the exception of posh establishments, service today tends to be less deferential and personal than in the past. Prices are held in check by stressing fast and efficient operations, usually at the expense of extra refinements and courtesies. When prices dip with quality, the trade-off is genuine, often being part of the marketing strategy. It is higher prices for less quality that has aroused censure of the postal service, railroads, and nursing homes. Such examples are fodder for the perception that management either is not aware of declining quality or does not care.

Future productivity gains will increasingly be sought in the service sector, and quality enhancement is sure to be the target.

15-10 SUMMARY

Quality assurance ranges over the complete input–transformation–output sequence of a production system. Acceptance sampling measures the quality of inputs. Quality programs urge workers to avoid mistakes in the transformation process. Control charting measures the performance of a process in meeting output objectives. Industry, government, and service organizations follow essentially the same quality assurance procedures, differing mostly in the way that output is inspected and measured.

Quality assurance

Production quality starts with a process capable of producing to design specifications. The design must meet the precision demanded by customers and be within the accuracy attainable by the process. The economics of precision versus accuracy relates the capital cost of achieving greater accuracy to the higher returns expected from supplying greater precision.

Relation of value to cost

Inspections are scheduled when they will do the most good. They are usually a function of the phase of product development: at the receiving point for raw materials and components; in the process before operations that are costly, irreversible, or masking; and at the end of the process before delivery. Inspectors typically follow the product in fixed-position and process layouts.

Inspecting

Formal inspection stations are associated with product layouts. Some products must be sent to central laboratories for testing. When there is a choice between laboratory and on-site inspections, the advantages of lab standardization are compared with the added costs of handling.

Acceptance sampling

The purpose of acceptance sampling is to recommend a specific action, it is not an attempt to estimate quality or to control quality directly. A lot-by-lot acceptance sampling plan for attributes specifies the sample size and the associated number of defectives that cannot be exceeded without rejecting the lot from which the sample was taken. The capability of the plan to distinguish between acceptable and unacceptable lots is revealed by its operating characteristic (OC) curve. The ideal OC curve is obtainable only from error-free, 100-percent inspection. The preferred plan balances the cost of taking larger samples against the greater precision obtainable from larger samples.

Sampling plans

Sampling plans are characterized by an acceptable quality level (AQL)—the grade considered adequate by the receiver—and by the lot tolerance percent defectives (LTPD)—the level of defectives a receiver will just tolerate. The AQL is associated with the producer's risk, α, and the LTPD with the consumer's risk, β.

Average outgoing quality (AOQ) is a measure of the quality level resulting from an acceptance plan when the rejected lots are subjected to 100-percent rectifying inspection and defectives in the sample are replaced by acceptable items. This percentage is calculated from the equation

$$AOQ = \frac{P_a \times p(N - n)}{N}$$

Statistical quality control

A control chart is a graphic aid for detecting quality variations in a production process. It has three main uses: to determine the capability of the process, to guide modifications, and to monitor output. Control limits help distinguish variations caused by chance from those due to assignable causes. An observation falls by chance outside the typical limits about three times in 1000.

Control charts

Control charts for variables record the average value of sample measurements ($\bar{x}$) and the range of measurements (R). Both the mean and range of a sample must be within control limits before the process represented is considered under control. One method of calculating the control limits for variables is to use predetermined factors based on sample size, n. The A factor is used for sample means, and the B and C factors are applied to sample ranges such as

$$UCL_x = \bar{\bar{x}} + A\bar{R} \qquad UCL_R = B\bar{R}$$
$$LCL_x = \bar{\bar{x}} - A\bar{R} \qquad LCL_R = C\bar{R}$$

When the sample size is not constant, different limits must be calculated for each n.

Control charts for attributes portray the percentage defective (p-chart) or the number of defects (c-chart). The p-chart control limits are based on the binomial distribution to provide

$$\text{Control limits}_{(p)} = \bar{p} \pm 3\sqrt{\frac{\bar{p}(100 - \bar{p})}{n}}$$

where $\bar{p}$ = average percentage defective. The control limits for the c-chart are derived from the Poisson distribution to give

$$\text{Control limits}_{(c)} = \bar{c} \pm 3\sqrt{\bar{c}}$$

where $\bar{c}$ = average number of defects per inspection unit. Limits for both charts are often developed from process samples taken during a base period after eliminating unrepresentative observations. Control limits should be revised when the process average changes.

Efforts to raise the quality of products and services are increasing. Employees can participate through quality (or QC) circles, a movement that started in Japan. Total quality control (TQC) is another program that has been used successfully in Japan. Many of the quality improvement methods applied in manufacturing can be adapted to the service sector.

Japanese quality control methods

15-11 REFERENCES

BESTERFIELD, D. H. *Quality Control: A Practical Approach.* Englewood Cliffs, N.J.: Prentice-Hall, 1978.

CHARBONNEAU, H. C., and G. L. WEBSTER. *Industrial Quality Control.* Englewood Cliffs, N.J.: Prentice-Hall, 1978.

CROSBY, P. B. *Quality Is Free.* New York: McGraw-Hill, 1979.

DODGE, N. F., and H. G. ROMIG. *Sampling Inspection Tables— Single and Double Sampling,* 2nd ed. New York: Wiley, 1957.

DUNCAN, A. J. *Quality Control and Industrial Statistics,* 4th ed. Homewood, Ill.: Irwin, 1974.

FEIGENBAUM, A. V. *Total Quality Control.* New York: McGraw-Hill, 1983.

GRANT, E. L., and R. S. LEAVENWORTH. *Statistical Quality Control,* 5th ed. New York: McGraw-Hill, 1980.

HALPERN, S. *The Assurance Sciences: An Introduction to Quality Control and Reliability.* Englewood Cliffs, N.J.: Prentice-Hall, 1978.

HARRIS, D. H., and F. B. CHANEY. *Human Factors in Quality Assurance.* New York: Wiley, 1969.

JURAN, J. M., and F. M. GRYNA, JR. *Quality Planning and Analysis,* 2nd ed. New York: McGraw-Hill, 1983.

Military Standard Sampling Procedures and Tables for Inspection by Attributes, MIL-STD-105C. Washington, D.C.: U.S. Government Printing Office, 1961.

SHEWHART, W. A. *Economic Control of Quality of Manufactured Product.* New York: Van Nostrand, 1931.

SINHA, M. N., and W. O. WILLBORN. *The Management of Quality Assurance.* New York: Wiley, 1985.

WADSWORTH, H. M., K. S. STEPHENS, and A. B. GODFREY. *Modern Methods for Quality Control and Improvement.* New York: Wiley, 1986.

15-12 SELF-TEST REVIEW

Answers to the following review questions are given in Appendix A.

1. T F *Quality assurance* is receiving less attention as concern for productivity improvement increases.

2. T F An efficient inspection procedure can ensure that a product meets

specifications even when the process that produces it is incapable of meeting *tolerances*.

3. T F *Costs and value* accelerate at increasing rates as the technical excellence of a product is increased.

4. T F *Inspections* should be scheduled immediately after an irreversible process to confirm that the process has been successful.

5. T F *Variables inspection* is usually applied to measure the quality of service organizations such as banks.

6. T F *Control sampling* is normally applied to incoming deliveries.

7. T F A *sample* is a single observation taken from a defined population.

8. T F A larger sample drawn from one *population* has a better chance than does a small sample to reveal the true characteristics of the population.

9. T F A *rectifying inspection* is a screening inspection.

10. T F The *acceptance number* in a sampling plan is the maximum number of defects a lot may contain and still be acceptable.

11. T F As an *OC curve* moves to the left and becomes steeper, the sampling plan's ability to discriminate good from bad lots increases.

12. T F A *100-percent inspection* always determines the true number of defectives in a lot.

13. T F The *mean range* is the average difference between the high and low observations for all of the samples.

14. T F *Upper and lower control* limits are always equally distant from the process mean.

15. T F A change in the *capability* of a process is usually considered to be more serious than is a change in the *level* of the process.

16. T F A point outside the control limits means that the process is *out of control*.

17. T F A point below the LCL of a *p-chart* does not need to be investigated when a lower percentage of defectives is desirable.

18. T F According to *ZD* philosophy, the lack of employee knowledge is more difficult to correct than is the lack of employee attention to quality.

19. T F In a *TQC program*, Japanese workers are expected to inspect all their own output and personally to do any necessary rework.

20. T F *Quality circles* originated in Japan and were introduced into the United States in about 1974.

21. T F The leader of a *quality control circle* is normally a supervisor.

22. T F *CEDAC* is a Computer Engineered Drawing with the Addition of Cards.

23. T F Quality circles are usually composed of *six* to *ten* members and meet *once* a week for about an *hour*.

24. T F Most quality control methods that have been effective in manufacturing can be adapted to the service sector.

15-13 DISCUSSION QUESTIONS

1. If a supervisor is indeed the interface between quality control goals of management and the achievement of those goals by workers, what are his or her responsibilities to both sides of the interface?

2. Relate the concepts of product *precision* and production *accuracy* to the appropriate functional areas represented in Figure 15.1.

3. The value line and cost line in Figure 15.2 tend to converge at higher levels of product refinement. How could the shape of the value line be changed to parallel the cost line more closely?

4. What are some potential advantages of treating the quality assurance staff as pseudoindependent operators? What practical measures could be used to establish this relatively independent status within the organization?

5. What is the relationship between a "zero defect" program and the more quantitative measures of quality control such as acceptance sampling and control charts?

6. What are the advantages and disadvantages of a central laboratory inspection facility with reference to the three basic plant layouts?

7. Discuss the considerations in selecting the inspection points within a process.

8. Why is statistical sampling generally more suitable for machine-produced products than for hand-crafted products? Under what conditions may sampling produce better results than 100-percent inspection?

9. Why is acceptance sampling not considered an attempt to control the quality of a process?

10. What is an *OC* curve? How can an *OC* curve be made more discriminating? Of what value is a more discriminating curve?

11. What is the relationship between *AQL* and α? Between *LTPD* and β?

12. Explain how the *AOQ* can decrease while the percentage of defectives increases. Refer to Figure 15.8.

13. Compare the economic and administrative considerations for single sampling, double sampling, and sequential sampling plans. What types of industrial situations could produce a preference for one of the plans?

14. Distinguish among natural variations in a process and variations due to an assignable cause. Give some examples of assignable causes.

15. What is the difference between statistical control limits and product tolerances? How do these two dimensions relate to product precision and production accuracy?

16. Why is it necessary to control both $\bar{x}$ and R variations for control of variables?

17. Why must separate charts be developed for each variable characteristic measured?

18. Distinguish between the types of inspection required for a *p*-chart and a *c*-chart.

19. Discuss why the control limits for a *p*-chart are based on the binomial distribution and those for a *c*-chart are based on the Poisson distribution.

20. Discuss the adequacy of today's production quality and possible reasons for current conditions.

21. What is the theory of runs, and how can it prove useful?

22. Comment on the administration and maintenance of control charts as they affect worker motivation.

23. Analyze each of the following quotations. State

two views for each statement: one supporting the quotation and one opposing or expressing a weakness in the logic. Which view agrees most closely with your personal opinion?

a. "We don't dare use any statistical techniques for quality control in our business because we can't afford to send out a single defective product to our customers."

b. "The only thing important to me in a sampling plan is the amount of the consumer's risk. I couldn't care less about protecting the producer. I'm the customer, and customers are always right."

c. "It's misleading to say 'quality is everybody's business.' You should say it is the manager's business to get everybody involved with quality. When a job belongs to everybody, it can easily become a job that nobody does."

24. Compare QC circles with zero defect programs.

25. Comment on the following views by J. F. Hird as expressed in "Japan's Q. C. Circles," *Industrial Engineering*, November 1972:

"The QC Circle also provides a door in the industrial structure, between labor and management, through which the experience of bringing about technical change is shared by both.

"While the Japanese are gathering their work force's talent, we seem to be going the other way. Hundreds of managements and labor unions are negotiating contracts designed to sustain their workers' material needs. At the same time, all kinds of supplementary schemes are going into contracts to give workers time off the job and to compensate them for it.

"Our workers are being provided with more and more ways to escape from work. Does this mean that both labor and management are admitting that, because we have separated a good portion of the work of the hands from the work of the mind, work as a whole is becoming less and less meaningful to the majority of our people? Are our unions and industrial leaders admitting that work no longer contains those qualities which make life on the job exciting? I am sure that this escapist view of work will do nothing to help us to improve the quality of work itself."

15-14 PROBLEMS AND CASES

1. In a preliminary study of a process containing five successive operations, the following data were obtained:

Operation	1	2	3	4	5
Percentage rejected	5.1	0.6	4.8	3.2	1.2
Cumulative manufacturing costs	$2.75	$3.98	$4.27	$4.85	$5.12

The standard unit of inspection is 100 products, and inspectors are paid $4 per hour. Inspection times in minutes per product are in order of the operations, 1.2, 0.6, 0.5, 2.0, and 1.8.

a. If two inspection stations are to be established, where in the sequence of operations should they be located? Why?

b. If three inspection stations are allowed, where should they be placed? Why?

2. a. Construct an *OC* curve for a single-sample, percentage-defective, lot-by-lot sampling plan where $n = 100$ and $c = 2$.

b. On the graph developed for part a, show by dotted lines the general shape of the curve for a 100-percent inspection plan and a plan where $n = 1000$ and $c = 20$.

c. What is the *AOQL* from part a? At what value of p does it occur?

d. If $\alpha = 0.05$ and $\beta = 0.10$ for the plan in part a, what are the *AQL* and the *LTPD*?

3. What size sample should be taken, and how many defectives can be allowed in the sample for a consumer's risk of 5 percent, a producer's risk of 10 percent, and *AQL* of 1 percent, and an *LTPD* of 9 percent? Determine the actual values of α and β for the sample plan selected.

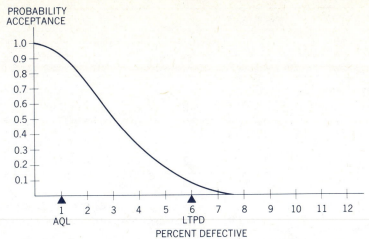

PROBABILITY
ACCEPTANCE

FIGURE 15.18 Operating characteristic curve with AQL = 0.01 and LTPD = 0.06.

4. Given a sampling plan defined by $n = 10$ and $c = 1$, what are the *AQL* and *LTPD* for $\alpha = 0.03$ and $\beta = 0.05$?

5. The operating characteristic curve shown in Figure 15.18 has been constructed for the sampling now used in your department for sampling incoming material.

a. What is the consumer's risk?

b. What is the producer's risk?

c. What is the probability of accepting a lot that contains 2 percent defectives?

d. What is the probability of rejecting a lot that contains a fraction defective of 11 percent?

6. Two identical cutting operations are depicted by the control charts in Figure 15.19. Twelve samples representing measurements of the same dimension

for the same product are shown in $\bar{x}$ and R charts for each operation.

a. Analyze the patterns for operator F on machine I. Is the process out of control? Explain.

b. Analyze the patterns for operator O on machine K. Is the process out of control? Explain.

c. Which of the two operations is likely to cause more trouble in the future? Why?

d. What physical conditions might cause the deviating pattern in each operation?

7. Ten samples, each containing five measurements, have been taken of two machines run by the same operator to produce the same part. The specification for the part is 0.750 ± 0.005 in. Measurements in the table below are given in *thousandths of an inch* in excess of 0.740 inch. That is, a measurement shown as 12 in the table indicates an actual dimension of $0.740 + 0.012 = 0.752$ inch.

		Machine 1						Machine 2			
Sample		Measurements				Sample		Measurements			
1	8	11	10	9	9	1	7	9	8	7	7
2	12	9	10	11	13	2	6	8	8	9	7

Sample	Machine 1 Measurements					Sample	Machine 2 Measurements				
3	7	10	8	12	8	3	9	7	8	9	9
4	7	6	10	8	9	4	8	6	9	7	8
5	10	7	11	12	10	5	8	10	9	10	10
6	11	6	9	10	12	6	9	8	8	10	9
7	11	14	10	9	11	7	11	10	9	10	10
8	9	10	12	11	9	8	10	11	10	11	10
9	13	7	8	12	10	9	11	12	13	12	12
10	12	11	10	12	4	10	11	14	12	12	13

a. Calculate the sample means and ranges.

b. Plot the sample means and ranges for the machines on two pairs of control charts. Indicate the control limits and the process mean on each chart.

c. Which machine has the better process capability? Why?

d. Can either machine meet the stated specifications? Explain.

8. Base-period data are as follows for the development of control charts:

a. What are the control limits and center line for the process average? Construct and discuss the $\bar{x}$ chart.

b. What are the control limits and center line for the process variation? Construct and discuss the range chart.

9. One characteristic of a product, x, is inspected. From 20 samples, each of 10 measurements, the following totals are obtained:

$$\sum \bar{x} = 139.8 \qquad \sum R = 23.9$$

The specification for the measured characteristic is

Sample Number	x_1	x_2	x_3	x_4	x_5	x_6
1	0.498	0.492	0.510	0.505	0.504	0.487
2	0.482	0.491	0.502	0.481	0.496	0.492
3	0.501	0.512	0.503	0.499	0.498	0.511
4	0.498	0.486	0.502	0.503	0.510	0.501
5	0.500	0.507	0.509	0.498	0.512	0.518
6	0.476	0.492	0.496	0.521	0.505	0.490
7	0.483	0.487	0.495	0.488	0.502	0.486
8	0.502	0.500	0.511	0.496	0.500	0.503
9	0.492	0.504	0.472	0.515	0.498	0.487
10	0.511	0.522	0.513	0.518	0.520	0.516
11	0.488	0.512	0.501	0.498	0.492	0.498
12	0.504	0.502	0.496	0.501	0.491	0.496
13	0.501	0.413	0.499	0.496	0.508	0.502
14	0.489	0.491	0.496	0.510	0.508	0.503
15	0.511	0.499	0.508	0.503	0.496	0.505

Cutting Operation 1: Machine I, Operator F

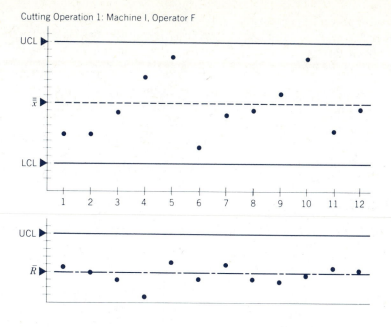

Cutting Operation 2: Machine K, Operator O

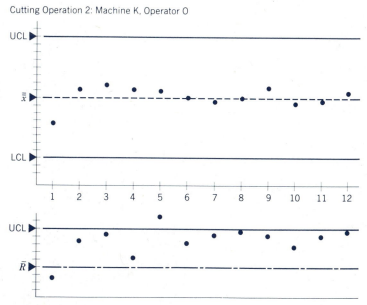

FIGURE 15.19 Pairs of x and R charts for cutting operations 1 and 2 in Problem 6.

7.10 ± 0.30. Assuming that the process is in control for the limits calculated from 20 samples, what conclusions can be drawn about the ability of the process to meet the specifications?

10. During the past year, an average of 32 units were found defective from 100-percent inspection of daily production. The output per day varied 2 percent at most from an average of 480 units. A review of the records revealed that the number of defectives occasionally went as high as 55 and as low as 4 or 5 per day. What help might be obtained from a control chart in improving process performance?

11. Construct the appropriate control chart for the process represented by the following data collected in the past month:

Sample Number	Sample Size	Number of Defectives
1	200	3
2	200	1
3	200	0
4	200	2
5	200	4
6	200	1
7	200	2
8	200	0
9	200	3
10	200	2
11	200	1
12	200	3
13	200	6
14	200	8
15	200	5
16	200	9
17	200	3
18	200	1
19	200	0
20	200	2
21	200	3
22	200	1

What can be concluded about the behavior of the process during the past month? What limits are appropriate for controlling next month's production?

12. A company that manufactures gyroscopes inspects all the produced devices. Each is classified as acceptable or unacceptable. Those rejected are reworked until they reach an acceptable state. What can be said about the 10 days' production described as follows when the process has previously produced an average percentage defect of 4 percent? Plot the control chart.

Day	Output	Number Rejected
1	80	3
2	121	6
3	57	2
4	141	7
5	132	6
6	154	9
7	171	13
8	101	4
9	88	3
10	71	1

(*Hint:* A straightforward approach to charting nonuniform sample sizes is to use the historical average, $\bar{p} = 4$, and to allow the control limits to vary according to $\pm 3\sqrt{4(100 - 4)}/\sqrt{n}$.)

13. The following number of defects were obtained from the inspection of 100-foot sections of high-quality, multichannel communication cable (read each line left to right to obtain the sequence of observations):

$$1\ 0\ 2\ 3\ 1\ 0\ 3\ 1\ 2\ 5\ 1\ 0\ 1\ 0\ 0$$
$$2\ 0\ 1\ 1\ 2\ 0\ 0\ 2\ 0\ 1\ 3\ 6\ 1\ 0\ 1$$

Do these data come from a controlled process?

14. *A Case of Quality Feedback*
A prominent manufacturer of aircraft equipment obtains "aircraft-quality" material (raw materials and

component parts) from many outside vendors. On all incoming material, this manufacturer maintains a quality rating on which a monthly report is sent to vendors. The rating formula is as follows:

$$\text{Rating} = 100 - \frac{100(\$ \text{ material review} + \$ \text{ source rejects} + \frac{1}{2} \$ \text{ vendor } CAD)}{\$ \text{ inspected}}$$

The item "$ material review" refers to dollars of material rejected at the manufacturer's receiving quality control check. The item "$ source rejects" means the reworking or replacement of parts rejected at the vendor's plant by the manufacturer's "source" inspector. CAD stands for "conditional acceptance deviation," involving borderline materials outside tolerance but accepted for specific use by the manufacturer.

When a monthly report for any vendor is construed by the manufacturer as unsatisfactory, the vendor is required to forward a statement showing action taken to improve the quality.

Comment on the general applicability, inherent fairness, and specific advantages, or the lack thereof, of this particular vendor quality–rating feedback.

15. *The Case of Quality at Wanabetcha Company*

The Wanabetcha Company recently discovered that the costs due to shipment of defective items had risen to an alarming level. To correct the situation, they decided to implement a *quality assurance program* (*QAP*). Previously, all inspection was done by workers on their own work. Because none of the present employees or managers had any formal education in quality assurance, they decided to compose the QAP team of recent college graduates.

When the team was formed, the company's president told them he expected the percentage of defective items being produced to be halved within one month. With this formidable introduction the team went to work.

Problems, however, began to plague the program immediately. Conflicts arose between the inspectors and the workers. Some of the older employees felt they were being insulted whenever a quality problem was traced to their work. This resentment often resulted in their work's deteriorating further

instead of improving. Other workers believed they were being wrongly accused of shoddy workmanship. Some even accused the inspectors of actually

making defects in their work so that they could claim they had found a problem spot and, hence, look good in the eyes of the QAP manager.

Monitoring reports after the first month showed that the quality level had actually worsened. Management felt that perhaps they had introduced the quality assurance program improperly.

a. What errors do you feel the Wanabetcha Company made in the implementation of QAP?

b. What remedial actions would you take to improve the present situation?

16. *The Case of a U.S. Company with a Japanese TQC Program*

In 1974 a division of Motorola was acquired by Matsushita Electric Industrial Company. It employs 1200 people in Franklin Park to produce color television receivers and microwave ovens. Five years after the changeover, in-process defects had dropped from 1.4 to 0.07 defects per set, and productivity had jumped almost 30 percent. The labor required to produce a color television receiver was cut in half. These gains were achieved by blending new equipment, technology, training, and managerial practices to revitalize an already skilled labor force.

Equipment. Automatic equipment developed in Japan for chassis assembly was used. Design changes reduced the number of required workers by 26 percent. Equipment and design engineers worked together to improve "producibility," making quality products easier to manufacture.

Technology. New assembly lines allowed workers to control their work flow individually. In place of a continuous, conveyor-paced line, operators were given foot levers to detour work to their station and to forward finished pieces to the next work stations. Closed-circuit television systems were installed to broadcast quality information to workers on the production line.

Training. The importance of quality was continually emphasized, placing responsibility on production workers, not inspectors. End-product inspection teams were replaced by a few in-process auditors who moved from one assembly line to another sampling quality. New employees receive both classroom and on-the-job training, up to five days of each, during which they learn about quality expectations, are judged on whether they can do the work adequately, and see whether they like the work conditions.

Managerial practices. Once a week all operations cease for 10 to 15 minutes while supervisors communicate with their crews. A supervisor typically talks with 45 workers about quality, productivity, absenteeism, scrap, and any other subjects that might come up. If the supervisors cannot answer a question, they make a note of it and come prepared with an answer at the next meeting.

Every six months, manufacturing and quality control people meet to set quality goals for different areas. Bar charts are kept to signal which areas are above, near, or below their targets. Special effort is concentrated on a particular production line, called the *model line,* to improve its performance. Workers on that line and support groups meet once a week to explore progress. Reasons for successes in the model line are identified and adapted to fit other lines.

A quality emphasis month is declared twice a year. Awareness is aroused by slogan and poster competitions, crossword puzzles with quality terms, and suggestion contests. Winners are entertained at a restaurant and given a modest award.

The underlying purpose of all activities is to create an environment that will be conducive to cooperation and encourage people to work together to identify problems and offer suggestions to solve them.

a. Discuss the relative contribution to quality improvement of the engineering/technology investments with the human resource–management changes made at the Motorola plant.

b. Quality control circles were not established at the plant by the Matsushita Company. Suggest reasons that the circles were not used, even though they enjoy great success in Japan.

PRODUCTIVITY

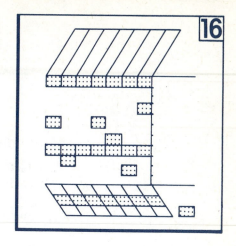

16

LEARNING OBJECTIVES

After studying this chapter you should

- realize the importance of productivity improvement to managers, workers, consumers, and national prosperity.
- recognize the relationship of productivity to economic competition and production systems management.
- understand how and why productivity is measured at different organizational levels by total indexes, partial indexes, and performance indicators, including models for interfirm, total firm, and work-unit measurement.
- be able to explain, construct, interpret, and implement an objectives matrix.
- appreciate how the awareness–improvement–maintenance (A.I.M.) process builds on the production planning–analysis–control phases to provide a coordinated approach to productivity improvement.

KEY WORDS

The following words characterize subjects presented in this chapter:

productivity	price recovery index
productivity ratio	cost effectiveness index
total productivity index	work-unit productivity indicators
price deflators	objectives matrix (Omax)
factor productivity index	productivity criteria
partial productivity indexes	performance indicators

level 3 and 10 scores
A.I.M. process
productivity audit
productivity sharing

Omax progress graph
interfirm productivity
comparison (IPC)

16-1 IMPORTANCE

Productivity is the quality or state of being productive. It is a concept that guides the management of a production system and measures its success. It is the quality that indicates how well labor, captial, materials, and energy are used. Productivity improvement is sought everywhere because it supports a higher standard of living, helps control inflation, and contributes to a stronger national economy. It is the implied theme throughout this book and is a subject of growing international concern.

Increasing productivity is a goal advocated by business, organized labor, and government. What worries U.S. leaders is the nation's poor productivity performance in recent years. Many reasons have been suggested for the declining productivity rate, but no single factor seems to deserve the full blame. Just as there are many causes, there are also many suggested remedies. The elusiveness of productivity stems mostly from its all-inclusive nature: a change in the productivity of a system results from the combined effect of all of the factors contributing to the system's performance.

The word *productivity* is bandied about so frequently that it assumes the proportions of a many-splendored cure-all. It can be yanked back into perspective by considering what it is not:

- *It is not a measure of production quantity.* It is the relationship of output to input; increasing production output may or may not improve productivity, depending on the inputs used to achieve that production increase.

- *It is not a measure of profitability.* It indicates the efficiency of operations and thereby suggests their profitability, but inefficient operations can occasionally be profitable if the product enjoys a favored market status.

- *It is not a guaranteed way to reduce inflation.* It may be a moderating factor, but it is only one among many economic factors that determine the general price trend.

- *It is not a technique to make workers work harder.* It is an approach that encourages workers to work together and to be more effective.

A definition appropriate to production managers is that *productivity is the measure of how specified resources are managed to accomplish timely objectives stated in terms of quantity and quality.*

The simplest statement of productivity is that it is just the ratio of output to input. An increase in the ratio, when properly adjusted for price changes, indicates greater production efficiency. It is thus a sensor in the production control feedback loop. A ratio lower than desired is a cue to initiate corrective

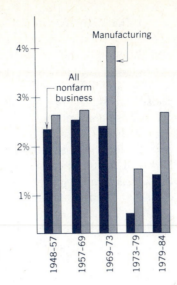

FIGURE 16.1 Productivity growth rates in the United States, 1948–1984.

action. This chapter presents ways to measure productivity, discusses what can be done to improve it, and examines the implications of productivity growth for productions systems.

16-2 PRODUCTIVITY PATTERNS

It may appear unreasonable to get excited about a sterile statement of, say, a 2-percent productivity increase for this year. The change from one year to the next in one organization is not immediately manifested in rising or falling prices and wages. Effects are more apparent and the impacts are more evident when they are examined over a span of years and compared with performance in other industries or countries. Then the ramifications of a productivity change are meaningful, and a 2-percent increase becomes a significant statistic.

Productivity in the United States since 1948 has gradually increased, as is shown in Figure 16.1. Annual gains have varied owing to business booms and busts, wars, economic policies, technology advances, as well as other factors. Dr. J. W. Kendrick, a prominent productivity analyst, reported that increases in productivity account for nearly one-half of the increases in real income for the United States in this century (the other half is due to increased resources, mainly labor and capital). About four-fifths of the growth in U.S. output since 1945 has resulted from the increased productivity of American workers. But if the output had advanced during the whole period at the 1974 to 1979 rate, output would have risen less than 70 percent as much. Looking at the same figures from a more personal angle, the lower rate would have caused about a one-third decrease in American living standards for the period since World War II.

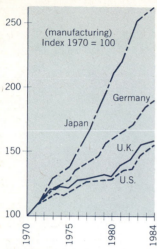

250 —
(manufacturing)
Index 1970 = 100

200

Germany

Japan

150

U.K.

U.S.

100

1970   1975   1980   1984

Comparison of manufacturing productivity trends in four countries, 1970–1984.

The United States experienced two periods of negative productivity rates: between 1974 and 1975 and in 1979, 1980, and 1982. In effect, the nation had virtually no growth in total productivity from 1974 to 1983. And in 1985 there was again no gain in productivity.

Deregulation in the 1980s generally improved productivity in the affected industries.

In 1984, R&D spending rebounded to close to the 1968 peak of 2.7 percent of GNP.

International implications

Throughout the world, productivity is taken to be a measure of the efficiency, and therefore the competitiveness, of a nation's industry. Losers in the competition face high inflation, persistent unemployment, a weak national currency, a worsening balance of trade, and a loss of markets to foreign competitors. As a loser during the last half of the 1970s, the United States has suffered through all of these ailments.

Although recent productivity performance in the United States leaves a lot to be desired, it is still in a favorable position with respect to most other countries. Part of the difference in the last decade's growth rates stems from the fact that the United States started from a much higher level of national production than did other countries and thus may have experienced diminishing rates of return that are associated with a larger scale of economic activity. Japan had the highest productivity growth rate, but its productivity is still slightly below the U.S. level. However, all northern nations except Ireland and the United Kingdom passed the United States in the early 1980s, and their lead is increasing.

Domestic implications

Growth in productivity makes a country more competitive in world markets and better able to sustain vigorous, noninflationary expansion. Over the long haul, it is the only way to raise living standards.

Manufacturing productivity in the United States climbed with the business upturn from the lows of the 1981–1982 recession. Much of the credit is owed to production managers who responded to foreign competition by raising the efficiency of their operations. The methods used by the threatened industries have been adopted by other companies, thereby increasing productivity throughout manufacturing.

The service sector has not done as well. Productivity in the nonfarm, nonmanufacturing sector (the shorter bars in Figure 16.1), which accounts for 70 percent of total output, grew at a rate of only 0.9 percent from 1980 through 1985. This is better than the pathetic 0.4 percent rate of the 1970s but is still less than half the rate of the 1950s and 1960s.

The slump in U.S. productivity is blamed by various experts on some or all of the following causes:

- *Government regulation.* Costs of complying with, or just reporting on, rules and regulations imposed by the government subtract from funds that might otherwise be put to more productive use.

- *Growth of services.* Increases in the service sector, which is generally characterized by lower productivity growth rates than in manufacturing has lowered the national average.

- *Decline in R&D.* National expenditure for research and development as a percentage of gross national product steadily declined from 1967 to 1981.

- *Work attitudes.* Some believe that positive attitudes toward work have declined. Boredom, a change in the composition of the work force, a shift from the work ethic, and personal or social goals in opposition to work goals are suggested as reasons.

- *Work restrictions.* Rules imposed by labor unions and management on what employees can and cannot do frequently stifle productivity by artificially restricting innovative approaches to work tasks.

- *Labor force.* The experience level of the work force was lowered in the 1970s by a 2.6-percent annual increase in workers. By 1985 the growth rate had dropped to 1.6 percent.

- *Management.* Much of the blame was directed at managers who maximized current gains at the expense of long-range prosperity.

- *And many others.* Energy cost fluctuations, stop-and-go government economic policies, lack of capital investment, inefficient technology transfer from research labs to factories, and the like.

High inflation masks inefficiencies in plant operations. Lower inflation rates in the mid-1980s focused attention on production.

These factors impinge to varying degrees on the many production systems that comprise the U.S. economy. Several factors are beyond the control of production managers, yet they must react to them. Ultimately, the difficulties cascade down to affect operating decisions. It is at this point that productivity emerges from concept to reality—finding the better way, adopting improved technology, reducing waste, selecting the most effective equipment and processes, motivating workers, and devising productive procedures.

EXAMPLE 16.1 Profits, prices, and productivity

A highly productive process is not always a highly profitable process, although there is usually a positive correlation with success. Exceptions occur when shortages or an inflationary economy masks inefficiencies, as when any price increase tends to be accepted and when the output of an efficient process is unwanted, regardless of attractive price and quality. The conventional relation between changes in profitability, price contribution, and productivity are shown in Figure 16.2.

The center column shows how profit variations are driven by changes in revenue and cost. The top and bottom rows, respectively, link changes in product quantity and price to revenue, and changes in resource quantity and price to cost. These are familiar relationships.

Less familiar are the outside columns that indicate that productivity change is a function of changes in production quantities and resource consumption and that price contribution depends on changes in the price of products and resources. Productivity is thus the ratio of output products to input resources. Price contribution measures how well resource price increases are compensated by higher prices for products sold. When a product's price goes up faster than the cost of the resources to produce it, the profit probably climbs, even without a productivity increase; this condition is inflationary pricing.

The same relationships can be observed in formulas. Starting from the traditional equation of profitability,

$$\text{Profitability} = \frac{\text{revenues}}{\text{expenses}}$$

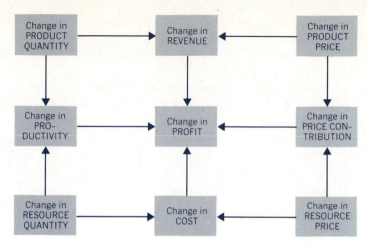

FIGURE 16.2 Interrelations of the prices and quantities of resources and products to profit. Adapted from John Parsons, "Productivity Profits and Prices," National Productivity Institute, Pretoria, S.A., 1984.

which can be expanded to

$$\text{Profitability} = \frac{\text{output quantity} \times \text{unit price}}{\text{input quantity} \times \text{unit cost}}$$

that converts to

$$\text{Profitability} = \frac{\text{output}}{\text{input}} \times \frac{\text{price}}{\text{cost}}$$

$$= \text{productivity} \times \text{contribution factor}$$

The last term in this equation is also called a *price recovery factor* to characterize an ability to pass input price increases to the customer as a higher output price.

Consideration of these formulations integrates operating efficiency with financial effectiveness, a natural perspective for engineering economists.

16-3 THE PRODUCTIVITY RATIO

The basic **productivity ratio,**

$$\text{Productivity} = \frac{\text{output}}{\text{input}}$$

can be applied to almost any human endeavor. As a measure of production efficiency, the ratio commonly takes the form of output per labor hours, with dollars or production units as the dimension of the numerator. But the ratio can be adapted to rate most production functions. For instance, the ratio could take the following forms:

$$\text{Productivity} = \frac{\text{papers processed}}{\text{labor hours}} \quad \text{or} \quad \frac{\text{materials handled}}{\text{labor hours}}$$

$$= \frac{\text{patient bed days}}{\text{staffing hours}} \quad \text{or} \quad \frac{\text{student credit-hours}}{\text{number of teachers}}$$

The preferred ratio is the one that best fits the mission, character, and resources of the organization.

An increase in productivity occurs when the output-to-input ratio rises from one period to the next. An increase in production does not necessarily lead to a productivity increase. For example, productivity rises when the output/input ratio goes from

$$\frac{120 \text{ (output)}}{100 \text{ (input)}} \quad \text{to} \quad \frac{150 \text{ (output)}}{100 \text{ (input)}}$$

If the second ratio had been 150/125, productivity would have remained the same when an increase in production was recorded. Thus the goal of increasing productivity is achieved from a continuously greater proportional surplus of output values over inputs.

What is done with a productivity increase affects opportunities for future gains. If a higher quantity of output is involved in the productivity rise, there has to be a demand for the additional output, or benefits from the gain cannot be realized. Similarly, continued gains cannot be sustained unless those responsible for the increased productivity share the resulting benefits. Therefore, distribution factors should be considered in concert with output and input.

Distribution and output

When production inputs are constant and output goes up, resources are being applied more efficiently. This condition is ideal for scarce commodities such as foodstuffs. Such productivity increases are listed as specific objectives in five-year plans for many developing countries and in policies favored by the United Nations. In instances in which consumption is ensured, the benefits from greater productivity are guaranteed, assuming that the distribution system can handle the larger load of outputs. However, higher output caused by redesigning the product could lead to an unwanted surplus if the new version failed to gain market acceptance. An equivalent situation happens in the service sector when an agency provides a new service that ups output values, which complement existing measures, but fails to satisfy consumers.

Concern about future scarcities of natural resources coupled with a growing world population leads directly to calls for higher worldwide productivity. Greater-than-proportional increases of output to input conserve resources and contribute to stable prices. When the gain from a productivity rise is shared with consumers in the form of constant or lowered prices, the consumers are more satisfied, and their increased purchasing power eventually supports the producer in the form of sustained demand and a stimulated national economy.

Distribution and input

Worker hours is one measure of input for productivity studies. A productivity rise with a constant input indicates that labor has been more effective. This greater effectiveness could have resulted from working faster, applying more efficient methods, reducing waste, organizing better, using improved tools, or

Productivity measurements are not exact readings; in essence, the total output of everything produced by an industry is divided by all the labor hours worked in that industry. Because service outputs are more difficult to quantify, productivity rates for service industries are even less precise.

Productivity gains in agriculture are indebted to large-scale government-sponsored research and extension programs (an agricultural experiment station was started in England in 1842). Interest in similar programs for industrial production is emerging.

any number of other labor–management tactics. A fair share of the surplus generated by a productivity rise should be distributed to the sources of the rise.

Workers are rewarded for productivity improvements by higher wages, bonuses, shared savings, and programs such as suggestion awards. These incentives are designed to encourage continued efforts toward maintaining and enlarging productivity gains. Rewards for higher *labor productivity* also increase the prosperity of the labor force, which ultimately supports further productivity efforts.

Investments in facilities and equipment are responsible for many spectacular leaps in productivity. Automated assembly lines, electronically controlled cutting machines, better-arranged and safer work places, equipment with higher capacities, and versatile robots have contributed to manufacturing efficiencies. These efficiences are attributed to *capital productivity*, and part of the resulting productivity gains should be set aside to finance continued investments.

Nonmonetary incentives can also increase productivity. These include employee training, more pleasant working conditions, enriched jobs, and other programs that improve workers' attitudes. These kinds of programs may be combined with direct incentive payments, but they usually rely on effective management practices to produce results. The costs of instituting the programs should be supported from productivity gains.

The various factors that contribute to "input" in the productivity ratio could be divided and subdivided into a large number of components. Subdividing is useful in analyzing costs, but the subfactors rarely exhibit independent effects. Their combined influence is similar to that found in agriculture: An additional unit of water added to a field might raise output by two bushels, another unit of fertilizer might increase it by three bushels, or an additional unit of labor might add four more bushels, to provide a total output increase of nine bushels if applied independently, but when all are applied together, the output could be increased by 16 bushels as a result of the interaction among the inputs that mutually support one another. A comparable industrial example is the combination of monetary incentives for workers, motivational programs by management, and capital investment to provide better facilities, which produces a *synergistic* relationship—the total effect is greater than the sum of the parts.

Productivity calculations provide a check for managers to evaluate the effects of new programs such as shortened workweeks.

EXAMPLE 16.2 Relationship of factors influencing productivity in a manufacturing system

The superficially simple ratio of output to input masks an intricate network of interrelated factors that complicate measurements and confuse improvement. Consider the technical factors in Figure 16.3 that influence the productivity of a production process (a corresponding diagram could be made of a government or service system). The primary *product* and *production* factor are roughly equivalent to output and input. A ratio of the amount of product to the magnitude of production is an index that relates re-

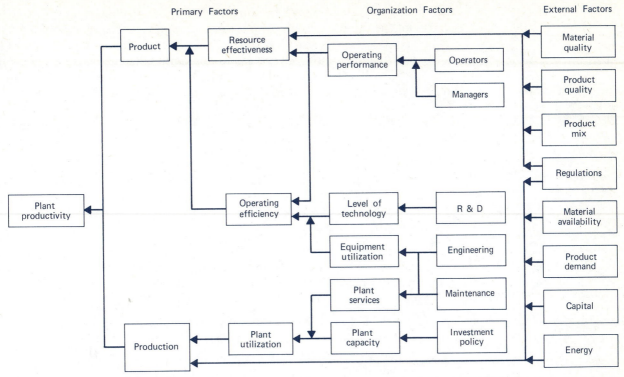

FIGURE 16.3 Technical factors that influence the productivity of a manufacturing process. Only the more prominent linkages between factors are indicated.

sources, operations, and facilities. But the relationships are not exclusive; factors that affect the quantity of output also influence the quantity of input.

The dilemma of *confounding*—the mingling together of factors—is a major obstacle in productivity studies. Confounding makes it difficult to trace the effect of a productivity change to a specific cause. For instance, an analyst might wish to investigate the influence of energy on productivity. The gross output/input ratio could be partitioned to output/energy consumed to provide a measure of the energy consumption per unit of product. This would show the relative consumption from one period to the next but would fail to reveal where and how the energy was consumed. In the preceding chapter, energy conservation is shown to be affected by operators, maintenance, type of equipment, design of facilities, and the like. Therefore, the energy productivity ratio may signal a need for concern, but it lacks the sensitivity to point out where improvements should be implemented.

Frustrations from confounded factors and distribution difficulties do not detract from the value of a productivity study; instead, they just confirm that productivity analysis must embrace the complete production system.

16-4 PRODUCTIVITY MEASUREMENT

For the reasons discussed in Example 16.2 and because organizations use productivity measures for different purposes, there is no standard formula for calculating a productivity index. Innumerable variations have been proposed to carry out the intent of the basic output/input ratio. Elaborate procedures take many pages to describe and a corresponding effort to conduct. Simplistic ones, such as airplanes produced per year to total annual labor hours, are generally too vague to be of much value. To be desirable, attributes of a productivity measurement instrument should

1. Measure both firm and operating units (ideally, the measure should aggregate so that the firm's total productivity is the sum of the productivities of its component parts).

2. Be understandable and reasonably easy to calculate—less complex formulas are usually better understood by employees and require less time for data collection (ideally, the formula should be compatible with the organization's accounting system).

3. Be accurate enough to present a realistic assessment—perfect accuracy is an unreasonable expectation and is not cost effective (ideally, the measure should provide a consistent gauge of significant operations from one period to the next).

4. Be insulated from changes in monetary values and external disruptions—factors that a firm cannot control should not distort the measure of performances it does control (ideally, an index tracks the efficiency of the utilization of key resources within the organization).

5. Stimulate motivation by associating measurements with achievable objectives—define criteria for productive performance that are modest enough to be an incentive but have standards that are a challenge for improvement (ideally, trade-off proportions among criteria would be clearly evident, as in time versus cost versus quantity versus quality).

6. Make the measurement system practical—data to fuel the system should be obtainable without excessive effort, and the mechanism for generating ratings should operate with a minimal administrative burden (ideally, the people whose performance is being measured should participate in designing the system, in collecting data, and in monitoring results, as well as in the recognition process that rewards superior productivity).

Total productivity index

A **total productivity index** is a single figure that expresses the efficiency of an entire organization. Its formulation includes an inclusive statement of the value of the product or service produced and a summary value for all of its inputs. Dollar dimensions are typically used for both the numerator and denominator to allow diverse products and resources to be expressed in equivalent terms. The familiar ratio of output/labor hours approaches a total productivity index only when the organization is extremely labor intensive.

A measure of total productivity is obtained from

$$\text{Total productivity index} = \frac{\text{product} + \text{service}}{\text{labor} + \text{materials} + \text{energy} + \text{capital}}$$

From this basic expression, adaptions can be made to represent more closely the functions of a particular organization. The intent of customizing the index is to reflect the firm's objectives. Consequently, many versions have been developed. For instance, one organization might believe that purchases of raw material represent someone else's productivity effort and should therefore be excluded from the user's input. Other firms with large material inputs might disagree that the exclusion is justified. Still another firm might have small and constant energy usage, suggesting that the energy input can be ignored in the model. An aluminum producer would, of course, feel differently.

One of many possible total productivity models has the following formula:

$$\text{Total productivity index} = \frac{\text{sales} + \text{inventory change} + \text{plant}}{\text{labor} + \text{material} + \text{services} + \text{depreciation} + \text{investment}}$$

The output and input factors are individually described and numerically demonstrated next.

SALES—NET SALES BILLED
FOR THE MEASURED YEAR

To make data comparable from one year to the next, a *deflation factor*, or a **price deflator,** must be applied to convert current dollars to constant dollars. Readily available price index tables can supply the current deflation figures. Assuming a base year of 1986 and that the measured year is 1987, a change in the general price trend of 13 percent between 1986 and 1987 would result in a general 1987 deflator of 100/113 = 0.885. Then sales of $133,333,000 in 1987 would amount to $133,333,000 × 0.885 = $118,000,000 in constant 1986 dollars.

Price deflators are published by the U.S. Department of Commerce, Bureau of Economic Analysis.

INVENTORY CHANGE—VARIATION IN THE VALUE OF
INVENTORY LEVELS BETWEEN THE BEGINNING AND
END OF THE MEASURED YEAR

Inventory includes work-in-process and finished goods. Each category of inventory could have its own deflation factor. Assuming that a deflator of 0.85 is applicable to all material categories and that the total value of all inventories on hand increased by $3.5 million during the year, the amount of inventory change has a constant dollar value of $3,500,000 × 0.85 = $2,975,000.

Plant improvements represent an increase in the worth of that production system in the total economy.

PLANT—VALUE ADDED INTERNALLY
TO THE ORGANIZATION

The capabilities of an organization are often increased by the output of its employees in ways that increase the worth of the production system but do not directly increase income through sales. Maintenance, R&D, and company-produced machinery are examples. The value contributed by plant improvement can be valued at the internal cost or market value but in either case must be deflated to constant dollars. Assume that the value of plant increased by 2.2 million constant dollars in 1987.

LABOR—WAGES AND BENEFITS PAID TO EMPLOYEES OR SET ASIDE IN RESERVE ACCOUNTS

Annual compensation for labor includes hourly wages, salaries, overtime, vacation and sick pay, insurance, profit sharing, social security tax, and bonuses. Assume that after deflating by the consumer price index, labor costs for 1987 were $35 million.

MATERIAL—RAW MATERIALS AND PURCHASED COMPONENTS CONSUMED IN PRODUCTION

After applying appropriate material deflators, assume that the net consumption for 1987 was $61 million.

SERVICES—EXTERNALLY PURCHASED ENERGY AND SUPPLIES PLUS RENTED EQUIPMENT AND CONTRACTED LABOR

Annual expenditures are deflated to constant dollars: 1987 services equals $2,500,000.

DEPRECIATION—DECREASE IN CAPITAL VALUE OF THE ASSETS THAT PROVIDE PRODUCTION CAPABILITY

Services and depreciation are sometimes considered to be an "exclusion" which is subtracted from output in the numerator of the ratio. This approach reasons that these factors represent the productivity of external operations.

Because depreciation charges are not responsive to current price trends, they are included at their current value and are assumed to be $3.2 million in 1987.

INVESTMENT—COST TO INVESTORS FOR FIXED AND WORKING CAPITAL IN THE PRODUCTION SYSTEM

Fixed capital includes land, buildings, machinery and equipment, and deferred charges. Working capital includes cash, notes, accounts receivable, inventories, and prepaid expenses. The investment input is then the rate of return that should be earned by all of the capital assets. If we assume a constant value of 15 percent for the annual cost of capital to the organization, total capital assets of $58 million, and a price deflator of 0.885, the constant dollar investment input for 1987 will be $58,000,000 \times 0.15 \times 0.885 = $7,700,000.

From these output and input factors, the total productivity index for 1987 is calculated as

EXAMPLE 16.3 Tablular format for calculating changes in total and factor productivity in manufacturing

Another version of the total productivity index is used in Figure 16.4 to analyze annual operations. The five factors in the index are deflated by their associated price indexes to provide comparable year-to-year figures. These figures are collected in different ratios to furnish insights into the organization's performance.

Labor productivity (9) relates net output (sales minus material and service expenses) to labor costs to indicate the efficiency of the work force.

Labor–capital productivity (10) is a measure of the efficiency of the combined labor and capital (represented by depreciation charges) input.

Total productivity (11) is the ratio of net sales to all inputs.

Many things can influence productivity measures in a given year. Conditions such as shortages of raw materials caused by international politics or bad weather cannot be controlled. A better perspective is obtained by viewing each year's performance

Productivity Analysis of Annual Operations

	1987			1988			
	Current Dollars (in 000s)	Price Index (1986 = 100)	Constant (1986) Dollars	Current Dollars (in 000s)	Price Index (1986 = 100)	Constant (1986) Dollars	Productivity Rise (1987 to 1988)
1. Net sales	$1831	108	$1695	$2293	122	$1880	
2. Labor	295	107	276	352	120	293	
3. Materials	880	105	838	1161	131	886	
4. Services	301	106	284	365	118	309	
5. Depreciation			96			122	
6. Total inputs (2 + 3 + 4 + 5)			1494			1610	
7. Net output (1 − 3 − 4)			573			685	
8. Labor-capital input (2 + 5)			372			415	
9. Labor productivity (7/2)			2.08			2.34	12.5%
10. Labor-capital productivity (7/8)			1.54			1.65	7.1%
11. Total productivity (1/6)			1.13			1.17	3.5%

FIGURE 16.4 Breakdown of input and output from the annual corporate financial statement used for productivity analyses. Current dollars are converted to the 1986 base-year values as (current dollars × 100)/(price index) = constant dollars. 1988 operations are compared to 1987 by the productivity rise as [1988 − 1987 constant dollars] (100%)/(1987 constant dollars). Note that total productivity increased at a lesser rate than other productivity ratios, a condition caused mostly by the steep rise in material costs.

in relation to previous years. By plotting ratios from successive accounting periods, trends may become apparent. Erratic movement from an established trend is important feedback information and should be checked to see whether the cause of the movement is controllable. A productivity ratio that moves sharply downward is an obvious cause for alarm. A major shift upward also deserves attention to determine what factors caused the improvement.

Total productivity index, 1987

$$= \frac{\$118,000,000 + \$2,975,000 + \$2,200,000}{\$35,000,000 + \$61,000,000 + \$2,500,000 + \$3,200,000 + \$7,700,000}$$

$$= \frac{\$123,175,000}{\$109,400,000} = 1.126$$

This figure has more meaning when it is compared with the index for the same organization from the preceding year. Assuming that the total productivity index for 1986 was 1.073,

$$\text{Productivity increase 1986 to 1987} = \frac{1.126 - 1.073}{1.073}$$

$$= 0.0494 \quad \text{or} \quad 4.94\%$$

Partial productivity indexes

A widely used firm-level productivity index is

$$\frac{\text{Value added}}{\text{Labor hours}}$$

where value added = raw materials − expenses paid − depreciation + inventory at beginning of period − inventory at end of period − adjustments.

Productivity measurement at the firm level or total organization level yields a figure that rates the application efficiency of all resources. It is akin to the return on equity figure which rates the utilization effectiveness of invested capital. These indexes are strategic yardsticks. Efficiency of individual operations and effectiveness of specific capital expenditures are lost in the inclusive indexes. A more valuable rating for the utilization efficiency of specific resources is obtained by calculating the **productivity index** for individual factors.

In a labor-intensive industry the productivity of workers is critical. The conventional method to determine labor productivity is to weight each product produced by its standard time and summarize the weighted values to obtain the total output; this is then the numerator that is divided by the total labor hours to establish the labor factor productivity index. A simpler approach uses payroll data. When a piece-rate pay system is in effect, the productivity of a department is measured by

$$\text{Labor–productivity index} = \frac{\text{actual pay}}{\text{standard pay}}$$

and the labor productivity of the entire plant is measured by summing the department labor–productivity indexes, each multiplied by the fraction of total plant production hours performed by each department.

The American Productivity Center was organized in 1977 by Dr. C. Jackson Grayson. It is a nonprofit, privately funded institute.

A more general approach to factor productivity is proposed by the American Productivity Center (APC). The APC measurement model starts with the calculation of **partial productivity indexes,** or measures, and combines the data from several factors to obtain a plant-level productivity measurement. The following basic relationship is designed to tie in with a firm's financial accounting system and to relate current price-weighted output and input data to a base period:

$$\text{Factor productivity index} = \frac{\dfrac{\text{current output quantities}}{\text{base output quantities}}}{\dfrac{\text{current input quantities}}{\text{base input quantities}}} = \frac{\dfrac{\Sigma O_2 P_1}{\Sigma O_1 P_1}}{\dfrac{\Sigma I_2 C_1}{\Sigma I_1 C_2}}$$

where subscripts 1 and 2 indicate, respectively, the base and current periods; O and I indicate, respectively, output and input quantities; and P and C indicate, respectively, output price and input cost.

Variations of the basic relationship are used to analyze operations from different vantage points. A **price recovery index** shows to what extent the firm has been able to absorb increases in costs of input to combat inflation:

$$\text{Price recovery index} = \frac{\Sigma O_2 P_2 / \Sigma O_2 P_1}{\Sigma I_2 C_2 / \Sigma I_2 C_1}$$

A **cost effectiveness index** reflects how costs for the current period compare with the cost relationship established for the base period (ideal costs).

$$\text{Cost effectiveness index} = \frac{\Sigma O_2 P_2 / \Sigma O_1 P_1}{\Sigma I_2 C_2 / \Sigma I_1 C_1}$$

Note that the cost effectiveness index is the product of the factor productivity index and the price recovery index. A sample application of the APC measurement models is given in Table 16.1.

Many desirable features for measuring productivity are incorporated into the APC model. Instead of using an average value for all the products produced, as is done in some models, the output is weighted according to value. For instance, an industrywide ratio of tons of coal to labor hours does not distinguish between the difference in value of bituminous and anthracite coal. Similarly, a product mix change from high-volume, low-price units to low-volume, high-value products would register as a decline in productivity unless quantity is price weighted.

It is equally important to weight inputs according to value. With a constant output, a decrease in labor hours made possible by increased capital expenditure for mechanization and greater energy use would register as a productivity increase when only labor hours are considered. However, the cost of capital and energy could be such that productivity actually declined, even as labor hours dropped. Labor input alone can vary as a result of a change in the mix of high-priced technicians and unskilled workers, even though the total hours remain constant. Without the leveling effect of price–cost weighting, changes in productivity ratios do not necessarily reflect actual accomplishments.

"Developments in Firm-Level Productivity Measurement," by J. Hamlin, AIIE Proceedings, Spring Conference, 1979.

Dr. M. E. Mundel reported that the Philippine Bureau of the Budget accused the Weather Bureau of a drop in productivity of 35 percent; it tracked 26 major typhoons in 1974 but only 17 in 1975. ("Measures of Productivity," *Industrial Engineering,* May 1976.)

TABLE 16.1

Calculation of Productivity, Price Recovery, and Cost Effectiveness Indexes

The following data describe the labor productivity for a product:

	Period 1		Period 2	
	Quantity, Q_1	Price, P_1	Quantity, Q_2	Price, P_2
Product	1000 units	$100 per unit	1200 units	$125 per unit
	Quantity, I_1	Cost, C_1	Quantity, I_2	Cost, C_2
L (1)	5000 hours	$8 per hour	5200 hours	$9 per hour
L (2)	1500 hours	$12 per hour	1000 hours	$14 per hour

The change in labor mix and costs in relation to the variations in the output quantity and price is analyzed according to the following ratios:

$$\text{Factor productivity index} = \frac{(1200)(\$100)/(1000)(\$100)}{\dfrac{(5200)(\$8) + (1000)(\$12)}{(5000)(\$8) + (1500)(\$12)}} = \frac{1.200}{0.924} = 1.3 \text{ or a 30\% increase}$$

$$\text{Price recovery index} = \frac{(1200)(\$125)/(1200)(\$100)}{\dfrac{(5200)(\$9) + (1000)(\$14)}{(5200)(\$8) + (1000)(\$12)}} = \frac{1.250}{1.134} = 1.1 \text{ or a 10\% increase}$$

$$\text{Cost effectiveness index} = \frac{(1200)(\$125)/(1000)(\$100)}{\dfrac{(5200)(\$9) + (1000)(\$14)}{(5000)(\$8) + (1500)(\$12)}} = \frac{1.500}{1.048} = 1.43 \text{ or a 43\% increase}$$

The indexes indicate that sales revenue is increasing faster than costs and that the productivity increase is more significant than the product price increase: $1.43 = 1.3 \times 1.1$.

Weighting factors obtained from production records are more appropriate than industrywide deflators. Actual price and cost changes between evaluation periods provide more accurate representations of potential trade-offs between inputs and outputs.

It is said that financial data tell you where you are and that productivity trends tell you where you are going. Productivity ratios do reveal the effectiveness of resource utilization, but they do not point out specific operations that need attention, nor do they reveal what actions are necessary to improve productivity. Future developments in productivity measurement should alleviate these weaknesses.

Work-unit productivity indicators

Productivity-improvement effort is focused on the people who actually produce the output, by the direct measurement of their performance. Productivity indicators differ from the standards set for accomplishing a given amount of work within a given time. Team indicators measure the characteristics of an operation or project carried out by a group of people. The composition of the groups may cross traditional organizational boundaries. The criterion for inclusion is that all members of the group are producing a definable end product or accomplishment—maintenance of certain production facilities, operation of a warehouse, an administrative function such as accounting, fabrication of a subassembly, the issuing of licenses, or completion of one stage of production. The size of the group naturally depends on the situation.

To be effective, productivity indicators must be readily understandable, easily measured, competently administered, and acceptable by those being measured. Most employees are apprehensive about having their output formally and critically evaluated. Being group instead of individual oriented, team measurements are usually considered less threatening. However, they still must be reasonable indicators of performance quality.

No single parameter can capture the gist of productivity. Because productivity results from a complex interaction of a number of parameters, many of which are interrelated, productivity indicators must be selected carefully. Omitting a significant indicator, or including an unimportant one, lowers the credibility of the whole measurement exercise. Examples of indicators applied to research and development activities, an exceedingly difficult area to measure, are presented in Example 16.4.

The worth of measuring—**work-unit productivity indicators**—is ultimately decided by the management action it provokes. Benign indifference undermines its value as surely as does disciplinary overreaction.

EXAMPLE 16.4 Productivity indicators for R&D

Hughes Aircraft Company sponsored a five-year study of the productivity of research and development.[1] According to the study, "*Efficiency, effectiveness,* and *value* are the key factors related to productivity. Work can be efficient but highly ineffective and of little or no value; conversely, work can be effective and valuable but grossly inefficient."

Productivity in R&D is an elusive subject because it involves so many variables: individual abilities and work habits, vague standards of performance, erratic schedules, changing goals, etc. However, some basic tenets were identified to foster productivity improvement:

• Improvements in any organization are "there for the asking," but few take advantage of them.

[1]*R & D Productivity* (Culver City, Calif.: Hughes Aircraft Co., 1978).

• There are underutilized resources in every organization and individual.

• Higher productivity results from greater involvement by management.

• Seemingly small individual improvements add up to significantly increased productivity, and this can frequently have a large impact on profits.

Quantitative productivity indicators suggested by the study include the following:

1. Profits generated per R&D dollar spent.
2. Dollar value of proposals won versus dollars spent to secure them.
3. Number of errors detected per square foot of drawing.
4. Number of changes per drawing per year.

5. Ratio of staff personnel to line personnel.

6. Secretarial support ratios—managerial, scientific, and engineering.

7. Absentee rate.

8. Employee voluntary turnover rate.

It was generally agreed that productivity evaluations should be conducted at frequent intervals by responsible line managers.

16.5 PERFORMANCE MEASUREMENT USING THE OBJECTIVES MATRIX

Many organizations resist measuring productivity for such reasons as distrust of accuracy, wariness of more paperwork, and lack of qualified measurers. These are flimsy excuses. Accuracy results from measuring the right things in the right way, when the right things are controllable factors and the proper way is conscientiously. Selecting appropriate, readily observable factors to measure and reporting the results through existing communication channels increases knowledge flow, not paper flow. Assuming that the current work force is conscientious, training and supervisory support will generate consistently correct numbers without bloating the bureaucracy.

A productivity ratio has many forms because organizations produce many different goods and services from many different resources. Larger organizations with a greater diversity of outputs and inputs obviously have more complicated ratios. Because the productivity of a large organization is determined by the performance of its constituent parts, improvement depends on the collective accomplishments of smaller organization units. These smaller units must, therefore, be included in the productivity measurement system.

I created the objectives matrix in the mid-1970s to measure the productivity of work units in hospitals. Since then, many versions have been developed. It is best known for the measurement of white-collar productivity.

A measurement method called the **objectives matrix (Omax)** is particularly appropriate to basic work units such as crews, departments, and staffs, although it can also represent a complete organization. It can be conveniently applied to knowledge-based activities that are considered difficult to measure, as well as skill-based work that can be metered by more conventional measures. It has been favorably received in manufacturing, service, and government sectors by both large and small organizations.

The theory behind the objectives matrix is that productivity is a function of several performance factors, each with distinct dimensions that vary among work units, and the most practical way to assess unit productivity is to measure the most influential factors. It is commensurately reasonable to assume that the productivity of larger organizational units can also be represented by an appropriate collection of performance factors. However, figures obtained from measuring the performance of the component units of the parent organization cannot be simply added together to indicate the parent's productivity. A sophisticated weighting system would be required. Therefore, a more practical procedure is to measure directly the performance factors that collectively indicate the productivity of each unit of interest in the organizational hierarchy, regardless of size.

A 600-bed hospital might have an Omax for the whole hospital, one for each major department, and several for work units within departments.

16-6 FORMAT AND FUNCTIONS OF THE OBJECTIVES MATRIX

Most of the desirable features for a productivity measurement system listed in Section 16-4 are accounted for in the objectives matrix. It employs a multidimensional format that accommodates employee involvement and managerial leadership to produce a single measurement that represents the productive performance of an organizational unit.

The structure of the matrix is shown in Figure 16.5, in which the different divisions are identified by letters. The functions associated with each component are briefly described in the figure and will be examined more closely next.

By definition, a matrix is "that within which something originates, takes form." In Omax, a profile of a work unit's total performance takes form.

Productivity criteria

Most employees know how their activities affect the productivity of their work unit. They realize which of the functions that they perform support the organization's output, as opposed to those that are incidental to production or are simply customary behavior. These significant functions are the criteria of productive performance. Defining these criteria would be valuable even if the matrix development went no further, because they clarify performance expectations and explore the diverse factors in achievement.

Different groups of workers share a set of work characteristics that distinguish their contribution to the organization's productivity. What contributes most to an office may not be as important to a machine shop. Activities of consequence to productivity differ between supervisors and their subordinates. Yet a few general categories encompass most of the factors. The six broad classifications given in Table 16.2 encompass most of the productivity criteria pertinent to nonmanagerial positions.

TABLE 16.2
Generic Classes of Productivity Criteria

Six Generic Categories That Define a Person's, Work Unit's, or Total Organization's Contribution to Productivity

1. *Quantity*—the number of items produced or a measure of the service provided (output maximization).
2. *Quality*—precise or inferential indicators of the quality of goods or services produced (customer satisfaction).
3. *Timeliness*—the extent to which activities or functions are completed on schedule (elimination of delays).
4. *Yield*—degree of efficiency of the transformation process (input minimization and waste avoidance).
5. *Utilization*—the effectiveness with which critical resources are utilized (availability of key people and machines).
6. *Group traits*—individual and organizational properties that contribute to productive performance (such as safety, turnover, and absenteeism).

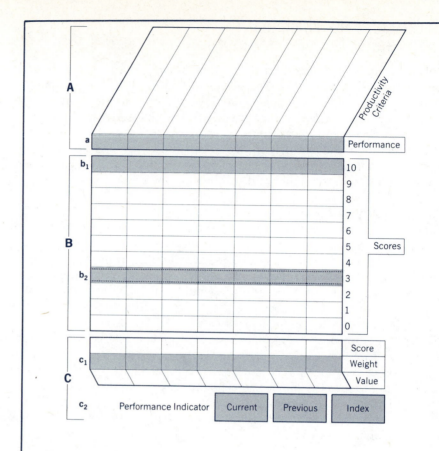

A *DEFINING*—The factors that determine the performance of a work unit are identified as *productivity criteria* and stated as ratios.

a A work unit's actual accomplishments during an assessment period are registered in the *performance* row.

B *QUANTIFYING*—The body of the matrix is composed of 11 levels of achievement, ranging from a *score* of 0 for unsatisfactory performance to 10 for superior accomplishment.

b_1 The prevailing level of performance when the matrix is initiated is considered to have a score of 3 for all criterion ratios.

b_2 Realistic performance objectives for the work unit to strive for during a prescribed period have a score of 10.

C *MONITORING*—A *performance indicator* is the sum of *values* obtained by multiplying each criterion's *score* by its *weight*. The *index* is the percentage difference between the *current* and *previous* performance indicators.

c_1 Importance weights are assigned to all criteria to show their relative impact on the work unit's productivity objectives.

c_2 An indication of work-unit productivity is provided by the rate of change of the performance indicator.

FIGURE 16.5 Basic structure of an objectives matrix with explanations of the functions served by the components identified by letters.

Actual entries in the generic classes given in Table 16.2 take the form of ratios. Two or three ratios may be required to account for one category, such as quality, or a category may be ignored if it does not affect productivity in a given situation. Because most white-collar and knowledge-worker positions do not have readily distinguishable outputs, their ratios are measurable modes of behavior that are known to affect their collective output. Typical ratios for various categories applicable to different work situations are given in Table 16.3.

Acceptance by employees, simplicity, and data availability are important considerations in selecting ratios.

Each organizational unit may have a different set of productivity criteria. Only those factors that are pertinent to the unit's mission should be included, but the collection should encompass all the necessary characteristics of productive performance. **Performance indicators,** or criteria, should be selected according to how well they represent the service performed by the unit. The ratios must be thoroughly defined for consistent measurement.

Performance scores

Performance scales in the body of the objectives matrix run from 0 to 10. There are thus 11 levels of accomplishment for each criterion: a single criterion occupies a column that stretches from the top to bottom of the matrix. Levels of accomplishment extend across the body of the matrix, as indicated by the rows marked from 0 through 10. The assignment of the results expected at each level is the crucial part of scaling, because the results set specific hurdles that reflect the accomplishment of a work unit's productivity objectives. The scale is anchored by designated numbers at three levels:

- *Level 0.* The lowest level recorded for the criterion ratio over a recent period of time, say, the last year, in which normal operating conditions existed; nominally the worst ratio reading that might be expected.

- *Level 3.* Operating results indicative of performance proficiency at the time the rating scale was established; current ratio reading at the time measuring is initiated.

The objectives matrix gets its names from the level 10 goal-setting process. The two most important lines in the matrix are the score 10 line and the weight line.

- *Level 10.* A realistic estimate of results that can be attained in the foreseeable future, say two years, with essentially the same resources that are now available; a stimulating productivity objective.

Level 0 and **3 scores** are clearly defined benchmarks. Level 10 is the challenge. An overly optimistic objective may later prove to be discouraging by its unattainability, and a conservative goal may inhibit motivation if it is too easily achieved.

The scoring columns for all the criteria are completed by determining the appropriate entries for the cells between the benchmark levels. These are often equal-interval scales. In the first column of the matrix in Figure 16.6, the ratio of equivalent units of output per hour increases linearly from 10 at level 3 to 17 at level 10 and decreases at the same rate from level 3 to 7 at level 0. Thus minigoals or hurdles are set for each score. For this quantity column, each hurdle is $(17 - 7)/10 = 1$ equivalent unit per hour.

Objective measures should be used as much as possible. Subjective scales are most commonly used to evaluate quality, especially the quality of services.

TABLE 16.3

Examples of Performance Ratios Frequently Encountered in Manufacturing Industries, Government and Service Sectors, and Work Groups in Any Organization

Ratios Associated with Manufacturing

$$\frac{\text{Number of units produced}}{\text{Labor hours}} \qquad \frac{\text{Hours of rework required}}{\text{Units produced}}$$

$$\frac{\text{Equivalent vehicles unloaded}}{\text{Labor hours}} \qquad \frac{\text{Weight handled or loaded}}{\text{Labor hours}}$$

$$\frac{\text{Number of defects}}{\text{Total units produced}} \qquad \frac{\text{Machine operating hours}}{\text{Total possible machine hours}}$$

$$\frac{\text{Pounds of waste}}{\text{Total pounds processed}} \qquad \frac{\text{Finished product weight}}{\text{Raw material weight}}$$

Ratios Associated with Services

$$\frac{\text{Number of customer complaints}}{\text{Orders delivered}} \qquad \frac{\text{Number of data entry errors}}{\text{Total data lines written}}$$

$$\frac{\text{Errors made on policies}}{\text{Number of new policies}} \qquad \frac{\text{Number of drawings}}{\text{Total drafting hours}}$$

$$\frac{\text{Number of pages completed}}{\text{Employee day}} \qquad \frac{\text{Number of late deliveries}}{\text{Total number of deliveries}}$$

$$\frac{\text{Number of customers}}{\text{Number of employees}} \qquad \frac{\text{Project planned cost}}{\text{Project actual cost}}$$

$$\frac{\text{Investigations completed}}{\text{Investigator hours}} \qquad \frac{\text{Department expenses}}{\text{Total hours worked}}$$

Ratios Associated with Work Groups

$$\frac{\text{Hours missed from irregular absences}}{\text{Total possible work hours}} \qquad \frac{\text{Overtime hours}}{\text{Regular hours}}$$

$$\frac{\text{Number of appointments missed}}{\text{Total appointments}} \qquad \frac{\text{Number of quits}}{\text{Average group size}}$$

$$\frac{\text{Hours lost to accidents}}{\text{Total paid hours}} \qquad \frac{\text{Number of calls handled}}{\text{Total operator hours}}$$

$$\frac{\text{Number of orders processed}}{\text{Number of department hours}} \qquad \frac{\text{Volume of case loads}}{\text{Employee days}}$$

In the customer-rating column, which is a subjective scale of quality, the entries conform to the scoring scale shown on the right side of the matrix. When a 0-to-10 subjective scale is used, each level of the scale should be carefully defined to ensure that the same score is given each time the same level of performance occurs.

Nonlinear scales are appropriate when it is progressively more difficult to

Equivalent units / Labor hours	Actual cost / Budgeted cost	On-time Percentage	Customer rating	Errors / Output	Injury cost / Hours worked	Hours worked / Hours paid	
							Performance
17	(—)	98	10	0.5	0.1	0.90	10
16	1.00	97	9	1.0	0.2	0.89	9
15	1.02	96	8	1.5	0.3	0.88	8
14	1.05	94	7	2.0	0.4	0.86	7
13	1.09	92	6	2.5	0.5	0.84	6
12	1.14	90	5	3.0	0.6	0.81	5 Scores
11	1.20	88	4	3.5	0.7	0.78	4
10	1.27	85	3	4.0	0.8	0.75	3
9	1.40	81	2	5.0	0.9	0.71	2
8	1.60	76	1	6.0	1.1	0.66	1
7	1.80	70	0	7.0	1.3	0.60	0
							Score
13	20	14	22	16	7	8	Weight
							Value

Performance Indicator | Current | Previous | Index

FIGURE 16.6 Sample objectives matrix based on frequently used productivity criteria. The middle column is an evaluation of quality derived from subjective opinions of customers. Such rating scales are developed according to the practices described in Section 5-8. Considerations in developing each of the criteria are as follows:

Equivalent units of output are developed when there is no readily countable product. A unit may be a package of services.

Actual versus budgeted costs can be applied to materials, energy, labor, or a collection of resources.

Actual versus expected times pertain to any output or function of the work unit.

The customers who provide the ratings may be outside buyers or in-house clients. Results of several surveys may be combined into one scale.

Errors can take the form of rejects, misdeeds, complaints, inaccuracies, or any type of mistake that wastes resources.

Safety cost, a factor in most manufacturing matrices, is often replaced by turnover cost in a service organization.

An available-hours ratio may be directed at absenteeism, overtime, or other personnel utilization factors.

Adapted from "Monitoring With A Matrix That Motivates As It Measures Performance" by James L. Riggs, *Industrial Engineering*, January, 1986.

reach equal-step hurdles. As shown in the second column of the matrix in Figure 16.6, the numerical distance from each intermediate score to the next grows smaller as scores get higher. This pattern recognizes the increasing difficulty of finding additional savings after each successive cost reduction is achieved.

The quality column—errors per equivalent unit of output—has both linear and nonlinear sections. The issue concerns less the structure of the scale than its appropriateness and how well it is understood by those whose performance it registers.

For any criterion, the scaling may differ among work units according to the size of the unit and the nature of the work. Even for identical units, the scales may vary according to their objectives. Divisions of the scale are of no great consequence because it is the rating earned from one period to the next that measures gains or losses in each performance criterion. The important consideration is that work-unit members agree that the scales are meaningful and fair.

Scores, weights, values and indicators

All the criteria of productive performance do not have equal effect on the overall productivity of the work unit. Assigned weights, 100 points distributed among the criteria, reflect management's perceived contribution of each criterion to the total organization's productivity objectives.

Weights are usually determined by the managers of the work unit represented by the matrix.

Weight assignment is not a trivial undertaking. It provides an opportunity to direct attention to activities that have the greatest potential for improving productivity. For instance, if reducing the amount of wasted raw material is a problem, the "waste-reduction" criterion would be weighted heavily. A weight for waste reduction twice as large as "output per hour" encourages material conservation, perhaps at the expense of lowering output to avoid scrap. Ambitious groups naturally concentrate on the criteria that tender the most recognition.

The final phase ties together criteria scores and weights to determine a performance index. Data for ratios are collected periodically—once a week, month, or quarter—depending on the use of the monitoring system. Results are entered on the *performance* line of the matrix and translated into scores according to the rating scale of each criterion. Because each scoring step is a "minigoal," performance must equal or exceed the number associated with a given score. That is, if the rating scales for scores of 3 and 4 were 100 and 110, respectively, a performance of 107 would yield a score of 3 because it is above the 100 hurdle but has not met the minigoal of 110 to earn the higher score. Scores are entered in the *score* line and are multiplied by the *weight* immediately below each score to complete the *value* row. The procedure is demonstrated in Figure 16.7.

The rule for scoring is to award the lower score when the number in the performance line lies between two scoring levels.

The sum of the numbers in the value row is entered in the first box below it. This is the *current performance indicator*. It is a single number that represents the composite performance of the work unit or organization being monitored. An index of performance is calculated by dividing the difference between the current and *previous performance indicator* by the previous indicator. This per-

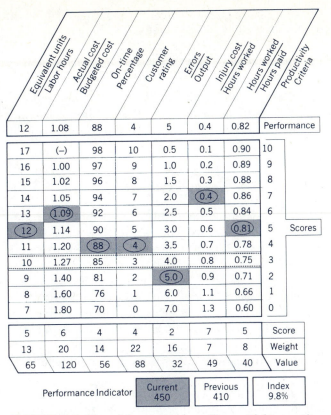

FIGURE 16.7 Use of the matrix from Figure 16.6 to monitor performance. Operating data in the performance row are converted to scores in the body of the matrix, as shown by the circled numbers, which are multiplied by their respective weights to give products that are summed to obtain the current performance indicator. The percentage change from the previous indicator is the index for the latest evaluation period.

centage is entered in the *index* box and represents the work unit's productivity during the evaluation period.

A series of performance scores must be collected before they can become truly useful. The first one is simply a score. What counts is the rate of change from one period to the next. The productivity index for a period is calculated as

$$\text{Productivity index} = \frac{\text{performance during the current period minus the previous period's indicator}}{\text{previous period's performance indicator}} \times 100\%$$

From the matrix in Figure 16.7, in which the new performance indicator shows a gain of 40 points over the reading obtained during the previous period, the improvement is displayed as

$$\text{Productivity index} = \frac{450 - 410}{410} \times 100\% = 9.8\%$$

The perspective obtainable by continuous indexing is destroyed by changes in the rating scale or weights. The maximum usefulness results from correctly setting up the matrix the first time and continuing to apply it for several periods.

16-7 OMAX APPLICATIONS

No measurement system can be implemented without careful planning and a commitment from upper management to make it work. Dedication is especially important to an Omax application because it relies on extensive input from the people in each organizational unit for which a matrix is constructed.

An exercise that encourages upper-level commitment is to have managers develop a master matrix for their plant or major division. This matrix is a ratio-oriented business plan. It gives dimensions to the business unit's mission. Level 10 states the goals, but unlike the matrices for the work units, the master matrix often uses ultimate objectives, such as achieving zero defects and zero late deliveries. These objectives serve as guidelines for matrix development lower in the organization.

It is usually advisable to introduce Omax gradually, but even a confined introduction should be thorough. Friendly work units should be selected for the trial. Union or employee representatives should be brought in early and kept informed.

The first work-unit meeting should be upbeat, stressing the good things that should result from improving productivity and instituting a measurement method that credits the people who produce. The promise of recognition is particularly appealing to staff and administrative personnel who feel that line units receive more than their legitimate share of credit because they are rewarded for meeting output goals, whereas workers whose output cannot be readily measured are ignored.

Considerations that expedite the construction of an objectives matrix include the following:

- Not all the ratios on the master matrix will be represented on a work unit's matrix. A master matrix naturally has more ratios than does a work unit because the larger organization has a more complex mission. A single generic criterion, such as quality, may require more than one ratio.

- The criteria should represent conditions and activities that are essentially controlled by the work unit. However, it should be realized that no group is completely independent. Each relies on inputs from the other units and suppliers, as well as being subject to such external factors as emergencies and production fluctuations.

- At least one criterion should represent the unit's customers. In service organizations, output quality is often measured by a customer evaluation. A 10-point scale for the evaluation should be carefully crafted in order to ensure consistent scoring. It helps to have customers involved in the scaling.

- The interrelation between criteria must be considered in setting level 10 objectives. A goal to reach a new high in units produced per hour may be reasonable only if quality expectations are relaxed to allow, say, 5 percent rejects. A lower output goal would have a correspondingly lower reject percentage.

- When conventional work standards are included in the matrix, the accepted standard of performance is given a score of 5.

After the members of a work unit have presented their completed matrix for management review and in turn have had the management-assigned weights justified to them, the improvement phase begins. Although the purpose of measuring performance is to improve it, measures alone do not reveal how to make improvements. With help in identifying and pursuing promising possibilities, work units can undertake significant projects. Sometimes they score impressively.

Improvements show up as higher scores in the matrix. They also boost morale and provide cost reductions or revenue increases. These achievements deserve recognition. There are several well-known gain-sharing plans, and a few companies have designed their own plans based on objectives matrix scores. A lump-sum award can be split among members of a work unit when their effort produces special economic gains—a promptly paid bonus spurs more activity.

Nonmonetary—meaning noncash—awards can take many forms, ranging from mementos to elaborate entertainment. All awards should be shared by the entire group represented by the matrix, in order to build cohesiveness and motivation.

The level of proficiency among work units cannot be compared directly according to their performance-indicator numbers unless identical matrices are appropriate. Otherwise, some units may set modest goals and may consequently have high performance indicators. Each unit should therefore be judged by the rate of increase (or decrease) of its performance indicator. This provides a customized productivity index for each unit.

Charts are frequently drawn in order to display current performance records and to monitor progress over several time periods. The latest matrix may be displayed by posting a copy in the work area, showing it on a TV monitor, or exhibiting it on large placards hanging from the factory ceiling. Computerized

Standardized objectives matrices may be developed for units that do essentially identical work. Motivational advantages are lost by standardization, but comparability is gained.

An example of a progress chart for Omax scores is shown in Figure 16.11.

trend charts may track the performance of individual criteria as well as of the composite indicator. Visual reporting provides potent feedback.

EXAMPLE 16.5 Construction and implementation of an objectives matrix for an engineering design unit

An objectives matrix is more than simply a scorecard. Although the format has tic-tac-toe simplicity, its application recipe includes ingredients of acclaimed management practices. Its development follows time-management advice to identify goals, assign priorities, construct to-do lists, and track progress. Its versatility is also a valued feature. A weighted indicator method is perhaps most attractive for nonmanufacturing activities, for which measurement is generally more difficult. When engineered standards are available, the ratio of *actual time to standard time* usually has a heavy weight. When standards are unavailable,

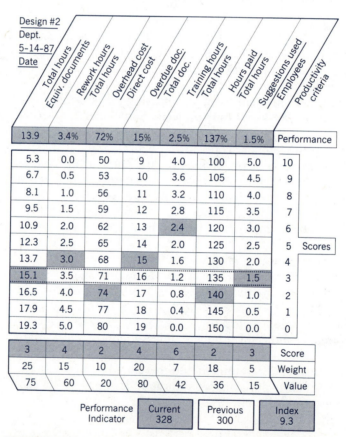

Design #2 Dept. 5-14-87 Date

Total hours / Equiv. documents	Rework hours / Total hours	Overhead cost / Direct cost	Overdue doc. / Total doc.	Training hours / Total hours	Hours paid / Total hours	Suggestions used / Employees	Productivity criteria
13.9	3.4%	72%	15%	2.5%	137%	1.5%	Performance

5.3	0.0	50	9	4.0	100	5.0	10
6.7	0.5	53	10	3.6	105	4.5	9
8.1	1.0	56	11	3.2	110	4.0	8
9.5	1.5	59	12	2.8	115	3.5	7
10.9	2.0	62	13	2.4	120	3.0	6
12.3	2.5	65	14	2.0	125	2.5	5 Scores
13.7	3.0	68	15	1.6	130	2.0	4
15.1	3.5	71	16	1.2	135	1.5	3
16.5	4.0	74	17	0.8	140	1.0	2
17.9	4.5	77	18	0.4	145	0.5	1
19.3	5.0	80	19	0.0	150	0.0	0

3	4	2	4	6	2	3	Score
25	15	10	20	7	18	5	Weight
75	60	20	80	42	36	15	Value

Performance Indicator — Current 328 — Previous 300 — Index 9.3

FIGURE 16.8 An Omax developed and implemented by an engineering design group that serves other departments in the production system.

the criteria that represent accomplishment of service functions are identified and weighted according to their impact on related functions.

A design team in an engineering department is represented by the matrix in Figure 16.8. The group provides drawings, diagrams, and associated computer printouts to other production departments. It is composed of knowledge- or skill-based workers who seldom generate easily identified products. Their output is information, and it is delivered in several ways.

The productivity criteria are typical for a service unit. They focus on the behaviors and activities that promote customer satisfaction. In this example the customers are the line and staff departments in the same manufacturing organization. The ratios that quantify the productivity criteria have the following interpretations:

$$\frac{\text{Total hours}}{\text{Equivalent documents}}$$

The numerator is hours chargeable to the production of documents. Because reports, drawings, and diagrams vary widely in complexity, a formula equates each to a standard document.

$$\frac{\text{Rework hours}}{\text{Total hours}}$$

The degree of rework required is a measure of quality. A comparable measure would be customer complaints per document.

$$\frac{\text{Overhead cost}}{\text{Direct cost}}$$

This ratio indicates the proportion of management and organizational support burden required for direct design operations. The intent is to reduce overhead expense.

$$\frac{\text{Overdue documents}}{\text{Total documents}}$$

Late deliveries are apparently a problem, or this ratio would not have been included.

$$\frac{\text{Hours paid}}{\text{Hours worked}}$$

The intent of this measurement is to reduce overtime and absenteeism. Excused hours for vacation and such are not included in the numerator. The denominator is the actual on-the-job hours for the period. If the unit were a profit center, the numerator might have been billable hours.

$$\frac{\text{Training hours}}{\text{Total hours}}$$

Training is included as a ratio because it is deemed important to the sustained performance of knowledge workers.

$$\frac{\text{Suggestions used}}{\text{Employees}}$$

This ratio suggests the interest and involvement of employees in their work. The goal is to obtain one usable suggestion per month from every 20 employees.

The scales for all the criteria except hours/document are given in percentages. For instance, the initial amount of time devoted to training was about one-half hour per 40-hour week $(0.5/40 = 1.25\%)$ which gave a rating of 1.2 at level 3. The performance objectives for a score of 10 are based on the group's improvement goals for the next 18 months, assuming that the mission and resources of the unit remain essentially unchanged.

The measurements in the *performance row* are for the first month of the improvement period. Members of the work unit who developed the ratios may also carry out the measurement process for monitoring their progress. The performance measures translate into the scores indicated by the shaded cells of the matrix and are entered in the *score row*.

Managers from the engineering department and the departments it serves establish the criterion weights that should remain applicable for the 18-month evaluation period. If conditions change, such as the acquisition of a new CAD system or the ad-

dition of new design responsibilities, the matrix will probably have to be revised, including new weightings.

As shown in the performance indicator boxes, the design group recorded a 9.3-percent gain in the first month. This gain results from the sum of values of the weighted scores (328) being compared with the 300 score which is, of course, the value of the *previous* performance indicator for every *first* measurement period in a new Omax application.

16-8 AWARENESS-IMPROVEMENT-MAINTENANCE (A.I.M.) PROCESS

In recent years there has been a great outflow of advice about what American industries should do to escape the productivity doldrums. "Follow the Japanese example" was a popular cry. Managers were urged to wander about, think further ahead, and to seek consensus decisions. Robots and a host of computer-aided devices were pushed as high-tech solutions. Zero defects and zero inventory were the proclaimed goals of production managers.

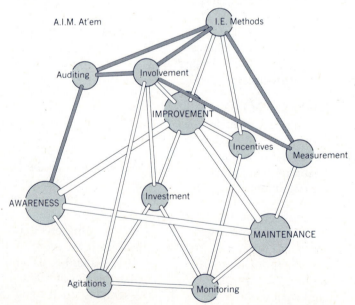

FIGURE 16.9 Relationship of phases and techniques featured in the Awareness-Improvement-Maintenance (A.I.M.) process.

Successes and failures have been recorded for all productivity-inducing tactics. Quality circles thrived in some companies and wilted in others. Incentive wages were introduced with mixed results. Smoother work flows and tighter material controls usually boosted output and reduced waste, but not always. Most quick-fix productivity promotions failed. These mixed results leave many managers wondering what to do.

A three-phase process designed to foster the successes is known by the acronym **A.I.M.** It recommends events to create *Awareness*, activities that cause *Improvement*, and measures for the *Maintenance* of gains. There is nothing new in the sequence. In fact, its phases should seem familiar because they have long been endorsed by management experts to improve operations. A.I.M. is distinguished, however, by its narrow focus on the fundamental output/input ratio. It attracts attention to the importance of using resources efficiently, seeks measurable gains, and implements methods to sustain progress.

The "A.I.M. atom" in Figure 16.9 provides a framework for evaluating various paths to productivity. Surrounding the *awareness–improvement–maintenance* nucleus are implementation elements with connecting bonds that indicate opportune relationships. For instance, the shaded bonds between *audit*, *involvement*, industrial engineering (*I.E. methods*), and *measurement* suggest that these activities are mutually supportive.

A.I.M. embodies the entire contents of this text. In effect, it is a restatement of the production phases shown in Figure 1.2.

First phase: Awareness

To become more productive, workers must want to do so. New tools and streamlined methods add nothing except cost unless they are used properly, and this demands effort by the users. Even a modest change requires willingness to adapt and prevail. The first step toward reform is to convince people that improving productivity will benefit them personally.

Overcoming the fear that advances in productivity necessarily lead to unemployment is essential. It is not enough to point out that jobs have historically increased in the more productive sectors of the economy. This observation is too impersonal. Employees should be assured that their own jobs are not at stake. They should also realize that productivity gains can be secured from actions other than labor reductions, such as reducing scrap and conserving energy. Then they are more likely to accept the other virtues of productivity growth.

Convincing employees that they should push productivity is the prime function of the awareness phase. A simple explanation of what should be done and the value of doing it may be enough to gain full backing. On other occasions, cynicism bred by unmet expectations from previous campaigns must be overcome. Because the composition and attitudes of work forces vary so widely, the seemingly simple act of starting a productivity push deserves careful planning. A clumsy kickoff can damage all subsequent moves. Two start-up tactics are suggested: agitation and auditing.

AGITATION

To agitate is to excite. Getting everyone excited about productivity is the epitome

A big-bang opener soon fizzles when insufficient resources are programmed for the continuation or if the next phase is delayed.

of awareness. Imagination is the only limit to discovering ways to attract attention. Possible tactics are mass meetings with famous speakers, contests, group gatherings to solicit pledges, retreats, committees formed to develop team competition, posters and published announcements, morning meetings for exercise and motivational talks, and so on. However, too much hoopla can cheapen the start-up, especially if it has been preceded by similar extravaganzas for safety or zero defect promotions.

A lavish kickoff can churn up short-lived excitement that dissolves before much is accomplished. Productivity growth relies on sustained effort. A campaign that gradually builds momentum is likely to endure longer, but it still needs enough visibility to be properly launched.

Clear evidence of management commitment is a powerful stimulant. Workplace innovations and installation of new equipment are ideal occasions on which to launch a productivity drive, because they confirm that the organization is willing to invest in the existing work place and work force.

A definite distinction must be apparent between a major productivity drive and regular cost-reduction and suggestion programs. The search for operating economies is supposed to be a continuous process in which everyone participates. In practice, however, the search is typically spearheaded by a staff group that generates periodic spasms of activity. These routines are easily distinguished from a concerted productivity drive when the drive's breadth, backing, and long-term schedule are acknowledged.

AUDITING

A **productivity audit** delivers a message about an organization's current state and its dedication to improving its productivity. An audit that surveys the work force can concurrently extract information about work conditions and practices while it invokes a communal spirit.

As opposed to managerial audits, which collect financial and operational data, or attitude surveys, which indicate how employees feel about their jobs, a productivity audit questions resource-utilization efficiency, identifies promising areas for improvement, and promotes employee involvement. An audit can be a one-page questionnaire or an exhaustive written examination followed by a battery of interviews, or anywhere between. Characteristics of a productivity diagnostic are discussed in Example 16.6.

EXAMPLE 16.6 Causes and effects of productivity auditing

A diagnostic audit examines the factors that retard or stimulate productivity in an organizational unit. It is ideally designed to examine the most serious issues affecting each unit. For most organizations, however, the cost of a customized audit is prohibitive, and so standardized audits are often used. An advantage of standardization is the opportunity to compare results from different units within an organ-

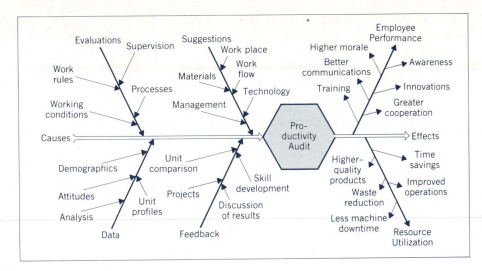

FIGURE 16.10 Contributors to high productivity (causes) and the results of productivity improvements (effects) displayed in a C&E diagram. This type of information is obtained from a productivity audit.

ization on an equivalent basis and to make similar comparisons among organizations. Relations among causes and effects are shown in Figure 16.10.

A productivity audit is not an employee attitude survey. An attitude survey is usually sponsored by the personnel or human-resources department to find out how employees feel about their wages, supervision, fellow employees, work situation, and other factors that affect job satisfaction; they are often asked to compare present conditions with previously experienced conditions and with those found in other organizations. Although a productivity audit may also seek employee perceptions regarding human relations and employment benefits that affect morale, and consequently may affect personal productivity, the main issues are work conditions, methods, facilities, and environment. An audit concentrates on factors that influence the quantity, timeliness, and quality of the output. The immediacy of an audit is apparent from the following sample of typical questions:

• *What are the most serious problems that inhibit production*—absenteeism, turnover, accidents, slowdowns,

equipment downtime, reject rates, poor quality of supplies or raw materials, and so forth?

• *What characteristics of operations contribute to lower-than-expected performance*—inattention, horseplay, lack of enthusiasm, make-work, no respect for quality, poor work flow, tardiness, wrong assignments, too many delays and disruptions, and so forth?

• *What opinions do employees have about working conditions*—adequacy of facilities (such as crowding, lighting, ventilation, and noise), condition of equipment, sufficient conveniences (such as cafeteria, parking, in-plant transportation, restrooms, lockers), availability of modern machines and tools, and the like?

Based on the answers to such questions, an organization should recognize its current strengths and weaknesses, be able to set reasonable objectives, and be in a position to decide how to move ahead. An audit also sets the stage for introducing productivity measurement and any other program that fosters productivity, by increasing the employees' awareness of the relation between productivity and prosperity, especially when management makes known the results.

Several features of the auditing process substantiate its value for elevating the awareness of productivity issues. When members of individual work units get together to complete a survey and to listen to the associated explanation of what is to be accomplished, unity is fostered. Everyone starts thinking about productivity. Feedback from the analyzed surveys and responses to suggested changes can prove that management is serious. Resources are being committed. If managers listen to what the employees are saying and encourage further participation, the stage is set for significant improvement.

Second phase: Improvement

The phases of awareness and improvement tend to blend together. Awareness of the benefits of productivity growth is reinforced as gains are recorded from improved operations. Similarly, early demonstrations of easily implemented improvements generate enthusiasm for more accomplishments. But such gains are not automatic. Even the most willing work force needs direction.

Four improvement paths were mapped in Figure 16.9: *investment, incentives, involvement,* and *I.E. methods.* Which paths to pursue most vigorously depends on the nature and condition of the organization. A prosperous manufacturer might gain most from a sophisticated new CAM system, whereas a struggling service company could neither afford nor benefit from such an equipment investment.

INVESTMENT

Spectacular turnarounds have been attributed to plant modernizations, notably the adoption of advanced technology. Stodgy companies supposedly rocket almost overnight to lofty levels of prosperity and productivity when they become computer assisted. Enough success stories have circulated to give credence to the claim that the future lies in high technology. There are also less-publicized disappointments.

The proper economic justification methods should be applied to all investment proposals. The *first rule* is to be objective. Such emotional considerations as puffed-up pride from buying the latest robot should not enter this evaluation unless their monetary value can be assessed. The *second rule* is to conduct the economic analysis correctly. To be complete, an analysis should embrace all associated cash flows over the life of the investment, including taxes, and take into account any significant uncertainties, such as questionable estimates of future operating costs or rate of inflation. A *third rule* is to consider an investment's affect on productivity.

Most organizations have reasonably good capital-budgeting procedures. Investment proposals are ranked by their relative profitability potential. But seldom are proposals awarded priority for their potentially strong impact on future productivity. The following adaptations to the investment screening process would favor productivity-oriented proposals:

- A percentage of capital funds available each year for internal investment could be set aside for productivity projects. Priority would be given to small-scale innovations and long-term or high-risk projects to improve resource utilization.

Economic sensitivity analyses are ideal for assessing the risk of promising but untested productivity-bettering investments. See Section 5-5.

- A section could be added to the customary request-for-expenditure form to require an explanation of the proposed investment's affect on such productivity considerations as conservation of materials, output quality, and employees' working conditions.

- An unceremonious application form could be developed for seed money that would allow workers to develop ideas for improving products or processes. Successful applications should be ceremoniously announced.

INCENTIVES

A sure way to attract the attention of employees is to wave dollar bills. Attaching cash payments to productivity gains, commonly called **productivity sharing,** is a powerful inducement for active participation. It is the most obvious way to convince employees that they have a stake in productivity improvement; they unmistakably will gain as the organization gains. Yet there are barriers of distrust to break through and touchy decisions to make about the nature of sharing— who shares, for what, and how much?

Monetary incentives for work units or structured employee teams can be clearly related to productivity gains, but fair amounts are difficult to determine. In many situations, simple and sincere praise for a job well done is more effective than a reward that could cause dissension, as long as the kudos are given atop an adequate pay level. Clever ways to recognize high achievers with nonmonetary awards have been derived, among which are the following:

- A table prominently reserved for a week in the cafeteria with free refreshments given at breaks.

- Reserved parking places in spots normally set aside for top executives.

- Trading stamps that can be collected to earn special vacations.

- Presentation of a memento by a visiting V.P. A personal award that can be displayed at a work station has a lasting impact.

- Any reward that can be shared by the group being honored, such as a group dinner, team holiday at a resort, or clothing emblazoned with the team logo, will build camaraderie.

These "nonmonetary" awards cost money, of course, but they do not distort the existing wage structure. Although productivity sharing may have Machiavellian—or even Pavlovian—overtones, it conspicuously recognizes productive performance. And that stimulates more productivity.

INVOLVEMENT

The idea of everyone working together and enjoying it is so innately appealing that nearly all managers try in their own fashion to install some form of worker participation. Management theorists have expounded the merits of a motivated work force in exquisite detail, complete with a bountiful smorgasbord of in-

Wages can act as disincentives for productivity when slow performance is rewarded by overtime pay or sloppy work earns the same pay as good work does.

volvement techniques and a shifting set of behavioral methodology to craft their own application mosaic. It is no surprise that varying degrees of success result, ranging from do-more-than-asked groups powered by high-octane morale to openly antagonistic groups that do just enough to survive.

Employees bring personal characteristics to work, where they merge with the social texture of the rest of the work force to shape an organization. Managers with diverse personalities and skills also blend into the organizational fabric. It is preposterous to expect a master set of motivation practices to fit all employees in all organizations, and it is equally unrealistic to believe that all managers possess the same abilities to execute the maneuvers. A governing precept is therefore to realize that each exercise to develop greater employee involvement must be crafted for that occasion.

For employee involvement to be fulfilling, it must give workers a degree of control over their destiny without abrogating management's power to manage resources—including workers. The balance is fragile.

Training to perfect currently needed skills and to develop crossover skills appeases both interests. The training sessions spice the work routine and build self-confidence. Workers with multiple skills allow more scheduling versatility and rapid response to emergencies. Perhaps the best productivity training is the development of general-purpose abilities that can be applied to all kinds of operations. Some of the more applicable types are

- *Creativity*—techniques to inspire new ideas and produce work-related innovations.

- *Time management*—awareness of how much time is routinely wasted and procedures for better utilization.

- *Economic justification*—appreciation of what data are required to justify an investment and how proposals should be evaluated.

- *Methods improvement*—development of a questioning attitude that detects inefficient operations and the application of motion studies and process charting to correct the inefficiencies.

Training nurtures involvement especially when it is conducted by supervisors and higher-level managers. A worker–manager interaction forges a bond of shared experiences and exposes both parties to the other's thinking. This additional channel of communication may deliver messages that would otherwise go unheeded, such as the existence of minor administrative irritations.

I.E. METHODS

"I.E. (industrial engineering) methods" is a catch-all term that represents a very large family of procedural techniques and managerial practices used to make operations more productive. These techniques and practices are collectively the arsenal that arms a productivity movement. They strike directly at poor quality, material and energy waste, and inefficient activities.

Involvement is hurt by archaic work rules and out-of-date bosses who equate worker participation with management giveaways.

Another beneficiary of general purpose training is the suggestion system. The yield is increased by employees who have the training to discover and write up potential improvements.

Industrial engineering is a profession devoted to productivity improvement. Although I.E.s still carry the stigma of "efficiency experts," being barely tolerated by some unions, their century-old heritage of scientific management is ideally suited to today's quest for greater quality and productivity. The services of an I.E. staff are greatly leveraged by educating all employees to be conscious of the continual need to improve operations and to be able to use basic I.E. methods to make the improvements.

Some of the simplest I.E. tools are the most productive. Large projects such as designing a total plant should be done by experienced engineers, but the redesign of a work area, in which the nature of the work has changed, can be done by operators who work in the area with little staff help. Many improvements result from the initiative of individual workers who detect a difficulty and on their own find a way to eliminate it. This process is more likely to get started and progress more smoothly when the workers are familiar with basic analysis techniques.

Three of the I.E. methods found to be most useful at all levels are flow process charts, motion economy, and time management.

Third phase: Maintenance

To maintain is to support and preserve from decline. Maintenance of a productivity-improvement process depends on measuring performance and monitoring progress to sustain motivation and momentum.

Sometimes it is more difficult to preserve gains than it was to achieve them. And it is still more difficult to continue gaining. The flush of enthusiasm that accompanies the kickoff of a productivity campaign almost guarantees some immediate advances. But as spirits wane, momentum will decline unless something is done to spark more interest.

Rewards, competition, threats, and other stimulants can be injected into a program to prolong its life, but they too become less effective as they are repeated. Their impact declines with each repetition because they lack continuity. Each time a reward or threat is issued, it is perceived as an isolated act, and it probably has been seen before, which means that it must have greater magnitude every time it is used to sustain its potency. The law of diminishing returns limits repetition.

MEASUREMENT

An improvement process is easier to maintain when the participants have a clear goal to aim at. It should be perceived by all concerned as worthwhile and attainable with reasonable effort. When the goal is a popular challenge, rewards and friendly competition are accepted as helpers. Conversely, when a goal is arbitrary and not embraced by workers, the stimulants conferred by management are mostly ineffectual.

Once a goal is met, a new campaign with a higher goal can be launched, or a new goal can be set and pursued with essentially the same process. For example, an emphasis on better customer service to boost productivity could be replaced by a goal to raise quality. Because it is obviously better to reach a goal and start over than to bog down in pursuit of an impossible goal, short-range, modest objectives are preferable. Success is the greatest motivator.

The multiple criteria with minigoals in Omax promote motivation.

The existence of a goal implies a measurement system. A goal just to do better is too vague to be of value. Specific levels of measurable achievement are necessary. The scaling system can then be used to monitor progress. Feedback to workers about their accomplishments delivers both a reading on current headway and message about expectations. An objectives matrix or a comparable measurement system is an essential ingredient of the A.I.M. process.

MONITORING

A productivity-improvement process without benchmarks is a race without a timer. To monitor is to watch or check for a special purpose and to regulate or control an operation. Productivity monitoring should provide feedback, *preferably in visual form,* that work units can use to regulate their efforts to improve operations.

Monitoring begins with the collection of data. Their accuracy, completeness, and timeliness is important. Unreliable data can destroy confidence and mislead improvement efforts. Without trustworthy data to pump through communication channels, the most perfectly designed feedback systems will be worthless.

Given dependable data, monitoring continues with the delivery of information in a usable form. Messages are most forcefully delivered by graphics. Graphic representations are to management what numbers are to mathematics— a language of abbreviation that displays the core of a situation concisely and understandably.

A performance chart based on data from an objectives matrix can be used to record a work unit's accomplishments. The one shown here, an **Omax progress graph,** plots two of the productivity criteria from the matrix given in Figure

Quality control charts are often posted by work units to display their achievements during a productivity campaign, because quality is inevitably a productivity criterion.

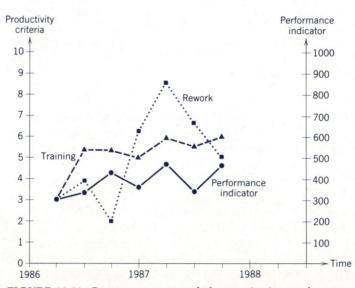

FIGURE 16.11 Omax progress graph for monitoring performance.

16.11, plus the unit's performance indicator. The left-hand scale that runs from 0 to 10 is for the productivity criteria, and the right-hand scale has values for the performance indicator.

Many different charting formats are available. It might be more revealing to calculate the rate of change of the performance indicator in each period and plot these percentages on a bar chart; this would be a *productivity index* chart. The type of visual display that is most likely to promote discussion and maintain interest should be selected.

The A.I.M. cycle

A.I.M. is thus a three-phase, technique-oriented process for improving productivity. It is a structured approach that can be adapted to fit different circumstances, by emphasizing certain techniques over others. For instance, a highly motivated work force might best be served initially by larger investments in equipment and more training to improve work methods, whereas a troubled work force might need an audit to uncover difficulties and project teams to increase involvement. Both groups would benefit from performance measurement, of course, but the methods of measuring might differ.

Because no improvement process retains its effectiveness forever in a production setting, users should anticipate successive cycles. As suggested in Figure 16.12, the A.I.M. process can be recycled indefinitely to prolong employees' commitment to more productive production systems.

Employees should realize that A.I.M. supports their interests and that the quality of their work life will rise commensurately with productivity gains.

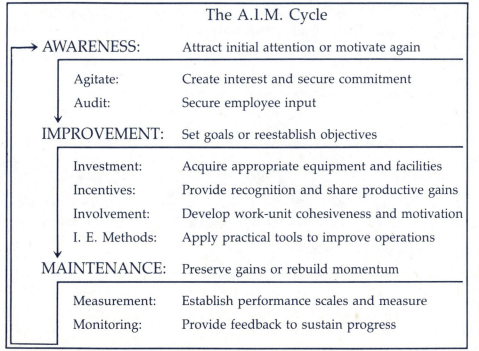

The A.I.M. Cycle

AWARENESS: Attract initial attention or motivate again

    Agitate: Create interest and secure commitment

    Audit: Secure employee input

IMPROVEMENT: Set goals or reestablish objectives

    Investment: Acquire appropriate equipment and facilities

    Incentives: Provide recognition and share productive gains

    Involvement: Develop work-unit cohesiveness and motivation

    I. E. Methods: Apply practical tools to improve operations

MAINTENANCE: Preserve gains or rebuild momentum

    Measurement: Establish performance scales and measure

    Monitoring: Provide feedback to sustain progress

FIGURE 16.12 The A.I.M. process is simply a selective collection of proven practices woven into a productivity-driven menu that organizations can repeatedly apply to make their operations more effective.

16-9 CLOSE-UPS AND UPDATES

All of the Close-ups and Updates sections in previous chapters have dealt with productivity concerns, usually featuring recent developments, many of which will surely shape the future of production systems. The same emphasis continues in this section, which examines a productivity-oriented data exchange and extrapolations of productivity trends.

Cooperating competitors use interfirm productivity comparisons

The comparison of selected productivity indicators among firms within a limited industrial sector is a potent tool for management. The mechanism is simple. Data regarding common operations are exchanged by firms through a third party, so as to ensure confidentiality. Each firm receives a report tabulating its performance in several categories and its standing in these categories relative to those of other participating companies. By comparing a firm's standings with its competitors' standings, managers locate their firm's weaknesses and strengths, enabling them to concentrate their attention and resources where most needed.

Interfirm productivity comparisons (IPC) are conducted in many countries. They were instituted by the United Kingdom Centre for Interfirm Comparison in 1959 as an activity of the British Productivity Council: studies were undertaken in more than 50 industries covering over 1000 firms. The Australian Department of Productivity arranged comparisons among 1500 firms in 60 industries. In contrast with the government-sponsored studies in other countries, trade associations run most of the interfirm comparisons in the United States. The number of firms participating in the data exchange is increasing steadily.

The relationships between financial and productivity ratios are outlined in Figure 16.13. From the first ratio, which relates profit to investment, progressive differentiations lead to economic indicators, cost indicators, operating indicators, and so on. The family of ratios shown in each indicator classification has many relatives that are not listed. Additional entries customize the comparison for specific industry sectors or service agencies.

The closer a ratio comes to representing the formative stages of the product, the more clues it reveals for productivity improvement. Each ratio in Figure 16.13 bears a message. For instance, a figure higher than the median ratio for

Ratio 4 Indicates disproportionately large manufacturing expense.

Ratio 9 Suggests excessive spoilage, lack of standardization, or inefficient purchasing.

Typifying associations in comparative measurements are Air Transport, North American Wholesale Grocers, and American Railroads.

An IFC for private electric utilities uses 103 different ratios. A more typical number is 20 to 30.

Cost ratios tell whether gross margins adequately cover costs; they do not tell whether an operation is as efficient as it could be or whether it is as efficient as the industry norm is.

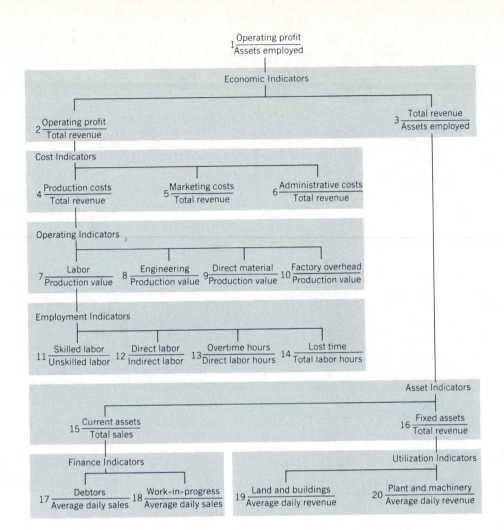

FIGURE 16.13 Family tree of typical ratios that can be used as productivity indicators for interfirm and intrafirm comparisons of financial and operating efficiencies. (Adapted from Interfirm Comparison, Australian Department of Productivity, Canberra, 1979.)

Ratio 10 Indicates higher-than-average energy costs, maintenance, or level of depreciation.

Ratio 14 Suggests the possibility of unpleasant working conditions, poor staff relations, unattractive factory location, or personality problems.

Ratio 18 Indicates problems in work flow, plant layout, length of the production cycle, or number of "rush" orders received.

Ratio 20 Indicates ownership of more capacity or more expensive plant and machinery than other companies have, which should be compensated by a significantly lower labor cost, ratio 7.

The same ratios are used in an intrafirm comparison as in an IPC, and the same ones appear in a total organization Omax.

A single corporation or government agency with several organizational divisions may benefit as much from internal as from external comparisons of productivity indicators. The ratios locate areas of excellence in different divisions and may explain the reasons for favorable readings. Just paying attention to financial ratios can obscure long-run effects on productivity, making a manager temporarily look good by sacrificing future productivity capability for current profits or, conversely, not giving a manager credit for laying the foundation for future productivity growth. Cost ratios show today's competitive standing. Productivity ratios foretell tomorrow's position.

Varying perspectives on production systems and productivity

Most people greet a claim of increased productivity with applause. They are convinced that productivity gains in this century are responsible for current high living standards and that more increases are needed to maintain standards and to control inflation. No one wants to argue against a proposition that gives more for less. But there are some who question whether we can afford to be more productive.

Are we really getting more for less, or does the gain depend on what is measured and the nimbleness of the measurer? And who actually benefits from a productivity increase? It may be difficult for a worker who has been automated out of work by a productivity push to appreciate the improvement. Other people never get the productivity message. They are still working harder, not smarter, and are probably enjoying it less. Worse yet, they are likely wasting resources. These varying perspectives deserve consideration because productivity is a concept that ties production systems into the total framework of society.

Investments made to protect the environment seldom contribute to greater productivity. Indeed, the opposite is more often true.

The concept of productivity improvement impinges on several of today's emotional issues. The old misgivings of labor that equated productivity gains with fewer jobs have been replaced by dire predictions from special-interest groups that the total cost to society for productivity improvements may exceed their benefits.

Working conditions contribute to the quality of life. Many clever approaches have been aimed at making jobs more attractive to their holders. One difficulty in redesigning work methods and conditions is to determine what the new designs actually accomplish. Surveys of workers' opinions and short-term production records are notoriously deceptive. What seems to work in one instance often fails to produce equal satisfaction in a different but comparable situation. Production output by itself is an inadequate measure of the success of work improvement efforts, but it is certainly a significant indicator.

An auxiliary productivity criterion could rate the effect of people-oriented production costs on production output. In the same way that consecutive productivity ratios are indexed to reflect a consistent level of quality, a method could be devised to give credit for advances in job "livability" and to reveal the inputs that fostered those advances. This criterion would show the extent of resources devoted to legislated employee protection programs and organizational attempts to motivate the work force. Although any such measurement can be faulted for lack of precision, it would at least link input cost to output results in a way that could disclose some of the black-box workings between cause and effect.

There is little doubt that today's production–consumption relationships will change in response to changing social expectations and future resource availabilties. As shown in Figure 16.14, interactions among government agencies, industrial systems, human populations, and the natural environment can be represented by flows that are subject to productivity assessments.

Consider, for instance, the factors that determine "quality of life" for the public. The availability of goods and the capacity to buy them are functions of working hours in facilities provided from investments. The time thus spent working is made more or less pleasant by accommodations owed to returns derived from output. The attractiveness of leisure time relies strongly on the services provided by government agencies supported by taxes. These agencies, in concert with the public and industry, attempt to manage the environment. Thus the single "quality of life" criterion embodies a host of factors, any of which can be recognized by a customized productivity ratio.

Most of the concepts supporting productivity in this chapter are implied in Figure 16.14. Changes in the flow levels among the four sectors would be reflected by variations in the socioeconomic productivity ratios, even though the measurements lack preciseness. The rate of change would signal the need for corrective actions to remedy unbalances or to modify objectives. Thus different production-system perspectives could identify deviating activities that warrant attention and could suggest the factor adjustments for rebalancing the total system.

These observations on production systems and society should bring to mind one more time the many systems concepts we have explored throughout this book: suboptimization, satisficing, capital budgeting, time–cost trade-offs, feedback, intangibles, sensitivity, and so forth. The collection of system management tools we have surveyed provides a respectable repertoire for making production decisions. Experience will sharpen the tools. It will also reveal more areas for profitable study and the need to keep abreast of new developments. The history of production has been a parade of change. The future promises to be even more dynamic than the past. Few undertakings offer more important, challenging, or diversified

It is estimated that U.S. output per labor hour must grow at a 4-percent rate, in real terms, during the next decade to just maintain current employment and living standards.

The ratios in Figure 16.14 are of an intrafirm-productivity nature. Many are already being collected. Others could be. Together, they would provide a valuable report on national prosperity prospects.

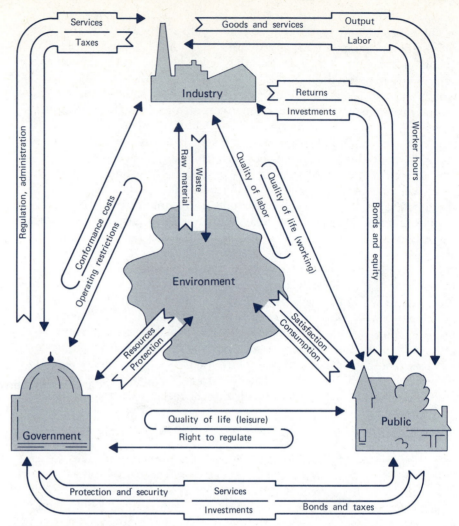

FIGURE 16.14 Productivity ratios for socioeconomic factors relating the public, government, industry, and environment.

opportunities than do the planning, analysis, and control of production systems.

16-10 SUMMARY

Importance of productivity

Productivity is a major concern throughout the world. Its rate of increase has declined alarmingly since the 1960s in the United States. The slump cannot be

674 PRODUCTIVITY

blamed on a single cause. Because numerous factors contributed to the slump, the proposals for improvement are many and varied.

The basic productivity ratio is output/input. The most common dimensions are units of product for output and labor hours for input. In evaluating the ratio it is advisable to consider the distribution of output and input, the confounding of factors, and the technical composition of the ratio.

Productivity measurement

Many versions of the productivity ratio have been proposed. A macrolevel index can be expressed as

$$\text{Total productivity index} = \frac{\text{product} + \text{service}}{\text{labor} + \text{materials} + \text{energy} + \text{capital}}$$

Productivity indexing

The APC model for factor productivity measurement is

$$\text{Factor productivity index} = \frac{\Sigma O_2 P_1 / \Sigma O_1 P_1}{\Sigma I_2 C_1 / \Sigma I_1 C_1}$$

where subscripts 1 and 2 indicate, respectively, the base and current periods; O equals output; I equals input; P equals output price; and C equals input cost. Related ratios evaluate price recovery and cost effectiveness.

A multidimensional method of measuring productive performance is known as the objectives matrix (Omax). It is generally applied at the work-unit level, although it can represent a total organization. The appeal of Omax is owed to its implicit potential for securing employee involvement, and applicability to knowledge or service workers.

Objectives matrix (Omax)

A matrix is constructed by identifying appropriate productivity criteria and ratios for a work unit; developing a scoring scale with 11 levels, in which level 3 is based on performance numbers in effect at the time the matrix is initiated and level 10 designates the unit's objectives; and assigning weights to the criteria to indicate their relative importance. Workers typically set the scoring scales by designating performance for level 10, and managers do the weighting by distributing 100 points among the criteria.

Omax construction

The matrix is used to monitor progress by measuring the performance in each criterion, converting the performance number to a score, multiplying each score by its weight to obtain a value, and then summing the values to get the performance indicator for the period. That period's productivity index is the percentage change in the performance indicator from the previous period.

Omax application

A.I.M. is a process that promotes productivity by increasing *awareness*, encouraging *improvement*, and providing measures for the *maintenance* of gains. It emphasizes the use of proven practices. Awareness can be raised by kickoff programs and a productivity audit. Productivity improvements can be generated by investing in new technology, providing incentives to employees, arousing greater employee involvement, and directly employing I.E. methods to improve operations. Motivation and momentum can be maintained by implementing performance measurement and using visual aids to monitor progress. Which practices to emphasize depend on each situation, but if they are selected and applied competently, they can achieve significant productivity advances.

Awareness-improvement-maintenance (A.I.M.) process

Interfirm productivity comparison (IPC)

Firms in the same industry learn the relative efficiency of their operations with respect to their competitors by participating in an interfirm productivity comparison (IPC).

Productivity and production systems

Productivity is a unifying theme between production systems and the rest of society. It involves the conservation of resources, protection of the environment, and quality of life. Productivity affects the environment as it interfaces with industry, government, and the public. All of these sectors of the economy rely on production-system concepts to integrate their interests and to fulfill their missions.

16-11 REFERENCES

DRUCKER, P. F. *Technology Management and Society.* New York: Harper & Row, 1970.

FABRICANT, S. *A Primer on Productivity.* New York: Random House, 1969.

KATZELL, M. E. *Productivity: The Measure and the Myth.* New York: AMACOM, 1975.

MCBEATH, G. *Productivity through People.* New York: Wiley, 1974.

MORRIS, W. T. *Work and Your Future.* Reston, Va.: Reston, 1975.

NORMAN, R. G., and S. BAHIRI. *Productivity Measurement and Incentives.* London: Butterworth, 1972.

PARSONS, J. *Productivity Profits and Prices.* Pretoria; National Productivity Institute, 1984.

PENDERGAST, C., ed. *Productivity: The Link to Economic and Social Progress.* Washington, D.C.: Work in America Institute, 1976.

RIGGS, J. L. "A.I.M. at Productivity." Corvallis, Ore.: Oregon Productivity Center, 1985.

———. *Productive Supervision.* McGraw-Hill, New York, 1985.

———. "Monitoring with a Matrix That Motivates As It Measures Performance," *Industrial Engineering,* 1, January 1986.

———, and G. H. FELIX. *Productivity by Objectives.* Englewood Cliffs, N.J.: Prentice-Hall, 1983.

———, and T. M. WEST. *Engineering Economics.* New York: McGraw-Hill, 1986.

ROSS, J. E. *Managing Productivity.* Reston, Va.: Reston, 1977.

SHEN, G. C. *Productivity Measurement and Analysis.* Tokyo: Asian Productivity Organization, 1985.

Work in America. Department of Health, Education and Welfare. Cambridge, Mass.: MIT University Press, 1973.

16-12 SELF-TEST REVIEW

Answers to the following review questions are given in Appendix A.

1. T F Increasing *productivity* will raise the standard of living and increase efficiency, quantity, and quality.

2. T F The *productivity ratio* increases when production (output) increases.

3. T F Rewarding increased *labor productivity* will help ensure the success of productivity improvement programs.

4. T F A *productivity measurement instrument* should be simplistic, so it is easy for employees to understand and calculate.

5. T F Constant dollars are used to determine the *total productivity index.*

6. T F *The price recovery index* relates the price of the product to the cost of inputs.

7. T F *Partial productivity indexes* point out specific operations that need attention.

8. T F *Work unit productivity indicators* measure only characteristics of groups who perform operations with definable end products or accomplishments.

9. T F The *objectives matrix* was created to measure the productivity of basic work units.

10. T F *Productivity criteria* are ratios that represent only factors pertinent to a work unit's output.

11. T F There are 11 *levels of accomplishment* in the scoring scale of an objective matrix.

12. T F *Weights* for the relative importance of criteria are usually determined in consultation with the employees who are being represented by the matrix.

13. T F If a *score* of 3 has a *performance level* of 12 units and a score of 4 requires 14 units, performance at the level of 13 units can earn a score of either 3 or 4.

14. T F *A.I.M.* is a three-phase process that stands for Awareness, Improvement, and Measurement.

15. T F A *productivity audit* is a survey that indicates how employees feel about their jobs, identifies promising areas for improvement, and promotes employee involvement.

16. T F When a firm chooses the *investment* path to improvement, it should apply proper economic justification methods to questionable proposals.

17. T F Monetary *incentives* may actually act as disincentives for productivity.

18. T F *Productivity sharing* occurs when employees receive cash payments for their productivity gains.

19. T F *I.E. methods* are procedural techniques and managerial practices that make operations more productive.

20. T F The objective of the *maintenance* phase is to continue improvement strategies that have been started.

21. T F The *Omax progress graph* plots data from the objectives matrix.

22. T F The ratios used in the *interfirm productivity comparisons* are similiar to the ones appearing in a total organization Omax.

23. T F *Cost ratios* tell whether an operation is as efficient as the industry norm.

24. T F *Environmental protection* efforts increase productivity.

16-13 DISCUSSION QUESTIONS

1. Distinguish between labor and capital productivity. Why should each share the productivity gains resulting from its contribution?

2. Why is it important to consider the distribution of benefits gained from a productivity rise?

3. Describe a situation in which a company or government agency has apparently offered a new service without increasing the production inputs, but the new offering serves no useful purpose (or at least consumers are ignoring it). Does this situation really constitute an increase in productivity?

4. Do you feel that stockholders should share the benefits produced from productivity increases by the companies in which they hold stock?

5. Evaluate the performance of the company described by the financial data and ratios in Figure 16.4. Explain what is indicated by each of the ratios. Which one do you feel is the best single indicator of performance? Why?

6. Answer the following reservations about supporting plans for a productivity-improvement program:

a. "Why should I cooperate with that new scheme to have us all switch positions on the line every hour? All they're trying to do is get us to work faster."

b. "I don't plan to tell anyone how I can do my job better. I might get some cash for the suggestion, but then I'd have to live with it forever. Now I can speed up whenever I need to meet the quota quickly and loaf when I want to."

c. "They say that if we follow all those new methods, we'll put out 20 percent more units and the job will be easier, too. Maybe my job will be easier, but I figure that before long they'll come back and say that one out of every five of us will be laid off."

7. How does confounding input factors impede productivity studies?

8. Decide in your opinion which three of the causes of lower productivity in the United States contributed most to the slump. Rank them, and state why you ranked them as you did.

9. Are the concepts of productivity improvement and intermediate technology compatible? Why?

10. Although most of the productivity ratios suggested in Figure 16.14 are not formally measured, many of them have surrogate measures that are familiar statistics in news releases. Discuss three significant surrogates that are available to evaluate relationships among the environment, industry, government, and the public.

11. Name and discuss the implementation considerations of the five changes in the numerator and denominator of the basic productivity equation (productivity = output/input) that indicate a productivity increase. (*Hint:* The most obvious change is to increase the numerator while maintaining the same input level, which is accomplished by more efficient utilization of input resources.)*

12. Productivity awareness is considered to be an important factor for improving productivity. Much like Adam Smith's invisible hand, pursuing productivity improvements as a selfish goal, only to better one's own economic health, collectively has a widespread benefit to everyone through the improved utilization of resources.

Construct a C&E diagram to explore what could be done to increase workers' awareness of the value of high productivity to their own and everyone else's prosperity. Use the following main ribs and complete the diagram with factor arrows to or from all ribs:

Causes. Students, universities, governments, workers, corporations, professional associations. (For each of these headings, indicate categories of activities or activists that could publicize and promote the awareness issue.)

Effects. Publications, nonprint media, programs, contracts. (For each of these headings that represent

*James L. Riggs and Thomas M. West, *Engineering Economics,* Third Edition. Copyright © 1986. Reprinted by permission of McGraw-Hill Book Company, New York.

the results of the identified causes, indicate on arrows the specific channels of information flow.)

13. A list of adjectives that describe desirable attributes of a measurement system is as follows. For each adjective, write a specific feature of measurement that is supposed to satisfy that requirement. For instance, the adjective *consistent* is achieved when a measurement system produces the same result, regardless of who applies it, and such an attribute results from having clear directions, a simple procedure, and specific descriptions of performance proficiency required at each level of the rating scale.**

a. feasible
b. accurate
c. participative
d. fair
e. understandable
f. motivational
g. timely
h. economical
i. friendly

14. The examples of Omax applications illustrated group monitoring. Explain how the objectives matrix can be applied to the appraisal of an individual person.

15. Why is the "previous" performance indicator always 300 on the first time that measurements are taken with an objectives matrix?

16. Construct a *T* chart comparing the pros and cons of each of the *I* paths in A.I.M. (investment, incentives, I.E. methods, and involvement).

17. Several obstacles and detours on the four paths to productivity improvement in A.I.M. (investment, incentives, I.E. methods, and involvement) have been blamed for the poor productivity performance in the United States during the 1970s and early 1980s. For each of the following possible inhibitors to productivity, discuss how it is being relieved currently or how you believe it could be reduced in the future:

a. Management decisions motivated by short-term gains rather than directed toward the long-term prosperity of the firm.
b. Decline in the work ethic among young employees.
c. Government regulations that mandate investments to improve environmental or social conditions that siphon money from investments that could boost productivity. (Consider both sides of this situation: What happens when social responsibilities are neglected and when social expenditures are so high that industries become unproductive and noncompetitive?)
d. Lack of concern about designing and building higher-quality products.
e. Inadequate support for basic research.
f. Influx of untrained or less-trained workers into the work force as a result of high birthrates from 1946 to 1964 and a larger proportion of women seeking employment during the 1970s.
g. Low rate of saving by Americans that has limited the amount of capital available for borrowing for industrial purposes.
h. Tax policies that provide little incentive for modernizing production facilities and investing in technologically advanced manufacturing processes.
i. Confrontational relationship between management and unions that restricts cooperation in productivity-improving projects.
j. Management resistance to employee participation plans that give workers more voice in job design and decisions about working conditions.
k. Work rules that enforce the inefficient use of resources, especially labor.
l. Shift of employment from manufacturing, in which productivity is high, to service industries that generally have lower productivity.
m. Lack of government support for ailing industries and an unwillingness by the government to protect new industries from foreign competition.
n. Reluctance of employees at all levels to have wage increases tied to measured increases in productivity.
o. General lack of awareness of the negative impact of declining productivity on the standard of living.

**James L. Riggs, *Production Supervision*, 1985. Reprinted by permission of Prentice-Hall, Inc., Englewood Cliffs, New Jersey.

1. Complete the following table to make the sales data equivalent through the use of quantity and price indexes. Entries in the bottom line (comparable sales) should be equal; the 1987 sales are corrected for price, and the 1988 sales are corrected for volume.

	1987	1988
Sales	$2,000,000	$4,800,000
Price per unit	$80	$120
Units sold	_____	_____
Price index	100	_____
Quantity index	100	_____
Comparable sales	_____	_____

2. Production of a product starts in the fabrication department and finishes in the assembly department. Last year 20,000 units were processed through the two departments, and this year the total is 22,080. Standard labor hours per unit in fabrication and assembly are 0.47 and 0.5, respectively. Actual labor hours last year were 9640 in fabrication and 9920 in assembly; corresponding actual labor hours this year are 10,064 and 10,820.

a. What are the partial productivity indexes for each department?

b. What is the labor productivity for the product?

3. Charley's Chickenpluckers buys chickens, processes them, and sells the cut meat to wholesalers. From the following data regarding Charley's production last period and this period, determine

a. The quantity productivity index.

b. The price-weighted productivity index.

c. The conclusions that can be drawn from the two indexes.

	Last Period		This Period	
	$/pound	Pounds	$/pound	Pounds
Input	1.39	22,000	1.24	24,200
Output	1.65	17,500	1.73	19,400

4. Toyco is a small, labor-intensive producer of toys. During the last month of 1987, the president of the company, encouraged by recent marketing success, acquired the assets of a small competitor that had been forced into bankruptcy. This doubled the production capacity and consequently doubled the work force. The resulting financial statistics are as follows and caused the president to proclaim the acquisition a complete success. Assume that the data are given in constant dollars.

	1987	1988
Net Sales	$520,000	$1,040,000
Cost of goods sold:		
Material	100,000	200,000
Labor	300,000	620,000
Other expense	50,000	100,000
Depreciation	10,000	20,000
Total	460,000	940,000
Gross income	60,000	100,000
Tax (50%)	30,000	50,000
Net income	30,000	50,000

a. Calculate the productivity ratios applicable to Toyco for 1987 and 1988 by using the method given in Example 16.3. Comment on the productivity changes from one year to the next.

b. Was the president's claim justified? Why?

c. What suggestion might you make to improve Toyco's productivity?

5. Last year, because of competition and inflation,

Zorzaz Corporation barely made a profit. Near the end of the year, the corporation had an opportunity to purchase rights to a patent that appeared likely to lower the cost of production and to increase the amount of sales in future years. A contract was made to give the patent owner 1 percent of net sales by Zorzaz for the next five years. Use of the patented process in equivalent manufacturing situations has reduced labor costs by 10 percent, with no additional capital expenditures for equipment. Financial data for the two years in constant dollars are as follows:

	Last Year	This Year
Net sales	$3,000,000	$3,300,000
Cost of goods sold:		
Material	1,000,000	1,050,000
Labor	1,400,000	1,500,000
Patent royalty		33,000
Other expense	400,000	410,000
Depreciation	150,000	150,000
Total	2,950,000	3,143,000
Gross income	50,000	157,000

a. Calculate the productivity ratios using the model from Example 16.3 for both years, and interpret the results.

b. What logic should guide the distribution of productivity gains expected in the future from the application of the patented process?

6. a. Calculate the base-period, factor productivity index for a mine that last year produced a million tons of coal valued at $50 per ton and a million and one-half tons valued at $80 per ton. To do so, 1.5 million labor hours at $12 per hour and 2.5 million labor hours at $16 per hour were expended.

b. If prices and costs both increase by 10 percent next year while the quantity and proportion of labor hours remain constant, how much must the output of the mine increase to produce a 5-percent productivity gain for the year? The proportion of the two types of coal remains unchanged.

7. Two products, J and K, are the output of a manufacturing company. Quantities sold and price per unit for two years of operation are shown in the accompanying table. Also shown are input hours and cost per hour for two labor classes, $L1$ and $L2$.

	Last Year		This Year	
	Quantity	$/Unit	Quantity	$/Unit
Product J	1,000 units	$200	1,800 units	$240
Product K	2,000 units	$150	1,500 units	$150
Labor, $L1$	5,000 hours	$8.00	7,500 hours	$8.50
Labor, $L2$	10,000 hours	$8.50	9,000 hours	$9.00

a. What is the labor productivity each year based *only* on quantity? Explain the weakness of this ratio.

b. Calculate the labor factory productivity index.

c. Calculate the labor price recovery index.

d. Calculate the labor cost effectiveness index.

e. Compare the three indexes, indicating the meaning of each and what the three together suggest about the company's operations.

8. Quantities and costs for three additional inputs that are used in the production of products J and K from problem 7 are given below.

a. Calculate the energy factory productivity index.

b. Calculate the energy price recovery index.

c. Calculate the energy cost effectiveness index.

d. Compare the energy indexes to the labor indexes calculated in Problem 7. What conclusions can be drawn?

e. Calculate the material factor productivity index. Compare this to the energy factor productivity index, and discuss the implications.

f. Calculate the capital factor productivity index, and discuss its meaning to the company.

g. Calculate the total productivity index using output for products J and K and inputs of labor, energy, material, and capital.

	Last Year		This Year	
	Quantity	*$/Unit*	*Quantity*	*$/Unit*
Energy	300,000 units	$0.09	280,000 units	$0.15
Material	100,000 units	$2.00	120,000 units	$2.10
Capital	$3,000,000	0.12*	$3,200,000	0.13*

*Percentage cost for invested capital.

9. Your organization has the productivity pattern, corrected to a base-year index, shown in the following table. Assume that net profit increases at a commensurate rate with total productivity gains each year.

a. Is the increase in total productivity from 1986 to 1987, at no increase in labor input, a sufficient reason to grant a wage increase to the workers? Why?

b. If the decrease in capital productivity in 1987 re-

sulted from a mandatory capital investment to conform to a new law restricting pollution, should this social benefit be recognized differently in the productivity measure? Why?

c. Who should benefit from the 4-percent gain in 1988? What percentage, if any, of the total profit for 1988 should be distributed to each of the following parties: (1) consumers, (2) workers, (3) owners or stockholders, and (4) the organization, as retained earnings? Why?

	1986	Change	1987	Change	1988
Net profit (1)	100	15%	115	13%	130
Labor input (2)	50	0	50	14%	57
Capital input (3)	40	25%	50	4%	52
Labor productivity (1 ÷ 2)	2.0	15%	2.3	0	2.3
Capital productivity (1 ÷ 3)	2.5	−9%	2.3	9%	2.5
Total productivity [1 ÷ (2 + 3)]	1.11	4%	1.15	4%	1.19

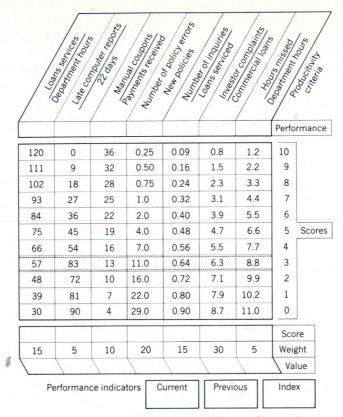

FIGURE 16.15 Objectives matrix for financial operations.

10. Performance records for the seven criteria used in the objectives matrix are shown in Figure 16.15. These results were obtained at the end of the first two evaluation periods for the financial operations.

	Ratio Results	
Criteria	*First Six Months*	*Next Six Months*
Loans serviced/department hours	63	67
Late computer reports/22 days	60	55
Manual coupons/payments received	14	12
Policy errors/number of new policies	6.8	3.9
Number of inquiries/loans services	0.73	0.88
Investor complaints/commercial loans	5.4	5.6
Hours missed/department hours	7.0	6.6

a. Calculate the performance indexes for both of the evaluation periods.

b. Calculate the productivity indexes for each of the periods and the productivity index for the first year.

c. Consider the performance in each criterion. Which criterion might be considered "out of control" on a control chart? Does any overall pattern of progress appear to be developing?

11. *The Case for Partitioned Productivity Gains*

The owner-manager of a custom bakery faces a delicate but pleasant problem: how to distribute rewards. A year ago the bakery was close to going out of business because of competition from the bakery-product section built into the new shopping centers in the suburbs. Fewer people visited the downtown site of the bakery, and those that did were purchasing less to take home. The turnaround in business originated at a meeting of all employees called by the manager last year.

The purpose of the meeting was to urge the employees to reduce costs further or face the possibility of having the shop close. As the meeting progressed and the employees became aware of how serious the situation was, they started offering many ideas that the manager had never considered before. Suggestions ranged from an easier way to make butter-flake rolls, to opening a coffee-roll sidewalk bar and a discount booth inside the shop for day-old products.

The manager shrewdly recognized the productivity-improvement potential of the suggestions and gave them full backing. Highlights of the resulting year's activities compared with those of the previous year and the price index data are as follows:

Total sales increased by 60 percent (including a 5-percent general increase owing to inflation in the overall economy) to $480,000.

Cost of materials used increased by 50 percent (including a 10-percent industrywide rise in material expenses) to $150,000.

Labor costs increased by 10 percent (the same rate as industrywide wage average) to $110,000.

Annual depreciation charges jumped by $15,000 (owing to investments made to facilitate innovations) to $65,000.

Other expenses increased by 40 percent (including in-

flation) to $140,000. The *increased output* with essentially the same number of labor hours worked was due to improved methods instituted by the bakers and partially to the new equipment. Sales increased more than did material costs because waste was reduced.

a. Calculate the percentage increase in different productivity measures to determine the effect of revised operations in the bakery. Interpret the significance of the different ratios.

b. How do you think the manager should distribute the benefits from the productivity rise in order to sustain the gain?

12. *The Case of PACE*

A novel method of appraising group performance uses the technique of work sampling described in Section 10.5. The method was devised by the Northrop Corporation and named *performance and cost evaluation* (PACE). The concepts of PACE are still interesting, even though the method has never been fully accepted by industry.

PACE is designed to measure the effectiveness of a team working on an assigned project. Measurement include four key factors: (1) number of persons assigned to the project, (2) number idle, (3) number absent from the project area, and (4) the group-effort rating. These factors are combined into a formula to provide a PACE index:

$$\text{PACE index} = \frac{(1) - (2) - (3)}{(1)} \times (4) \times 100\%$$

where the group-effort rating (4) has a base point of 1.0 for a normal tempo of work.

The values for the four factors are obtained by having trained observers make head counts in the work area at random intervals. Guidelines are established to define each state, such as gossiping versus "talking shop," and to rate the effectiveness of group activities. As in most personnel appraisal methods, PACE depends on subjective evaluations by the observers, but it provides a systematic framework for the evaluations. A chart of the PACE index and related factors allows managers to appraise the trend of the observations, as shown in Figure 16.16.

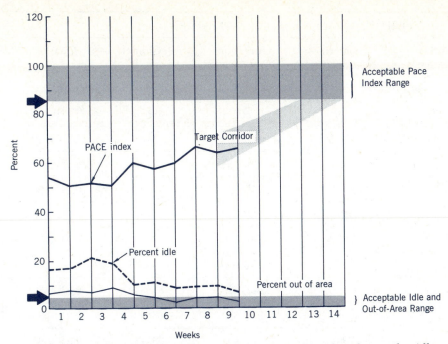

FIGURE 16.16 PACE chart. Weekly averages of the PACE index, the number idle, and the out-of-area percentages are plotted to determine the trend of group performance. The upper shaded area of the chart shows the acceptable range for a mature PACE index (85 to 100 percent), and the target corridor defines acceptable progress for initial PACE index values. The lower 5-percent band represents the acceptable range for the idle time and out-of-area percentages. Whenever a weekly reading falls outside the shaded zones in an established PACE program, management is expected to investigate to find the cause.

a. What reasons can you give for the apparent failure of PACE to earn acceptance as a performance appraisal technique? Compare its use with that of the productivity ratio as a measure of production performance.

b. How might PACE be adapted to overcome the obstacles mentioned in part a., in anticipation of applying the revised version to evaluate performance at a university?

13. *The Case of Increasing Productivity through Teamwork*

After several weeks of planning and indoctrination, a company initiated a unit productivity-improvement program. Teams were formed to represent natural divisions of the production process according to the nature of their value-added contributions. Organizational structures were established to provide unit training and to respond to team suggestions.

During weekly meetings the teams developed their own objectives and planned ways to achieve their goals. Concurrently, performance indicators were identified, and rating scales were established. Then management's "productivity council" proposed a weighting plan for each team. An abbreviated version of one team's performance indicator table is shown in Figure 16.17.

a. Based on the data in Figure 16.17, construct an objectives matrix using the standard format.

b. The unit that developed the performance efficiency indicators in Figure 16.17 has measured its

PERFORMANCE EFFICIENCY INDICATORS

Point Scale	Delivery Days late Total orders	Facilities Hours downtime Line operating hours	Material Pounds salvaged 100 units	Operations Completed units Period	Personnel Overtime hours Period	Quality Rework Units Produced	Point Scale
10	0	0	9	800	0	0	10
9	0–0.2	0–0.02	8.9–8.8	799–780	1–8	0–0.10	9
8	0.2–0.5	0.03–0.04	8.7–8.6	799–755	9–16	0.11–0.20	8
7	0.5–1	0.05–0.06	8.5–8.3	756–725	17–24	0.21–0.30	7
6	1–2	0.07–0.10	8.2–7.9	726–690	25–36	0.31–0.40	6
5	2–3	0.11–0.13	7.7–7.4	691–650	37–48	0.41–0.50	5
4	3–4	0.14–0.17	7.3–6.8	651–610	49–60	0.51–0.60	4
3	4–5	0.18–0.21	6.7–6.1	611–570	61–84	0.61–0.70	3
2	5–6	0.22–0.25	6.0–5.3	571–530	85–108	0.71–0.80	2
1	6–7	0.26–0.30	5.2–4.4	531–490	109–132	0.81–0.90	1
0	Over 7	Over 0.30	Under 4.4	Under 490	Over 132	Over 0.90	0
Effectiveness weights	0.05	0.10	0.05	0.30	0.20	0.30	

FIGURE 16.17 Performance indicator table for one team. Columns list the points earned for each level of accomplishment for six performance-efficiency criteria. The weighting scale for each criterion is shown in the bottom row.

performance recently with the following results:

Delivery. average days late per delivery = 1.2

Facilities. 11 hours downtime per 168 operating hours = 0.07

Material. 61 pounds salvaged for 748 finished units = 8.2

Operations. 748 units completed

Personnel. 15 overtime hours used

Quality. rework required on 88 of 748 units = 0.12

Determine the performance indicator for the period.

c. Assume that the performance indicator calculated in Problem 13b is based on the most recent measurements and that the measurements of operations taken six months previously by the same unit produced a performance indicator of 695. What is the productivity index for the period?

14. *The Case of to Get You Gotta Give*
A company with 200 employees produces over 100 different types of envelopes and mail containers. In an attempt to increase productivity, the management instituted a sharing program based on a company-wide objectives matrix, as shown in Figure 16.18.

Five criteria were selected for the total plant monitoring:

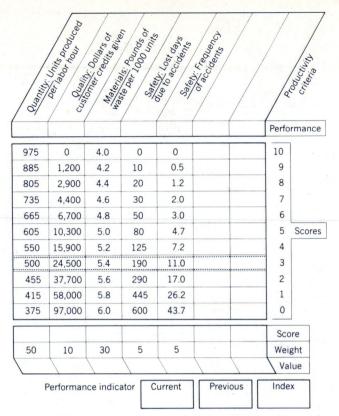

FIGURE 16.18 Productivity-sharing plan displayed in an objectives matrix format.

a. *Quantity*—Number of units produced divided by the total number of straight-time and overtime hours for the entire work force.

b. *Quality*—All credits given to customers during the period as the result of plant quality problems or errors, based on a production level of 4 million units per week.

c. *Materials*—Equivalent pounds of wasted raw materials, mostly papers of comparable value, per 1000 units produced.

d. *Safety*—Number of days lost during the period for 200 employees owing to industrial accidents.

e. *Safety*—Number of doctor-related accidents during the period times 200,000 divided by total hours worked.

Goals were set and scale levels developed to measure improvement above existing performance, designated as level 3. Dollar values for cost savings resulting from a change in each scoring level of all criteria were calculated. Based on these values, the company agreed to split the savings 50-50 with employees, paid as a bonus every six months.

Employees endorsed the plan wholeheartedly and were rewarded with significant bonuses when productivity improved by 30 percent during the first year. Quantity, quality, and materials increased by a point each, and the other criteria held steady at level 3. During the next six-month period, orders were heavy and the enthusiastic employees boosted quantity to 610 units per hour, but quality costs rose to $39,270 in credits and waste increased to 5.35 pounds per 1000 units. Safety did not change.

a. What was the rate of change of productivity during the period?

b. Management began to worry about the effect of declining quality on the reputation of their products. Perhaps the penalty for lower quality was not large enough. Should they consider changing the matrix? How might it be changed? What effect might a change have on employee morale? What do you suggest?

15. *The Case of Fencing with Figures*

Jonesy is the only manager in the Fancy Fence Company. He is also the owner. There are 12 employees: a bookkeeper/secretary, an estimator, and 10 fence builders. All the fences are constructed of cedar, but some of them are very fancy; there are 19 basic designs. By far the largest cost of operations is for materials. Next comes labor and then travel.

Most orders are received by phone. The estimator calls on a prospective customer and makes a bid to do the job, or if a standard fence is ordered, a bid is quoted over the phone. The estimator is very good. Jonesy does the work scheduling, material ordering, and general administration.

Jonesy has scheduled a company meeting for next Friday at two o'clock. He plans to set up an objectives matrix to monitor performance. He is also thinking about giving rewards for improvements that are justified by productivity increases. Before going into the meeting, he wants to prepare notes and a tentative matrix. He asks your help. You agree.

a. Prepare at least five ratios that could be used as criteria of performance. Consider how output can be measured when standard hours per foot of fence are not known and the times to build fences of different designs vary greatly.

b. Place weights on the criteria you have developed. Note why certain criteria are accorded more importance than are others.

c. What types of recognition for improved productivity do you suggest? Why? Should the estimator and secretary be included? What level of recognition should trigger a reward?

APPENDIX A
SOLUTIONS

A-1 ANSWERS TO SELF-TEST REVIEW QUESTIONS

CHAPTER 1 PRODUCTION OPERATIONS AND OPPORTUNITIES

1. True.

2. False. Taylor initiated scientific management.

3. False. Automation has had very little impact on employment.

4. False. Computers are associated with "soft" automation.

5. True.

6. True.

7. False. Consensus, not quantitative, decision making is the mode.

8. True.

9. False. A look-alike replica is a physical model.

10. True.

11. True.

12. False. There is not a set sequence, but the given order is often applicable.

13. True.

14. True.

15. False. CAD is not inexpensive but is more versatile than manual drafting.

CHAPTER 2 PRODUCTS AND SERVICES: DESIGN AND DEVELOPMENT

1. True.

2. False. The final phase is to update, which could be to modify rather than to terminate.

3. True.

4. False. Brainstorming is a group ideation technique.

5. False. *Cause* ribs point toward the problem statement, but *effect* ribs point away from it.

6. False. The only time reference is the implementation date.

7. True.

8. True.

9. True.

10. False. Each working unit reports to two bosses, a project manager and a functional manager.

11. False. Patent rights expire after 17 years.

12. True.

13. True.

14. False. Feedback provides information about a system's operations, but it does not control the operation.

15. True.

16. False. Some functions have far greater effect on performance than others do and, therefore, deserve closer attention.

17. False. Physical flow of material occurs between about half the functional areas. Information flows throughout the production process.

18. True.

19. True.

20. False. Design and manufacturing engineers work together in the simultaneous engineering approach.

21. True.

CHAPTER 3 FORECASTING

1. False. Considerable free information is available, but customer opinions are obtained only at raw cost.

2. True.

3. True.

4. True.

5. False. Cyclic variations are not necessarily repeating patterns.

6. False. Normal equations are solved to determine a and b.

7. True.

8. False. Only the $Y = ab^x$ exponential equation takes the logarithmic form from which its formula can be determined by the least squares method.

9. True.

10. False. Alpha can vary from zero to $+1$.

11. True.

12. True.

13. False. The coefficient of determination equals r^2.

14. False. There is no correlation when $r = 0$.

15. True.

16. True.

17. False. To anticipate changes, trend spotting relies on a qualitative assessment of clues from the printed media.

18. True.

CHAPTER 4   SYSTEM ECONOMICS

1. True.

2. False. Tactics are associated with operational efficiency.

3. False. Organizational suboptimization happens when returns from optimizing the performance of a lower unit causes greater losses for its parent unit.

4. True.

5. False. The break-even point is the lower limit for profitable production.

6. True.

7. False. Lower variable cost decreases the break-even point when other factors are fixed.

8. False. Dumping occurs only when it charges different prices for the same product in different markets.

9. True.

10. False. Maximum profit occurs when marginal revenue and cost are equal.

11. False. Marginal and total average costs are equal at minimum average total cost.

12. True.

13. True.

14. True.

15. True.

16. False. The distinctive feature of a DDS is its ability to assist the investigation of alternative courses of action.

17. False. A DBMS is a software program for accessing and modifying various databases that store collected data.

18. True.

19. False. The BEA provides no policy advice.

CHAPTER 5   OPERATIONS ECONOMY

1. False. A future sum is worth less than an equal amount received now.

2. False. Nominal interest is the rate that results from multiplying the per-period rate by the number of periods in a year.

3. True.

4. True.

5. True.

6. False. A comparison is always made to the next lowest *acceptable* level.

7. False. Physical depreciation is caused by normal use.

8. True.

9. False. Depreciation charges are not actual cash flows.

10. True.

11. True.

12. False. The present worth of a proposal caused by deviations of different economic factors are recorded in a sensitivity graph.

13. True.

14. False. The difference in expected values with and without knowledge of which outcomes will occur is the value of perfect information.

15. True.

16. False. The payback method does not use discounted cash flow.

17. False. B–C ratios use present-worth calculations.

18. False. Neither an ordinal nor an interval scale has a natural zero, but a ratio scale does have one.

19. True.

20. True.

21. True.

22. False. Office productivity has not kept pace with the average U.S. productivity growth. Downsizing is primarily aimed at cost reduction.

CHAPTER 6 ALLOCATION OF RESOURCES

1. False. Linear programming originated in the 1940s.

2. True.

3. True.

4. True.

5. False. The objective function is an equation that should be maximized (profit) or minimized (cost). Resource constraints are expressed as inequations.

6. False. Only two products can be conveniently graphed. Any number of resource restrictions can be accommodated.

7. True.

8. False. The isoprofit line farthest from the origin identifies the optimal mix.

9. False. The number of rows and columns need not be equal.

10. True.

11. False. VAM is used to obtain an initial feasible solution.

12. False. The "stone" may be occupied by a real or an epsilon quantity.

13. True.

14. True.

15. False. The scope of aggregate planning is one month to a year, and physical facilities are assumed to be fixed during the period.

16. False. R&D budgets are not production inputs, but hours worked are.

17. True.

18. False. The classification comes from the type of work done, not the level of education.

19. True.

20. False. Priority should be given to the more important tasks.

21. False. Views can be discussed, but no arguments are allowed.

22. True.

23. True.

CHAPTER 7   RESOURCE SCHEDULING

1. True.

2. False. CPM originated in 1956, but it was designed for construction and production scheduling. PERT is used for research and development work.

3. True.

4. True.

5. False. *ES* plus the activity duration equals *EF*.

6. False. The critical path sets the shortest possible project duration, but it is also the longest correctly ordered chain of activities.

7. True.

8. False. Computers do not draw networks very well and are rarely used for that purpose. But computers can calculate boundary times and draw bar charts very quickly.

9. True.

10. True.

11. False. A job shop is a place where individually different but similar jobs are worked on in the same facility.

12. True.

13. True.

14. False. Matrix organizational structures are associated with project management.

15. False. Personnel are usually "borrowed" for internal projects, and line managers are often reluctant to loan skilled workers.

CHAPTER 8   HUMAN FACTORS

1. False. Identification is the process of grouping differentiated sensations.

2. True.

3. True.

4. False. The second aim of screening is to create a favorable impression for the job applicant. Orientation starts on the first day of the job.

5. True.

6. False. Regular rewards are often taken for granted. More incentive is provided by an immediate reward for a specific accomplishment.

7. False. Increasing the diversity of work is job enlargement. Job enrichment stresses greater work responsibility.

8. True.

9. False. OSHA is the nationally legislated Occupational Safety and Health Act.

10. True.

11. True.

12. False. Decentralization has more motivation potential.

13. False. No single span of control fits every situation. The ideal supervisor-to-subordinate ratio depends on several situation variables.

14. True.

15. True.

CHAPTER 9   WORK ENVIRONMENT

1. False. No site factor is invariably the most important. Transportation is a major consideration, but labor considerations are often more crucial.

2. True.

3. False. Assembly lines are associated with a product layout.

4. True.

5. True.

6. False. From-to charts display distances between departments, number of trips, or cost for material handling. A link analysis explores communications.

7. False. Vowels plus the letter x rate the importance of closeness.

8. True.

9. True.

10. False. Glare is the most harmful effect of illumination.

11. False. The decibel scale is logarithmic.

12. True.

13. True.

14. True.

15. False. A work station should be able to accommodate a range of physical sizes, say 40 percent of the user population.

CHAPTER 10   METHODS AND MEASUREMENT

1. True.

2. False. Operations might be eliminated, combined, or rearranged—not duplicated.

3. True.

4. False. Contribution is the difference between output value and the sum of labor plus machine burden cost.

5. False. Motion economy principles suggest the most efficient body movements.

6. True.

7. True.

8. False. Normal time equals selected time multiplied by a rating factor.

9. True.

10. True.

11. False. MTM is a predetermined time system.

12. True.

13. True.

14. False. Time-based wages require less clerical work than do incentive wages.

15. False. The Scanlon and Rucker plans were started in the 1930s.

16. False. Comparable worth is determined from equivalent qualifications for jobs.

17. True.

CHAPTER 11 MACHINES AND MAINTENANCE

1. False. General-purpose machine tools are the most common.

2. True.

3. False. Johnson's rule sequences n jobs through just two machines.

4. False. Assembly lines in a product layout are the candidates for line balancing.

5. False. Group replacement is most feasible for identical low-cost items that are increasingly prone to fail as they age.

6. True.

7. False. Preventive maintenance is unglamourous, important, and very difficult to do well.

8. False. Twice the number of servers does not necessarily cut waiting time in half.

9. True.

10. True.

11. False. Simulation indicates a preferred solution, but it is not necessarily the best possible solution.

12. True.

13. False. Manufacturing automation protocol (MAP) is the standardized communication system.

14. True.

15. True.

16. True.

CHAPTER 12 MATERIAL AND INVENTORY MANAGEMENT

1. True.

2. False. Hedging is the practice of buying contracts for future deliveries.

3. False. The difference between ordering and delivering is the lead time.

4. True.

5. False. Cash discounts are earned by prompt payments. Trade discounts are determined by the classification of the buyer.

6. True.

7. False. Running out of stock is an opportunity cost.

8. False. Order quantities are associated with independent demand.

9. True.

10. True.

11. True.

12. False. The order size depends on usage during the fixed interval in a periodic interval system. Fixed-order sizes are used in the periodic system.

13. True.

14. True.

15. False. Four elements of AS/RS are racks, S/R machines, conveyors, and controls.

16. True.

CHAPTER 13 SYSTEMS SYNTHESIS: MRP, MRP II, AND JIT

1. True.

2. True.

3. False. The Wagner–Whitin algorithm is more demanding, but it provides an optimal solution, whereas the Silver–Meal heuristic does not.

4. True.

5. False. Well under half the MRP applications are successful.

6. False. The master production schedule gives planned production quantities.

7. False. An end-item has independent demand.

8. True.

9. True.

10. True.

11. False. The boulders represent production problems.

12. False. MRP II, not JIT, provides a database for simulation.

13. False. Kanban is essentially a manual system.

14. True.

15. False. JIT is associated with thinking workers who have multiple skills.

16. True.

CHAPTER 14 QUANTITY CONTROL

1. True.

2. False. Order control is more complex than is flow control.

3. False. Backward scheduling is used when a deadline is known.

4. True.

5. False. Work orders are issued as part of the dispatching function.

6. False. A critical ratio is an index of the relative priority of jobs.

7. True.

8. False. Float is symbolized by dashed lines on time charts, but there is no CPM network symbol for float.

9. True.

10. False. An objective chart shows the completion schedule for all products. A program plan shows the operations needed for one product.

11. True.

12. True.

13. False. The symbol a stands for the time required to produce the first unit.

14. True.

15. True.

16. True.

CHAPTER 15 QUALITY ASSURANCE

1. False. Quality and productivity are two sides of the same coin. Concerns for both are increasing.

2. False. A process must be able to produce products within tolerance limits in order to meet specifications *and* be economically viable.

3. False. Value increases at a decreasing rate as technical excellence is increased.

4. False. Inspection should precede irreversible processes, so as to reduce costs and to detect errors before they are hidden.

5. False. Attributes inspection is appropriate to most service functions.

6. False. Acceptance sampling is used for both incoming and outgoing shipments.

7. True.

8. True.

9. True.

10. False. The acceptance number is the number of allowable defectives in the sample, not the total lot.

11. True.

12. False. Bored and tired inspectors may make mistakes in 100-percent screening.

13. True.

14. False. The distance from the mean to the UCL may be larger than the distance to the LCL when the LCL is zero.

15. True.

16. False. A point can fall outside the control limits by chance, but the chance is remote.

17. False. Even desirable samples should be investigated to see whether there is a reason that the results are so good.

18. False. Lack of attention is more serious and more difficult to combat.

19. True.

20. True.

21. True.

22. False. CEDAC is a Cause-and-Effect Diagram with the Addition of Cards.

23. True.

24. True.

CHAPTER 16   PRODUCTIVITY

1. False. Increasing productivity does not necessarily increase quantity and quality.

2. False. An increase in production does not necessarily lead to a productivity increase.

3. True.

4. False. Simplistic instruments are usually too vague to be of much value.

5. True.

6. True.

7. False. Productivity ratios reveal the effectiveness of resource utilization.

8. True.

9. True.

10. True.

11. True.

12. False. Weights are usually determined by the managers of the work unit represented by the matrix.

13. False. The score should be 3, as the performance level for 3 has been surpassed, but a score of 4 has not yet been earned.

14. False. A.I.M. is the acronym for awareness, improvement, and maintenance.

15. False. A productivity audit questions resource-utilization efficiency instead of how employees feel about their jobs.

16. False. Economic justification should be applied to all investment proposals.

17. True.

18. True.

19. True.

20. False. The maintenance phase measures performance and monitors progress to sustain motivation and momentum.

21. True.

22. False. The ratios are the same.

23. False. Cost ratios tell whether gross margins adequately cover costs.

24. False. Investments made to protect the environment seldom contribute to greater productivity.

A-2 SOLUTIONS TO SELECTED PROBLEMS

CHAPTER 3

3.2 a. Y_{Jan} = \$247,000 b. $r = 0$ c. range: 1482 $\pm$ 31.24 d. F_{Jan} = \$247,000 e. F_{Jan} = 248

3.4 a. Y_{f9} = 162.355 b. confidence interval: $Y_f \pm 5.332$

3.6 a. F_{winter} = 114.5, F_{spring} = 67.9, F_{summer} = 76.9, F_{fall} = 58.2 b. F_{winter} = 110.6, F_{spring} = 69.6, F_{summer} = 73.6 F_{fall} = 54.0

CHAPTER 4

4.2 $12,000

4.4 a. second modification   b. 11,500 units

4.6 50 percent

4.8 a. 53 and 187 units   b. 110 units

4.10 a. $P = 0.4931 + (7.6269)\dfrac{1}{D}$ b. $P = 7.5577$ c. $P = \$7,627.39$

CHAPTER 5

5.2 $64,358

5.4 4.2 percent

5.6 $2,580.77

5.8 plan 1

5.10 71.4 percent

5.12 choose new model

5.14 a. buy one B machine   b. buy one B machine

5.16 annual receipts

5.18 alternative B

5.20 site 1

CHAPTER 6

6.2 a. $P1$ uses $MC2$, $P2$ uses $MC3$, $P3$ uses $MC1$, $P4$ uses $MC4$ b. $P4$ uses the new machine

6.4 a. $L = 10,000$, $S = 0$, $Z = \$15,000$ b. any point on the line segment MP

6.6 b. Problem A total transportation cost $= \$242$, Problem B total transportation cost $= \$14,722$

6.8 total transportation cost $= \$5,620$

6.10 a. total transportation cost $= \$2,465$ b. $0.392

6.12 a. choose plan 2   b. $659,550

CHAPTER 7

7.2 B-1, G-2, F-2, C-0, H-0, I-2

7.4 a. durations: 40, 90, 140, 190 . . . b. 45.55

7.7 $LF = 39$

7.8 12

7.10 19.67 weeks

7.12 c. 0.9987   d. 0.9252

7.14 a. 19 days   b. 15 days   c. 1 day—$50, 2 days—$125, 3 days—$210

7.17 a and b. critical path: 1-3-4-6-7   c. critical path: 1-3-5 and 1-5, probability $= 0.5$

7.18 duration $= 18$ hours, total cost $= \$6,760$

CHAPTER 9

9.2 $x > 0.9936$

9.3 a. minimum travel distance $= 2110$ ft

9.6 b. transportation distance $= 165,900$ ft

CHAPTER 10

10.3 project B

10.5 $14.24

10.7 $\pm 20\%$

10.8 74

10.10 a. 76   b. 7.32%

10.12 a. time $= 36(x - 6) + [60 + (x - 6)30]\dfrac{Z}{200}$ b. 151.8 sec

10.14 s $= 98\%$, range: $13.8 - 26.2$

10.16 a. 64%   b. s $= \pm 17.02\%$ c. range: $31.2 - 96.5$ d. day 6

10.18 a. increase   b. nothing   c. $11.50

CHAPTER 11

11.2 B-F-C-J-E-H-D-A-I-G and B-F-J-C-E-H-D-A-I-G

11.4 19

11.6 a. cycle time $= 45$, total idle time $= 30$ b. cycle time $= 61$, total idle time $= 20$

11.8 group replacement is preferred

11.10 5

11.12 alternative 1

11.14 a. 0.00122   b. no

11.16 0.00122

11.18 6 cabs − 0.93, 5 cabs − 0.05, 4 cabs − 0.02, 3 cabs − 0.00

CHAPTER 12

12.2 $1079.80

12.4 5.78 weeks

12.6 500-ton size discount preferred

12.8 a. 5 bottles   b. $1,326, 117 bottles

12.10 510 units

12.12 c. 1. 10 units 2. 22.36 units

CHAPTER 13

13.2 a. $643.20   b. $501.20   c. $501.20

CHAPTER 14

14.2 a. 711, 712, 713, 714   b. Job 711: $CR = 12$, $P = 3$; Job 712: $CR = 2$, $P = 2$; Job 713: $CR = 0.67$, $P = 1$; Job 714: $CR = 10$, $P = 4$

14.10 a. 90.12%   b. $Y_2 = 4.818$ hr, $Y_3 = 4.446$ hr, $Y_4 = 4.23$ hr   c. 9 units

14.12 77.97%

CHAPTER 15

15.2 c. AOQL = 0.013534, occurs between $P = 0.020$ and $P = 0.025$ d. AQL = 0.8%, LTPD = 5.5%

15.4 AQL = 2.75%, LTPD = 47.50%

15.5 a. 0.10   b. 0.05   c. 0.75   d. approximately 1.00

15.8 a. UCL = 0.0539, $\bar{\bar{x}} = 0.4992$, LCL = 0   b. UCL = 0.0539, $\bar{R} = 0.0269$, LCL = 0

15.9 Process needs centering.

15.12 under control, $\bar{P} = 4$

CHAPTER 16

16.2 a. Fabrication PPI = 1.06, Assembly PPI = 1.01   b. 1.034

16.4 a. Labor productivity: −2.90%, Factor productivity: −3.16% Total productivity: −2.10%   b. no, productivity declined

16.6 a. Labor productivity = 2.931   b. Total output = 2.625 million tons

16.8   a. *EPI* = 1.254   b. *EPRI* = 0.674   c. *ECEI* = 0.944   e. *MPI* = 1.097   f. *CFPI* = 1.097   g. *TPI* = 1.060

16.10 a. first-period performance indicator = 325, second-period performance indicator = 310   b. first-period PI = 8.3%, second-period PI = 3.3%, first-year PI = −4.6% c. policy errors

16.13 b. 730   c. 5.04%

16.14 a. −2.56%

APPENDIX B
STATISTICAL TABLES*

TABLE B.1
Areas of a standard normal distribution

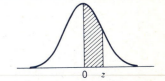

An entry in the table is the proportion under the entire curve, which is between $z = 0$ and a positive value of z. Areas for negative values of z are obtained by symmetry.

z	0.00	0.01	0.02	0.03	0.04	0.05	0.06	0.07	0.08	0.09
0.0	0.0000	0.0040	0.0080	0.0120	0.0160	0.0199	0.0239	0.0279	0.0319	0.0359
0.1	0.0398	0.0438	0.0478	0.0517	0.0557	0.0596	0.0636	0.0675	0.0714	0.0753
0.2	0.0793	0.0832	0.0871	0.0910	0.0948	0.0987	0.1026	0.1064	0.1103	0.1141
0.3	0.1179	0.1217	0.1255	0.1293	0.1331	0.1368	0.1406	0.1443	0.1480	0.1517
0.4	0.1554	0.1591	0.1628	0.1664	0.1700	0.1736	0.1772	0.1808	0.1844	0.1879
0.5	0.1915	0.1950	0.1985	0.2019	0.2054	0.2088	0.2123	0.2157	0.2190	0.2234
0.6	0.2257	0.2291	0.2324	0.2357	0.2389	0.2422	0.2454	0.2486	0.2517	0.2549
0.7	0.2580	0.2611	0.2642	0.2673	0.2703	0.2734	0.2764	0.2794	0.2823	0.2852
0.8	0.2881	0.2910	0.2939	0.2967	0.2995	0.3023	0.3051	0.3078	0.3106	0.3133
0.9	0.3159	0.3186	0.3212	0.3238	0.3264	0.3289	0.3315	0.3340	0.3365	0.3389
1.0	0.3413	0.3438	0.3461	0.3485	0.3508	0.3531	0.3554	0.3577	0.3599	0.3621
1.1	0.3643	0.3665	0.3686	0.3708	0.3729	0.3749	0.3770	0.3790	0.3810	0.3830
1.2	0.3849	0.3869	0.3888	0.3907	0.3925	0.3944	0.3962	0.3980	0.3997	0.4015
1.3	0.4032	0.4049	0.4066	0.4082	0.4099	0.4115	0.4131	0.4147	0.4162	0.4177
1.4	0.4192	0.4207	0.4222	0.4236	0.4251	0.4265	0.4279	0.4292	0.4306	0.4319
1.5	0.4332	0.4345	0.4357	0.4370	0.4382	0.4394	0.4406	0.4418	0.4429	0.4441
1.6	0.4452	0.4463	0.4474	0.4484	0.4495	0.4505	0.4515	0.4525	0.4535	0.4545
1.7	0.4554	0.4564	0.4573	0.4582	0.4591	0.4599	0.4608	0.4616	0.4625	0.4633
1.8	0.4641	0.4649	0.4656	0.4664	0.4671	0.4678	0.4686	0.4693	0.4699	0.4706
1.9	0.4713	0.4719	0.4726	0.4732	0.4738	0.4744	0.4750	0.4756	0.4761	0.4767
2.0	0.4772	0.4778	0.4783	0.4788	0.4793	0.4798	0.4803	0.4808	0.4812	0.4817
2.1	0.4821	0.4826	0.4830	0.4834	0.4838	0.4842	0.4846	0.4850	0.4854	0.4857
2.2	0.4861	0.4864	0.4868	0.4871	0.4875	0.4878	0.4881	0.4884	0.4887	0.4890
2.3	0.4893	0.4896	0.4898	0.4901	0.4904	0.4906	0.4909	0.4911	0.4913	0.4916
2.4	0.4918	0.4920	0.4922	0.4925	0.4927	0.4929	0.4931	0.4932	0.4934	0.4936
2.5	0.4938	0.4940	0.4941	0.4943	0.4945	0.4946	0.4948	0.4949	0.4951	0.4952
2.6	0.4953	0.4955	0.4956	0.4957	0.4959	0.4960	0.4961	0.4962	0.4963	0.4964
2.7	0.4965	0.4966	0.4967	0.4968	0.4969	0.4970	0.4971	0.4972	0.4973	0.4974
2.8	0.4974	0.4975	0.4976	0.4977	0.4977	0.4978	0.4979	0.4979	0.4980	0.4981
2.9	0.4981	0.4982	0.4982	0.4983	0.4984	0.4984	0.4985	0.4985	0.4986	0.4986
3.0	0.4987	0.4987	0.4987	0.4988	0.4988	0.4989	0.4989	0.4989	0.4990	0.4990

*Source: The tables in this appendix were taken with permission from Paul G. Hoel, *Elementary Statistics*, 2nd ed., New York: Wiley, 1966.

TABLE B.2
Student's t distribution

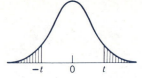

The first column lists the number of degrees of freedom (v). The headings of the other columns give probabilities (P) for t to exceed numerically the entry value.

v \ P	0.50	0.25	0.10	0.05	0.025	0.01	0.005
1	1.00000	2.4142	6.3138	12.706	25.452	63.657	127.32
2	0.81650	1.6036	2.9200	4.3027	6.2053	9.9248	14.089
3	0.76489	1.4226	2.3534	3.1825	4.1765	5.8409	7.4533
4	0.74070	1.3444	2.1318	2.7764	3.4954	4.6041	5.5976
5	0.72669	1.3009	2.0150	2.5706	3.1634	4.0321	4.7733
6	0.71756	1.2733	1.9432	2.4469	2.9687	3.7074	4.3168
7	0.71114	1.2543	1.8946	2.3646	2.8412	3.4995	4.0293
8	0.70639	1.2403	1.8595	2.3060	2.7515	3.3554	3.8325
9	0.70272	1.2297	1.8331	2.2622	2.6850	3.2498	2.6897
10	0.69981	1.2213	1.8125	2.2281	2.6338	3.1693	3.5814
11	0.69745	1.2145	1.7959	2.2010	2.5931	3.1058	3.4966
12	0.69548	1.2089	1.7823	2.1788	2.5600	3.0545	3.4284
13	0.69384	1.2041	1.7709	2.1604	2.5326	3.0123	3.3725
14	0.69242	1.2001	1.7613	2.1448	2.5096	2.9768	3.3257
15	0.69120	1.1967	1.7530	2.1315	2.4899	2.9467	3.2860
16	0.69013	1.1937	1.7459	2.1199	2.4729	2.9208	3.2520
17	0.68919	1.1910	1.7396	2.1098	2.4581	2.8982	3.2225
18	0.68837	1.1887	1.7341	2.1009	2.4450	2.8784	3.1966
19	0.68763	1.1866	1.7291	2.0930	2.4334	2.8669	3.1737
20	0.68696	1.1848	1.7247	2.0860	2.4231	2.8453	3.1534
21	0.68635	1.1831	1.7207	2.0796	2.4138	2.8314	3.1352
22	0.68580	1.1816	1.7171	2.0739	2.4055	2.8188	3.1188
23	0.68531	1.1802	1.7139	2.0687	2.3979	2.8073	3.1040
24	0.68485	1.1789	1.7109	2.0639	2.3910	2.7969	3.0905
25	0.68443	1.1777	1.7081	2.0595	2.3846	2.7874	3.0782
26	0.68405	1.1766	1.7056	2.0555	2.3788	2.7787	3.0669
27	0.68370	1.1757	1.7033	2.0518	2.3734	2.7707	3.0565
28	0.68335	1.1748	1.7011	2.0484	2.3685	2.7633	3.0469
29	0.68304	1.1739	1.6991	2.0452	2.3683	2.7564	3.0380
30	0.68276	1.1731	1.6973	2.0423	2.3596	2.7500	3.0298
40	0.68066	1.1673	1.6839	2.0211	2.3289	2.7045	2.9712
60	0.67862	1.1616	1.6707	2.0003	2.2991	2.6603	2.9146
120	0.67656	1.1559	1.6577	1.9799	2.2699	2.6174	2.8599
∞	0.67449	1.1503	1.6449	1.9600	2.2414	2.5758	2.8070

APPENDIX C
TABLE OF RANDOM NUMBERS*

03 99 11 04 61	93 71 61 68 94	66 08 32 46 53	84 60 95 82 32	88 61 81 91 61
38 55 59 55 54	32 88 65 97 80	08 35 56 08 60	29 73 54 77 62	71 29 92 38 53
17 54 67 37 04	92 05 24 62 15	55 12 12 92 81	59 07 60 79 36	27 95 45 89 09
32 64 35 28 61	95 81 90 68 31	00 91 19 89 36	76 35 59 37 79	80 86 30 05 14
69 57 26 87 77	39 51 03 59 05	14 06 04 06 19	29 54 96 96 16	33 56 46 07 80
24 12 26 65 91	27 69 90 64 94	14 84 54 66 72	61 95 87 71 00	90 89 97 57 54
61 19 63 02 31	92 96 26 17 73	41 83 95 53 82	17 26 77 09 43	78 03 87 02 67
30 53 22 17 04	10 27 41 22 02	39 68 52 33 09	10 06 16 88 29	55 98 66 64 85
03 78 89 75 99	75 86 72 07 17	74 41 65 31 66	35 20 83 33 74	87 53 90 88 23
48 22 86 33 79	85 78 34 76 19	53 15 26 74 33	35 66 35 29 72	16 81 86 03 11
60 36 59 46 53	35 07 53 39 49	42 61 42 92 97	01 91 32 83 16	98 95 37 32 31
83 79 94 24 02	56 62 33 44 42	34 99 44 13 74	70 07 11 47 36	09 95 81 80 65
32 96 00 74 05	36 40 98 32 32	99 38 54 16 00	11 13 30 75 86	15 91 70 62 53
19 32 25 38 45	57 62 05 26 06	66 49 76 86 46	78 13 86 65 59	19 64 09 94 13
11 22 09 47 47	07 39 93 74 08	48 50 92 39 29	27 48 24 54 76	85 24 43 51 49
31 75 15 72 60	68 98 00 53 39	15 47 04 83 55	88 65 12 25 96	03 15 21 91 21
88 49 29 93 82	14 45 40 45 04	20 09 49 89 77	74 84 39 34 13	22 10 97 85 08
30 93 44 77 44	07 48 18 38 28	73 78 80 65 33	28 59 72 04 05	94 20 52 03 80
22 88 84 88 93	27 49 99 87 48	60 53 04 51 28	74 02 28 46 17	82 03 71 02 68
78 21 21 69 93	35 90 29 13 86	44 37 21 54 86	65 74 11 40 14	87 48 13 72 20
41 84 98 45 47	46 85 05 23 26	34 67 75 83 00	74 91 06 43 45	19 32 58 15 49
46 35 23 30 49	69 24 89 34 60	45 30 50 75 21	61 31 83 18 55	14 41 37 09 51
11 08 79 62 94	14 01 33 17 92	59 74 76 72 77	76 50 33 45 13	39 66 37 75 44
52 70 10 83 37	56 30 38 73 15	16 52 06 96 76	11 65 49 98 93	02 18 16 81 61
57 27 53 68 98	81 30 44 85 85	68 65 22 73 76	92 85 25 58 66	88 44 80 35 84
20 85 77 31 56	70 28 42 43 26	79 37 59 52 20	01 15 96 32 67	10 62 24 83 91
15 63 38 49 24	90 41 59 36 14	33 52 12 66 65	55 82 34 76 41	86 22 53 17 04
92 69 44 82 97	39 90 40 21 15	59 58 94 90 67	66 82 14 15 75	49 76 70 40 37
77 61 31 90 19	88 15 20 00 80	20 55 49 14 09	96 27 74 82 57	50 81 69 76 16
38 68 83 24 86	45 13 46 35 45	59 40 47 20 59	43 94 75 16 80	43 85 25 96 93
25 16 30 18 89	70 01 41 50 21	41 29 06 73 12	71 85 71 59 57	68 97 11 14 93
65 25 10 76 29	37 23 93 32 95	05 87 00 11 19	92 78 42 63 40	18 47 76 56 22
36 81 54 36 25	18 63 73 75 09	82 44 49 90 05	04 92 17 37 01	14 70 79 39 97
64 39 71 16 92	05 32 78 21 62	20 24 78 17 59	45 19 72 53 32	33 74 52 25 67
04 51 52 56 24	95 09 66 79 46	48 46 08 55 58	15 19 11 87 82	16 93 03 33 61
83 76 16 08 72	43 25 38 41 45	60 33 32 59 83	01 29 14 13 49	20 36 80 71 26
14 38 70 63 45	80 85 40 92 79	43 52 90 63 18	38 38 47 47 61	41 19 63 74 80
51 32 19 22 46	80 08 87 70 74	88 72 25 67 36	66 16 44 94 31	66 91 93 16 78
72 47 20 00 08	80 89 01 80 02	94 81 33 19 00	54 15 58 34 36	35 35 25 41 31
05 46 65 53 06	93 12 81 84 64	74 45 79 05 61	72 84 81 18 34	79 98 26 84 16
39 52 87 24 84	82 47 42 55 93	48 54 53 52 47	18 61 91 36 74	18 61 11 92 41
81 61 61 87 11	53 34 24 42 76	75 12 21 17 24	74 62 77 37 07	58 31 91 59 97
07 58 61 61 20	82 64 12 28 20	92 90 41 31 41	32 39 21 97 63	61 19 96 79 40
90 76 70 42 35	13 57 41 72 00	69 90 26 37 42	78 46 42 25 01	18 62 79 08 72
40 18 82 81 93	29 59 38 86 27	94 97 21 15 98	62 09 53 67 87	00 44 15 89 97
34 41 48 21 57	86 88 75 50 87	19 15 20 00 23	12 30 28 07 83	32 62 46 86 91
63 43 97 53 63	44 98 91 68 22	36 02 40 08 67	76 37 84 16 05	65 96 17 34 88
67 04 90 90 70	93 39 94 55 47	94 45 87 42 84	05 04 14 98 07	20 28 83 40 60
79 49 50 41 46	52 16 29 02 86	54 15 83 42 43	46 97 83 54 82	59 36 29 59 38
91 70 43 05 52	04 73 72 10 31	75 05 19 30 29	47 66 56 43 82	99 78 29 34 78

*The random numbers in this appendix were developed by the Rand Corporation.

APPENDIX D
TABLE OF
LOGARITHMS

N	0	1	2	3	4	5	6	7	8	9
10	0000	0043	0086	0128	0170	0212	0253	0294	0334	0374
11	0414	0453	0492	0531	0569	0607	0645	0682	0719	0755
12	0792	0828	0864	0899	0934	0969	1004	1038	1072	1106
13	1139	1173	1206	1239	1271	1303	1335	1367	1399	1430
14	1461	1492	1523	1553	1584	1614	1644	1673	1703	1732
15	1761	1790	1818	1847	1875	1903	1931	1959	1987	2014
16	2041	2068	2095	2122	2148	2175	2201	2227	2253	2279
17	2304	2330	2355	2380	2405	2430	2455	2480	2504	2529
18	2553	2577	2601	2625	2648	2672	2695	2718	2742	2765
19	2788	2810	2833	2856	2878	2900	2923	2945	2967	2989
20	3010	3032	3054	3075	3096	3118	3139	3160	3181	3201
21	3222	3243	3263	3284	3304	3324	3345	3365	3385	3404
22	3424	3444	3464	3483	3502	3522	3541	3560	3579	3598
23	3617	3636	3655	3674	3692	3711	3729	3747	3766	3784
24	3802	3820	3838	3856	3874	3892	3909	3927	3945	3962
25	3979	3997	4014	4031	4048	4065	4082	4099	4116	4133
26	4150	4166	4183	4200	4216	4232	4249	4265	4281	4298
27	4314	4330	4346	4362	4378	4393	4409	4425	4440	4456
28	4472	4487	4502	4518	4533	4548	4564	4579	4594	4609
29	4624	4639	4654	4669	4683	4698	4713	4728	4742	4757
30	4771	4786	4800	4814	4829	4843	4857	4871	4886	4900
31	4914	4928	4942	4955	4969	4983	4997	5011	5024	5038
32	5051	5065	5079	5092	5105	5119	5132	5145	5159	5172
33	5185	5198	5211	5224	5237	5250	5263	5276	5289	5302
34	5315	5328	5340	5353	5366	5378	5391	5403	5416	5428
35	5441	5453	5465	5478	5490	5502	5514	5527	5539	5551
36	5563	5575	5587	5599	5611	5623	5635	5647	5658	5670
37	5682	5694	5705	5717	5729	5740	5752	5763	5775	5786
38	5798	5809	5821	5832	5843	5855	5866	5877	5888	5899
39	5911	5922	5933	5944	5955	5966	5977	5988	5999	6010
N	0	1	2	3	4	5	6	7	8	9

N	0	1	2	3	4	5	6	7	8	9
40	6021	6031	6042	6053	6064	6075	6085	6096	6107	6117
41	6128	6138	6149	6160	6170	6180	6191	6201	6212	6222
42	6232	6243	6253	6263	6274	6284	6294	6304	6314	6325
43	6335	6345	6355	6365	6375	6385	6395	6405	6415	6425
44	6435	6444	6454	6464	6474	6484	6493	6503	6513	6522
45	6532	6542	6551	6561	6571	6580	6590	6599	6609	6618
46	6628	6637	6646	6656	6665	6675	6684	6693	6702	6712
47	6721	6730	6739	6749	6758	6767	6776	6785	6794	6803
48	6812	6821	6830	6839	6848	6857	6866	6875	6884	6893
49	6902	6911	6920	6928	6937	6946	6955	6964	6972	6981
50	6990	6998	7007	7016	7024	7033	7042	7050	7059	7067
51	7076	7084	7093	7101	7110	7118	7126	7135	7143	7152
52	7160	7168	7177	7185	7193	7202	7210	7218	7226	7235
53	7243	7251	7259	7267	7275	7284	7292	7300	7308	7316
54	7324	7332	7340	7348	7356	7364	7372	7380	7388	7396
55	7404	7412	7419	7427	7435	7443	7451	7459	7466	7474
56	7482	7490	7497	7505	7513	7520	7528	7536	7543	7551
57	7559	7566	7574	7582	7589	7597	7604	7612	7619	7627
58	7634	7642	7649	7657	7664	7672	7679	7686	7694	7701
59	7709	7716	7723	7731	7738	7745	7752	7760	7767	7774
60	7782	7789	7796	7803	7810	7818	7825	7832	7839	7846
61	7853	7860	7868	7875	7882	7889	7896	7903	7910	7917
62	7924	7931	7938	7945	7952	7959	7966	7973	7980	7987
63	7993	8000	8007	8014	8021	8028	8035	8041	8048	8055
64	8062	8069	8075	8082	8089	8096	8102	8109	8116	8122
65	8192	8136	8142	8149	8156	8162	8169	8176	8182	8189
66	8195	8202	8209	8215	8222	8228	8235	8241	8248	8254
67	8261	8267	8274	8280	8287	8293	8299	8306	8312	8319
68	8325	8331	8338	8344	8351	8357	8363	8370	8376	8382
69	8388	8395	8401	8407	8414	8420	9426	8432	8439	8445
70	8451	8457	8463	8470	8476	8482	8488	8494	8500	8506
71	8513	8519	8525	8531	8537	8543	8549	8555	8561	8567
72	8573	8579	8585	8591	8597	8603	8609	8615	8621	8627
73	8633	8639	8645	8651	8657	8663	8669	8675	8681	8686
74	8692	8698	8704	8710	8716	8722	8727	8733	8739	8745
75	8751	8756	8762	8768	8774	8779	8785	8791	8797	8802
76	8808	8814	8820	8825	8831	8837	8842	8848	8854	8859
77	8865	8871	8876	8882	8887	8893	8899	8904	8910	8915
78	8921	8927	8932	8938	8943	8949	8954	8960	8965	8971
79	8976	8982	8987	8993	8998	9004	9009	9015	9020	9025
N	0	1	2	3	4	5	6	7	8	9

N	0	1	2	3	4	5	6	7	8	9
80	9031	9036	9042	9047	9053	9058	9063	9069	9074	9079
81	9085	9090	9096	9101	9106	9112	9117	9122	9128	9133
82	9138	9143	9149	9154	9159	9165	9170	9175	9180	9186
83	9191	9196	9201	9206	9212	9217	9222	9227	9232	9238
84	9243	9248	9253	9258	9263	9269	9274	9279	9284	9289
85	9294	9299	9304	9309	9315	9320	9325	9330	9335	9340
86	9345	9350	9355	9360	9365	9370	9375	9380	9385	9390
87	9395	9400	9405	9410	9415	9420	9425	9430	9435	9440
88	9445	9450	9455	9460	9465	9469	9474	9479	9484	9489
89	9494	9499	9504	9509	9513	9518	9523	9528	9533	9538
90	9542	9547	9552	9557	9562	9566	9571	9576	9581	9586
91	9590	9595	9600	9605	9609	9614	9619	9624	9628	9636
92	9638	9643	9647	9652	9657	9661	9666	9671	9675	9680
93	9685	9689	9694	9699	9703	9708	9713	9717	9722	9727
94	9731	9736	9741	9745	9750	9754	9759	9763	9768	9773
95	9777	9782	9786	9791	9795	9800	9805	9809	9814	9818
96	9823	9827	9832	9836	9841	9845	9850	9854	9859	9863
97	9868	9872	9877	9881	9886	9890	9894	9899	9903	9908
98	9912	9917	9921	9926	9930	9934	9939	9943	9948	9952
99	9956	9961	9965	9969	9974	9978	9983	9987	9991	9996
N	0	1	2	3	4	5	6	7	8	9

½%

n	To find F, given P: $(1 + i)^n$	To find P, given F: $\dfrac{1}{(1 + i)^n}$	To find A, given F: $\dfrac{i}{(1 + i)^n - 1}$	To find A, given P: $\dfrac{i(1 + i)^n}{(1 + i)^n - 1}$	To find F, given A: $\dfrac{(1 + i)^n - 1}{i}$	To find P, given A: $\dfrac{(1 + i)^n - 1}{i(1 + i)^n}$	n
	$(f/p,½,n)$	$(p/f,½,n)$	$(a/f,½,n)$	$(a/p,½,n)$	$(f/a,½,n)$	$(p/a,½,n)$	
1	1.005	0.9950	1.00000	1.00500	1.000	0.995	1
2	1.010	0.9901	0.49875	0.50375	2.005	1.985	2
3	1.015	0.9851	0.33167	0.33667	3.015	2.970	3
4	1.020	0.9802	0.24183	0.25313	4.030	3.950	4
5	1.025	0.9754	0.19801	0.20301	5.050	4.926	5
6	1.030	0.9705	0.16460	0.16960	6.076	5.896	6
7	1.036	0.9657	0.14073	0.14573	7.106	6.862	7
8	1.041	0.9609	0.12283	0.12783	8.141	7.823	8
9	1.046	0.9561	0.10891	0.11391	9.182	8.779	9
10	1.051	0.9513	0.09777	0.10277	10.288	9.730	10
11	1.056	0.9466	0.08866	0.09366	11.279	10.677	11
12	1.062	0.9419	0.08107	0.08607	12.336	11.619	12
13	1.067	0.9372	0.07464	0.07964	13.397	12.556	13
14	1.072	0.9326	0.06914	0.07414	14.464	13.489	14
15	1.078	0.9279	0.06436	0.06936	15.537	14.417	15
16	1.083	0.9233	0.06019	0.06519	16.614	15.340	16
17	1.088	0.9187	0.05615	0.06151	17.697	16.259	17
18	1.094	0.9141	0.05323	0.05823	18.786	17.173	18
19	1.099	0.9096	0.05030	0.05530	19.880	18.082	19
20	1.105	0.9051	0.04767	0.05267	20.979	18.987	20
21	1.110	0.9006	0.04528	0.05028	22.084	19.888	21
22	1.116	0.8961	0.04311	0.04811	23.194	20.784	22
23	1.122	0.8916	0.04113	0.04613	24.310	21.676	23
24	1.127	0.8872	0.03932	0.04432	25.432	22.563	24
25	1.133	0.8828	0.03767	0.04265	26.559	23.446	25
26	1.138	0.8784	0.03611	0.04111	27.692	24.324	26
27	1.144	0.8740	0.03469	0.03969	28.830	25.198	27
28	1.150	0.8697	0.03336	0.03836	29.975	26.068	28
29	1.156	0.8653	0.03213	0.03713	31.124	26.933	29
30	1.161	0.8610	0.03098	0.03598	32.280	27.794	30
31	1.167	0.8567	0.02990	0.03490	33.441	28.651	31
32	1.173	0.8525	0.02889	0.03389	34.609	29.503	32
33	1.179	0.8482	0.02795	0.03295	35.782	30.352	33
34	1.185	0.8440	0.02706	0.03206	36.961	31.196	34
35	1.191	0.8398	0.02622	0.03122	38.145	32.035	35
40	1.221	0.8191	0.02265	0.02765	44.159	36.172	40
45	1.252	0.7990	0.01987	0.02487	50.324	40.207	45
50	1.283	0.7793	0.01765	0.02265	56.645	44.143	50
55	1.316	0.7601	0.01548	0.02084	63.126	47.981	55
60	1.349	0.7414	0.01433	0.01933	69.770	51.726	60
65	1.383	0.7231	0.01306	0.01806	76.582	55.377	65
70	1.418	0.7053	0.01197	0.01697	83.566	58.939	70
75	1.454	0.6879	0.01102	0.01602	90.727	62.414	75
80	1.490	0.6710	0.01020	0.01520	98.068	65.802	80
85	1.528	0.6545	0.00947	0.01447	105.594	69.108	85
90	1.567	0.6383	0.00883	0.01383	113.311	72.331	90
95	1.606	0.6226	0.00825	0.01325	121.222	75.476	95
100	1.647	0.6073	0.00773	0.01273	129.334	78.543	100

	To find F, given P: $(1 + i)^n$	To find P, given F: $\dfrac{1}{(1 + i)^n}$	To find A, given F: $\dfrac{i}{(1 + i)^n - 1}$	To find A, given P: $\dfrac{i(1 + i)^n}{(1 + i)^n - 1}$	To find F, given A: $\dfrac{(1 + i)^n - 1}{i}$	To find P, given A: $\dfrac{(1 + i)^n - 1}{i(1 + i)^n}$	
n	$(f/p,1,n)$	$(p/f,1,n)$	$(a/f,1,n)$	$(a/p,1,n)$	$(f/a,1,n)$	$(p/a,1,n)$	n
1	1.010	0.9901	1.00000	1.01000	1.000	0.990	1
2	1.020	0.9803	0.49751	0.50751	2.010	1.970	2
3	1.030	0.9706	0.33002	0.34002	3.030	2.941	3
4	1.041	0.9610	0.24628	0.25628	4.060	3.902	4
5	1.051	0.9515	0.19604	0.20604	5.101	4.853	5
6	1.062	0.9420	0.16255	0.17255	6.152	5.795	6
7	1.072	0.9327	0.13863	0.14863	7.214	6.728	7
8	1.083	0.9235	0.12069	0.13069	8.286	7.652	8
9	1.094	0.9143	0.10674	0.11674	9.369	8.566	9
10	1.105	0.9053	0.09558	0.10558	10.462	9.471	10
11	1.116	0.8963	0.08645	0.09645	11.567	10.368	11
12	1.127	0.8874	0.07885	0.08885	12.683	11.255	12
13	1.138	0.8787	0.07241	0.08241	13.809	12.134	13
14	1.149	0.8700	0.06690	0.07690	14.947	13.004	14
15	1.161	0.8613	0.06212	0.07212	16.097	13.865	15
16	1.173	0.8528	0.05794	0.06794	17.258	14.718	16
17	1.184	0.8444	0.05426	0.06426	18.430	15.562	17
18	1.196	0.8360	0.05098	0.06098	19.615	16.398	18
19	1.208	0.8277	0.04805	0.05805	20.811	17.226	19
20	1.220	0.8195	0.04542	0.05542	22.019	18.046	20
21	1.232	0.8114	0.04303	0.05303	23.239	18.857	21
22	1.245	0.8034	0.04086	0.05086	24.472	19.660	22
23	1.257	0.7954	0.03889	0.04889	25.716	20.456	23
24	1.270	0.7876	0.03707	0.04707	26.973	21.243	24
25	1.282	0.7798	0.03541	0.04541	28.243	22.023	25
26	1.295	0.7720	0.03387	0.04387	29.526	22.795	26
27	1.308	0.7644	0.03245	0.04245	30.821	23.560	27
28	1.321	0.7568	0.03112	0.04112	32.129	24.316	28
29	1.335	0.7493	0.02990	0.03990	33.450	25.066	29
30	1.348	0.7419	0.02875	0.03875	34.785	25.808	30
31	1.361	0.7346	0.02768	0.03768	36.133	26.542	31
32	1.375	0.7273	0.02667	0.03667	37.494	27.270	32
33	1.391	0.7201	0.02573	0.03573	38.869	27.990	33
34	1.403	0.7130	0.02484	0.03484	40.258	28.703	34
35	1.417	0.7059	0.02400	0.03400	41.660	29.409	35
40	1.489	0.6717	0.02046	0.03046	48.886	32.835	40
45	1.565	0.6391	0.01771	0.02771	56.481	36.095	45
50	1.645	0.6080	0.01551	0.02551	64.463	39.196	50
55	1.729	0.5785	0.01373	0.02373	72.852	42.147	55
60	1.817	0.5504	0.01224	0.02224	81.670	44.955	60
65	1.909	0.5237	0.01100	0.02100	90.937	47.627	65
70	2.007	0.4983	0.00993	0.01993	100.676	50.169	70
75	2.109	0.4741	0.00902	0.01902	110.913	52.587	75
80	2.217	0.4511	0.00822	0.01822	121.672	54.888	80
85	2.330	0.4292	0.00752	0.01752	132.979	57.078	85
90	2.449	0.4084	0.00690	0.01690	144.863	59.161	90
95	2.574	0.3886	0.00636	0.01636	157.354	61.143	95
100	2.705	0.3697	0.00587	0.01587	170.481	63.029	100

	To find F, given P: $(1 + i)^n$	To find P, given F: $\dfrac{1}{(1 + i)^n}$	To find A, given F: $\dfrac{i}{(1 + i)^n - 1}$	To find A, given P: $\dfrac{i(1 + i)^n}{(1 + i)^n - 1}$	To find F, given A: $\dfrac{(1 + i)^n - 1}{i}$	To find P, given A: $\dfrac{(1 + i)^n - 1}{i(1 + i)^n}$	
n	$(f/p,1½,n)$	$(p/f,1½,n)$	$(a/f,1½,n)$	$(a/p,1½,n)$	$(f/a,1½,n)$	$(p/a,1½,n)$	n
1	1.015	0.9852	1.00000	1.01500	1.000	0.985	1
2	1.030	0.9707	0.49628	0.51128	2.015	1.956	2
3	1.046	0.9563	0.32838	0.34338	3.045	2.912	3
4	1.061	0.9422	0.24444	0.25944	4.091	3.854	4
5	1.077	0.9283	0.19409	0.20909	5.152	4.783	5
6	1.093	0.9145	0.16053	0.17553	6.230	5.697	6
7	1.110	0.9010	0.13656	0.15156	7.323	6.598	7
8	1.126	0.8877	0.11858	0.13358	8.433	7.486	8
9	1.143	0.8746	0.10461	0.11961	9.559	8.361	9
10	1.161	0.8617	0.09343	0.10843	10.703	9.222	10
11	1.178	0.8489	0.08429	0.09930	11.863	10.071	11
12	1.196	0.8364	0.07668	0.09168	13.041	10.908	12
13	1.214	0.8240	0.07024	0.08524	14.237	11.732	13
14	1.232	0.8118	0.06472	0.07972	15.450	12.543	14
15	1.250	0.7999	0.05994	0.07494	16.682	13.343	15
16	1.269	0.7880	0.05577	0.07077	17.932	14.131	16
17	1.288	0.7764	0.05208	0.06708	19.201	14.908	17
18	1.307	0.7649	0.04881	0.06381	20.489	15.673	18
19	1.327	0.7536	0.04588	0.06088	21.797	16.426	19
20	1.347	0.7425	0.04325	0.05825	23.124	17.169	20
21	1.367	0.7315	0.04087	0.05587	24.471	17.900	21
22	1.388	0.7207	0.03870	0.05370	25.838	19.621	22
23	1.408	0.7100	0.03673	0.05173	27.225	19.331	23
24	1.430	0.6995	0.03492	0.04992	28.634	20.030	24
25	1.451	0.6892	0.03325	0.04826	30.063	20.720	25
26	1.473	0.6790	0.03173	0.04673	31.514	21.399	26
27	1.495	0.6690	0.03032	0.04532	32.987	22.068	27
28	1.517	0.6591	0.02900	0.04400	34.481	22.727	28
29	1.540	0.6494	0.02778	0.04278	35.999	23.376	29
30	1.563	0.6398	0.02664	0.04164	37.539	24.016	30
31	1.587	0.6303	0.02557	0.04057	39.102	24.646	31
32	1.610	0.6210	0.02458	0.03958	40.688	25.267	32
33	1.634	0.6118	0.02364	0.03864	42.229	25.879	33
34	1.659	0.6028	0.02276	0.03776	43.933	26.482	34
35	1.684	0.5939	0.02193	0.03693	45.592	27.076	35
40	1.814	0.5513	0.01834	0.03343	54.268	29.916	40
45	1.954	0.5117	0.01572	0.03072	63.614	32.552	45
50	2.105	0.4750	0.01357	0.02857	73.683	35.000	50
55	2.268	0.4409	0.01183	0.02683	84.530	37.271	55
60	2.443	0.4093	0.01039	0.02539	96.215	39.380	60
65	2.632	0.3799	0.00919	0.02419	108.803	41.338	65
70	2.835	0.3527	0.00817	0.02317	122.364	43.155	70
75	3.055	0.3274	0.00730	0.02230	136.973	44.842	75
80	3.291	0.3039	0.00655	0.02155	152.711	46.407	80
85	3.545	0.2821	0.00589	0.02089	169.665	47.861	85
90	3.819	0.2619	0.00532	0.02032	187.930	49.210	90
95	4.114	0.2431	0.00482	0.01982	207.606	50.462	95
100	4.432	0.2256	0.00437	0.01937	228.803	51.625	100

	To find F, given P: $(1 + i)^n$	To find P, given F: $\dfrac{1}{(1 + i)^n}$	To find A, given F: $\dfrac{i}{(1 + i)^n - 1}$	To find A, given P: $\dfrac{i(1 + i)^n}{(1 + i)^n - 1}$	To find F, given A: $\dfrac{(1 + i)^n - 1}{i}$	To find P, given A: $\dfrac{(1 + i)^n - 1}{i(1 + i)^n}$	
n	$(f/p,2,n)$	$(p/f,2,n)$	$(a/f,2,n)$	$(a/p,2,n)$	$(f/a,2,n)$	$(p/a,2,n)$	n
1	1.020	0.9804	1.00000	1.02000	1.000	0.980	1
2	1.040	0.9612	0.49505	0.51505	2.020	1.942	2
3	1.061	0.9423	0.32675	0.34675	3.060	2.884	3
4	1.082	0.9238	0.24262	0.26262	4.122	3.808	4
5	1.104	0.9057	0.19216	0.21216	5.204	4.713	5
6	1.126	0.8880	0.15853	0.17853	6.308	5.601	6
7	1.149	0.8706	0.13451	0.15451	7.434	6.472	7
8	1.172	0.8535	0.11651	0.13651	8.583	7.325	8
9	1.195	0.8368	0.10252	0.12252	9.755	8.162	9
10	1.219	0.8203	0.09133	0.11133	10.905	8.983	10
11	1.243	0.8043	0.08216	0.10218	12.169	9.787	11
12	1.268	0.7885	0.07456	0.09456	13.412	10.575	12
13	1.294	0.7730	0.06812	0.08812	14.680	11.348	13
14	1.319	0.7579	0.06260	0.08260	15.974	12.106	14
15	1.346	0.7430	0.05783	0.07783	17.293	12.849	15
16	1.373	0.7284	0.05365	0.07365	18.639	13.578	16
17	1.400	0.7142	0.04997	0.06997	20.012	14.292	17
18	1.428	0.7002	0.04670	0.06670	21.412	14.992	18
19	1.457	0.6864	0.04378	0.06378	22.841	15.678	19
20	1.486	0.6730	0.04116	0.06116	24.297	16.351	20
21	1.516	0.6598	0.03878	0.05878	25.783	17.011	21
22	1.546	0.6468	0.03663	0.05663	27.299	17.658	22
23	1.577	0.6342	0.03467	0.05467	28.845	18.292	23
24	1.608	0.6217	0.03287	0.05287	30.422	18.914	24
25	1.641	0.6095	0.03122	0.05122	32.030	19.523	25
26	1.673	0.5976	0.02970	0.04970	33.671	20.121	26
27	1.707	0.5859	0.02829	0.04829	35.344	20.707	27
28	1.741	0.5744	0.02699	0.04699	37.051	21.281	28
29	1.776	0.5631	0.02578	0.04578	38.792	21.844	29
30	1.811	0.5521	0.02465	0.04465	40.568	22.396	30
31	1.848	0.5412	0.02360	0.04360	42.379	22.938	31
32	1.885	0.5306	0.02261	0.04261	44.227	23.468	32
33	1.922	0.5202	0.02169	0.04169	46.112	23.989	33
34	1.961	0.5100	0.02082	0.04082	48.034	24.499	34
35	2.000	0.5000	0.02000	0.04000	49.994	24.999	35
40	2.208	0.4529	0.01656	0.03656	60.402	27.355	40
45	2.438	0.4102	0.01391	0.03391	71.893	29.490	45
50	2.692	0.3715	0.01182	0.03182	84.579	31.424	50
55	2.972	0.3365	0.01014	0.03014	98.587	33.175	55
60	3.281	0.3048	0.00877	0.02877	114.052	34.761	60
65	3.623	0.2761	0.00763	0.02763	131.126	36.197	65
70	4.000	0.2500	0.00667	0.02667	149.978	37.499	70
75	4.416	0.2265	0.00586	0.02586	170.792	38.677	75
80	4.875	0.2051	0.00516	0.02516	193.722	39.745	80
85	5.383	0.1858	0.00456	0.02456	219.144	40.711	85
90	5.943	0.1683	0.00405	0.02405	247.157	41.587	90
95	6.562	0.1524	0.00360	0.02360	278.085	42.380	95
100	7.245	0.1380	0.00320	0.02320	312.232	43.098	100

n	To find F, given P: $(1 + i)^n$	To find P, given F: $\dfrac{1}{(1 + i)^n}$	To find A, given F: $\dfrac{i}{(1 + i)^n - 1}$	To find A, given P: $\dfrac{i(1 + i)^n}{(1 + i)^n - 1}$	To find F, given A: $\dfrac{(1 + i)^n - 1}{i}$	To find P, given A: $\dfrac{(1 + i)^n - 1}{i(1 + i)^n}$	n
	(f/p,2½,n)	(p/f,2½,n)	(a/f,2½,n)	(a/p,2½,n)	(f/a,2½,n)	(p/a,2½,n)	
1	1.025	0.9756	1.00000	1.02500	1.000	0.976	1
2	1.051	0.9518	0.49383	0.51883	2.025	1.927	2
3	1.077	0.9386	0.32514	0.35014	3.076	2.856	3
4	1.104	0.9060	0.24082	0.26582	4.153	3.762	4
5	1.131	0.8839	0.19025	0.21525	5.256	4.646	5
6	1.160	0.8623	0.15655	0.18155	6.388	5.508	6
7	1.189	0.8413	0.13250	0.15750	7.547	6.349	7
8	1.218	0.8207	0.11447	0.13947	8.736	7.170	8
9	1.249	0.8007	0.10046	0.12546	9.955	7.971	9
10	1.280	0.7812	0.08926	0.11426	11.203	8.752	10
11	1.312	0.7621	0.08011	0.10511	12.483	9.514	11
12	1.345	0.7436	0.07249	0.09749	13.796	10.258	12
13	1.379	0.7254	0.06605	0.09105	15.140	10.983	13
14	1.413	0.7077	0.06054	0.08554	16.519	11.691	14
15	1.448	0.6905	0.05577	0.08077	17.932	12.381	15
16	1.485	0.6736	0.05160	0.07660	19.380	13.055	16
17	1.522	0.6572	0.04793	0.07293	20.865	13.712	17
18	1.560	0.6412	0.04467	0.06967	22.386	14.353	18
19	1.599	0.6255	0.04176	0.06676	23.946	14.979	19
20	1.639	0.6103	0.03915	0.06415	25.545	15.589	20
21	1.680	0.5954	0.03679	0.06179	27.183	16.185	21
22	1.722	0.5809	0.03465	0.05965	28.863	16.765	22
23	1.765	0.5667	0.03270	0.05770	30.584	17.332	23
24	1.809	0.5529	0.03091	0.05591	32.349	17.885	24
25	1.854	0.5394	0.02928	0.05428	34.158	18.424	25
26	1.900	0.5262	0.02777	0.05277	36.012	18.951	26
27	1.948	0.5134	0.02638	0.05138	37.912	19.464	27
28	1.996	0.5009	0.02509	0.05009	39.860	19.965	28
29	2.046	0.4887	0.02389	0.04889	41.856	20.454	29
30	2.098	0.4767	0.02278	0.04778	43.903	20.930	30
31	2.150	0.4651	0.02174	0.04674	46.000	21.395	31
32	2.204	0.4538	0.02077	0.04577	48.150	21.849	32
33	2.259	0.4427	0.01986	0.04486	50.354	22.292	33
34	2.315	0.4319	0.01901	0.04401	52.613	22.724	34
35	2.373	0.4214	0.01821	0.04321	54.928	23.145	35
40	2.685	0.3724	0.01484	0.03984	67.403	25.103	40
45	3.038	0.3292	0.01227	0.03727	81.516	26.833	45
50	3.437	0.2909	0.01026	0.03526	97.484	28.362	50
55	3.889	0.2572	0.00865	0.03365	115.551	29.714	55
60	4.400	0.2273	0.00735	0.03235	135.992	30.909	60
65	4.978	0.2009	0.00628	0.03128	159.118	31.965	65
70	5.632	0.1776	0.00540	0.03040	185.284	32.898	70
75	6.372	0.1569	0.00465	0.02965	214.888	33.723	75
80	7.210	0.1387	0.00403	0.02903	248.383	34.452	80
85	8.157	0.1226	0.00349	0.02849	286.279	35.096	85
90	9.229	0.1084	0.00304	0.02804	329.154	35.666	90
95	10.442	0.0958	0.00265	0.02765	377.664	36.169	95
100	11.814	0.0846	0.00231	0.02731	432.549	36.614	100

3%

n	To find F, given P: $(1 + i)^n$	To find P, given F: $\dfrac{1}{(1 + i)^n}$	To find A, given F: $\dfrac{i}{(1 + i)^n - 1}$	To find A, given P: $\dfrac{i(1 + i)^n}{(1 + i)^n - 1}$	To find F, given A: $\dfrac{(1 + i)^n - 1}{i}$	To find P, given A: $\dfrac{(1 + i)^n - 1}{i(1 + i)^n}$	n
	$(f/p,3,n)$	$(p/f,3,n)$	$(a/f,3,n)$	$(a/p,3,n)$	$(f/a,3,n)$	$(p/a,3,n)$	
1	1.030	0.9709	1.00000	1.03000	1.000	0.971	1
2	1.061	0.9426	0.49261	0.52261	2.030	1.913	2
3	1.093	0.9151	0.32353	0.35353	3.091	2.829	3
4	1.126	0.8885	0.23903	0.26903	4.184	3.717	4
5	1.159	0.8626	0.18835	0.21835	5.309	4.580	5
6	1.194	0.8375	0.15460	0.18460	6.468	5.417	6
7	1.230	0.8131	0.13051	0.16051	7.662	6.230	7
8	1.267	0.7894	0.11246	0.14246	8.892	7.020	8
9	1.305	0.7664	0.09843	0.12843	10.159	7.786	9
10	1.344	0.7441	0.08723	0.11723	11.464	8.530	10
11	1.384	0.7224	0.07808	0.10808	12.808	9.253	11
12	1.426	0.7014	0.07046	0.10046	14.192	9.954	12
13	1.469	0.6810	0.06403	0.09403	15.618	10.635	13
14	1.513	0.6611	0.05853	0.08853	17.086	11.296	14
15	1.558	0.6419	0.05377	0.08377	18.599	11.938	15
16	1.605	0.6232	0.04961	0.07961	20.157	12.561	16
17	1.653	0.6050	0.04595	0.07595	21.762	13.166	17
18	1.702	0.5874	0.04271	0.07271	23.414	13.754	18
19	1.754	0.5703	0.03981	0.06981	25.117	14.324	19
20	1.806	0.5537	0.03722	0.06722	26.870	14.877	20
21	1.860	0.5375	0.03487	0.06487	28.676	15.415	21
22	1.916	0.5219	0.03275	0.06275	30.537	15.937	22
23	1.974	0.5067	0.03081	0.06081	32.453	16.444	23
24	2.033	0.4919	0.02905	0.05905	34.426	16.936	24
25	2.094	0.4776	0.02743	0.05743	36.459	17.413	25
26	2.157	0.4637	0.02594	0.05594	38.553	17.877	26
27	2.221	0.4502	0.02456	0.05456	40.710	18.327	27
28	2.288	0.4371	0.02329	0.05329	42.931	18.764	28
29	2.357	0.4243	0.02211	0.05211	45.219	19.188	29
30	2.427	0.4120	0.02102	0.05102	47.575	19.600	30
31	2.500	0.4000	0.02000	0.05000	50.003	20.000	31
32	2.575	0.3883	0.01905	0.04905	52.503	20.389	32
33	2.652	0.3770	0.01816	0.04816	55.078	20.766	33
34	2.732	0.3660	0.01732	0.04732	57.730	21.132	34
35	2.814	0.3554	0.01654	0.04654	60.462	21.487	35
40	3.262	0.3066	0.01326	0.04326	75.401	23.115	40
45	3.782	0.2644	0.01079	0.04079	92.720	24.519	45
50	4.384	0.2281	0.00887	0.03887	112.797	25.730	50
55	5.082	0.1968	0.00735	0.03735	136.072	26.774	55
60	5.892	0.1697	0.00613	0.03613	163.053	27.676	60
65	6.830	0.1464	0.00515	0.03515	194.333	28.453	65
70	7.918	0.1263	0.00434	0.03434	230.594	29.123	70
75	9.179	0.1089	0.00367	0.03367	272.631	29.702	75
80	10.641	0.0940	0.00311	0.03311	321.363	30.201	80
85	12.336	0.0811	0.00265	0.03265	377.857	30.631	85
90	14.300	0.0699	0.00226	0.03226	443.349	31.002	90
95	16.578	0.0603	0.00193	0.03193	519.272	31.323	95
100	19.219	0.0520	0.00165	0.03165	607.288	31.599	100

	To find F, given P: $(1 + i)^n$	To find P, given F: $\dfrac{1}{(1 + i)^n}$	To find A, given F: $\dfrac{i}{(1 + i)^n - 1}$	To find A, given P: $\dfrac{i(1 + i)^n}{(1 + i)^n - 1}$	To find F, given A: $\dfrac{(1 + i)^n - 1}{i}$	To find P, given A: $\dfrac{(1 + i)^n - 1}{i(1 + i)^n}$	
n	$(f/p,4,n)$	$(p/f,4,n)$	$(a/f,4,n)$	$(a/p,4,n)$	$(f/a,4,n)$	$(p/a,4,n)$	n
1	1.040	0.9615	1.00000	1.04000	1.000	0.962	1
2	1.082	0.9246	0.49020	0.53020	2.040	1.886	2
3	1.125	0.8890	0.32035	0.36035	3.122	2.775	3
4	1.170	0.8548	0.23549	0.27549	4.246	3.630	4
5	1.217	0.8219	0.18463	0.22463	5.416	4.452	5
6	1.265	0.7903	0.15076	0.19076	6.633	5.242	6
7	1.316	0.7599	0.12661	0.16661	7.898	6.002	7
8	1.369	0.7307	0.10853	0.14853	9.214	6.733	8
9	1.423	0.7026	0.09449	0.13449	10.583	7.435	9
10	1.480	0.6756	0.08329	0.12329	12.006	8.111	10
11	1.539	0.6496	0.07415	0.11415	13.486	8.760	11
12	1.601	0.6246	0.06655	0.10655	15.026	9.385	12
13	1.665	0.6006	0.06014	0.10014	16.627	9.986	13
14	1.732	0.5775	0.05467	0.09467	18.292	10.563	14
15	1.801	0.5553	0.04994	0.08994	20.024	11.118	15
16	1.873	0.5339	0.04582	0.08582	21.825	11.652	16
17	1.948	0.5134	0.04220	0.08220	23.698	12.166	17
18	2.026	0.4936	0.03899	0.07899	25.645	12.659	18
19	2.107	0.4746	0.03614	0.07614	27.671	13.134	19
20	2.191	0.4564	0.03358	0.07358	29.778	13.590	20
21	2.279	0.4388	0.03128	0.07128	31.969	14.029	21
22	2.370	0.4220	0.02920	0.06920	34.248	14.451	22
23	2.465	0.4057	0.02731	0.06731	36.618	14.857	23
24	2.563	0.3901	0.02559	0.06559	39.083	15.247	24
25	2.666	0.3751	0.02401	0.06401	41.646	15.622	25
26	2.772	0.3607	0.02257	0.06257	44.312	15.983	26
27	2.883	0.3468	0.02124	0.06124	47.084	16.330	27
28	2.999	0.3335	0.02001	0.06001	49.968	16.663	28
29	3.119	0.3207	0.01888	0.05888	52.966	16.984	29
30	3.243	0.3083	0.01783	0.05783	56.085	17.292	30
31	3.373	0.2965	0.01686	0.05686	59.328	17.588	31
32	3.508	0.2851	0.01595	0.05595	62.701	17.874	32
33	3.648	0.2741	0.01510	0.05510	66.210	18.148	33
34	3.794	0.2636	0.01431	0.05431	69.858	18.411	34
35	3.946	0.2534	0.01358	0.05358	73.652	18.665	35
40	4.801	0.2083	0.01052	0.05052	95.026	19.793	40
45	5.841	0.1712	0.00826	0.04826	121.029	20.720	45
50	7.107	0.1407	0.00655	0.04655	152.667	21.482	50
55	8.646	0.1157	0.00523	0.04523	191.159	22.109	55
60	10.520	0.0951	0.00420	0.04420	237.991	22.623	60
65	12.799	0.0781	0.00339	0.04339	294.968	23.047	65
70	15.572	0.0642	0.00275	0.04275	364.290	23.395	70
75	18.945	0.0528	0.00223	0.04223	448.631	23.680	75
80	23.050	0.0434	0.00181	0.04181	551.245	23.915	80
85	28.044	0.0357	0.00148	0.04148	676.090	24.109	85
90	34.119	0.0293	0.00121	0.04121	827.983	24.267	90
95	41.511	0.0241	0.00099	0.04099	1012.785	24.398	95
100	50.505	0.0198	0.00081	0.04081	1237.624	24.505	100

5%

n	To find F, given P: $(1 + i)^n$	To find P, given F: $\dfrac{1}{(1 + i)^n}$	To find A, given F: $\dfrac{i}{(1 + i)^n - 1}$	To find A, given P: $\dfrac{i(1 + i)^n}{(1 + i)^n - 1}$	To find F, given A: $\dfrac{(1 + i)^n - 1}{i}$	To find P, given A: $\dfrac{(1 + i)^n - 1}{i(1 + i)^n}$	n
	(f/p,5,n)	(p/f,5,n)	(a/f,5,n)	(a/p,5,n)	(f/a,5,n)	(p/a,5,n)	
1	1.050	0.9524	1.00000	1.05000	1.000	0.952	1
2	1.103	0.9070	0.48780	0.53780	2.050	1.859	2
3	1.158	0.8638	0.31721	0.36721	3.153	2.723	3
4	1.216	0.8227	0.23201	0.28201	4.310	3.546	4
5	1.276	0.7835	0.18097	0.23097	5.526	4.329	5
6	1.340	0.7462	0.14702	0.19702	6.802	5.076	6
7	1.407	0.7107	0.12282	0.17282	8.142	5.786	7
8	1.477	0.6768	0.10472	0.15472	9.549	6.463	8
9	1.551	0.6446	0.09069	0.14069	11.027	7.108	9
10	1.629	0.6139	0.07950	0.12950	12.578	7.722	10
11	1.710	0.5847	0.07039	0.12039	14.207	8.306	11
12	1.796	0.5568	0.06283	0.11283	15.917	8.863	12
13	1.886	0.5303	0.05646	0.10646	17.713	9.394	13
14	1.980	0.5051	0.05102	0.10102	19.599	9.899	14
15	2.079	0.4810	0.04634	0.09634	21.579	10.380	15
16	2.183	0.4581	0.04227	0.09277	23.657	10.838	16
17	2.292	0.4363	0.03870	0.08870	25.840	11.274	17
18	2.407	0.4155	0.03555	0.08555	28.132	11.690	18
19	2.527	0.3957	0.03275	0.08275	30.539	12.085	19
20	2.653	0.3769	0.03024	0.08024	33.066	12.462	20
21	2.786	0.3589	0.02800	0.07800	35.719	12.821	21
22	2.925	0.3418	0.02597	0.07597	38.505	13.163	22
23	3.072	0.3256	0.02414	0.07414	41.430	13.489	23
24	3.225	0.3101	0.02247	0.07247	44.502	13.799	24
25	3.386	0.2953	0.02095	0.07095	47.727	14.094	25
26	3.556	0.2812	0.01956	0.06966	51.113	14.375	26
27	3.733	0.2678	0.01829	0.06829	54.669	14.643	27
28	3.920	0.2551	0.01712	0.06712	58.403	14.898	28
29	4.116	0.2429	0.01605	0.06605	62.323	15.141	29
30	4.322	0.2314	0.01505	0.06505	66.439	15.372	30
31	4.538	0.2204	0.01413	0.06413	70.761	15.593	31
32	4.765	0.2099	0.01328	0.06328	75.299	15.803	32
33	5.003	0.1999	0.01249	0.06249	80.064	16.003	33
34	5.253	0.1904	0.01176	0.06176	85.067	16.193	34
35	5.516	0.1813	0.01107	0.06107	90.320	16.374	35
40	7.040	0.1420	0.00828	0.05828	120.800	17.159	40
45	8.985	0.1113	0.00626	0.05626	159.700	17.774	45
50	11.467	0.0872	0.00478	0.05478	209.348	18.256	50
55	14.636	0.0683	0.00367	0.05367	272.713	18.633	55
60	18.679	0.0535	0.00283	0.05283	353.584	18.929	60
65	23.840	0.0419	0.00219	0.05219	456.798	19.161	65
70	30.426	0.0329	0.00170	0.05170	588.529	19.343	70
75	38.833	0.0258	0.00132	0.05132	756.654	19.485	75
80	49.561	0.0202	0.00103	0.05103	971.229	19.596	80
85	63.254	0.0158	0.00080	0.05080	1245.087	19.684	85
90	80.730	0.0124	0.00063	0.05063	1594.607	19.752	90
95	103.035	0.0097	0.00049	0.05049	2040.694	19.806	95
100	131.501	0.0076	0.00038	0.05038	2610.025	19.848	100

	To find F, given P: $(1 + i)^n$	To find P, given F: $\dfrac{1}{(1 + i)^n}$	To find A, given F: $\dfrac{i}{(1 + i)^n - 1}$	To find A, given P: $\dfrac{i(1 + i)^n}{(1 + i)^n - 1}$	To find F, given A: $\dfrac{(1 + i)^n - 1}{i}$	To find P, given A: $\dfrac{(1 + i)^n - 1}{i(1 + i)^n}$	
n	$(f/p,6,n)$	$(p/f,6,n)$	$(a/f,6,n)$	$(a/p,6,n)$	$(f/a,6,n)$	$(p/a,6,n)$	n
1	1.060	0.9434	1.00000	1.06000	1.000	0.943	1
2	1.124	0.8900	0.48544	0.54544	2.060	1.833	2
3	1.191	0.8396	0.31411	0.37411	3.184	2.673	3
4	1.262	0.7921	0.22859	0.28859	4.375	3.465	4
5	1.338	0.7473	0.17740	0.23740	5.637	4.212	5
6	1.419	0.7050	0.14336	0.20336	6.975	4.917	6
7	1.504	0.6651	0.11914	0.17914	8.394	5.582	7
8	1.594	0.6274	0.10104	0.16104	9.897	6.210	8
9	1.689	0.5919	0.08702	0.14702	11.491	6.802	9
10	1.791	0.5584	0.07587	0.13587	13.181	7.360	10
11	1.898	0.5268	0.06679	0.12679	14.972	7.887	11
12	2.012	0.4970	0.05928	0.11928	16.870	8.384	12
13	2.133	0.4688	0.05296	0.11296	18.882	8.853	13
14	2.261	0.4423	0.04758	0.10758	21.015	9.295	14
15	2.397	0.4173	0.04296	0.10296	23.276	9.712	15
16	2.540	0.3936	0.03895	0.09895	25.673	10.106	16
17	2.693	0.3714	0.03544	0.09544	28.213	10.477	17
18	2.854	0.3503	0.03236	0.09236	30.906	10.828	18
19	3.026	0.3305	0.02962	0.08962	33.760	11.158	19
20	3.207	0.3118	0.02718	0.08718	36.786	11.470	20
21	3.400	0.2942	0.02500	0.08500	39.993	11.764	21
22	3.604	0.2775	0.02305	0.08305	43.392	12.042	22
23	3.820	0.2618	0.02128	0.08128	46.996	12.303	23
24	4.049	0.2470	0.01968	0.07968	50.816	12.550	24
25	4.292	0.2330	0.01823	0.07823	54.865	12.783	25
26	4.549	0.2198	0.01690	0.07690	59.156	13.003	26
27	4.822	0.2074	0.01570	0.07570	63.706	13.211	27
28	5.112	0.1956	0.01459	0.07459	68.528	13.406	28
29	5.418	0.1846	0.01358	0.07358	73.640	13.591	29
30	5.743	0.1741	0.01265	0.07265	79.058	13.765	30
31	6.088	0.1643	0.01179	0.07179	84.802	13.929	31
32	6.543	0.1550	0.01100	0.07100	90.890	14.084	32
33	6.841	0.1462	0.01027	0.07027	97.343	14.230	33
34	7.251	0.1379	0.00960	0.06960	104.184	14.368	34
35	7.686	0.1301	0.00897	0.06897	111.435	14.498	35
40	10.286	0.0972	0.00646	0.06646	154.762	15.046	40
45	13.765	0.0727	0.00470	0.06470	212.744	15.456	45
50	18.420	0.0543	0.00344	0.06344	290.336	15.762	50
55	24.650	0.0406	0.00254	0.06254	394.172	15.991	55
60	32.988	0.0303	0.00188	0.06188	533.128	16.161	60
65	45.145	0.0227	0.00139	0.06139	719.083	16.289	65
70	59.076	0.0169	0.00103	0.06103	967.932	16.385	70
75	79.057	0.0126	0.00077	0.06077	1300.949	16.456	75
80	105.796	0.0095	0.00057	0.06057	1746.600	16.509	80
85	141.579	0.0071	0.00043	0.06043	2342.982	16.549	85
90	189.465	0.0053	0.00032	0.06032	3141.075	16.579	90
95	253.546	0.0039	0.00024	0.06024	4209.104	16.601	95
100	339.302	0.0029	0.00018	0.06018	5638.368	16.618	100

7%

n	To find F, given P: $(1 + i)^n$	To find P, given F: $\dfrac{1}{(1 + i)^n}$	To find A, given F: $\dfrac{i}{(1 + i)^n - 1}$	To find A, given P: $\dfrac{i(1 + i)^n}{(1 + i)^n - 1}$	To find F, given A: $\dfrac{(1 + i)^n - 1}{i}$	To find P, given A: $\dfrac{(1 + i)^n - 1}{i(1 + i)^n}$	n
	(f/p,7,n)	(p/f,7,n)	(a/f,7,n)	(a/p,7,n)	(f/a,7,n)	(p/a,7,n)	
1	1.070	0.9346	1.00000	1.07000	1.000	0.935	1
2	1.145	0.8734	0.48309	0.55309	2.070	1.808	2
3	1.225	0.8163	0.31105	0.38105	3.215	2.624	3
4	1.311	0.7629	0.22523	0.29523	4.440	3.387	4
5	1.403	0.7130	0.17389	0.24389	5.751	4.100	5
6	1.501	0.6663	0.13980	0.20980	7.153	4.767	6
7	1.606	0.6227	0.11555	0.18555	8.654	5.389	7
8	1.718	0.5820	0.09747	0.16747	10.260	5.971	8
9	1.838	0.5439	0.08349	0.15349	11.978	6.515	9
10	1.967	0.5083	0.07238	0.14238	13.816	7.024	10
11	2.105	0.4751	0.06336	0.13336	15.784	7.499	11
12	2.252	0.4440	0.05590	0.12590	17.888	7.943	12
13	2.410	0.4150	0.04965	0.11965	20.141	8.358	13
14	2.579	0.3878	0.04434	0.11434	22.550	8.745	14
15	2.759	0.3624	0.03979	0.10979	25.129	9.108	15
16	2.952	0.3387	0.03586	0.10586	27.888	9.447	16
17	3.159	0.3166	0.03243	0.10243	30.840	9.763	17
18	3.380	0.2959	0.02941	0.09941	33.999	10.059	18
19	3.617	0.2765	0.02675	0.09675	37.379	10.363	19
20	3.870	0.2584	0.02439	0.09439	40.995	10.594	20
21	4.141	0.2415	0.02229	0.09229	44.865	10.836	21
22	4.430	0.2257	0.02041	0.09041	49.006	11.061	22
23	4.741	0.2109	0.01871	0.08871	53.436	11.272	23
24	5.072	0.1971	0.01719	0.08719	58.177	11.469	24
25	5.427	0.1842	0.01581	0.08581	63.249	11.654	25
26	5.807	0.1722	0.01456	0.08456	68.676	11.826	26
27	6.214	0.1609	0.01343	0.08343	74.484	11.987	27
28	6.649	0.1504	0.01239	0.08239	80.698	12.137	28
29	7.114	0.1406	0.01145	0.08145	87.347	12.278	29
30	7.612	0.1314	0.01059	0.08059	94.461	12.409	30
31	8.145	0.1228	0.00980	0.07980	102.073	12.532	31
32	8.715	0.1147	0.00907	0.07907	110.218	12.647	32
33	9.325	0.1072	0.00841	0.07841	118.923	12.754	33
34	9.978	0.1002	0.00780	0.07780	128.259	12.854	34
35	10.677	0.0937	0.00723	0.07723	138.237	12.948	35
40	14.974	0.0668	0.00501	0.07501	199.635	13.332	40
45	21.002	0.0476	0.00350	0.07350	285.749	13.606	45
50	29.457	0.0339	0.00246	0.07246	406.529	13.801	50
55	41.315	0.0242	0.00174	0.07174	575.929	13.940	55
60	57.946	0.0173	0.00123	0.07123	813.520	14.039	60
65	81.273	0.0123	0.00087	0.07087	1146.755	14.110	65
70	113.989	0.0088	0.00062	0.07062	1614.134	14.160	70
75	159.876	0.0063	0.00044	0.07044	2269.657	14.196	75
80	224.234	0.0045	0.00031	0.07031	3189.063	14.222	80
85	314.500	0.0032	0.00022	0.07022	4478.576	14.240	85
90	441.103	0.0023	0.00016	0.07016	6287.185	14.253	90
95	618.670	0.0016	0.00011	0.07011	8823.854	14.263	95
100	867.716	0.0012	0.00008	0.07008	12381.662	14.269	100

n	To find F, given P: $(1 + i)^n$ (f/p,8,n)	To find P, given F: $\dfrac{1}{(1 + i)^n}$ (p/f,8,n)	To find A, given F: $\dfrac{i}{(1 + i)^n - 1}$ (a/f,8,n)	To find A, given P: $\dfrac{i(1 + i)^n}{(1 + i)^n - 1}$ (a/p,8,n)	To find F, given A: $\dfrac{(1 + i)^n - 1}{i}$ (f/a,8,n)	To find P, given A: $\dfrac{(1 + i)^n - 1}{i(1 + i)^n}$ (p/a,8,n)	n
1	1.080	0.9259	1.00000	1.08000	1.000	0.926	1
2	1.166	0.8573	0.48077	0.56077	2.080	1.783	2
3	1.260	0.7938	0.30803	0.38803	3.246	2.577	3
4	1.360	0.7350	0.22192	0.30192	4.506	3.312	4
5	1.469	0.6806	0.17046	0.25046	5.867	3.933	5
6	1.587	0.6302	0.13632	0.21632	7.336	4.623	6
7	1.714	0.5835	0.11207	0.19207	8.923	5.206	7
8	1.851	0.5403	0.09401	0.17401	10.637	5.747	8
9	1.999	0.5002	0.08008	0.16008	12.488	6.247	9
10	2.159	0.4632	0.06903	0.14903	14.487	6.710	10
11	2.332	0.4289	0.06008	0.14008	16.645	7.139	11
12	2.518	0.3971	0.05270	0.13270	18.977	7.536	12
13	2.720	0.3677	0.04652	0.12652	21.495	7.904	13
14	2.937	0.3405	0.04130	0.12130	24.215	8.244	14
15	3.172	0.3152	0.03683	0.11683	27.152	8.559	15
16	3.426	0.2919	0.03298	0.11298	30.324	8.851	16
17	3.700	0.2703	0.02963	0.10963	33.750	9.122	17
18	3.996	0.2502	0.02670	0.10670	37.450	9.372	18
19	4.316	0.2317	0.02413	0.10413	41.446	9.604	19
20	4.661	0.2145	0.02185	0.10185	45.762	9.818	20
21	5.034	0.1987	0.01983	0.09983	50.423	10.017	21
22	5.437	0.1839	0.01803	0.09803	55.457	10.201	22
23	5.781	0.1703	0.01642	0.09642	60.893	10.371	23
24	6.341	0.1577	0.01498	0.09498	66.765	10.529	24
25	6.848	0.1460	0.01368	0.09368	73.106	10.675	25
26	7.396	0.1352	0.01251	0.09251	79.954	10.810	26
27	7.988	0.1252	0.01145	0.09145	87.351	10.935	27
28	8.627	0.1159	0.01049	0.09049	95.339	11.051	28
29	9.317	0.1073	0.00962	0.08962	103.966	11.158	29
30	10.063	0.0994	0.00883	0.08883	113.283	11.258	30
31	10.868	0.0920	0.00811	0.08811	123.346	11.350	31
32	11.737	0.0852	0.00745	0.08745	134.214	11.435	32
33	12.676	0.0789	0.00685	0.08685	145.951	11.514	33
34	13.690	0.0730	0.00630	0.08630	158.627	11.587	34
35	14.785	0.0676	0.00580	0.08580	173.317	11.655	35
40	21.725	0.0460	0.00386	0.08386	259.057	11.925	40
45	31.920	0.0313	0.00259	0.08259	386.506	12.108	45
50	46.902	0.0213	0.00174	0.08174	573.770	12.233	50
55	68.914	0.0145	0.00118	0.08118	848.923	12.319	55
60	101.257	0.0099	0.00080	0.08080	1253.213	12.377	60
65	148.780	0.0067	0.00054	0.08054	1847.248	12.416	65
70	218.606	0.0046	0.00037	0.08037	2720.080	12.443	70
75	321.205	0.0031	0.00025	0.08025	4002.557	12.461	75
80	471.955	0.0021	0.00017	0.08017	5886.935	12.474	80
85	693.456	0.0014	0.00012	0.08012	8655.706	12.482	85
90	1018.915	0.0010	0.00008	0.08008	12723.939	12.488	90
95	1497.121	0.0007	0.00005	0.08005	18701.507	12.492	95
100	2199.761	0.0005	0.00004	0.08004	27484.516	12.494	100

	To find F, given P: $(1 + i)^n$	To find P, given F: $\dfrac{1}{(1 + i)^n}$	To find A, given F: $\dfrac{i}{(1 + i)^n - 1}$	To find A, given P: $\dfrac{i(1 + i)^n}{(1 + i)^n - 1}$	To find F, given A: $\dfrac{(1 + i)^n - 1}{i}$	To find P, given A: $\dfrac{(1 + i)^n - 1}{i(1 + i)^n}$	
n	$(f/p,9,n)$	$(p/f,9,n)$	$(a/f,9,n)$	$(a/p,9,n)$	$(f/a,9,n)$	$(p/a,9,n)$	n
1	1.090	0.9174	1.00000	1.09000	1.000	0.917	1
2	1.188	0.8417	0.47847	0.56847	2.090	1.759	2
3	1.295	0.7722	0.30505	0.39505	3.278	2.531	3
4	1.412	0.7084	0.21867	0.30867	4.573	3.240	4
5	1.539	0.6499	0.16709	0.25709	5.985	3.890	5
6	1.677	0.5963	0.13292	0.22292	7.523	4.486	6
7	1.828	0.5470	0.10869	0.19869	9.200	5.033	7
8	1.993	0.5019	0.09067	0.18067	11.028	5.535	8
9	2.172	0.4604	0.07680	0.16680	13.021	5.995	9
10	2.367	0.4224	0.06582	0.15582	15.193	6.418	10
11	2.580	0.3875	0.05695	0.14695	17.560	6.805	11
12	2.813	0.3555	0.04965	0.13965	20.141	7.161	12
13	3.066	0.3262	0.04357	0.13357	22.953	7.487	13
14	3.342	0.2992	0.03843	0.12843	26.019	7.786	14
15	3.642	0.2745	0.03406	0.12406	29.361	8.061	15
16	3.970	0.2519	0.03030	0.12030	33.003	8.313	16
17	4.328	0.2311	0.02705	0.11705	36.974	8.544	17
18	4.717	0.2120	0.02421	0.11421	41.301	8.756	18
19	5.142	0.1945	0.02173	0.11173	46.018	8.950	19
20	5.604	0.1784	0.01955	0.10955	51.160	9.129	20
21	6.109	0.1637	0.01762	0.10762	56.765	9.292	21
22	6.659	0.1502	0.01590	0.10590	62.873	9.442	22
23	7.258	0.1378	0.01438	0.10438	69.532	9.580	23
24	7.911	0.1264	0.01302	0.10302	76.790	9.707	24
25	8.623	0.1160	0.01180	0.10181	84.701	9.823	25
26	9.399	0.1064	0.01072	0.10072	93.324	9.929	26
27	10.245	0.0976	0.00973	0.09973	102.723	10.027	27
28	11.167	0.0895	0.00885	0.09885	112.968	10.116	28
29	12.172	0.0822	0.00806	0.09806	124.135	10.198	29
30	13.268	0.0754	0.00734	0.09734	136.308	10.274	30
31	14.462	0.0691	0.00669	0.09669	149.575	10.343	31
32	15.763	0.0634	0.00610	0.09610	164.037	10.406	32
33	17.182	0.0582	0.00556	0.09556	179.800	10.464	33
34	18.728	0.0534	0.00508	0.09508	196.982	10.518	34
35	20.414	0.0490	0.00464	0.09464	215.711	10.567	35
40	31.409	0.0318	0.00296	0.09296	337.882	10.757	40
45	48.327	0.0207	0.00190	0.09190	525.859	10.881	45
50	74.358	0.0134	0.00123	0.09123	815.084	10.962	50
55	114.408	0.0087	0.00079	0.09079	1260.092	11.014	55
60	176.031	0.0057	0.00051	0.09051	1944.792	11.048	60
65	270.864	0.0037	0.00033	0.09033	2998.288	11.070	65
70	416.730	0.0024	0.00022	0.09022	4619.223	11.084	70
75	641.191	0.0016	0.00014	0.09014	7113.232	11.094	75
80	986.552	0.0010	0.00009	0.09009	10950.556	11.100	80
85	1517.948	0.0007	0.00006	0.09006	16854.444	11.104	85
90	2335.501	0.0004	0.00004	0.09004	25939.000	11.106	90
95	3593.513	0.0003	0.00003	0.09003	39917.378	11.108	95
100	5529.089	0.0002	0.00002	0.09002	61422.544	11.109	100

	To find F, given P: $(1 + i)^n$	To find P, given F: $\dfrac{1}{(1 + i)^n}$	To find A, given F: $\dfrac{i}{(1 + i)^n - 1}$	To find A, given P: $\dfrac{i(1 + i)^n}{(1 + i)^n - 1}$	To find F, given A: $\dfrac{(1 + i)^n - 1}{i}$	To find P, given A: $\dfrac{(1 + i)^n - 1}{i(1 + i)^n}$	
n	(f/p,10,n)	(p/f,10,n)	(a/f,10,n)	(a/p,10,n)	(f/a,10,n)	(p/a,10,n)	n
1	1.100	0.9091	1.00000	1.10000	1.000	0.909	1
2	1.210	0.8264	0.47619	0.57619	2.100	1.736	2
3	1.331	0.7513	0.30211	0.40211	3.310	2.487	3
4	1.464	0.6830	0.21547	0.31547	4.641	3.170	4
5	1.611	0.6209	0.16380	0.26380	6.105	3.791	5
6	1.772	0.5645	0.12961	0.22961	7.716	4.355	6
7	1.949	0.5132	0.10541	0.20541	9.487	4.868	7
8	2.144	0.4665	0.08744	0.18744	11.436	5.335	8
9	2.358	0.4241	0.07364	0.17364	13.579	5.759	9
10	2.594	0.3855	0.06275	0.16275	15.937	6.144	10
11	2.853	0.3505	0.05396	0.15396	18.531	6.495	11
12	3.138	0.3186	0.04676	0.14676	21.384	6.814	12
13	3.452	0.2897	0.04078	0.14078	24.523	7.103	13
14	3.797	0.2633	0.03575	0.13575	27.975	7.367	14
15	4.177	0.2394	0.03147	0.13147	31.772	7.606	15
16	4.595	0.2176	0.02782	0.12782	35.950	7.824	16
17	5.054	0.1978	0.02466	0.12466	40.545	8.022	17
18	5.560	0.1799	0.02193	0.12193	45.599	8.201	18
19	6.116	0.1635	0.01955	0.11955	51.159	8.363	19
20	6.727	0.1486	0.01746	0.11746	57.275	8.514	20
21	7.400	0.1351	0.01562	0.11562	64.002	8.649	21
22	8.140	0.1228	0.01401	0.11401	71.403	8.772	22
23	8.954	0.1117	0.01257	0.11257	79.543	8.883	23
24	9.850	0.1015	0.01130	0.11130	88.497	8.985	24
25	10.835	0.0923	0.01017	0.11017	98.347	9.077	25
26	11.918	0.0839	0.00916	0.10916	109.182	9.161	26
27	13.110	0.0763	0.00826	0.10826	121.100	9.237	27
28	14.421	0.0693	0.00745	0.10745	134.210	9.307	28
29	15.863	0.0630	0.00673	0.10673	148.631	9.370	29
30	17.449	0.0573	0.00608	0.10608	164.494	9.427	30
31	19.194	0.0521	0.00550	0.10550	181.943	9.479	31
32	21.114	0.0474	0.00497	0.10497	201.138	9.526	32
33	23.225	0.0431	0.00450	0.10450	222.252	9.569	33
34	25.548	0.0391	0.00407	0.10407	245.477	9.609	34
35	28.102	0.0356	0.00369	0.10369	271.024	9.644	35
40	45.259	0.0221	0.00226	0.10226	442.593	9.779	40
45	72.890	0.0137	0.00139	0.10139	718.905	9.863	45
50	117.391	0.0085	0.00086	0.10086	1163.909	9.915	50
55	189.059	0.0053	0.00053	0.10053	1880.591	9.947	55
60	304.482	0.0033	0.00033	0.10033	3034.816	9.967	60
65	490.371	0.0020	0.00020	0.10020	4893.707	9.980	65
70	789.747	0.0013	0.00013	0.10013	7887.470	9.987	70
75	1271.895	0.0008	0.00008	0.10008	12708.954	9.992	75
80	2048.400	0.0005	0.00005	0.10005	20474.002	9.995	80
85	3298.969	0.0003	0.00003	0.10003	32979.690	9.997	85
90	5313.023	0.0002	0.00002	0.10002	53120.226	9.998	90
95	8556.676	0.0001	0.00001	0.10001	85556.760	9.999	95
100	13780.612	0.0001	0.00001	0.10001	137796.123	9.999	100

12%

n	To find F, given P: $(1 + i)^n$	To find P, given F: $\dfrac{1}{(1 + i)^n}$	To find A, given F: $\dfrac{i}{(1 + i)^n - 1}$	To find A, given P: $\dfrac{i(1 + i)^n}{(1 + i)^n - 1}$	To find F, given A: $\dfrac{(1 + i)^n - 1}{i}$	To find P, given A: $\dfrac{(1 + i)^n - 1}{i(1 + i)^n}$	n
	$(f/p,12,n)$	$(p/f,12,n)$	$(a/f,12,n)$	$(a/p,12,n)$	$(f/a,12,n)$	$(p/a,12,n)$	
1	1.120	0.8929	1.00000	1.12000	1.000	0.893	1
2	1.254	0.7972	0.47170	0.59170	2.120	1.690	2
3	1.405	0.7118	0.29635	0.41635	3.374	2.402	3
4	1.574	0.6355	0.20923	0.32923	4.779	3.037	4
5	1.762	0.5674	0.15741	0.27741	6.353	3.605	5
6	1.974	0.5066	0.12323	0.24323	8.115	4.111	6
7	2.211	0.4523	0.09912	0.21912	10.089	4.564	7
8	2.476	0.4039	0.08130	0.20130	12.300	4.968	8
9	2.773	0.3606	0.06768	0.18768	14.776	5.328	9
10	3.106	0.3220	0.05698	0.17698	17.549	5.650	10
11	3.479	0.2875	0.04842	0.16842	20.655	5.938	11
12	3.896	0.2567	0.04144	0.16144	24.133	6.194	12
13	4.363	0.2292	0.03568	0.15568	28.029	6.424	13
14	4.887	0.2046	0.03087	0.15087	32.393	6.628	14
15	5.474	0.1827	0.02682	0.14682	37.280	6.811	15
16	6.130	0.1631	0.02339	0.14339	42.753	6.974	16
17	6.866	0.1456	0.02046	0.14046	48.884	7.120	17
18	7.690	0.1300	0.01794	0.13794	55.750	7.250	18
19	8.613	0.1161	0.01576	0.13576	63.440	7.366	19
20	9.646	0.1037	0.01388	0.13388	72.052	7.469	20
21	10.804	0.0926	0.01224	0.13224	81.699	7.562	21
22	12.100	0.0826	0.01081	0.13081	92.503	7.645	22
23	13.552	0.0738	0.00956	0.12956	104.603	7.718	23
24	15.179	0.0659	0.00846	0.12846	118.155	7.784	24
25	17.000	0.0588	0.00750	0.12750	133.334	7.843	25
26	19.040	0.0525	0.00665	0.12665	150.334	7.896	26
27	21.325	0.0469	0.00590	0.12590	169.374	7.943	27
28	23.884	0.0419	0.00524	0.12524	190.699	7.984	28
29	26.750	0.0374	0.00466	0.12466	214.582	8.022	29
30	29.960	0.0334	0.00414	0.12414	241.333	8.055	30
31	33.555	0.0298	0.00369	0.12369	271.292	8.085	31
32	37.582	0.0266	0.00328	0.12328	304.847	8.112	32
33	42.091	0.0238	0.00292	0.12292	342.429	8.135	33
34	47.142	0.0212	0.00260	0.12260	384.520	8.157	34
35	52.800	0.0189	0.00232	0.12232	431.663	8.176	35
40	93.051	0.0107	0.00130	0.12130	767.091	8.244	40
45	163.988	0.0061	0.00074	0.12074	1358.230	8.283	45
50	289.002	0.0035	0.00042	0.12042	2400.018	8.305	50

n	To find F, given P: $(1 + i)^n$	To find P, given F: $\dfrac{1}{(1 + i)^n}$	To find A, given F: $\dfrac{i}{(1 + i)^n - 1}$	To find A, given P: $\dfrac{i(1 + i)^n}{(1 + i)^n - 1}$	To find F, given A: $\dfrac{(1 + i)^n - 1}{i}$	To find P, given A: $\dfrac{(1 + i)^n - 1}{i(1 + i)^n}$	n
	$(f/p,15,n)$	$(p/f,15,n)$	$(a/f,15,n)$	$(a/p,15,n)$	$(f/a,15,n)$	$(p/a,15,n)$	
1	1.150	0.8696	1.00000	1.15000	1.000	0.870	1
2	1.322	0.7561	0.46512	0.61512	2.150	1.626	2
3	1.521	0.6575	0.28798	0.43798	3.472	2.283	3
4	1.749	0.5718	0.20027	0.35027	4.993	2.855	4
5	2.011	0.4972	0.14832	0.29832	6.742	3.352	5
6	2.313	0.4323	0.11424	0.26424	8.754	3.784	6
7	2.660	0.3759	0.09036	0.24036	11.067	4.160	7
8	3.059	0.3269	0.07285	0.22285	13 727	4.487	8
9	3.518	0.2843	0.05957	0.20957	16.786	4.772	9
10	4.046	0.2472	0.04925	0.19925	20.304	5.019	10
11	4.652	0.2149	0.04107	0.19107	24.349	5.234	11
12	5.350	0.1869	0.03448	0.18448	29.002	5.421	12
13	6.153	0.1625	0.02911	0.17911	34.352	5.583	13
14	7.076	0.1413	0.02469	0.17469	40.505	5.724	14
15	8.137	0.1229	0.02102	0.17102	47.580	5.847	15
16	9.358	0.1069	0.01795	0.16795	55.717	5.954	16
17	10.761	0.0929	0.01537	0.16537	65.075	6.047	17
18	12.375	0.0808	0.01319	0.16319	75.836	6.128	18
19	14.232	0.0703	0.01134	0.16134	88.212	6.198	19
20	16.367	0.0611	0.00976	0.15976	102.444	6.259	20
21	18.821	0.0531	0.00842	0.15842	118.810	6.312	21
22	21.645	0.0462	0.00727	0.15727	137.631	6.359	22
23	24.891	0.0402	0.00628	0.15628	159.276	6.399	23
24	28.625	0.0349	0.00543	0.15543	184.168	6.434	24
25	32.919	0.0304	0.00470	0.15470	212.793	6.464	25
26	37.857	0.0264	0.00407	0.15407	245.711	6.491	26
27	43.535	0.0230	0.00353	0.15353	283.569	6.514	27
28	50.066	0.0200	0.00306	0.15306	327.104	6.534	28
29	57.575	0.0174	0.00265	0.15265	377.170	6.551	29
30	66.212	0.0151	0.00230	0.15230	434.745	6.566	30
31	76.143	0.0131	0.00200	0.15200	500.956	6.579	31
32	87.565	0.0114	0.00173	0.15173	577.099	6.591	32
33	100.700	0.0099	0.00150	0.15150	664.664	6.600	33
34	115.805	0.0086	0.00131	0.15131	765.364	6.609	34
35	133.176	0.0075	0.00113	0.15113	881.170	6.617	35
40	267.863	0.0037	0.00056	0.15056	1779.090	6.642	40
45	538.769	0.0019	0.00028	0.15028	3585.128	6.654	45
50	1083.657	0.0009	0.00014	0.15014	7217.716	6.661	50

	To find F, given P: $(1 + i)^n$	To find P, given F: $\dfrac{1}{(1 + i)^n}$	To find A, given F: $\dfrac{i}{(1 + i)^n - 1}$	To find A, given P: $\dfrac{i(1 + i)^n}{(1 + i)^n - 1}$	To find F, given A: $\dfrac{(1 + i)^n - 1}{i}$	To find P, given A: $\dfrac{(1 + i)^n - 1}{i(1 + i)^n}$	
n	(f/p,20,n)	(p/f,20,n)	(a/f,20,n)	(a/p,20,n)	(f/a,20,n)	(p/a,20,n)	n
1	1.200	0.8333	1.00000	1.20000	1.000	0.833	1
2	1.440	0.6944	0.45455	0.65455	2.200	1.528	2
3	1.728	0.5787	0.27473	0.47473	3.640	2.106	3
4	2.074	0.4823	0.18629	0.38629	5.368	2.598	4
5	2.488	0.4019	0.13438	0.33438	7.442	2.991	5
6	2.986	0.3349	0.10071	0.30071	9.930	3.326	6
7	3.583	0.2791	0.07742	0.27742	12.916	3.605	7
8	4.300	0.2326	0.06061	0.26061	16.499	3.837	8
9	5.160	0.1938	0.04808	0.24808	20.799	4.031	9
10	6.192	0.1615	0.03852	0.23852	25.959	4.192	10
11	7.430	0.1346	0.03110	0.23110	32.150	4.327	11
12	8.916	0.1122	0.02526	0.22526	39.581	4.439	12
13	10.699	0.0935	0.02062	0.22062	48.497	4.533	13
14	12.839	0.0779	0.01689	0.21689	59.196	4.611	14
15	15.407	0.0649	0.01388	0.21388	72.035	4.675	15
16	18.488	0.0541	0.01144	0.21144	87.442	4.730	16
17	22.186	0.0451	0.00944	0.20944	105.931	4.775	17
18	26.623	0.0376	0.00781	0.20781	128.117	4.812	18
19	31.948	0.0313	0.00646	0.20646	154.740	4.843	19
20	38.338	0.0261	0.00536	0.05236	186.688	4.870	20
21	46.005	0.0217	0.00444	0.20444	225.025	4.891	21
22	55.206	0.0181	0.00369	0.20369	271.031	4.909	22
23	66.247	0.0151	0.00307	0.20307	326.237	4.925	23
24	79.497	0.0126	0.00255	0.20255	392.484	4.937	24
25	95.396	0.0105	0.00212	0.20212	471.981	4.948	25
26	114.475	0.0087	0.00176	0.20176	567.377	4.956	26
27	137.371	0.0073	0.00147	0.20147	681.853	4.964	27
28	164.845	0.0061	0.00122	0.20122	819.223	4.970	28
29	197.813	0.0051	0.00102	0.20102	984.068	4.975	29
30	237.376	0.0042	0.00085	0.20085	1181.881	4.979	30
31	284.851	0.0035	0.00070	0.20070	1419.257	4.982	31
32	341.822	0.0029	0.00059	0.20059	1704.108	4.985	32
33	410.186	0.0024	0.00049	0.20049	2045.930	4.988	33
34	492.223	0.0020	0.00041	0.20041	2456.116	4.990	34
35	590.668	0.0017	0.00034	0.20034	2948.339	4.992	35
40	1469.772	0.0007	0.00014	0.20014	7343.858	4.997	40
45	3657.258	0.0003	0.00005	0.20005	18281.331	4.999	45
50	9100.427	0.0001	0.00002	0.20002	45497.191	4.999	50

APPENDIX F
SIMPLEX METHOD

F-1 PROBLEM STATEMENT

The simplex method is a procedure to solve linear programming problems of greater complexity than those we dealt with in Chapter 6. It is a collection of rules applied in a relatively mechanical manner to obtain sequentially improved solutions to a problem with linear relationships. The rules lend themselves nicely to computer programming, and the tediousness of iterative solution cycles makes the computer a most appreciated assistant.

We examine the rules and computational steps of the simplex method by applying them to the problem previously encountered in Section 6-4. The conditions of the problem are shown in the following reprint of Table 6.3.

| | Hours Per Unit | | |
	Product A	Product B	Hours Available
Machine center J	3	2	42
Machine center K	2	2	30
Machine center L	2	4	48
Contribution per unit	$12	$8	

The objective function in the simplex method is stated in the same form as is the graphical solution—maximize the profit given by the equation

$$Z = 12A + 8B$$

where

$$A = \text{units produced of product } A$$

$$B = \text{units produced of product } B$$

Maximum profit is restrained by the limited amount of time available in each machine center. Instead of stating these machine constraints as inequations, as was done in the graphical method, they are converted to restriction equations by inserting *slack variables* $S1$, $S2$, and $S3$, respectively, for machine centers J, K, and L:

$$3A + 2B + S1 = 42 \qquad \text{for machine center } J$$

$$2A + 2B + S2 = 30 \qquad \text{for machine center } K$$

$$2A + 4B + S3 = 48 \qquad \text{for machine center } L$$

When the entire capacity is used, the slack variables equal zero; otherwise they equal the unused capacity.

F-2 SIMPLEX TABLEAU

The problem data are organized in the format shown in Figure F.1. The top two rows show the variables involved and the contribution of each to the objective function: profit. In the *stub* (entries to the left of the vertical line), the amount and contribution of the variables in successive solutions are recorded. In the initial tableau, the objective column contains only zeros because the slack variables have zero coefficients in the objective equation. By considering the vertical line as an equals sign, the restriction equations are defined. For instance, the initial condition for machine center J (top line beginning with

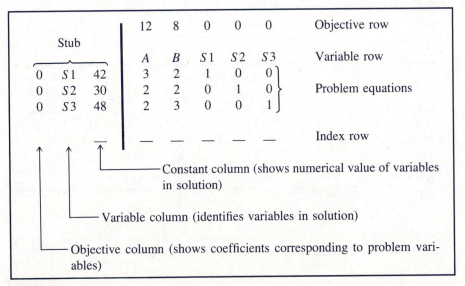

FIGURE F.1

42 in the "problem equations" portion of the tableau) is

$$42 = 3A + 2B + 1S1 + 0S2 + 0S3$$

which reduces to

$$42 = 3A + 2B + S1$$

as was given previously in the machine center J restriction.

The tableau is completed by calculating values for the index row to evaluate the initial feasible solution. The formula for each entry in the index row is

Index number $= \Sigma$ [(number in its column) $\times$ (corresponding row number in objective column)] $-$ (number in objective row heading its column)

The calculations required for all entries in the initial index row are as follows:

42×0	3×0	2×0	1×0	0×0	0×0
$+30 \times 0$	$+2 \times 0$	$+2 \times 0$	$+0 \times 0$	$+1 \times 0$	$+0 \times 0$
$+48 \times 0$	$+2 \times 0$	$+4 \times 0$	$+0 \times 0$	$+0 \times 0$	$+1 \times 0$
$-\quad 0$	$-\quad 12$	$-\quad 8$	$-\quad 0$	$-\quad 0$	$-\quad 0$
0	$-\quad 12$	$-\quad 8$	0	0	0

Constant column   Index row

Thus the initial index row is simply the objective row with each entry preceded by a minus sign. The trivial solution it represents gives zero profit from producing zero products; the variables are equal to the constants to yield $A = 0$, $B = 0$, $S1 = 42$, $S2 = 30$, $S3 = 48$, and, therefore,

$$Z = 0 = 12(0) + 8(0) + 0(42) + 0(30) + 0(48)$$

F-3 OPTIMALITY TEST

The initial feasible solution and each improved solution are tested for optimality by reference to the following criteria:

1. The solution will be optimal if all entries in the index row are zero or positive.

2. An optimal solution with an infinitely large Z is indicated when all values in a column with a negative index number are zero or negative.

3. The solution can be improved when one or more values in a column with a negative index number are positive.

Our initial solution falls under criterion 3; there are positive values in both columns containing negative index numbers, -12 and -8. These negative index numbers serve almost the same mission as do negative opportunity costs in the MODI and a stepping-stone check for optimality in a transportation matrix.

F-4 REVISION OF A NONOPTIMAL SOLUTION

A *key column* is signified by the column containing the largest negative number in the index row of a nonoptimal solution. The column variable associated with the key column is introduced into a revised solution as a row variable. The row it replaces is the *key row* and is identified by dividing each number in the constant column by the corresponding nonzero, positive number in the key column. The *lowest* ratio obtained marks the key row. For the example problem, the key column has -12 in the index row; the quotients formed by dividing the constant column values by the corresponding key column values are

$$\text{First row: } \tfrac{42}{3} = 14 \longleftarrow \text{Key row}$$

$$\text{Second row: } \tfrac{30}{2} = 15$$

$$\text{Third row: } \tfrac{48}{2} = 24$$

which lead to the conditions shown in Figure F.2.

The intersection of the key row and column shows the *key number*. This key number is the dividend used to calculate all the new entries in the key row except those for the objective and variable columns. Thus the values for the key row of the second tableau are the quotients of existing entries divided by the key number as follows:

Objective column	Variable column	Constant column		A	B	$S\,1$	$S\,2$	$S\,3$
12	A	$\tfrac{42}{3} = 14$		$\tfrac{3}{3} = 1$	$\tfrac{2}{3}$	$\tfrac{1}{3}$	$\tfrac{0}{3} = 0$	$\tfrac{0}{3} = 0$

Entries shown in the objective and variable columns are taken, respectively, from the key column values of the objective and variable rows, 12 and A.

The remaining values in the revised tableau, other than those in the objective and variable columns that are just repeated from the previous tableau, are calculated from the formula

$$\text{New value} = \text{old number} - \frac{\left(\begin{array}{c}\text{associated number} \\ \text{in key row}\end{array}\right) \times \left(\begin{array}{c}\text{corresponding number} \\ \text{in key column}\end{array}\right)}{\text{key number}}$$

				12	8	0	0	0	
Key number				A	B	$S1$	$S2$	$S3$	Key row (outgoing variable)
0	$S1$	42		3	2	1	0	0	
0	$S2$	30		2	2	0	1	0	
0	$S3$	48		2	4	0	0	1	
		0		-12	-8	0	0	0	←Index row

Key column (incoming variable)

FIGURE F.2 First tableau.

			12	8	0	0	0
			A	B	$S1$	$S2$	$S3$
12	A	14	1	$\frac{2}{3}$	$\frac{1}{3}$	0	0
0	$S1$	2	0	$\frac{2}{3}$	$-\frac{2}{3}$	1	0
0	$S2$	20	0	$\frac{8}{3}$	$-\frac{2}{3}$	0	1
		168	0	0	4	0	0

FIGURE F.3 Second tableau.

For number 30 in the constant column,

$$\text{New value} = 30 - \frac{42 \times 2}{3} = 30 - 28 = 2$$

and for number 2 in the key column, A,

$$\text{New value} = 2 - \frac{3 \times 2}{3} = 2 - 2 = 0$$

(Both number 2's in the column have the same new value.) By continuing the new value calculations and determining the new index row numbers for the optimality check, the second tableau is completed (Figure F.3).

F-5 OPTIMAL SOLUTION

The solution in the second tableau is optimal, as defined by criterion 1 of the optimality test. The solution is shown in the stub. Variables entered in the variable column have the values shown in the constant column. In the given problem, there is only one variable, A, with a value of 14. The value of the objective function, 168, is shown in the index row of the constant column. When returning to the original objective function of the problem, the solution is written:

$$Z = 12A + 8B$$

$$168 = 12(14) + 8(0)$$

An optimal solution means that no other solution will *better* satisfy the objective function, but other solutions may satisfy it equally as well. This condition exists in our sample problem. The presence of equally optimal solutions is indicated whenever there is a zero in the index row in the column of a variable that is not in the variable column. In the example, the variable B does not occur in the variable column, and there is a zero in the index row in column B.

F-6 DEGENERACY

A degenerate condition occurs when two or more rows tie for the smallest nonnegative quotient while the key row is being selected. The tie affords the possibility of making the wrong choice for the key row. An incorrect selection could lead to problem cycling, a situation in which the disappearance and reappearance of variables defy attaining an optimal solution.

A degenerate condition is resolved by dividing the values in the tied rows by the key column number. Quotients obtained are compared in a left-to-right sequence among the columns of the slack variables first and then among the other columns. The key row is identified by the first column containing unequal ratios. The row with the lowest ratio marks the variable to replace.

F-7 SUMMARY AND EXTENSION OF SIMPLEX SOLUTION PROCEDURES

The procedure previously described and summarized as follows will satisfy the constraints of a linear problem and maximize the objective function:

1. Make a precise statement of the objective function, and state all constraints as equalities through the use of slack variables.

2. Establish an initial feasible solution by placing only slack variables in the variable column.

3. Calculate index row values, and test for optimality.

4. Identify the key column in a nonoptimal solution.

5. Find the key row, and thereby establish the key number.

6. Calculate new entries in the key row.

7. Calculate new values for the remaining numbers in the revised tableau (those in the objective and variable columns remain unchanged).

8. Repeat Steps 3 through 7 until an optimal solution is obtained.

Additional rules are necessary to minimize an objective function. There are also rules available to convert the *primal* problem (original statement of the problem) to a *dual* problem. The conversion essentially amounts to reversing the rows and columns, and it thereby reduces the amount of calculations for problems with many rows and few columns. Such refinements are beyond the scope of this introductory presentation, but they are available in many books devoted primarily to linear programming.

More information than just the optimal solution is available in the final tableau. The index row shows *shadow prices*, the projected loss that would occur if variables not in the optimal solution were actually used. In the example problem, we already observed a zero cost for including product B, an alternative optimal solution. We can also observe that there would be a $4 decrease in profit if one unit of the machine center represented by $S1$ (machine center J) were left out of the problem. Further familiarity with the simplex process allows analysts to evaluate the sensitivity of solutions to changes in the objective function and constraints. Such interpretations are extremely useful for considering intangibles or other influencing factors that cannot be included formally in the simplex tableau.

INDEX